Grzimek's
Animal Life Encyclopedia

Second Edition

••••

Grzimek's
Animal Life Encyclopedia

Second Edition

●●●●

Volume 1
Lower Metazoans and Lesser Deuterostomes

Dennis A. Thoney, Advisory Editor
Neil Schlager, Editor

Joseph E. Trumpey, Chief Scientific Illustrator

Michael Hutchins, Series Editor
In association with the American Zoo and Aquarium Association

Detroit • New York • San Diego • San Francisco • Cleveland • New Haven, Conn. • Waterville, Maine • London • Munich

Grzimek's Animal Life Encyclopedia, Second Edition

Volume 1: Lower Metazoans and Lesser Deuterostomes
Produced by Schlager Group Inc.
Neil Schlager, Editor
Vanessa Torrado-Caputo, Associate Editor

Project Editor
Melissa C. McDade

Editorial
Stacey Blachford, Deirdre S. Blanchfield, Madeline Harris, Christine Jeryan, Kate Kretschmann, Mark Springer

Indexing Services
Synapse, the Knowledge Link Corporation

Permissions
Margaret Chamberlain

Imaging and Multimedia
Mary K. Grimes, Lezlie Light, Christine O'Bryan, Barbara Yarrow, Robyn V. Young

Product Design
Tracey Rowens, Jennifer Wahi

Manufacturing
Wendy Blurton, Dorothy Maki, Evi Seoud, Mary Beth Trimper

For permission to use material from this product, submit your request via Web at http://www.gale-edit.com/permissions, or you may download our Permissions Request form and submit your request by fax or mail to: The Gale Group, Inc., Permissions Department, 27500 Drake Road, Farmington Hills, MI, 48331-3535, Permissions hotline: 248-699-8074 or 800-877-4253, ext. 8006, Fax: 248-699-8074 or 800-762-4058.

Cover photo of sea anemone by AP/Wide World Photos/University of Wisconsin-Superior. Back cover photos of sea anemone by AP/Wide World Photos/University of Wisconsin-Superior; land snail, lionfish, golden frog, and green python by JLM Visuals; red-legged locust © 2001 Susan Sam; hornbill by Margaret F. Kinnaird; and tiger by Jeff Lepore/Photo Researchers. All reproduced by permission.

While every effort has been made to ensure the reliability of the information presented in this publication, The Gale Group, Inc. does not guarantee the accuracy of the data contained herein. The Gale Group, Inc. accepts no payment for listing; and inclusion in the publication of any organization, agency, institution, publication, service, or individual does not imply endorsement of the editors and publisher. Errors brought to the attention of the publisher and verified to the satisfaction of the publisher will be corrected in future editions.

ISBN 0-7876-5362-4 (vols. 1–17 set)
    0-7876-5777-8 (vol. 1)

This title is also available as an e-book.
ISBN 0-7876-7750-7 (17-vol set)
Contact your Gale sales representative for ordering information.

LIBRARY OF CONGRESS CATALOGING-IN-PUBLICATION DATA

Grzimek, Bernhard.
    [Tierleben. English]
    Grzimek's animal life encyclopedia.— 2nd ed.
        v. cm.
Includes bibliographical references.
Contents: v. 1. Lower metazoans and lesser deuterosomes / Neil Schlager, editor — v. 2. Protostomes / Neil Schlager, editor — v. 3. Insects / Neil Schlager, editor — v. 4-5. Fishes I-II / Neil Schlager, editor — v. 6. Amphibians / Neil Schlager, editor — v. 7. Reptiles / Neil Schlager, editor — v. 8-11. Birds I-IV / Donna Olendorf, editor — v. 12-16. Mammals I-V / Melissa C. McDade, editor — v. 17. Cumulative index / Melissa C. McDade, editor.
ISBN 0-7876-5362-4 (set hardcover : alk. paper)
    1. Zoology—Encyclopedias. I. Title: Animal life encyclopedia. II. Schlager, Neil, 1966- III. Olendorf, Donna IV. McDade, Melissa C. V. American Zoo and Aquarium Association. VI. Title.
QL7 .G7813 2004

590′.3—dc21
2002003351

Printed in Canada
10 9 8 7 6 5 4 3 2 1

Recommended citation: *Grzimek's Animal Life Encyclopedia,* 2nd edition. Volume 1, *Lower Metazoans and Lesser Deuterostomes,* edited by Michael Hutchins, Dennis A. Thoney, and Neil Schlager. Farmington Hills, MI: Gale Group, 2003.

AEH-0047

Contents

Contents

• • • • •

Foreword

Earth is teeming with life. No one knows exactly how many distinct organisms inhabit our planet, but more than 5 million different species of animals and plants could exist, ranging from microscopic algae and bacteria to gigantic elephants, redwood trees and blue whales. Yet, throughout this wonderful tapestry of living creatures, there runs a single thread: Deoxyribonucleic acid or DNA. The existence of DNA, an elegant, twisted organic molecule that is the building block of all life, is perhaps the best evidence that all living organisms on this planet share a common ancestry. Our ancient connection to the living world may drive our curiosity, and perhaps also explain our seemingly insatiable desire for information about animals and nature. Noted zoologist, E. O. Wilson, recently coined the term "biophilia" to describe this phenomenon. The term is derived from the Greek *bios* meaning "life" and *philos* meaning "love." Wilson argues that we are human because of our innate affinity to and interest in the other organisms with which we share our planet. They are, as he says, "the matrix in which the human mind originated and is permanently rooted." To put it simply and metaphorically, our love for nature flows in our blood and is deeply engrained in both our psyche and cultural traditions.

Our own personal awakenings to the natural world are as diverse as humanity itself. I spent my early childhood in rural Iowa where nature was an integral part of my life. My father and I spent many hours collecting, identifying and studying local insects, amphibians and reptiles. These experiences had a significant impact on my early intellectual and even spiritual development. One event I can recall most vividly. I had collected a cocoon in a field near my home in early spring. The large, silky capsule was attached to a stick. I brought the cocoon back to my room and placed it in a jar on top of my dresser. I remember waking one morning and, there, perched on the tip of the stick was a large moth, slowly moving its delicate, light green wings in the early morning sunlight. It took my breath away. To my inexperienced eyes, it was one of the most beautiful things I had ever seen. I knew it was a moth, but did not know which species. Upon closer examination, I noticed two moon-like markings on the wings and also noted that the wings had long "tails", much like the ubiquitous tiger swallow-tail butterflies that visited the lilac bush in our backyard. Not wanting to suffer my ignorance any longer, I reached immediately for my *Golden Guide to North American Insects* and searched through the section on moths and butterflies. It was a luna moth! My heart was pounding with the excitement of new knowledge as I ran to share the discovery with my parents.

I consider myself very fortunate to have made a living as a professional biologist and conservationist for the past 20 years. I've traveled to over 30 countries and six continents to study and photograph wildlife or to attend related conferences and meetings. Yet, each time I encounter a new and unusual animal or habitat my heart still races with the same excitement of my youth. If this is biophilia, then I certainly possess it, and it is my hope that others will experience it too. I am therefore extremely proud to have served as the series editor for the Gale Group's rewrite of *Grzimek's Animal Life Encyclopedia*, one of the best known and widely used reference works on the animal world. *Grzimek's* is a celebration of animals, a snapshot of our current knowledge of the Earth's incredible range of biological diversity. Although many other animal encyclopedias exist, *Grzimek's Animal Life Encyclopedia* remains unparalleled in its size and in the breadth of topics and organisms it covers.

The revision of these volumes could not come at a more opportune time. In fact, there is a desperate need for a deeper understanding and appreciation of our natural world. Many species are classified as threatened or endangered, and the situation is expected to get much worse before it gets better. Species extinction has always been part of the evolutionary history of life; some organisms adapt to changing circumstances and some do not. However, the current rate of species loss is now estimated to be 1,000–10,000 times the normal "background" rate of extinction since life began on Earth some 4 billion years ago. The primary factor responsible for this decline in biological diversity is the exponential growth of human populations, combined with peoples' unsustainable appetite for natural resources, such as land, water, minerals, oil, and timber. The world's human population now exceeds 6 billion, and even though the average birth rate has begun to decline, most demographers believe that the global human population will reach 8–10 billion in the next 50 years. Much of this projected growth will occur in developing countries in Central and South America, Asia and Africa—regions that are rich in unique biological diversity.

Finding solutions to conservation challenges will not be easy in today's human-dominated world. A growing number of people live in urban settings and are becoming increasingly isolated from nature. They "hunt" in supermarkets and malls, live in apartments and houses, spend their time watching television and searching the World Wide Web. Children and adults must be taught to value biological diversity and the habitats that support it. Education is of prime importance now while we still have time to respond to the impending crisis. There still exist in many parts of the world large numbers of biological "hotspots"—places that are relatively unaffected by humans and which still contain a rich store of their original animal and plant life. These living repositories, along with selected populations of animals and plants held in professionally managed zoos, aquariums and botanical gardens, could provide the basis for restoring the planet's biological wealth and ecological health. This encyclopedia and the collective knowledge it represents can assist in educating people about animals and their ecological and cultural significance. Perhaps it will also assist others in making deeper connections to nature and spreading biophilia. Information on the conservation status, threats and efforts to preserve various species have been integrated into this revision. We have also included information on the cultural significance of animals, including their roles in art and religion.

It was over 30 years ago that Dr. Bernhard Grzimek, then director of the Frankfurt Zoo in Frankfurt, Germany, edited the first edition of *Grzimek's Animal Life Encyclopedia*. Dr. Grzimek was among the world's best known zoo directors and conservationists. He was a prolific author, publishing nine books. Among his contributions were: *Serengeti Shall Not Die*, *Rhinos Belong to Everybody* and *He and I and the Elephants*. Dr. Grzimek's career was remarkable. He was one of the first modern zoo or aquarium directors to understand the importance of zoo involvement in *in situ* conservation, that is, of their role in preserving wildlife in nature. During his tenure, Frankfurt Zoo became one of the leading western advocates and supporters of wildlife conservation in East Africa. Dr. Grzimek served as a Trustee of the National Parks Board of Uganda and Tanzania and assisted in the development of several protected areas. The film he made with his son Michael, *Serengeti Shall Not Die*, won the 1959 Oscar for best documentary.

Professor Grzimek has recently been criticized by some for his failure to consider the human element in wildlife conservation. He once wrote: "A national park must remain a primordial wilderness to be effective. No men, not even native ones, should live inside its borders." Such ideas, although considered politically incorrect by many, may in retrospect actually prove to be true. Human populations throughout Africa continue to grow exponentially, forcing wildlife into small islands of natural habitat surrounded by a sea of humanity. The illegal commercial bushmeat trade—the hunting of endangered wild animals for large scale human consumption—is pushing many species, including our closest relatives, the gorillas, bonobos and chimpanzees, to the brink of extinction. The trade is driven by widespread poverty and lack of economic alternatives. In order for some species to survive it will be necessary, as Grzimek suggested, to establish and enforce

a system of protected areas where wildlife can roam free from exploitation of any kind.

While it is clear that modern conservation must take the needs of both wildlife and people into consideration, what will the quality of human life be if the collective impact of short-term economic decisions is allowed to drive wildlife populations into irreversible extinction? Many rural populations living in areas of high biodiversity are dependent on wild animals as their major source of protein. In addition, wildlife tourism is the primary source of foreign currency in many developing countries and is critical to their financial and social stability. When this source of protein and income is gone, what will become of the local people? The loss of species is not only a conservation disaster; it also has the potential to be a human tragedy of immense proportions. Protected areas, such as national parks, and regulated hunting in areas outside of parks are the only solutions. What critics do not realize is that the fate of wildlife and people in developing countries is closely intertwined. Forests and savannas emptied of wildlife will result in hungry, desperate people, and will, in the long-term lead to extreme poverty and social instability. Dr. Grzimek's early contributions to conservation should be recognized, not only as benefiting wildlife, but as benefiting local people as well.

Dr. Grzimek's hope in publishing his *Animal Life Encyclopedia* was that it would "...disseminate knowledge of the animals and love for them", so that future generations would "...have an opportunity to live together with the great diversity of these magnificent creatures." As stated above, our goals in producing this updated and revised edition are similar. However, our challenges in producing this encyclopedia were more formidable. The volume of knowledge to be summarized is certainly much greater in the twenty-first century than it was in the 1970's and 80's. Scientists, both professional and amateur, have learned and published a great deal about the animal kingdom in the past three decades, and our understanding of biological and ecological theory has also progressed. Perhaps our greatest hurdle in producing this revision was to include the new information, while at the same time retaining some of the characteristics that have made *Grzimek's Animal Life Encyclopedia* so popular. We have therefore strived to retain the series' narrative style, while giving the information more organizational structure. Unlike the original *Grzimek's*, this updated version organizes information under specific topic areas, such as reproduction, behavior, ecology and so forth. In addition, the basic organizational structure is generally consistent from one volume to the next, regardless of the animal groups covered. This should make it easier for users to locate information more quickly and efficiently. Like the original Grzimek's, we have done our best to avoid any overly technical language that would make the work difficult to understand by non-biologists. When certain technical expressions were necessary, we have included explanations or clarifications.

Considering the vast array of knowledge that such a work represents, it would be impossible for any one zoologist to have completed these volumes. We have therefore sought specialists from various disciplines to write the sections with which they are most familiar. As with the original *Grzimek's*, we have engaged the best scholars available to serve as topic editors, writers, and consultants. There were some complaints about inaccuracies in the original English version that may have been due to mistakes or misinterpretation during the complicated translation process. However, unlike the original *Grzimek's*, which was translated from German, this revision has been completely re-written by English-speaking scientists. This work was truly a cooperative endeavor, and I thank all of those dedicated individuals who have written, edited, consulted, drawn, photographed, or contributed to its production in any way. The names of the topic editors, authors, and illustrators are presented in the list of contributors in each individual volume.

The overall structure of this reference work is based on the classification of animals into naturally related groups, a discipline known as taxonomy or biosystematics. Taxonomy is the science through which various organisms are discovered, identified, described, named, classified and catalogued. It should be noted that in preparing this volume we adopted what might be termed a conservative approach, relying primarily on traditional animal classification schemes. Taxonomy has always been a volatile field, with frequent arguments over the naming of or evolutionary relationships between various organisms. The advent of DNA fingerprinting and other advanced biochemical techniques has revolutionized the field and, not unexpectedly, has produced both advances and confusion. In producing these volumes, we have consulted with specialists to obtain the most up-to-date information possible, but knowing that new findings may result in changes at any time. When scientific controversy over the classification of a particular animal or group of animals existed, we did our best to point this out in the text.

Readers should note that it was impossible to include as much detail on some animal groups as was provided on others. For example, the marine and freshwater fish, with vast numbers of orders, families, and species, did not receive as detailed a treatment as did the birds and mammals. Due to practical and financial considerations, the publishers could provide only so much space for each animal group. In such cases, it was impossible to provide more than a broad overview and to feature a few selected examples for the purposes of illustration. To help compensate, we have provided a few key bibliographic references in each section to aid those interested in learning more. This is a common limitation in all reference works, but *Grzimek's Encyclopedia of Animal Life* is still the most comprehensive work of its kind.

I am indebted to the Gale Group, Inc. and Senior Editor Donna Olendorf for selecting me as Series Editor for this project. It was an honor to follow in the footsteps of Dr. Grzimek and to play a key role in the revision that still bears his name. *Grzimek's Animal Life Encyclopedia* is being published by the Gale Group, Inc. in affiliation with my employer, the American Zoo and Aquarium Association (AZA), and I would like to thank AZA Executive Director, Sydney J. Butler; AZA Past-President Ted Beattie (John G. Shedd Aquarium, Chicago, IL); and current AZA President, John Lewis (John Ball Zoological Garden, Grand Rapids, MI), for approving my participation. I would also like to thank AZA Conservation and Science Department Program Assistant, Michael Souza, for his assistance during the project. The AZA is a professional membership association, representing 215 accredited zoological parks and aquariums in North America. As Director/ William Conway Chair, AZA Department of Conservation and Science, I feel that I am a philosophical descendant of Dr. Grzimek, whose many works I have collected and read. The zoo and aquarium profession has come a long way since the 1970s, due, in part, to innovative thinkers such as Dr. Grzimek. I hope this latest revision of his work will continue his extraordinary legacy.

Silver Spring, Maryland, 2001
Michael Hutchins
Series Editor

· · · · ·

How to use this book

Grzimek's Animal Life Encyclopedia is an internationally prominent scientific reference compilation, first published in German in the late 1960s, under the editorship of zoologist Bernhard Grzimek (1909–1987). In a cooperative effort between Gale and the American Zoo and Aquarium Association, the series has been completely revised and updated for the first time in over 30 years. Gale expanded the series from 13 to 17 volumes, commissioned new color paintings, and updated the information so as to make the set easier to use. The order of revisions is:

Volumes 8–11: Birds I–IV
Volume 6: Amphibians
Volume 7: Reptiles
Volumes 4–5: Fishes I–II
Volumes 12–16: Mammals I–V
Volume 3: Insects
Volume 2: Protostomes
Volume 1: Lower Metazoans and Lesser Deuterostomes
Volume 17: Cumulative Index

Organized by taxonomy

The overall structure of this reference work is based on the classification of animals into naturally related groups, a discipline known as taxonomy—the science in which various organisms are discovered, identified, described, named, classified, and catalogued. Starting with the simplest life forms, the lower metazoans and lesser deuterostomes, in volume 1, the series progresses through the more complex classes of animals, culminating with the mammals in volumes 12–16. Volume 17 is a stand-alone cumulative index.

Organization of chapters within each volume reinforces the taxonomic hierarchy. In the case of the volume on Lower Metazoans and Lesser Deuterostomes, introductory chapters describe general characteristics of all organisms in these groups, followed by taxonomic chapters dedicated to Phylum or Class. Species accounts appear at the end of the taxonomic chapters. To help the reader grasp the scientific arrangement, each type of chapter has a distinctive color and symbol:

■ = Phylum Chapter (lavender background)

◆ = Class Chapter (peach background)

Introductory chapters have a loose structure, reminiscent of the first edition. Chapters on taxonomic groups, by contrast, are highly structured, following a prescribed format of standard rubrics that make information easy to find. These chapters typically include:

Opening section
  Scientific name
  Common name
  Phylum
  Class (if applicable)
  Number of families
  Thumbnail description

Main chapter
  Evolution and systematics
  Physical characteristics
  Distribution
  Habitat
  Behavior
  Feeding ecology and diet
  Reproductive biology
  Conservation status
  Significance to humans

Species accounts
  Common name
  Scientific name
  Order
  Family
  Taxonomy
  Other common names
  Physical characteristics
  Distribution
  Habitat
  Behavior
  Feeding ecology and diet
  Reproductive biology
  Conservation status
  Significance to humans

Resources
  Books
  Periodicals
  Organizations
  Other

Color graphics enhance understanding

Grzimek's features approximately 3,000 color photos, including nearly 110 in the Lower Metazoans and Lesser Deuterostomes volume; 3,500 total color maps, including approximately 130 in the Lower Metazoans and Lesser Deuterostomes volume; and approximately 5,500 total color illustrations, including approximately 350 in the Lower Metazoans and Lesser Deuterostomes volume. Each featured species of animal is accompanied by both a distribution map and an illustration.

All maps in *Grzimek's* were created specifically for the project by XNR Productions. Distribution information was provided by expert contributors and, if necessary, further researched at the University of Michigan Zoological Museum library. Maps are intended to show broad distribution, not definitive ranges.

All the color illustrations in *Grzimek's* were created specifically for the project by Michigan Science Art. Expert contributors recommended the species to be illustrated and provided feedback to the artists, who supplemented this information with authoritative references and animal specimens from the University of Michigan Zoological Museum library. In addition to illustrations of species, *Grzimek's* features drawings that illustrate characteristic traits and behaviors.

About the contributors

Virtually all of the chapters were written by scientists who are specialists on specific subjects and/or taxonomic groups. Dennis A. Thoney reviewed the completed chapters to insure consistency and accuracy.

Standards employed

In preparing the volume on Lower Metazoans and Lesser Deuterostomes, the editors relied primarily on the taxonomic structure outlined in *Invertebrates*, edited by R. C. Brusca, and G. J. Brusca (1990). Systematics is a dynamic discipline in that new species are being discovered continuously, and new techniques (e.g., DNA sequencing) frequently result in changes in the hypothesized evolutionary relationships among various organisms. Consequently, controversy often exists regarding classification of a particular animal or group of animals; such differences are mentioned in the text.

Grzimek's has been designed with ready reference in mind, and the editors have standardized information wherever feasible. For **Conservation Status**, *Grzimek's* follows the IUCN Red List system, developed by its Species Survival Commission. The Red List provides the world's most comprehensive inventory of the global conservation status of plants and animals. Using a set of criteria to evaluate extinction risk, the IUCN recognizes the following categories: Extinct, Extinct in the Wild, Critically Endangered, Endangered, Vulnerable, Conservation Dependent, Near Threatened, Least Concern, and Data Deficient. For a complete explanation of each category, visit the IUCN Web page at <http://www.iucn.org /themes/ssc/redlists/categor.htm>.

In addition to IUCN ratings, chapters may contain other conservation information, such as a species' inclusion on one of three Convention on International Trade in Endangered Species (CITES) appendices. Adopted in 1975, CITES is a global treaty whose focus is the protection of plant and animal species from unregulated international trade.

In the species accounts throughout the volume, the editors have attempted to provide common names not only in English but also in French, German, Spanish, and local dialects..

Grzimek's provides the following standard information on lineage in the **Taxonomy** rubric of each Species account: [First described as] *Actinia xanthogrammica* [by] Brandt, [in] 1835, [based on a specimen from] Sitka, Alaska. The person's name and date refer to earliest identification of a species. If the species was originally described with a different scientific name, the researcher's name and the date are in parentheses.

Readers should note that within chapters, species accounts are organized alphabetically by order name, then by family, and then by genus and species.

Anatomical illustrations

While the encyclopedia attempts to minimize scientific jargon, readers will encounter numerous technical terms related to anatomy and physiology throughout the volume. To assist readers in placing physiological terms in their proper context, we have created a number of detailed anatomical drawings that are found within the particular taxonomic chapters to which they relate. Readers are urged to make heavy use of these drawings. In addition, many anatomical terms are defined in the **Glossary** at the back of the book.

Appendices and index

In addition to the main text and the aforementioned **Glossary**, the volume contains numerous other elements. **For further reading** directs readers to additional sources of information about lower metazoans and lesser deuterostomes. Valuable contact information for **Organizations** is also included in an appendix. An exhaustive **Lower Metazoans and Lesser Deuterostomes order list** records all orders of lower metazoans and lesser deuterostomes as recognized by the editors and contributors of the volume. And a full-color **Geologic time scale** helps readers understand prehistoric time periods. Additionally, the volume contains a **Subject index.**

Acknowledgements

Gale would like to thank several individuals for their important contributions to the volume. Dr. Dennis A. Thoney, topic editor for the Lower Metazoans and Lesser Deuterostomes volume, oversaw all phases of the volume, including creation of the topic list, chapter review, and compilation of the appendices. Neil Schlager, project manager for the Lower Metazoans and Lesser Deuterostomes volume, and Vanessa

Torrado-Caputo, associate editor at Schlager Group, coordinated the writing and editing of the text. Dr. Michael Hutchins, chief consulting editor for the series, and Michael Souza, program assistant, Department of Conservation and Science at the American Zoo and Aquarium Association, provided valuable input and research support.

• • • • •

Advisory boards

Series advisor

Michael Hutchins, PhD
Director of Conservation and Science/
William Conway Chair
American Zoo and Aquarium Association
Silver Spring, Maryland

Subject advisors

Volume 1: Lower Metazoans and Lesser Deuterostomes

Dennis A. Thoney, PhD
Director, Marine Laboratory & Facilities
Humboldt State University
Arcata, California

Volume 2: Protostomes

Sean F. Craig, PhD
Assistant Professor, Department of Biological Sciences
Humboldt State University
Arcata, California

Dennis A. Thoney, PhD
Director, Marine Laboratory & Facilities
Humboldt State University
Arcata, California

Volume 3: Insects

Arthur V. Evans, DSc
Research Associate, Department of Entomology
Smithsonian Institution
Washington, DC

Rosser W. Garrison, PhD
Research Associate, Department of Entomology
Natural History Museum
Los Angeles, California

Volumes 4–5: Fishes I– II

Paul V. Loiselle, PhD
Curator, Freshwater Fishes
New York Aquarium
Brooklyn, New York

Dennis A. Thoney, PhD
Director, Marine Laboratory & Facilities
Humboldt State University
Arcata, California

Volume 6: Amphibians

William E. Duellman, PhD
Curator of Herpetology Emeritus
Natural History Museum and Biodiversity
Research Center
University of Kansas
Lawrence, Kansas

Volume 7: Reptiles

James B. Murphy, DSc
Smithsonian Research Associate
Department of Herpetology
National Zoological Park
Washington, DC

Volumes 8–11: Birds I–IV

Walter J. Bock, PhD
Permanent secretary, International
Ornithological Congress
Professor of Evolutionary Biology
Department of Biological Sciences,
Columbia University
New York, New York

Jerome A. Jackson, PhD
Program Director, Whitaker Center for Science,
Mathematics, and Technology Education
Florida Gulf Coast University
Ft. Myers, Florida

Volumes 12–16: Mammals I–V

Valerius Geist, PhD
Professor Emeritus of Environmental Science
University of Calgary
Calgary, Alberta
Canada

Contributing writers

Charles I. Abramson, PhD
Oklahoma State University
Stillwater, Oklahoma

Andrey Adrianov, PhD
Russian Academy of Sciences
Far East Branch
Institute of Marine Biology
Vladivostok, Russia

Eliane Pintor Arruda, MSc
Universidade Estadual de Campinas
Campinas, Brazil

William Arthur Atkins
Atkins Research and Consulting
Normal, Illinois

A. N. Baker, PhD
Tasmacetus Inc.
Kerikeri, New Zealand

Ferdinando Boero
Università di Lecce
Lecce, Italy

Jean Bouillon
Oignies, Viroinval, Belgium

Graham Budd, PhD
Institute of Earth Sciences
Uppsala, Sweden

Isaure de Buron, PhD
College of Charleston
Charleston, South Carolina

David Bruce Conn, PhD
Berry College
Mount Berry, Georgia

Vincent A. Connors, PhD
University of South Carolina
at Spartanburg
Spartanburg, South Carolina

Kevin F. Fitzgerald, BS
Independent Science Writer

Scott C. France, PhD
College of Charleston
Charleston, South Carolina

Steven Mark Freeman, PhD
ABP Marine Environmental
Research Ltd.
Southampton, United Kingdom

Hidetaka Furuya, PhD
Osaka University
Osaka, Japan

Nobuhiro Fusetani, PhD
The University of Tokyo
Tokyo, Japan

Boyko B. Georgiev, PhD
Parasite Biodiversity Group
Bulgarian Academy of Sciences
Sofia, Bulgaria

Ben Hanelt, PhD
Louisiana State University
Baton Rouge, Louisiana

Iben Heiner, MSc
Zoological Museum
University of Copenhagen
Copenhagen, Denmark

Rick Hochberg, PhD
Smithsonian Marine Station at
Fort Pierce
Fort Pierce, Florida

Tohru Iseto, PhD
The Kyoto University Museum
Kyoto, Japan

Graham Kearn, PhD, DSc, FLS
University of East Anglia
Norwich, United Kingdom

Alexander M. Kerr, PhD
University of Guam
Mangilao, Guam

Reinhardt Møbjerg Kristensen,
PhD
Zoological Museum
University of Copenhagen
Copenhagen, Denmark

Sally P. Leys, PhD
The University of Alberta
Edmonton, Alberta, Canada

Fábio Sá MacCord, MSc
Universidade Estadual
de Campinas
Campinas, Brazil

George I. Matsumoto, PhD
Monterey Bay Aquarium
Research Institute
Monterey, California

Leslie Ann Mertz, PhD
Wayne State University
Detroit, Michigan

Surindar Paracer, PhD
Worcester State College
Worcester, Massachusetts

Annelies C. Pierrot-Bults, PhD
Institution and Zoological
Museum Amsterdam
University of Amsterdam
Amsterdam, The Netherlands

Jennifer E. Purcell, PhD
Shannon Point Marine Center
Anacortes, Washington

Henry M. Reiswig, PhD
Royal British Columbia Museum
University of Victoria
Victoria, British Columbia, Canada

Alexandra Eliane Rizzo, PhD
Universidade Estadual de Campinas
Campinas, Brazil

Karen Sanamyan, PhD
Russian Academy of Sciences
Far East Branch
Kamchatka Branch of Pacific Institute
of Geography
Kamchatka, Russia

Michael S. Schaadt, MS
Cabrillo Marine Aquarium
San Pedro, California

Rob Sherlock, PhD
Monterey Bay Aquarium
Research Institute
Monterey, California

Mattias Sköld, PhD
Institute of Marine Research
National Board of Fisheries
Lysekil, Sweden

Martin Vinther Sørensen, PhD
Zoological Museum
University of Copenhagen
Copenhagen, Denmark

Thomas Günther Stach, PhD
Royal Swedish Academy
of Sciences
Kristinebergs Marina
Forskningsstation
Fiskebäckskil, Sweden

Wolfgang Sterrer, PhD
Bermuda National History Museum
Flatts, Bermuda

Per A. Sundberg, PhD
Göteborg University
Göteborg, Sweden

Dennis A. Thoney, PhD
Humboldt State University
Arcata, California

Les Watling, PhD
University of Maine
Darling Marine Center
Walpole, Maine

Wallie H. de Weerdt, PhD
Zoological Museum Amsterdam
University of Amsterdam
Amsterdam, The Netherlands

Gert Wörheide, PhD
Geoscience Centre
Göttingen, Germany

Contributing illustrators

Drawings by Michigan Science Art

Joseph E. Trumpey, Director, AB, MFA
Science Illustration, School of Art and Design,
University of Michigan

Wendy Baker, ADN, BFA

Ryan Burkhalter, BFA, MFA

Brian Cressman, BFA, MFA

Emily S. Damstra, BFA, MFA

Maggie Dongvillo, BFA

Barbara Duperron, BFA, MFA

Jarrod Erdody, BA, MFA

Dan Erickson, BA, MS

Patricia Ferrer, AB, BFA, MFA

George Starr Hammond, BA, MS, PhD

Gillian Harris, BA

Jonathan Higgins, BFA, MFA

Amanda Humphrey, BFA

Emilia Kwiatkowski, BS, BFA

Jacqueline Mahannah, BFA, MFA

John Megahan, BA, BS, MS

Michelle L. Meneghini, BFA, MFA

Katie Nealis, BFA

Laura E. Pabst, BFA

Amanda Smith, BFA, MFA

Christina St.Clair, BFA

Bruce D. Worden, BFA

Kristen Workman, BFA, MFA

Thanks are due to the University of Michigan, Museum of Zoology, which provided specimens that served as models for the images.

Maps by XNR Productions

Paul Exner, Chief Cartographer
XNR Productions, Madison, WI

Tanya Buckingham

Jon Daugherity

Laura Exner

Andy Grosvold

Cory Johnson

Paula Robbins

• • • • •

Topic overviews

What are lower metazoans
and lesser deuterostomes?

Evolution and systematics

Reproduction, development, and life history

Ecology

Symbiosis

Behavior

Lower metazoans,
lesser deuterostomes, and humans

Conservation

What are lower metazoans and lesser deuterostomes?

Introduction

The Metazoa, today taken as a synonym of Animalia, comprises a large grouping of organisms that may be characterized as being multicellular and heterotrophic, i.e., they do not synthesize their own food, but obtain it from external sources. While there has been some debate in the past, it now seems overwhelmingly likely that the metazoans are monophyletic, and thus that all living examples are descended from an animal that lived some time in the Proterozoic (probably at least 600 million years ago [mya]). Even this statement is controversial, however. Molecular evidence suggests that metazoans had emerged up to one billion years ago or more; but the fossil record is most reasonably read as implying a much later origin, with definitive metazoans not appearing before 600 mya, and perhaps even later. Unfortunately, most "lower" metazoans today lack substantial hard parts such as mineralized shells, so their fossil record is correspondingly very poor. Telling the true time of appearance of such animals just from the fossil record may therefore well be inaccurate.

Animal evolution

Because of their long history and enormous adaptability, animals are organized in a remarkable number of different ways, ranging from simple sponges with only a few cell types through to the vertebrates with their complex nervous and immune systems. Indeed, it is possible to arrange animals in a broad series, from organisms that do not possess true tissues (e.g., sponges), through organisms with tissues but no organs (e.g., cnidarians), into the bilaterally symmetrical animals (the Bilateria) with both. The Bilateria typically also possess a central nervous system and muscles; some of them are segmented; and some possess a body cavity called a coelom.

The evolution of the animals has long been a contentious issue that has generated a huge number of theories. Nevertheless, at the heart of the issue is whether or not the organizational gradient that can be erected tells anything at all about animal evolution, or whether it merely reflects different adaptive needs of each organism. To put it more simply: are all the simple animals more basal than the more complex ones, or is animal evolution less tidy than that? The traditional assumption has been that organization is indeed a re-

flection of animal relationships and evolution, although the more thoughtful authors have refrained from definitively stating this. In this view, it makes sense to talk about animal evolution being a more or less stately progress from simple to complex; thus, one can label the simple animals, which are at the bottom of the tree, the "lower" metazoans, and the more advanced ones, the "higher" metazoans. To be more precise, animals without coeloms or segments are typically thought of as being "lower." These sorts of organisms show a variety of functional adaptations. Sponges and cnidarians typically have some sort of central fluid-filled cavity, which is critical to many roles including support, nutrition, excretion, and reproduction. Small bilaterians, on the other hand, have typically no need of any such system, as they are small enough to

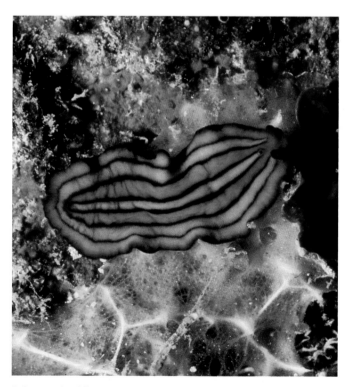

A three striped flatworm (*Pseudoceros tristriatus*) showing aposematic coloration. (Photo by ©A. Flowers & L. Newman/Photo Researchers, Inc. Reproduced by permission.)

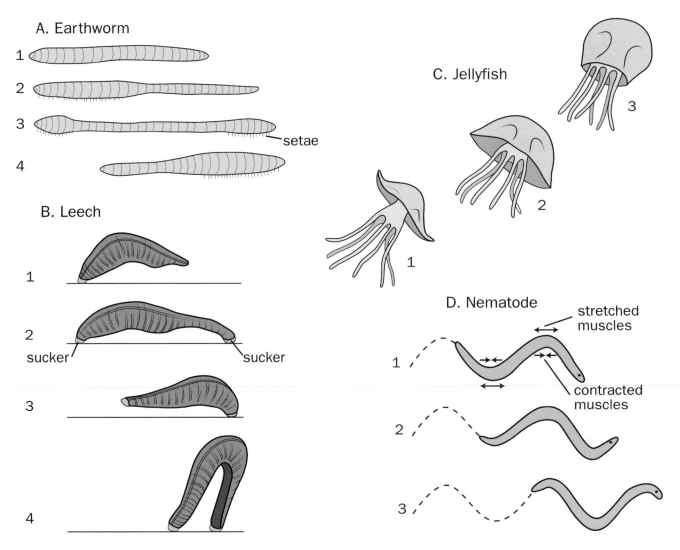

Locomotion in different animals: A. Earthworm (a protostome); B. Leech (a protostome); C. Jellyfish; D. Nematode. (Illustration by Patricia Ferrer)

allow diffusion directly to and from the body tissues. Although lower metazoans by the definition here lack a true coelom, which can be used as a hydrostatic skeleton, such a tack is replaced either by the use of a rather solid array of muscles, or by some other type of body cavity, such as the so-called pseudocoelom. It should be stressed that this "lower" terminology is a remnant of certain types of eighteenth century views of the world that are in many ways entirely inappropriate to the modern evolutionary ways of thinking about animals. Furthermore, modern systematic practice forbids the use of taxonomic units that are defined by exclusion—lower metazoans defined by being everything except the higher metazoans, which are usually taken as deuterostomes, arthropods, mollusks, and perhaps annelids, together with their close allies. Nevertheless, the central issue of the relationship of overall form to evolution remains unsolved, and, indeed, has been in hot contention since the last years of the twentieth century.

The debate has become sharp because of the introduction of entirely new sources of data that have bearing on the problem, i.e., evidence from molecules. Analysis of the nucleic acids allows a view of animal evolution that is completely, or largely, independent from that provided by classical morphological studies, and the results have sometimes been surprising. The sponges, with their relatively poorly organized morphology, remain basal within the tree, followed by the cnidarians (jellyfish, corals, and allies) and the ctenophores (the comb jellies), although the exact relationships between these three is contentious. All other animals fall into the Bilateria, but the relationships within this group remain highly debated. On a strictly "progressionist" view, the most basal bilaterians would be the flatworms, followed by animals that possess a body cavity that is not fully bounded by mesodermally derived epithelium (i.e., the coelom), followed by the coelomates themselves. However, this view, prevalent among scientists in Great Britain and the United States until a few years ago, was always rejected by many zoologists in Ger-

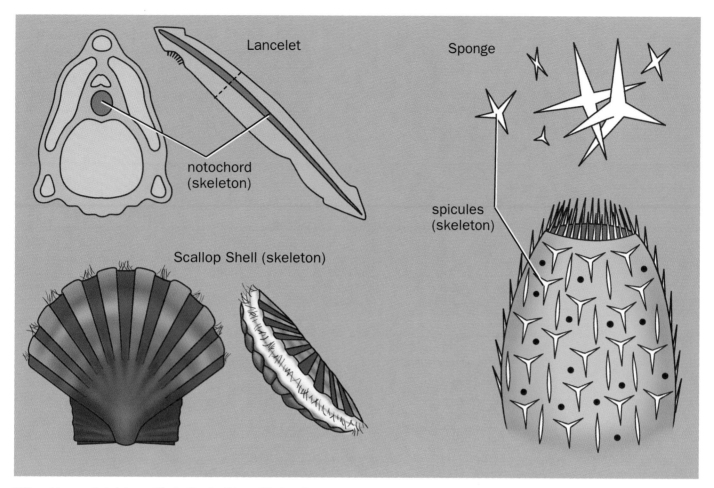

Different types of skeletons. (Illustration by Kristen Workman)

many and elsewhere. In their view, the basal metazoans already possessed a coelom and segments, and the bilaterians that lack these features have therefore lost them secondarily. Therefore, there are at least two radically differing views of what a "lower metazoan" is, at least as applied to the bilaterians: one, a relatively simple organism with no through-gut, blood vascular system, or coelom, and another with all of these features present.

The advent of molecular systematics has had a dramatic effect on the view of animal relationships, especially within the Bilateria. Two important features stand out. First, the group of lower metazoans referred to collectively as pseudocoelomates, possessing a body cavity that does not fall under the definition of the coelom, is seen to be a highly heteregeneous group. The most surprising aspect of their reassignment has been the proposal that some of them, notably the nematodes and priapulids, are close relatives to the arthropods (forming the Ecdysozoa), displacing the annelids that are traditionally placed in this position. Secondly, at least some of the flatworms have been largely displaced from the base of the tree to form a group loosely related to the annelids and mollusks in a remnant of one of the old branches of the bilaterians, the protostomes. Both these developments lend support to the idea that

the ancestral bilaterians were rather complex. Nevertheless, at least some of the flatworms, the acoels, have now been reinstated at the base of the tree. Are their simple features primitive for all of the bilaterians, or have they simply lost their

Sea squirts in the South Pacific. (Photo by ©Nancy Sefton/Photo Researchers, Inc. Reproduced by permission.)

A northern sea star (*Asterias vulgaris*) feeding on sea urchin. (Photo by ©Andrew J. Martinez/Photo Researchers, Inc. Reproduced by permission.)

laterians. Genes that control the layout of the complex morphology, including segments, eyes, and the development of the heart, are all highly conserved between the "higher" metazoans such as the arthropods and chordates. The implication is that these genes are present in the lower bilaterians (and indeed, this has been shown in some cases), and that they may originally have functioned as they do today—implying the very deep origin of the structures these genes now regulate. Such conclusions are controversial, and have by no means been accepted by all, especially morphologists.

Summary

The lower metazoans comprise animals that are thought to lie relatively close to the evolutionary root of animals as a whole. While they all have their own specializations, it is possible, and to be hoped for, that they retain some features that were characteristic of the very earliest stages of metazoan and bilaterian evolution. Although it is widely accepted that the sponges are basal within the animals, broadly followed by the ctenophores and cnidarians, the relationships of the bilaterians remain highly controversial. Molecular systematics places the bilaterians into three groups: the Deuterostomia (including, among others, echinoderms and chordates); the Ecdysozoa (principally arthropods and nematodes), and the Trochozoa (annelids, mollusks, most flatworms, and several other small groups, including the brachiopods). At least one group of the flatworms may still be basal within the bilaterians, but the implications this position may have for the evolution of the lower metazoans are presently unclear. The fundamental questions about bilaterian evolution, including the origin of the coelom and segmentation, remain unanswered, and, indeed, as vigorously contested as ever.

complex features, as have the other flatworms by implication of their higher position? One further line of evidence has come from the shared developmental mechanisms between the bi-

Resources

Books

Nielson, C. *Animal Evolution: Interrelationships of the Living Phyla.* Oxford: Oxford University Press, 2001.

Periodicals

Budd, G. E., and S. Jensen. "A Critical Reappraisal of the Fossil Record of the Bilaterian Phyla." *Biological Reviews* 75 (2000): 253–295.

Graham Budd, PhD

Evolution and systematics

Origin of lower Metazoa

The lower Metazoa comprises a diverse assemblage of animal phyla traditionally considered primitive by most biologists in the nineteenth and twentieth centuries. Despite claims about their primitive appearance (e.g., simple anatomy, small size), the lower Metazoa collectively displays some of the greatest morphological and developmental diversity within the Animalia. There is also mounting evidence to suggest that the primitiveness of many lower metazoans is a secondary phenomenon, i.e., the lack of complexity is interpreted as a loss of derived features of more complex ancestors. Moreover, many animals once thought to have simple anatomy are now regarded as morphologically complex by way of advanced techniques to explore their organ systems (electron microscopy, immunofluorescence, and molecular probes). Still, debate exists over the evolutionary relationships of many of the lower Metazoa, and a proper examination of metazoan origins is necessary to discern the basis of these arguments.

The evolutionary origin of the Metazoa has long been a source of major controversy, and several theories have been proposed to explain the phylogenetic jump from unicellular organism to multicellular animal. Ernst Haeckel (1874) was the first biologist to speculate that animals share a common ancestry with unicellular protists, and his subsequent theories on metazoan origins revolve around the perceived similarities between unicellular protists and the process of animal embryology. In Haeckel's 1866 penultimate work, in which he formulated his fundamental biogenetic law, he posits that embryonic development (ontogeny) recapitulates evolution (phylogeny), or rather, that the development of individuals (from zygote to larva) is a stepwise progression through adult ancestral forms. While the idea of recapitulation (one-to-one correspondence between ontogeny and phylogeny) has in the ensuing period been rejected, developmental characters still play an important role in reconstructing metazoan evolution. Two of Haeckel's theories on the origin of the Metazoa have relied almost exclusively on developmental characters and the biogenetic law: the cellularization theory and the colonial theory.

In Haeckel's cellularization theory, a ciliate-like ancestor became multicellular through a process of cellularization, i.e., cytokinesis that results in the synchronous formation of many cells. This organism, variously regarded as planula larva-like (phylum Cnidaria) or even an acoel flatworm (Platyhelminthes, Acoela), served as a model of the metazoan ancestor and was subsequently championed in numerous theories on metazoan evolution (e.g., planula or planuloid-acoeloid theory of von Graff in the 1880s and Hyman in the 1950s; the syncytial or ciliate-acoel theory of Hadzi in 1953 and Steinböck in 1963). The implications of these theories are threefold: first, the cnidarian planula larva or adult acoel flatworm represents the most primitive metazoan; second, that acoel flatworms are the link between Cnidaria and the rest of the animal kingdom (Bilateria); and third, that the acoelomate body organization is primitive within the Bilateria. Recent findings using ultrastructural and molecular sequence data have essentially disproved these theories.

Haeckel's second theory, and one that has received more widespread acceptance, is the colonial theory (1874), which states that the most primitive metazoans were derived from a colonial amoeba-like organism (synamoebium) that later became ciliated, similar in appearance to modern-day *Volvox* (phylum Chlorophyta). While protists such as *Volvox* are colonial flagellates, and not amoeba-like (phylum Rhizopoda) or ciliated (phylum Ciliophora), they nonetheless have a characteristic hollow, spherical appearance analogous to the blastula stage of metazoan embryos. Haeckel coined this metazoan precursor a "blastaea." Following his biogenetic law, Haeckel suggested the blastaea evolved into a "gastraea," a two-layered sac with two germ layers, corresponding to the gastrula stage of animal embryogeny. This theory is important for two reasons: first, it introduced comparative embryology into phylogenetic discussions that would later have prominence in the creation of Protostomia and Deuterostomia; and second, it presupposed the ancestral metazoan to have had primary radial symmetry, as is the case for some primitive metazoans (e.g., Cnidaria). Adding further support to this theory was the discovery of choanoflagellates (phylum Choanoflagellata), a group of protists that possesses a collar complex (a forwardly directed flagellum surrounded by an inverted cone-shaped collar of 30–40 retractile microvilli). The collar complex is interpreted as a synapomorphy uniting Choanoflagellata and Metazoa (=Animalia) because a homologous structure is present in basal metazoans (e.g., the choanocyte layer of sponges). In 1880, the British protozo-

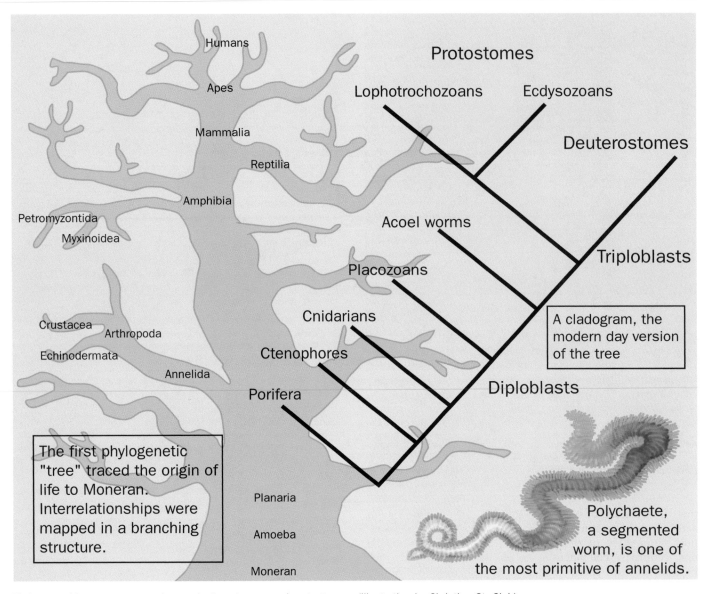

Humans
Apes
Mammalia
Reptilia
Amphibia
Petromyzontida
Myxinoidea
Crustacea
Arthropoda
Echinodermata
Annelida

Protostomes
Lophotrochozoans
Ecdysozoans
Deuterostomes
Acoel worms
Placozoans
Triploblasts
Cnidarians
Ctenophores

A cladogram, the modern day version of the tree

Porifera

Diploblasts

The first phylogenetic "tree" traced the origin of life to Moneran. Interrelationships were mapped in a branching structure.

Planaria
Amoeba
Moneran

Polychaete, a segmented worm, is one of the most primitive of annelids.

Phylogeny of lower metazoans, lesser deuterostomes, and protostomes. (Illustration by Christina St. Clair)

ologist Saville-Kent went so far as to consider the sponges (phylum Porifera) as colonial protists derived directly from choanoflagellates. The existence of colonial choanoflagellates such as *Proterospongia* (Greek: *protero* = earlier or former) has added fuel to this theory, and since this time, it has remained conventional wisdom that Choanoflagellata is the most likely sister-group of Metazoa.

The early radiation of the Metazoa still remains a major puzzle in animal evolution. The earliest known metazoan fossils were once thought to extend only as far back as the Cambrian Period, 540 million years ago (mya). The Cambrian marks a time when most of the major animal phyla first appear in the fossil record. The period of time over which much of this diversification appeared, approximately 30 million years, is relatively rapid compared to the known age of metazoan life (low estimate: 600 million years based on paleon-

tology; high estimate: 1,100 based on molecular clock). Consequently, it is often referred to as the Cambrian explosion. However, intriguing evidence suggests that there was a cryptic Precambrian metazoan history. Some of the most recent finds (Xiao et al., 1998) include the smallest known fossils, in the form of embryos and early larval stages, dating back to the Neoproterozoic (570 mya). These fossils clearly indicate the existence of a fauna prior to the Cambrian explosion, and perhaps allude to the presence of small, planktonic or interstitial organisms that did not readily fossilize. While molecular and fossil evidence verifies Precambrian existence, most of the larger animal phyla clearly diversified during the Cambrian. Some of the earliest molecular studies (Field et al., 1988; Christen et al., 1991) also provided evidence to suggest that the Metazoa was diphyletic, and that two subkingdoms, Radiata and Bilateria, arose independently from flagellated protozoa in the Precambrian. Both terms allude to the ani-

Artist's rendition of Precambrian life forms based on fossils from the Ediacara Hills of South Australia. Prominent in the diorama are jellyfish (*Ediacaria flindersi*), *Mawsonites spriggi*, *Kimberella quadrata*, sea pens (*Charniodiscus arboreus*), pink paddles (*Charniodiscus oppositus*), flat worms (*Dickensonia costata*), and algae. (Photo by ©Chase Studios, Inc/Photo Researchers, Inc. Reproduced by permission.)

mals' primary axis of symmetry, and though the terms remain active in the literature, there is conclusive evidence to support monophyly of the Metazoa; synapomorphies include multicellularity, cell junctions, collagen, gametic meiosis with haploid egg and sperm, sperm with acrosome, and fundamentally radial cleavage. These ground-pattern characteristics are still found in basal and derived metazoans, despite up to one billion years of animal evolution. As of recent times, most phyla are considered monophyletic, i.e., a clade consisting of an ancestral species and all its descendants and all sharing a characteristic combination of specific anatomical features (Nielsen, 2003).

Systematics of lower Metazoa

Molecular, morphological, and paleontological evidence suggests the most primitive phyla are found within two paraphyletic groups, Parazoa and Radiata. The Parazoa includes two phyla, Porifera and Placozoa, which display the most primitive grade of organization and are at the base of the evolutionary tree. The Radiata contains the radiate phyla, Cnidaria and Ctenophora, the first true metazoans prior to the evolution of the bilateral body plan in the Precambrian. Some enigmatic groups with questionable affinities to either Radiata or Bilateria include the Monoblastozoa, Orthonectida, and Rhombozoa, often included under the name Mesozoa. The

largest assemblage of animals is found within the Bilateria, a monophylum consisting of two large clades, Protostomia and Deuterostomia. Branching order within these clades is unresolved, though molecular evidence indicates the Protostomia may be subdivided into Ecdysozoa (molting animals) and Lophotrochozoa (ciliated animals). Lophotrochozoa contains several phyla, including the Lophophorata, a superphylum of Brachiopoda, Ectoprocta and Phoronida, still considered by many authors to be deuterostomes. Additionally, subordinate clades within the Protostomia define certain groups of lower Metazoa. These groups go by a variety of names, including Aschelminthes, Nemathelminthes, Gnathifera, Cycloneuralia, Introverta, and Cephalorhyncha. Nemathelminthes and Aschelminthes are older terms proposed in 1859 and 1910, respectively, to consume a variety of acoelomate and pseudocoelomate taxa; neither group is considered monophyletic.

The monophylum Gnathifera was erected in 1995 to include three phyla, Gnathostomulida, Rotifera, and Acanthocephala. The Micrognathozoa, discovered in 2002, is now included in this clade, and some biologists also place the Chaetognatha within Gnathifera. Cycloneuralia includes six acoelomate and pseudocoelomate taxa: Gastrotricha, Nematoda, Nematomorpha, Kinorhyncha, Loricifera, and Priapulida. The Introverta makes up a subset of Cycloneuralia, and Cephalorhyncha is yet a smaller subset of Introverta. Phyla of uncertain taxonomic affinity include the Entoprocta and

Cycliophora. Some biologists consider entoprocts closely related to ectoprocts or a sister taxon of Cycliophora. Within Deuterostomia, subordinate clades include Lophophorata (also considered protostomes), Neorenalia, and Cyrotreta. The animal, *Xenoturbella bocki*, is perhaps the most enigmatic bilaterian to date, being variously regarded as a species of protobranch bivalve (Mollusca), member of the Acoela (Platyhelminthes), relative of acorn worms (Enteropneusta) and Holothuroidea (Echinodermata), or as the most basal bilaterian; *Xenoturbella's* affinities remain undetermined.

The Parazoa (Greek: *para* = beside, *zoa* = animals, Metazoa) includes animals with the most primitive grade of organization: they lack definable nervous systems, true tissues, or organ systems, and epithelia do not rest on a basement membrane. Sponges (phylum Porifera) are the most basal parazoans. According to some researchers, sponges are little more than colonial choanoflagellates, but arguments can be made to classify sponges as more advanced than colonial, heterotrophic eurkaryotes such as *Proterospongia*. Sponges contain many more specialized cell types (including sex cells similar to other metazoans), have at least a rudimentary ability to form tissue-like epithelia via septate junctions (between choanocytes), and undergo an embryonic process of tissue formation (though probably not homologous to gastrulation of eumetazoa). The Porifera is a well-defined monophylum characterized by two synapomorphies, cellular totipotency and acquiferous systems of choanocyte chambers. Nearly all sponges are benthic, sessile suspension-feeders with the exception of *Asbestopluma*, a carnivorous demosponge. More than 5,000 living species of sponge are known globally, and many more fossil species exist, some dating as far back as 600 mya. Four sponge-like fossil groups are well known to have produced massive reef-like structures back in the early Paleozoic (550–480 mya): archaeocyathans, stromatoporids, sphinctozoans, and chaetetids.

Unlike Porifera, the phylum Placozoa has no known fossil record. Discovered in 1883, *Trichoplax adhaerens*, the only known member of Placozoa, is best characterized as a minute (0.07–0.11 in [2–3 mm]), flattened bag of epithelial cells with dorso-ventral polarity and no anterior-posterior polarity or symmetry. The outer epithelium is composed of monociliate cells (one cilium per cell) interconnected with belt desmosomes and perhaps septate junctions. The narrow space between epithelial layers is filled with a gel-like mesenchyme. Asexual and sexual reproduction is known, but details of sexual reproduction are incomplete. The existence of striated ciliary rootlets in the monociliate cells suggests an evolutionary link to eumetazoans, and the lack of autapomorphies defining the placozoan body plan suggests a placozoan-like ancestor may have given rise to the Eumetazoa.

Eumetazoa (Greek: *eu* = good) refers to the remaining animal groups, approximately 30 phyla. Eumetazoans share several synapomorphies, including a synaptic nervous system, gap junctions between cells, a basal lamina, striated myofibrils, organized gonads, definable germ layers (ectoderm, endoderm), and body symmetry. The Cnidaria is the most basal eumetazoan phylum and contains approximately 10,000 extant species and several thousand fossil species. The fossil record extends back to the Precambrian (540 mya), though many fossils, especially those of Ediacara and Burgess Shale, have questionable affinities to the Cnidaria. The phylum is defined by at least three synapomorphies: cnidae (nematocysts), planula larva, and coelenteron with mouth surrounded by tentacles. Other synapomorphies might characterize the group, depending on one's view of the interrelationships. The phylum is traditionally divided into four classes: Hydrozoa, Scyphozoa, Cubozoa, and Anthozoa. Body morphology is highly variable within the classes but is generally a variation on one or two basic body plans: polyp and medusa. Polyps are generally sessile, benthic forms with a ring of tentacles around the mouth, although some polyps have become colonial and taken up a pelagic existence (e.g., *Velella velella*). The medusa, or jellyfish, is often a flattened disc-shaped animal that lives in the water column, though some species have adopted a polypoid form and become benthic (*Haliclystus*). The polyp may be a sexual form that produces gametes (e.g., Anthozoa) or an asexual form that produces medusae (e.g., Cubozoa, Schyphozoa, Hydrozoa). Only polyps within Anthozoa form large reefs in mostly tropical seas. Interrelationships within the phylum are contentious.

Two prominent theories on the origin of Cnidaria have become established: the medusa theory and the polyp theory. The medusa theory postulates that the cnidarian ancestor was a planula-like animal that first gave rise to the medusoid form, subsequently followed by evolution of the polyp stage. Such a scenario places the trachyline hydrozoans as basal cnidarians, and therefore the medusoid body form would be a synapomorphy of the phylum. The polyp theory states that the polypoid form is ancestral, and therefore the Anthozoa is the most basal group within the phylum. In this scenario, Anthozoa is a paraphyletic taxon and the medusoid form is a synapomorphy of more derived clades within the Cnidaria. These theories have played important roles in the understanding of metazoan evolution, particularly regarding the origin of flatworms, the origin of the coelom, and the origin of mesodermal muscles (Rieger, 1986).

The triploblasts, or Triploblastica (Greek: *triplo* = three, *blast* = bud), refer to phyla that possess three embryonic germ layers (ectoderm, mesoderm, endoderm) formed during gastrulation, though mesoderm is probably not homologous between protostomes and deuterostomes. Prior to 1985, the phylum Ctenophora was generally considered a diploblastic phylum because true mesoderm (entomesoderm) was not found; however, subsequent studies have revealed the presence of mesoderm during embryogeny. Ctenophora is considered the most primitive of the triploblastic phyla because it is hypothesized to have originated prior to the evolution of the bilateral body plan. Ctenophora is a small phylum of approximately 80 gelatinous animals known as comb jellies and sea gooseberries. Their paleontological history extends back to the Cambrian. The phylum is well defined by the presence of adhesive colloblasts, eight rows of ciliary plates, a unique apical sense organ, and a cydippid larval stage. Interrelationships among the seven inclusive orders are not well resolved, and there is disagreement around which form of ctenophore is most primitive, the tentaculate or atentactulate form. The ctenophore body plan has played a historical role in various theories on metazoan evolution (e.g., ctenophore-polyclad theory, ctenophore-trochophore theory).

The Bilateria (Greek: *bi* = two, *latus* = side) is a well-defined monophylum characterized by a bilaterally symmetric body plan, a pronounced antero-posterior axis, and cephalization (brain). The phylum Platyhelminthes has figured prominently in nearly all discussions of the origin and relationships of the Bilateria. Originally classified within the taxon Vermes by Linnaeus in his 1735 book, *Systema Naturae*, the Platyhelminthes has since had a contentious history. In 1851 Vogt was first to isolate the phylum (with the nemerteans) into a single taxon called Platyelmia, later changed to Platyelminthes (now Platyhelminthes) by Gregenbaur in 1859. Platyhelminthes was subsequently erected to phylum-level status containing four classes: Turbellaria, Nemertea, Trematoda, and Cestoda. Minot (1876) later dropped Nemertea from this taxon. As of recent times, the Platyhelminthes contains close to 20,000 species of free-living and parasitic flatworms, with very little fossil record. The phylum is divided into three monophyletic clades: Acoelomorpha, Catenulida, and Rhabditophora. The terms Acoelomorpha and Rhabditophora are recent systematic additions to the platyhelminth literature (Ehlers, 1985). Acoelomorphs include the subclades (orders) Acoela and Nemertodermatida, and Rhabditophora includes all remaining groups (free living and parasitic), with the exception of Catenulida.

Historically, the Catenulida, both acoelomorph taxa, and all free-living groups of Rhabditophora (approximately nine orders) made up class Turbellaria; however, since recently, the Turbellaria is considered paraphyletic and there is little agreement on its precise composition because of the lack of well-defined synapomorphies. Turbellarian worms are best characterized as lower platyhelminths with a ciliated epidermis and most use cilia for locomotion. The remaining parasitic classes within the Rhabditophora are often referred to together as Neodermata (Latin: *neo* = new, *derm* = skin), in reference to their tegument. Class Trematoda historically contained all parasitic flukes (endo- and ectoparasites), but has subsequently been changed to include only the digenetic and aspidogastrean flukes. Class Monogenea contains only monogenetic flukes, and class Cestoda contains the tapeworms.

Relationships among all major clades and classes are unknown and highly speculative. Morphological and molecular data argue for a major restructuring of the Platyhelminthes, particularly in the removal of the taxon name Turbellaria, recognition of its (potentially) evolutionary independent clades, and a search for synapomorphies to unite the major taxa. Arguments based on molecular data, mostly sequence data in the form of 18S rDNA, assert a polyphyletic Platyhelminthes, with complete removal of the Acoelomorpha and its transference to a position at the base of the Bilateria. Morphologists recognize the uniqueness of the Acoelomorpha, but contend removal from the phylum is not yet warranted. They emphasize the search for shared, derived characters to unite the three main clades, and assume a cladistic stance that accentuates the similarities that unite taxa (synapomorphies), not the differences that separate them. Potential synapomorphies exist at different levels within the phylum. According to

Rieger and Ladurner (2001), the mode of epidermal replacement during growth or general maintenance appears to be unique to the Platyhelminthes. New epidermal cells arise from neoblasts below the body wall, instead of from dividing epidermal cells as occurs in all other metazoans. Tyler (2001) mentions the unregionalized sac-like gut as a potential autapomorphy of the phylum, and the structure of the hermaphroditic reproductive system and presence of biflagellate spermatozoa may be synapomorphies of Acoelomorpha and Rhabditophora. Within the Rhabditophora, the exclusively parasitic classes, Trematoda, Monogenea, and Cestoda, are related by the shared presence of a syncytial tegument that arises through the developmental replacement of the larval epidermis. Relationships among the smaller clades of free-living and symbiotic "turbellarians" are still a work in progress. The phylogenetic position of the Platyhelminthes remains unknown and will likely prevail as the main topic of discussion among biologists interested in the origins and early radiation of the Metazoa. Enigmatic animals such as *Xenoturbella* and the turbellariomorph annelid, *Lobatocerebrum*, along with enteropneust models proposed by Tyler (2001), may also figure prominently in understanding platyhelminth evolution.

The phylum Nemertea is a small group of approximately 900 predatory worms occupying marine, freshwater, and terrestrial habitats. There is no reliable fossil record. Nemerteans have been known since 1758, but it was not until 1876 that they were recognized as a group distinct from the turbellarian flatworms. The monophyletic nature of Nemertea is confirmed by the presence of their unique proboscis apparatus, but other synapomorphies may exist depending on their accepted theory of origin. Two classes exist, Anopla and Enopla, based on the structure of the proboscis and the position of the proboscis pore. The Anopla is generally considered primitive to the Enopla and may be a paraphyletic taxon since the defining features are symplesiomorphies. Relationships to other metazoans are less well known. Spiral cleavage and the presence of a true coelom place the nemerteans within the Protostomia, but questions have been raised as to the homology of the nemertean coelom (rhynchocoel, bloodvessel system) with other protostomes. The traditional placement of Nemertea close to the Platyhelminthes is often favored because both groups are functionally acoelomate, though no synapomorphies are evident.

The clade Gnathifera contains four phyla defined by the ultrastructure of jaw-like elements in the mouth, though the orientation and number of elements are phylum specific. Historically, Rotifera is the best-known group of gnathiferans, containing approximately 2,000 species of exclusively microscopic, pseudocoelomate animals with a ciliated wheel-like organ on their head (Latin: *rota* = wheel, *fera* = bearer). The fossil record is sparse and only dates back to the Eocene epoch (54 mya) where animals are preserved in Dominican amber. The phylum is monophyletic and characterized by the ciliated corona, a retrocerebral organ, and unique jaw structure. Some researchers prefer to include the phylum Acanthocephala within the Rotifera based on the shared presence of an intracytoplasmic lamina and anterior flagellum on the sperm. The name Syndermata is often used to refer to both phyla together. Acanthocephala consists of approximately

Middle Silurian crinoids (*Calliocrinus cronutus*) based on fossils from the Racine Dolomite of the Chicago-Milwaukee area. Although they resemble flowers, crinoids are animals related to starfish. With a net of delicate, feathery arms splayed out in the current, crinoids were able to trap microscopic food particles. The food was transported down the arms into the mouth, centrally located on the bulb-like calyx. The long stem-like column served to lift the animal above the sea floor, exposing it to currents for feeding. (Photo by ©Chase Studios, Inc/Photo Researchers, Inc. Reproduced by permission.)

jaw structure. A fossil record is lacking. Micrognathozoa is the most recent group of gnathiferans to be discovered, described by Reinhardt Kristensen and Peter Funch in 2000. The group consists of microscopic, ciliated, jawed worms found in a cold spring at Disko Island, Greenland. Only a single species has been described, *Limnognathia maerski*. The authors chose to treat Micrognathozoa as a class within the phylum Gnathifera, relegating the phyla Rotifera, Acanthocephala, and Gnathostomulida to classes within Gnathifera. Gnathostomulida is the sister group to the clade containing Micrognathozoa, Acanthocephala, and Rotifera.

Cycloneuralia is a morphologically defined clade containing Gastrotricha, Nematoda, Nematomorpha, Priapula, Kinorhyncha, and Loricifera. Molecular evidence indicates the clade is paraphyletic, but several potential synapomorphies exist: terminal mouth, radial pharynx, and tripartite peripharyngeal brain. The phylum Gastrotricha is the most primitive group of cycloneuralians, and the one often considered having the weakest affiliation with the clade. Gastrotrichs are microscopic (0.003–0.11 in [0.1–3 mm]) ciliated worms that bear cuticular adhesive organs. There is no fossil record. Synapomorphies defining the small phylum of approximately 600 species include a multilayered cuticle that covers the cilia, tube-like dual-gland adhesive organs, the unique structure of their sensory cilia, and helicoidal muscles that surround the digestive tract. The two orders within the phylum are monophyletic. Relationships of gastrotrichs to specific clades within the Protostomia are uncertain; the possession of both a cuticle and external cilia places gastrotrichs in an intermediate position between Ecdysozoa and Lophotrochozoa.

Once considered closely related to Gastrotricha, phylum Nematoda (also Nemata) is the largest phylum of cycloneuralians, containing greater than 20,000 described species. Together with its sister group Nematomorpha, these two phyla form a basal group within the Introverta, a subclade of Cycloneuralia defined by the loss of external cilia, a molted cuticle, and an introvert (an anterior part of the head than can be invaginated). The Nematoda is monophyletic and well-defined by the presence of sensory organs called amphids and a unique form of excretory system (renette cells). The presence of an introvert is not often considered part of the ground pattern of the phylum (only known in *Kinonchulus*), so relationships to other Introverta are tenuous based on this character. The fossil record is sparse and dates back only to the Carboniferous Period (354–290 mya). Molecular evidence suggests the class of predominantly marine, free-living nematodes, Adenophorea, may be paraphyletic as it includes the ancestors of Secernentea, the class of terrestrial parasites. The closely related phylum, Nematomorpha, contains approximately 325 species of exclusively parasitic worms, often called horsehair worms. The phylum is defined by the presence of a unique larval stage that parasitizes arthropods, and an adult body devoid of a gut. The fossil record is sparse and extends only back to the Eocene (15–45 mya). Nematomorphs share several characteristics with nematodes that suggest a sister-group relationship: collagenous cuticle without microvilli, absence of circular muscles, and ectodermal longitudinal cords. According to some zoologists, nematomorphs may have evolved from the mermithoid nematodes that they resemble.

1,100 species of obligate intestinal parasites of vertebrates. Traditionally, acanthocephalans were closely aligned with turbellarians or Aschelminthes, but recent studies support the sister-group relationship of Acanthocephala and Rotifera. If considered part of Gnathifera, acanthocephalans must be considered highly derived, having lost the main gnathiferan characteristic (jaws) in their evolution towards parasitism.

Autapomorphies of Acanthocephala include a hook-bearing proboscis with unique ligament system and the absence of a gut. The phylum is traditionally divided into three classes, though relationships among the classes are poorly known. Gnathostomulids were first described in 1956 by Peter Ax as turbellarians, but later raised to phylum status in 1969. The phylum is small, containing approximately 100 species divided into two monophyletic classes and characterized by the unique

The Cephalorhyncha, a sub-clade of Introverta within the Cycloneuralia, contains the phyla Priapula, Kinorhyncha, and Loricifera. These phyla are united by the shared presence of chitin in their cuticles, of rings of scalids on their introverts, flosculi, and two rings of introvert retractor muscles attached through their brains. Phylum Priapula is the most primitive of the cephalorhynchs, contains fewer than 20 extant species, and has an extensive fossil record. Recent species range in size from minute to inches (millimeters to centimeters) and are exclusively benthic predators. Autapomorphies of Priapula include a large body cavity containing amoebocytes and erthryocytes, and perhaps the unique structure of their urogenital system. The fossil record for priapulans is unique within the Cycloneuralia because it extends back to the Cambrian: *Ottoia prolifica* is the most abundant fossil species and closely resembles the extant *Halicryptus spinulosus*. The two remaining phyla, Kinorhyncha and Loricifera, are closely united by the structure of their introvert, which actually protrudes out rather than evaginates out (turns inside out by eversion of an inner surface) as in priapulans. The Kinorhyncha is a small phylum of approximately 150 species, all microscopic with a body composed of 13 segments, and with an internally segmented musculature and nervous system. Despite the tough cuticle, no fossilized kinorhynchs have been found. Reinhardt Kristensen, the co-discoverer of Micrognathozoa and Cycliophora, first described Loriciferans in 1983. The Loricifera is a small phylum of 100 species characterized by scalids (spines) with intrinsic muscles and a lifecycle with a Higgins larva. No fossils are known.

The phyla Entoprocta and Cycliophora comprise a monophylum, according to cladistic studies. Phylum Entoprocta consists of 150 extant species and few fossilized forms that extend back to the Upper Jurassic (200 mya). Species are either individual or colonies of globular polyps with ciliated tentacles on a short stalk. Entoprocts are functionally acoelomate but may have small pseudocoelic sinuses. Embryonic cleavage is clearly of the spiral pattern and results in a trochophore larva, placing entoprocts within the Protostomia. Traditionally, entoprocts were allied with ectoprocts in the phylum Bryozoa, but a lack of synapomorphies led to their eventual separation in 1921. Danish biologist Claus Nielsen (2001) still supports a close relationship between the taxa and suggests several possible synapomorphies: myoepithelial cells in the apical organ, similar metamorphosis, and the ultrastructure of larval eyes. Cladistic studies have pointed out a potential relationship to Cycliophora, a recently discovered phylum of symbiotic micrometazoans that live on the mouthparts of lobsters (Funch and Kristensen, 1995). The phylum is characterized by a highly complicated lifecycle involving asexual female stages, sexual female stages, sexual dwarf males, and a variety of larval forms (pandora larvae, chordoid larvae). The chordoid larva is thought to be homologous with the typical protostome trochophore larva, and may have developed neotenically from entoprocts, according to some researchers. A single synapomorphy unites the two phyla: mushroom-shaped extensions from the basal lamina into the epidermis.

The Deuterostomia is a morphologically well-defined group with a long fossil history. Traditionally, the Deuterostomia is characterized by embryological criteria such as radial cleavage (a symplesiomorphy), blastopore becomes the anus, enterocoelous archenteric mesoderm, and a tripartite coelom. The deuterostome clade generally includes the Chaetognatha, Lophophorata (Brachiopoda, Ectoprocta, Phoronida), Hemichordata (Enteropneusta, Pterobranchia), Echinodermata, and Chordata. Several phyla, namely Chaetognatha and the lophophorate groups, share characteristics with both protostomes and deuterostomes, and their phylogenetic affinities are muddled. The advent of molecular phylogenetics in the 1990s also raised several questions regarding traditional deuterostome relationships. The hemichordates, echinoderms, and chordates make up a characteristic clade of derived deuterostomes defined by at least five synapomorphies, depending on views about their evolutionary origins: glomerular complex, epithelia that binds iodine and secretes iodothyrosine, separate gonoducts, pharyngeal gill slits, and a dorsally concentrated nervous system. The Hemichordata is a small phylum of a few hundred species, with a fossil record dating back to the Cambrian, and generally considered monophyletic by the presence of an anterior evagination of the gut called the stomochord. Brusca and Brusca (2002) consider the glomerulur complex, an excretory organ comprising a heart and blood vessel network in the prosome of the proboscis, an additional synapomorphy, but Nielsen (2001) regards it as a synapomorphy of Neorenalia (= Deuterostomia: Hemichordata + Echinodermata + Chordata). Nielsen also considers the Hemichordata to be polyphyletic with pterobranchs forming the sister group of Echinodermata. Alternatively, molecular studies suggest that enteropneusts and echinoderms are sister groups, or echinoderms are the closest sister group to the chordates.

Despite their extensive fossil record extending to the Cambrian, echinoderms and chordates have been notoriously difficult in terms of identifying a closest sister group. Echinoderms are defined primarily by the endoskeletal system of plates with a stereom structure, and external ciliary grooves for suspension feeding. Two extinct groups, the Carpoids and Helicoplacoids, are basal groups with the phylum. According to some authors, the Carpoids represent the ancestor to most fossil and modern echinoderms, but the lack of radial symmetry and evidence of a water vascular system make their presence within the phylum debatable. The Helicoplacoids, however, had triradial symmetry and water vascular systems with open ambulacral grooves, and are so considered basal echinoderms. The six remaining classes are united by the characteristic pentaradial symmetry, with mouth and anus on oral surface, and attachment to the substratum by the aboral surface.

According to the paleontologist R. P. S. Jefferies (1986), a group of extinct echinoderms called calcichordates gave rise to the chordate lineage. Jefferies claims to have identified chordate-like structures such as gill slits, a brain, notochord, and dorsal nerve cord in fossils of calcichordates. This calcichordate theory is still hotly contested, although many paleontologists have rejected the theory because they do not accept Jefferies' identification of the chordate features in these extinct echinoderms. The traditional branching of the chordate line is generally well accepted with Urochordata arising first, followed by Cephalochordata and Vertebrata. The cephalochordate-vertebrate line is hypothesized to have evolved via paedomorphosis from a urochordate-like ancestor. Some zoologists even prefer Garstang's theory (1928) that vertebrates evolved directly from a larval urochordate by paedomorphosis. However, lately, the model cephalochordate, *Branchiostoma* (Amphioxus), is considered to be the closest living relative of Vertebrata. Synapomorphies uniting Cephalochordata and Vertebrata include segmental muscles (myomeres), a ventral pulsating blood vessel (homologous with the vertebrate heart), an intestinal diverticulum (embryonic precursor of the vertebrate liver), and separate dorsal and ventral roots of the spinal cord.

Resources

Books

Brusca, R. C., and G. J. Brusca. *Invertebrates*. 2nd ed. New York: Sinauer Associates Inc., 2003.

Jefferies, R. P. S. *The Ancestry of the Vertebrates*. London: British Museum (Natural History), 1986.

Nielsen, C. *Animal Evolution. Interrelationships of the Living Phyla*. 2nd ed. New York: Oxford University Press, 2001.

Tyler, S. "The Early Worm: Origins and Relationships of Lower Flatworms." In *Interrelationships of the Platyhelminthes*, edited by D. T. J. Littlewood and R. A. Bray. London: Taylor and Francis, 2001.

Periodicals

Christen, R., A. Ratto, A. Baroin, R. Perasso, K. G. Grell, and A. Adouette. "An Analysis of the Origin of Metazoans, Using Comparisons of Partial Sequences of the 28S RNA, Reveals an Early Emergence of Triploblasts." *EMBO* 10 (1991): 499–503.

Field, K. G., et al. "Molecular Phylogeny of the Animal Kingdom." *Science* 239 (1988): 748–753.

Funch, P., and R. M. Kristensen. "Cycliophora Is a New Phylum with Affinities to Entoprocta and Ectoprocta." *Nature* 378 (1995): 711–714.

Garstang, W. "The Morphology of the Tunicata, and Its Bearing on the Phylogeny of the Chordata." *Quarterly Journal of Microscopical Science* 72 (1928): 51–187.

Haeckel, E. "The Gastrea-Theory, The Phylogenetic Classification of the Animal Kingdom and the Homology of the Germ Lamellae." *Quarterly Journal of Microscopical Science* 14 (1866): 142–165, 223–247.

Nielsen, C. "Defining Phyla: Morphological and Molecular Clues to Metazoan Evolution." *Evolution & Development* 5 (2003): 386–393.

Rieger, R. M. "Uber den Ursprung der Bilateria: die Bedeutung der Ultrastrukturforschung fur ein neues Verstehen der Metazoenevolution." *Verhandlungen der Deutschen Zoologischen Gesellschaft* 79 (1986): 31–50.

Rieger, R. M., and P. Ladurner. "Searching for the Stem Species of the Bilateria." *Belgian Journal of Zoology* 131 (2001): 27–34.

Xiao S., Y. Zhang, and A. H. Knoll. "Three Dimensional Preservation of Algae and Animal Embryos in a Neoproterozoic Phosphorite." *Nature* 391 (1998): 553–558.

Rick Hochberg, PhD

Reproduction, development, and life history

What is a life history?

Each animal species can be viewed as a collection of individuals, all sharing a common pool of genes. The genes determine all of the animal's characteristics, and are made of the complex molecule known as DNA. The genes are exchanged among the members of the species through a variety of mechanisms, all of which are related in some way to the process of reproduction. For this reason, the ability to reproduce is often considered to be one of the most important defining characteristics of life. But reproduction, like most basic biological functions, occurs primarily at the level of a single individual that may interact with another individual to exchange genetic information through the transfer of DNA. This individual parent animal reproduces to form another new individual offspring. The offspring, however, begins its individual life as a single cell rather than as a multicellular animal like its parent. If it is to reproduce, it must usually become multicellular as well, and must ultimately mature to be anatomically similar to its parent. This entire cycle of development is generally known as the life history of an individual animal. In other words, the life history of any individual animal comprises the processes of reproduction and development; in turn, each of these processes is broken down into other more specific processes, each designed to accomplish a particular function.

Asexual reproduction

Fundamentally, reproduction is the copying of an individual animal's DNA combined with transferring the copy into a newly formed individual. In some cases, the copy of the DNA is nearly exact, and the offspring develop from a single parent. This is typically the case in what is known as asexual reproduction. In its truest form, asexual reproduction involves simple cell division, or mitosis, without any reorganization of DNA fragments during the process. Asexual reproduction also involves only one parent, and thus involves no new combinations of genes resulting from mixing complementary genes from two parents. Asexual reproduction has the advantage of allowing a very fast rate of reproduction, with a resulting rapid increase in the population of a given species. The primary disadvantage of asexual reproduction is that it does not permit much genetic variability; as a result, the population as a whole

is relatively unable to adapt to diverse or changing environmental conditions.

Sexual reproduction

In most cases, the DNA copy is not exact, so that the genetic makeup of the newly formed offspring differs from that of its parent. This programmed variability is accomplished primarily by sexual reproduction. The genetic process that defines sexual reproduction occurs only during a very brief specialized phase of cell division, and only within cells belonging to the germ cell line. A germ cell is defined as a cell belonging to a cellular lineage that, at some point, will deviate from normal cell division (i.e., mitosis, which results in exact duplicate copies of DNA) to engage in meiosis. Meiosis is often known as reductional division, because it results in reducing the number of chromosomes by half in preparation for an exchange with the complementary chromosome of a mating partner that restores the full set. The most important aspect of meiosis, however, occurs long before this reduction of chromosome number, which occurs late in meiosis. Early in meiosis, gene segments are actually recombined by the exchange of DNA sequences, so that the individual chromosomes are transformed into genetically unique combinations of genes. This process is known as crossing over; it is the defining event that distinguishes sexual from asexual reproduction. In this way, germ cells become genetically unique while they are still in the parent animal, long before they become fully formed sperm or oocytes, and even longer before they fuse with gametes from another parent. At the later point of gamete fusion, the recombined genes in the chromosomes from two parents will fertilize each other to produce new chromosome combinations. Thus, the genetic recombination that takes place during meiosis, combined with the reaggregation of the chromosomes that occurs during fertilization, results in the distinctive offspring that characterize most animals. It is important, however, to recognize that sexual reproduction does not require either fertilization or two parents. If meiosis takes place, the reproduction is sexual. In some lower metazoans, a single parent animal may produce gametes by meiosis, which can then develop into fully formed offspring by various processes known collectively as sexual parthenogenesis.

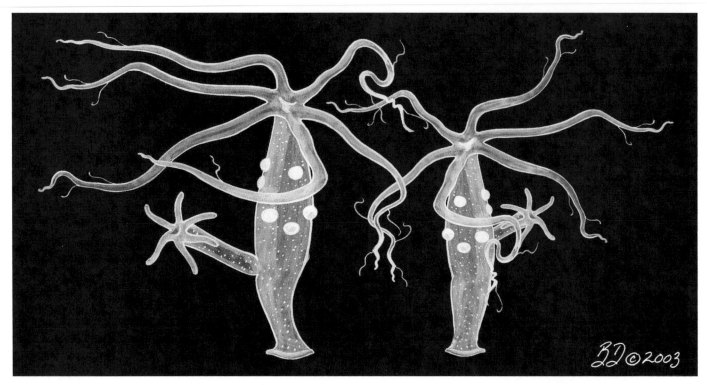

Hydras reproduce by budding. (Illustration by Barbara Duperron)

Reproductive duality

Many animals, particularly the lower metazoans, actually use both asexual and sexual reproduction at various times. This reproductive duality gives them the advantages of both modes. It is rare for both asexual and sexual reproduction to occur simultaneously, however. In many lower metazoans, the asexual and sexual processes are cyclical, occurring in different seasons. Examples include many marine sponges (phylum Porifera), in which a single individual may alternate between reproductive modes. Another example may be seen among the freshwater rotifers (phylum Rotifera), in which the two reproductive modes will be restricted to sequential generations that are otherwise anatomically similar. In other lower metazoan groups, asexual and sexual processes are segregated into two distinct stages in the life history, with radically different anatomical and behavioral traits, or even different habitats. The best examples of these stages are found among the hydroids and jellyfishes (phylum Cnidaria), and among the parasitic flatworms (phylum Platyhelminthes). Since the life history stages responsible for asexual and sexual phases are so distinctive in these groups, and occur in a regularly alternating pattern, this general reproductive strategy is often called alternation of generations.

Gametogenesis

During sexual reproduction, each parent animal must form specialized cells known as gametes, which are genetically recombined and in which the chromosome number is reduced by half from a diploid double set to a haploid single set. Both processes occur during meiosis, which is the first stage of ga-
mete formation, or gametogenesis. In virtually all animals that reproduce sexually, the gametes occur in two morphologically distinct forms corresponding to male and female. These distinctions in form and structure are related to the specific functions of each gamete. The differences become apparent during the latter stages of spermatogenesis (for male gametes) and oogenesis (for female gametes).

After spermatogenic meiosis, the morphological transformation of the male gamete generally includes development of a small motile sperm. The sperm's function is to move toward and ultimately meet the female gamete, beginning a sequence of events that ends in the fusion of the two gametes. After the gametes fuse, the sperm's role is essentially complete except for the final genetic contribution from its half of the new offspring's genome. Thus, the primary structures developed by the sperm are concerned with movement and with engaging the female gamete and its coatings. The specific locomotory structures of sperm vary among the lower metazoans, ranging from pseudopodia (temporary extensions of cell material) resembling amoebas in the roundworms (phylum Nemata) to flagella (long whiplike projections) in most other groups.

After oogenetic meiosis, the morphological transformation of the female gamete generally includes development of a large oocyte that does not move around. The oocyte's functions are far more numerous than those of the sperm. For most lower metazoan groups, they include equal or greater participation in the process of gamete fusion. Following fusion, however, the oocyte must provide the coordination of and materials for all the early stages of embryo and larval de-

A West Indies sea egg (*Tripneustes ventricosus*) releasing sperm into water. (Photo by ©Andrew J. Martinez/Photo Researchers, Inc. Reproduced by permission.)

velopment. To carry out these functions, the oocyte must build up large stores of energy-rich nutrients (e.g., carbohydrates in the form of glycogen, and lipid or proteinaceous yolk); phospholipid stores for membrane synthesis; extra nucleotides or redundant DNA; transcripts of RNA for protein synthesis; extra regulatory and structural proteins; and occasionally materials for eventual eggshell formation. In most cases, the oocyte must fulfill all these functions by itself. Some animals, however, use other germ cells to assist the oocyte. Perhaps the best example of this assistance is found among the parasitic flatworms (phylum Platyhelminthes). These organisms have special vitelline (resembling yolk) germ cells that never become gametes but instead supply the gametes with needed materials. The oogenic stage that is most often involved in fertilization is the oocyte; however, there are some exceptions to this generalization. The ambiguous term "egg" is often applied to oocytes and other fertilizable stages of female gametes. "Egg" may, however, also refer to fully formed embryos or juveniles within various embryonic coverings, so the word should be avoided in most instances.

Spermatogenesis and oogenesis most often occur in different individual animals known as males and females respectively. This differentiation of sexes is known as gonochorism. Alternatively, it is quite common for the same individual to produce both sperm and oocytes. This condition is known as hermaphroditism; it may involve either simultaneous or sequential production of sperm and oocytes. Oogenesis and spermatogenesis may occur in different gonads, namely ovaries and testes, or may occur in a single hermaphroditic gonad. Such lower metazoans as sponges (phylum Porifera) may lack distinct gonads, with gametes developing in normally somatic regions of the body. Whether through gonochorism or hermaphroditism, gametogenesis may occur throughout the adult life of the animal, as in parasitic flatworms (phylum Platyhelminthes). The more common pattern among the lower metazoans, however, is one of seasonal reproduction.

Gamete exchange

Gametes come together in a variety of ways among the animals in the lower metazoan groups. Sperm are generally motile and engage in oocyte-seeking behavior of some sort. The small size and short-term motility of the sperm, however, mean that their efforts are effective for only a very short time; thus sperm motility is only effective for meeting the oocyte within very small spaces. Bringing the sperm and oocyte into these small spaces depends on the behavior of the parent animal, which must engage in some form of gamete exchange.

Adult animals fall into two broad categories in relation to gamete exchange: spawners and copulators. Lower metazoans demonstrate a broad range of variations within both of these categories. The vast majority of spawners release their gametes directly into the surrounding water, an activity known as broadcast spawning. Broadcast spawning is the most common method of gamete exchange in free-living marine invertebrates, but is rare among most freshwater groups. In the very lowest phyla (e.g., Porifera, Placozoa, Cnidaria) spawning is the only method of gamete exchange. In some groups, only the males spawn while the females take up the sperm while retaining their oocytes for internal fertilization. In others, both sperm and oocytes are spawned, resulting in external fertilization. For broadcast spawning with external fertilization to be successful, the parent animal must use some strategy to increase the chances of the gametes coming together. The most common strategy involves simultaneous spawning, in which all gametes are released at the same time by all members of a population. Other strategies include producing gametes with similar densities, adhesive properties, or other features that cause them to settle out into the same general parts of the water column or substrate.

Copulation occurs in many groups, and varies considerably among the lower metazoans. In all cases, there is some mechanism for transferring sperm directly from the male to the female. This transfer may occur by direct injection or by transfer of sperm packets known as spermatophores. Most copulators have specialized genital structures for transferring the sperm. Most often these structures inject the sperm directly into the female's reproductive system, but they may also inject them through the body wall, as in the hypodermic traumatic insemination seen in some turbellarian flatworms (phylum Platyhelminthes).

In many animals, the males and females are morphologically distinct, thus exhibiting sexual dimorphism. Sexual dimorphism is, however, more common among higher animals. The males and females of many lower metazoans are sexually monomorphic, and thus are distinguishable only microscopically by the nature of their gametes. Generally speaking, most broadcast spawners are sexually monomorphic, while copulators tend to be sexually dimorphic. The dimorphism may relate only to differences in copulatory structures or genitalia, but in some cases it may relate to differences in other habits or roles of the two sexes. Marked somatic dimorphism is much more common among higher animals, but does occur in some lower metazoans; examples include the schistosomatid flukes (phylum Platyhelminthes) and certain roundworms (phylum Nemata). In both of these cases, the dimorphism reflects differences in size as well as form.

Fertilization

After the gametes come together, they must fuse with one another to form a diploid zygote, which is the genetically complete new animal in a unicellular (single-celled) form. Fertilization is not a single event; rather, it is a complex series of events in which both sperm and oocyte actively participate. It begins with simple recognition among the sperm and oocytes of a given species, and concludes with the fusion of the haploid pronucleus of the sperm and the haploid pronucleus of the oocyte. Between these two events, there is considerable variation among the lower metazoans. In some cases, the oocytes have a covering that must be penetrated by the sperm. In most cases, the sperm must initiate the actual process of fusion of the cell membranes of the two gametes; however, it is the oocyte that generally is most active in directing the fusion and actually incorporating the sperm into its cytoplasm. The actual fusion of the two cells, known as syngamy, is accomplished largely by the oocyte. After syngamy occurs, pronuclear fusion may follow rapidly or be delayed up to several days, depending on the species. At this point, the new genetically unique and genetically complete animal, in the form of a single-celled zygote, is ready to complete the development of its body.

Embryogenesis

The development of a fully functional animal body begins with the transformation of the single-celled zygote into a multicellular embryo. Part of the definition of an animal (i.e., a metazoan) is that it possesses true multicellularity, which it acquires during the process of embryogenesis. True multicellularity is defined not only as the possession of multiple cells in the body, but also a specific division of labor among those cells. In particular, the body of an animal must have somatic (body) cells separated from the reproductive (germ) cells. But, beyond that the somatic cells must be differentiated into different functional groups. During the process of embryogenesis, the new individual first develops numerous cells, then differentiates them according to space, structure, and function. Among the lower metazoans, we find a range of structural complexity, from simple double-layered groups of cells in the phylum Placozoa to complex organ systems in such phyla as the Platyhelminthes, Nemertea, and Nemata. Many other lower metazoan phyla fall at various places along this spectrum.

The first stage of embryogenesis is cleavage, which creates the simplest form of multicellularity. Cleavage is simply a series of mitotic cell divisions that take the new individual from a unicellular zygote to a multicellular mass of cells generally known as a morula. Depending on the phylum and species, this cell mass may consist of a few dozen to hundreds of cells. The actual pattern of cleavage varies in terms of the cells' spatial relationships to one another, the plane of the cell axis at which cleavage occurs, and the degree to which the cytoplasm is divided. The subject of cleavage is complex, but the type of cleavage is important in defining a group's evolutionary relationships to other groups. There is also great variation in the degree of cellular differentiation at this stage. In most lower metazoans, however, the cells do differentiate during cleavage into large macromeres and small micromeres; some groups also have intermediate-sized mesomeres. In most cases, these classes of cells are destined to become specific layers within the later embryo. In turn, each layer will give rise to certain tissues of the adult body. Prior to forming layers, most embryos go through a stage of minor reorganization known as blastulation, which provides the spatial framework in which the actual layering takes place.

The vast majority of animals develop either two or three layers of cells, known as germ layers. These germ layers develop out of the dramatic reorganization of the cells that were formed during cleavage and slightly reorganized during blastulation. The stage of radical reorganization is called gastrulation. Possession of two or fewer germ layers is found only in certain lower metazoan phyla. A very few simple phyla (e.g., the Placozoa, Monoblastozoa, Orthonectida, and Rhombozoa) do not have specific germ layers. Sponges (phylum Porifera) are often interpreted as having no germ layers, but some biologists regard them as having two. Members of the phyla Cnidaria (jellyfishes, corals, etc.) and Ctenophora posses two well-developed germ layers, and thus are described as diploblastic. All other lower metazoans, and indeed all other animal phyla, possess three well-developed germ layers, and are correspondingly described as triploblastic. The actual mechanisms of layering vary widely among different phyla; they range from the migration of cells into the interior of the cell mass to the infolding of an entire hemisphere of the spherical blastula. In all cases, however, the embryo is left with an endoderm layer on the inside and an ectoderm layer on the outside. In triploblastic phyla, a layer of mesoderm forms between the other two. The endoderm typically develops into the digestive system of the adult animal, while the ectoderm develops into the epidermis and nervous system. The mesoderm, if one is present, gives rise to such structures as excretory systems and the tissues that line body cavities. Other organ systems develop from various layers, depending on the phylum, and most often involve contributions from two germ layers working cooperatively.

At the end of gastrulation the embryo is fully formed; embryogenesis is complete. The new organism now looks a bit like an animal with a skin and a gut; all the basic layers are present, ready to differentiate further into a more definitive animal having the recognizable characteristics of its taxonomic group.

Most higher animals, above the platyhelminthes, can be divided into two groups based primarily on embryonic features. These two major branches are known as the protostomes and the deuterostomes. Protostomes undergo determinate cleavage or mosaic development, in contrast to the indeterminate cleavage or regulative development of deuterostomes. The determinate cleavage of protostomes results from a plane of cell division, usually visible after the second division, that cuts diagonally across the original zygote axis, thus compartmentalizing different regulative and nutritive chemicals in each of the resulting cells. This is referred to as spiral cleavage, since the cells dividing diagonally appear under the microscope to spiral around the original axis. In contrast, the indeterminate cleavage of deuterostomes results from a planes of cell division that cut alternatively longitudinally along the zygote axis, then transversely across the axis, thus leaving each resulting tier of cells with similar regulative and nutritive chemicals. This is referred to as radial cleavage, since the cells dividing at alternating parallel and right angles to the original axis appear under the microscope to radiate in parallel planes from that axis. The most important thing is not whether the resulting cell masses appear to spiral or to radiate, but that only the spiraling cells of the protostomes

show determination of specific germ layers as early as the first cell division, and almost universally by the third. Thus, at the very earliest stages of cleavages, specific cells of protostomes have already been determined to a fate of forming a specific one of the three germ layers. During gastrulation, the embryo is left with an opening to the outside called the blastopore, which will develop into an opening into the gut in the adult animal. The precise nature of the opening is the second major feature differentiating deuterostomes from protostomes. In protostomes, the blastopore becomes the adult mouth, whereas in deuterostomes, the blastopore becomes the anus. Shortly or immediately after gastrulation is complete, higher animals form their body cavity, the coelom. By definition, the true coelom is always a body cavity within mesodermal tissue. The mechanism by which the coelom is formed is the third primary distinction between protostomes and deuterostomes. In most deuterostomes, the coelom forms by outpocketing from the original archenteron, a process known as enterocoely since the coelomic cavities are thus derived directly from embryonic enteric cavities. In protostomes, the coelom forms from a split in the previously solid mass of mesodermal cells, a process thus known as schizocoely. There are some exceptions to this rule, but it applies well to most.

Postembryonic development

Most animal embryos look rather similar to one another up through the stage of gastrulation. It is during the important stages of postembryonic development, however, that the characteristic features of specific phyla, classes, and orders finally emerge. For this reason, much of postembryonic development is said to involve morphogenesis, or the establishment of the animal's definitive body form. Along with the completion of form comes the establishment of function, so that the end result is a fully functional animal. For some species, this fully functional individual will be a juvenile, which resembles an adult in form but lacks a mature reproductive system. Development that proceeds from embryo to juvenile with no intervening stage is known as direct development. Direct development occurs in some lower metazoans, including the nematodes, gnathostomulids, rotifers, and gastrotrichs. In contrast, most lower metazoans undergo indirect development, in which a larval stage is inserted between the embryo and the juvenile or adult.

A larva is a fully functional animal, generally feeding and moving about independently. The larval form of a given species is generally as characteristic of the species as the adult, and may complete some critical parts of the life history strategy for its species. The most common task of larvae is long-distance migration in order to colonize new environments for the species. This phenomenon, known as planktonic dispersal, is especially critical in marine species whose adult forms have limited or no mobility, such as corals, sponges, ribbonworms, and polyclad flatworms. Larvae may also make use of food resources that differ from those needed by the parent, thus avoiding competition within the species. Because of the critical and distinctive attributes of larvae, their formation is frequently referred to as larvigenesis, and represents a discrete (separate) stage of postembryonic morphogenesis.

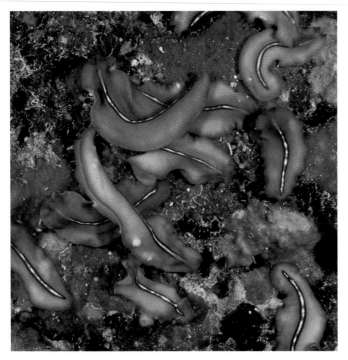

Mating flatworms (*Pseudoceros bifurcus*) with insemination marks. (Photo by ©A. Flowers & L. Newman/Photo Researchers, Inc. Reproduced by permission.)

By definition, larvae and adults are dissimilar in structure and function. For this reason, the transition from larva to adult requires radical changes in the morphological, behavioral, and physiological characteristics of the animal. This transformation between successive postembryonic forms is known as metamorphosis. As with cleavage and gastrulation, the exact mechanism of metamorphosis varies widely among different lower metazoan phyla. In all, it involves the loss of some specifically larval structures and the development of new adult structures from groups of undifferentiated cells.

Sexual maturation

After the embryo or larva is transformed into a juvenile form, all that remains for the individual is maturation of its sexual reproductive systems, accompanied or followed by meiotic maturation of the gametes to form fully functional and fertilizable sperm and oocytes. The mechanisms for maturation vary widely among animal phyla, but are especially diverse among the lower metazoans. Such phyla as placozoans and sponges have no discernible gonads; the gametes simply form out of previously undifferentiated somatic cells. In such others as the diploblastic cnidarians, the gametes develop from specialized cells within one of the germ layers, but no other gonadal tissues are present. At the other end of the scale of complexity, such animals as roundworms and flatworms develop elaborate gonads and ducts; these structures include discrete testes with sperm ducts, discrete ovaries with oviducts, and even numerous types of specialized reproductive glands. In some respects, the parasitic flatworms have the most structurally sophisticated reproductive systems in the animal king-

dom, even though as a group they are considered primitive organisms.

Another variation that occurs among the lower metazoa relates to reproductive seasonality. While some groups, such as tapeworms and flukes, reproduce continually and thus retain all of their sex organs, such others as hydroids and turbellarian flatworms reproduce only temporarily, but repeatedly. These organisms often lose their reproductive organs during their nonreproductive periods and redevelop them during the next mating season.

Structural maturation of the sex organs and gametes is often accompanied by certain changes in behavior as well as the development of special structures that are not gonadal but are nonetheless related to reproduction. The most obvious behavioral changes involve mate-seeking and copulatory behaviors. Structural changes include the development of special genitalia for coupling. The latter includes the copulatory cirrus (flexible penis) of parasitic flatworms and the copulatory spicules of many roundworms. Other related behavior may include various forms of brooding or other parental care strategies. While parental care is not generally as common or as well developed among lower metazoans as in higher animals, some examples do exist. Among the lower metazoans, most of these are more structural than behavioral. Examples include the maintenance of amphiblastula larvae within specially adapted radial canals of calcareous sponges; the development of special egg-enclosure organs in several tapeworms; the retention of fully-formed juveniles within the uterus of some roundworms; and the retention of successive generations as colonies in many hydroids and other cnidarians.

Phylum summaries

Brief summaries of the primary reproductive and developmental strategies of each lower metazoan and deuterostome phyla follow. The variations are within each phylum are great, however, and the short summaries below are intended only to situate each phylum within the overall context of reproductive strategies and processes discussed earlier. Interested readers should consult the references listed for further details and analyses.

Phylum Placozoa

The limited information on *Trichoplax adhaerens*, the only known species of Placozoa, indicates that the organism has no capacity for sexual reproduction. Asexual reproduction, however, is very effective and diverse, occurring in three distinct modes. Simple fission, or division of the simple body by cell separation, is the most common mode. More rarely, two types of budding occur. In one form, hollow swarmers bud off the parent organism and may swim to remote locations to develop further. In other individuals, the attached buds may stretch and attach themselves to adjacent substrates before detaching from the parent. In both cases, cellular rearrangements similar to gastrulation occur.

Phylum Monoblastozoa

This phylum, represented only by the single genus *Salinella*, has been observed and described by only one author, and some researchers question its existence. The original description is vague, but describes asexual reproduction by a sort of transverse fission similar to that of placozoans.

Phylum Porifera

Sponges may be either gonochoristic or hermaphroditic, but most undergo some form of sexual reproduction. The exact origin of germ cells varies somewhat among species, but most sperm and oocytes develop from undifferentiated cells known as archeocytes in the central mesohyl (connective tissue) layer. In some species, sperm may develop through transformation of the flagellated collar cells that line the sponge's chambers and create the water currents responsible for the exchange of all materials within the sponge. Males are broadcast spawners, but most sponges undergo internal fertilization. Fertilization is followed by internal brooding of larvae in many sponges, including most marine calcareous sponges and the spongillid family of freshwater sponges. Whether sponges have true germ layers is often debated, since some cells can transform into any cell type even in the adult; however, cellular rearrangements comparable to gastrulation take place at the end of embryogenesis. Sponges are perhaps the most efficient phylum in the animal kingdom for asexual reproduction. They employ a number of different strategies ranging from simple fragmentation of the adult body to formation of specialized gemmules (reproductive buds), the latter being more common in the overwintering stages of freshwater species.

Phylum Cnidaria

Sexual reproduction is well developed throughout the phylum. Gonochoristic and hermaphroditic species are known to occur; however, the true jellyfishes (class Scyphozoa) and colonial hydroids (class Hydrozoa) are primarily gonochoristic. Germ cells develop in either the ectoderm or endoderm, depending on the class, but always originate from undifferentiated cells known as interstitial cells. Most species of cnidarians are broadcast spawners, but in some, such as the freshwater *Hydra*, the oocyte may be retained for internal fertilization. In most species, embryonic development leads to a planula larva that settles to the substrate for metamorphosis into the adult cnidarian. Asexual reproduction is very common and takes many different forms. Many species undergo alternation of generations, with asexually produced medusae (free-swimming jellyfish) alternating with sexually produced polyps attached to the substrate. Sexual and asexual mechanisms of reproduction may occur in either stage, however, depending on the species.

Phylum Ctenophora

Comb jellies are primarily hermaphroditic, with only a few gonochoristic species. They have simple gonads resembling those of the closely related cnidarians. Most are broadcast spawners, but some are fertilized internally and may even brood their larvae. Embryogenesis results in a cydippid larva that swims freely during its metamorphosis into an adult. Asexual reproduction is not known to occur in this phylum.

Phylum Rhombozoa

The dicyemid mesozoans are all parasites, and alternate between sexual stages in the adult host and asexual stages in the juvenile host. The sexual forms are hermaphroditic. These animals are structurally simple, barely qualifying as truly multicellular. They lack layers comparable to the germ layers of other animal phyla.

Phylum Orthonectida

Orthonectids are parasites that alternate between sexual and asexual stages within their host animal. Asexually produced plasmodia may develop into sexual forms, most of which are gonochoristic, with a few hermaphroditic species. Copulation is followed by internal fertilization, and the larva ultimately leaves the parent to seek a new host.

Phylum Platyhelminthes

The vast majority of flatworms are hermaphroditic, but some gonochoristic forms occur, including the medically important schistosome flukes. The reproductive systems of predominantly free-living turbellarians are simple and transient (temporary), whereas those of the parasitic tapeworms and flukes are complex and permanent, with many specialized organs. Copulation is the rule for reproduction in this phylum, followed by internal fertilization and either internal or external development. Fertilization generally involves incorporation of the full sperm into the oocyte. Internal development often takes place within specialized structures for maternal care of the larvae. Cleavage and embryogenesis occur in patterns unique to this phylum, especially among the tapeworms and flukes; there are many different forms of larvae and patterns of metamorphosis in this group. In addition, the tapeworms and flukes engage in regular alternation of sexual and asexual generations, perhaps producing the greatest number of progeny in the animal kingdom.

Phylum Nemertea

Nemerteans are primarily gonochoristic (except for the few freshwater and terrestrial species), with large but simple gonads. Most species are marine, and reproduce by broadcast spawning followed by external fertilization and embryonic development. They undergo spiral cleavage, and thus are generally considered to be related to the protostomes. Postembryonic development leads to formation of a pilidium larva in most nemerteans. Some species may reproduce asexually by fragmentation, but this pattern is uncommon.

Phylum Nemata (Nematoda)

Sexual reproduction is the rule among the roundworms; most species are gonochoristic with some sexual dimorphism. Some hermaphroditic species do exist. Reproductive systems are tubular; in copulation, the male introduces amoeboid sperm into the vagina of the female. The embryogenetic process begins with an unusual form of bilateral cleavage, which ends with the direct development of a juvenile form (often incorrectly called a larva) that is structurally like a miniature adult. There are five juvenile molts before the adult form is reached. Asexual reproduction is extremely rare, and only involves the mitotic division of female germ cells.

Phylum Nematomorpha

Horsehair worms are exclusively sexual and gonochoristic. The gametes develop in long strands attached to support cells.

Adult nematomorphs copulate, often in large masses; fertilization is either external or internal, depending on the species. Little is known about embryogenesis in this phylum, but it culminates in a distinctive free-swimming larva that must invade an arthropod host before it can transform itself into a juvenile. The juvenile in turn must leave the host before maturing into an adult. Asexual reproduction is unknown in this phylum.

Phylum Priapulida

Priapulids are all gonochoristic, with no known mode of asexual reproduction. Their reproductive systems are poorly described, but are similar to those of nematomorphans in being formed as strands of oocytes attached to a common stalk. Most priapulids are broadcast spawners, with external fertilization and embryogenesis. Postembryonic development involves a distinctive larva that undergoes metamorphosis to become an adult.

Phylum Acanthocephala

Thorny-headed worms are gonochoristic, with complex reproductive systems, copulation, and internal fertilization. They develop through an intricate series of stages, including an acanthor larva and a cystacanth juvenile. The various stages of development occur inside different hosts of these parasitic animals. Asexual reproduction does not occur in this phylum.

Phylum Rotifera

As a group, rotifers exhibit a variety of reproductive strategies, with the three classes distinguished by either hermaphroditic or gonochoristic sexuality. Some species, especially in freshwater habitats, alternate between generations produced by parthenogenesis, in which no fertilization occurs, and typical generations produced by copulation and fertilization. The gonads of these tiny animals consist of only a few gametes enclosed by a thin sac. Embryogenesis culminates in direct development of a juvenile that quickly matures into an adult.

Phylum Gastrotricha

Gastrotrichs are generally hermaphroditic, with sperm and oocytes generally forming within the same gonad. Adults reciprocally inseminate each other during copulation. Fertilization is internal, but embryonic development is external. Postembryonic development is direct, and there is no known example of asexual reproduction in this phylum.

Phylum Loricifera

The loriciferans are exclusively sexual and gonochoristic in their reproduction. Little is known about their embryonic development, but it ends with the formation of a distinctive Higgins larva, or perhaps juvenile, that is similar to the adult.

Phylum Kinorhyncha

All known kinorhynchs are sexual and gonochoristic. Copulation, fertilization, and embryonic development are poorly known. Postembryonic development appears to be direct.

Phylum Gnathostomulida

Gnathostomulids are primarily hermaphroditic, and none are known to reproduce asexually. Simple reproductive systems, copulation, and internal fertilization characterize this group. The spiral cleavage is similar to that of protostome animals, and development progresses directly into a juvenile and then an adult form.

Phylum Chaetognatha

Adult arrow worms are hermaphroditic, with well-developed male and female gonads in separate body cavities. Fertilization is internal, but development of the embryos is external, though some species brood their young. Cleavage is radial, and the coeloblastula undergoes gastrulation similar to that of echinoderms and other deuterostomes. Asexual reproduction is not known to occur.

Phylum Hemichordata

Acorn worms are gonochoristic, and gametes are spawned into the open seawater. Fertilization and embryonic development, beginning with radial cleavage, occur in the plankton, and are similar to the patterns of echinoderms. Embryos develop directly or through a distinctive tornaria larva. Some species may reproduce asexually by fragmentation.

Phylum Echinodermata

Starfishes, sea urchins, sea lillies, and their relatives are extremely diverse, and exhibit a variety of reproductive and development modes. In all, however, typical deuterostome development is the rule, beginning with radial cleavage. Larval forms are varied, and tend toward different forms in different classes. Both sexual and asexual reproduction occur within the phylum, but broadcast spawning and planktonic development are the most common patterns.

Phylum Chordata

The invertebrate chordates, including tunicates, lancelets, and their relatives, generally reproduce by spawning and development planktonically in the seawater. Sexual and asexual reproduction may occur in the urochordates, but cephalochordates only undergo sexual processes. Cleavage is generally radial, but may be more mosaic than that of other deuterostomes.

Phylum Entoprocta

Various entoprocts may be either gonochoristic or hermaphroditic, depending on the species. A few may reproduce asexually by budding. Males generally spawn into open water, but the sperm are usually taken up by females for internal fertilization. Cleavage is spiral, possibly indicating some relationship to protostomes.

Phylum Ectoprocta (Bryozoa)

Most bryozoans are hermaphroditic. As colonial animals, all reproduce by asexual means as well. Males spawn sperm, which females take up for internal fertilization. Cleavage is radial, and most species have planktonic larvae.

Phylum Brachiopoda

Lampshells are primarily gonochoristic. Fertilization is variable, but cleavage is always radial. Depending on the class, they may undergo planktonic development through a larva, or may develop directly. Asexual reproduction has not been described for the group.

Phylum Phoronida

The vast majority of phoronids are hermaphroditic, with female and male gonads functioning simultaneously. Fertilization is usually internal. Cleavage is radial, followed by planktonic development in most species, generally through a distinctive actinotroch larva. Asexual reproduction by budding or fission occurs in a few species.

Phylum Cycliophora

Cycliophorans alternate between sexual and asexual stages. They are gonochoristic, and the male attaches to the female for insemination followed by internal fertilization and development. The modified trochophore larva is somewhat similar to that of some protostome groups.

References

Books

Conn, David Bruce. *Atlas of Invertebrate Reproduction and Development*. 2nd ed. New York: Wiley-Liss, 2000.

Gilbert, Scott F., and Anne M. Raunio, eds. *Embryology: Constructing the Organism*. Sunderland, MA: Sinauer Associates, 1997.

Strathmann, Megumi F., ed. *Reproduction and Development of Marine Invertebrates of the Northern Pacific Coast*. Seattle: University of Washington Press, 1987.

Wilson, W. H., Stephen A. Sticker, and George L. Shinn, eds. *Reproduction and Development of Marine Invertebrates*. Baltimore: Johns Hopkins University Press, 1994.

Young, Craig M., M. A. Sewell, and Mary E. Rice, eds. *Atlas of Marine Invertebrate Larvae*. San Diego: Academic Press, 2002.

Periodicals

Fautin, Daphne Gail. "Reproduction of Cnidaria." *Canadian Journal of Zoology* 80 (2002): 1735–1754.

Leys, Sally P., and Bernard M. Degnan. "Embryogenesis and Metamorphosis in a Haplosclerid Demosponge: Gastrulation and Transdifferentiation of Larval Ciliated Cells to Choanocytes." *Invertebrate Biology* 121 (2002): 171–189.

Slyusarev, George S., and Marco Ferraguti. "Sperm Structure of *Rhopalura littoralis* (Orthonectida)." *Invertebrate Biology* 121 (2002): 91–94.

Swiderski, Zdislaw, and David Bruce Conn. "Ultrastructural Aspects of Fertilization in *Proteocephalus longicollis*, *Inermicapsifer madagascariensis*, and *Mesocestoides lineatus* (Platyhelminthes: Cestoda)." *Acta Parasitologica* 44 (1999): 19–30.

Weiss, Mitchell J. "Widespread Hermaphroditism in Freshwater Gastrotrichs." *Invertebrate Biology* 120 (2001): 308–341.

David Bruce Conn, PhD

Ecology

Introduction

The ecology of invertebrates consists of all the external factors acting upon that organism. These factors may be either physical or biological. The physical or abiotic environment consists of the nonliving aspects of an organism's surroundings, including temperature; salinity; pH (a measurement of acidity or alkalinity); exposure to sunlight; ocean currents; wave action; and the type and size of sediment particles. The biotic environment consists of living organisms and the ways in which they interact with one another.

Invertebrate species have colonized all types of aquatic habitats. For example, sponges of the class Calcarea are restricted to firm substrates. They are also restricted by physical factors that affect their skeletons, limiting their habitats to shallow zones. Hexactinellida sponges colonize soft surfaces; they prefer to live in deep water. Demosponges can live on such different substrates as rock, unstable shell, sand, and mud; in some cases they burrow into calcareous material. They are found in a variety of underwater habitats ranging from upper intertidal to hadal depths (below 20,000 ft or 6,100 m). The ecological dominance of the Demospongiae reflects their diversity in form, structure, reproductive capabilities and physiological adaptation. Cnidarians and ctenophores are mostly marine; however, a few groups have successfully made their way into freshwater habitats.

Most lower metazoans are either sessile polyps, which means that they are attached at the base to the surface that they live on, or planktonic carnivores. Some, however, employ suspension feeding and many species harbor symbiotic intracellular algae that supply them with energy. Hydroids, scyphozoans, and anthozoans live in seas around the globe, from polar to tropical oceans. Most lower metazoans, however, live in coastal waters.

Physical factors

Light

Sunlight has an important role in both terrestrial and marine environments, powering the process of photosynthesis that provides energy either directly or indirectly to nearly all forms of life on earth. The diel, or 24-hour cyclical migra-

tions of epipelagic species, are at least in part active responses to changing light levels. Epipelagic refers to the upper levels of the ocean that are penetrated by enough sunlight for photosynthesis to occur. *Aurelia aurita* approaches the surface during the day, at both midday and midnight, or only at night, and becomes scattered throughout the water column at night or during the sunlit days. Diel migrations probably do not occur in the bathypelagic zone (about 3,280–6,562 ft or 1–2 km); migration in the mesopelagic zone (about 656–3,280 ft or 200–1,000 m) depends on the levels of available light in that zone. Sunlight is necessary for vision as well as photosynthesis. Many animals rely on their vision to capture prey, avoid predation, and communicate with one another.

Bioluminescence is a type of visible light produced by marine animals such as scyphozoans, hydrozoans, ctenophores, squids, thaliaceans, and fishes. It may be used for counterillumination or as ventral camouflage. Another possibility is that bioluminescence is a useful defense mechanism against potential predators.

Turbidity

Turbidity refers to the cloudiness of sea water caused by the suspension of sediment particles and organic matter. High concentrations of suspended particles in the water over offshore coral reefs are considered a stress factor for coral colonies because they reduce the amount of light for photosynthesis and smother coral tissues. Nevertheless, many reefs with large growths of coral are found in relatively turbid waters, such as the fringe reefs around the inshore continental islands in the Great Barrier Reef lagoon. This finding suggests that turbid water is not necessarily harmful to coral. Fine, suspended particles provide a large surface area for colonization by microorganisms that produce organic nutrients. By limiting light penetration, turbid water also limits the distribution of both benthic algae and phytoplankton, which are at the base of the web.

Temperature

All lower metazoans are ectothermic (sometimes referred to as "cold-blooded"), which means that they retain the same temperture as their surroundings. Because of this restraint; invertebrate physiology has evolved to operate in a specific temperature range for each species. Most organisms can tolerate only a narrow range of temperatures; changes above or below this critical range disrupt their metabolism, resulting in a lowered rate of reproduction, injury, or even death. Since temperatures change less rapidly in the open sea than in shallow waters, species in shallow waters can tolerate a wider range of temperature than deep water species. Temperature often influences the distribution, reproduction, and morphology (form and structure) of these organisms. Colonies of *Obelia geniculata* and *Silicularia bilabiata* living in cold water develop long, branching hydrocauli (stalks), whereas colonies of these species living in warm waters have short stems with few branches. Gametogenesis in the hydrozoan *Sertularella miuresis* begins when the temperature reaches 50°F (10°C) and stops when it reaches 64°F (18°C). *Coryne tubulosa* reproduces asexually at around 57°F (14°C), but produces medusae when the temperature falls to 35°F (2°C). The acclimation temperature of *Chrysaora quinquecirrha* polyps is about 51°F (10.50°C), but the upper lethal temperature dose, defined as the temperature at which 50% of the test animals die (LD-50), is 95°F (35°C).

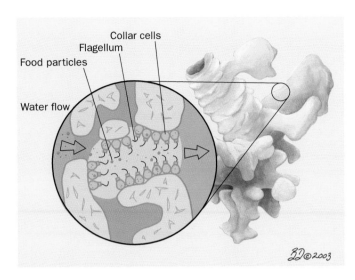

A sponge feeding. (Illustration by Barbara Duperron)

Salinity

Salinity, or the level of salt content in seawater, can affect invertebrates. Species that have evolved to live in freshwater can rarely live in salt water, and few marine species can tolerate low salinities or freshwater. This becomes quite apparent when one studies species richness (number of species) as one moves down a river into an estuary. Species richness (number of species) is relatively high in freshwater, then decreases considerably as salinity increases to about 5ppt, where most freshwater species cannot exist. Species richness then increases with salinity as more low-salinity-tolerant species are encountered. Species richness is at its greatest at the mouth of the estuary, where fully marine species occur with estuarine species.

Salinity can effect the morphology of organisms. For example, the shape, number, and size of tentacles of *Cordylophora caspia* polyps is affected by salinity. The scyphozoan medusae of *Rhopiena esculenta* can survive at levels of salt concentration as low as 8 parts per thousand (ppt), the scyphistomae to 10 ppt and the planulae to 12 ppt. The estromatolites of *Phylloriza peronlesueri*, however, form in hypersaline (very salty) waters. A rise in the salt content of the Baltic Sea allowed *A. aurita* and *C. capillata* to expand into northern waters, and allowed *Rhizostoma pulmo* and *A. aurita* to move from the Azov Sea into the Black Sea. The hydroid *Laomedea flexuosa* increases its production of gonozooids when the seawater concentration is around 30–40 ppt; at higher concentrations, however, the colonies begin to degenerate. The cephalochordate *Branchiostoma nigeriense* becomes opaque above salinities of 13 ppt.

Ocean currents and turbulence

Moving water is essential to lower metazoans because it supplies food and dissolved gases; prevents the accumulation of sediment; and disperses waste products, medusae, and larvae. *Aglaophenia picardi* resorbs the tissues of its hydrocauli into the hydrorhiza when the surrounding water is relatively stagnant, but regenerates them when the water begins to move more rapidly. The speed and direction of current flow affect the form and size of some lower metazoans. The size of hydroids is usually inversely related to the speed of water movement; large specimens are found in calm water and smaller specimens in rougher water. *Aglaophenia pluma* develops unbranched hydrocauli about 0.6 in (1.5 cm) tall in shallow, turbulent water, but produces branched hydrocauli as high as 19.6 in (50 cm) in deeper water with bidirectional currents. Planar (flat) forms such as *A. pluma*, *Plumularia setacea*, and *Eudendrium rameum* are most abundant where the current tends to flow in one direction, while radial or arborescent (treelike) forms such as *Lytocarpia myriophyllum*, *Nemertesia antenna*, and *E. racemosum* flourish in bi- or multidirectional currents. The distribution of species that inhabit coral reefs and display highly specific patterns of tolerance is greatly affected by water movement. Morphological differences in hydroids and anthozoans are also regarded as indicators of distinct patterns in water movement.

Water depth

The majority of organisms are not able to survive in great depths (below 3,281 ft or 1,000 m). In general, the number of invertebrates is highest in shallow water communities and decreases as water depth increases. However, species diversity may be quite high at great depths on the abyssal plain where the environment has been extremely stable for millennia. Though diversity can be high, biomass may be low in these deep benthic habitats, because the lack of light prevents any primary production. Therefore, these habitats are usually limited by food and depend on organic input from sunlit seas above.

The lack of mixing and primary production result in oxygen-minimum layers in the ocean, and many species are either adapted to lower oxygen concentrations or avoid these areas. The scyphozoans *Periphylla periphylla* and *Nausithoe rubra* show high levels of the anaerobic enzyme lactate dehydrogenase, probably as an adaptation to moving at depths between 1,312 and 4,921 ft (400–1,500 m), which has minimal levels of oxygen.

Aerial view of the reef complex of Heron and Wistari Reefs at low tide, showing an extensive reef system with a deep channel and coral cay, southern Great Barrier Reef, Australia. (Photo by A. Flowers & L. Newman. Reproduced by permission.)

Environmental contaminants

Pollution may result from contamination by sewage, hydrocarbons, polyvinyl biphenals (i.e., PCBs), pesticides (e.g., DDT), and heavy metals such as cadmium, copper, lead, mercury, and zinc. Experiments have revealed that exposure to pollutants can lead to sublethal effects in hydroids, including changes in the curvature or branching of the hydrorhiza; loss of hydranths; stimulation of gonozooid production; or changes in the rate of growth. Low concentrations of metal ions such as copper and mercury may inhibit growth regulation in hydroids while increasing the growth rate in *Laomedea flexuosa* and *Clavopsella michaeli*. In Elefsis Bay, a polluted area of Greece, populations of *Aurelia aurita* have multiplied to rates of more than 1,500 medusae per 10 m^3. Certain species of *Rhizostoma* have survived in parts of Madras Harbor that have been polluted by diesel oil; however, the presence of crude petroleum in the waters of Alaska has caused a reduction or cessation in the strobilation in the polyps of *A. aurita*, and the production of ephyra and polyps with both morphological and behavioral abnormalities. *Pelagia noctiluca*, a scyphozoan from the Mediterranean Sea, acquires high concentrations of cadmium, lead, mercury and zinc. Individuals of the species *Chrysaora quinquecirrha* have been found to have highly concentrated levels of the herbicide pendimethalin in their tentacles; they show no change in behavior at concen-

trations of the pesticide that are lethal to fishes such as perch. The dumping of raw sewage in may tropical areas of the world destroys coral reefs by increasing turbidity that prevents light penetration, increasing sediment loads that smother corals, and increasing nutrient loads that encourage algae growth that can out-compete corals.

Species interactions

Competition occurs when organisms require the same limited resources, such as food, living space, or mates; or when two groups of organisms try to occupy an ecological niche in the same location at the same time. Competition may either be interspecific (between different species) or intraspecific (within the same species). Hydroids, which have a stoloniferous growth pattern, demonstrate two different growth strategies. The first, a guerrilla strategy, is characterized by extensive hydrorhizal growth with little branching and sparsely spaced hydrocauli or polyps; this pattern is exhibited by *L. flexuosa*. Guerilla behavior is an opportunist-style strategy for reducing interspecific competition for space and the possibility of overgrowth by other organisms. The second strategy, phalanx; results in highly branched hydrorhizae with dense hydranths carried on large hydrocauli. The phalanx strategy is exemplified by *Podocoryne carnea* and *Hydractinia echinata*, which grow on the shells of hermit crabs. Intraspecific variation in the allocation of resources that lead to hydrorhizal growth can be observed in the colonies of *H. echinata*. This species usually shows either little outward hydrorhizal growth combined with a high rate of reproduction, or extensive hydrorhizal growth combined with a lower rate of reproduction and correspondingly greater competitive ability. Contact with another colony leads the hydrorhiza to produce an abnormally large number of stolons (shoots or runners) armed with nematocysts, which sting and kill the tissue of other hydroids. Guerrilla growth strategies have adaptive value in situations where there is relatively little space available, as on shells occupied by other hydroids, while the phalanx strategy is more advantageous for expanding the colony to shells that are inhabited by hydroids.

Competition for space is of prime importance in the coral reef ecosystem. Most of the aggressive species are small and have slow growth rates, while the less aggressive coral species have faster growth rates and are able to outpace their competitors through rapid growth. The ability to maintain either rapid growth or aggressively dominative practices, but never both, explains why no single species of coral is able to dominate a coral reef. One possible outcome of competition is the extinction of the less successful competitor.

A niche can be subdivided into two or more small niches with minimal overlap, allowing competing organisms to share a resource. Examples of resource partitioning may be found on coral reefs. Small ecological niches can be occupied by similar species if the anatomy, feeding behavior, and territory of each species are only slightly different from those of another. The hydrozoans *Hydractinia* (retained gonophores), *Stylactis* (medusoids), and *Podocoryne* (medusae) have similar morphologies but different reproductive strategies, which allows them to occupy similar niches on the shells of hermit crabs. The competitive ability of colonies may also depend on their size. *Podocoryne carnea* hydroids show a greater selective advantage in aggressiveness in relation to *H. echinata* in interspecific competition. In instances of intraspecific competition among different colonies of *P. carnea*, however, the colony most likely to lose out is the one with the slowest rate of growth.

Feeding mechanisms and behavior

Suspension feeding

Lower metazoans demonstrate a remarkable variety of feeding mechanisms. Most sponges are suspension feeders that subsist on such fine particles as bacterioplankton and dissolved organic matter. Sponges acquire food and oxygen from water that flows through them; this flow is actively generated by sponges beating their flagella (microscopic whiplike structures). This process also acts as a means of waste removal for sponges. The movement of water through sponges is aided by ambient currents passing over raised excurrent (providing outward passage) openings, which creates an area of low pressure above these openings. Sponges are also capable of regulating the amount of flow through their bodies by narrowing or partly closing off various openings. The volume of water passing through a sponge can be enormous—as much as 20,000 times its volume over a 24-hour period.

Sponges are size-selective particle feeders. Their aquiferous systems create a series of "sieves" of varying mesh size. The largest diameter of incurrent openings is usually around 0.002 in (50 μm), which keeps larger particles from entering the aquiferous system. A few species have larger incurrent pores, reaching diameters of 0.006–0.0069 in (150–175 μm). Some sponges trap roughly 90% of all bacteria in the water they filter. Other sponges also take significant amounts of dissolved organic matter into their aquiferous systems. In some demosponges, 80% of the organic matter that is filtered through their aquiferous system is too small to be seen by light microscopy. The other 20% is composed primarily of bacteria and dinoflagellates. Other sponges harbor symbionts such as green algae, dinoflagellates (zooxanthellae), or cyanobacteria, which also provide them with nutrients.

Predation

Many invertebrates are predators that feed on protozoans, other invertebrates, and fishes. The discovery of several Mediterranean species of sponges that capture and digest entire animals came as a surprise to marine biologists. These species of the family Cladorhizidae have no choanocytes or aquiferous systems, but anatomic and biological analysis revealed the presence of spiky filaments with raised hook-shaped spicules. These carnivorous sponges capture and hold small crustaceans with their spicules, which act like Velcro® tape when they come in contact with the crustaceans' exoskeletons. Once captive, the crustaceans cannot free themselves. They struggle for several hours, which indicates that the spicules do not produce any paralyzing or toxic secretions. Cells then migrate around the helpless prey, and digestion takes place outside the cell walls.

Most cnidarians are carnivorous, using cnidocytes on their tentacles to capture prey. Polyps, the sessile stage of cnidarians, are generally believed to be passive predators, feeding on animals that blunder into their tentacles. Some cnidarian medusae possess sensory structures resembling primitive eyes; they are active predators. Many corals and anemones feed by suspending thin strands or sheets of mucus over the surface of their colony. The sticky mucus collects fine particles of nourishment from the water; cilia present on the organisms drive the food-laden mucus into the mouths of coral or anemones. Many species have developed complementary adaptations such as ectodermal ciliary currents on their tentacles, oral discs or columns, or the ability to position themselves strategically within the water's flow pattern.

In sea anemones, the presence of nearby food evokes behavior that has two phases: a prefeeding, and a feeding response that leads to the ingestion of prey. The prefeeding response consists of the expansion of the oral disc, the movement of its tentacles, and both the extension and swaying of the column. This prefeeding behavior increases the chances of catching nearby food. The feeding response, which takes place after the prey has made contact with the anemone's tentacles or oral disc, includes the discharge of nematocysts and ingesting movements. Sea anemones are able to detect prey from the prey's emission of small dissolved molecules of

Flatworms (class Turbellaria) feed on smaller animals such as this water louse, or on dead animals. They help to keep the ecosystem in balance. (Photo by Animals Animals ©G. I. Bernard, OSF. Reproduced by permission.)

amino acids, tripeptides, and vitamins. The cnidae, which ensure the capture of the prey, are spirocysts and nematocysts. The numerous spirocysts on the tentacles of sea anemones appear to have an adhesive function. The contact of solid food with the tentacles leads to a massive discharge of spirocysts, which hold the prey while the nematocysts inject their toxin. Ingestion is directed by the chemical and mechanical stimuli produced by the immobilized prey. After ingestion, the prey is enclosed by filaments whose cnidoglandular tracts contain nematocysts but no spirocysts. The penetrating filaments of these nematocysts inject more toxin. The prey is then subject to the action of secretory cells that ensure its extracellular digestion.

The prey of sedentary cnidarians is composed of small motile animals such as zooplanktonic larvae, isopods, amphipods, and polychaetes. Sea anemones found closer to shore may complete their diet with larger sessile prey dislodged by wave action or foraging predators. The size of the prey is generally small considering the diameter of the sea anemone, and many species may be considered microphagous. Of the nine common species of Caribbean reef sea anemones, seven are planktivores. *Condylactis gigantea* and *Stoichactis helianthus*, which eat macroscopic prey such as gastropods and echinoids, probably depend on heavy wave action in the reef to supply them with prey.

Comb jellies are entirely predatory in their habitats. The long tentacles of ctenophores have muscular cores with an epidermal cover that contains colloblasts, or adhesive cells. The tentacles trail passively through the water or are twirled about by various circular movements of the body. Upon contact with the prey, the colloblasts burst and discharge a strong sticky material. In ctenophores, which bear very short tentacles (orders Lobata and Cestida), small zooplankton are trapped in

mucus on the body surface and then carried to the mouth by ciliary currents (along ciliated auricular grooves in ctenophore lobates and ciliated oral grooves in cestids). Under conditions of starvation in an aquarium, lobate adults often swim vertically through the water only to descend with their lobes extended; in this position the width of the lobes may be as much as 116% of the animal's length. The lobe width decreases to about 79% of the body length after food is placed in the aquarium and the animal begins to feed. Once the digestive tract is full, these adults still continue to feed by entangling their prey in mucus, which produces a bolus or clump near the mouth. Quite frequently they will either spit out this ball of food or completely empty their digestive tract and continue to feed. This behavior pattern can continue for several hours until the concentration of food is reduced to the point at which all the prey that have been captured can be ingested.

Most pelagic ctenophores and cnidarians feed primarily on copepods. There are a few intriguing examples of pelagic cnidarians that feed exclusively on one type of prey. The siphonophore *Hippopodius hippopus* feeds only on ostracods; the hydromedusae *Bougainvillia principis* feeds mainly on barnacle nauplii; and *Proboscidactyla flavicirrata* eats only veligers (mollusk larvae). Some cnidarians and ctenophores feed specifically on gelatinous prey or fishes. Ctenophores of the genus *Beroe* offer well-known examples of selective feeding on gelatinous prey: *B. cucumis* feeds exclusively on the ctenophore *Bolinopsis vitrea*, and *B. gracilis* feeds only on *Pleurobrachia pileus*. When beroids prey on animals larger than themselves, they appear to attach themselves to the prey and suck its tissues into their mouths. Beroids lack tentacles; however, they do possess some 3,000 macrocilia that are hexagonally arranged and form a ciliary band around the inside of the mouth that beats inward, and forcing tissue from the prey into the beroid's pharynx. Gelatinous species that include high proportions of soft-bodied prey in their diets often eat fish eggs and larvae when they are available.

The diets of gelatinous predators generally show some selectivity, and are dependent on factors including the prey's size; the width and spacing of the predator's tentacles; the predator's swimming behavior and speed; water flow; and the prey's ability to escape. In general, species that catch large prey have few and widely spaced tentacles, while those that feed on small prey have numerous closely spaced tentacles. Most gelatinous predators move while feeding, which allows them to make use of water currents that will bring prey toward their tentacles. Cnidarians that are ambush predators are able to catch large, fast-moving prey; cruising predators prefer small, slow-moving prey. For example, siphonophores, which are ambush predators, tend to select large and relatively swift prey, while *Aurelia aurita*, which is a cruising predator, selects slow-moving organisms. Swimming offers the advantage of allowing predators to scan larger volumes of water; however, it also has the disadvantage of alerting the prey to the predator's presence. Some predators deal with this disadvantage by remaining stationary while they are "fishing."

Swimming is not the only way for prey to escape. Bivalve veligers will close themselves off when they are disturbed by the medusae of *Chrysaora quinquecirrha*; 99% of captured veligers are ingested alive.

Ecological significance of predation

The feeding rate of predators can be expressed in terms of clearance rate (volume of water × predator −1 × time −1), or in terms of numbers or biomass of prey captured (predator −1 × time −1) and are often combined with estimates of predator and prey densities in order to estimate the effects of predation on prey populations. One characteristic of gelatinous predators is that clearance rates tend to be constant even at extremely high prey densities. In species with few tentacles (e.g., *Pleurobrachia* spp.), however, ingestion is limited by prey handling. Saturation feeding at typical prey densities rarely occurs in situ. Gelatinous predators feeding on small prey seldom fill themselves at natural prey densities. They may not reduce populations, but they do keep them in check. Predators that consume ichthyoplankton often appear in areas of intense spawning activity and are major causes of fish egg and larva mortality. In some areas large schyphomedusae consume high numbers of commercially important fish larvae, and they also compete with fishes for food. Swarms of jellyfish may be so dense that they clog and damage fishing nets. Many hydromedusae species, including *Porpita* spp., *Velella* spp., and *Physalia* spp., also occur in huge concentrations, particularly in tropical seas.

Some species have such a significant ecological effect that they are considered "keystone species." If these species disappear or appear in an environment, the entire habitat can shift dramatically. For example, active predation by sea stars can significantly affect prey populations. The crown-of-thorns sea stars (*Acanthaster* spp.) feed on coral polyps in tropical reef habitats worldwide. Occasionally, population explosions of *A. planci* occur that can have devastating effects on coral reefs. Thousands of square miles (kilometers) of bare coral skeletons can result. Species composition and diversity of other inhabitants are affected secondarily as algae and other organisms colonize the reef. Organisms that depend on live coral for survival either leave the area or die.

Aggressive and defensive behavior

Gastropods, pycnogonids, sea stars, sea urchins, fishes, and sea turtles are predators that feed on invertebrates. Moreover, jellyfishes and comb jellies can be predators of other cnidarians and ctenophores. Sea turtles, especially *Dermochelys coriacea*, feed on scyphomedusae such as *A. aurita*. In addition, birds may add scyphozoans to their diets. Some benthic animals like the nudibranchs may feed on the scyphistomae of *Cyanea capillata* and *A. aurita*. A single nudibranch can consume as many as 200 polyps per day; however, not all nudibranchs are able to eat scyphozoans. As in the pelagic environment, scyphozoans in their benthic stages may eat one another; for example, the scyphistomae of *A. aurita* eat the planulae larvae of *C. capillata* as well as the larvae of their own species. In addition to natural predation, scyphozoans may be affected by fisheries.

In general, these species defend themselves against predators by the production of physical structures or the emission of various chemicals. Sponges are the most diverse source of marine natural products; some of these compounds offer potential pharmacological benefits. Compounds isolated from sponges vary widely in structural complexity; they include sterols, terpenoids, amino acid derivatives, saponins, and macrolides. The toxins produced by temperate and tropical sponges have been shown to deter predation by fishes, asteroids, and gastropods. The organic component of sponges consists primarily of NaOH-soluble and insoluble protein, NaOH being the chemical formula for sodium hydroxide, or lye. About 56% of Antarctic sponges are toxic. *Leucetta leptorhapsis* and *Mycale acerata* are highly toxic; however, the asteroid *Perknaster fucus antarcticus* is a specialist and is able to feed on *M. acerata* without succumbing to its toxins. This finding suggests that some asteroids have evolved physiological mechanisms that neutralize or sequester (compartmentalize) sponge toxins. In many cases these secondary metabolites are not necessarily toxic, but may make the sponge distasteful to predators and may be more effective in deterring predation. These compounds not only help the sponges avoid predators, but also prevent infection by microorganisms. In addition, they allow the sponges to compete for space with other sessile invertebrates such as ectoprocts, ascidians, corals and even other sponges. Clinoid sponges are among the most common and destructive endolithic (living embedded in rock surfaces) borers on coral reefs worldwide. *Cliona*, *Anthosigmella*, and *Spheciospongia* of the order Hadromerida; and *Siphonodictyon*, of the order Haplosclerida are siliceous sponges known to bore into hard substrates. Such sponges are able to excavate galleries in calcareous material by removing small fragments of the mineral by specialized archaeocyte cells. The cells secrete chemicals that dissolve the calcareous substrate. When infested corals are split open, clinoid sponges appear as brown, yellow, or orange patches lining the corroded interiors of the coral skeleton.

The sponge *Cinachyra antarctica* has distinctively long spicule tufts that emerge from the spiral conules on their surface. This species, found throughout Antarctica at depths of 59–2,496 ft (18–761 m), uses its spicules to protect itself from predators. When spicules are removed, *C. antarctica* is made vulnerable to predators.

Defensive and feeding activities are closely associated in most cnidarians; the tentacles of most anemones and jellyfishes serve both purposes. In some cases, however, both functions are performed by separate structures. Sea anemones and corals have developed several specialize structures used to defend against territorial invasion. Three of these structures, namely acrorhagi (special fighting tentacles), catch tentacles, and sweeper tentacles, are modified feeding tentacles. The mesentenic filament is another modified defensive structure. The acrorhagi are located at the margin of the anemone's body column. When these anemones make physical contact with one another, usually with their tentacles, the acrorhagi expand and apply themselves on the target organism. The ectodermal tissue of the acrorhagus lifts away from its underlying mesenclyme (cellular jellylike material) while the acrorhagus discharges its nemactocysts, and the ectoderm then clings to the target organism. This process is called peeling. As a result of continued discharges from the nematocysts, the victim's tissue beneath the acrorhagial peel becomes necrotic and dies. Catch tentacles (in sea anemones) and sweeper tentacles (in scleractinian corals) develop from feeding tentacles that undergo a morphological change when the organism comes into contact with appropriate other species. In response to weeks of contact, the feeding tentacles alter their form, structure, and complement of cnidae. Catch and sweeper tentacles do not adhere to potential food objects; when they are touched by prey, they actually retract. Sweeper tentacles emerge at night; as their name implies, they flail about or undulate. They can reach 5–10 times the length of feeding tentacles. The coral *Montastrea cavernosa* is a mildly aggressive coral, capable of destroying the tissue of a variety of subordinate coral species with its mesenterial filaments. But its own mesenterial filaments can be destroyed by *M. annularis* when both are placed together. These species have sweeper tentacles, which are multipurpose structures with the capacity to regulate the distance between colonies, thus functioning as organs of competition.

Although the ability of hydroids to resist predation is often attributed to their nematocysts and associated toxins, the chemical compounds make them much less attractive to a potential hydroid predator. Some species of hydroids secrete chemicals that deter feeding by the pinfish *Lagodon rhomboides*. After having been treated with potassium chloride, which forces them to discharge their nematocysts, both *Halocordyle disticha* and *Tubularia crocea* become palatable; this suggests that these species rely on nematocysts to defend themselves against predators. However, species such as *Corydendrium parasiticum*, *Eudendrium carneum*, *Hydractinia symbiologicarpus*, and *Tridentata marginata*, remain unpalatable after their nematocysts have been discharged.

Resources

Books

Arai, N. A. *A Functional Biology of Scyphozoa*. London: Chapman and Hall, 1997.

Bergquist, P. R. *Sponges*. Berkeley: University of California Press, 1978.

Nybbaken, J. W. *Marine Biology: An Ecological Approach*. New York: Harper and Row, 1982.

Periodicals

Fautin, D. G. "Reproduction of Cnidaria." *Canadian Journal of Zoology* 80 (2002): 1735–1754.

Gili, J. M., and R. G. Hughes. "The Ecology of Marine Benthic Hydroids." *Oceanography and Marine Biology: An Annual Review* 33 (1995): 351–426.

Harbison, G. R., and R. L. Miller. "Not All Ctenophores Are Hermaphrodites. Studies on the Systematics, Distribution, Sexuality and Development of Two Species of *Ocyropsis*." *Marine Biology* 90 (1986): 413–424.

Kass-Simon, G., and A. A. Scappaticci, Jr. "The Behavioral and Developmental Physiology of Nematocysts." *Canadian Journal of Zoology* 80 (2002): 1772–1794.

McClintock, J. B. "Investigation of the Relationship Between Invertebrate Predation and Biochemical Composition, Energy Content, Spicule Armament and Toxicity of Benthic Sponges at McMurdo Sound, Antarctica." *Marine Biology* 94 (1987): 479–487.

Purcell, J. E. "Pelagic Cnidarians and Ctenophores as Predators: Selective Predation, Feeding Rates and Effects on Prey Populations." *Annales de l'Institut Océanographique* 73, no. 2 (1997): 125–137.

Reeve, M. R., and M. A. Walter. "Nutrionatal Ecology of Ctenophores — A Review of Recent Research." *Advances in Marine Biology* 15 (1978): 249–287.

Stachowicz, J. J., and N. Lindquist. "Hydroid Defenses Against Predators: The Importance of Secondary Metabolites Versus Nematocysts." *Oecologia* 124 (2000): 80–288.

Van-Praet, M. "Nutrition of Sea Anemones." *Advances in Marine Biology* 22 (1985): 65–99.

Other

"Spiky sponge *Cinachyra antarctica*." Underwater Field Guide to Ross Island and McMurdo Sound, Antarctica. Scripps Institution of Oceanography Library. <http://scilib.ucsd.edu/sio/nsf/fguide/porifcra16.html>

Alexandra Elaine Rizzo, PhD
Eliane Pintor Arruda, MSc
Dennis A. Thoney, PhD

Symbiosis

Introduction

Symbiosis is an association between two or more different species of organisms. The association may be permanent, the organisms never being separated, or it may be long lasting. Life is complex and often involves a delicate balancing act between hosts and symbionts, in associations that range from parasitism to mutualism. In the long history of life on Earth, symbionts have evolved many protective strategies in their attempts to overcome a host's defenses, including molecular camouflage, deception, mimicry, and subversion. By studying symbiosis one gains a wider evolutionary perspective on the extent and nature of biological interactions between species.

Symbiosis and modern biology

The recognition of symbiotic relationships has had a revolutionary impact on modern biological thought. The idea that mitochondria and chloroplasts are transformed by symbiotic bacteria provides a common thread to the biological world and raises hope of finding other symbiotic wonders among life's diversity. Plants and animals have acquired new metabolic capabilities through symbioses with bacteria and fungi. Mammalian herbivores and termites digest cellulose with the help of microbial symbionts. The luminescent bacteria contained in the specialized light organs of some fishes and squids produce marine bioluminescence. Diverse animal life around deep-sea vents is based on symbiosis with bacteria that oxidize hydrogen sulfide and chemosynthetically fix carbon dioxide into carbohydrates. Associations between fungi and algae have resulted in unique morphological structures called lichens. Early land plants formed associations with mycorrhizal fungi, which greatly facilitated their phosphorous uptake and thus played a significant role in the plants' ability to colonize terrestrial habitats. Evolutionary changes in organisms and their gene pools are not restricted to nuclear events and sexual mechanisms. Horizontal gene transfer between species has been documented in all forms of life. Bacterial cells possess plasmids and viruses that transfer new genetic properties from one cell to another. Many virulence factors in pathogenic bacteria are expressed through plasmid-borne genes. Similarly, bacteria become resistant to antibiotics when they incorporate plasmids with genes for antibiotic resistance. Horizontal gene transfer has been suggested in the evolution of flowers, fruits, and storage structures from gall-forming insects and viruses. The role of viruses as genetic engineers is gaining importance in evolutionary biology. The *Rhizobium*-legume symbiotic relationship is an excellent example of how host cells and bacterial symbionts within root nodules undergo transformation, which allows the bacterial cells to fix nitrogen-converting atmospheric nitrogen into a chemical form that can be taken up by plants. Within the host cells, Rhizobia acting as bacteroids behave as temporary cell organelles that fix nitrogen. Intragenomic conflict is an evolutionary force. The evolution of sex was a form of genomic conflict management. Uniparental inheritance of cytoplasmic genes, mating types, and many features of sexual behavior may have evolved as a result of evolutionary conflict. The two-sex model that is widespread throughout the diversity of life may have been the result of ancient intracellular symbiosis. The Red Queen hypothesis suggests that harmful parasites and virulent pathogens exert selection pressure on their hosts so that sexual selection is maintained. Parasites, pathogens, and their hosts are involved in a microevolutionary "arms race" and in time, the symbionts' offense and the host defenses produce cycles of coadaptations.

Types of symbioses

The term "symbiosis" was, in a broad sense, originally intended by Anton de Bary in 1879 to refer to different organisms living together. Proposals to change this definition and redefine symbiosis, such as equating it to mutualism, have led to confusion. Various types of symbioses, whether beneficial or harmful, are described by the terms commensalism, mutualism, and parasitism.

The term "commensalism" was first used by P. J. van Beneden in 1876 for associations in which one animal shared food caught by another animal. An example of a commensalistic symbiosis is the relationship between silverfish and army ants. The silverfish live with the army ants, participate in their raids, and share their prey. They neither harm nor benefit the ants.

In mutualistic symbiosis, both partners benefit from the relationship. The extent to which each symbiont benefits, however, may vary and is generally difficult to assess. The complex interactions that take place between the symbionts may involve a reciprocal exchange of nutrients. For example, in the symbioses of algae and invertebrates (such as corals, anemones, and flatworms), the algae provides the animals with organic compounds that are products of photosynthesis, while the animals provide the algae with waste products such as nitrogenous compounds and carbon dioxide, which the algae use in photosynthesis. Unfortunately, in many academic circles, the terms symbiosis, mutualism, and cooperation have similar meanings and are often used interchangeably. Mutualism has also been widely used to describe intraspecific cooperative behavior in various animal species. The study of cooperation has enjoyed a resurgence during the past several decades. The evolution of cooperation via byproduct mutualism is generally found in the context of interspecific associations.

Parasitism is a form of symbiosis in which one symbiont benefits at the expense of its host. Parasitic symbioses affect the host in different ways. Some parasites are so pathogenic that they produce disease in the host shortly after parasitism begins. In other associations, the host and parasite have coevolved into a controlled parasitism in which the death of the host cells is highly regulated. Associations among many species are not clear and are more difficult to define categorically. For instance, when in their larval form, flukes might be considered parasites to snails because they harm their host; but, adult flukes have a commensal relationship with snails because when present in the alimentary tract of invertebrates they only share digesting food.

Classification of symbioses

Many scientists have attempted to standardize the many conflicting terms that have been used to describe different symbioses, including:

- Ectosymbiosis: The partners remain external to each other, such as in lichens.

- Endosymbiosis: The smaller symbionts are inside the host, but remain extracellular. Most of the time endosymbionts are in the digestive tract, or inside particular organs.

- Endocytobiosis is intracellular symbiosis. Symbiosome membranes are the host cell's vacuoles that enclose the symbiont.

- Obligate symbionts are so highly adapted to a symbiotic experience that they cannot live outside of it.

- Facultative symbionts, however, can live in a free-living condition.

Some examples of symbiosis in lower metazoans and tunicates

Commensalistic associations

Sharing of food and the provision of shelter are two main features of commensalistic relationships. Many species that display commensalistic relationships inhabit the internal spaces of sponges, clams, and sea cucumbers. The symbionts are often smaller and more streamlined than their free-living relatives and show evidence of long-term associations. For example, there are crab and shrimp species that live in the mantle cavities of bivalve mollusks; the pearl fish, *Corpus*, shows both structural and behavioral modifications that adapted in order to live in the cloacal spaces of sea cucumbers. These adaptations include a dramatic shift of the anal opening to just beneath the head, and the loss of both scales and the pelvic fins. In tropical water the hat-pin urchin *Diadema*, with its long needle-like spines, provides protection to fish such as *Aeoliscus* (the shrimpfish) and *Diademichthys* (the clingfish). These elongated fish species hide among the host's spines, which are constantly moving, by orienting themselves parallel to the spines. Another common example of commensalism is the relationship that exists between fishes and jellyfish. Fishes of the family Nomeidae congregate among the tentacles of jellyfish for protection. The anemonefishes keep the surface of their host anemones free of debris and may also lure fishes into the tentacles, thus providing food to the host.

Sponges

Marine sponges contain a variety of endosymbionts, including bacteria, dinoflagellates, diatoms, and cryptomonads. Symbionts are especially common among tropical sponges. Many sponges contain endosymbiotic cyanobacteria that are intercellular (in sponge tissue). The sponge obtains nutrients from the digestion of bacteria or from the excretion of compounds such as glycerol and nitrogen from bacteria. In turn the bacteria receives nutrients and a place to live.

Phytosynthetic associations

Green hydra-*Chlorella* symbiosis

Hydra are common inhabitants of freshwater lakes and ponds, where they feed on small animals. *Hydra viridis* contains the green alga *Chlorella*. Algae reproduces asexually within the gastrodermal cells and a single hydra may contain about 150,000 algal cells. Under normal conditions, symbiotic algae are not digested by hydra. There are two reasons for this: first, the cell wall of algae contains sporopollenin, a protein that resists digestive enzymes; second, vacuoles containing algae do not fuse with lysosomes, the organelles that contain digestive enzymes and normally fuse with food particles. But if a digestive cell takes in more algal cells than normal, the extra cells are either digested or ejected. A bilateral movement of nutrients takes place between the symbionts. Algae supplies the animals with photosynthetic products such as maltose. At an acidic pH level, almost 60% of the carbon fixed by the algae is excreted as maltose, but at a neutral pH level, very little maltose is excreted. The rapidly hydrolyzed maltose is converted to glucose, and then glycogen is produced. Algae also provide the animal with oxygen, which they produce during photosynthesis. Hydra provides the algae with nutrients, including precursors of proteins and nucleic acids, and a protected place to live. As digestion is avoided and the host cells are able to regulate algal reproduction, the sym-

A sponge (*Suberites ficus*) gets more food by travelling with a hermit crab (*Pagurus beringanus*). The hermit crab profits from the sponge's protective cover. (Photo by ©Tom McHugh/Photo Researchers, Inc. Reproduced by permission.)

biosis that exists between *H. viridis* and *Chlorella* appears to be a finely tuned, nonpathogenic equilibrium.

Marine algal-invertebrate symbioses

Many marine invertebrates, such as sea anemones, coral, and flatworms have formed mutualistic symbioses with the photosynthetic algae known as the dinoflagellates. Their chloroplasts have efficient light-harvesting complexes that include chlorophyll a, chlorophyll c, and large amounts of xanthophylls. A common dinoflagellate of marine invertebrates is *Symbiodinium microadriaticum*, and this is greatly modified when it lives inside animal cells. The algal cell wall becomes thinner, loses the groove and flagella, and divides only by binary fission. In the host animal the algae excrete large amounts of glycerol, in addition to glucose, alanine, and organic acids. When the algae are isolated from animals and grown in culture, they stop excreting these substances.

Sea anemones and jellyfish

The sea anemone *Anthopleura xanthogrammica*, contains two types of symbiotic algae: zoochlorellae and zooxanthellae. The relative proportion of each algal symbiont in the animal depends on the water temperature. The anemones position themselves in ways to increase the exposure of their symbionts to light.

Cassiopea xamachana is a jellyfish that has been used to study how an invertebrate selects its algal symbionts. The lifecycle of *Cassiopea* includes a sexual medusoid stage, which contains algae that does not swim freely, but rather lies upside down in shallow waters, a behavioral adaptation that allows the algae in its tentacles to receive maximum daylight for photosynthesis, and gives the animal its common name, the upside-down jellyfish.

Anemone-clownfish symbiosis

Fishes of the genera *Amphiprion*, *Dascyllus*, and *Premnas*, commonly called clownfish, form mutualistic associations with giant sea anemones that live in coral reefs throughout the Pacific Ocean. The association is obligatory for the fish, but facultative for the anemones. The anemones eat prey that have been paralyzed by means of poisonous nematocysts discharged from specialized cells in their tentacles. The clownfishes are immune to the stinging nematocysts and can nestle among tentacles without harm. Some clownfish go through a period of acclimation before they become immune to the anemones' poison. Symbiosis with the anemone changes the mucous coating around the fish and the fish is no longer recognized as prey by the anemone. Clownfishes are brightly colored and marked, and attract larger fish to the anemone. These fish, if they come too close, are stung by the tentacles and eaten by the anemone. The clownfish share in the meal. A similar relationship exists between the Portugese man-of-war (*Physalia physalia*) and the horse mackerel (*Trachurus trachurus*). The bright blue and silver color of the fish, as well as its small size, attract prey for the man-of-war.

Reef-building corals

The symbiotic association between *Symbiodinium*-reef-building corals (Scleractinia) is of great importance in marine tropical ecosystems and has been the subject of many studies. Coral reefs support large communities of organisms. Coral polyps excrete a calcium carbonate shell around their body. As the polyp dies, the shells harden, and new polyps grow over them. After many years of this process, coral reefs are formed. Symbiotic dinoflagellates live inside nutrient-rich cells of the gasterodermis of the coral polyp. In some corals more than 90% of photosynthate may be released by the symbiont to its host cell. The algae supply the coral with oxygen, carbon, and nitrogen compounds. The animal obtains vitamins, trace elements, and other essential compounds from the digestion of old algal symbionts. Animal waste products such as ammonia are converted by the algae into amino acids, which are translocated to the animals. Such a recycling of nitrogen is an important feature in the nitrogen-poor habitats of coral. Coral bleaching is caused by the loss of symbiotic algae from the host and may be caused by environmental stresses such as global warming, pollution, and increased ultraviolet radiation.

Green flatworms

Convoluta roscoffensis is a small marine flatworm that lives in the intertidal zones of beaches in the Channel Islands of the United Kingdom and in western France. The worms are 0.08–0.16 in (2–4 mm) long and deep green in color from the algae they contain. During high tide, the worms are buried in the sand, but at low tide, during daylight, they move up to the surface. During this time the green algal symbiont, *Tetraselmis convolutae*, photosynthesizes until the next high tide. The *Convoluta*-algal symbiotic relationship is an early example of detailed studies (1910) that attracted public attention to the broader significance of symbiosis in nature.

Tunicates

Some marine tunicates contain an unusual photosynthetic symbiont, *Prochloron*, that has characteristics of both cyano-

Coleman's shrimp (*Periclimenes colemani*) lives exclusively with the venomous fire urchin (*Asthenosoma varium*). They are found often as a male/female pair on a bare patch that they have created by removing the urchin's tube feet and spines. They can move through the venomous spines without being harmed. (Photo by ©David Hall/Photo Researchers, Inc. Reproduced by permission.)

bacteria and green algae. In some tunicates the symbionts lie within a cellulose matrix that surrounds the outer surface of the animal, whereas in other tunicates symbionts are loosely attached to the cloacal wall. The larvae of some tunicates have specialized pouches that carry *Prochloron* cells that they obtain from the parent.

Symbiosis and animal parasitism

Scientists estimate that up to 50% of all animal species are parasitic symbionts. Some phyla such as the Platyhelminthes, Nematoda, and Arthropoda contain large a number of parasitic species. Hosts and parasites have coevolved together and under natural conditions many have become mutually tolerant. Host organisms can live independently, but, in most cases, the parasite's association with its host is obligatory. Animal parasites affect the health of humans and domesticated animals throughout the world. In most warm climates parasitic infections from flukes, nematodes, and arthropods greatly diminish the quality of life for people.

Helminths are widely distributed parasites of vertebrates. Infections caused by helminths such as schistosoma, hook-

worms, and filarial nematodes are a major cause of sickness of humans inhabiting the tropics. Helminths have complex lifecycles. They live for a long period within host animals, and they often possess a remarkable ability to evade the host's defense mechanisms. The prevalence of helminthic infections in some areas is high; however, only a few hosts develop disease. Helminths do not multiply in humans, and therefore the severity of the disease depends on the extent of the original infection. However, some helminthes may accumulate after repeated infection of a host.

Some fluke symbioses

Flukes (including the liver fluke, lung fluke, and human blood fluke) are obligate endoparasites of vertebrates as adults. After mating, the female fluke produces eggs into the host environment which are then are passed out of the host with feces or urine. There is a series of larval stages that multiply asexually in snails, serving as the first intermediate hosts. A larval stage (cercaria) with a characteristic tail emerges from the snail and either penetrates a vertebrate host immediately, encysts on to vegetation, or is eaten by a second intermediate host such as a crab or a fish, which may then be ingested by a vertebrate. *Fasciola hepatica*, the sheep liver fluke, commonly inhabits the bile duct, liver, or gallbladder of cattle, horses, pigs, and other farm animals. The Chinese liver fluke, *Clonorchis sinensis*, is an important parasite of humans and other fish-eating mammals in Far Eastern countries. Fish farming in east Asia is a major source of fluke infections in people. In other areas, dogs and cats serve as reservoir hosts of *C. sinensis*. *Paragonimus westermani*, the lung fluke, infects humans, cats, dogs, and rats. Occurences of this fluke are extremely prevalent in the people of China, the Philippines, Thailand, and other Asian countries.

Next to malaria, schistosomiasis is the most important parasitic disease in the world, affecting more than 200 million people in more than 75 countries. Schistosomes are blood flukes, and they reside in the mesenteric blood vessels of humans. Adult flukes are elongated and wormlike, and the female fluke is permanently held in the ventral groove of the male fluke. The presence of blood fluke eggs in various host tissues triggers an immune response, causing the affected person to show symptoms of disease that include enlargement of the liver and spleen, bladder calcification, and kidney disorders. Three important blood flukes that infect humans are *Schistosoma mansoni*, *S. japonicum*, and *S. haematobium*. *Schsitosoma mansoni* has been known to cause intestinal bilharziasis among people in South America, Central America, and the Middle East. Urinary schistosomiasis is caused by *S. haematobium* and is thought to occur in about 40 million people in Africa and the Middle East.

Cestodes: Tapeworms

Tapeworms represent the ultimate example of biological adaptation in order to live in another organism. All tapeworms are obligate symbionts of vertebrates and arthropods. Sexually mature tapeworms live in the intestines of vertebrates; in their larval stages they develop in the visceral organs of an alternate host, which may be a vertebrate or an arthropod. Serious diseases are caused by the progressive larval stages that take place in the muscles and nervous tissue of the vertebrate host. Some scientists view the adult tapeworms in the alimentary canal as endocommensals living in a nutrient-rich environment. Tapeworms lack a digestive system and absorb all their nutrients through their tegument, which is remarkably similar to that of flukes. *Diphyllobothrium latum*, the fish tapeworm, is a common inhabitant of the alimentary canal of fish-eating mammals, birds, and fishes. In temperate climates, people who eat raw fish often carry *D. latum*. The fish tapeworm is well known for its ability to absorb vitamin B[12], thereby causing the host to be deficient in a vitamin that is essential for the development of red blood cells. Humans become infected with pork tapeworm (*Taenia solium*) when they eat undercooked meat. Humans acquire the beef tapeworm, *Taenia saginata*, by eating raw or undercooked beef. *Hymenolepis diminuta*, the rat tapeworm, has been a favorite experimental subject to investigate the nutrition, biochemistry, immunology, and developmental biology of tapeworms.

Nematodes

Roundworms are second only to insects as the most abundant animals on earth. Most nematodes are free living. They occur in freshwater, marine, and soil habitats, feeding on microorganisms and decaying organic matter. Many nematodes are adapted for a parasitic lifestyle in plants, fungi, and ani-

Young brittle stars on sea lettuce. (Photo by Animals Animals ©G. I. Bernard, OSF. Reproduced by permission.)

mals. Scientists estimate that every kind of animal is inhabited by at least one parasitic nematode. Many nematodes live in the alimentary canals of their hosts, while others parasitize organs such as the eyes, liver, and lungs, often causing destruction to the host tissue. *Ascaris lumbricoides* is one of the largest intestinal nematodes present in humans and is prevalent throughout warmer climates. The two most important hookworms are *Ancylostoma duodenale*—the oriental hookworm of China, Japan, Asia, North Africa, the Caribbean Islands, and South America, and *Necator americanus*—which is primarily in South and Central America but also present in Africa and Asia. An estimated one billion people who live in the warmer climates of the world are infected with these nematodes, but most are asymptomatic. *Trichinella spiralis*, one of the most common parasites of vertebrates, has been studied extensively by physicians, experimental biologists, and ecologists.

Filarial nematodes are obligate parasites with complex life-cycles involving humans, other vertebrates, and arthropod vectors. *Wuchereria bancrofti* and *Brugia malayi* occur in the lymphatic system and cause the elephantiasis disease. The filarial worm, *Onchocerca volvulus*, causes skin tumors and blindness and is prevalent among the people of Africa and the Middle East. A typical filarial lifecycle begins when humans acquire the parasite from the bite of an infected bloodsucking insect. Once in the bloodstream or lymphatic system the nematode larvae become sexually mature. The mature female gives birth to larval stage microfilariae, which infest the biting insects that continue the lifecycle. Symptoms of filarial diseases are the result of host immune response and the physical blockage of ducts.

Nematode-insect symbioses

Insects are the dominant form of life on earth, and nematodes have successfully evolved symbioses with many of them. Nematodes that are symbionts of insects have intricate lifecycles that are synchronized with those of their hosts. *Heterorhabditis* and *Steinernema* are nematodes that parasitize insects and transmit bacteria that kill the hosts. The possibility of developing a biological control for mosquitoes has heightened interest in the mermithid nematode, *Romanomermis*, which kills mosquito larvae.

Most nematodes that attack plants are obligate parasites and include root-knot nematodes (*Meloidogyne* spp.) and cyst nematodes (*Globodera* and *Heterodera* spp.) that cause destructive infections in crop plants. There are more than 2,000 species of plant-parasitic nematodes, but few species are pests. Cell proliferation, giant cell formation, suppression of cell division, and cell wall breakdown are some of the host responses to nematode parasitism.

Bursaphaloenchus exylophilus is a nematode that lives in weakened or dead pine trees. A beetle that may carry up to 15,000 juvenile nematodes to a new location spreads the nematode, which feed on wood tissue and are suspected of killing pine trees. The relationship between the plant-parasitic nematode and the bark beetle is thought to be an example of mutualistic symbiosis.

Resources

Books

Cheng, Thomas. *General Parasitology*, 2nd edition. Orlando, FL: Academic Press, 1986.

Douglas, Angela. *Symbiotic Interactions.* Oxford: Oxford University Press, 1994.

Kennedy, M. W., and W. Harnett, eds. *Parasitic Nematodes: Molecular Biology, Biochemistry and Immunology.* Oxon, U.K.: CABI Publishing, 2001.

Margulis, Lynn. *Symbiosis in Cell Evolution*, 2nd edition. New York: W. H. Freeman, 1993.

Paracer, Surindar, and Vernon Ahmadjian. *An Introduction to Biological Associations*, 2nd edition. New York: Oxford University Press, 2000.

Organizations

International Society of Endocytobiology.
Web site: <http://www.endocytobiology.org/>

International Symbiosis Society.
Web site: <http://www.ma.psu.edu/lkh1/iss/>

Surindar Paracer, PhD

Behavior

Behavioral characteristics of invertebrates

This chapter will familiarize you with issues and examples related to the behavior of invertebrates. The large number and sheer diversity of invertebrates requires a restriction in the types of behaviors (and species) that can be discussed. The behaviors selected were based in part on their importance to the survival of an individual organism. Since there is little known about the behavior of many of the lower invertebrate and deuterstome phyla, examples of insects and other protostomes were used to illustrate the various kinds of behavior mentioned.

All animals are metazoans and are characterized by being multicellular. The principles of behavior discovered in unicellular organisms are fundamentally the same in multicellular organisms. The similarities that exist between both forms of organisms are fascinating because the evolution of multicellularity has led organisms to develop a fantastic array of complex activities and modifiable behavior. Consider, for instance, the behavior of the marine sponge *Sycon gelatinosum* (phylum Porifera). During the larval stage it is a free-swimming animal that moves toward light at the water's surface. As it develops, it lives near the substrate and eventually becomes a sessile adult. The adult sponge is no longer a free-swimming animal and is entirely incapable of active movement. Reactive behavior is elicited from the adult sponge when individual cells are stimulated directly. The resulting responses are localized, slow, and uncoordinated. The inability to hunt food items such as bacteria, plankton, and detritus requires the sponge to develop specially designed feeding structures that bring food to it.

Contrast the poorly coordinated behavior of sponges with the more active behavior of animals in the phylum Cnidaria, which includes multicellular marine animals such as jellyfish, corals, sea anemones, and the freshwater *Hydras*. In cnidarians, cells performing the same function are grouped into tissue. The creation of tissue allows cnidarians to behave in more complex ways than sponges. *Hydras*, for example, coordinate their tentacles to grasp prey, contract their entire bodies in response to strong mechanical stimulation such as predatory attacks, and move a single tentacle in response to a non-threatening organism or passing shadow.

A great advance in behavior is seen in worms of the phylum Platyhelminthes. This phylum contains animals such as planarians, flukes, and tapeworms. In these animals we find the first evidence of characteristics critical for the development of complex behavior. These characteristics include bilateral symmetry, the appearance of a brain, polarized neurons, and definitive anterior and posterior ends—with the anterior end containing a head, and eyes. The advances present in this phylum make complex orientation possible. The first examples of complex learning are also present in flatworms. An example of this development is the flatworm's ability to to discriminate between two signals—one of which leads to a biologically relevant stimulus. This organism's ability to modify its behavior based on the possible consequences of an encounter in order to avoid dangerous situations moves an existing reflex into a new context. Although primitive types of behavior modification are possible in members of the phyla Porifera and Cnidaria, they are not as complex as those found in flatworms, nor are they retained for as as long as they are in worms.

The advances first seen in members of the phylum Platyhelminthes and elaborated by worms in the phylum Annelida (e.g., polychetes, earthworms, leeches) and reach their apex in members of the phyla Mollusca (e.g., snails, clams, squid, octopus) and Arthropoda (e.g., spiders, crabs, crayfish, lobsters, honey bees, wasps, ants). For example, the cephalopods' neural development, problem solving capability, sensory apparatus, and ability to modify behavior is unsurpassed among the invertebrates. Among the arthropods, social insects such as the honey bee and ant have astonishing examples of defensive, social, and learned behavior patterns. What is behavior, and who studies it? Behavior is not easy to define and various definitions exist. For example, physiologists might describe the "behavior of a neuron," but some comparative psychologists would find this objectionable. Generally speaking, behavior is defined as "what organisms do." Behavior is the action an animal takes in order to adjust, manipulate, and interact with its environment. Actions such as moving, grooming, and feeding can be referred to as maintenance behavior. Action that influences members of the same and/or different species can be called communication behavior. Behavior that is modifiable is known as learned behavior. In general, each of these three types of behavior defines or "orientates" the animal in space.

Various disciplines have contributed to the study of invertebrate behavior. These disciplines include comparative psychology, ethology, physiology, ecology, and entomology.

A stalked jellyfish (*Haliclystus auricula*) attaches to kelp and eelgrass near British Columbia. (Photo by ©Neil G. McDaniel/Photo Researchers, Inc. Reproduced by permission.)

Scientists engaged in the study of behavior often do so from an interdisciplinary approach in which psychologists, ethologists, physiologists, and entomologists all work alongside each other. Comparative psychologists have a special interest in searching for similarities and differences in behavior.

Orientation behavior

Over the years various classification systems have been developed to describe orientation, and the terminology used within these systems is confusing. Popular systems in which orientation can be described are kinesis and taxis. Orientation behavior represents an example of the type of behavior referred to as maintenance behavior.

Kinesis

The simplest response through which invertebrates find a suitable location to live is referred to as kinesis. The response is not directed toward or away from a stimulus, but nevertheless places the animal in an optimum location. Changes in activity, rate of movement, and/or turning is non-directional and directly related to the intensity of the stimulus from moment to moment. Kinetic responses often occur when the source of the stimulus cannot be sensed at a distance. Several types of kinesis are recognized including: barokinesis, hygrokinesis, orthokinesis, photckinesis, thigmokinesis, and klinokinesis. Theories regarding kinesis are made more con-

fusing when describing orientation using two kineses. For example, an animal that changes its rate of movement under illumination is said to exhibit "photo-orthokinesis." The words negative and positive also can be added to these terms in order to further adapt their meaning; an animal that is active under little or no illumination is said to exhibit "photo-negative kinesis."

Examples of kinesis:

BAROKINESIS

Various classes of invertebrates react to pressure changes, including increased locomotion because of changes in barometric pressure. Larvae of the crab genus *Carcinus* swim toward the water's surface when pressure increases. Copepods, adult and larval polychaetes, and jellyfish medusae are other examples of animals that increase their activity in response to pressure changes.

HYGROKINESIS

Increased locomotion in reaction to changes in humidity is referred to as hygrokinesis. Some species of nematoda are stimulated to move due to conditions of low humidity and are less active when there are high degrees of humidity. The increase in activity under dry conditions increases the chances of finding a suitable damp environment. Increasing locomotion based on fluctuations in humidity levels is important among terrestrial invertebrates (e.g., planaria) because, with the exception of insects, very few have developed methods of conserving significant amounts of water.

PHOTOKINESIS

Increased locomotion resulting from changes in levels of light is called photokinesis. Flatworms (e.g., *Dugesia dorotocephala*, *Dugesia tigrina*) all increase their activity depending on the intensity of illumination that surrounds them. Other examples of organisms that exhibit photokinesis include gill and skin fluke larvae (monogenea), jellyfishes, and rotifers. Not all invertebrates respond to increases in illumination. When conducting studies on the effect of light on activity levels it is important to separate the role of illumination from the temperature increases produced by light.

THIGMOKINESIS

This form of kinesis is defined as increased locomotion in response to changes in contact with the immediate physical environment. Some invertebrates are more active in open spaces than in closed spaces. Examples of closed spaces include cracks and crevices. For example, the contraction of longitudinal muscles in nematodes produces a whiplike undulatory motion that relies on environmental substrata for the body to push against; when they are placed in fluid without substrata they thrash around.

OTHER FORMS OF KINESIS

Orthokinesis refers to changes in the speed or frequency of movement in reaction to changes in the intensity of a stimulus. Stimuli that produces a change in direction (such as turning) is known as klinokinesis. Movement influenced by gravity is known as geokinesis, and changes of movement caused by water currents is known as rheokinesis. Kinetic responses also can be in reaction to chemical and temperature stimuli (chemokinesis and thermokinesis, respectively). The oncomiricidia (larvae) of the Monogenea have been shown to change speed and direction in reaction to gravity, current, and light stimuli.

Taxis

Taxis is a more complex response through which invertebrates find a suitable location to live. The response is directed toward or away from a stimulus to place the animal in an optimum location to inhabit. Changes in activity, rate of movement, and/or direction are related to the intensity of the stimulus gradient from moment to moment. Taxis differs from kinesis in that taxic responses allow invertebrates to engage in specific activity as opposed to general activity, relative to a stimulus source.

Taxis can be characterized by:

1. Whether the animal moves toward or away from a stimulus.

2. The way in which the animal moves.

3. The complexity of the sensory structures used to detect the stimulus.

An invertebrate with a single visual receptor can determine the direction of a light source simply by moving the receptor (such as turning its head) and sampling the stimulus gradient produced by the light. If the animal is attracted to light, its receptor becomes more active the closer it moves toward the light source. The majority of invertebrates have at least two receptors; the second receptor allows the animal to make simultaneous comparisons from each side of its body from moment to moment as it moves through a stimulus gradient.

Several different types of taxis are recognized including, phototaxis, klinotaxis, phototropotaxis, and phototelotaxis. Moreover, movements toward the source of stimulation are called positive, and movements away from the source are called negative. For example, movements toward a source of light is called "positive phototaxis," while movement away from light is referred to as "negative phototaxis."

Examples of various form of taxis include:

PHOTOTAXIS

An animal that moves toward (positive phototaxis) or away (negative phototaxis) from light is exhibiting phototaxis. Movement is parallel to the direction of light. Examples of animals that exhibit this type of behavior are jellyfishes, oncomiricidia (monogenea larvae), and some echinoderms (sea stars and sea urchins). Planarians are negatively phototoxic in that they seek less illuminated areas.

KLINOTAXIS

A change in directed movement based on successive comparisons of a stimulus is referred to as klinotaxis. The larvae of many flies, including the common house fly, *Musca domes-*

tica, find the location of a light source by moving their head left and right in order to compare the relative intensity of a stimulus. Derivatives of this behavior also occur to many lower invertebrates, including planaria and echinoderms.

PHOTOTROPOTAXIS

Animals that display phototropotaxis undergo movement toward a source of illumination based on a comparison of information gathered by their eyes. The animal orientates toward the direction of light (assuming it exhibits positive phototropotaxis) and moves in a direction that keeps the eyes equally stimulated. Phototropotaxis is best demonstrated experimentally in what is known as a "two-light experiment." In this design, an animal is placed between two light sources. Phototropotaxis is indicated if the animal follows a path between the two light sources thereby stimulating the photoreceptors equally. Phototropotaxis also can be detected by preventing visual information reaching one eye. This test can be done by removing or painting an eye. Phototropotaxis is indicated if the animal begins to engage in "circus movements." For example, when the honey bee (*Apis mellifera*) is blinded in one eye it will perform "circus movements" (sideways movements) toward a light source. Positive phototropotaxis can be detected if the animals continuously turns toward a light source; if an animal is negatively phototropotaxic it will continuously turn away from a light source (i.e., it will turn so that its blind side faces the light).

PHOTOTELOTAXIS

Movement directed toward one of two sources of illumination is known as phototelotaxis. This form of behavior is dependent upon the type of sophistication present in the visual receptors. If the eye is capable of forming an image that will allow the animal to identify the source of illumination, the animal can move directly toward the source without the need for comparing two sources of illumination. Phototelotaxis is found in invertebrates possessing compound eyes (arthropods such as insects and crustaceans) including hermit crabs, isopods, and mysid crustaceans. There are no known examples of this behavior among the lower invertebrate and invertebrate deuterostomes.

GEOTAXIS

Geotaxis refers to movement along lines of gravitational force. As with all forms of taxic behavior, the direction of movement can be either positive or negative. Geotaxis is observed on surfaces (especially inclines), in water, air, sand, or mud. The most pronounced examples of geotaxis are found in invertebrates that live in sand or mud. Many examples of geotatic behavior occur in animals with statocysts, although there are examples where statocysts are not involved. The statocyst is a heavy object (statolith) located in a fluid-filled chamber used to detect gravitational forces. When the statocyst is moved, the statolith induces movement by activating various sensory and motor systems that return the animal back to its normal balance. Geotaxis is most readily studied in invertebrates by having an animal crawl on a vertical glass plate that is gently rotated or inclined. In order to test burrowing animals such as polychaetes the animal can be sandwiched

A common or edible sea urchin (*Echinus esculentus*) on the move, with tube-feet extended. (Photo by Jane Burton. Bruce Coleman, Inc. Reproduced by permission.)

between two glass plates filled with sand, or on a rotating table or centrifuge. Examples of geotactic behavior can be found in medusa (e.g., *Cotylorhiza tuberculata*), planaria (e.g., *Convoluta roscoffensis*), and polychaetes (e.g., *Arenicola grubei*). There are many cases of invertebrates that exhibit geotatic behavior without statocysts, including *Helix, Limax,* the sea anemone *Cerianthus,* monogeneans, starfishes, and sea urchins.

RHEOTAXIS

Rheotactic behavior involves movement directed by water flow, and can be found in most classes of invertebrates that inhabit water. Examples of organisms that display rheotaxis include anemones, planarians, monogeneans, and many protostomes (gastropods, crustaceans, and both nymphs and larvae of insects).

OTHER FORMS OF TAXIS

Thigmotaxis is defined as movement when direction is determined by a stimulus making contact with an animal's body. Turbellarians are positively thigmotactic on their ventral sides, and negatively thigmotactic on their dorsal sides, which keeps their ventral side against the substrate. Movement influenced by air currents is known as Anemotaxis. Taxic responses also are created by chemical and temperature stimuli (referred to as chemotaxis and thermotaxis respectively).

Learned behavior

Learned behavior is another class of behavior exhibited by invertebrates. The reasons for studying learning in invertebrates are varied and include gaining further knowledge of how biochemistry and physiology affect the process of learn-

ing, searching for similarities and differences within and between phyla, and using learning paradigms to explore applied and basic research questions (e.g., how pesticides influence the foraging behavior of the honey bee).

The term learning, like the term behavior, has several definitions. When reviewing studies of learning, the reader should be aware that definitions may vary from researcher to researcher. For example, a researcher may consider behavior controlled by its consequences (i.e., behavior that is rewarded or punished) as an example of operant behavior, while others believe that it depends upon the type of behavior being modified. Moreover, some believe that any association between stimuli represents an example of Pavlovian conditioning, while others believe that the "conditioned stimulus" must never elicit a trained response prior to the process of association.

We will define learning as a relatively permanent change in behavior potential that comes as a result of experience. This definition contains several important principles. First, learning is inferred from behavior. Second, learning is the result of experience. Third, temporary fluctuations are not considered learning; rather, the change in behavior identified as learned must persist as such behavior is appropriate. This definition excludes changes in behavior produced as the result of physical development, aging, fatigue, adaptation, or circadian rhythms. To better understand the process of learning in invertebrates, many behavioral scientists have divided the categories of learning into non-associative and associative.

Non-associative learning

This form of behavior modification involves an association developing from one event, as when the repeated presentation of a stimulus leads to an alteration of the frequency or speed of a response. Non-associative learning is considered to be the most basic of the learning processes and forms the building blocks of higher types of learning in metazoans. The organism does not learn to do anything new or better; rather the innate response to a situation or to a particular stimulus is modified. Many basic demonstrations of non-associative learning are available in scientific literature, but there is little sustained work on the many parameters that influence such learning (e.g., time between stimulus presentations, intensity of stimulation, number of repeated trials). There are two types of non-associative learning: habituation and sensitization.

HABITUATION

Habituation refers to a reduction in the response elicited by a stimulus as it is repeated. For a decline in responsiveness to be considered an instance non-associative learning, it must be determined that any decline related to sensory and motor fatigue do not exert an influence.

Studies of habituation show that it has several characteristics, including the following:

1. The more rapid the rate of stimulation is, the faster habituation occurs.

2. The weaker the stimulus is, the faster habituation occurs.

3. Habituation to one stimulus will produce habituation to similar stimuli.

4. Withholding the stimulus for a long period of time will lead to the recovery of the response.

SENSITIZATION

Sensitization refers to the augmentation of a response to a stimulus. In essence, it is the opposite of habituation and refers to an increase in the frequency or probability of a response. Studies of sensitization show that this process has several defining characteristics, including the following:

1. The stronger the stimulus is, the greater the probability that sensitization will be produced.

2. Sensitization to one stimulus will produce sensitization to similar stimuli.

3. Repeated presentations of the sensitizing stimulus tend to diminish its effect.

Associative learning

A form of behavior modification involving the association of two or more events, such as between two stimuli, or between a stimulus and a response is referred to as associative learning. This form of learning allows a participant to aqcuire the ability to perform a new task, or improve on their ability to perform a task. Associative learning differs from non-associative learning by the number and kind of events that are learned and how the events are learned. Another difference between the two forms of learning is that non-associative learning is considered to be a more fundamental mechanism for behavior modification than those mechanisms present in associative learning; examples of these differences can easily be found in the animal kingdom. Habituation and sensitization are present in all invertebrates, but classical and instrumental conditioning seems to occur first in flatworms (phylum Platyhelminthes). In addition, the available evidence suggests that the behavioral and cellular mechanisms uncovered for non-associative learning may serve as building blocks for the type of complex behavior characteristic of associative learning. The term associative learning is reserved for a wide variety of classical, instrumental, and operant procedures in which responses are associated with stimuli, consequences, and other responses.

CLASSICAL CONDITIONING

Classical conditioning refers to the modification of behavior in which an originally neutral stimulus—known as a conditioned stimulus (CS)—is paired with a second stimulus that elicits a particular response—known as the unconditioned stimulus (US). The response which the US elicits is known as the unconditioned response (UR). A participant exposed to repeated pairings of the CS and the US will often respond to the originally neutral stimulus as it did to the US. Studies of classical conditioning show that it has several characteristics, including the following:

1. The more intense the CS is, the greater the effectiveness of the training.

2. The more intense the US is, the greater the effectiveness of the training.

3. The shorter the interval is between the CS and the US, the greater the effectiveness of the training.

4. The more pairings there are of the CS and the US, the greater the effectiveness of the training.

5. When the US no longer follows the CS, the conditioned response gradually becomes weaker over time and eventually stops occurring.

6. When a conditioned response has been established to a particular CS, stimuli similar to the CS may elicit the response.

INSTRUMENTAL AND OPERANT CONDITIONING

Instrumental and operant conditioning refer to the modification of behavior involving an organism's responses and the consequences of those responses. In order to gain further understanding of this concept it may be helpful to conceptualize an operant and instrumental conditioning experiment as a classical conditioning experiment in which the sequence of stimuli and reward is controlled by the behavior of the participant. Studies of instrumental and operant conditioning show that they have several characteristics, including the following:

1. The greater the amount and quality of the reward, the faster the acquisition is.

2. The greater the interval of time between response and reward, the slower the acquisition.

3. The greater the motivation, the more vigorous the response.

4. When reward no longer follows the response, the response gradually becomes weaker over time and eventually stops occurring.

Non-associative and/or associative learning has been demonstrated in all the invertebrates in which it has been investigated, including planarians and many protostomes (polychaetes, earthworms, leeches, water fleas, acorn barnacles, crabs, crayfish, lobsters, cockroaches, fruit flies, ants, honey bees, pond snails, freshwater snails, land snails, slug, sea hare, and octopus). While there is no general agreement, most behavioral scientists familiar with the literature would suggest that the most sophisticated examples of learning occur in several of the protostome taxa (crustaceans, social insects, gastropod mollusks, and cephalopods). Many of the organisms in these groups can solve complex and simple discrimination tasks, learn to use an existing reflex in a new context, and learn to control their behavior by the consequences of their actions.

Defensive behavior

Defensive behavior represents a class of behavior referred to as communication behavior. Metazoans must defend themselves against an impressive array of predators. To survive against an attack, various strategies have evolved. These

Forbes' common sea star (*Asteruas firbesii*) regenerating arms that have been lost. This ability is a type of defense mechanism, enabling the sea star to stay alive even if a predator takes an arm. (Photo by ©Andrew J. Martinez/Photo Researchers, Inc. Reproduced by permission.)

strategies include active mimicry, flash and startle displays, and chemical/physical defense.

Physical and chemical defense

A common behavior exhibited by invertebrates in response to danger is the adoption of a threatening posture. For example, when the long-spined sea urchin (*Diadema* sp.) is threatened it will point its spines toward the predator or threat. Aquatic organisms found in the order Decapoda, such as the cuttlefish and squid, defend themselves by discharging an ink that temporarily disorientates the predator, allowing the organisms time to escape. Some decapods (such as the octopus) in the order Octopoda have a similar ink defense system. At least one case has been observed in which *Octopus vulgaris* was recorded holding stones in its tentacles as a defensive shield against a moray eel. When sea cucumbers are threatened they expel their intestines to confuse a predator and allow them to escape.

In general, organisms during early ontogenetic development approach low-intensity stimulation and withdraw from high-intensity stimulation (e.g., light intensity). Sessile invertebrates like anemones, corals, and tunicates will contract or withdraw to protect their most vulnerable body parts. Mobile invertebrates can usually escape an aggressor's high intensity stimulation by engaging in kinesis and/or taxis such as crawling, swimming, flying, or jumping. Such behavior is easily observed in the cephalopod *Onychoteuthis* (protostome) popularly known as the "flying squid." The flying squid can escape aggressors by emitting strong water bursts from its mantle, which propels the animal into the air where finlike structures allow it to glide for a brief period of time. Most other non-

sessile invertebrates—like flatworms, echinoderms, and arrow worms—crawl away to hide under a rock, or change direction and swim away to escape predatory stimuli.

MIMICRY

There are various forms of mimicry and only a few can be mentioned here. Some of the more well-known invertebrates that engage in mimicry are butterflies. The species *Zeltus amasa maximianus* (Lycaenidae) presents a "false head" to aggressors that is made more attractive to the predator by the motion of its wings. False-head mimicry requires not only morphological adaptations but also an ability to engage in behavior patterns that force the predator to focus its attention on the false structure. By presenting a predator with a convincing false target the probability of surviving an attack is increased. A similar strategy is also common in caterpillars. Species of *Lirimiris* (Notodontidae) actually inflate a head-like sac that is found posteriorly. The resulting fictitious appendage draws the attention of the predator away from the actual head and toward the comparatively tough posterior end. An interesting version of false-head mimicry exists in crab spiders (*Phrynarachne* spp.), and longhorn beetles (*Aethomerus* spp.), which both mimic the appearance of bird feces, and the Anaea butterfly caterpillar (Nymphalidae), which mimics the appearance of dried leaf tips. Mimicry and false mimicry—where animals mimic another animal—are not well developed among the invertebrate taxa.

STARTLE DISPLAYS AND FLASH COLORATION

When some invertebrates are stimulated by an aggressor they quickly modify their posture in an attempt to make it appear larger, and at the same time their body will quickly present a "flash" of color. This type of behavior has evolved mostly in protostomes. However, many combjellies and jellyfish can produce flashes of bioluminescent light that deters or confuses predators.

Migratory behavior

Migration is a second example of communication behavior. Migratory behavior refers to the movement of entire populations. For invertebrates such movement can range from one or two meters to hundreds of meters. Some well known examples of migratory behavior can be found among insects such as the monarch butterfly, *Danaus plexippus* (Danaidae) and the locust *Schistocerca gregaria* (Tettigoniidae).

During migration, activities such as foraging for food and engaging in mate selection are reduced or suspended altogether. The separation of movement from vegetative activities such as feeding, defense, and reproduction is one criterion used to determine if migratory behavior is occurring.

Migratory behavior is usually confined to animals living in temporary habitats. The ability to leave a particular habitat is important for those animals that feed on vegetation or plankton that is seasonal or limited, and that live in unstable environments. Leaving aversive conditions related to crowding or food shortages is one hypothesis that explains migratory behavior in invertebrates. Examples of lower invertebrate migrations are few. While it is known that several species of jellyfish often congregate in groups of thousands, the mechanism that brings them together is largely unknown.

Many species that do not or are not capable of migration (i.e., sessile forms) may encyst, or produce encysted formations or eggs that can withstand seasonal variation in food and other environment conditions. Many sponges, flatworms, rotifers, nematodes, and gastrotrichs can produce resistant eggs or other forms that are capable of withstanding temporary environmental fluctuation.

Resources

Books

Abramson, C. I. *Invertebrate Learning: A Laboratory Manual and Source Book.* Washington, DC: American Psychological Association, 1990.

———. *A Primer of Invertebrate Learning: The Behavioral Perspective.* Washington, DC: American Psychological Association, 1994.

Abramson, C. I., and I. S. Aquino. *A Scanning Electron Microscopy Atlas of the Africanized Honey Bee (Apis mellifera L.): Photographs for the General Public.* Campina Grande, PB, Brazil: Arte Express, 2002.

Abramson, C. I., Z. P. Shuranova, and Y. M. Burmistrov, eds. *Russian Contributions to Invertebrate Behavior.* Westport, CT: Praeger, 1996.

Brusca, R. C., and G. J. Brusca. *Invertebrates.* 2nd ed. Sunderland, MA: Sinauer Associates, 2003.

Fraenkel, G. S., and D. L. Gunn. *The Orientation of Animals: Kineses, Taxes, and Compass Reactions.* New York: Dover Publications, Inc, 1961.

Lutz, P. E. *Invertebrate Zoology.* Menlo Park, CA: Benjamin/Cummings Publishing Company, Inc, 1986.

Matthews, R. W., and J. R. Matthews. *Insect Behavior.* New York: John Wiley and Sons, 1978.

Preston-Mafham, R., and K. Preston-Mafham. *The Encyclopedia of Land Invertebrate Behavior.* Cambridge, MA: The MIT Press, 1993.

Romoser, W. S., and J. G. Stoffolano, Jr. *The Science of Entomology.* 3rd ed. Dubuque, IA: W. C. Brown Publishers, 1994.

Charles I. Abramson, PhD
Dennis A. Thoney, PhD

Lower metazoans, lesser deuterostomes, and humans

Marine organisms from all marine phyla have been a source of food since humans first began to explore marine environments. In stark contrast to terrestrial plants and animals that have been widely used for remedies of human disorders—the World Health Organization (WHO) estimates that 60–80% of the world's population relies primarily on plants for their basic health care—only a small number of marine species have been used for medicines, mostly due to human's limited access to marine resources.

Perhaps the earliest industrial application of components of marine organisms is Tyrian purple or royal purple, a brilliant dye produced from marine gastropods (protostome) of the superfamily Muricacea, which dates back to 1600 B.C. Phoenicians processed the dye from the mucus in the hypobrachial glands of such mollusks as *Murex trunculus* at many dyeing factories located on the coast of Lebanon. However, only 0.03 oz (1 g) of the dye could be obtained from 10,000 animals and was worth as much as 0.35–0.7 oz (10–20 g) of gold. Actually, the mucus contained in the hypobranchial glands is colorless, which turns purple by enzymatic, oxidation, and photochemical reactions during processing.

Regarding the medicinal application of marine organisms, though, there were only a few examples, including using the red alga, *Digenea simplex*, to treat ascariasis in Asian countries. No one had seriously thought about this issue until the early 1950s when a professor at Yale University discovered unusual nucleosides, spongouridine and spongothymidine, from the Caribbean sponge, *Tectitethya crypta* (=*Chryptotethya crypta*); these nucleosides are composed of arabinose in place of ribose or deoxyribose found in those of RNA or DNA. This discovery was unprecedented and impressed researchers with the uniqueness of marine metabolites. Nearly 15 years later, Ara-A (arabinosyl adenine) and Ara-C (arabinosyl cytosine), antiviral and anticancer drugs, respectively, were developed from these sponge-derived nucleosides. Obviously, this development was the most significant driving force for organizing the symposium entitled "Drugs from the Sea," which was held at University of Rhode Island in 1967. The 1969 discovery of a large amount of prostaglandins from the Caribbean gorgonian, *Plexaura homomala* further stimulated research into marine metabolites. These achievements oc-

curred largely due to scuba diving, which allowed researchers not only to observe but also to collect exotic marine creatures.

A new research field of marine natural products was created in the early 1970s, and many researchers from academia as well as industry embarked on exploring medicinally active compounds from marine organisms. This resulted in the isolation of more than 10,000 new compounds over the next 30 years, including a number of structurally unusual and biologically interesting compounds. Clinical trials on more than 30 marine natural products and their derivatives have since been conducted.

More than any other organism, sponges have been the most actively exploited sources for drugs, because they con-

In Greece sponges are gathered from the ocean and sold, often to tourists. This practice is common around the world. (Photo by ©Margot Granitsas/Photo Researchers, Inc. Reproduced by permission.)

tain not only metabolites of unique structures, but also of potent biological activities. It is widely believed that symbiotic bacteria or other microbes are responsible for the production of biologically active sponge metabolites. In fact, sponges contain numerous compounds reminiscent of microbial metabolites (e.g., sponge peptides often embrace D-amino acids and unusual amino acids). Increasing evidence for the involvement of bacteria in production of unusual sponge metabolites has been obtained; highly unusual antifungal cyclic peptides, theopalauamides, were found to be contained in the new δ-proteobacterium *Entotheonella palauensis* isolated from the Palauan sponge, *Theonella swinhoei.*

Among sponge-derived compounds, halichondrin B, an unusual polyether macrolide originally discovered from the Japanese marine sponge, *Halichondria okadai*, is highly promising as an anticancer agent. However, its low contents (10% based on wet weight) and its complex structure are the most serious obstacles for the development of drugs, as is the case for other marine-derived drug candidates. Fortunately, a New Zealand sponge, *Lissodendoryx* sp., inhabiting depths of 279–344 ft (85–105 m) off the Kaikoura Peninsula, South Island, was found to contain larger amounts of halichondrin B and analogues. To guarantee sponge supplies required for clinical trials, aquaculture of this sponge has been initiated. Small pieces of the sponge were cultured on lantern arrays in shallow waters; small explants grew rapidly, 50-fold in six weeks in some areas. Although the bath sponges have been cultured in the Mediterranean and other regions for more than 80 years, this is the first sponge aquaculture for the production of drugs.

Another promising sponge-derived anticancer agent is discodermolide, discovered from *Discodermia dissolute* collected at depths of 656 ft (200 m) off the Bahamas. It causes anticancer activity by stabilizing microtubules, as in the case of the bestselling anticancer drug, Taxol, which is extracted from the Pacific yew tree. Again, large amounts cannot be supplied by extracting the sponge that inhabits deep sea, but its relatively simple structure indicates the possibility of chemical synthesis.

Cnidaria usually harbor symbiotic dinoflagellates that are thought to be responsible for the synthesis of cnidarian metabolites such as terpenoids. Diterpenoid glycosides called pseudopterosins, which are isolated from the Caribbean sea whip, *Pseudopterogorgia elisabethae*, have been added to skincare creams because of their anti-inflammatory properties.

Opisthobranch mollusks (protostome) are unique animals that have abandoned their protective shells in the course of evolution. Instead, they have developed chemical defenses, which are extracted from their prey organisms. For example, the Spanish dancer, *Hexabranchus sanguineus*, extracts powerful antifeedants, trisoxazole-containing macrolides from sponges. Similarly, the sea hare, *Dolabella auricularia*, accumulates numbers of bioactive compounds from cyanobacteria such as *Lyngbya majuscula*, among which dolastatin 10, an unusual linear peptide, is highly promising as an anticancer drug (under Phase II clinical trials in 2003). *Elysia rufescens*, a Hawaiian sacoglossan mollusk, extracts a cyclic peptide, kahalalide F, from a green alga, *Bryopsis* sp. (actually derived

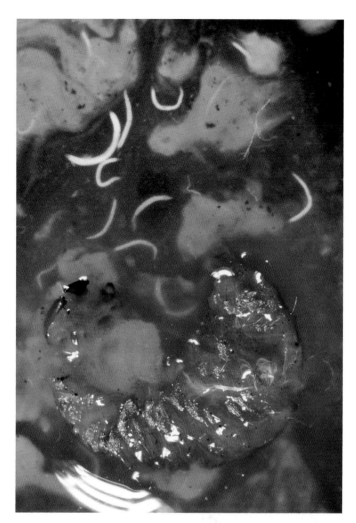

A vine weevil larva (*Otiorhynchus sulcatus*) infected with a parasitic nematode of the genus *Heterorhabditis* shows the contents of its body cavity containing immature nematodes. This nematode is watered into the soil in commercial greenhouses and market gardens to contol vine weevils. (Photo by H. S. I. (Nigel Cattlin/Photo Researchers, Inc. Reproduced by permission.)

from an epiphytic cyanobacteria, most likely *Lyngbya* spp.). The peptide entered Phase I clinical trials in 2002 as an anticancer drug.

Cone snails (protostome), comprising 500 species, hunt fishes, mollusks, and worms using venomous harpoons, which is a very rare adaptation for marine animals. Unexpectedly, the venom glands contain hundreds of small biologically active peptides. For example, *Conus geographus*, a fish-hunting species that occasionally causes death in humans, contains peptides, tabbed conotoxins, that act on Na+ channels, Ca2+ channels, acetylcholine receptors, and others. Consequently, conotoxins are considered potential drugs for the treatment of neurological disorders. In fact, Ω-conotoxin MVIIA derived from *C. magnus*, an N-type Ca2+ channel blocker, is promising as a painkiller for cancer and HIV patients; it is 50 times more potent than morphine.

Colonial ascidians often contain compounds of highly unusual structures with potent biological activities, which perhaps is due to the presence of microbial symbions such as prochlorons. *Ecteinascidia turbinate*, which grows on mangrove roots in the Caribbean, was first reported to be highly antitumoral in 1967, but the active components were not unveiled until 1990. Interestingly, ecteinascidin 743, an active component, is closely related to saframycins, antibiotics isolated from terrestrial actinomycetes. This alkaloid is shortly expected to become the first marine anticancer drug. It should be noted that echteinascidins are likely produced by symbiotic microbe(s). The Mediterranean tunicate, *Aplidium albicans*, contains a depsipeptide, aplydine (dehydrodidemnin B), which showed good results in clinical trials as a anticancer drug.

Cephalostatin 1, a highly unusual dimeric steroid discovered from the hemichordate, *Cephalodiscus gilchristi*, collected at depths of 197–262 ft (60–80 m) off East Africa, proved to inhibit the growth of P388 murine leukemia cells at incredibly low concentrations. Interestingly, closely related ritterazines were isolated from the Japanese ascidian, *Ritterella tokioka*, thus indicating the involvement of symbiotic microbes in synthesis of this unique dimeric steroid.

Increasing numbers of marine natural products have been found to have promising properties for the treatment of human medical disorders. Obviously, marine organisms, particularly benthic invertebrates, are an important source of drugs.

Resources

Books

Baslow, Morris H. *Marine Pharmacology*. Baltimore: Williams and Wilkins, 1969.

Fusetani, Nobuhiro, ed. *Drugs from the Sea*. Basel: Karger, 2000.

Pietra, Francesco. *Biodiversity and Natural Product Diversity*. London: Pergamon, 2002.

Scheuer, Paul J. *Chemistry of Marine Natural Products*. New York: Academic Press, 1973.

Periodicals

Faulkner, D. John. "Marine Natural Products." *Natural Product Reports* 19 (2002): 1–48.

Newman, David J., Gordon M. Cragg, and Kenneth M. Snader. "The Influence of Natural Products upon Drug Discovery." *Natural Product Reports* 17 (2000): 215–234.

Nobuhiro Fusetani, PhD

Conservation

Need for conservation

The conservation of habitat and species poses an enormous challenge for humans in the twenty-first century. As the earth's population grows, human use of natural resources increases. Municipalities, reservoirs, aqueducts and roads impinge upon natural habitat, fragmenting it into smaller pieces. As people burn fossil fuels, remove minerals, harvest commercial species, or convert wild land into farms, they necessarily compel other species to adapt to or emigrate from shrinking habitat or other changes in the environment. As humans consume more of the world's natural resources, they are faced with environmental concerns that range from acid rain, carbon emissions, and ozone depletion to starvation and the emergence of previously unknown diseases. The loss of a species—unlike pollution, unlike hunger, unlike global warming—cannot be reversed. Once a species is gone, it is gone forever.

In a seminal scientific paper first published in 1968, Garret Hardin pointed out that such areas as the open ocean that are not considered private property inevitably suffer from overuse resulting from their being common to all. Hardin argued that because they belong to no one in particular, they do not benefit from anyone's stewardship; no one wants to make personal sacrifices in order to increase the profits of others. He called this phenomenon the "tragedy of the commons." Hardin's article began as a retiring president's address and later appeared in the prestigious journal *Science*. Subsequently his paper was discussed in academic contexts ranging from engineering to political science; his 600 reprints were exhausted in a matter of months. Hardin began his original essay with this citation from a 1964 article published in *Scientific American* by J. B. Wiesner and H. F. York: "Both sides in the arms race are confronted by the dilemma of steadily increasing military power and steadily decreasing national security. It is our considered professional judgment that this dilemma has no technical solution. If the great powers continue to look for solutions in the area of science and technology only, the result will be to worsen the situation." Today, this citation remains uniquely pertinent.

Wiesner and York's observation led to Hardin's incendiary suggestion that there are problems for which science has no solution. Scientists do their work under the explicit or implicit assumption that the questions they address have technical and logical answers. To suggest otherwise is to strike at a collective Achilles heel. Wiesner and York were speaking of the potential for nuclear war in their original article, but Hardin thought their words were equally relevant to ecological concerns. He applied their reasoning to the social goal first proposed by the nineteenth-century English philosopher Jeremy Bentham: acquiring the greatest good for the greatest number of people. In essence, Hardin was talking about human population control; he wanted to know if humans could have their cake and eat it too.

Deductively and simply he showed that the optimum population is less than the maximum, although he admitted that the latter term is difficult to quantify. Humans want the maximum good for the maximum number of people, but differ among themselves in their definitions of "the good." Hardin used the term "tragedy" in its original Greek sense of downfall caused by a remorseless and inexorable fate rather than its modern connotation of personal unhappiness. He used the example of a herdsman who shares a pasture open to all herdsmen in a specific community. Since all who are using the pasture want to graze as many sheep as they can on it, the pasture eventually reaches its carrying capacity. Thereafter, when any herdsman adds another sheep to his flock, the result is that that herdsman has a net gain of almost +1. Because of the shared overgrazing, all others realize a net loss of some fraction of 1. The mathematics and illustration are simple but the results inarguable. Certainly the tragedy of the commons led to the near extirpation of the American bison, many cetaceans, tigers, rhinoceros, and the list continues. Dugongs and dodo birds were less fortunate. Moreover, the example of the commons can be readily extended to the use of rivers and streams for agriculture, hydroelectric power, or fishing (upstream = benefit, downstream = impoverished). It also may be applied to any contemporary marine fishery.

Hardin's logic extends further, but in the opposite direction, when he addresses such issues as the pollution of air or water and the disposal of wastes in leaky landfills or at sea. The tragedy of the commons includes putting certain things into the natural world as well as taking them out; thus it encompasses the release of carbon dioxide and other greenhouse gases by industrialized nations that cause warming worldwide. When the herdsman states his case to a board of inquiry, few could argue with his rationale; he merely wants to add one more animal to his herd. Tragically, that is the conclusion reached by each and every herdsman. Hardin was speaking directly to the issue of human overpopulation that exacerbates the tragedy of the commons, but implicit throughout his message is the need for conservation.

Rate of extinctions

The condition of rarity generally precedes extinction of a species, even though passenger pigeons numbered in the millions less than a century before they died out entirely. Endangered species tend to have the smallest populations and are hence the most likely to die out in the short term. While extinctions can occur for a host of reasons, they can be broadly categorized into two types: systematic pressures and random events. The former include such human activities as habitat destruction, overharvesting, and behaviors that affect the rate of climate change—all of these are factors that can systematically push a species toward extinction. Random pressures, however, are stochastic and may be less obvious. They include such catastrophic events as fires and floods; disease; or demographic fluctuations caused by genetic drift, bottlenecking or inbreeding depression, often acting in combination.

Of course, extinctions are not new events. Fossils provide a clear record of many terrestrial and marine species that once lived on earth and do so no longer. The naturalist Alfred Russell Wallace wrote over a hundred years ago that "...we live in a zoologically impoverished world, from which all the hugest, and fiercest, and strangest forms have recently disappeared...." For Wallace, "recent" meant the Quaternary period, about 12,500 to 11,000 years ago; and the animals he referred to varied according to the continent on which they lived. In the global ocean and in Africa and Asia, there were relatively few Quaternary extinctions; in the Americas, Madagascar, and Australia, however, extinction was nothing short of cataclysmic in its reach. Almost three-quarters of the genera that weighed over 97 lb (44 kg) died out from North America. Australia fared even worse; the continent lost every terrestrial vertebrate species larger than a human being. These included carnivorous kangaroos, a horned tortoise the size of a small automobile, and a monitor lizard almost 22.9 ft (7 m) long. Many small mammals, reptiles, and flightless birds also died out. In fact, eggshells from the giant flightless bird *Genyornis newtoni* indicate that it went extinct simultaneously in three disparate locations. These extinctions occurred during a time in the Quaternary when climate change was relatively mild. They also happened to coincide with the coming of humans to Australia.

Of course, species extinctions also predate human influence and have occurred throughout history as the consequence of climate change, natural selection, and evolution—a fact erroneously used as evidence that conservation is unnecessary. Unfortunately, a consensus exists among many scientists that the earth's species are vanishing at an alarmingly fast rate when compared to background levels of extinction. Sobering warnings come from esteemed scientists like Paul Dayton, Paul Ehrlich, Jane Lubchenco, Stuart Pimm, Michael Soulé, and E. O. Wilson, who have each worked for decades in their respective fields. We dismiss their opinions at grave risk. E. O. Wilson, professor and curator of entomology at Harvard University's Museum of Comparative Zoology, predicts that as many as 20% of the species alive today will be extinct by the year 2030 if conservation measures are not implemented.

Conservation biology

Conservation biology is an interdisciplinary science that attempts to integrate the fields of biology, ecology, economics, and conservation. Scientists Michael Soulé and Bruce Wilcox held the First International Conference on Conservation Biology in the United States in 1978 to address such problems as extinction and habitat loss. Those who attended were ecologists and population biologists already studying these issues. The discipline of conservation biology was founded on the principle that advances in population biology could be applied to conservation issues and put into practice by managers of protected areas. Organizations like the World Conservation Union's (IUCN) Species Survival Commission (SSC) had emerged from the improved quantification of biological diversity as well as the need for conservation. As of 2003, the SSC is a clearinghouse for information on the health and status of species worldwide. Toward that end, the SSC publishes a Red List that uses set criteria (population size, distributional range, rates of decline) to assess and manage extinction risk. As of the year 2000, the Red List included all known birds and mammals on its inventory of 18,000 species. Of these, 11,000 are designated as threatened. The reader should bear in mind that there are about 1.5 million species of insects alone that have been described in the scientific literature. Plants have evolved together with insects, and their species numbers are similarly diverse. Conservative estimates place the total number of eukaryotic (multicelled) species at somewhere near 7 million, so the compilers of the Red List have much information to gather.

Need for biodiversity

When a species is commercially important and goes extinct, its loss can be assigned a dollar value. Atlantic cod, American mahogany, and great auks are listed by the IUCN Red List as Vulnerable, Endangered and Extinct species respectively. All are or were harvested commercially, and all could be ascribed some worth by the industries that made use of them. Such large organisms as whales, tigers, bison, and manatees are mammalian megafauna that, while not commercially important, hold symbolic or aesthetic values for humans. But what about the lower metazoans, the species described in this volume of *Grzimek's*? Most people would not know that an ophiuoroid is related to a sea star or that ascidians in their early life possess a structure very like a human notochord, let alone value these organisms. "Priapulan" is harder to pronounce than to describe, but few people will have even seen the word. Loriceferans were described for the first time in 1983 as resembling tiny "ambulatory pineapples." People might realize that sea cucumbers do not improve either the taste or the appearance of garden salads, but if one refers to them as holothuroids, most will respond with blank stares. When the poor name recognition of some of the lower metazoans is combined with their lack of commercial value, the fact that some of their populations are close to the vanishing point makes their value seem questionable.

The loss of any species, however, goes beyond its monetary worth. Certainly the morality of allowing species, including many that have been evolving longer than humans have, to suffer extinction as the result of human interference is indefensible. But a more scientific and impartial reason to prevent extinction exists: biodiversity. First coined by E. O. Wilson in 1986, the term "biodiversity" refers to the sum of all diversity, all the variability in a given area that is genetic, conferred by other species, or inherent in the ecosystem itself. Simply put, biodiversity is the natural variability among living organisms and everything that fosters that variability.

From a strictly anthropocentric standpoint, humans have benefited directly and greatly from biological diversity. Penicillin comes from the mold *Penicillium* sp. The tree *Calophyllum lanigerum* was found to produce a substance that inhibits replication in the AIDS virus. Aequorin, collected from the jellyfish *Aequorea victoria*, is a common fluorescent marker used in medicine and microbiology. Studies of the venom of a South American pit viper led to the discovery of the angiotensin system that regulates human blood pressure. Venom from marine cone snails has given rise to a synthetic analgesic and is used to keep nerve cells alive following ischemia. The compound cytarabine is more effective at inducing remission in one form of leukemia than any other drug. The polymerase chain reaction (PCR), a technique that revolutionized the field of microbiology, was made possible because of an enzyme discovered in a bacterium in the hot springs of Yellowstone, Wyoming. PCR enables us to perform rapid DNA testing of criminal suspects. It allows microbiologists to modify the genomes of bacteria, insert specific genes into them, and ultimately produce genetic modifications of other plants and animals. The thermophilic bacteria found in those hot springs are more similar to bacteria found at hydrothermal vents in the deep sea than to common bacteria like *Escherichia coli*; scientists have classified them in their own kingdom, Archaea.

As we find new organisms or look more closely at familiar ones, we discover more human uses for those organisms. This fact underscores the very tangible benefits to humans of conservation—even if it is based on aesthetic values, as was the inception of Yellowstone National Park over a century ago. Moreover, conservation offers emphatic demonstrations of the value of biological surveys because scientists continue to find new organisms even at the most general level of classification, the animal kingdom.

Yet even species that have no present or apparent commercial use benefit humans in ways that are taken for granted. Processes as fundamental as natural selection and evolution depend on genetic variability. Plants require nitrogen to grow. Although nitrogen is the largest single component of the air humans breathe, most plants cannot use it in its stable atmospheric form N_2. Some bacteria and blue-green algae help to "fix" atmospheric nitrogen into forms that can be used by plants. Different species of bacteria form different types of nitrogenous byproducts. Some bacteria are even endosymbiotic and live in the root tissues of legumes. As organisms die

and then decompose into their elemental components—mostly water, carbon, nitrogen and some minerals—which are eventually recycled, the very mechanics of nutrient and energy transfer are dependent upon a diversity of plant and animal species. Humans do not ascribe a monetary value to such ecosystem services as decomposition, carbon, or nitrogen cycling. They are, however, invaluable because life as we know it would not exist without them.

Risk assessment

On a practical level, it is difficult to assess which species are most at risk. Because biodiversity encompasses the range of variation, from individuals to populations to habitats themselves, it is virtually impossible to enumerate and quantify. Nevertheless, it is precisely that complexity that is worth conserving. In the absence of long-term data and exact numbers for such ecological parameters as abundance and diversity, conservation efforts usually focus on surrogate species and assume that the protection of some species will include others. Generally, surrogate species fall under three categories: flagship species, umbrella species, and biodiversity indicators. The first group includes well-known or well-publicized species that appeal to the general public. Umbrella species are those that require such large tracts of habitat that their protection envelops other species; for example, large mammalian carnivores. The third surrogate category includes sets of species or taxa whose presence indicates a rich variety of other species. For example, a species of bee may indicate several species of plants it pollinates and they, in turn, may provide habitat to a number of species of birds, invertebrates or mammals. In this particular example, the bee functions as a "keystone" species, since its activities support the well-being of others.

Species are often interdependent in ways that are less apparent. For example, white wartyback mussels (*Plethobasus cicatricosus*) are dependent on a fish host in their larval phase. When rivers in which the mussels and fishes lived were dammed, the fishes could not survive in the colder river water released from those dams and died out. Thus, although white wartyback mussels can still be found, the population cannot reproduce and is functionally extinct, a "living ghost." Such rare species as white wartyback mussels run a higher risk of extinction than common species.

Endemic species are restricted to a given area and are rare by definition; that is, they are found nowhere else. Endemism is common among island flora and fauna because they have evolved more or less in isolation from other populations. Over time, natural selection acts to mold such species into what are often very specific niches; they adapt themselves to very narrowly defined habitats. The many shapes of bills found on the finches of the Galapagos Islands may serve as a common example. Endemic species are commonly used surrogates in conservation strategies.

Habitat conservation

Organisms have evolved to live in certain habitats. It stands to reason, then, that the most severe problem in protecting global biodiversity concerns habitat destruction and fragmentation. Habitats are generally defined by myriad physical parameters like temperature, rainfall, elevation, topography, salinity, soil type and many others. For tuna, habitat might be generally defined as the pelagic ocean within a certain temperature range. For tiny tardigrades, or moss bears, habitat may be forest moss, lichen, beach sand or arctic tundra. For gnathostomulids, the interstices between silt-sized sand grains in the deep sea comprise their habitat. Flightless birds and many herptiles (reptiles and amphibians) may not require much physical space but may be highly specific regarding the space they can inhabit because of such factors as predators, food availability, or breeding sites. An organism's habitat is thus defined by a combination of physical and biological factors.

Biodiversity is not equally distributed but varies from ocean to tidepool, entire rainforest to one strangler fig tree, from temperate zones to tropical. Ecologists and conservationists understand that biodiversity is unrelated to aesthetics or the sweeping vistas of national parks. In fact, habitats with some of the highest numbers of species ("species-rich") include such places as streams, wetlands, coastal mangroves, rainforests, sloughs and estuaries—places often targeted by farmers and developers.

Hotspots are found in areas where the habitats of several rare species overlap. Because so much diversity is concentrated within relatively small areas they are, in fact, hot spots for extinction. The tropics have an inordinate number of hotspots because both species diversity and endemism increase as one travels from the poles to the equator. By way of example, the southeastern Appalachian mountains and all of southern California have many endangered species, but the vast majority of endangered species in the United States occur in the much smaller area of the Hawaiian Islands.

As of 2003, habitat has been already greatly reduced within biodiversity hotspots. This fact is of great concern to scientists. Worldwide, approximately two-thirds of all eukaryotic (multicellular) species occur in humid tropical forests. As the twenty-first century begins, these same forests are being cleared at a rate of approximately 386,100.5 mi^2 (1 million km^2) every 5–10 years. Logging and burning account for several times that loss. From the time a species' habitat begins to decline to the time when the population of that species also declines there will usually be a lag. Thus, the loss of habitat may initially cause only a few extinctions as small pockets of habitat remain; then, however, as habitat is lost entirely or so fragmented that some species cannot survive, individuals die out and the number of extinctions rises. The concern here in this "fewer extinctions now, more later" scenario is that by the time scientists have quantified the loss, it will be too late to stop it.

Consequently, designating and protecting hotspots may be a logical first step in conservation. Myers and colleagues calculate that a staggering 46% of all plant diversity involves endemics. They also estimate that 30–40% of all terrestrial vertebrates could be protected in 24 hotspots. Moreover, those 24 hotspots would span only about 2% of the earth's surface. Protecting hotspots makes economic sense as well. Although the expense of conservation per unit area varies

hugely, from less than a cent in United States currency to over a million dollars per 0.4 mi^2 (1 km^2), costs are generally lower in less developed countries. More importantly, these same countries often have the most left to conserve. Furthermore, as habitat becomes fragmented through development or exploitation, the costs of mitigation and conservation rise. Pimm and Raven agree that the selective protection of hotspots is necessary but caution that it is insufficient for long-term conservation of biodiversity. Usually conservation biologists favor increasing the size of protected areas as a means of including more organisms and as safeguarding against some of the random causes of extinction discussed previously. In some cases less diverse but cheaper habitats can be purchased, ultimately conserving the same number of species on larger or more numerous tracts of land.

Marine conservation and shifting baselines

The conservation strategies outlined above are based largely on terrestrial research. This bias may result in part from the fact that the marine environment was long considered less at risk for environmental degradation because of its sheer immensity and the "unlimited bounty" of the seas. Many marine invertebrates and vertebrates have planktonic larvae capable of spreading over great distances in ocean currents; such species would be less likely to suffer from genetic bottlenecks, inbreeding, or the risk associated with endemic status. Yet in fact the ocean's bounty is limited. Many scientists are concerned that we may be seeing those limits breached.

Overfishing refers to fishing practiced unsustainably. Overfished species are species that have had their numbers so depleted that stocks may never recover. In most coastal ecosystems, manatees, dugongs, sea cows, monk seals, crocodiles, swordfish, codfish, sharks, and rays are functionally or formally extinct. The data compiled by these authors are impressive and cause for foreboding. In order to detect ecological trends a baseline for comparison is necessary, as is a distinction between natural and human-caused changes. As a starting point for such a baseline, Jackson and Kirby gathered paleoecological data, beginning about 125,000 years ago, together with archeological data from human coastal settlements from about 100,000 years ago. To augment these data, they made use of historical records from documents, charts and journals from the fifteenth century. Ecological records from scientific literature spanning the twentieth century completed their data set. Based on these data the authors found that from the onset of overfishing, lag times of only decades to centuries preceded large-scale changes in ecological communities.

In some cases, such time lags existed because other species filled in the gap left by the overfished species. For example, sea otters were all but eliminated from the northern Pacific by fur traders in the 1800s. Because the voracious appetite of sea otters kept sea urchin populations in check, sea urchin numbers increased and decimated kelp forests. Sea otters off the southern California coast were similarly wiped out at approximately the same time as those in the northern Pacific; however, the California kelp forests did not begin to disappear until the 1950s. Diversity across trophic levels was the reason for the lag time between overharvest and threshold response. Predatory fishes like sheephead also ate sea urchins, and spiny lobsters and abalone competed with urchins for kelp. Together these animals effectively "shifted over" to occupy a portion of the ecological niche formerly occupied by the sea otters. Since sea urchins have become a popular fishery, some well-developed kelp forests have returned. Unfortunately, they now have only a vestige of their former complexity. Kelp forests off southern California now lack the trophic diversity of sea otters and such predatory fishes as sheephead, black seabass, and white seabass that they once possessed. Unregulated fisheries exist for echinoids like sea cucumbers, for crabs, and for small snails. The numbers of abalone—greens, reds, whites, and blacks—have dwindled from overharvest and disease. The once diverse southern California kelp forest community has been reduced to a community of primary producers. Thus overfishing affects not only the target species, but dramatically alters ecosystem diversity when "keystone" species are removed.

Nor are coastal ecosystems the only habitats affected. Pelagic longlines are the most widespread fishing gear used in the ocean and threaten many open ocean species. Some species form legitimate fisheries; such others as sea turtles constitute by-catch. Because many pelagic animals have such extensive habitat and evolved to swim great distances often at high speeds, they can be extremely difficult to study. Nevertheless, we know that billfishes, tuna, and sea turtles are in need of conservation because they have been subjected to such intense exploitation. Relatively less is known about open ocean sharks, but with the exception of makos their numbers are estimated to have declined by at least half in less than two decades. Such large animals as these generally bear fewer young, reproduce less often, and do so at an older age. In ecological terms, they have a low intrinsic rate of increase. This low rate means that their ability to rebound if fishing pressures are decreased is slim, and that conservation efforts will have to be long-term if their numbers are to increase.

Jellyfish, medusae, ctenophores and siphonophores are invertebrate predators that typically feed on the same prey as larval and adult fishes do. Concern exists that jellies may be sliding over to fill the void left by declining pelagic predators. Although Carr and his colleagues studied fishes, their field experiments provide insight as to how such "cascading negative consequences" occur. They found that by removing important predators (groupers or jacks) and other highly competitive fish (territorial damselfishes), as is often done by fishermen as well as by the aquarium trade, species interactions changed. As species interactions decreased, population fluctuations increased. As fish populations grew more unstable, the likelihood of extinction increased locally and regionally. In part, jellies are able to shift and fill the ecological "holes" left by declining fish stocks because they can reproduce quickly and in great numbers. *Mnemiopsis leidyi*, a tiny comb jelly, was introduced into the Black Sea, probably when a grain ship pumped out her ballast. The *Mnemiopsis* population was able to take advantage of habitat conditions that were unfavorable to potential competitors; as a result, *M. leidyi* populations peaked in the late 1980s and 1990s. Over this same time *Mnemiopsis* consumed most of the zooplankton production that had previously

A diver with a camera looking at a coral reef near Lizard Island in the Great Barrier Reef. (Photo by A. Flowers & L. Newman. Reproduced by permission.)

been taken up by commercial fisheries. Consequently commercial fisheries in the Black Sea went virtually extinct themselves. This example illustrates both the shift in keystone species as well as the devastating effect that introduced species can have at the level of an entire ecosystem.

In addition, fisheries suffer because the fishing industry is a powerful political lobby. Ample scientific evidence exists that illustrates the need for change in fisheries management. The industry itself, however, continually clamors for additional proof, unable or unwilling to listen to the nails being driven into its own coffin. If a dearth of evidence exists, it is any that can demonstrate that fisheries are harvested sustainably. In part, fisheries may be slow to acknowledge problems because, as the numbers of fish decrease, technological development has advanced in the fishing industry—such as satellite imagery that can be downloaded to computers; sensitive sonar that can locate fish schools; large fleets of fast boats—disguising diminishing stocks by more efficiently harvesting what remains.

Aquaculture is commonly believed to relieve pressure on fisheries. It is true that between 1986 and 1996 the global production of fishes by aquaculture more than doubled. Aqua-

The calm mangroves of the Florida Keys are a protective network of canals for manatees and a diverse range of marine flora and fauna. (Photo by A. Flowers & L. Newman. Reproduced by permission.)

culture alone, however, is not an answer; by reducing wild fish supplies for seedstock collection or for feed, aquaculture can in many ways be detrimental. Hundreds of thousands of Atlantic salmon raised in pens in the Pacific have escaped to locations where they can hybridize and genetically weaken native stocks. Atlantic salmon escaped from Atlantic pens can also interbreed with wild stocks and interfere with the latter's ability to find their spawning grounds, which is a trait that is passed on genetically. Other problems posed by aquaculture include the spread of diseases amongst pens and to wild stocks as well as the discharge of untreated effluent and nitrogenous wastes that result in eutrophication.

Marine reserves are controlled-take areas that have been helpful in restoring depleted fish and invertebrate populations. Because open ocean species as well as fishing fleets move around, however, the effectiveness of reserves to help species living in the open ocean is equivocal. Reserves, although helpful, are not enough; effective conservation will require intelligent consumer choices. The Monterey Bay Aquarium is making an attempt to educate the public regarding sustainable harvesting of fishes. Toward this end the aquarium offers a free "Seafood Watch Card" that takes into account the sustainability of the fishery as well as the ecology of the species; then rates that species accordingly. Informed consumers can make more intelligent purchasing

decisions and use their buying power to encourage conservation and improved management of fisheries.

Global warming

Aside from being responsible for causing the most species extinction worldwide, tropical deforestation creates 20–30 percent of global carbon emissions, with the burning of fossil fuels accounting for most of the remainder. Although it is commonly thought that carbon dioxide emissions are to blame for the rapid warming trends observed in recent decades, such other noncarbon greenhouse gases as chlorofluorocarbons and nitrous oxides contribute as well. Aerosols are tiny particles emitted into the air as pollution, which we see as visible smog or haze. Aerosols alter the brightness of clouds and increase solar heating in the atmosphere. These changes serve in turn to weaken the earth's hydrologic cycle, reducing rainfall and fresh water supplies.

Biological indicators for global warming abound. Species from butterflies to marine invertebrates show a tendency to migrate northward; by so doing, they act as indicators of climate change. Like extinction, global warming predates human influence; however, the global effects that humans can and do have on the earth's climate are facts that cannot be dismissed. As with concerns over extinction, global warming is not an unfounded notion propounded by aging hippies or ecoterrorists. Rather, it is a measurable phenomenon that concerns scientists around the world. Warming has been shown to increase the spread of infectious diseases; and in concert with the overharvesting of resources terrestrial, freshwater or marine, temperature change can have synergistic affects that lead to more extinctions and loss of biodiversity. Warming causes bleaching in coral reefs as well as some sea anemones. Coral reefs fringe no less than a sixth of the world's coastlines. They are species-rich and more biologically diverse than any other shallow-water marine ecosystem.

An obvious way to decrease human contribution to global warming is to restrict the release of greenhouse gases. Now, as we contemplate using the deep sea to store excess carbon dioxide, scientists are asked to assess the risks. Certainly sequestering carbon in the deep sea seems a logical way to decrease atmospheric input and concomitant warming. This approach, however, is only a bandage solution because it fails to address causative agents. Moreover, many of the lower metazoans discussed in this chapter live in the deep sea. Deep sea organisms generally have slow metabolism and difficulty dealing with even minor changes in the acidity or alkalinity of sea water. Dumping carbon dioxide into the deep sea causes pH changes of a little to a lot depending on proximity. It impairs the metabolism of deep sea animals and weakens their exoskeletons, either of which causes increased mortality. As a pool of carbon dioxide forms in the deep sea, such destructive changes are not limited to the benthic fauna living on top but extend to the benthic infauna as well—the habitat occupied by gnathostomulids, loriciferans and their metazoan kin.

Humans claim to cherish our natural environment, yet each year we lose between 14,000 and 40,000 species from tropical forests alone. Between one-third to one-half of the land surface has been transformed by our species; we use more than half of the accessible freshwater. Gone with those species may be life-saving medicines, models for research, or services to the ecosystem that sustain our quality of life. Not that of future generations, but our own. Through it all we must remember the importance of biodiversity and conservation. Conservation is not a luxury; rather, it has been a luxury for humankind to progress as far as the twenty-first century without putting proper emphasis on conservation. Designating reserves can no longer be an opportunistic action performed at the whim of politicians with financial ties to business and industry. Can conservation be a priority for the twenty-first century? Humans are too knowledgeable for excuses and too skilled to do nothing. What is biodiversity worth? What price conservation? Are these questions that science can answer? When it does, are we willing to listen?

Resources

Books

Chaplin, Stephen J., Ross A. Gerrard, Hal M. Watson, Lawrence L. Master, and Stephanie R. Flack. "The Geography of Imperilment." In *Precious Heritage: The Status of Biodiversity in the United States*, edited by Bruce A. Stein, L. S. Kutner, and J. S. Adams. New York: Oxford University Press, 2000.

Cunningham, W. P., and M. A. Cunningham. *Principles of Environmental Science: Inquiry and Applications*, 1st ed. New York: McGraw-Hill, 2002.

Erwin, Terry L. "Biodiversity at Its Utmost: Tropical Forest Beetles." In *Biodiversity II: Understanding and Protecting Our Biological Resources*, edited by M. L. Reaka-Kudla, D. E. Wilson, and E. O. Wilson. Washington, DC: Joseph Henry Press, 1997.

Lovejoy, T. E. "Biodiversity: What Is It?" In *Biodiversity II: Understanding and Protecting Our Biological Resources*, edited by M. L. Reaka-Kudla, D. E. Wilson, and E. O. Wilson. Washington, DC: Joseph Henry Press, 1997.

Master, Lawrence L., Bruce A. Stein, Lynn S. Kutner, and Geoffrey A. Hammerson. "Vanishing Assets: Conservation Status of U. S. Species." In *Precious Heritage: The Status of Biodiversity in the United States*, edited by Bruce A. Stein, L. S. Kutner, and J. S. Adams. New York: Oxford University Press, 2000.

Quammen, D. *The Song of the Dodo: Island Biogeography in an Age of Extinctions*. New York: Scribner, 1996.

Reaka-Kudla, M. L. "The Global Biodiversity of Coral Reefs: A Comparison with Rain Forests." In *Biodiversity II:*

Understanding and Protecting Our Biological Resources, edited by M. L. Reaka-Kudla, D. E. Wilson, and E. O. Wilson. Washington, DC: Joseph Henry Press, 1997.

Stork, Nigel E. "Measuring Global Biodiversity and Its Decline." In *Biodiversity II: Understanding and Protecting Our Biological Resources*, edited by M. L. Reaka-Kudla, D. E. Wilson, and E. O. Wilson. Washington, DC: Joseph Henry Press, 1997.

Wallace, Alfred Russell. *The Geographical Distribution of Animals, with a Study of the Relations of Living and Extinct Faunas as Elucidating Past Changes of the Earth's Surface*. New York: Harper, 1876.

Wilson, E. O. *The Diversity of Life*. New York: Harvard University Press, 1992.

Zaitsev, Yu, and V. Mamaev. *Marine Biological Diversity in the Black Sea: A Study of Change and Decline*. New York: United Nations Publications, 1997.

Periodicals

Allison, Gary W., Jane Lubchenco, and Mark H. Carr. "Marine Reserves Are Necessary But Not Sufficient for Marine Conservation." *Ecological Applications* 8 (Supplement 1): S79–S92.

Andelman, Sandy J., and William F. Fagan. "Umbrellas and Flagships: Efficient Conservation Surrogates or Expensive Mistakes?" *PNAS* 97, no. 11 (2000): 5954–5959.

Ando, Amy, Jeffrey Camm, Stephen Polasky, and Andrew Solow. "Species Distributions, Land Values, and Efficient Conservation." *Science* 279, no. 5359 (1998): 2126–2128.

Resources

Balmford, Andrew, Kevin J. Gaston, Simon Blyth, Alex James, and Val Kapos. "Global Variation in Terrestrial Conservation Costs, Conservation Benefits, and Unmet Conservation Needs." *PNAS* 100, no. 3 (2003): 1046–1050.

Barry, James P., C. Baxter, R. Sagarin, and S. Gilman. "Climate-Related, Long-Term Faunal Changes in a Californian Rocky Intertidal Community." *Science* 267 (1995): 672–675.

Baum, Julia K., Ransom A. Myers, Daniel G. Kehler, Boris Worm, Shelton J. Harley, and Penny A. Doherty. "Collapse and Conservation of Shark Populations in the Northwest Atlantic." *Science* 2999, no. 5605 (2003): 389–392.

Bellwood, D. R., and T. P. Hughes. "Regional-Scale Assembly Rules and Biodiversity of Coral Reefs." *Science* 292, no. 5521 (2001): 1532–1534.

Botsford, Louis W., Juan Carlos Castilla, and Charles H. Peterson. "The Management of Fisheries and Marine Ecosystems." *Science* 277, no. 5323 (1997): 509–515.

Brewer, Peter G., Edward T. Peltzer, Gernot Friederich, Izuo Aya, and Kenji Yamane. "Experiments on the Ocean Sequestration of Fossil Fuel CO_2: pH Measurements and Hydrate Formation." *Marine Chemistry* 72, nos. 2–4 (2000): 83–93.

Brussard, Peter F., and Paul R. Ehrlich. "The Challenges of Conservation Biology." *Ecological Applications* 2, no. 1 (1992): 1–2.

Carr, Mark H., Todd W. Anderson, and Mark A. Hixon. "Biodiversity, Population Regulation, and the Stability of Coral-Reef Fish Communities." *PNAS* 99, no. 17 (2002): 11241–11245.

Carroll, Carlos, R. F. Noss, and Paul C. Paquet. "Carnivores as Focal Species for Conservation Planning in the Rocky Mountain Region." *Ecological Applications* 11, no. 4 (2001): 961–980.

Chivian, Eric. "Environment and Health; 7. Species Loss and Ecosystem Disruption—The Implications for Human Health." *CMAJ* 164, no. 1 (2001): 66–69.

Dayton, Paul K., Mia J. Tegner, Peter B. Edwards, and Kristin L. Riser. "Sliding Baselines, Ghosts, and Reduced Expectations in Kelp Forest Communities." *Ecological Applications* 8, no. 2 (1998): 309–322.

Delcourt, Paul A., and Hazel R. Delcourt. "Paleoecological Insights on Conservation of Biodiversity: A Focus on Species, Ecosystems, and Landscapes." *Ecological Applications* 8, no. 4 (1998): 921–934.

Dobson, A. P., A. D. Bradshaw, and A. J. M. Baker. "Hopes for the Future: Restoration Ecology and Conservation Biology." *Science* 277, no. 5325 (1997): 515–522.

Dobson, A. P., J. P. Rodriguez, W. M. Roberts, and D. S. Wilcove. "Geographic Distribution of Endangered Species in the United States." *Science* 275, no. 5299 (1997): 550–553.

Earn, David J. D., Simon A. Levin, and Pejman Rohani. "Coherence and Conservation." *Science* 290, no. 5495 (2000): 1360–1364.

Flannery, Timothy F. "Paleontology: Debating Extinction." *Science* 283, no. 5399 (1999): 182–183.

Goldstein, Bob, and Mark Blaxter. "Tardigrades." *Current Biology* 12, no. 14 (2002): R475.

Goldstein, Paul Z. "Functional Ecosystems and Biodiversity Buzzwords." *Conservation Biology* 13, no. 2 (1999): 247–255.

Hansen, James, Makiko Sato, Reto Ruedy, Andrew Lacis, and Valdar Oinas. "Global Warming in the Twenty-First Century: An Alternative Scenario." *PNAS* 97, no. 18 (2000): 9875–9880.

Hardin, Garrett. "Essays on Science and Society: Extensions of 'The Tragedy of the Commons.'" *Science* 280, no. 5364 (1998): 682–683.

———. "The Tragedy of the Commons." *Science* 162, no. 3859 (1968): 1243–1248.

Harvell, C. Drew, Charles E. Mitchell, Jessica R. Ward, Sonia Altizer, Andrew P. Dobson, Richard S. Ostfield, and Michael D. Samuel. "Climate Warming and Disease Risks for Terrestrial and Marine Biota." *Science* 296, no. 5576 (2002): 2158–2162.

Humphries, Christopher J., Paul H. Williams, and Richard I. Vane-Wright. "Measuring Biodiversity Value for Conservation." *Annual Review of Ecology and Systematics* 26 (1995): 93–111.

Jackson, Jeremy B. C., et al. "Historical Overfishing and the Recent Collapse of Coastal Ecosystems." *Science* 293, no. 5530 (2001): 629–637.

Jones, K., F. Hibbert, and M. Keenan. "Glowing Jellyfish, Luminescence, and a Molecule Called Coelenterazine." *Trends in Biotechnology* 17, no. 12 (1999): 477–481.

Karpov, K. A., P. L. Haaker, D. Albin, I. K. Taniguchi, and D. Kushner. "The Red Abalone, *Haliotis rufescens*, in California: Importance of Depth Refuge to Abalone Management." *Journal of Shellfish Research* 17, no. 3 (1998): 863–870.

Keith, David A. "An Evaluation and Modification of World Conservation Union Red List Criteria for Classification of Extinction Risk in Vascular Plants." *Conservation Biology* 12, no. 5 (1998): 1076–1090.

Kerr, Jeremy T. "Species Richness, Endemism, and the Choice of Areas for Conservation." *Conservation Biology* 11, no. 5 (1997): 1094–1100.

Kremen, C., J. O. Niles, M. G. Dalton, G. C. Daily, P. R. Ehrlich, J. P. Fay, D. Grewal, and R. P Guillery. "Economic Incentives for Rain Forest Conservation Across Scales." *Science* 288, no. 5472 (2000): 1828–1832.

Lafferty, K. D., M. D. Behrens, G. E. Davis, P. L. Haaker, D. J. Kushner, D. V. Richards, I. K. Taniguchi, and M. J. Tegner. "Habitat of Endangered White Abalone, *Haliotis sorenseni*." *Biological Conservation*, in press.

Levitus, Sydney, John I. Antonov, Julian Wang, Thomas L. Delworth, Keith W. Dixon, and Anthony J. Broccoli. "Anthropogenic Warming of Earth's Climate System." *Science* 292, no. 5515 (2001): 267–270.

Lubchenco, Jane. "Entering the Century of the Environment: A New Social Contract for Science." *Science* 279, no. 5350 (1998): 491–497.

Margules, C. R., and R. L. Pressey. "Systematic Conservation Planning." *Nature (London)* 405 (2000): 243–253.

Resources

McKinney, Michael L. "Extinction Vulnerability and Selectivity: Combining Ecological and Paleontological Views." *Annual Review of Ecology and Systematics* 28 (1997): 495–516.

Miller, Gifford H., John W. Magee, Beverly J. Johnson, Marilyn L. Fogel, Nigel A. Spooner, Malcolm T. McCulloch, and Linda K. Ayliffe. "Pleistocene Extinction of *Genyornis newtoni*: Human Impact on Australian Megafauna." *Science* 283, no. 5399 (1999): 205–208.

Mills, Claudia E. "Jellyfish Blooms: Are Populations Increasing Globally in Response to Changing Ocean Conditions?" *Hydrobiologia* 451 (2001): 55–68.

Mittermeier, Russell A., Norman Myers, Jorgen B. Thomsen, Gustavo A. B. da Fonseca, and Silvio Olivieri. "Biodiversity Hotspots and Major Tropical Wilderness Areas: Approaches to Setting Conservation Priorities." *Conservation Biology* 12, no. 3 (1998): 516–520.

Myers, Norman, Russell A. Mittermeier, Cristina G. Mittermeier, Gustavo A. B. da Fonseca, and Jennifer Kents. "Biodiversity Hotspots for Conservation Priorities." *Nature (London)* 403 (2000): 853–858.

Naylor, Rosamond L., Rebecca J. Goldburg, Jurgenne H. Primavera, Nils Kautsky, Malcolm C. M. Beveridge, Jason Clay, Carl Folke, Jane Lubchenco, Harold Mooney, and Max Troell. "Effect of Aquaculture on World Fish Supplies." *Nature (London)* 405 (2000): 1017–1024.

Naylor, Rosamond L., Rebecca J. Goldburg, Harold Mooney, Malcolm Beveridge, Jason Clay, Carl Folke, Nils Kautsky, Jane Lubchenco, Jurgenne Primavera, and Meryl Williams. "Ecology: Nature's Subsidies to Shrimp and Salmon Farming." *Science* 282, no. 5390 (198): 883–884.

Parmesan, Camille, Nils Ryrholm, Constanti Stefanescu, Jane K. Hill, Chris Thomas, Thomas H. Descimon, Brian Huntley, Lauri Kaila, Jaakko Kullberg, Toomas Tammaru, W. John Tennent, Jeremy A. Thomas, and Martin Warren. "Poleward Shifts in Geographical Ranges of Butterfly Species Associated with Regional Warming." *Nature (London)* 399 (1999): 579–583.

Pimm, Stuart L., and Peter Raven. "Biodiversity: Extinction by Numbers." *Nature (London)* 403 (2000): 843–845.

Ramanathan, V., P. J. Crutzen, J. T. Kiehl, and D. Rosenfeld. "Aerosols, Climate, and the Hydrological Cycle." *Science* 294, no. 5549 (2001): 2119–2124.

Ricciardi, Anthony, and Joseph B. Rasmussen. "Extinction Rates of North American Freshwater Fauna." *Conservation Biology* 13, no. 5 (1999): 1220–1222.

Roberts, Callum M., Colin J. McClean, John E. N. Veron, Julie P. Hawkins, Gerald R. Allen, Don E. McAllister, Cristina G. Mittermeier, Frederick W. Schueler, Mark Spalding, Fred Wells, Carly Vynne, and Timothy B. Werner. "Marine Biodiversity Hotspots and Conservation Priorities for Tropical Reefs." *Science* 295, no. 5558 (2002): 1280–1284.

Saunders, Brian K., and Gisele Muller-Parker. "The Effects of Temperature and Light on Two Algal Populations in the Temperate Sea Anemone *Anthopleura elegantissima* (Brandt, 1835)." *Journal of Experimental Marine Biology and Ecology* 211, no. 2 (1997): 213–224.

Saunders, Denis A., Richard J. Hobbs, and Chris R. Margules. "Biological Consequences of Ecosystem Fragmentation: A Review." *Conservation Biology* 5, no. 1 (1991): 18–32.

Seibel, Brad A., and Patrick J. Walsh. "Biological Impacts of Deep-Sea Carbon Dioxide Injection Inferred from Indices of Physiological Performance." *Journal of Experimental Biology* 204, no 4 (2003): 641–650.

———. "Carbon Cycle: Enhanced: Potential Impacts of CO2 Injection on Deep-Sea Biota." *Science* 294, no. 5541 (2001): 319–320.

Shaffer, Mark L. "Minimum Population Sizes for Species Conservation." *BioScience* 31, no. 2.

Simberloff, Daniel. "Flagships, Umbrellas, and Keystones: Is Single-Species Management Passé in the Landscape Era?" *Biological Conservation* 83, no. 3 (1998): 247–257.

Soulé, M. E., and R. F. Noss. "Rewilding and Biodiversity: Complementary Goals for Continental Conservation." *Wild Earth Journal* (Fall, 1998): 19–28.

Tamburri, Mario N., Edward D. Pelzer, Gernot E. Friederich, Izuo Aya, Kenji Yamane, and Peter G. Brewer. "A Field Study of the Effects of CO2 Ocean Disposal on Mobile Deep-Sea Animals." *Marine Chemistry* 72, nos. 2–4 (2000): 95–101.

Tietjen, John H. "Ecology and Distribution of Deep-Sea Meiobenthos off North Carolina." *Deep Sea Research and Oceanographic Abstracts* 18, no. 10 (1971): 941–944.

van Jaarsveld, Albert S., Stefanie Freitag, Steven L. Chown, Caron Muller, Stephanie Koch, Heath Hull, Chuck Bellamy, Martin Krüger, Sebastian Endrödy-Younga, Mervyn W. Mansell, and Clarke H. Scholtz. "Biodiversity Assessment and Conservation Strategies." *Science* 279, no. 5359 (1998): 2106–2108.

Vetriani, Costantino, Holger W. Jannasch, Barbara J. MacGregor, David A. Stahl, and Anna-Louise Reysenbach. "Population Structure and Phylogenetic Characterization of Marine Benthic Archaea in Deep-Sea Sediments." *Applied and Environmental Microbiology* 65, no. 10 (1999): 4375–4384.

Wiesner, J. B., and H. F. York. *Scientific American* 211, no. 4 (1964): 27.

Other

Global Warming: Early Warning Signs. <http//www .climatehotmap.org>

IUCN Red List of Threatened Species. <http://www .redlist.org>

Seafood Watch Regional Card. Monterey Bay Aquarium. http://www.mbayaq.org/cr/cr_seafoodwatch/sfw_regional .asp>

Union of Concerned Scientists. <http://www.ucsusa.org>

Organizations

IUCN Species Survival Commission. Rue Mauverney 28, Gland, 1196 Switzerland. Phone: +41 (22) 999 0000. Fax: +41 (22) 999 0002. E-mail: ssc@iucn.org Web site: <http:// www.iucn.org>

Rob Sherlock, PhD

Calcarea

(Calcareous sponges)

Phylum Porifera

Class Calcarea

Number of families 22

Thumbnail description
Marine sponges with calcareous skeletal elements (spicules)

Photo: A calcareous sponge in a cryptic reef environment in waters near the Little Cayman Islands, at a depth of 82 ft (25 m). (Photo by ©Gregory G. Dimijian, M. D./Photo Researchers, Inc. Reproduced by permission.)

Evolution and systematics

The fossil record of unambiguously identified Calcarea is relatively poor and fragmented. Most calcareous sponges in the fossil record were classified as either stromatoporoids, chaetetids, archaeocyaths, inozoans, pharetronids, or sphinctozoans. They are common in the Paleozoic and Mesozoic, however, rare in the Cenozoic. It is now established that many of these forms actually belong to several groups of demosponges because of the possession of primary siliceous spicules, and only few to Calcarea (pharetronids and some sphinctozoans). Identification of "true" calcareous sponges in the fossil record is difficult because fossil remains often lack diagnostic spicules at all. Heteractinida, characterized by a spiculate (consisting of six-rayed heteractinid octactines, polyactines) and aspiculate calcitic skeleton, are now regarded as an extinct order of Calcarea, restricted to the Paleozoic. The oldest probable calcareous sponge with affinities to modern subclass Calcaronea (*Gravestockia pharetroniensis* Reitner, 1992) was described from the lower Cambrain of South Australia. The assignment of many records of so-called "Pharetronida," calcareous sponges with a rigid calcareous skeleton, to subclasses Calcaronea or Calcinea is difficult if they do not possess characteristic spicules to allow precise assignment. However, most Pharetronids probably belong to subclass Calcaronea. The majority of modern spiculate calcareans would be found as dissociated spicules in the fossil record; there is only one record from the middle Jurassic at King's Sutton, Northamptonshire, where the form and arrangement of a calcareous sponge was preserved (*Leucandra walfordi* Hinde, 1893).

Calcarea are regarded as one of four classes of the phylum Porifera (three extant [Demospongiae, Hexactinellida, Cal-

carea] and one fossil [Archaeocyatha]), distinctive in possessing a spicule skeleton composed exclusively of calcium carbonate and being the only poriferan taxon realizing all three stages of development of the aquiferous system (asconoid-syconoid-leuconoid). There is still dispute about the true phylogenetic relationships of the three extant sponge classes, including also the relationship of the class Calcarea to other (higher) diploblastic taxa like Ctenophora and Cnidaria. Two competing hypotheses group a) Hexactinellida + Demospongiae more closely together based on the possession of silicious spicules ("Silicea") in contrast to Calcarea ("Calcispongia") and b) Demospongiae more closely with Calcarea based on the possession of a cellular pinacoderm ("Cellularia"/"Pinacophora") to the exclusion of Hexactinellida, which possess a cyncitial tissue structure ("Symplasma"). Both proposals, however, assume poriferan monophyly. More recently, several authors have suggested from ribosomal DNA sequence data that Calcarea might be more closely related to the phyla Ctenophora/Cnidaria than to the other two extant classes of Porifera, rendering phylum Porifera paraphyletic. Class Calcarea was elevated to phylum status ("Calcispongia," a term that was already used in the mid-nineteenth century) (Zrzavy, et al., 1998; Borchiellini, et al., 2001), but as yet without robust statistical support (e.g., Medina, et al., 2001). However, this proposal is not followed in the most comprehensive systematic treatment of sponges to date, the *Systema Porifera* (Hooper and Van Soest, 2002) and the issue of sponge paraphyly is at the time of writing (2003) far from being resolved. Therefore, it should be regarded as still contentious until further corroboratory data, such as a molecular multi-locus approach, is presented. However, new chemotaxonomic data from lipid biomarkers (Thiel, et al., 2002) support a closer relationship of Hexactinellida and Demosponges. Although this

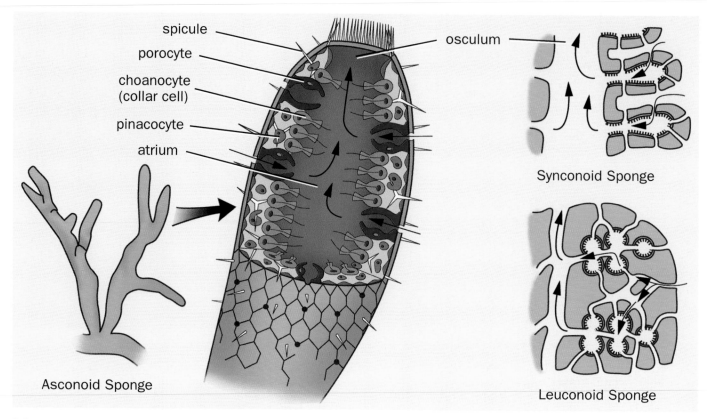

Calcareous sponge anatomy. (Illustration by Kristen Workman)

confirms that Calcarea are chemotaxonomically different from "Silicosponges" or "Silicea" (Demospongiae + Hexactinellida), it does not necessarily imply sponge paraphyly.

Number of classes and families: 1 class (Calcarea); 2 subclasses (Calcinea, Calcaronea); 5 orders (2 in Calcinea: Clathrinida, Murrayonida; 3 in Calcaronea: Leucosoleniida, Lithonida, Baeriida); 22 families; 75 genera; about 500 described species.

Physical characteristics

Calcareous sponges are mostly small and inconspicuous; they occur in a variety of forms, as single tubes, sometimes vase shaped, a mass of small tubes ("cormus"), a bushy arrangement of single tubes, or sometimes massive without any apparent symmetry. Three types of aquiferous system are realized in Calcarea: asconoid, all internal cavities are lined by choanocytes (flagellated cells) without folding of the choanoderm; syconoid, simple folding of the choanoderm; and leuconoid, choanocytes are arranged in discrete "choanocyte chambers."

Calcareous sponges range from minute size an inch or less (few millimeters), to about a maximum of about 12 in (30 cm) (*Pericharax heteroraphis*). They are mostly colorless (whitish to beige), sometimes bright yellow (*Leucetta chagosensis*), dark greenish-brown (*Pericharax heteroraphis*), or fluorescent red/orange (*Leucetta microraphis*, sometimes).

Calcareous sponges have a skeleton that is made of calcium carbonate (calcite), composed of free diactines, triactines, tetracines, and/or polyactine spicules, to which a solid basal calcitic skeleton may be added, with either cemented basal spicules or which is fully embedded in an enveloping calcareous cement. Calcareans are viviparous and have blastula larvae.

Distribution

Calcareous sponges are found globally in all oceans, from intertidal to the deep sea, but not the abyss.

Habitat

Calcareous sponges live in diverse habitats. In tropical coral reefs, they dwell mainly in shaded and/or cryptic habitats and prefer calmer waters.

Behavior

Not applicable; calcareous sponges are sessile filter feeders.

Feeding ecology and diet

Calcareous sponges are sessile filter feeders, whose main diet is dissolved organic matter and small particulate matter (bacteria) filtered from seawater by pumping activity.

Clathrina sponges are usually dull colored and less than 0.16 in (4 mm) long. These were seen in Papua, New Guinea. (Photo by Ron and Valerie Taylor. Bruce Coleman, Inc. Reproduced by permission.)

This calcareous sponge *Pericharax* sp. has been eaten by nudibranchs *Notodoris*. (Photo by Bill Wood. Bruce Coleman, Inc. Reproduced by permission.)

Reproductive biology

Calcareous sponges have internal fertilization, with egg size ranging from 25 to 100 µm. They are sexual and viviparous, with some species probably asexual by budding.

Conservation status

No species are listed by the IUCN.

Significance to humans

There is no known significance of calcareous sponges to humans.

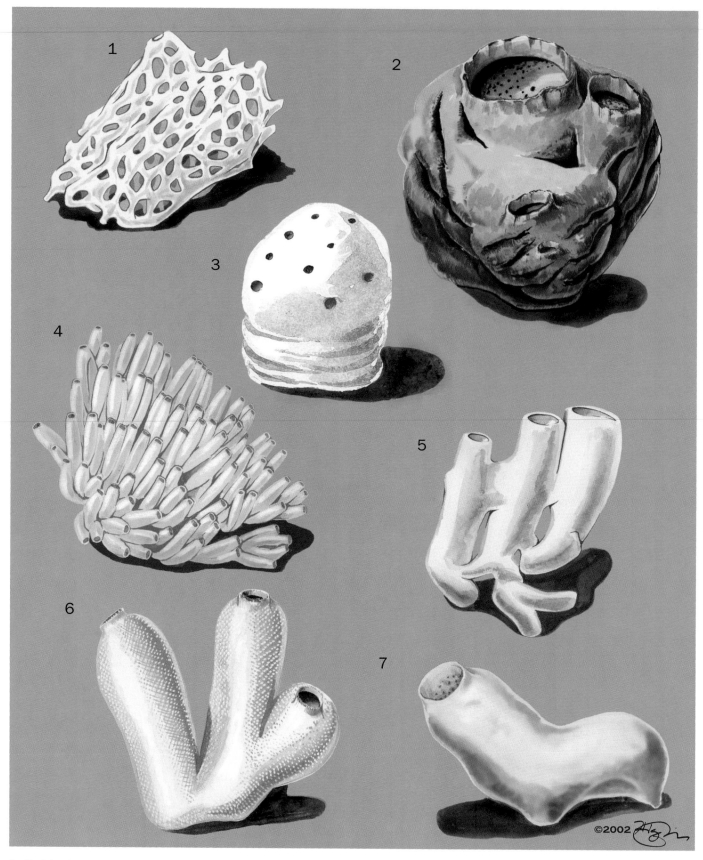

1. *Clathrina heronensis*; 2. *Pericharax heteroraphis*; 3. *Petrobiona masselina*; 4. *Soleneiscus radovani*; 5. *Grantiopsis heroni*; 6. *Sycon capricorn*; 7. Lemon-sponge (*Leucetta chagosensis*). (Illustration by Jonathan Higgins)

Species accounts

No common name
Clathrina heronensis

ORDER
Clathrinida

FAMILY
Clathrinidae

TAXONOMY
Clathrina heronensis Wörheide & Hooper, 1999, Heron Island, at Wistari Channel, Great Barrier Reef.

OTHER COMMON NAMES
None known.

PHYSICAL CHARACTERISTICS
Mass of loosely anastomosing tubes, approximately 0.04 in (1 mm) diameter, with fairly large space between tubes, whole sponge 1.2 × 0.8 in (3 × 2 cm), flat. Life color white. No visible oscules nor distinct exhalant system. With soft, compressible, and delicate texture, smooth surface. Skeleton consists solely of a layer of irregular triactines. Triactines tangentially orientated, actines sometimes overlap. No differentiation or zonation of skeleton, appears to be uniform throughout cormus (sponge body). Asconoid grade of aquiferous system. One type of triactines, with a more-or-less blunt tip, actines measuring 80–130 × 8–12μm.

DISTRIBUTION
Currently only known from the Great Barrier Reef, Australia. Putative member of the cosmopolitan species group *Clathrina coriacea*.

HABITAT
Cryptic, under rubble at reef crest, intertidal.

BEHAVIOR
Sessile.

FEEDING ECOLOGY AND DIET
Sessile filter feeder.

REPRODUCTIVE BIOLOGY
Viviparous, not much known.

CONSERVATION STATUS
Not listed by the IUCN.

SIGNIFICANCE TO HUMANS
None known. ◆

Lemon-sponge
Leucetta chagosensis

ORDER
Clathrinida

FAMILY
Leucettidae

TAXONOMY
Leucetta chagosensis Dendy, 1913, Chagos Archipelago, Indian Ocean.

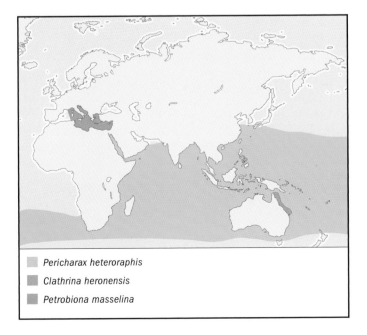

Pericharax heteroraphis
Clathrina heronensis
Petrobiona masselina

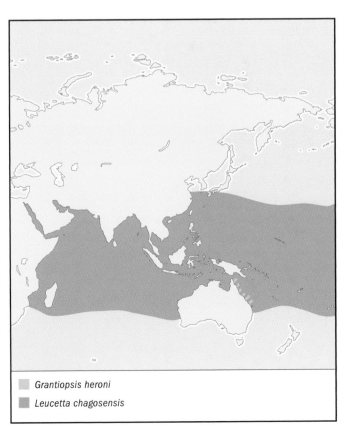

Grantiopsis heroni
Leucetta chagosensis

OTHER COMMON NAMES
None known.

PHYSICAL CHARACTERISTICS
Massive, globular, slightly elongated-globular to pyriform or elongate growth form. Specimens range from 0.39–2.0 in (1–5 cm) in size. Bright yellow color in life (while alive). Globular specimens with one prominent osculum with a naked "lip." Elongated specimens with a few oscules with a naked "lip" on the ridge of the sponge body. Firm and smooth, with slightly translucent surface. With a distinct thin cortex, up to 50 μm thick, sustained by tangentially arranged small regular triactines (three rayed "mercedes-star" spicules), with some large triactines also scattered tangentially in the cortex. Small sub-dermal cavities present (50–150 μm diameter). Choanocyte-chamber free zone of sub-dermal cavities with a thickness of up to 250 μm. Large regular triactines also found here. A peculiar, special, small sagittal triactine only found in the oscular rim ("lip"). Dense irregular meshwork of small regular triactines form choanoskeleton. Larger triactines occasionally irregularly scattered in the choanosome in small numbers. Small tetractines (four-rayed spicules) concentrated around exhalant canals. Leuconoid aquiferous system. Spicules small: larger regular triactines about 250–500 × 25–50 μm, smaller regular triactines 100–200 × 10–25 μm; regular tetractines of the excurrent canals 90–125 × 10–20 μm.

DISTRIBUTION
Probably circum Indo-Pacific, from (southern) Red Sea to French Polynesia in most tropical coral reefs. Recorded from Indo-West Pacific, in Western Australia (Houtman Abrolhos, Fremantle), Queensland (Great Barrier Reef); also Indian Ocean (Chagos) and western Pacific (New Caledonia, Fiji, Vanuatu, French Polynesia).

HABITAT
In crevices and under overhangs of coral bommies, also abundant in illuminated reef habitats.

BEHAVIOR
Sessile.

FEEDING ECOLOGY AND DIET
Sessile filter feeder.

REPRODUCTIVE BIOLOGY
Viviparous, not much known.

CONSERVATION STATUS
Not listed by the IUCN.

SIGNIFICANCE TO HUMANS
None known. ◆

No common name
Pericharax heteroraphis

ORDER
Clathrinida

FAMILY
Leucettidae

TAXONOMY
Pericharax carteri heteroraphis Poléjaeff, 1883, Tristan da Cunha (South Atlantic).

OTHER COMMON NAMES
None known.

PHYSICAL CHARACTERISTICS
Massive, bulbous, rarely clavate growth form. Small and young specimens pyriform, with external surface not folded; older and larger specimens with characteristic irregularly folded external surface. Maximum size of about 12 in (30 cm) in height. Yellow greenish to dark greenish brown color in life, one large terminal osculum always present, mostly with a "lip." Firm, harsh texture, surface smooth but brittle, large triactines protruding through the surface readily visible. With a thin but distinct, up to 50 μm thick, cortex consisting of tangentially aligned small, characteristic tripod-like triactines, but small and large triactines also present. Sub-dermal cavities present in most areas below the cortex (50–200 μm diameter), devoid of choanocyte chambers. Sub-dermal cavity zone up to 400 μm thick. Dense irregular meshwork of mainly regular small triactines forms the choanosomal skeleton. Some small tetractines also. Many large "giant" triactines irregularly dispersed throughout the choanosome. Small tetractines concentrated at, but not restricted to, excurrent water canals. Leuconoid aquiferous system. Large regular triactines 500–1,600 × 70–200 μm; smaller regular triactines 120–200 × 15–25 μm; tetractines 90–180 × 10–20 μm; tripod-like cortical sagittal triactines 45–130 × 7–15 μm.

DISTRIBUTION
Widely distributed, allegedly nearly cosmopolitan. From tropical coral reefs (e.g., Great Barrier Reef) to the Subantarctic, also south Atlantic, Indian Ocean, and Indo-Malayan region.

HABITAT
Widely distributed in exposed and semi-shaded habitats, sometimes under overhangs.

BEHAVIOR
Sessile.

FEEDING ECOLOGY AND DIET
Sessile filter feeder.

REPRODUCTIVE BIOLOGY
Viviparous. Not much known.

CONSERVATION STATUS
Not listed by the IUCN.

SIGNIFICANCE TO HUMANS
None known. ◆

No common name
Soleneiscus radovani

ORDER
Clathrinida

FAMILY
Soleneiscidae

TAXONOMY
Soleneiscus radovani Wörheide & Hooper, 1999, south side of Wistari Reef, Great Barrier Reef, 56 ft (17 m) depth.

OTHER COMMON NAMES
None known.

Soleneiscus radovani

Sycon capricorn

PHYSICAL CHARACTERISTICS
Aborescent, bushy, with single, delicate tubes branching dichotomous and polychotomous from a few central tubes. Bright yellow color. Central, proximal tube larger than the distal tubes; tubes ramify only in the lower part of the sponge "bush." Distal parts of tubes mostly longer than ramified parts. Single tubes approximately 0.08 in (2 mm) in diameter. Size of the sponge "bush" is less than 4 in (10 cm). One naked osculum on top of each tube. Soft and delicate texture, easily torn. Sagittal tetractines only make up skeleton. Tangentially arranged facial plane of tetractines forms wall of tubes, with the longer ray of basal triradiate system pointing in direction of central tube (growth axis). The curved and free actines of tetractines protrude into tube, tips of the free actines bent in direction of osculum. Asconoid grade of aquiferous system. Only one spicule type is present. Non-curved, longer unpaired actine of basal triradiate system approximately 120–200 μm × 8–12 μm, the paired shorter (curved) actines of pseudosagittal basal plane approximately 85–130 × 8–12 μm.

DISTRIBUTION
Currently only described from the Great Barrier Reef, probably wider distribution in the tropical western Pacific.

HABITAT
Small patches of coral, under overhangs.

BEHAVIOR
Sessile.

FEEDING ECOLOGY AND DIET
Filter feeder.

REPRODUCTIVE BIOLOGY
Viviparous, details not known.

CONSERVATION STATUS
Not listed by the IUCN.

SIGNIFICANCE TO HUMANS
None known. ◆

No common name
Sycon capricorn

ORDER
Leucosoleniida

FAMILY
Sycettidae

TAXONOMY
Sycon capricorn Wörheide & Hooper, 2003, south side of Heron Island, Great Barrier Reef.

OTHER COMMON NAMES
None known.

PHYSICAL CHARACTERISTICS
Tubular and mostly branching growth form. If small, only a single tube with a few smaller branches arising from a narrower base. Otherwise bushy, with multiple dichotomous or occasionally polychotomous branching digits. Apical oscule of each branch always fringed. Digitations with a diameter of 0.2–0.4 in (0.5–1 cm). Beige color in life. Smooth, soft, tessellated texture. No defined cortex (i.e., with tangentially arranged spicules). Ectosomal skeleton (forming the external surface) consists of characteristic "tufts" of bundled free actines of t-shaped sagittal triactines and two types of microdiactines. Tufts, with a diameter of 100–150 μm, form minute tessellation of external sponge surface, located over radially arranged choanocyte chambers. Long, thin diactines around the osculum, sustaining the oscular fringe, with a size of 200–600 × 7–15 μm. Choanosomal skeleton composed of two types of sagittal t-shaped triactines in articulated arrangement. First type (shorter unpaired actines) builds walls of tubular choanocyte chambers, with size of unpaired actines 69–170 × 5–15 μm, the paired actines are 40–95 μm long. Second type only found in the distal parts of the choanocyte chamber-tubes, with size of unpaired actines 150–290 × 4–15 μm, length paired actines 40–80 μm. Their longer, unpaired actines contributes to ectosomal tessellation, with brush-like tufts of microdiactines arranged around them. Sinuous fusiform diactines of spicule tufts with a size of 70–180 × 3–12 μm, and smaller microdiactines with a characteristic "ball"-type thickening, with a size of 55–75 × 7–12 μm. Atrial skeleton of sagittal triactines and tetractines, their elongated free unpaired/apical actines protruding into atrium, with a size of 45–230 × 6–15 μm. Syconoid grade of aquiferous system.

DISTRIBUTION
Currently only known from the Great Barrier Reef, Australia. Member of the allegedly more widespread *Sycon gelationosum/arborea* species group.

HABITAT
Cryptic, mostly in caves and under overhangs, rarely in the open.

BEHAVIOR
Sessile.

FEEDING ECOLOGY AND DIET
Sessile filter feeder.

REPRODUCTIVE BIOLOGY
Viviparous, amphiblastula larva.

CONSERVATION STATUS
Not listed by the IUCN.

SIGNIFICANCE TO HUMANS
None known. ◆

No common name
Grantiopsis heroni

ORDER
Leucosoleniida

FAMILY
Lelapiidae

TAXONOMY
Grantiopsis heroni Wörheide & Hooper, 2003, northern side of Wistari Reef, Great Barrier Reef.

OTHER COMMON NAMES
None known.

PHYSICAL CHARACTERISTICS
Mass of branching and anastomosing tubes, about 2 × 1.2 × 0.8 in (5 × 3 × 2 cm) in size. Tubes anastomose in proximal part of cormus at the base, but become unified and only partly branching in distal part. With a terminal osculum of about 0.20 in (5 mm) diameter. White color in life, soft texture, no surface ornamentation. With a distinct cortex of tangentially arranged thick sagittal triactines, with more-or-less cylindrical actines and an unpaired angle of up to 160°. Paired actines with a size of 220–350 µm, unpaired actines 210–370 µm, maximum thickness 15–40 µm. Ectosomal membrane (exopinacoderm) supported by perpendicularly arranged micro-diactines, restricted to the ectosomal (distal) part of cortex, 40–100 µm long, often finely spined. Sometimes small subcortical cavities, with diameter of 50–100 µm. Inarticulated choanoskeleton only supported by two (non-articulated) spicules, apical actines of sub-atrial sagittal tetractines and nail-shaped triactines (with totally reduced paired actines). With a size of 140–350 × 4–12 µm. Nail-shaped triactines pointing with "unpaired" actines towards cortex, unpaired actines of the sub-atrial triactines do not extend to the cortex. Subatrial sagittal tetractines delimit choanosome towards atrial skeleton with their regular basal triradiate system, their elongated apical actine forms proximal part of choanoskeleton. Actines of basal triradiate system 30–80 µm long, longer apical actine 130–330 µm long. The atrial skeleton is supported by sagittal tetractines with their curved free actines pointing towards and into the atrium. Spicule with a "plough-like" shape. Apical actine, with a size of 65–120 µm, bent towards unpaired actine of basal triradiate system. Basal triradiate actines are 30–75 µm long, with maximum thickness of 2–12 µm. Syconoid grade of aquiferous system.

DISTRIBUTION
Currently only known from the Great Barrier Reef, Australia. Putative member of the allegedly more widespread *Grantiopsis cylindrica* species group.

HABITAT
Overhangs, swim throughs and crevices between coral bommies at the reef edge.

BEHAVIOR
Sessile.

FEEDING ECOLOGY AND DIET
Sessile filter feeder.

REPRODUCTIVE BIOLOGY
Viviparous.

CONSERVATION STATUS
Not listed by the IUCN.

SIGNIFICANCE TO HUMANS
None known. ◆

No common name
Petrobiona masselina

ORDER
Lithonida

FAMILY
Petrobionidae

TAXONOMY
Petrobiona masselina Vacelet & Levi, 1958, Mediterranean.

OTHER COMMON NAMES
None known.

PHYSICAL CHARACTERISTICS
Massive, subspherical or multi-lobate growth form, encrusting in high energy habitats or dead stalk in calm habitats. Maximum size of up to 2.4 in (6 cm) in diameter when encrusting, if not, living "head" of 0.39–0.47 in (1.0–1.2 cm) in diameter with a stalk 0.79 in (2 cm) long. Stony texture, white color, smooth surface. In subspherical specimens apical oscules, 600–800 µm in diameter. Living tissue only located at the surface and between crests of basal skeleton, with a choanosome 600 µm thick. Spicules are sagittal triactines (with actines 25–200 × 6–40 µm), tuning-fork (diapason) triactines (basal actines 30–70 × 5–8.5 µm, apical actines 20–50 × 4–7 µm), two size categories of tetractines (pugioles) (apical actines 40–130 × 22–28 µm and 16–40 × 5.5–8.5 µm, actines of basal triradiate system 8–100 × 10–28 µm and 30–70 × 5.5–8.5 µm), spined microdiactines 30–60 × 2–3 µm. Elongate, irregular skeletal elements, with a radial orientation of the crystals giving them a pseudo-spherulitic appearance, form solid calcareous basal skeleton of Mg-calcite, with crests and depressions on the surface. Some spicules entrapped in basal skeleton, randomly arranged. Leuconoid grade of aquiferous system.

DISTRIBUTION
Mediterranean: eastern basin (Adriatic, Ionian Sea, Crete, Malta, Tunisia), western part of the eastern basin (not recorded west of the Rhone delta and Algeria).

HABITAT
Common near the entrance of dark caves, more rarely on the under surface of stones, 1.6–82 ft (0.5–25 m) depth.

BEHAVIOR
Sessile.

FEEDING ECOLOGY AND DIET
Sessile filter feeder.

REPRODUCTIVE BIOLOGY
Viviparous. Amphiblastula larva.

CONSERVATION STATUS
Not listed by the IUCN.

SIGNIFICANCE TO HUMANS
None known. ◆

Resources

Books

Borojevic, R., N. Boury-Esnault, M. Manuel, and J. Vacelet. "Order Baerida." In *Systema Porifera*, edited by J. N. A. Hooper and R. W. M. Van Soest. New York: Plenum, 2002.

———. "Order Clathrinida Hartman, 1958." In *Systema Porifera*, edited by J. N. A. Hooper and R. W. M. Van Soest. New York: Plenum, 2002.

———. "Order Leucosolenida Hartman, 1958." In *Systema Porifera*, edited by J. N. A. Hooper and R. W. M. Van Soest. New York: Plenum, 2002.

Fell, P. E. "Porifera." In *Reproductive Biology of Invertebrates. Volume I: Oogenesis, Oviposition and Oosorption*, edited by K. G. Adiyodi, and R. G. Adiyodi. New York: John Wiley and Sons, 1983.

———. "Porifera." In *Reproductive Biology of Invertebrates. Volume IV: Fertilization, Development, and Parental Care (Part A)*, edited by K. G. Adiyodi, and R. G. Adiyodi. New York: John Wiley and Sons, 1989.

———. "Porifera." In *Reproductive biology of invertebrates. Volume 6 Part A: Asexual propagation and reproductive strategies*, edited by K. G. Adiyodi, and R. G. Adiyodi. Chichester, New York: John Wiley and Sons, 1993.

———. "Poriferans, the Sponges." In *Embryology: constructing the organism*, edited by S. F. Gilbert, and A. M. Raunio. Sunderland, MA: Sinauer Associates, Inc., Publishers, 1997.

Hooper, J. N. A., and R. W. M. Van Soest, eds. *Systema Porifera. Guide to the Supraspecific Classification of Sponges and Spongiomorphs (Porifera)*. New York: Plenum, 2002.

Lévi, C., ed. *Sponges of the New Caledonian Lagoon*. Paris: Orstom, 1998.

Manuel, M., R. Borojevic, N. Boury-Esnault, and J. Vacelet. "Class Calcarea Bowerbank, 1864." In *Systema Porifera*, edited by J. N. A. Hooper and R. W. M. Van Soest. New York: Plenum, 2002.

Pickett, J. "Fossil Calcarea. An overview." In *Systema Porifera*, edited by J. N. A. Hooper and R. W. M. Van Soest. New York: Plenum, 2002.

Vacelet, J., R. Borojevic, N. Boury-Esnault, and M. Manuel. "Order Lithonida Vacelet, 1981, recent." In *Systema Porifera*, edited by J. N. A. Hooper and R. W. M. Van Soest. New York: Plenum, 2002.

———. "Order Murrayonida Vacelet, 1981." In *Systema Porifera*, edited by J. N. A. Hooper and R. W. M. Van Soest. New York: Plenum, 2002.

Periodicals

Borchiellini, C., M. Manuel, E. Alivon, N. Boury-Esnault, J. Vacelet, and Y. Le Parco. "Sponge paraphyly and the origin of Metazoa." *Journal of Evolutionary Biology* 14, no. 1 (2001): 171–179.

Medina, M., A. G. Collins, J. D. Silberman, and M. L. Sogin. "Evaluating hypotheses of basal animal phylogeny using complete sequences of large and small subunit rRNA." *Proceedings of the National Academy of Science of the USA* 98, no.17 (2001): 9707–9712.

Thiel, V., M. Blumenberg, J. Hefter, T. Pape, S. A. Pomponi, J. Reed, J. Reitner, G. Wörheide, and W. Michaelis. "A chemical view on the most ancestral Metazoa—biomarker chemotaxonomy of hexactinellid sponges." *Naturwissenschaften* 89 (2002): 60–66.

Wörheide, G., and J. N. A. Hooper. "Calcarea from the Great Barrier Reef. 1: Cryptic Calcinea from Heron Island and Wistari Reef (Capricorn-Bunker Group)." *Memoirs of the Queensland Museum* 43, no. 2 (1999): 859–891.

———. "New species of Calcaronea (Porifera: Calcarea) from cryptic habitats of the southern Great Barrier Reef (Heron Island and Wistari Reef, Capricorn-Bunker Group, Australia)." *Journal of Natural History* 37 (2003): 1–47.

Zrzavy, J., S. Mihulka, P. Kepka, A. Bezdek, and D. Tietz. "Phylogeny of the Metazoa based on morphological and 18S ribosomal DNA evidence." *Cladistics* 14, no. 3 (1998): 249–285.

Other

Atlas of Sponges. (2 July 2003). <http://www.ulb.ac.be/sciences/biodic/EImAnatepon.html>.

Gert Wörheide, PhD

Hexactinellida
(Glass sponges)

Phylum Porifera

Class Hexactinellida

Number of families 17

Thumbnail description
Deepwater marine sponges with a glass
skeleton, and typically six rays; unusual
because of their multinucleate tissues and
ability to conduct electrical signals in the
absence of nerves

Photo: A glass sponge (*Aphrocallistes vastus*)
living in deep waters off of British Columbia.
(Photo by ©Neil G. McDaniel/Photo
Researchers, Inc. Reproduced by permission.)

Evolution and systematics

Hexactinellids (glass sponges) are deepwater marine sponges
that have skeletons of siliceous (glass) spicules with a distinctive triaxonic (cubic three-rayed) symmetry. Unlike the other
two main classes of sponges (Calcarea and Demospongiae),
glass sponges lack either a calcareous or organic skeleton. Furthermore, glass sponges are highly unusual in that their major tissue component is a giant "syncytium" (see below) that
ramifies throughout the entire body, stretching like a cobweb
over the glass skeleton. As their skeletons are both made of
glass rather than calcium, early classification schemes grouped
hexactinellids with the demosponges; however, at present
hexactinellids are separated from cellular sponges (the Calcarea and Demospongiae) in the subphylum Symplasma because of their unique (syncytial) structure. Nevertheless, recent
molecular evidence suggests that whereas modern hexactinellids are descended from the most ancient multicellular animals, they are more closely related to demosponges than either
group is to the calcareous sponges or other metazoans. There
are approximately 500 species of hexactinellids in two subclasses containing five orders, 17 families, and 118 genera.

Hexactinellids have left the oldest fossil record of multicellular animals on Earth. Their triaxonic spicules are known
from the Late Proterozoic of Mongolia and China. The group
thrived during the Middle Cambrian and reached its maximum radiation and diversity during the Cretaceous, when
hexactinellids formed vast reefs in the Tethys Sea. Their fossilized skeletons now make up the stony outcrops upon which
many castles are built from southern Spain through France,
Germany, and Poland to Romania.

Physical characteristics

Within each of the two subclasses of hexactinellids are
sponges with loose skeletons—spicules held together by living tissue—and sponges with fused skeletons. Sponges of both
designs are essentially vase-shaped, with a large central or
atrial cavity, usually with one opening, the osculum. The
species with fused skeletons often have mittenlike or finger-
like protrusions of the body wall, and some form platelike
structures; these species have oscula on each of the projections. The tissue of hexactinellids generally is creamy yellow
to white. Some animals are quite clean, so that the whiteness
of their tissue looms out of the darkness of their deepwater
habitats. Others tend to accumulate particulate matter on the
outside and can look quite dirty. The diverse species of hexactinellids vary in length from 0.2 to 5 ft (0.5 cm to 1.5 m);
many of the largest hexactinellids are as wide as they are tall.

The body wall is composed of three parts: both the inner
and outer peripheral trabecular networks, and the feeding region, which is called the choanosome. Large incurrent and
excurrent canals meet at the choanosome, where the fine,
branchlike endings of the incurrent canals contact oval, flagellated chambers that create the feeding current through the
sponge.

While sponges of other groups are constructed of single
cells, each with a single nucleus, the greater part of the soft tissue in a hexactinellid consists of the trabecular reticulum, which
contains thousands of nuclei and cytoplasm that is free to move
as it is unimpeded by membrane barriers (the "syncytial" condition). The trabecular reticulum hangs from the skeleton in
thin strands, resembling a cobweb, and stretches from the out-
ermost layer, termed the dermal membrane, to the innermost
layer, the atrial membrane. Single cells, specialized for particular functions, are also present. Cells are attached to one another and the trabecular reticulum by a unique type of
attachment structure referred to as the "plugged junction,"
which is called so because it is slotted into the neck of cytoplasm between two regions, resembling a plug. It is not an
extension of the lipo-protein cell membrane, but is a multi-

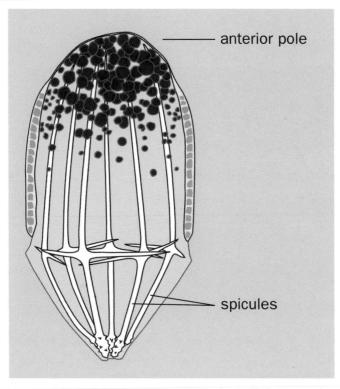

The larva of a hexactinellid sponge. (Illustration by Emily Damstra)

laminar proteinaceous structure. While it undoubtedly serves as a partial barrier to the free exchange of materials, electric currents can flow through it and transport vesicles are able to move through pores in its structure. Between cellular and syncytial components of the sponge there is a very thin collagen layer, the mesohyl. This layer is believed to be too thin for cells to migrate within, as is the case with other sponges. Instead, transport of nutrients and other materials appears to occur along vast networks of microtubules within the multinucleate tissue.

Distribution

Hexactinellid sponges are known at depths from 30 to 22,200 ft (9.14–6,770 m) in all oceans. There are no records of this class in freshwater. The fossil record suggests that their historical range was similar.

Habitat

The vast majority of hexactinellids live at depths greater than 1,000 ft (304.8 m). In a few coastal locations, however, such as Antarctica, the northeastern Pacific, New Zealand, and some caves in the Mediterranean, species are found at depths accessible by scuba divers. These habitats have in common cold water (35–52°F, or 2–11°C), relatively high levels of dissolved silica, and low light intensity. Although many hexactinellids require a firm substratum, such as rocks, for attachment, others grow on fused skeletons of dead sponges, and still others live over soft sediments. The latter group, though not numerous, support themselves on struts made of bundles of long spicules that project down into the sediment.

Behavior

Sponges are not noted for their complex behavior. Nevertheless, hexactinellids can respond to mechanical or electrical stimuli by instantly shutting down the feeding current. The explanation for this unusual ability lies in their possession of a trabecular reticulum, which acts like a nervous system, conveying impulses to all parts of the body. Sponges of other groups lack any such system and show no evidence of an ability to conduct electrical impulses. Electrical signals traveling at 0.07 in per second (0.26 cm per second) have been recorded from slabs of the body wall of *R. dawsoni*. It is presumed that when the signals reach the flagellated chambers, the pumping stops. No rhythmic pattern has been found in the cessation of pumping. It is thought that since glass sponges lack motile cells that would otherwise remove unwanted material from the sponge, shutting down the feeding current may prevent the clogging of the canal system with large amounts of debris.

Feeding ecology and diet

Like the majority of sponges, glass sponges are thought to filter food from the water that they pump through the choanosome. Two in situ studies confirm this to be true for sponges that lack debris on their outer surfaces. Studies comparing the content of inhaled and exhaled water showed that both *Aphrocallistes vastus* and *Sericolophus hawaiicus* retain particulate matter—mostly bacteria—and rely little on dissolved organic carbon. The results of one study that compared such water samples from a sponge covered in debris (*Rhabdocalyptus dawsoni*) suggest that particulate matter is not retained and that the sponge instead relies for nutrition entirely on the uptake of dissolved organic carbon. It is thought that the organisms coating the sponge produce sufficient organic carbon for

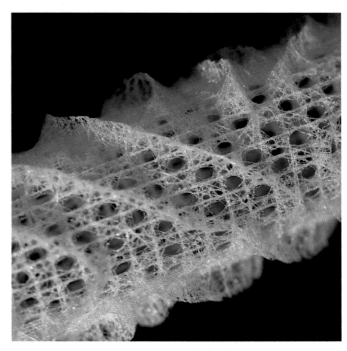

A glass sponge (*Aphrocallistes vastus*) skeleton. (Photo by ©Ken M. Highfill/Photo Researchers, Inc. Reproduced by permission.)

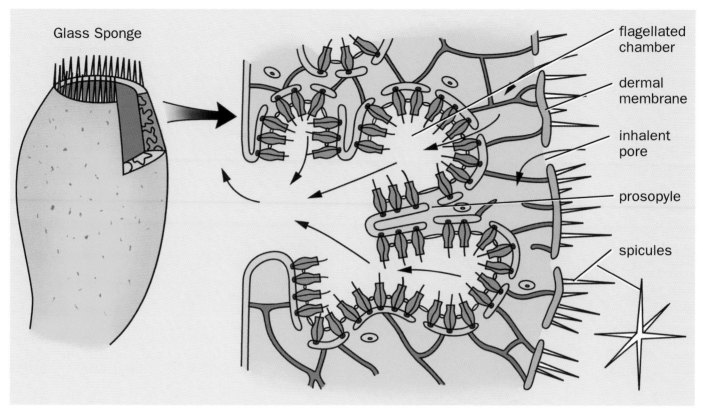

Glass sponge anatomy. (Illustration by Kristen Workman)

themselves and their host. Nevertheless, *R. dawsoni* can phagocytose both bacteria and latex beads in laboratory preparations.

Laboratory experiments with *R. dawsoni* and *Oopsacas minuta* (also a rosellid sponge, but one with a clean exterior) have shown that uptake of particulates occurs in the trabecular syncytium near to and in the flagellated chambers, not in choanocytes, as is normally the case in cellular sponges. In hexactinellid sponges the structure equivalent to a choanocyte, the collar body, lacks a nucleus, and in most species examined so far the collar body is enveloped by extensions of the trabecular syncytium (the primary and secondary reticula), which do most of the particulate capture and uptake. The siliceous skeleton may protect hexactinellids from many predators, but at least one asteroid species is not deterred. *Pteraster tesselatus* frequently can be found digesting the soft tissues off the skeleton of *R. dawsoni*.

Reproductive biology

It is generally thought that most hexactinellids lack a seasonal reproductive period because of their deepwater habitat. Nevertheless, because of the difficulty in collecting and preserving these sponges there is little information on reproduction in most deepwater populations. Our knowledge of their development comes from studies done on only a handful of species. Hexactinellid sponges are viviparous. Eggs arise from cells within groups of archaeocytes, a type of pleuripotent cell found in all sponges. The first cell divisions that occur are

equal and result in the formation of a hollow ball of cells (a blastula). Gastrulation (the formation of two cells layers during embryonic development) is said to occur by delamination. The larvae are top-shaped with a girdle, band, or cilia around their middle. The tissues are already syncytial, although it is not yet known how the multinucleate tissue arises. When the larva matures, it is released through the osculum. The species *Oopsacas minuta*, which is reproductive throughout the year, produces the only known live larvae. When studied in a laboratory setting larvae swim slowly to the surface of dishes in left-handed rotations (clockwise, as seen from the anterior pole); although they can swim for several days, they begin to settle and transform into juvenile sponges within 12 hours of release from the parent sponge.

Conservation status

In general, most hexactinellid sponges inhabit areas well out of the reach of human activity. However, on the continental shelf of the northeastern Pacific in British Columbia, reefs of hexactinellid sponges several city blocks in area have been damaged by trawlers. New legislation for the establishment of marine protected areas around these reefs is under development. No species of hexactinellid is listed by the IUCN.

Significance to humans

Euplectella aspergillum, which harbors a pair of crustaceans within its enclosed atrial cavity for life, is commonly given to newlyweds in Japan as a symbol of bonding.

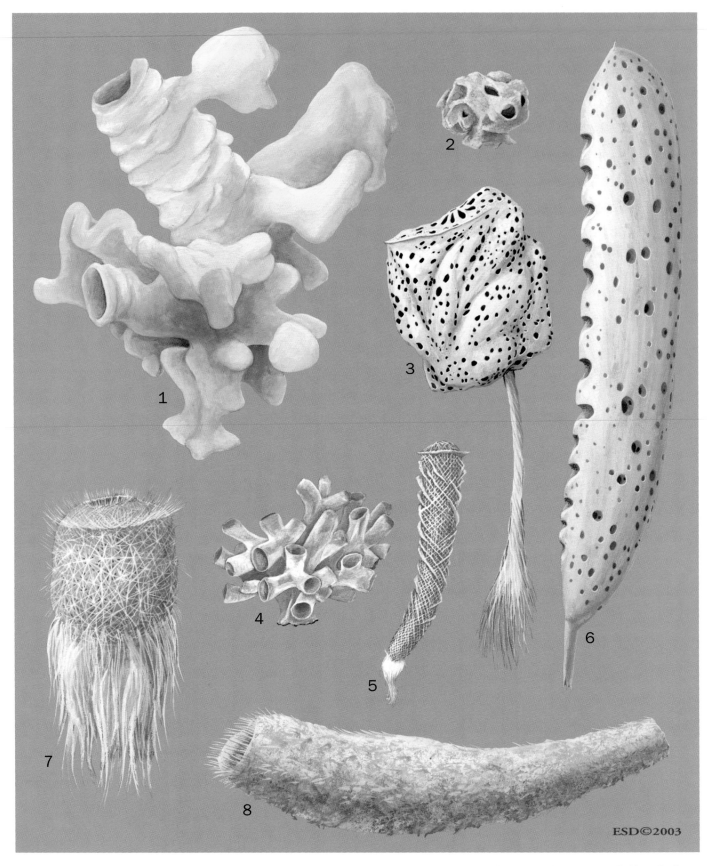

1. Cloud sponge (*Aphrocallistes vastus*); 2. *Neoaulocystis grayi*; 3. Glass-rope sponge (*Hyalonema sieboldi*); 4. *Farrea occa*; 5. Venus's flower basket (*Euplectella aspergillum*); 6. *Monorhaphis chuni*; 7. Bird's nest sponge (*Pheronema carpenteri*); 8. Sharp-lipped boot sponge (*Rhabdocalyptus dawsoni*). (Illustration by Emily Damstra)

Species accounts

Glass-rope sponge
Hyalonema sieboldi

ORDER
Amphidiscosida

FAMILY
Hyalonematidae

TAXONOMY
Hyalonema sieboldi Gray, 1835, Japan.

OTHER COMMON NAMES
English: Glass plant; Japanese: Hoshi-gai.

PHYSICAL CHARACTERISTICS
Truncated oval body, 3–6 in (75–155 mm) long by 2.2–6 in (55–155 mm) wide, which is borne on a compact, twisted bundle, 16–26 in (400–650 mm) long, of 200–300 coarse, siliceous root spicules. Flat body top is covered by perforated sieve plate. Spicules are never fused.

DISTRIBUTION
Known only from entrance of Tokyo Bay, Japan.

HABITAT
Restricted to soft-bottom habitats; anchored in sediments by long root tuft. No reliable depth record is known.

BEHAVIOR
Nothing known.

FEEDING ECOLOGY AND DIET
Presumably filter feeds.

REPRODUCTIVE BIOLOGY
Nothing is known.

CONSERVATION STATUS
Not listed by the IUCN.

SIGNIFICANCE TO HUMANS
This hexactinellid, the first to be described, is displayed in museums worldwide. ◆

No common name
Monorhaphis chuni

ORDER
Amphidiscosida

FAMILY
Monorhaphididae

TAXONOMY
Monorhaphis chuni Schulze, 1904, eastern Indian Ocean north of Madagascar.

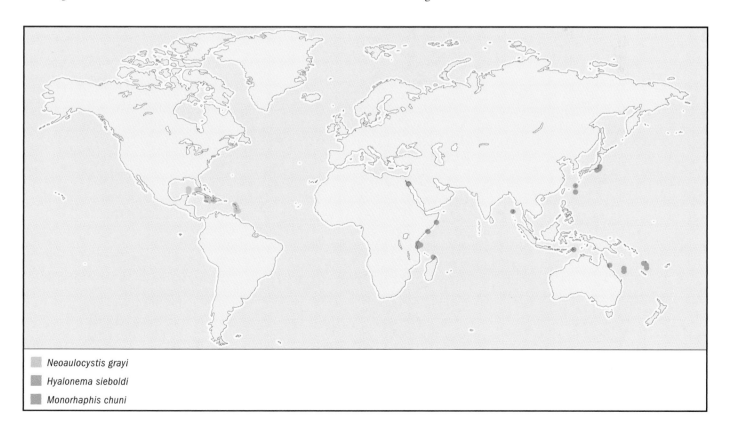

◻ *Neoaulocystis grayi*
◼ *Hyalonema sieboldi*
◼ *Monorhaphis chuni*

OTHER COMMON NAMES
None known.

PHYSICAL CHARACTERISTICS
Cylindrical body, 41 in (105 cm) long and 6 in (15 cm) wide, supported on a single large anchor or basal spicule; the largest siliceous structure formed by any organism (excluding humans), which may reach 9.8 ft (3 m) in length and 0.4 in (10 mm) in width.

DISTRIBUTION
Indo-West Pacific from eastern Africa to New Caledonia.

HABITAT
Lives suspended over muddy bottoms on long anchor spicule at depths of 1,604–6,553 ft (489–1,998 m).

BEHAVIOR
Nothing is known.

FEEDING ECOLOGY AND DIET
Presumably filter feeds.

REPRODUCTIVE BIOLOGY
Nothing is known.

CONSERVATION STATUS
Not listed by the IUCN.

SIGNIFICANCE TO HUMANS
None known. ◆

Bird's nest sponge
Pheronema carpenteri

ORDER
Amphidiscosida

FAMILY
Pheronematidae

TAXONOMY
Holtenia carpenteri Thomson, 1869, northeastern Atlantic off northern Scotland.

OTHER COMMON NAMES
Portuguese: Ninhos de mer.

PHYSICAL CHARACTERISTICS
Ranges in shape from tall and narrow with a deep interior cavity (resembling a barrel), to a squat shape that is broader than it is tall with a shallow depression on top (resembling a cake). Grows to 10 in (25 cm) tall and 8 in (20 cm) wide. Thick, cavernous body wall tapers to single sharp-edged upper opening, which leads to narrow internal atrial cavity. Thin, hairlike siliceous spicules project from annulus around upper third of body and over entire lower third of the body, where they serve as attachment anchors in soft mud. Spicules are never fused.

DISTRIBUTION
Northeastern Atlantic from Iceland to northern Africa, including the Mediterranean Sea. Reports from Brazil and eastern Africa are dubious.

Pheronema carpenteri

Euplectella aspergillum

Rhabdocalyptus dawsoni

HABITAT
On soft, muddy bottoms at depths of 1,007–7,118 ft
(307–2,170 m).

BEHAVIOR
Nothing is known.

FEEDING ECOLOGY AND DIET
Presumably filter feeds.

REPRODUCTIVE BIOLOGY
Nothing is known.

CONSERVATION STATUS
Not listed by the IUCN.

SIGNIFICANCE TO HUMANS
The reasons for the surprisingly high population density and
biomass of this species at bathyal depths, usually poor in animal abundance, has been a focus of oceanographic research. ◆

Cloud sponge
Aphrocallistes vastus

ORDER
Hexactinosida

FAMILY
Aphrocallistidae

TAXONOMY
Aphrocallistes vastus Schulze, 1886, Sagami Bay, Japan.

OTHER COMMON NAMES
None known.

PHYSICAL CHARACTERISTICS
An expanding hollow cone, usually with mittenlike external
projections. Grows to more than 3 ft (1 m). Rigid siliceous
skeleton of body wall is 0.4 in (1 cm) thick and composed of
honeycomb array of small tubes, 0.25 in (1 mm) in diameter,
oriented perpendicular to the surface and passing through the
entire body wall.

DISTRIBUTION
Northern Pacific Ocean west to Japan and east to Central
America.

HABITAT
Attached by cementation to hard bottom at depths of
16–10,500 ft (5–3,000 m).

BEHAVIOR
Nothing is known.

FEEDING ECOLOGY AND DIET
Filter feeds on bacteria easily visible with normal laboratory mi-
croscopes and smaller, microscopically invisible particles (col-
loids, viruses).

REPRODUCTIVE BIOLOGY
Nothing is known.

CONSERVATION STATUS
Not listed by the IUCN.

SIGNIFICANCE TO HUMANS
One of few hexactinellids living within scuba range along the
coast of British Columbia, Canada, and thus available for *in*

Farrea occa

Aphrocallistes vastus

situ and laboratory study. One of three hexactinellids that build unique living glass sponge reefs, which in 2002 were proposed for designation as marine protected areas by the Canadian Parks and Wilderness Society. ◆

No common name
Farrea occa

ORDER
Hexactinosida

FAMILY
Farreidae

TAXONOMY
Farrea occa Bowerbank, 1862, Comoro Islands, Indian Ocean.

OTHER COMMON NAMES
None known.

PHYSICAL CHARACTERISTICS
Network of thin-walled, branching and fusing tubules, 0.2–0.8 in (0.5–2 cm) in diameter. Skeletal framework of tubule wall consists of single layer of fused six-ray spicules, forming a rectangular lattice. Loose spicules include pin-shaped forms.

DISTRIBUTION
Cosmopolitan between latitudes of 48° north and 35° south.

HABITAT
Attached by cementation to hard bottom at depths of 282–6,235 ft (86–1,901 m).

BEHAVIOR
Nothing is known.

FEEDING ECOLOGY AND DIET
Presumably filter feeds.

REPRODUCTIVE BIOLOGY
Nothing is known.

CONSERVATION STATUS
Not listed by the IUCN.

SIGNIFICANCE TO HUMANS
One of three hexactinellids that build unique living glass sponge reefs on the coast of British Columbia, Canada, which in 2002 were proposed for designation as marine protected areas by the Canadian Parks and Wilderness Society. ◆

No common name
Neoaulocystis grayi

ORDER
Lychniscosida

FAMILY
Aulocystidae

TAXONOMY
Myliusia grayi Bowerbank, 1869, Saint Vincent, West Indies.

OTHER COMMON NAMES
None known.

PHYSICAL CHARACTERISTICS
Hemispherical; grows to 4.7 in (12 cm) in diameter. Composed of a network of branching and fusing tubules, 0.10–0.24 in (3–6 mm) in diameter. Skeletal frame of tubule wall consists of two to five layers of six-ray spicules fused into a rigid lattice. Intersections of this framework have 12 supporting struts and are called lantern nodes.

DISTRIBUTION
Gulf of Mexico and West Indies.

HABITAT
Attached by cementation to hard bottom at depths of 348–4,536 ft (106–1,383 m).

BEHAVIOR
Nothing is known.

FEEDING ECOLOGY AND DIET
Presumably filter feeds.

REPRODUCTIVE BIOLOGY
Nothing is known.

CONSERVATION STATUS
Not listed by the IUCN.

SIGNIFICANCE TO HUMANS
The first and most well known of the few surviving members of the order Lychniscosida, a once dominant reef-building group during the Jurassic and Cretaceous periods. ◆

Venus's flower basket
Euplectella aspergillum

ORDER
Lyssacinosida

FAMILY
Euplectellidae

TAXONOMY
Euplectella aspergillum Owen, 1841, Philippines.

OTHER COMMON NAMES
Spanish: Regadera de filipinas; Norwegian: Venuskurv.

PHYSICAL CHARACTERISTICS
Thin-walled tubular body up to 9.5 in (240 mm) long by 2 in (50 mm) wide, with numerous holes through sides and an upper terminal, colander-like sieve plate. External ridges occur obliquely on the sides and as a circular cuff around edge of sieve plate. Skeletal spicules are fused into rigid network in mature specimens. Attached to soft bottom by root of thin, hairlike glass strands, ending in microscopic anchors.

DISTRIBUTION
Indo-West Pacific from the Philippines to eastern Africa.

HABITAT
Lives on soft, muddy bottoms at depths of 144–1,520 ft (44–463 m).

BEHAVIOR
Nothing is known.

FEEDING ECOLOGY AND DIET
Presumably filter feeds.

REPRODUCTIVE BIOLOGY
Nothing is known.

CONSERVATION STATUS
Not listed by the IUCN.

SIGNIFICANCE TO HUMANS
In Japan and the Philippines it traditionally is given as a marriage gift symbolizing fidelity, because a pair of crustaceans often live imprisoned inside the hollow sponge. This demand and its desirability as a beautiful curio have supported a Philippine fishery for hundreds of years. ◆

Sharp-lipped boot sponge
Rhabdocalyptus dawsoni

ORDER
Lyssacinosida

FAMILY
Rossellidae

TAXONOMY
Bathydorus dawsoni Lambe, 1893, Vancouver Island, British Columbia, Canada. The recent transfer of this species to the genus *Adanthascus* is not recognized here.

OTHER COMMON NAMES
None known.

PHYSICAL CHARACTERISTICS
Soft, thick-walled tube, often J shape; grows to more than 33 in (1 m) in length. Has large terminal hole (osculum) that opens into wide, deep atrial cavity. External surface is shaggy, owing to a thick veil of five-ray spicules projecting to 0.4 in (1 cm) fouled by dense community of small animals and bacteria. Attaches to hard bottom by grappling spicules; direct cementation is not reported. Spicules are never fused.

DISTRIBUTION
Northern Pacific Ocean from southern California to the Bering Sea.

HABITAT
Lives at depths of 33–1,433 ft (10–437 m) on rock surfaces and adjacent silt bottoms, to which it presumably falls after detachment from original site.

BEHAVIOR
Stops water pumping for short periods at irregular intervals when undisturbed or immediately when subjected to mechanical or electrical stimulus. Sheds outer spicule veil seasonally.

FEEDING ECOLOGY AND DIET
Captures and presumably digests small algae and bacteria in the laboratory; field study suggests that it captures primarily dissolved organic matter.

REPRODUCTIVE BIOLOGY
Presumed embryos found in adult tissues year-round but are more common in late summer; mature sperm and larvae are unknown.

CONSERVATION STATUS
Not listed by the IUCN.

SIGNIFICANCE TO HUMANS
This is the most intensely studied hexactinellid, because of its occurrence within scuba range on the coast of British Columbia, Canada, and its high rate of survival in the laboratory. ◆

Resources

Books
Reiswig, Henry M. "Class Hexactinellida Schmidt, 1870." In *Systema Porifera, A Guide to the Classification of Sponges.* Vol. 2, *Calcarea, Hexactinellida, Sphinctozoa, Archaeocyatha, Unrecognizable Taxa, and Index of Higher Taxa,* edited by John N. A. Hooper and Rob W. M. Van Soest. New York: Kluwer Academic/Plenum Publishers, 2002.

Periodicals
Boury-Esnault, N., S. Efremova, C. Bézac, and J. Vacelet. "Reproduction of a Hexactinellid Sponge: First Description of Gastrulation by Cellular Delamination in the Porifera." *Invertebrate Reproduction and Development* 35, no. 3 (1999): 187–201.

Conway, K. W., M. Krautter, J. V. Barrie, and M. Neuweiler. "Hexactinellid Sponge Reefs on the Canadian Continental Shelf: A Unique 'Living Fossil.'" *Geoscience Canada* 28, no. 2 (2001): 71–78.

Lawn, I. D., G. O. Mackie, and G. Silver. "Conduction System in a Sponge." *Science* 211 (1981): 1169–1171.

Leys, Sally P. "The Choanosome of Hexactinellid Sponges" *Invertebrate Biology* 118 (1999): 221–235.

———. "Cytoskeletal Architecture and Organelle Transport in Giant Syncytia Formed by Fusion of Hexactinellid Sponge Tissues." *Biological Bulletin* 188 (1995): 241–254.

Leys, Sally P., and N. R. J. Lauzon. "Hexactinellid Sponge Ecology: Growth Rates and Seasonality in Deep Water Sponges." *Journal of Experimental Marine Biology and Ecology* 230 (1998): 111–129.

Leys, Sally P., and G. O. Mackie. "Electrical Recording from a Glass Sponge." *Nature* 387 (1997): 29–30.

Mackie, G. O., and C. L. Singla. "Studies on Hexactinellid Sponges. I. Histology of *Rhabdocalyptus dawsoni* (Lambe, 1873)." *Philosophical Transactions of the Royal Society of London B* 301 (1983): 365–400.

Mackie, G. O., Lawn, I. D., and M. Pavans de Ceccatty. "Studies on Hexactinellid Sponges. II. Excitability, Conduction and Coordination of Responses in *Rhabdocalyptus dawsoni* (Lambe, 1873)." *Philosophical*

Resources

Transactions of the Royal Society of London B 301 (1983): 401–418.

Perez, T. "La Rétention de Particules par une Éponge Hexactinellide, *Oopsacas minuta* (Leucopsacasidae): Le Rôle du Réticulum." *Comptes Rendus de l'Academie des Sciences de Paris* 4 (1996): 1–29.

Reiswig, Henry M. "Histology of Hexactinellida (Porifera)." *Colloques Internationaux de Centre Natnional de Recherche Scientifique* 291 (1979): 173–180.

Reiswig, Henry M., and G. O. Mackie. "Studies on Hexactinellid Sponges. III. The Taxonomic Status of Hexactinellida within the Porifera." *Philosophical Transactions of the Royal Society of London B* 301 (1983): 419–428.

Reitner, J., and D. Mehl. "Early Paleozoic Diversification of Sponges: New Data and Evidences." *Geologisch-Paläontologische Mitteilungen Innsbrück* 20 (1995): 335–347.

Schultz, F. E. "Report on the Hexactinellida Collected by H.M.S. Challenger during the Years 1873–1876." *Report on Scientific Research of the Challenger, Zooolgy* 21 (1887): 1–513.

———. "On the Structure and Arrangement of the Soft Parts in *Euplectella aspergillum*." *Royal Society of Edinburgh Transactions* 29 (1880): 661–673.

Wyeth, R. C., Leys, S. P., and G. O. Mackie. "Use of Sandwich Cultures for the Study of Feeding in the Hexactinellid Sponge *Rhabdocalyptus dawsoni* (Lambe, 1892)." *Acta Zoologica* 77, no. 3 (1998): 227–232.

Other

The Sponge Project. "Recent Hexactinellid Sponge Reefs on the Continental Shelf of British Columbia, Canada." April 22, 2003 [June 11, 2003]. <http://www.porifera.org/a/cifl.htm>.

Sally P. Leys, PhD
Henry M. Reiswig, PhD

Demospongiae
(Demosponges)

Phylum Porifera

Class Demospongiae

Number of families 80

Thumbnail description
Soft, elastic, but also tough, friable, or hard, frequently brightly colored sponges; varying in shape from encrusting, massive, tubes, or branches to cups or vases; the body reinforced by spongin, siliceous (containing silica) spicules, or a combination of both

Photo: A row pore rope sponge (*Aplysina cauliformis*) seen near the Cayman Islands. (Photo by ©Andrew J. Martinez/Photo Researchers, Inc. Reproduced by permission.)

Evolution and systematics

The demosponges originated in the Cambrian period and form the largest class of the phylum Porifera, containing about 85% of all described Holocene species. The class Demospongiae is divided into three subclasses:

1. Subclass Homoscleromorpha, with one order, Homosclerophorida; one family; and about 60 species.

2. Subclass Tetractinomorpha, with four orders, Astrophorida (also known as Choristida), Chondrosida, Hadromerida, and Spirophorida; 22 families; and several hundred species.

3. Subclass Ceractinomorpha, with nine orders, Agelasida, Dendroceratida, Dictyoceratida, Halichondrida, Halisarcida, Haplosclerida, Poecilosclerida, Verongida, and Verticillitida; 57 families; and several thousand species.

The names of these subclasses have been in use for several decades. As of 2002, however, with the publication of *Systema Porifera*, several changes in classification have been made and definitions refined. These changes have made the subclasses more homogeneous, though still not completely so.

The subclass Homoscleromorpha is a small and well-defined group of sponges with or without a skeleton, characterized by

viviparous reproduction and a unique incubated cinctoblastula type of larva. If skeletal elements are present, they are relatively small, consisting of tetraxonic (four-rayed) siliceous spicules without a clear distinction between megascleres (large spicules) and microscleres (small spicules). The Tetractinomorpha have monaxonic (single-rayed) spicules in addition to large tetraxonic spicules; asterose (star-shaped) microscleres; a skeleton that is usually radial or axially compressed; predominantly oviparous reproduction and parenchymellar (solid) or blastular (hollow) larvae. Ceractinomorpha is the largest and most diverse subclass, with a wide variety of monactine megascleres and various kinds of microscleres, with the exception of asterose forms. In general, sponges in this subclass have skeletons made of spongin and spicules in different proportions, with a variety of skeletal structures. Their reproduction is predominantly viviparous and their larvae are parenchymellar.

The former class of Sclerospongiae, which was proposed in 1970 ("sclerosponges"), together with the former order Ceratoporellida, formed a polyphyletic (descended from more than one line of ancestors) group of coralline sponges that included several Holocene species as well as fossil sponges. The Sclerospongiae are hard, stony sponges with a rigid calcareous basal skeleton in addition to an otherwise "normal" demosponge type of skeleton and spicule complement. Since 1985 the class name Sclerospongiae has been discarded and its fam-

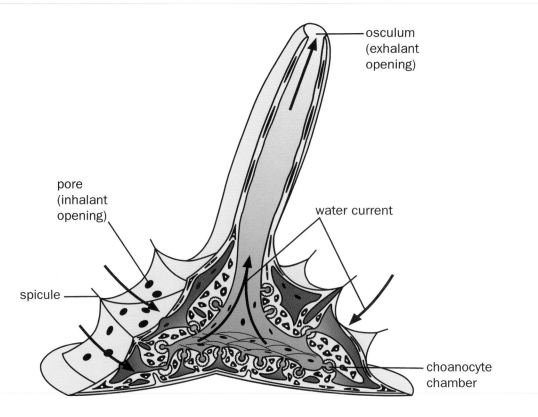

Demosponge anatomy. (Illustration by Laura Pabst)

ilies reassigned to different orders on the basis of characteristics reflecting common ancestry.

Another polyphyletic group is the former order Lithistida, which included many fossil and several Holocene species characterized by a special type of spicules called desmas. Most species in this group were deep-water sponges. The evolutionary history of these sponges is still far from resolved; some appear to be related to the Astrophorida and others to the Hadromerida. Most taxa (categories) in this group, however, have been classified as an artificial fifth order (Lithistida) in the subclass Tetractinomorpha.

Axinellida, another polyphyletic group, is no longer defined as an order. Its families have been reassigned to various orders of Tetractinomorpha and Ceractinomorpha.

Verticillitida consists of the fossil family Verticillitidae. It belongs to an unrelated assemblage of mainly calcified fossil sponges with chambered structures known as Sphinctozoa. One Holocene genus, *Vaceletia*, which has one known polymorphic species and possibly other "living fossil" species, has been assigned to this order.

Physical characteristics

The demosponges as a group display a wide variety of shapes, colors, textures, skeletal architectures, and spicule morphology. There are species that are capable of hollowing out limestone, penetrating deep inside rocks, coral heads, and

shells. Most demosponges have skeletons made of siliceous spicules, spongin fibers, or a combination of both; one group, however, has no skeleton at all. The architecture varies widely among the different groups; it may be reticulate (netlike), confused, radial (spreading outward from a common center), plumose (feathery) or axially compressed. The spicules are usually divided into two size categories (megascleres and microscleres) with a distinct morphology.

The order Poecilosclerida is the largest and most diverse order, with 25 families and several thousand species. Although this group displays a wide variety in form and skeletal architecture, it has a unique feature— chelae, which are meniscoid (crescent-shaped) microscleres with a curved shaft and recurved, winglike or broadly rounded structures at each end. These chelae are extremely diverse, and new ultrastructural characteristics are still being discovered.

The order Haplosclerida comprises 13 families and hundreds of species. All freshwater sponges belong to this order as the suborder Spongillina. They are frequently cushion-shaped; however, encrusting, branching, tubular, vase-, and fan-shaped forms are also quite common. Their coloring is not very intense; most sponges in this order come in delicate shades of purple, lavender, light brown or blue. Most freshwater sponges are green. They are rather soft and easily squeezed except for species of the suborder Petrosina, which are firm and cannot be compressed. All haplosclerids have a netlike skeleton of smooth, single-rayed, one-pointed short megascleres bound together by different amounts of spongin. Most

A giant barrel sponge (*Xestospongia muta*) in the Lighthouse Reef, Belize. (Photo by ©Andrew J. Martinez/Photo Researchers, Inc. Reproduced by permission.)

marine haplosclerids have no microscleres. Where microscleres are present, they are very simple in structure and none are unique to the order. The spicules of the Spongillina are more elaborate, with smooth or variably ornamented megascleres and several kinds of microscleres. The simple structure of the spicules, combined with a very high degree of variability in skeletal architecture in some species, make the marine Haplosclerida among the most difficult sponges to identify.

The Dendroceratida, Dictyoceratida, and Verongida, also known as Keratosa, are sponges with a skeleton made up only of spongin without spicules. All commercial bath sponges belong to the Dictyoceratida. Taken together, these orders contain 10 families and about 450 species. The sponges are often rather tough and flexible; in one family, the Spongiidae, both the surface and the spongin fibers may be heavily coated with foreign spicules and detritus. Species of the order Verongida are easily noticed tube-, fan-, or vase-shaped sponges, frequently colored a deep sulphur yellow. When these sponges are damaged or exposed to air, their color changes rapidly to a deep purple or black.

Distribution

The Astrophorida, Chondrosida, Hadromerida, Halichondrida, both the marine and freshwater Haplosclerida, the Homoscleromorpha, Poecilosclerida and most Spirophorida have a worldwide distribution. The Agelasida, Dictyoceratida, and the sclerosponges, however, are found mostly in the tropics. The Verticillitida; the spirophorid family Spirasigmidae; and two families of the Verongida, the Pseudoceratinidae and the Aplysinellidae, are restricted to the Indian and Pacific Oceans; while the Halisarcida, the dendroceratid family Dictyodendrillidae, and the dictyoceratid family Thorectidae are not found in the polar regions.

Habitat

Most demosponges occur in all habitats at all depths. The Homoscleromorpha, Chondrosida, Agelasida, Dendroceratida, Halisarcida, and most Dictyoceratida occur mainly in the shallower parts of the oceans. The sclerosponges prefer cryptic (hidden) habitats.

Behavior

Most demosponges are immobile animals attached at the base to a substrate, or surface on which they live. Some species, however, successfully compete with corals and other sponges for space by releasing toxic chemicals.

Feeding ecology and diet

Like all other sponges, the Demospongiae are filter-feeders. One genus consists of carnivorous species that engulf and digest small crustaceans.

Reproductive biology

Some demosponges are hermaphroditic while others have distinct sexes. Their reproduction may be viviparous, oviparous, or asexual. Asexual reproduction occurs by means of budding, fragmentation, or the production of resistant globular bodies called gemmulae. Demosponge larvae are partly or completely ciliated, usually somewhat elongated blastulae (hollow larvae) or parenchymellae (solid larvae) about 300 µm long. The larvae swim or crawl around for a few hours or days at most, after which they settle on a substrate and metamorphose into an adult sponge.

Conservation status

In response to the overfishing of commercial sponges, patrimonial interest, and rare and remarkable characteristics of certain sponges, eight Mediterranean sponges are protected under the Bern Convention of 1998, and an additional seven species are protected in Italy.

Significance to humans

Several species are of pharmacological interest because of the production of bioactive compounds with antiviral (spongothymidine) and antibacterial (polybrominated diphenyl ethers) properties. Mediterranean and Caribbean horny sponges have commercial value as bath sponges.

A *Niphates digitalis* releasing sperm. (Photo by ©Nancy Sefton/Photo Researchers, Inc. Reproduced by permission.)

1. Yellow boring sponge (*Cliona celata*), boring stage; 2. Barrel sponge (*Xestospongia testudinaria*); 3. *Carteriospongia foliascens*; 4. Stove-pipe sponge (*Aplysina archeri*); 5. Eyed finger sponge (*Haliclona oculata*); 6. Carnivorous sponge (*Asbestopluma hypogea*); 7. Bath sponge (*Spongia officinalis*); 8. Freshwater sponge (*Spongilla lacustris*). (Illustration by Michelle Meneghini)

Species accounts

Eyed finger sponge
Haliclona oculata

ORDER
Haplosclerida

FAMILY
Chalinidae

TAXONOMY
Haliclona oculata Pallas, 1766, British Isles.

OTHER COMMON NAMES
Dutch: Geweispons.

PHYSICAL CHARACTERISTICS
Clusters of thin, commonly somewhat flattened branches, up to 12 in (30 cm) high, arising from a common stalk and attached to the substrate with a small pedicel or foot. In places with strong water currents, branches may fuse to the point of becoming flabellate, or fan-shaped. Oscules (small mouthlike openings) are small and circular, regularly distributed along the narrower sides of the branches. Sponges have a soft, velvety consistency and are light-brown or pinkish-brown in color.

DISTRIBUTION
Arctic-boreal.

HABITAT
Infralittoral to about 328 ft (100 m), on shores with rocky or sandy bottoms.

BEHAVIOR
Little is known besides feeding ecology and reproductive biology.

FEEDING ECOLOGY AND DIET
Filter-feeder, like all other sponges.

REPRODUCTIVE BIOLOGY
Viviparous. White oval larvae are produced from July to November. Asexual reproduction occurs occasionally by means of gemmules attached to the base of the stalk.

CONSERVATION STATUS
Not threatened.

SIGNIFICANCE TO HUMANS
An elegant, attractive sponge; as such, a pleasure for snorklers and divers to collect. ◆

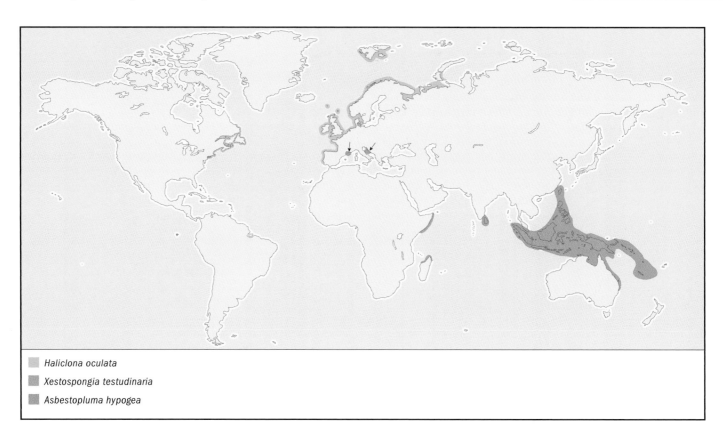

Haliclona oculata
Xestospongia testudinaria
Asbestopluma hypogea

Barrel sponge
Xestospongia testudinaria

ORDER
Haplusclenda

FAMILY
Petrosiidae

TAXONOMY
Xestospongia testudinaria Lamarck, 1815, Cape Denison, Queensland, Australia.

OTHER COMMON NAMES
English: Great vase sponge, volcano sponge; German: Grosser Vasenschwamm.

PHYSICAL CHARACTERISTICS
A large erect reddish brown, barrel- or cup-shaped, thick-walled sponge as much as 4.92 ft (1.5 m) high, with prominent ridges or knobs at the surface. The upper edge of the cup is irregularly indented; the cup itself forms a conspicuous central cavity occupying as much as a third of the total height of the sponge. The sponge is firm and slightly compressible in consistency.

DISTRIBUTION
Western and Central Indian Ocean, Indo-Malesia, northeastern Australia, New Caledonia.

HABITAT
Reefs and lagoons, on rock or dead coral substrates.

BEHAVIOR
Little is known besides feeding ecology and reproductive biology.

FEEDING ECOLOGY AND DIET
Filter-feeder like all other sponges.

REPRODUCTIVE BIOLOGY
Individual sponges are of separate sexes. Mass release of the gametes takes place in September ('smoking' sponges), after which fertilization occurs in the sea water.

CONSERVATION STATUS
Not threatened.

SIGNIFICANCE TO HUMANS
A pleasure for snorklers and divers to find or collect. ◆

Stove-pipe sponge
Aplysina archeri

ORDER
Verongida

FAMILY
Aplysinidae

TAXONOMY
Aplysina archeri Higgin, 1875, Yucatán, Mexico.

OTHER COMMON NAMES
Italian: Spugna a tuba di stufa.

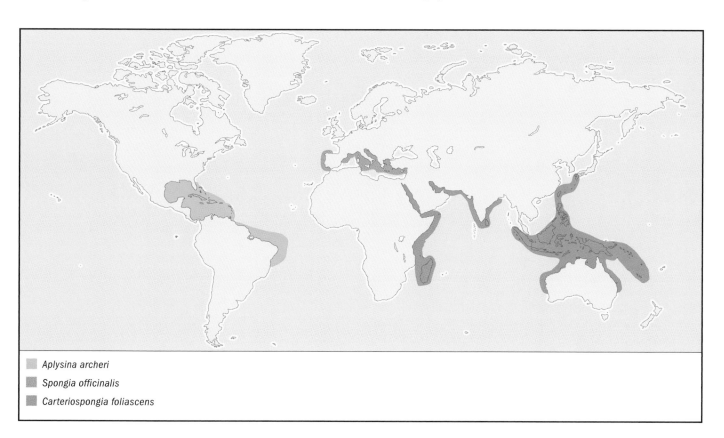

■ *Aplysina archeri*
■ *Spongia officinalis*
■ *Carteriospongia foliascens*

PHYSICAL CHARACTERISTICS
Large pink or purplish gray tubes, growing in groups of several isolated tubes arising from a common base, gradually tapering towards a terminal thick-walled vent. Individual tubes as large as 5 ft (1.5 m) high and 3 in (8 cm) thick. Surface finely conulose (cone-shaped), commonly with a pattern of rounded disc-shaped elevations.

DISTRIBUTION
Caribbean.

HABITAT
Grows in reef localities, 6.5-130 ft (2–40 m) in depth.

BEHAVIOR
Little is known besides feeding ecology and reproductive biology.

FEEDING ECOLOGY AND DIET
Filter-feeder, like all other sponges.

REPRODUCTIVE BIOLOGY
Oviparous, separate sexes, simultaneous spawning during a short period.

CONSERVATION STATUS
Not threatened.

SIGNIFICANCE TO HUMANS
An attractive sponge to snorklers and divers. ◆

Yellow boring sponge
Cliona celata

ORDER
Hadromerida

FAMILY
Clionaidae

TAXONOMY
Cliona celata Grant, 1826, Firth of Forth, Scotland.

OTHER COMMON NAMES
Spanish: Esponja perforadora

PHYSICAL CHARACTERISTICS
An excavating, bright yellow sponge occurring in two different forms: a boring stage appearing as low, rounded papillae sticking out from such limestone substrates as calcareous rocks and shells; and a so-called gamma stage that consists of massive lobes as large as 3 ft (1 m) across and 20 in (50 cm) high. Lobes have raised rounded ridges and small round nodules spread over the surface.

DISTRIBUTION
Northeastern Atlantic from Norway south to the Gulf of Guinea; Mediterranean; eastern coast of North America from Newfoundland south to North Carolina.

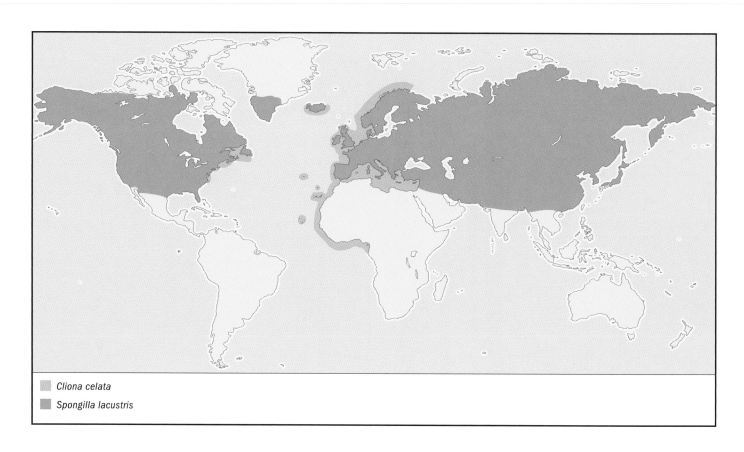

Cliona celata
Spongilla lacustris

HABITAT
Infralittoral zone, 0-650 ft (0–200 m).

BEHAVIOR
Little is known besides feeding ecology and reproductive biology.

FEEDING ECOLOGY AND DIET
Filter feeder, like all other sponges.

REPRODUCTIVE BIOLOGY
Hermaphroditic; oocytes and sperm develop between June and November. Occasionally reproduces by means of asexual buds.

CONSERVATION STATUS
Not threatened.

SIGNIFICANCE TO HUMANS
A pest to oyster growers. ◆

Bath sponge
Spongia officinalis

ORDER
Dictyoceratida

FAMILY
Spongiidae

TAXONOMY
Spongia officinalis Linnaeus, 1759, Mediterranean.

OTHER COMMON NAMES
German: Meerschwamm; Greek: Fino, Matapas; Italian: Spugna da bagno.

PHYSICAL CHARACTERISTICS
Globular-massive sponges, usually over 4 in (10 cm) in diameter, varying in color from white to black depending on environmental circumstances, with a finely conulose (cone-shaped) surface and spongy-elastic consistency.

DISTRIBUTION
Atlantic coasts of Spain; Mediterranean.

HABITAT
On rocks and in caves from the shoreline to the edge of the continental shelf.

BEHAVIOR
Little is known besides feeding ecology and reproductive biology.

FEEDING ECOLOGY AND DIET
Filter-feeder, like all other sponges.

REPRODUCTIVE BIOLOGY
Viviparous; separate sexes. The parenchymellar larvae are large (to 500 µm) and elliptical in shape with short cilia over most of the body.

CONSERVATION STATUS
As of 1986 populations declined as a result of an epidemic disease; protected under the Bern Convention 1998.

SIGNIFICANCE TO HUMANS
Regarded as the finest quality bath sponge in Europe. ◆

No common name
Carteriospongia foliascens

ORDER
Dictyoceratida

FAMILY
Thorectidae

TAXONOMY
Carteriospongia foliascens Pallas, 1766, India.

OTHER COMMON NAMES
German: Blattschwamm.

PHYSICAL CHARACTERISTICS
Lamellate (thinly layered) or foliose (leaflike) greyish-blue sponges, the surface heavily coated with foreign debris, with a characteristic pattern of mounds or ridges. Consistency coarse and flexible.

DISTRIBUTION
Indo-Pacific, Red Sea.

HABITAT
Shallow waters around reefs.

BEHAVIOR
Little is known besides feeding ecology and reproductive biology.

FEEDING ECOLOGY AND DIET
Filter-feeder, like all other sponges.

REPRODUCTIVE BIOLOGY
Viviparous.

CONSERVATION STATUS
Not threatened.

SIGNIFICANCE TO HUMANS
Produces cytotoxic chemicals named sesterterpenoids with biomedical properties. ◆

Carnivorous sponge
Asbestopluma hypogea

ORDER
Poecilosclerida

FAMILY
Cladorhizidae

TAXONOMY
Asbestopluma hypogea Vacelet & Boury-Esnault, 1996, La Ciotat, Mediterranean.

OTHER COMMON NAMES
French: Éponge carnivore.

PHYSICAL CHARACTERISTICS
Small white sponge with an ovoid body as large as 0.25 in (6.5 mm) high and 0.04 in (1.2 mm) thick, attached to the substrate with a long, thin stalk up to 0.5 in (14 mm) long and .007 in (0.18 mm) in diameter. The body bears 30–60 filaments up to 2.3 in (60 mm) long and 50–80 µm in diameter. The filaments are sticky and shaggy because of their dense cover of tiny raised hook-shaped microscleres (anisochelae).

DISTRIBUTION
Known from two caves near Marseille, France, and one in La Croatia, (Mediterranean).

HABITAT
Rocky surfaces in caves where water is trapped all year long and thus has a constant low temperature of 55-58° F (13–14.7° C), 50-195 ft (15–60 m) away from the entrance, at a depth of 55-75 ft (17–23 m).

BEHAVIOR
Their carnivorous feeding behavior makes these sponges unique.

FEEDING ECOLOGY AND DIET
These sponges have no aquiferous system or choanocyte chambers. Their diet consists of small crustaceans that they capture in their filaments. The prey is completely surrounded by new filaments and digested within a few days.

REPRODUCTIVE BIOLOGY
Viviparous; however, little is known as of 2003 regarding the exact time of reproduction and structure of the embryos.

CONSERVATION STATUS
Known only from La Ciotat; protected under the Bern Convention of 1998 due to its patrimonial interest.

SIGNIFICANCE TO HUMANS
None known. ◆

Freshwater sponge
Spongilla lacustris

ORDER
Haplusclenida

FAMILY
Spongillidae

TAXONOMY
Spongilla lacustris Linnaeus, 1759, Lake of Småland.

OTHER COMMON NAMES
French: Éponge d'eau douce; German: Süsswasserschwamm.

PHYSICAL CHARACTERISTICS
Encrusting, branched, arborescent or massive sponges with a fragile, soft consistency and whitish or green color, with irregularly scattered and inconspicuous oscula. Surface uneven and roughened by tiny spines. Gemmules subspherical to oval, occurring in dense clusters or irregularly scattered in the skeletal network.

DISTRIBUTION
Palaearctic.

HABITAT
In standing and running fresh water.

BEHAVIOR
Little is known besides feeding ecology and reproductive biology.

FEEDING ECOLOGY AND DIET
Filter-feeder, like all other sponges.

REPRODUCTIVE BIOLOGY
Overwinters as gemmules, the dormant stage. Viviparous, with sexual reproduction during the summer.

CONSERVATION STATUS
Not threatened.

SIGNIFICANCE TO HUMANS
None known. ◆

Resources

Books

Bergquist, Patricia R. *Sponges.* London: Hutchinson; Berkeley and Los Angeles: University of California Press, 1978.

Hooper, John N.A., and Rob W. M. van Soest, eds. *Systema Porifera: A Guide to the Classification of Sponges.* New York: Kluwer Academic/Plenum Publishers, 2002.

Hooper, John N. A., and Felix Wiedenmayer. "Porifera." In *Zoological Catalogue of Australia.* Vol. 12, edited by A. Wells. Melbourne, Australia: CSIRO, 1994.

Moss, David, and Graham Ackers, eds. *The UCS Sponge Guide.* Ross-on-Wye: The Underwater Conservation Society, 1982.

Other

van Soest, Rob W. M., Bernard Picton, and Christine Morrow. *Sponges of the North East Atlantic.* [CD-ROM] World Biodiversity Database CD-ROM Series. Windows version 1.0. Amsterdam: Biodiversity Center of ETI, Multimedia Interactive Software, 2000.

Wallie H. de Weerdt, PhD

Placozoa
(Placozoans)

Phylum Placozoa
Number of families 1

Thumbnail description
A large, flat, amoeba-like creature up to 0.078 in (2 mm) in diameter; the grayish white body of the organism consists of several thousand cells that form two epithelia (thin layers of tissue composed of closely apposed cells) but are not organized into tissues and organs

---

Photo: Dorsal view of *Trichoplax adhaerens*. (Photo by Hidetaka Furuya. Reproduced by permission.)

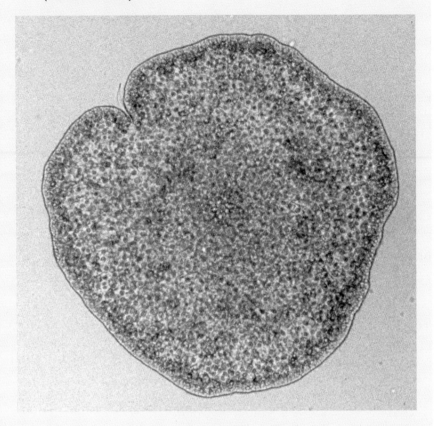

Evolution and systematics

Phylum Placozoa includes only one species, *Trichoplax adhaerens*. Because this phylum is so small, it has no classes or orders assigned to it. There is one family, the Trichoplacidae. Little is known about this organism because it has never been observed in the wild; it has been studied from samples cultured in laboratory aquaria around the world. *Trichoplax adhaerens* was first discovered in the aquarium of the Graz Zoological Institute in Austria in 1883. A second species, *Trichoplax reptans*, was found in 1896. This species, however, has never been seen since it was first described, causing some researchers to doubt its existence.

The placozoans were formerly assigned to the phylum Mesozoa together with the dicyemids, orthonectids, and *Salinella* on the basis of their simple body organization. It became evident, however, that placozoans are not like other mesozoans and do not fit into any other metazoan phylum. As a result, phylum Placozoa was established in 1971. Since that date, specimens of *T. adhaerens* have been found in aquaria around the world. It is not known whether placozoans are cosmopolitan (widely distributed around the world); they are, however, so cryptic (hidden) that their diversity may be much greater than we realize.

Some zoologists have inferred that placozoans may represent the earliest form of animal life. Others regard them as a modified form of the planula stage of a cnidarian. Recent molecular phylogenetic studies suggest that the placozoans are closely related to cnidarians. If this finding is confirmed, it would imply that the placozoans are a secondary simplification of more complex ancestors that possessed fully differentiated tissues and organs, including muscles and nerves.

The taxonomy for this species is: *Trichoplax adhaerens* Schulze, 1883, aquarium of the Zoological Institute in Graz, Austria.

Physical characteristics

Placozoans are extremely simple multicellular animals, with no anterior-posterior (front to rear) polarity or bilateral (right to left) symmetry. They have only dorsoventral (upper surface to lower surface) polarity. The bodies of placozoans consist of several thousand cells but only four cell types. The dorsal epithelium is thin and loosely constructed of cover cells that bear a single flagellum (microscopic whip-like appendage), and contain droplets of fatty material that refract light. The ventral epithelium is composed of a thicker, denser layer of nonflagellated gland cells and columnar cylindrical

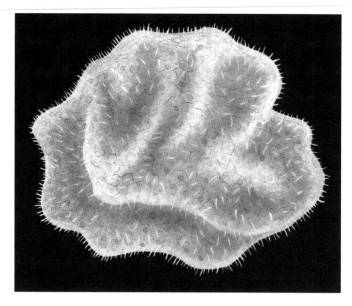

Trichoplax adhaerens. (Illustration by Dan Erickson)

cells with a single flagellum. A fluid-filled space lies between the two epithelia. It contains mesenchyme, a network of loosely organized cells known as fiber cells. The fiber cell is a syncytium, or mass of protoplasm containing many nuclei. Fiber cells are star-shaped, and connected to one another and to both epithelia by branched extensions. Microtubules and microfilaments can be seen within the cytoplasm of the fiber cells; they may be associated with the contraction of cells. Each fiber cell has a synapse-like structure at the point of its attachment to neighboring cells. The fiber cells are thought to function as both muscle and nerve cells. An additional feature of the fiber cells is the presence of symbiotic bacteria within the space of the endoplasmic reticulum.

Distribution

Placozoans are distributed in aquaria located in the littoral zones of tropical and subtropical seas.

Habitat

As of 2003, nothing is known about the biology of placozoans under natural conditions. They may occur on the surface of underwater rocks and benthic marine organisms with shells.

Behavior

Placozoans move around by waving or beating their cilia (tiny hairlike projections), and their outer shape changes continuously. Feeding behavior depends on the amount of available food. When food concentration is low, the organisms move rapidly with frequent random changes in shape. At high concentrations, however, they flatten themselves and move around less.

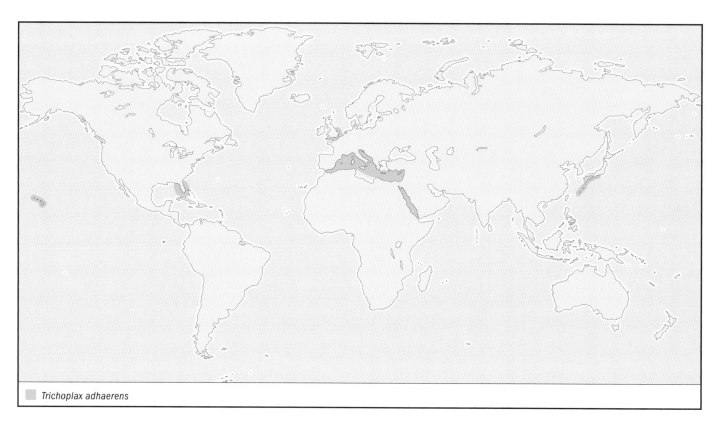

Trichoplax adhaerens

Feeding ecology and diet

In aquaria and laboratory cultures, placozoans have been observed to creep over the substrate by beating the cilia on their ventral surface. They often lift themselves off the substrate, forming a digestive bag in which they can digest their food more efficiently. They are able to digest detritus, living protozoans, and algae by extracellar digestion. The ventral epithelium absorbs the digested material by phagocytosis.

Reproductive biology

Placozoans appear to undergo asexual reproduction in three ways: fission, in which the body divides in half; fragmentation, in which small parts separate from the body; and budding. Sexual reproduction has occasionally been observed in laboratory vessels containing two different clones of placozoans.

Conservation status

This species is not listed by the IUCN.

Significance to humans

There is no known significance to humans.

Resources

Books

Grell, Karl G., and A. Ruthmann. "Placozoa." In *Microscopic Anatomy of Invertebrates*, vol. 2, edited by Frederick W. Harrison and J. A. Westfall. New York: John Wiley and Sons, 1991.

Periodicals

Grell, Karl G. "Einbildung und Forschung von *Trichoplax adhaerens* F. E. Schulze (Placozoa)." *Zeitschrift für Morphologie der Tiere* 73 (1972): 297–314.

Grell, Karl G., and G. Benwitz. "Die Ultrastruktur von *Trichoplax adhaerens* F. E. Schulze." *Cytobiologie* 4 (1971): 216–240.

Kim, Jihee, W. Kim, and C. W. Cunningham. "A New Perspective on Lower Metazoan Relationships from 18S rDNA Sequences." *Molecular Biology and Evolution* 16, no. 3 (1999): 423–427.

Hidetaka Furuya, PhD

Monoblastozoa
(Salinella)

Phylum Monoblastozoa

Number of families 1

Thumbnail description
A small organism consisting of a single cell layer and lacking tissues and organs

Illustration: *Salinella salve.* (Illustration by Dan Erickson)

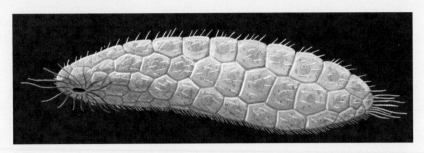

Evolution and systematics

Phylum Monoblastozoa includes only one species, *Salinella salve.* Because this phylum is so small, it has no classes or orders assigned to it. *Salinella salve* was found in 1892 in a culture of 2% saline solution made from material taken from the salt beds of Cordova, Argentina. According to Frenzel, the author of the only published record of the organism, the body of *Salinella* consists of a single cell layer and lacks tissues and organs. *Salinella*, however, has never been observed since its discovery. As a result, some zoologists doubt that it really exists.

Salinella has been classified within the phylum Mesozoa, together with the dicyemids, orthonectids, and *Trichoplax.* The body of *Salinella*, however, differs from those of any other metazoans in lacking internal cells. Its body organization is clearly different from the usual pattern of relegating some cells to the interior of the organism to form endoderm. In this regard, *Salinella* appears to be more closely related to unicellular (one-celled) organisms than to living multicellular animals. If contemporary researchers are able to find new specimens of *Salinella* and study them in detail, they might find that the organism represents an intermediate stage between unicellular and multicellular animals.

The taxonomy for this species is: *Salinella salve* Frenzel, 1892, in a culture made from materials taken from the salt beds of Cordoba, Argentina.

Physical characteristics

The body of *Salinella* consists of about a hundred cells and a single cell layer enclosing a digestive cavity. The cavity is open at both ends, with the openings functioning as a mouth and anus respectively. Distinct bristles can be seen around the mouth and anus. The organism's dorsal surface carries a sparse collection of bristles. The ventral surface is somewhat flattened but is heavily ciliated. The cell walls facing the inner cavity are also heavily ciliated.

Distribution

Salinella salve has been identified only in Argentina.

Habitat

As of 2003, nothing is known about the biology of *Salinella* under natural conditions.

Behavior

Salinella is reported to move by ciliary gliding in the manner of ciliates and small flatworms.

Feeding ecology and diet

Salinella appears to feed by ingesting organic detritus in its internal cavity. Undigested materials are carried to the anus by the movement of the cilia.

Salinella salve

Reproductive biology

Asexual reproduction occurs by transverse fission; however, another mode of reproduction was frequently seen in culture. *Salinella* appears to form a cyst by the conjugation or coupling of two individuals. Although the details of the process are unknown, a unicellular individual that possibly came from the cyst was found in the culture. It is not known whether sexual reproduction takes place within the cyst.

Conservation status

The species is not listed by the IUCN.

Significance to humans

There is no known significance to humans.

Resources

Periodicals

Frenzel, Johannes. "Untersuchungen über die mikroskopische Fauna Argentiniens." *Archiv für Naturgeschichte* 58 (1892): 66–96.

Hidetaka Furuya, PhD

Rhombozoa
(Rhombozoans)

Phylum Rhomobozoa

Number of families 3

Thumbnail description
Characteristic parasites of the kidney of benthic cephalopod mollusks; the body consists of only 8–40 cells, which are fewer in number than in any other metazoans and are organized very simply

Photo: *Dicyemodeca deca* within its host *Octopus dofleini.* (Photo by Hidetaka Furuya. Reproduced by permission.)

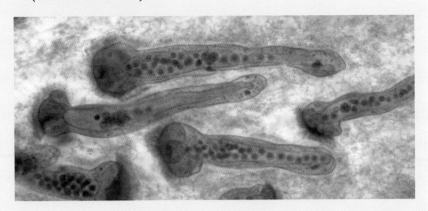

Evolution and systematics

The phylum Rhombozoa includes three families, eight genera, and approximately 150 species. The three families are Conocyemidae, Dicyemidae, and Kantharellidae.

Édouard van Beneden proposed the name Mesozoa for these organisms as an intermediate between Protozoa and Metazoa in body organization. Later zoologists considered rhombozoans degenerated from metazoans such as trematodes because of adaptation for the parasitic lifestyle. Results of phylogenetic studies with nucleotide sequences of 18S recombinant DNA and Hox gene have suggested rhombozoans are spiralians (protostomes) such as turbellarians, nemerteans, annelids, and mollusks. Many scientists now place this group within the phylum Dicyemida.

Physical characteristics

Rhombozoans have two phases of body organization—the vermiform stages of vermiform embryo and adult (nematogen, rhombogen) and the infusoriform embryo. The body of vermiform rhombozoans consists of a central cylindrical cell called the axial cell and a layer of 8–30 ciliated external cells called peripheral cells. The number of peripheral cells is fixed and species specific. At the anterior region, 4–10 peripheral cells form the calotte, the cilia of which are shorter and denser than in more posterior peripheral cells. The calotte comprises two tiers of cells—propolar cells and metapolar cells. Calotte shape varies with species.

Infusoriform embryos consist of 37 or 39 cells, which are more differentiated than those of vermiform organisms. Inside the embryo are four large cells called urn cells, each containing a germinal cell that may give rise to the next generation. At the anterior region of the embryo is a pair of unique cells called apical cells, each containing a refringent body composed of magnesium inositol hexaphosphate. At the posterior region, external cells are ciliated.

Distribution

Temperate and subtropical continental waters. Okhotsk Sea; Sea of Japan; northern, eastern, and western Pacific Ocean; New Zealand; Australia. Mediterranean Sea; northern, eastern, western Atlantic Ocean; Gulf of Mexico; Antarctic Ocean.

Habitat

Vermiform stages of rhombozoans are restricted to the renal sac of cephalopods. In decapods, rhombozoans are also found in the renopancreatic coelom. Some rhombozoans are found in the pericardium of decapods. On the surface of the renal appendage, vermiform rhombozoans insert their heads into renal tubules and folds.

Behavior

Vermiforms and infusoriform embryos swim with their cilia. There appears to be positive thigmotaxis to renal appendages in the vermiform stages.

Feeding ecology and diet

The surface of the rhombozoan body possesses numerous cilia and has a folded structure believed to contribute to absorption of nutrients from the urine of hosts.

Reproductive biology

The vermiform embryo develops asexually from an agamete (axoblast) and grows into an adult. A high population density in the host renal sac can cause a shift from an asexual mode to a sexual mode of reproduction. The functionally hermaphroditic gonad, the infusorigen, forms at high population density. Mature spermatozoa without tails fertilize the pri-

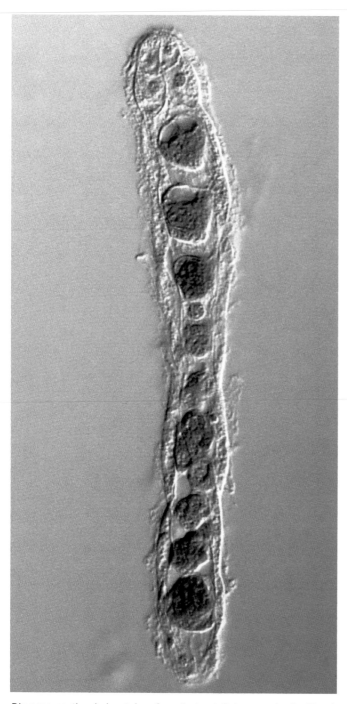

Dicyema acuticephalum taken from its host *Octopus vulgaris*. (Specimens in McConnaughey collection, Santa Barbara Museum of Natural History. Photo by Hidetaka Furuya. Reproduced by permission.)

mary oocytes. A fertilized egg develops into an infusoriform embryo. It remains to be understood how infusoriform larvae infect the new host and develop into vermiforms.

Conservation status

No species is listed by the IUCN.

Significance to humans

None known.

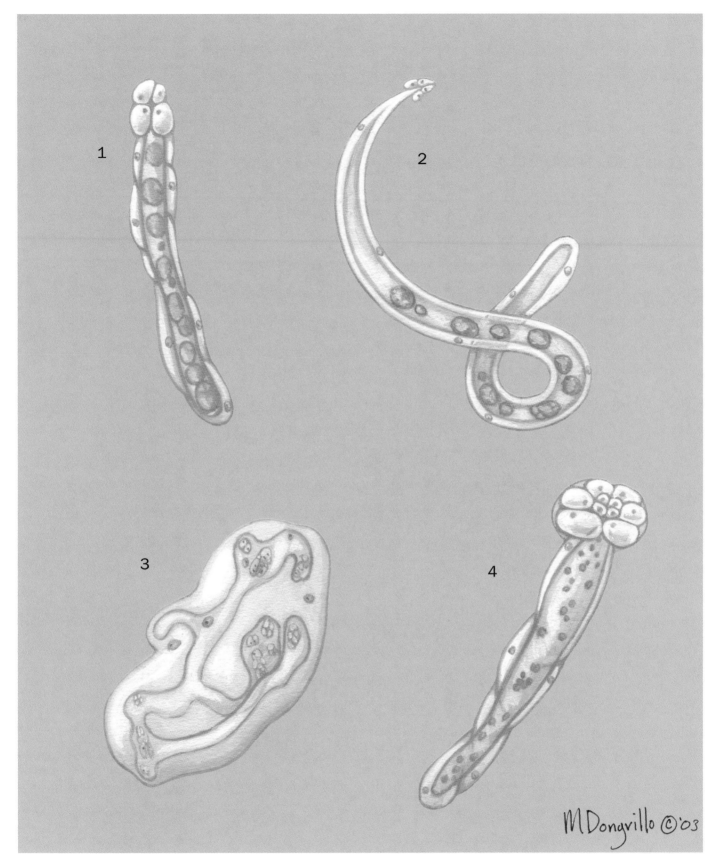

1. *Dicyema acuticephalum*; 2. *Dicyemennea antarcticensis*; 3. *Microcyema vespa*; 4. *Dicyemodeca deca*. (Illustration by Marguette Dongvillo)

Species accounts

No common name
Dicyema acuticephalum

FAMILY
Dicyemidae

TAXONOMY
Dicyema acuticephalum Nouvel, 1947, Japan.

OTHER COMMON NAMES
None known.

PHYSICAL CHARACTERISTICS
Dicyema acuticephalum is characterized by the presence of four propolar and metapolar cells. In this genus, cells of the propolar tier are opposite cells in the metapolar tier. Adults are up to 0.03 in (800 μm) long. Vermiform embryos are approximately 0.002 in (50 μm) long. There are 16–18 peripheral cells. The calotte is bell shaped. Infusoriform embryos consist of 37 cells and are approximately 0.001 in (30 μm) long.

DISTRIBUTION
Japan.

HABITAT
Renal sacs of *Octopus vulgaris.*

BEHAVIOR
Vermiform rhombozoans and infusoriform embryos swim with their cilia. There appears to be positive thigmotaxis to renal appendages in the vermiform stages.

FEEDING ECOLOGY AND DIET
Absorption of nutrients from the urine of hosts.

REPRODUCTIVE BIOLOGY
Adults have a single hermaphroditic gonad, the infusorigen. Approximately nine egg-line cells (oogonia primary oocytes) and 16 sperm-line cells (spermatogonia, primary spermatocytes, secondary spermatocytes, sperm) usually are found in an infusorigen.

CONSERVATION STATUS
Not listed by the IUCN.

SIGNIFICANCE TO HUMANS
None known. ◆

▢ *Dicyema acuticephalum*

▢ *Dicyemodeca deca*

▢ *Microcyema vespa*

No common name
Dicyemennea antarcticensis

FAMILY
Dicyemidae

TAXONOMY
Dicyemennea antarcticensis Short and Hochberg, 1970, Antarctic Ocean.

OTHER COMMON NAMES
None known.

PHYSICAL CHARACTERISTICS
Dicyemennea antarcticensis is characterized by the presence of four propolar and five metapolar cells. The body length of adults is up to 0.2 in (5,500 µm). The body length of vermiform embryos is approximately 0.01 in (300 µm). There are 34–36 peripheral cells. The calotte is triangular in the side view and pointed to rounded in the anterior view. Infusoriform embryos consist of 39 cells. The body length of infusoriform embryos is approximately 0.002 in (51 µm).

DISTRIBUTION
Antarctic Ocean.

HABITAT
Renal sacs of the Antarctic octopus (*Pareledone turqueti*).

BEHAVIOR
Vermiform rhombozoans and infusoriform embryos swim with their cilia. There appears to be positive thigmotaxis to renal appendages in the vermiform stages.

FEEDING ECOLOGY AND DIET
Absorption of nutrients from the urine of hosts.

REPRODUCTIVE BIOLOGY
Adults have a single hermaphroditic gonad, the infusorigen. Approximately 10 egg-line cells (oogonia primary oocytes) and four sperm-line cells (spermatogonia, primary spermatocytes, secondary spermatocytes, sperm) usually are found in an infusorigen.

CONSERVATION STATUS
Not listed by the IUCN.

SIGNIFICANCE TO HUMANS
None known. ◆

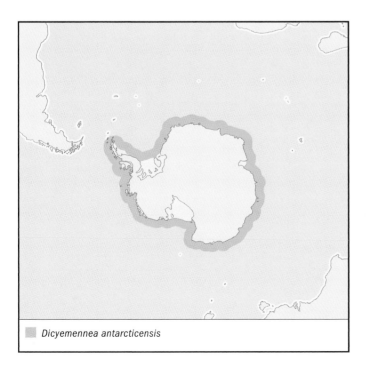

Dicyemennea antarcticensis

No common name
Dicyemodeca deca

FAMILY
Dicyemidae

TAXONOMY
Dicyemodeca deca McConnaughey, 1957, northeastern Pacific Ocean (Washington, United States).

OTHER COMMON NAMES
None known.

PHYSICAL CHARACTERISTICS
Dicyemodeca deca is characterized by the presence of four propolar and six metapolar cells. The body length of adults does not exceed 0.04 in (1,000 µm). The body length of vermiform embryos is approximately 0.002 in (40–60 µm). There are 23 or 24 peripheral cells. The calotte is disk shaped. Infusoriform embryos consist of 35 cells and are approximately 0.001 in (33 µm) long.

DISTRIBUTION
Northeastern Pacific Ocean.

HABITAT
Renal sacs of *Octopus dofleini*.

BEHAVIOR
Vermiform rhombozoans and infusoriform embryos swim with their cilia. There appears to be positive thigmotaxis to renal appendages in the vermiform stages.

FEEDING ECOLOGY AND DIET
Absorption of nutrients from the urine of hosts.

REPRODUCTIVE BIOLOGY
Adults have two infusorigens. Approximately 16 egg-line cells (oogonia primary oocytes) and 15 sperm-line cells (spermatogonia, primary spermatocytes, secondary spermatocytes, sperms) usually are found in an infusorigen.

CONSERVATION STATUS
Not listed by the IUCN.

SIGNIFICANCE TO HUMANS
None known. ◆

No common name
Microcyema vespa

FAMILY
Conocyemidae

TAXONOMY
Microcyema vespa van Beneden, 1882, English Channel.

OTHER COMMON NAMES
None known.

PHYSICAL CHARACTERISTICS
Microcyema vespa is characterized by an irregular shape in adults; they lack external cilia and have a syncytiumal peripheral cell. The body length of adults is up to 0.03 in (800 μm). The body length of vermiform embryos is approximately 0.001 in (25 μm). There are 10 peripheral cells.

In vermiform embryos the calotte is a syncytium containing six nuclei. Infusoriform embryos consist of 39 cells. The body length of infusoriform embryos is approximately 0.001 in (26 μm).

DISTRIBUTION
English Channel, Mediterranean Sea.

HABITAT
Renal sacs of *Sepia officinalis.*

BEHAVIOR
Vermiform embryos and infusoriform embryos swim with their cilia. There appears to be positive thigmotaxis to renal appendages in the vermiform stages.

FEEDING ECOLOGY AND DIET
Absorption of nutrients from the urine of hosts.

REPRODUCTIVE BIOLOGY
Adults have approximately 10 infusorigens. Approximately 9 egg-line cells (oogonia primary oocytes) and 8 sperm-line cells (spermatogonia, primary spermatocytes, secondary spermatocytes, sperms) usually are found in an infusorigen.

CONSERVATION STATUS
Not listed by the IUCN.

SIGNIFICANCE TO HUMANS
None known. ◆

Resources

Books

Hochberg, F. G. "Diseases Caused by Protistans and Mesozoans." In *Diseases of Marine Animals.* Vol. III, edited by Otto Kinne. Hamburg: Biologische Anstalt Helgoland, 1990.

Periodicals

Furuya, Hidetaka. "Fourteen New Species of Dicyemid Mesozoans from Six Japanese Cephalopods, with Comments on Host Specificity." *Species Diversity* 4 (1999): 257–319.

McConnaughey, Bayard H. "Two New Mesozoa from the Pacific Northwest." *Journal of Parasitology* 43 (1957): 358–61.

Nouvel, Henri. "Les Dicyémides. Systématique, générations, vermiformes, infusorigène et sexualité." *Archives de Biologie, Paris* 58 (1947): 59–220.

Short, Robert B., and F. G. Hochberg. "A New Species of *Dicyemennea* (Mesozoa: Dicyemidae) from Near the Antarctic Peninsula." *Journal of Parasitology* 56 (1970): 517–22.

van Beneden, Édouard. "Contribution à l'histoire des Dicyémides." *Archives de Biologie, Paris* 31 (1882): 195–228.

Hidetaka Furuya, PhD

■
Orthonectida
(Orthonectidans)

Phylum Orthonectida
Number of families 1

Thumbnail description
Minute dioecious and dimorphic or hermaphroditic parasites found in tissues of a wide variety of marine invertebrate phyla.

Photo: A female *Rhopalura ophiocomae.* (Specimens in Nouvel collection, Santa Barbara Museum of Natural History. Photo by Hidetaka Furuya. Reproduced by permission.)

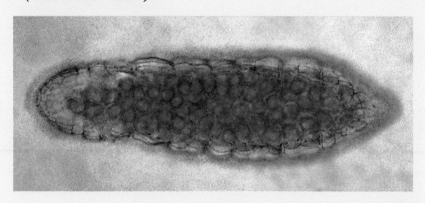

Evolution and systematics

The phylum Orthonectida encompasses one family, four genera, and 20 species. Twenty nominal species have been described and variously placed in four genera. Genera are separated on the basis of body shape and the number and arrangement of egg cells in the axial region of the females. Species are separated on the basis of body size; the number, size, and arrangement of rings of ciliated and unciliated jacket cells; the presence and location of pigment and refringent granules in jacket cells; and host specificity. One family, the Rhopaluridae, contains all four genera: *Rhopalura, Intoshia, Ciliocincta,* and *Stoecharthrum.*

The orthonectids were previously placed in the Mesozoa because of their simple body organization, but recent phylogenetic studies suggest the orthonectids are closer to myxozoans and nematodes. Many scientists now classify Orthonectida at the order level. A more traditional classification; with Orthonectida at the phylum level, is followed here.

Physical characteristics

Depending on the species, orthonectids are either dioecious and dimorphic, or hermaphroditic. The adults are minute, ranging in length from 0.002 to 0.031 in (50 to 800 µm). The body of the adult consists of a jacket of ciliated and unciliated somatic cells arranged in rings around an internal axial mass. Contractile muscle cells differentiate to pack the gonad with longitudinal, circular, and oblique orientations.

Distribution

Orthonectids occur in coastal regions of the English Channel, the Dover Strait, the Strait of Kattegat, the Barents Sea, the White Sea, the northwestern Pacific Ocean (Japan), and the northeastern Pacific Ocean (United States).

Habitat

Orthonectids are found in tissues of organisms in the marine invertebrate phyla Platyhelminthes, Nemertea, Annelida, Mollusca, Echinodermata, Bryozoa, and Urochordata. In the host, infective orthonectid germinal cells penetrate a host cell, and embryos develop and grow into adults within the cytoplasm of the host cell (the "plasmodium").

Behavior

The name Orthonectida means "straight swimming," because orthonectids swim in a straight line, but they generally swim in a spiral motion.

Feeding ecology and diet

Orthonectids may absorb nutrients within the host's cytoplasm.

Reproductive biology

During mating, males make brief contact with females when sperm are released. Fertilization is internal in females. Embryos form about 22 hours after the first cleavage of eggs. When the embryos are fully developed, the female ruptures and dies, releasing ciliated larvae that disperse and enter a new host.

Conservation status

No species of orthonectid is listed by the IUCN.

Significance to humans

None known.

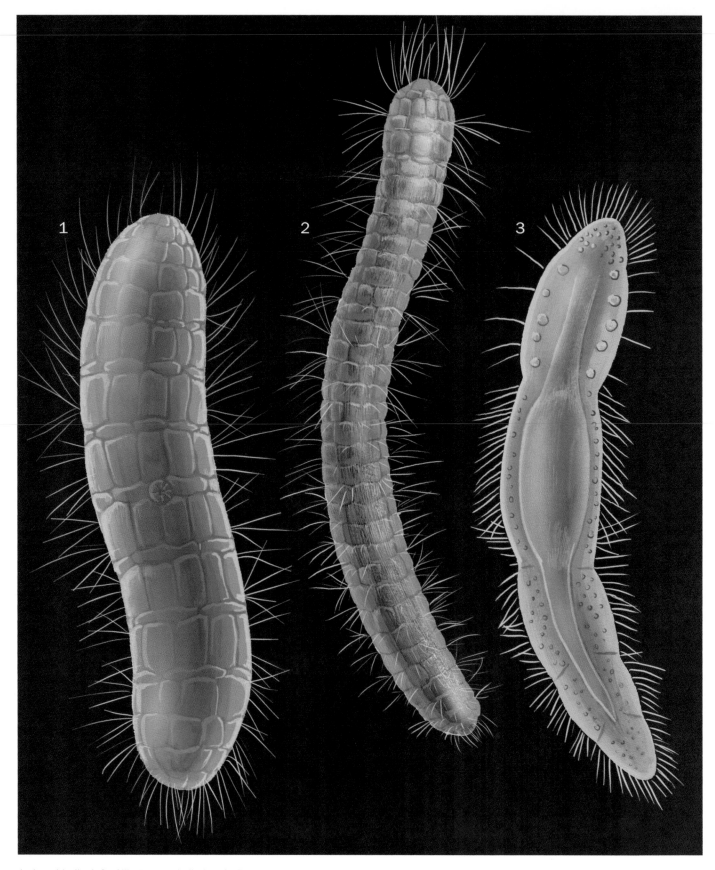

1. *Intoshia linei*; 2. *Ciliocincta sabellariae*; 3. *Rhopalura ophiocomae*. (Illustration by John Megahan)

Species accounts

No common name
Rhopalura ophiocomae

FAMILY
Rhopaluridae

TAXONOMY
Rhopalura ophiocomae Giard, 1877, Wimereux, France.

OTHER COMMON NAMES
None known.

PHYSICAL CHARACTERISTICS
Reaches length of 0.01 in (260 µm) in females, 0.005 in (130 µm) in males. This genus is characterized by sexual dimorphism, the demarcation of the body into regions being much sharper in males than in females. Males possess conspicuous crystalline inclusions in some epidermal cells. In females, the numerous oocytes form a compact mass that occupies most of the body; in males, the sperm mass is located in the middle third of the body.

DISTRIBUTION
Widely distributed. The usual host species, *Amphipholis squamata*, occurs off the coasts of France, Great Britain, and Italy, and Washington and California in the United States.

HABITAT
The perivisceral coelom are closely associated with the walls of the genital bursae or the gut of a parasitized brittle star.

BEHAVIOR
Generally swims by a spiraling motion.

FEEDING ECOLOGY AND DIET
May absorb nutrients in the host cytoplasm.

REPRODUCTIVE BIOLOGY
In males, the genital pore is located between epidermal rings 12 and 14; in females, in ring 19. The male and female may bring their genital pores together long enough for sperm transfer to be effected.

CONSERVATION STATUS
Not listed by the IUCN.

SIGNIFICANCE TO HUMANS
None known. ◆

No common name
Ciliocincta sabellariae

FAMILY
Rhopaluridae

TAXONOMY
Ciliocincta sabellariae Kozloff, 1965, San Juan Archipelago, Washington, United States.

OTHER COMMON NAMES
None known.

PHYSICAL CHARACTERISTICS
Body length of females up to 0.01 in (270 µm); usually 38 or 39 rings of epidermal cells. Males up to 0.005 in (130 µm); usually 30 rings of epidermal cells. This genus is characterized by slight sexual dimorphism. Males are smaller than females, but the arrangement of epidermal cells and the pattern of ciliation are similar in both sexes. Males have no crystalline inclusions in epidermal cells comparable to those of male *Rhopalura*. In females, oocytes are in a single series.

DISTRIBUTION
The host is the sabellid polychaete *Sabellaria cementarium*, which occurs in the San Juan Archipelago, off the coast of the state of Washington, in the United States.

HABITAT
Epidermal tissues of the body wall and dorsal cirri of the parasitized polychaete.

BEHAVIOR
Generally swims in a spiral motion.

FEEDING ECOLOGY AND DIET
May absorb nutrients in the host cytoplasm.

REPRODUCTIVE BIOLOGY
In males, the genital pore is located in the boundary between epidermal rings 15 and 16; in females, in ring 14. The male and female may bring their genital pores together long enough for sperm transfer to be effected.

CONSERVATION STATUS
Not listed by the IUCN.

SIGNIFICANCE TO HUMANS
None known. ◆

No common name
Intoshia linei

FAMILY
Rhopaluridae

TAXONOMY
Intoshia linei Giard, 1877, Wimereux, France.

OTHER COMMON NAMES
None known.

PHYSICAL CHARACTERISTICS
Body length up to 0.006 in (160 µm); usually 24 rings of epidermal cells. In males, the body of the largest individual is not over 0.0016 in (41 µm) in length. In this genus, more than half the rings of epidermal cells are completely ciliated; the rest lack cilia. Sexual dimorphism usually occurs. Males are much smaller than females, are ovoid, and have two genital pores. Males have no crystalline inclusions in epidermal cells compa-

Intoshia linei

Rhoparula ophiocomae

Ciliocincta sabellariae

rable to those in male *Rhopalura.* In females, the oocytes fill most of the axial mass, and are packed together so they seem to form two or three rows.

DISTRIBUTION
The hosts are the nemerteans *Lineus viridus, L. sanguineus,* and *L. rubber,* which occur on the English Channel coast of France at Wimereux and Roscoff.

HABITAT
Tissues of a parasitized nemertean.

BEHAVIOR
Generally swims in a spiraling motion.

FEEDING ECOLOGY AND DIET
May absorb nutrients in the host cytoplasm.

REPRODUCTIVE BIOLOGY
In males, there are two genital pores, located in epidural rings 8 and 16; in females, in ring 12. The male and female may bring their genital pores together long enough for sperm transfer to be effected.

CONSERVATION STATUS
Not listed by the IUCN.

SIGNIFICANCE TO HUMANS
None known. ◆

Resources

Books
Kozloff, Eugene N. "Phyla Placozoa, Dicyemida, and Orthonectida." In *Invertebrates.* Philadelphia: Saunders College Publishing, 1990.

Periodical
Kozloff, Eugene N. "The Genera of the Phylum Orthonectida." *Cahiers de Biologie Marine* 33 (1992): 377–406.

Pawlowski, J., J. Montoya-Burgos, J. F. Fahrni, J. Wüest, and L. Zaninetti. "Origin of the Mesozoa Inferred from 18S rRNA Gene Sequences." *Molecular Biology and Evolution* 13 (1996): 1128–1132.

Hidetaka Furuya, PhD

Anthozoa

(Anemones and corals)

Phylum Cnidaria

Class Anthozoa

Number of families 130

Thumbnail description
Exclusively polypoid cnidarians. Tubular body with hollow tentacles around the mouth; has a pharynx that opens into a digestive cavity subdivided by infoldings of the gut wall. May be solitary or colonial, with or without an internal or external skeleton.

Photo: Jeweled anenome (*Corynactis californica*) (©Shedd Aquarium. Photo by Patrice Ceisel. Reproduced by permission.)

Evolution and systematics

Anthozoa is the largest class in the phylum Cnidaria, with over 6,000 extant species divided among nine orders and classified in two subclasses. The nine orders are as follows:

Subclass Octocorallia (= Alcyonaria)

- Order Pennatulacea, the sea pens and sea pansies; 16 families

- Order Helioporacea, the blue corals; two families

- Order Alcyonacea, the soft corals, sea fans, and sea whips; 29 families

Subclass Hexacorallia (= Zoantharia)

- Order Actiniaria, the sea anemones; 42 families

- Order Scleractinia (= Madreporaria), the true (stony or hard) corals; 25 families

- Order Corallimorpharia, the mushroom (false) corals (also known as mushroom anemones or disc anemones); 4 families

- Order Zoanthidea (= Zoanthinaria), the zoanthids; four families

- Order Antipatharia, the black (thorny) corals and wire corals; five families

- Order Ceriantharia, the tube anemones; three families

Some authors classify the Ceriantharia and Antipatharia together in a third subclass, the Ceriantipatharia, but genetic evidence does not support this grouping.

Molecular evidence places the Anthozoa as the earliest branch of the phylum Cnidaria. Since anthozoans exist exclusively as benthic (sea bottom) polyps, whereas the remaining cnidarians have life cycles that alternate between benthic polyp and pelagic medusoid stages, this evidence indicates that the polyp stage is the ancestral condition among Cnidaria. Within the Hexacorallia, additional evidence suggests that actiniarians, and more recently corallimorpharians, may have evolved from a scleractinian ancestor, accompanied by the loss of a skeleton.

Fossils resembling anthozoans, including sea pens, are known from as early as the Precambrian eon (>540 million years ago [mya]), but it wasn't until the Ordovician period (about 465 mya) that stony coral fossils became common. Two groups of corals dominant among Ordovician fossils—the Tabulata and Rugosa—became extinct at the end of the Permian period (about 248 mya). The earliest scleractinian fossils come from the middle Triassic period (about 230 mya).

Physical characteristics

Anthozoans are polypoid cnidarians. As in all cnidarians, the body wall is composed of two cell layers—the outer ectodermis (or epidermis) and the inner gastrodermis—separated by a layer of gelatinous material known as mesoglea or mesenchyme. Tentacles bearing stinging cells surround the mouth, which is the only opening to the digestive system. A polyp is essentially a tubular sac, with the mouth and tentacles on a flattened upper surface called the oral disk. The mouth leads to a pharynx, a short tube projecting into the closed gut (the coelenteron or gastrovascular cavity). The pharynx typically has one or more ciliated grooves (siphonoglyphs) that funnel water into the coelenteron. The coelenteron is subdivided into chambers by vertical septa (or mesenteries), infoldings of the gut wall that may or may not

Frogspawn coral (*Euphyllia cristata*). (©Shedd Aquarium. Photo by Edward G. Lines, Jr. Reproduced by permission.)

attach to the pharynx. Below the pharynx, the free edges of the septa are thickened to form septal, or mesenterial, filaments that contain cells involved in digestion, including nematocysts. The tentacles are hollow and continuous with the coelenteron.

Octocorals have eight pinnate (parts arranged on each side of a common axis) tentacles and eight septa; hexacorals have septa and tentacles—usually simple—in multiples of six. Cerianthids have two circlets of tentacles, one around the mouth and the other around the edge of the oral disc. There are a number of other details of polyp anatomy that are used to distinguish the orders of Anthozoa.

A unique characteristic of cnidarians is the cnida, a complex intracellular capsule containing an eversible hollow tubule that can be released to sting or trap prey. Eversible means that the structure can turn inside out. There are three basic types of cnidae, and all can be found in the class Anthozoa. Nematocysts, which contain toxins and are typically armed with spines for penetrating the tissues of other organ-

isms, are possessed by all anthozoans. Spirocysts are sticky rather than penetrating and are found only in the hexacorallians. Ptychocysts are unique to the cerianthid tube anemones and are used to construct their tubes.

Species in the orders Scleractinia, Octocorallia, and Antipatharia produce skeletons. In the Scleractinia, the living tissue essentially lies above an external skeleton made of calcium carbonate secreted by the ectodermis. Scleractinian skeletons can take on a variety of shapes, generally described as massive, columnar, encrusting, branching, leaflike or platelike. It is these skeletons that form the framework of tropical coral reefs. Antipatharians secrete an internal horny skeleton that is flexible, black in color, and equipped with thorns on its surface. Octocorals secrete an internal skeleton composed of a protein called gorgonin, calcium carbonate, or a combination of both. Octocorallian and antipatharian skeletons are usually branching treelike or whiplike forms. Unlike other octocorals, helioporaceans produce a massive skeleton that is blue in color and resembles those found in the stony corals. Oc-

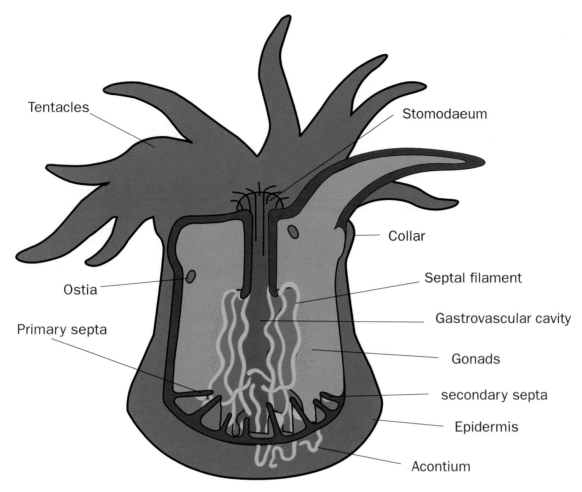

Anthozoa anatomy. (Illustration by Christina St. Clair)

tocorals also secrete small calcareous sclerites (hardened plates) of a variety of shapes and colors that are embedded in the mesenchyme and that may give a spiny or scaly appearance to the colony. Octocorallian soft corals lack a supporting internal skeleton and can inflate or deflate the fleshy colony by funneling water into or out of their polyps. With the exception of the zoanthid genus *Gerardia*, species in the remaining anthozoan orders do not secrete a skeleton. A few actiniarians secrete a chitinous tube, and one genus is able to form a chitinous coiled shell, similar in shape to a snail shell, that is inhabited by a hermit crab. Cerianthids build soft, felt-like tubes from fired ptychocysts.

Anthozoans may be either solitary or colonial. In colonial species, the polyps are united by living tissue, the co-enenchyme, and their gastrovascular cavities are joined by canals or tubes. Actiniarians and ceriantharians are exclusively solitary, and the octocorals and antipatharians are exclusively colonial, but the remaining orders have both types of morphologies. Solitary polyps are commonly 0.5–2 in (1–5 cm) in diameter at the oral disk, but the largest species grow to 3 ft (1 m) across. Polyps of colonial species are typically much smaller (<0.4 in (5 mm)), but the colonies themselves can be

quite large. Octocorallian and antipatharian colonies may grow >8 ft (2.5 m) tall, and some scleractinian corals may reach a size of 19.5 ft (6 m) in height and width. Colonial anthozoans may reach a great age. For example, many octocorals are 100 or more years old, and the deep-sea zoanthid *Gerardia* has been estimated to be 1800 years old.

Distribution

Anthozoans are found worldwide in all oceans.

Habitat

Anthozoans are restricted to marine habitats, but can be found from the intertidal zone to over 19,500 ft (6,000 m) deep. Solitary forms may be attached to a hard substrate or burrowed into soft bottom mud or sand. Colonial forms grow as an encrusting or stoloniferous form on other substrates, or build massive skeletons; tree-like colonial forms are attached by the base of the main stem. Sea pen colonies are anchored into soft bottoms by the base of the primary polyp. Reef-building corals are most typically found in clear, shallow, warm tropical waters, although a few species are known from the cold, dark deep sea.

Crimson anemone (*Cribinopsis fernaldi*). (©Shedd Aquarium. Photo by Patrice Ceisel. Reproduced by permission.)

Behavior

One of the most spectacular of anthozoan behaviors is the synchronous release of sperm and eggs by many colonies over a wide area of coral reef. Mass spawning events have been observed in octocorallian and zoanthid species, but are best known in the scleractinian corals of the Great Barrier Reef, Australia. In some cases, many species release gametes on the same night, and there is considerable evidence that hybridization between species may occur. Synchrony is achieved by timing gamete release with the lunar cycle, and many corals spawn shortly after sunset following a full moon. Eggs and sperm may be released separately into the water column; or, in some hermaphroditic species, may be combined into gamete bundles that float to the surface and break apart there. During these mass spawning events, huge slicks of gametes and developing larvae can be observed on the water surface, attracting a variety of predators that feed on the spawned gametes.

Anthozoans engage in aggressive interactions to defend space from neighboring individuals of the same or different species. A number of specialized structures may be used to repel encroachers. Sea anemones may possess acrorhagi, inflated saclike structures bearing nematocysts that ring the collar below the tentacles. Acrorhagi can be elongated to come into contact with an intruder, whereupon they cause tissue death. Scleractinian, octocorallian, and antipatharian corals may develop specialized tentacles ("sweeper tentacles") after prolonged contact with foreign species. These tentacles are five to ten times longer than normal feeding tentacles and have a greater number of stinging nematocysts. The tentacles search and sweep an area around the polyp and cause tissue death in neighboring species upon contact. Similar structures found in some sea anemones are called "catch tentacles." Some scleractinian corals and corallimorpharians also may extrude their septal filaments to digest the neighboring species' tissues.

Several octocorallian species can produce light when contact with the colony is made. This bioluminescence may be in the form of a bright green flash from a polyp, or a wave of light across the colony as polyps flash sequentially from the point of contact. It is likely the light is meant to startle visual predators.

Feeding ecology and diet

Most anthozoans are suspension feeders whose diet consists of small planktonic invertebrates, phytoplankton, bacterioplankton, or other suspended organic matter. Their methods of prey capture are generally, though not exclusively, passive. Drifting prey may be captured when it comes in contact with the extended tentacles of anthozoan polyps. Prey capture may also involve the firing of cnidae. Many scleractinians produce a slimy mucus that covers the polyp and traps floating and sinking food particles. The mucus is moved around by cilia (small hairlike projections), and eventually enters the mouth. Large sea anemones may feed on crabs, bivalves and fishes, while ptychodactarian anemones have been observed preying upon gorgonian octocorals. In 1997, colonies of the soft coral *Gersemia antarctica* were first reported to bend over and feed in the soft sediments of the Antarctic. Researchers proposed that this feeding strategy may be employed where suspended organic material is in low supply, as it is in the deep sea. Some anthozoans also may absorb dissolved organic matter directly from the seawater into their cells.

Another source of nutrition for many anthozoans comes from symbiotic photosynthetic algae living within cells of the gastrodermis. Like other plants, these algae produce energy-rich organic molecules through photosynthesis; between 20% and 95% of this production is transferred to the host. These

Bubble coral (*Pleurogyra sinuosa*). (©Shedd Aquarium. Photo by Edward G. Lines, Jr. Reproduced by permission.)

symbionts are usually dinoflagellates, called "zooxanthellae"; however, green algae symbionts ("zoochlorellae") are also known. Most shallow-water, reef-building corals contain zooxanthellae, and they also may be found in tropical gorgonians, anemones, and zoanthids. The zooxanthellae are additionally thought to increase the rate at which scleractinian reef corals produce their calcium carbonate skeleton.

The predators of anthozoans include nudibranchs, sea stars, crabs, polychaetes, and fishes.

Reproductive biology

Anthozoans display a wide range of reproductive strategies. Asexual clones may be produced in a variety of ways. Polyps may undergo fission in either a longitudinal or transverse direction. Many sea anemones produce clones by pedal laceration, wherein pieces of the pedal disk tear off or break free and develop into new individuals. The growth of anthozoan colonies may be considered a mode of asexual reproduction. After a free-living larva settles, it metamorphoses into a polyp that repeatedly divides to give rise to additional polyps, all of which remain connected by living tissue. In some species, budded polyps may be released from the parent colony, and these then settle and develop a new colony. Anthozoans, particularly colonial species, also may reproduce by fragmentation. For many scleractinian corals, damage caused by storms or strong wave action may produce fragments that lead to new colonies.

Anthozoans may be gonochoristic (having separate sexes) or hermaphroditic. In colonial species, gonochoristic colonies are composed entirely of male or female polyps, whereas hermaphroditic colonies may have both male and female polyps ("monoecious") or hermaphroditic polyps. Polyps lack well-defined gonads; rather, the gametes accumulate in the gastrodermis of some or all mesenteries. The gametes are typically shed into the gastrovascular cavity and are either released through the mouth for external fertilization ("broadcast spawners"), or eggs are retained for internal fertilization and the embryos released through the mouth at a later time ("brooders"). Brooding species may hold the embryos internally within the gastrovascular cavity or externally in a coat of mucus on the polyp's surface. Anthozoan embryos develop into ciliated planula larvae that may or may not feed, and that can stay in the water column for days to weeks. Contrary to initial assumptions, some species also can produce larvae asexually.

Conservation status

Some of the largest concentrations of anthozoans, in terms of both numbers and species diversity, occur on coral reefs. Such human activities as fishing, coastal development, terrestrial runoff, and marine pollution have had dramatic negative impacts on coral reefs, as have coral diseases that have increased in frequency and severity over the last decade of the twentieth century. Despite a ban in most countries, cyanide is still commonly used to collect reef fishes for the aquarium trade. The cyanide stuns the target fishes, allowing

for easy collection, but kills many of the anthozoans and other invertebrates living on the reef. In 1998, the World Resources Institute estimated that more than half of the world's coral reefs are threatened by these and other human activities. As of 2002, however, only two anthozoan species were listed on the International Union for Conservation of Nature and Natural Resources (IUCN) Red List of Threatened Species: the broad sea fan, *Eunicella verrucosa*, and the starlet sea anemone, *Nematostella vectensis*. All scleractinian corals, antipatharian black corals, and octocorallian blue corals and organ-pipe corals, are listed on Appendix II of the Convention on International Trade in Endangered Species (CITES). This listing means that "trade must be controlled in order to avoid utilization incompatible with their survival." A number of other treaties have been established to protect coral reef organisms, such as the International Coral Reef Initiative and the U.S. Coral Reef Initiative. Most recently, there has been concern over the impact of deep-sea fisheries trawling on slow-growing deep-water corals. Norway and Australia have created conservation areas to protect reefs within their territorial waters.

Significance to humans

Coral reefs, which are largely a framework of scleractinian skeletons glued together by sponges and other organisms, are a major tourist destination and source of recreation. Corals provide a habitat for a variety of organisms that humans use for food, including fishes, urchins, mollusks and crustaceans. It is estimated that approximately 50% of U.S. federally-managed fisheries depend on coral reefs for part of their life cycle, at an annual worth of over $100 million. Anthozoans of all orders (except Antipatharia) are sold in the aquarium trade, and octocorallian and antipatharian skeletons are used to make coral jewelry. Scleractinian skeletons are used as building material and in bone grafts, as the structure of the coral skeleton is similar to that of human bone. Black

Sea anemone seen near Cozumel, Mexico. (Photo by AP/Wide World Photos/University of Wisconsin-Superior. Reproduced by permission.)

coral skeleton was once thought to have medicinal properties, and the name "*Antipathes*" is a Latin word that means "against disease." A variety of natural products have been isolated from anthozoans for commercial use, from suntan lotions to antifoulants. In particular, octocorals produce a range of bioactive compounds; some of these have been harvested for molecular biological and pharmaceutical applications, including anticancer and anti-inflammatory agents.

1. Cauliflower coral (*Pocillopora damicornis*); 2. Red coral (*Corallium rubrum*); 3. Red soft tree coral (*Dendronephthya hemprichi*); 4. American tube dwelling anemme (*Ceriantheopsis americanus*); 5. Sea pansy (*Renilla reniformis*); 6. *Goniastrea aspera*; 7. Rubber coral (*Palythoa caesia*); 8. Mushroom coral (*Fungia scutaria*); 9. Black coral (*Antipathella fiordensis*). (Illustration by Emily Damstra)

1. Giant green anemone (*Anthopleura xanthogrammica*); 2. Elephant ear polyps (*Amplexidiscus fenestrafer*); 3. Magnificent sea anemone (*Heteractis magnifica*); 4. Deep water reef coral (*Lophelia pertusa*); 5. Frilled anemone (*Metridium senile*); 6. Starlet sea anemone (*Nematostella vectensis*); 7. Elkhorn coral (*Acropora palmata*); 8. *Acropora millepora*; 9. Close-up of *A. palmata*. (Illustration by Emily Damstra)

Species accounts

Giant green anemone
Anthopleura xanthogrammica

ORDER
Actiniaria

FAMILY
Actiniidae

TAXONOMY
Actinia xanthogrammica Brandt, 1835, Sitka, Alaska, United States.

OTHER COMMON NAMES
Portuguese: Anémona-verde-gigante.

PHYSICAL CHARACTERISTICS
Large, flat oral disk up to 9.8 in (25 cm) diameter; column densely covered with hollow adhesive wartlike protuberances known as verrucae; tentacles and disk are emerald green, column is olive or brownish.

DISTRIBUTION
Western coast of North America from Alaska south to Baja California.

HABITAT
Low intertidal to shallow subtidal zones on exposed coastlines where it is subject to strong wave action; it often forms carpets of individuals in surge channels.

BEHAVIOR
Nothing is known.

Renilla reniformis
Anthopleura xanthogrammica
Ceriantheopsis americanus

FEEDING ECOLOGY AND DIET
Feeds on sea urchins, crabs, and mussels dislodged by floating debris. One study found that *A. xanthogrammica* benefits when urchins fleeing from predatory seastars fall into the anemone's tentacles. Mussels that are detached by wave action also are eaten. Also derives nutrition from symbiotic association with zooxanthellae and zoochlorellae.

REPRODUCTIVE BIOLOGY
Gonochoristic; reaches sexual maturity in 5–10 years; planktotrophic larvae feed on algae. No asexual reproduction known.

CONSERVATION STATUS
Not listed by IUCN.

SIGNIFICANCE TO HUMANS
Produces toxins known as anthopleurins that stimulate heart muscle and that were considered for medical use. ◆

Starlet sea anemone
Nematostella vectensis

ORDER
Actiniaria

FAMILY
Edwardsiidae

TAXONOMY
Nematostella vectensis Stephenson, 1935, Isle of Wight, England.

OTHER COMMON NAMES
English: Athenarian burrowing anemone; dwarf mud anemone.

PHYSICAL CHARACTERISTICS
Tiny worm-like anemone, rarely more than 0.6 in (15 mm) in length, with 9–18 relatively long (up to 0.4 in [10 mm]) tentacles arranged in two rings; column is smooth with a rounded base called a physa; largely translucent with white bands on the tentacles.

DISTRIBUTION
Atlantic and Pacific coasts of North America; southern and eastern coasts of England. It is thought to have been introduced to England from North America.

HABITAT
Intertidal to shallow subtidal zones; burrows in mud of estuaries and salt marshes. Tolerates a broad range of salinities (8.96–51.54 ppt) and temperatures (30 to 82°F [-1 to 28°C]).

BEHAVIOR
When disturbed, the anemone can completely withdraw into its burrow. It may also move completely out of the burrow and climb onto algae and aquatic vegetation.

FEEDING ECOLOGY AND DIET
Feeds mainly on snails; however, copepods, ostracods, insects, and nematodes also have been found in the coelenteron.

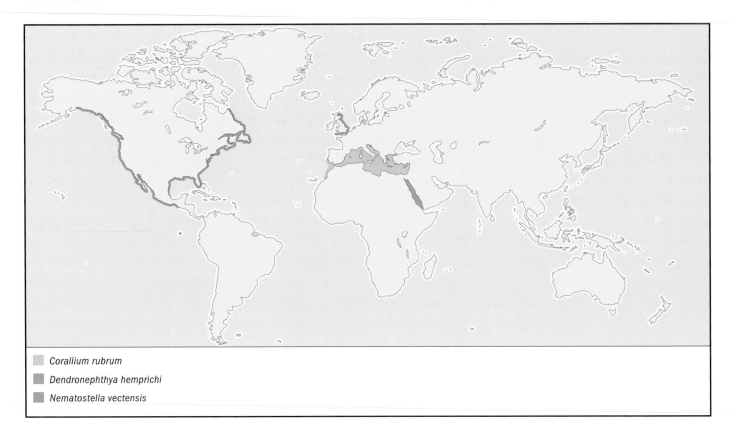

Corallium rubrum

Dendronephthya hemprichi

Nematostella vectensis

REPRODUCTIVE BIOLOGY

Gonochoristic, broadcast spawner; females release gelatinous egg masses that contain as few as five or as many as 2000 ova and hundreds to thousands of nematosomes, which are spherical, flagellated bodies containing nematocysts and are unique to this species. Planktotrophic larvae may settle in 7 days. Asexual reproduction may be more common than sexual reproduction. Several unisex populations have been discovered in North America, while no males have been observed in England. The starlet sea anemone is one of only five anemone species known to reproduce asexually by transverse fission and is the only anemone known to release gelatinous masses of eggs.

CONSERVATION STATUS

Although as many as five million individuals have been found in a single pond, this species is listed as vulnerable on the IUCN Red List. It is considered rare and endangered in the United Kingdom, largely because of its restricted habitat.

SIGNIFICANCE TO HUMANS

Used in laboratory studies of developmental genetics. ◆

Frilled anemone
Metridium senile

ORDER
Actiniaria

FAMILY
Metridiidae

TAXONOMY
Priapus senilis Linnaeus, 1761, Baltic Sea.

OTHER COMMON NAMES

English: Plumose anemone (Britain); French: Anémone plumeuse; German: Seenelke; Norwegian: Sjønellik.

PHYSICAL CHARACTERISTICS

Tall, to 11.8 in (30 cm), with hundreds to thousands of small, slender tentacles on a lobed crown giving a feathery or plumelike appearance; column smooth with a distinct collar below tentacles; with numerous threadlike acontia arising from bases of septa that can be discharged through column pores known as cinclides; color varies from white to brownish-orange.

DISTRIBUTION

Circumpolar, boreo-Arctic; found as far south as New Jersey, United States, in the western Atlantic; Bay of Biscay in the eastern Atlantic; southern California, United States, in the eastern Pacific; and South Korea in the western Pacific. Introduced populations have been found in South Africa and the Adriatic Sea.

HABITAT

Attached to rock, shell, wood, and other hard substrates from the intertidal zone to 540 ft (166 m) deep; tolerates temperatures between 32–80°F (0–27°C).

BEHAVIOR

Adjusts the length of the body column according to current flow. Uses catch tentacles equipped with specialized nematocysts to attack other species in competition for space; tips of the catch tentacles remain attached to the victim.

FEEDING ECOLOGY AND DIET

Passive suspension feeder that traps prey in mucus-coated tentacles; these particles are carried to the mouth by ciliary action.

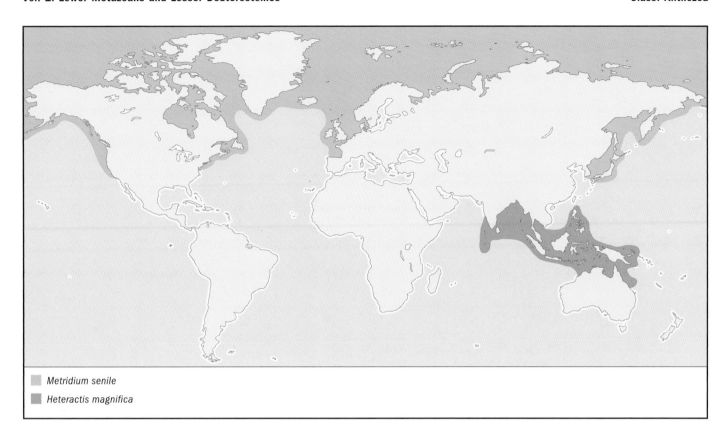

Metridium senile

Heteractis magnifica

Prey items include copepods and polychaetes; and decapod, bivalve, and gastropod larvae.

REPRODUCTIVE BIOLOGY
Gonochoristic, broadcast spawner; planktotrophic larvae feed primarily on dinoflagellates, but also on copepods, chaetognaths and other cnidarian larvae. Asexual reproduction by pedal laceration and fission.

CONSERVATION STATUS
Not listed by the IUCN or CITES.

SIGNIFICANCE TO HUMANS
None known. ◆

Magnificent sea anemone
Heteractis magnifica

ORDER
Actiniaria

FAMILY
Stichodactylidae

TAXONOMY
Actinia magnifica Quoy and Gaimard, 1833, Vanikoro, Santa Cruz Islands, New Hebrides.

OTHER COMMON NAMES
English: Bulb-tip anemone, maroon anemone, Ritteri anemone; German: Prachtanemone; Local dialects: Ramburambu (Indonesia), Burin (Malaysia)

PHYSICAL CHARACTERISTICS
Oral disc up to 39 in (1 m) in diameter; column as much as 8 in (20 cm) tall and wide at the base; tentacles 3 in (7.5 cm) long, of uniform thickness with blunt or slightly bulbous tips. Brightly colored column may be avocado green, magenta, cerulean blue, or white, with yellow, green, or white tentacles.

DISTRIBUTION
Tropical Indo-Pacific seas.

HABITAT
From less than 3–65 ft (1–20 m) deep on or near coral reefs, often perched on prominent structures in strong currents.

BEHAVIOR
Host to clownfish (*Amphiprion*) and damselfish (*Dascyllus*) that chase off potential predators. The fishes are protected from their predators by the anemone's nematocysts, but are themselves unaffected because of a coating of mucus that they pick up from the anemone.

FEEDING ECOLOGY AND DIET
Feeds on plankton that it traps from the water using nematocysts and mucus on the oral disc; digestion of prey begins prior to ingestion. Also derives nutrition from symbiotic zooxanthellae harbored within cells lining the inside of the tentacles and oral disc.

REPRODUCTIVE BIOLOGY
Capable of asexual reproduction by longitudinal fission. Sexual reproduction is presumed but data are lacking.

CONSERVATION STATUS
Not listed by the IUCN or CITES.

SIGNIFICANCE TO HUMANS
Reportedly one of the most frequently photographed anemones because of its coloration, shallow-water location, and tolerance of disturbance. Sold in the aquarium trade. ◆

Red coral
Corallium rubrum

ORDER
Alcyonacea

FAMILY
Coralliidae

TAXONOMY
Madrepora rubra Linnaeus, 1758, "Mari Numidico ad la Calle".

OTHER COMMON NAMES
English: Precious coral; French: Corail rouge; German: Rote Edelkoralle; Italian: Corallo rosso; Spanish: Coral rojo.

PHYSICAL CHARACTERISTICS
Arborescent colonies with red calcium carbonate skeleton up to 19.5 in (50 cm) tall; feeding polyps with eight pinnate tentacles can be completely withdrawn into tissue; also possesses tiny non-feeding polyps (siphonozooids) that lack tentacles and contain the gonads; red calcareous spicules embedded in mesoglea.

DISTRIBUTION
Central and western Mediterranean Sea, with a few records from the eastern Mediterranean; Atlantic coast of southern Portugal and northern Africa.

HABITAT
Cave walls, vertical cliffs, and overhangs from a depth of 32–812 ft (10–250 m).

BEHAVIOR
Nothing is known.

FEEDING ECOLOGY AND DIET
Captures small zooplankton from water column with tentacles. Also may directly absorb dissolved organic carbon from the water through epidermis.

REPRODUCTIVE BIOLOGY
Gonochoristic; male colonies release sperm that swim and are carried by currents to female colonies; eggs are fertilized internally and brooded for up to 30 days. Nonfeeding larvae are released from June through October and swim for up to 15 days before settling. Limited capacity for asexual reproduction.

CONSERVATION STATUS
Not listed by the IUCN or CITES. Population sizes are large but dominated by smaller, younger colonies, as larger colonies have been preferentially harvested. Commercial yields of precious coral declined from 100 tons (91,000 kg) in 1875 to 40 tons (36,000 kg) in 1989 and 27 tons (25,000 kg) in 1996. Underwent mass mortality in the northwestern Mediterranean in 1999 associated with abnormally high water temperatures.

SIGNIFICANCE TO HUMANS
The skeleton is highly valued for coral jewelry and has been harvested since Paleolithic times. It is commonly used by

Navajo and other Native American craftspeople in the southwestern United States for inlays in fine handmade jewelry. The ancient Phoenicians, Egyptians and Romans used coral for trade. According to Greek legend, it confers such magical powers as overcoming evil, protecting crops, warding off epilepsy, defending ships against lightning, and eliminating hatred from the home. Powdered coral skeleton is sold as herbal or homeopathic medicine to be used as an antacid, astringent, emmenagogue, nervine tonic, laxative, diuretic, emetic, or antibilious agent. The traditional Indian Ayurvedic herbal medicine presents the following applications for *C. rubrum* skeleton: "Coughs, wasting, asthma, low fever, urinary diseases, carbuncles, scrofula, spermatorrhoea, gonorrhea and other genital inflammation with mucus discharge, nerve headaches, giddiness, vertigo, chronic bronchitis, pulmonary tuberculosis, vomiting, dyspepsia, bilious headache, weakness, and debility. It is added to tooth powders as an astringent. "

The term "coral" was originally applied to this species, and only subsequently to black corals, stony corals, and soft corals. ◆

Red soft tree coral
Dendronephthya hemprichi

ORDER
Alcyonacea

FAMILY
Nephtheidae

TAXONOMY
Spongodes hemprichi Klunzinger, 1877, Red Sea.

OTHER COMMON NAMES
German: Hemprichs Schleierbäumchen; Italian: Alcionario rosa.

PHYSICAL CHARACTERISTICS
Highly-branched, fleshy, arborescent (treelike) colony lacking an axial skeleton; large embedded sclerites are visible in branches and are conspicuous in polyps, producing a spiky appearance; polyps are not retractile and are mainly clustered in bundles at the end of branches; translucent, colors vary from red to pink, orange, or violet.

DISTRIBUTION
Red Sea.

HABITAT
Found only in strong currents, often on vertical surfaces. Dominant organism on artificial reefs and oil platforms in the Red Sea.

BEHAVIOR
Nothing is known.

FEEDING ECOLOGY AND DIET
Suspension feeder almost exclusively on phytoplankton, a mode of nutrition considered rare among corals. Very small numbers of zooplankton also have been found in the gut. Lacks zooxanthellae.

REPRODUCTIVE BIOLOGY
Gonochoristic, broadcast spawner releases gametes year round. Eggs may remain attached to the mouth by a thread of mucus for several minutes after spawning; when released, the eggs

sink to the bottom. Larvae can metamorphose as soon as 27 hours or as long as 65 days after fertilization, and can live as long as 100 days. *Dendronephthya hemprichi* has a unique method of asexual reproduction: clusters of 4–12 polyps grow from the surface of the colony and autotomize (self-amputate) within two days; the detached polyp fragments have root-like processes that allow for rapid attachment to substrates. There may be hundreds of these fragments in a single colony, allowing for rapid clonal reproduction.

CONSERVATION STATUS
Not listed by IUCN.

SIGNIFICANCE TO HUMANS
Dendronephthya corals are very popular attractions for divers and underwater photographers. ◆

Black coral
Antipathella fiordensis

ORDER
Antipatharia

FAMILY
Myriopathidae

TAXONOMY
Antipathes fiordensis Grange, 1990, Doubtful Sound, New Zealand.

OTHER COMMON NAMES
None known.

PHYSICAL CHARACTERISTICS
Densely branched tree-like colonies grow to over 16 ft (5 m) tall. Tiny polyps, arranged in rows, are white with six tentacles surrounding a mouth that is raised on an oral cone. Proteinaceous black skeleton is covered with spines.

DISTRIBUTION
Endemic to southwestern New Zealand.

HABITAT
Attached to the walls of fjords from 13 to over 325 ft (4 to over 100 m) in depth (but most abundant between 32–114 ft [10–35 m]). This habitat range is unusually shallow for black coral, which are typically found in deeper waters.

BEHAVIOR
Produces sweeper tentacles, which are up to eight times longer, and more densely covered with nematocysts, than normal tentacles. Sweeper tentacles are used in aggressive interactions with other cnidarians in competition for space.

FEEDING ECOLOGY AND DIET
Diet consists of zooplankton, primarily copepods, which are captured by direct contact with the tentacles.

REPRODUCTIVE BIOLOGY
Gonochoristic; broadcast spawns annually during the summer to produce free-swimming planula larvae.

CONSERVATION STATUS
Endemic to the fjords of Fiordland in southwestern New Zealand. Population size has been estimated at more than seven million colonies. Black coral is a protected species in New Zealand, and all black corals are listed on CITES Appendix II.

SIGNIFICANCE TO HUMANS
The skeleton of black coral is used to make jewelry; however, no known fishery exists for *A. fiordensis* as of 2003. ◆

American tube dwelling anemone
Ceriantheopsis americanus

ORDER
Ceriantharia

FAMILY
Cerianthidae

TAXONOMY
Cerianthus americanus Agassiz in Verrill, 1864, Charleston, South Carolina, United States.

OTHER COMMON NAMES
English: Burrowing mud anemone, North American tube anemone, tube sea anemone.

PHYSICAL CHARACTERISTICS
Solitary, elongate, worm-like anemone up to 14 in (36cm) long and 0.79 in (2 cm) wide. Two rings of thin tentacles: up to 125 long ones around the outer edge of the oral disk, and an inner ring of short ones surrounding the mouth. Produces a soft, slippery, felt-like gray tube formed of ptychocysts, mucous, and debris. Colors range from brown or tan to maroon or purple.

DISTRIBUTION
Eastern coast of United States from Cape Cod south into the Gulf of Mexico; Caribbean; West Indies.

HABITAT
Intertidal zones up to 227 ft (70 m) deep; builds tube in vertical burrow as much as 18 in (45 cm) deep in muddy or sandy bottom; only the oral end emerges from the tube.

BEHAVIOR
Will withdraw rapidly to bottom of tube when disturbed.

FEEDING ECOLOGY AND DIET
Spreads tentacles over the surface of the sand or mud to obtain food. Captures its prey with the longer outer tentacles and transfers it to the shorter tentacles surrounding the mouth.

REPRODUCTIVE BIOLOGY
Hermaphroditic; fertilized eggs have been spawned in the laboratory, suggesting self-fertilization is possible.

CONSERVATION STATUS
Not threatened.

SIGNIFICANCE TO HUMANS
Sold in the aquarium trade. ◆

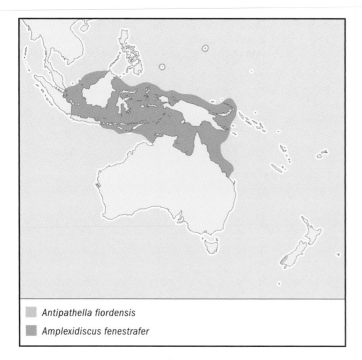

Antipathella fiordensis
Amplexidiscus fenestrafer

FEEDING ECOLOGY AND DIET
Captures prey by envelopment; the center of the oral disc is drawn down while the edges constrict to form a bowl that closes up like a drawstring within 3 sec, enclosing prey and up to 1 gal (4 l) of water. The prey, which may include zooplankton, shrimp and fishes, continues to swim or crawl within the enclosed space for 10–20 min after envelopment, after which it is swallowed. It has been suggested that *A. fenestrafer* mimics anemones that host anemonefish, thereby attracting anemonefish upon which it can feed. It also derives nutrition from symbiotic zooxanthellae harbored within cells lining the inside of the tentacles.

REPRODUCTIVE BIOLOGY
Capable of asexual reproduction by longitudinal fission, pedal laceration, and budding from the column. Data on sexual reproduction are lacking, although individuals with functional male gonads have been identified.

CONSERVATION STATUS
Not listed by the IUCN or CITES.

SIGNIFICANCE TO HUMANS
Sold in the aquarium trade. ◆

Elephant ear polyps
Amplexidiscus fenestrafer

ORDER
Corallimorpharia

FAMILY
Actinodiscidae

TAXONOMY
Amplexidiscus fenestrafer Dunn and Hamner, 1980, Lizard Island, Great Barrier Reef, Australia.

OTHER COMMON NAMES
English: Giant elephant ear mushroom anemone, disk anemone; French: Anémone disque; German: Grosses Elefantenohr; Italian: Anemone orecchio d'elefante.

PHYSICAL CHARACTERISTICS
Short, column up to 2 in (5 cm) tall. Oral disc typically 8–10 in (20–25 cm), but as much as 18 in (45 cm), in diameter; mouth atop a raised cone. 15–20 tentacles around the margin, and many short, conical tentacles arranged in radial rows on the oral disc. A 0.4 in (10 mm) wide tentacle-free "window" rings the oral disc about 0.4 in (10 mm) from the margin. Color is a dull gray to greenish-brown.

DISTRIBUTION
Australia, Indonesia, Palau, and Guam.

HABITAT
In subtidal zones on coral reefs, usually in shallow, quiet water on the inshore side of the reef; attached to vertical walls or coral.

BEHAVIOR
Nothing is known.

Sea pansy
Renilla reniformis

ORDER
Pennatulacea

FAMILY
Renillidae

TAXONOMY
Pennatula reniformis Pallas, 1766, "Mare Americum."

OTHER COMMON NAMES
English: Atlantic coral.

PHYSICAL CHARACTERISTICS
Colonial; a large primary polyp up to 3 in (7.5 cm) long and wide has a heart-shaped frond arising from a fleshy stalk. Smaller polyps are embedded in the upper surface of the frond: typical octocoral feeding polyps and tiny nonfeeding polyps lacking tentacles. Primary polyp appears purple because of colored sclerites in its tissue. The smaller embedded polyps are transparent.

DISTRIBUTION
Western Atlantic from North Carolina, United States, south to Tierra del Fuego, Argentina.

HABITAT
Stalk anchored in sand with frond lying flat on the surface, from low intertidal to shallow subtidal zones.

BEHAVIOR
Produces bioluminescent bright green waves of light that run across the surface of the colony when disturbed at night. Small tentacle-less polyps act as water pumps, allowing colony to quickly deflate to half its normal size or expand.

FEEDING ECOLOGY AND DIET
Polyps secrete a sticky net of mucus that can trap small zooplankton.

REPRODUCTIVE BIOLOGY
Nothing is known.

CONSERVATION STATUS
Not listed by IUCN.

SIGNIFICANCE TO HUMANS
Bioluminescence is created by a protein called "Green Fluorescent Protein" (GFP). The GFP gene has been isolated and is sold commercially for use in molecular biological studies of gene expression in mammals. Scientists have isolated from the sea pansy unique diterpene lipids known as renillafoulins that prevent fouling organisms (e.g., barnacles) from settling on boats and other manufactured marine structures without killing them. ◆

No common name
Acropora millepora

ORDER
Scleractinia

FAMILY
Acroporidae

TAXONOMY
Heteropora millepora Ehrenberg, 1834, Indian Ocean.

OTHER COMMON NAMES
None known.

PHYSICAL CHARACTERISTICS
Colonial; thick finger-like interlocking branches with upright branchlets form squat colonies. Branch surfaces appear scaly. Tissue color ranges from green with orange tips to bright salmon pink or blue.

DISTRIBUTION
Tropical and subtropical Indo-Pacific waters, from Sri Lanka and Thailand east to the Marshall Islands and Tonga, including Australia.

HABITAT
Common on reef flats, lagoons, and upper reef slopes less than 10 ft (3 m) deep.

BEHAVIOR
Mass spawning is timed to occur between 4 and 8 days following a full moon.

FEEDING ECOLOGY AND DIET
Feeds on minute zooplankton and derives nutrition from zooxanthellae harbored within cells lining the digestive cavity.

REPRODUCTIVE BIOLOGY
Hermaphroditic polyps release bundles containing both sperm and eggs that float to the sea surface, where they break open and fertilization occurs. Planula larva may be competent to settle 4–5 days after fertilization. Asexual reproduction occurs by fragmentation of colony branches.

CONSERVATION STATUS
All scleractinian corals are listed in CITES Appendix II.

SIGNIFICANCE TO HUMANS
Common reef-builder of Indo-Pacific coral reefs. ◆

Acropora millepora

Acropora palmata

Lophelia pertusa

Elkhorn coral
Acropora palmata

ORDER
Scleractinia

FAMILY
Acroporidae

TAXONOMY
Madrepora palmata Lamarck, 1816, "American Ocean."

OTHER COMMON NAMES
Spanish: Cuerno de alce.

PHYSICAL CHARACTERISTICS
Colonial; tree-like colonies up to 13 ft (4 m) across and 6.5 ft
(2 m) tall, with thick branches broadly flattened near tips to resemble moose or elk antlers; tubular coral cups protrude from
branch surface. Tissue tan or pale-brown, with tips of branches
white.

DISTRIBUTION
Caribbean; Florida Keys, Bahamas, West Indies to Brazil.

HABITAT
Subtidal zones to 65 ft (20 m); densely aggregated thickets
common on windward reef slopes exposed to heavy wave action.

BEHAVIOR
Nothing is known.

FEEDING ECOLOGY AND DIET
Feed on minute zooplankton and derive nutrition from
zooxanthellae harbored within cells lining the digestive cavity.

REPRODUCTIVE BIOLOGY
Hermaphroditic polyps broadcast spawn once a year in August
or September. Sperm and eggs are packaged together in bundles that float to the sea surface, where they break open and
fertilization occurs. Planula larva are ready to settle 4–5 days
after fertilization. Asexual reproduction occurs by fragmentation of colony branches. Damage to colonies caused by hurricanes is in part responsible for the large stands of *A. palmata* as
broken branches reattach and produce new colonies.

CONSERVATION STATUS
All scleractinian corals are listed in CITES Appendix II. Population declines across the Caribbean led to a designation as
Candidate Species for listing as Threatened or Endangered under the U.S. Endangered Species Act in 1999. Population declines have been attributed to disease outbreaks (white band
disease), compounded locally by hurricanes, increased predation, bleaching, and other factors. Poor water quality because
of land-derived pollutants, sewage, and sediment may stress
elkhorn coral and increase its susceptibility to disease.

SIGNIFICANCE TO HUMANS
Major structural component of Caribbean coral reefs. Provides
essential habitat for fishes and other reef invertebrates. Cuts
and scratches resulting from contact with elkhorn coral are reportedly slow to heal. ◆

Deep water reef coral
Lophelia pertusa

ORDER
Scleractinia

FAMILY
Caryophyllidae

TAXONOMY
Madrepora pertusa Linnaeus, 1758, type of locality not stated,
but probably the fjords of Norway.

OTHER COMMON NAMES
English: Spider hazards, spiders' nests (Nova Scotia); Norwegian: Glasskorall, øyekorall.

PHYSICAL CHARACTERISTICS
Colonial; irregularly branched to form bushy or tree-like
colonies up to 2 m tall; brittle tubular branches about 0.5 in
(1–1.5 cm) thick; white or pink.

DISTRIBUTION
Most records are from the North Atlantic, but also known
from the South Atlantic, northwestern Pacific; Indian Ocean;
and waters south of New Zealand.

HABITAT
Cold water (39–53°F [4–12°C]) and deep sea, from 162 to
>9750 ft (50 to >3000 m); hard substrates on slopes of conti-
nental margins and midoceanic islands. Colonies combine to
build reefs and mounds as large as 650 ft (200 m) high, 0.6 mi
(1 km) wide, and 3 mi (5 km) long. Studies have found *L. pertusa* associated with methane seeps, although this may simply
be because these features represent topographic highs where
ocean currents speed up, and not because *L. pertusa* is feeding
on methane.

BEHAVIOR
Nothing is known.

FEEDING ECOLOGY AND DIET
Feeds voraciously on zooplankton, including copepods.

REPRODUCTIVE BIOLOGY
Details of sexual reproduction are unknown. Asexual reproduction of new colonies occurs when the fragile branches break
and fragments continue to grow.

CONSERVATION STATUS
Commercial deep-sea fish trawls are likely causing mechanical
damage to *Lophelia* reefs. It is unknown what proportion of the
coral fragments survive trawling damage. *Lophelia* reefs have
received protected status in Norway. All scleractinian corals are
listed in CITES Appendix II.

SIGNIFICANCE TO HUMANS
Provides a habitat for a diverse community of invertebrates and
fishes. ◆

No common name
Goniastrea aspera

ORDER
Scleractinia

FAMILY
Faviidae

TAXONOMY
Goniastrea aspera Verrill, 1905, Hong Kong.

OTHER COMMON NAMES
None known.

PHYSICAL CHARACTERISTICS
Massive boulder-like or encrusting colonies; surface has a honeycomb-like appearance. Individual colonies may merge to form plateaus over 16 ft (5 m) across in intertidal zones. Color is usually pale brown; polyps have cream-colored centers.

DISTRIBUTION
Indo-Pacific; Red Sea, Indian Ocean, tropical and subtropical Pacific east to French Polynesia.

HABITAT
Intertidal to shallow subtidal zones; can withstand pounding surf and several hours of exposure to tropical sun and air.

BEHAVIOR
Nothing is known.

FEEDING ECOLOGY AND DIET
Derive nutrition from symbiotic zooxanthellae harbored within cells lining the digestive cavity, and may feed on minute zooplankton.

REPRODUCTIVE BIOLOGY
Hermaphroditic broadcast spawner; may release buoyant sperm and egg bundles, or eggs may be released moments before sperm. In the latter case, some eggs may be retained, with fertilization and development taking place within the polyp. Spawned larvae may settle within five days, but brooded larvae may remain within polyp for 18 days. Asexual reproduction by colony fission possible, but is not common.

CONSERVATION STATUS
All scleractinian corals are listed in CITES Appendix II. Studies have shown that low concentrations of copper or nickel negatively affect reproductive success. A similar effect is seen with relatively small increases in nutrients, as would occur from eutrophication (algae overgrowth associated with excessive amounts of nutrients in the water) resulting from human activity.

SIGNIFICANCE TO HUMANS
None known. ◆

Mushroom coral
Fungia scutaria

ORDER
Scleractinia

FAMILY
Fungiidae

TAXONOMY
Fungia scutaria Lamarck, 1801, no locality given.

OTHER COMMON NAMES
English: Plate coral; German: Pilzkoralle, Rasiermesserkoralle; Hawaiian: ko'a-kohe

PHYSICAL CHARACTERISTICS
Solitary and free-living (unattached) as an adult; oval to elongate skeleton from 1–7 in (2.5–18 cm) long resembles the underside of a toadstool mushroom; numerous small tentacles

■ *Goniastrea aspera*
■ *Fungia scutaria*

arise from lobes formed by underlying skeleton. Usually brown with irregular pink or violet patches; tentacular lobes often bright green.

DISTRIBUTION
Red Sea, Indian Ocean south to Madagascar, west into Pacific Ocean to Society Islands, north to Hawaiian and Line Islands.

HABITAT
On upper reef slopes exposed to strong wave action or around coral knolls at depths from 1.5–16 ft (0.5–5 m).

BEHAVIOR
In the Red Sea, spawning is synchronized to occur in the late afternoon 1–4 days after a full moon. Free-living, unattached adults that are overturned by wave action can right themselves by taking in water through the mouth to expand one-half of the body. This process may take several hours.

FEEDING ECOLOGY AND DIET
May feed on particulate organic material from water column. Derives nutrition from symbiotic zooxanthellae harbored within cells lining the digestive cavity.

REPRODUCTIVE BIOLOGY
Gonochoristic, broadcast spawners; males are smaller than females. Larvae acquire zooxanthellae in the water column. Larvae attach to the bottom and grow a short, calcareous stalk that eventually breaks, releasing the wide upper disk that becomes the free-living adult. The attached stalk may continue to grow and break off new polyps. Asexual reproduction by budding from adult tissue is also seen.

CONSERVATION STATUS
All scleractinian corals are listed in CITES Appendix II.

SIGNIFICANCE TO HUMANS
Skeleton was used by early Hawaiians as an abrasive for polishing wooden canoes and for removing bristles from pig skins before cooking. ◆

Cauliflower coral
Pocillopora damicornis

ORDER
Scleractinia

FAMILY
Pocilloporidae

TAXONOMY
Millepora damicornis Linnaeus, 1758, "Oceanus Asiatico."

OTHER COMMON NAMES
English: Bird's nest coral, lace coral; German: Buschkoralle.

PHYSICAL CHARACTERISTICS
Colonial; the colony is a compact clump, up to several meters across, formed of branches; the surface is dotted with verrucae (wartlike bumps) that intergrade with the branches. Growth form varies with environmental conditions and geographic location. Tissue color is pale brown, greenish, or pink.

DISTRIBUTION
Throughout the Indo-Pacific, western and eastern Australia, north to Japan and Hawaii, and east to Central America, Mexico and Ecuador.

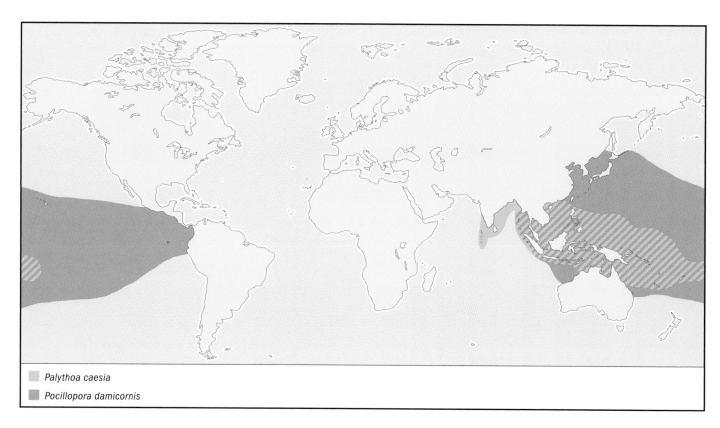

Palythoa caesia
Pocillopora damicornis

HABITAT
Common in all shallow-water habitats, from wharf piles and mangrove swamps to exposed reef fronts; rarely grows deeper than 30 ft (9 m).

BEHAVIOR
Polyp tentacles are usually extended only at night.

FEEDING ECOLOGY AND DIET
Feeds on minute zooplankton and derives nutrition from zooxanthellae harbored within cells lining the digestive cavity.

REPRODUCTIVE BIOLOGY
Polyps are hermaphroditic. Unlike most corals, eggs are fertilized internally and brooded (although in the eastern Pacific, *P. damicornis* is a broadcast spawner). Asexual production of larvae also has been reported. Larvae released from polyps can delay settlement for 100 days or longer, and therefore have the potential to travel great distances. Larvae acquire zooxanthellae from parent.

CONSERVATION STATUS
All scleractinian corals are listed in CITES Appendix II.

SIGNIFICANCE TO HUMANS
Pocilloporids are among the most successful colonizers of coral reefs following disturbance, and are an important frame-building species. ◆

Rubber coral
Palythoa caesia

ORDER
Zoanthidea

FAMILY
Sphenopidae

TAXONOMY
Palythoa caesia Dana, 1846, "Feejee Islands."

OTHER COMMON NAMES
English: *Palythoa* sea mat

PHYSICAL CHARACTERISTICS
Mat-like colonies appear rubbery in texture; typically form ovoid blobs 2–4 in (5–10 cm) in diameter. Polyps are completely embedded in thick coenenchyme encrusted with sediment; short, simple tentacles with knobbed tips. Color ranges from dark brown to tan.

DISTRIBUTION
Indo-Pacific.

HABITAT
Intertidal or subtidal zones on coral reefs and reef crests. In areas of high abundance, may form mats covering virtually all of the available substrate for tens to hundreds of square meters.

BEHAVIOR
Nothing is known.

FEEDING ECOLOGY AND DIET
Contains symbiotic zooxanthellae in its tissues. In laboratory studies, *P. caesia* feeds on zooplankton, algae, and bacteria.

REPRODUCTIVE BIOLOGY
Broadcast spawner. Eggs and larvae lack zooxanthellae, so symbionts must be acquired after settlement.

CONSERVATION STATUS
Not listed by the IUCN or CITES.

SIGNIFICANCE TO HUMANS
Palytoxins, which are found in the mucus and gonads of *Palythoa caesia*, are among the most potent toxins known in nature and can be painful and dangerous to humans if absorbed through open wounds. Ancient Hawaiians coated their spear tips with these toxins. ◆

Resources

Books

Davies, P. S. "Anthozoan Endosymbiosis." In *Proceedings of the Sixth International Conference on Coelenterate Biology*, edited by J. C. den Hartog. Leiden: National Natuurhistorisch Museum, 1997.

Dayton, P. K., K. W. England, and E. A. Robson. "An Unusual Sea Anemone, *Dactylanthus antarcticus* (Clubb, 1908) (Order Ptychodactiaria), on Gorgonians in Chilean Fjords." In *Proceedings of the Sixth International Conference on Coelenterate Biology*, edited by J. C. den Hartog. Leiden: National Natuurhistorisch Museum, 1997.

Fabricius, Katharina, and Philip Alderslade. *Soft Corals and Sea Fans*. Townsville: Australian Institute of Marine Science, 2001.

Friese, U. Erich. *Sea Anemones...As a Hobby*. Neptune City, NJ: T. F. H. Publications, Inc., 1993.

Martin, Vicki J. "Cnidarians, the Jellyfish and Hydras." In *Embryology: Constructing the Organism*, edited by Scott F.

Gilbert and Anne M. Raunio. Sunderland, MA: Sinauer Associates, Inc., 1997.

Shick, J. Malcolm. *A Functional Biology of Sea Anemones*. New York: Chapman & Hall, 1991.

Veron, John E. N. *Corals of the World*. Townsville: Australian Institute of Marine Science, 2000.

Wallace, Carden C. *Staghorn Corals of the World*. Collingwood: CSIRO, 1999.

Watling, Les, and Michael Risk, eds. *Biology of Cold Water Corals*. Vol. 471, *Hydrobiologia*. Dordrecht: Kluwer Academic Publishers, 2002.

Periodicals

Allemand, Denis. "The Biology and Skeletogenesis of the Mediterranean Red Coral." *Precious Corals and Octocoral Research* 2 (1993): 19–39.

Berntson, Ewann A., Scott C. France, and Lauren S. Mullineaux. "Phylogenetic Relationships within the Class

Anthozoa (Phylum Cnidaria) Based on Nuclear 18s RDNA Sequences." *Molecular Phylogenetics and Evolution* 13, no. 2 (1999): 417–433.

Bruckner, Andrew W. "Tracking the Trade in Ornamental Coral Reef Organisms: The Importance of CITES and Its Limitations." *Aquarium Sciences and Conservation* 3 (2001): 79–94.

Cappola, Valerie A., and Daphne G. Fautin. "All Three Species of Ptychodactiaria Belong to Order Actiniaria (Cnidaria: Anthozoa)." *Journal of the Marine Biological Association of the United Kingdom* 80 (2000): 995–1003.

Druffel, Ellen R. M., Sheila Griffin, Amy Witter, Erle Nelson, John Southon, Michaele Kashgarian, and John Vogel. "*Gerardia*: Bristlecone Pine of the Deep-Sea?" *Geochimica et Cosmochimicha Acta* 59, no. 23 (1995): 5031–5036.

Fabricius, Katharina E., Yehuda Benayahu, and Amatzia Genin. "Herbivory in Asymbiotic Soft Corals." *Science* 268 (1995): 90–92.

Fautin, Daphne G. "Reproduction of Cnidaria." *Canadian Journal of Zoology* 80 (2002): 1735–1754.

Frank, Uri, and Ofer Mokady. "Coral Biodiversity and Evolution: Recent Molecular Contributions." *Canadian Journal of Zoology* 80 (2002): 1723–1734.

Garrabou, J., T. Perez, S. Sartoretto, and J. G. Harmelin. "Mass Mortality Event in Red Coral *Corallium rubrum* Populations in the Provence Region (France, NW Mediterranean)." *Marine Ecology Progress Series* 217 (2001): 263–272.

Grigg, Richard W. "Precious Coral Fisheries of Hawaii and the U.S. Pacific Islands." *Marine Fisheries Reviews* 55, no. 2 (1993): 50–60.

Hall-Spencer, Jason, Valerie Allain, and Jan H. Fosså. "Trawling Damage to Northeast Atlantic Ancient Coral Reefs." *Proceedings of the Royal Society of London, Series B: Biological Sciences* 269, no. 1490 (2002): 507–511.

Hand, Cadet, and Kevin R. Uhlinger. "The Unique, Widely Distributed, Estuarine Sea Anemone, *Nematostella vectensis* Stephenson: A Review, New Facts, and Questions." *Estuaries* 17, no. 2 (1994): 501–508.

Hatta, Masayuki, Hironobu Fukami, Wenqia Wang, Makoto Omori, Kazuyuki Shimoike, Takeshi Hayashibara, Yasuo Ina, and Tsutomu Sugiyama. "Reproductive and Genetic Evidence for a Reticulate Evolutionary History of Mass-Spawning Corals." *Molecular Biology and Evolution* 16, no. 11 (1999): 1607–1613.

Hodgson, G. "A Global Assessment of Human Effects on Coral Reefs." *Marine Pollution Bulletin* 38, no. 5 (1999): 345–355.

Parker, N.R., P.V. Mladenov, and K.R. Grange. "Reproductive Biology of the Antipatharian Black Coral Antipathes Fiordensis in Doubtful Sound, Fiordland, New Zealand." *Marine Biology* 130 (1997): 11–22.

Pearson, C.V.M., A. D. Rogers, and M. Sheader. "The Genetic Structure of the Rare Lagoonal Sea Anemone, *Nematostella vectensis* Stephenson (Cnidaria; Anthozoa) in the United Kingdom Based on RAPD Analysis." *Molecular Ecology* 11 (2002): 2285–2293.

Rogers, Alex D. "The Biology of *Lophelia pertusa* (Linnaeus 1758) and Other Deep-Water Reef-Forming Corals and Impacts from Human Activities." *International Review of Hydrobiology* 84, no. 4 (1999): 315–406.

Ryland, John S. "Reproduction in Zoanthidea (Anthozoa: Hexacorallia)." *Invertebrate Reproduction and Development* 31, no. 1–3 (1997): 177–188.

Santangelo, Giovanni, and Marco Abbiati. "Red Coral: Conservation and Management of an Over-Exploited Mediterranean Species." *Aquatic Conservation: Marine and Freshwater Ecosystems* 11 (2001): 253–259.

Slattery, Marc, James B. McClintock, and Sam S. Bowser. "Deposit Feeding: A Novel Mode of Nutrition in the Antarctic Colonial Soft Coral *Gersemia antarctica*." *Marine Ecology Progress Series* 149 (1997): 299–304.

Other

Biogeoinformatics of Hexacorallia. 19 June 2003 [11 July 2003]. <http://www.kgs.ku.edu/Hexacoral/>.

NOAA's Coral Reef Information System. National Oceanic and Atmospheric Administration. 15 June 2003 [11 July 2003]. <www.coris.noaa.gov>.

Octocoral Research Center. 2001 [11 July 2003]. <http://www.calacademy.org/research/izg/orc_home.html>.

Tree of Life Web Project: Anthozoa. 4 Oct. 2002 [11 July 2003]. <http://tolweb.org/tree?group=Anthozoa&contgroup;=Cnidaria>.

Scott C. France, PhD

Hydrozoa
(Hydroids)

Phylum Cnidaria

Class Hydrozoa

Number of families 114

Thumbnail description

Invertebrates with a body plan that is comprised of a medusa with velum, a muscular projection from the subumbrellar margin that partially closes the subumbrellar cavity, and polyps; life cycles always involve the presence of a planula larva

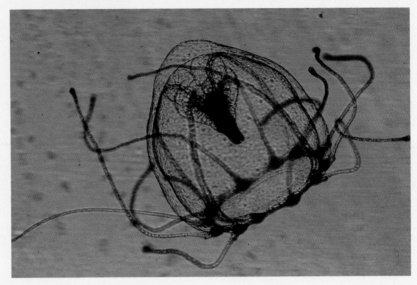

Photo: *A* newly released medusa of *Turritopsis nutricula.* (Photo by S. Piraino [University of Lecce]. Reproduced by permission.)

Evolution and systematics

Considered a superclass by some experts, this group has representatives since the Precambrian (velellids). Because of the rarity of skeletal structures, however, the fossil record is fragmentary. Recent interpretations of the medusary nodule (the structure that, from a polyp, gives rise to a medusa by a special type of budding) suggest that the Hydrozoa are triploblastic. The subumbrellar cavity of the hydromedusae, and the layer of striated muscle that lines it, are originated by a morphogenetic pattern very similar to schizocoely, with a third layer of tissue that is formed between ecto- and endoderm and that becomes hollow. The subumbrellar cavity, therefore, is interpreted as a coelom that later becomes open, with the origin of the velar opening. Molecular evidence is now available, showing, for instance, that the genes coding for mesodermic structures in the Bilateria is also present in the Hydrozoa. The polyp stage, thus, is diploblastic, whereas the medusae are triploblastic. If this view is confirmed by further evidence, the transition from diploblasts to triploblasts occurs every time a medusa is budded from a polyp, and one of the great mysteries of metazoan evolution is solved.

Hydrozoa (characterized by medusae with direct development or produced by lateral budding from polyps) are comprised in the subphylum Medusozoa, including also the Scyphozoa (characterized by medusa production by strobilation from polyps), and the Cubozoa (characterized by medusa production by complete metamorphosis of the polyp).

The classes of the superclass Hydrozoa include Automedusa, with the subclasses Actinulidae (one order, two families), Narcomedusae (one order, three families), and Trachymedusae (one order, five families); and Hydroidomedusa, with the subclasses Anthomedusae (two orders, 49 fam-

ilies), Laingiomedusae (one order, one family), Leptomedusae (two orders, 34 families), Limnomedusae (one order, three families), Siphonophorae (three orders, 16 families), and Polypodiozoa (one order, one family).

Physical characteristics

The Automedusa class is represented only by medusae; there are no polyp stages. Development is usually direct, sexes are separated; each fertilized egg leads to a planula that develops into a single medusa, except in some Narcomedusae, in which parasitic stages issued from the egg may give rise to several medusae by asexual budding. Medusae are not formed through a medusary nodule: the subumbrellar cavity and velum are formed by folding and deepening of the oral embryonic ectoderm, being only analogous to the subumbrellar cavity and velum of the Hydroidomedusa. The primary marginal tentacles are always formed before the subumbrellar cavity and the gastro vascular system. The marginal tentacles are deprived of tentacular bulbs. Sensory organs are in the form of ecto-endodermal statocysts, with an endodermal axis, growing out from the circular canal, with sensory cells characterized by numerous kinocilium-lacking rootlets, surrounded by stereocilia, innervated by the upper nerve ring; statoliths are of endodermal origin. Asexual reproduction is present only in "actinula"-like larvae and adults of Narcomedusae. The Actinulidae are all members of the interstitial fauna; they look like actinula-larvae, and the statocysts are the most distinctive medusan feature of these highly specialized medusae. The Narcomedusae have a flattened exumbrella, with lobed margin, incised by deep grooves. Usually with no radial canals, gametes are carried on the wide manubrium. Their intermediate tentaculated post-embryonic stages are juvenile medusae

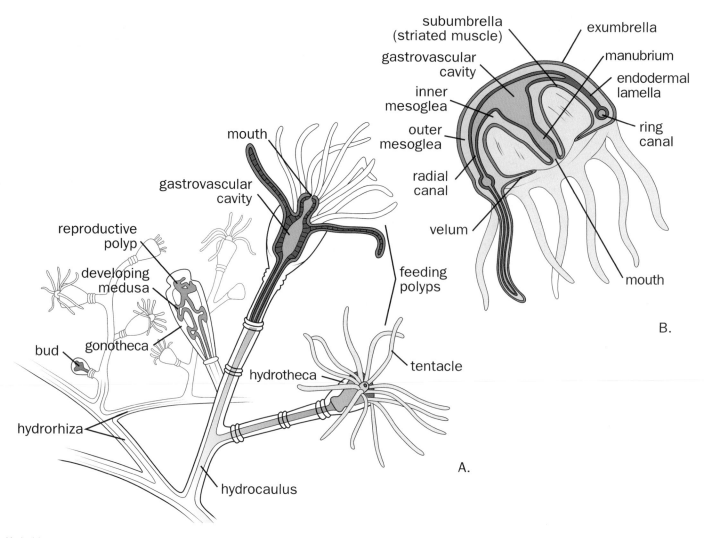

Hydroid anatomy. A. Colony; B. Medusa. (Illustration by Patricia Ferrer)

and are inappropriately called "actinulae," like the juvenile polyps of some Anthomedusae. The Trachymedusae have a bell-shaped umbrella, with circular and radial canals. Gametes ripen on the radial canals. The manubrium is often on a peduncle.

The Hydroidomedusa class is represented by a succession of three stages during indirect development. The planula is a ciliated motile gastrula; it typically develops into a benthic, modular, larval stage, polyp or hydroid (except in the Porpitidae, *Margelopsis* and *Pelagohydra*, where the hydroid is floating). Hydroids can be solitary, but generally form modular colonies by simple budding. Polyps can be specialized for different functions (defensive dactylozooids, reproductive gonozooids, nutritive gastrozooids, etc.). Polyps give rise, by asexual budding, to planktonic, free-swimming, and solitary hydromedusae, representing the sexual adult. The sense organs of pelagic hydroidomedusae, when present, are ocelli (Anthomedusae, some Leptomedusae), or statocysts (Leptomedusae, Limnomedusae); sometimes cordyli of unknown function are also present (Lep-

tomedusae). The Siphonophores have no visible sense organs. Medusae are often reduced to sporosacs (fixed gonophores), so that hydroids, by paedomorphosis, secondarily become the sexual stages. The Hydroidomedusa may also form pelagic, highly polymorphic colonies (Siphonophores). Medusa budding occurs via a medusary nodule or entocodon, forming a coelom-like cavity, the subumbrellar cavity, lined by striated muscle cells; primary marginal tentacles always develop after the subumbrellar cavity and the gastro vascular system. Both embryonic and larval stages, the planula and the polyp, are typically diploblastic; the adult sexual stages, the hydromedusae, acquire a triploblastic kind of organization during embryonic development (medusary nodule formation). Hydroidomedusae are frequently seasonal; the hydroid stage may develop several types of resting stages (frustules, propagules, cysts, stolon system) to overcome unfavorable ecological conditions.

The hydroids of the subclass Anthomedusae do not have a protective perisarc sheath around the polyps and are said athecates; they are usually colonial (but the most famous hy-

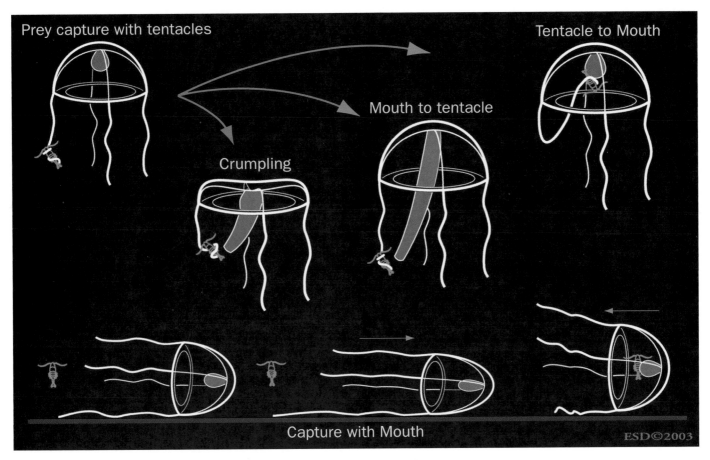

Hydroids capture prey with their tentacles and mouth. (Illustration by Emily Damstra)

droid, *Hydra*, is a solitary paedomorphic anthomedusa). The colonies can be either monomorphic or polymorphic, while the structure of the tentacles is in two forms: filiform tentacles do not present particular aggregations of cnidocysts, whereas capitate tentacles have cnidocyst knobs. The medusae are typically bell shaped; their gonads (aggregates of gametes commonly referred to as gonads) are confined on manubrium, sometimes extending on the most proximal parts of the radial canals. Their marginal sense organs, if present, are ocelli; the marginal tentacles are peripheral, hollow, or solid, usually with tentacular bulbs; sexual reproduction occurs through a complex planula. The hydroids of the subclass Laingiomedusae are unknown. The medusae have an almost hemispherical umbrella, with lobed margin, divided by peronial grooves or similar structures. The radial canals are four; there is no typical circular canal but a solid core of endodermal cells around the umbrella margin. The tentacles are solid, inserted above the exumbrellar margin, on the exumbrella. The manubrium is simple, quadrangular, tubular, or conical; the mouth opening is simple, quadrangular to circular; gametes are in four masses on the manubrium or as epidermal lining of interradial pockets of the manubrium. The hydroids of the subclass Leptomedusae are thecate: all parts of the colonies are typically protected by a chitinous perisarcal structure. The hydranth is protected by a hydrotheca, the nematophore by a nematotheca, and the gonophore by a gonotheca. Rarely, hydranths are naked. The medusae are typically with hemispherical or flattened umbrella; the masses of gametes are confined to radial canals, exceptionally extending onto the proximal part of manubrium; when present, the marginal sense organs are ectodermal velar statocysts, rarely cordyli, occasionally adaxial ocelli. The marginal tentacles are peripheral and hollow (except in *Obelia*), with tentacular bulbs. Sexual reproduction occurs through a complex planula. The hydroids of the subclass Limnomedusae are very simple; solitary or colonial; small, sessile; with or without tentacles; often close to planula structure and budding planula-like structures or frustules; there are no perisarcal thecae, but cysts and stolons are covered by chitin. The medusae usually have gamete masses along the radial canals or, exceptionally, on the manubrium. The marginal tentacles are peripheral, hollow, without a true basal bulb; their base is usually with a parenchymatic endodermal core embedded in umbrellar mesoglea. The marginal sense organs are internal, enclosed ecto-endodermal statocysts that are embedded in the mesoglea near the ring canal or in the velum. Exceptionally, medusae can be reduced medusoids. Sexual reproduction leads to simple planulae, without embryonic glandular cells.

The subclass Siphonophorae comprises generally pelagic, free-swimming, or floating species, forming highly polymor-

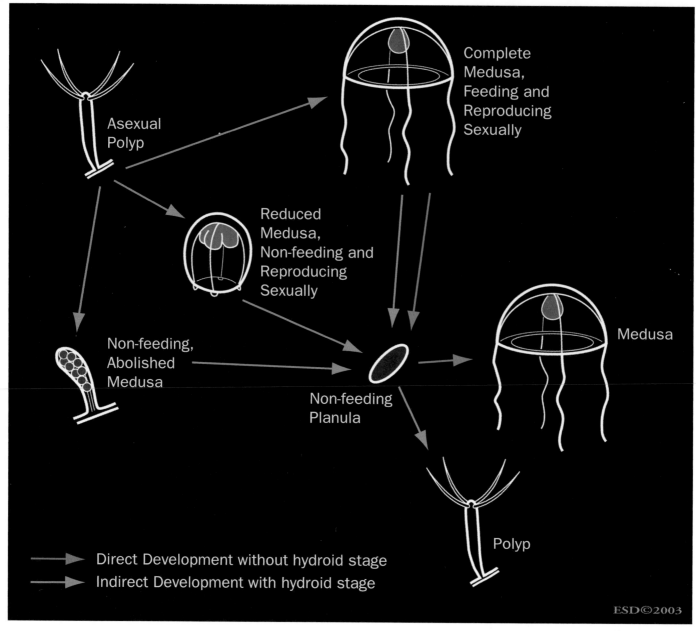

Hydroid life cycles vary by species. Here is shown an example of a "typical" life cycle. (Illustration by Emily Damstra)

phic modular colonies of polypoid and medusoid zooids attached to a stem or stolon supported by a floating and swimming system or nectosome (pneumatophores and nectophores).

The Polypodiozoa class is represented by a single species, *Polypodium hydriforme* Ussow, 1885, the only known metazoan adapted to intracellular parasitism. *Polypodium* has a unique lifecycle, having a succession of a free-living stage and of an intracellular parasitic stage of some Acipenseridae and Polyodontidae eggs. The earliest parasitic stage known is a binucleate cell observed in previtellogenetic fish oocytes. Further parasitic development leads to a didermic stolonal structure, with inverted germ layers, growing at the expense of the egg's yolk and forming numerous inverted buds. Before becoming free at fish spawning, eversion takes place and the germ layers take their normal position (ectoderm out, endoderm inside). Once liberated, the stolon becomes fragmented into individual buds, each originating a free-creeping globular stage that can multiply by longitudinal fission. These stages can move and feed, having an oral mouth-cone and tentacles. Germ cells are endodermal; the females have two kinds of gonads, each with a gonoduct opening in the gastral cavity; the gonads of the males form gametophores carrying cnidocysts. The free-living stage presumably represents the sexual medusae, the parasitic stages being considered as polypoid.

Seen here growing under a ledge near Lizard Island, North Great Barrier Reef, the hydrocoral *Distichopora violacea* is vivid purple in color. (Photo by A. Flowers & L. Newman. Reproduced by permission.)

They differ, however, from all other Hydrozoa by unique features: bilateral symmetry; presence of gonoducts; aberrant gametogenesis; unique structure of cnidocil apparatus; inversion of germ layers during parasitic life; and the complete separation of epidermal and muscle cells.

The umbrella of Hydroido- and Automedusae generally measures between 0.02 in (0.5 mm) and 2 in (50 mm), but in numerous species the size may be greater, reaching 3.9–7.8 in (100–200 mm) (*Aequorea*) and even, exceptionally, 15.7 in (400 mm) of diameter (*Rhacostoma atlanticum*). The colonies of the Hydroidomedusae usually have a reduced size. Most of them do not exceed a few inches (centimeters to a few decimeters) (i.e., *Cladocarpus lignosus* is 27.5 in [70 cm]); the hydranths are usually very tenuous, not exceeding a few inches (millimeters), but there are exceptions (i.e., *Hydrocoryne miurensis*: 2.3 in [6 cm]; *Corymorpha nutans*: 4.7 in [12 cm]; *Monocoryne gigantea*: 15.7 in [40 cm]; *Candelabrum penola*: 33.4 in [85 cm]; and *Branchiocerianthus imperator*: more than 6.5 ft [2 m]). The smallest polyps are those of the Microhydrulidae (Limnomedusae): they are reduced to a spherical or irregular body, ranging from 20 to 480 μm. The longest siphonophore is *Apolemia uvaria*, with colonies reaching 98.4 ft (30 m). The medusae of the hydrozoa are usually diaphanous, as are their polyps. Colored species, however, are frequent. Pigments derive from the diet or are produced directly. The most common color is reddish, deriving from crustaceans; other colors for medusae can be green, white, or orange; whereas polyps can be, according to the species, reddish, pink, white, or blue.

The medusae have typical and easily recognizable body architecture; the main feature that distinguishes them from other medusae is the velum. For this reason, they have also been called Craspedotae (with velum), as opposed to the medusae of the Scypozoa and Cubozoa, which were called Acraspedae (without velum). The polyps are quite varied in architecture, ranging from coral-like colonies (*Millepora* and the stylasterids) to gigantic polyps resembling those of the Anthozoa (*Branchiocerianthus*) to microscopic polyps reduced to a simple ball of tissue (*Microhydrula*).

Distribution

The Hydrozoa are cosmopolitan, and can be found in all water masses of the world, both in marine and in fresh waters. The Hydrozoa are known since the beginning of the modern study of animals; many species are Linnean and were described in the eighteenth century.

Habitat

The medusae of the Hydrozoa and the siphonophores are mostly planktonic; they are seasonal in occurrence and can be present in swarms, transported by the currents. Some medusae and some siphonophores, however, can be benthic. The polyp stages are usually benthic and live attached to the bottom, even though some species can be planktonic, such as the well-known *Velella velella*.

The Hydrozoa occur in all aquatic habitats, from anchialine caves to deep-sea trenches, from lakes and ponds to rocky coasts and the interstices among sand grains. The polyp stages of many species live exclusively on certain types of substrate, usually other organisms such as fishes, tunicates, polychaetes, bryozoans, mollusks, crustaceans, sponges, cnidarians, algae, sea-grasses, etc., with which they have symbiotic relationships ranging from simple epibiosis to commensalisms, mutualism, and parasitism.

Hydrozoa are mostly carnivores, using their habitat to acquire favorable positions to catch their prey. Planktonic stages are transported by currents, but can also move within water masses, searching for food. The position of benthic forms is decided by the planula, at the moment of settlement. Colonies are positioned at locations that will ensure a supply of new, fresh water around the hydranths, enhancing the transport of potential prey.

Behavior

The medusae are sharply individual; they can be gathered by winds and currents to form extensive swarms, but it is not known if they have any kind of social interaction while being

Stinging hydroids (*Lytocarpus philippinus*). (Photo by A. Flowers & L. Newman. Reproduced by permission.)

The species *Turritopsis nutricula* in its hydroid stage. (Photo by S. Piraino [University of Lecce]. Reproduced by permission.)

in close contact. The colonies of Hydrozoa, especially polymorphic ones, have been compared to superorganisms because of the complexity of the functions that they perform through the different types of zooids. The zooids of a colony usually derive from a single planula, thus being interconnected members of a single clone. It is possible, however, that different colonies merge their tissues or that different planulae aggregate to form coalescent colonies. In these cases, different individuals can be in such close connection that they become a single individual, possibly one of the extreme forms of social organization.

Most Hydrozoa have separate sexes. Fertilization is usually internal, with no copulation. Males spawn in the water and the sperm actively swim toward the eggs that are still on the maternal body (either a medusa or a polyp colony). The Hydrozoa are the first animals in which sperm attractants have been demonstrated, with species-specific attraction of sperms by the eggs. For many medusan species, both males and females spawn in the water, where fertilization occurs. Also in this case, however, sperm attractants facilitate gamete encounter.

The members of the same colony perform coordinated behaviors that surely involve communication. In *Thecocodium brieni*, for instance, the dactylozooids catch the prey with their tentacles, while the gastrozooids perceive that a prey is available and stretch towards the dactylozooids, detaching the prey from their tentacles and ingesting it. This division of labor, involving great coordination, is frequent in polymorphic colonies.

Planktonic animals do not show particular territoriality, being, by definition, transported by currents. It is probable, however, that medusae actively prevent being in too close contact to avoid competitive interactions while foraging. Territoriality is very strong in sessile organisms, where competition for the occupation of space is very evident. The arrangement of dactylozooids on the edge of many colonies is related to the defense of the territory from overgrowth by nearby animals. The feeding polyps can eat the settling larvae of potentially competing species, thus preventing competition for space.

When unfed, both polyps and medusae are always in search for food, with species-specific activity patterns. When the coelenteron is full of food, tentacles are usually contracted and do not catch prey, showing some control of cnidocyst discharge. Many medusae perform daily migrations through the water column.

Feeding ecology and diet

Hydrozoa are usually thought to feed on planktonic crustaceans such as copepods. Under laboratory conditions, they mostly survive with a diet based on *Artemia nauplii*. The medusae are mostly voracious carnivores and, when feeding on fish eggs and larvae, can be considered as being at the apex of trophic chains. The polyps are more varied in their food preferences. Recent investigations showed that the few species studied thoroughly feed on a great variety of prey, ranging from gelatinous plankton for the medusae to phytoplankton for the polyps. Some species have symbiotic zooxanthellae and are functionally photosynthetic.

Some medusae can remain immobile in the water, with their tentacles outstretched across the water column, performing ambush predation, whereas others can move across the water to contact prey, performing cruising predation. Polyps can simply extend their tentacles to catch passing prey, but they can also use special sense organs to perceive approaching prey and grab it actively, or they can form currents by moving their tentacles to direct food particles toward the mouth. Symbiotic species are very specialized in their feeding behavior; the most extreme cases are *Halocoryne epizoica* that feed on bryozoan tentacles, and *Polypodium hydriforme* that feed on sturgeon eggs from the inside, being the only intracellular parasitic metazoan.

Hydrozoa use cnidocysts as the main organelles to catch their food. The superclass has the richest variety of cnidocyst types of the whole phylum, with a fine range of adaptations to catch from tiny prey like the single cells of phytoplankton to the crustaceans and larvae of the zooplankton to the animals that live in the sand and mud, such as nematodes. Many species have very restricted diets, being specialized for just one type of food.

Both polyps and medusae are mostly carnivorous, feeding on almost all animals of proper size. In no other metazoan group (with the possible exception of parasitic trematodes) is the lifecycle of such paramount importance in defining the properties of a given species, and this is valid also for the type of prey. Tiny medusae that feed on fish eggs and larvae, sometimes impairing the success of recruitment, can be the most voracious predators of fish. The propensity of feeding on almost all types of larvae (both for polyps and for medusae) includes the Hydrozoa within the predators of almost all Metazoa with an indirect lifecycle.

Reproductive biology

Nothing is known regarding the courtship and mating. The eggs are brought in masses on female medusae or in female gonophores. According to the species, the eggs can be small and in great numbers, or they can be large and few, even a single one per gonophore.

In the Automedusae, with no polyp stages, there is little or no asexual reproduction in larval medusae (known as actinulae in the Narcomedusae) so that each fertilization event leads to one or a few adult medusae. In the Hydroidomedusae, the widespread asexual reproduction of the polyp stage can be considered as a polyembryony or as a larval amplification, as it happens in some parasites such as trematodes. Each fertilization event, therefore, leads to a single planula that produces a polyp colony that, in its turn, will produce many adult medusae. Since the lifespan of polyp colonies can last many years, each planula can lead to the production of hundreds or thousands medusae.

The so-called planula larva is nothing more than a gastrula, thus being more an embryo than a larva. Since gametes are shed before or soon after fertilization, embryonic development takes place outside the maternal body. Planulae can be solid (stereogastrula) or hollow (coeloblastula); usually the species with medusae in the lifecycle have hollow planulae that live part of their life in the water column, swimming with cilia or flagella to reach the settling sites. The species with suppressed medusae usually produce solid planulae that fall to the bottom and settle near the parent colony. Adults, by definition, are the sexually reproductive morphs in a lifecycle; if a medusa is present, it is the adult in the lifecycle. The polyp stage, in this framework, is a specialized (and perennial) larva that produces a great number of adults throughout its long life. In many species, however, the medusa stage can be reduced or even suppressed, so that the larvae, by paedomorphosis, become the sexually mature morphs of the lifecycle. Almost half of the species of Hydroidomedusae have suppressed or reduced medusae; the group is, therefore, the most paedomorphic of the whole animal kingdom.

Some medusae (e.g., Eleutheria) have special brood pouches where they safeguard developing medusae. Some hydroids have gonothecae with apical brood chambers that retain the planulae for a certain period.

Many species are sharply seasonal, being active only in narrow periods of time. The medusae can be present for a few weeks or months, completely disappearing from the water column and being represented by the correspondent polyps in the benthos. Polyp colonies can regress to resting hydrorhizae for long periods, reactivating at the onset of favorable conditions. Planulae can become encysted and remain dormant just as the resting hydrorhizae, being covered by a chitinous sheath.

Conservation status

No species are listed by the IUCN Red List. For most Hydrozoa, the distribution and the abundance are not known and only the few remaining specialists know about their presence. Many species are endemic simply because they are not searched for and only the areas of activity of specialists are covered. The only groups mentioned in regional or national red lists are the calcified, coral-like Milleporidae and Stylasteridae, which are also listed by CITES. Inclusion of these taxa on conservation lists is linked to the assessment of species value according to appearance.

Milleporidae and Stylasteridae have been subjected to trade, as are some hydroids (known as "white weeds" in the North Sea). Their decline is linked to habitat degradation.

Significance to humans

The Hydrozoa are mostly inconspicuous, both in the polyp and in the medusa stage, and are generally overlooked. The famous treatise by Tremblay, describing transplants in Hydra, inspired the fantastic novel Frankenstein by Mary Shelley. The illustrations of some monographs, especially those on medusae by Haeckel, are renowned for their beauty. The modern music composer, Frank Zappa, wrote a song on a hydromedusa that has been named after him: Phialella zappai.

"White weeds" (the colonies of hydroids of the genera Hydrallmania and Sertularia) had been used as decoration before the sharp reduction of their populations. Some Hydrozoa are used as laboratory animals for experimental biology; Hydra is the most popular one, but others include Aequorea victoria (for the production of the labeling enzyme aequorein), Hydractinia spp., Laomedea spp., and Tubularia spp.

Hydroids are important members of fouling communities, inhibiting the functioning of power plants by clogging their pipes and reducing the velocity of ships by settling on their hulls. Some species have been reported as pests in aquaculture, feeding on the larvae of the reared species or on their food. Polypodium hydriforme is a threat to the production of caviar, being a parasite of sturgeon eggs.

Some species of medusae (e.g., Gonionemus) can inflict severe stings on humans, as do some hydroid colonies such as the species of Millepora (fire corals) and some aglaopheniids. When present in swarms, even small medusae like those of Clytia can inflict slight stings on swimmers. The most important threat to human activities is the predation of some medusae (e.g., Aequorea victoria) and floating hydroids (e.g., Clytia gracilis) on the eggs and larvae of commercially exploited fish. This kind of predation can reduce the success of fish recruitment, reducing the yield of fisheries.

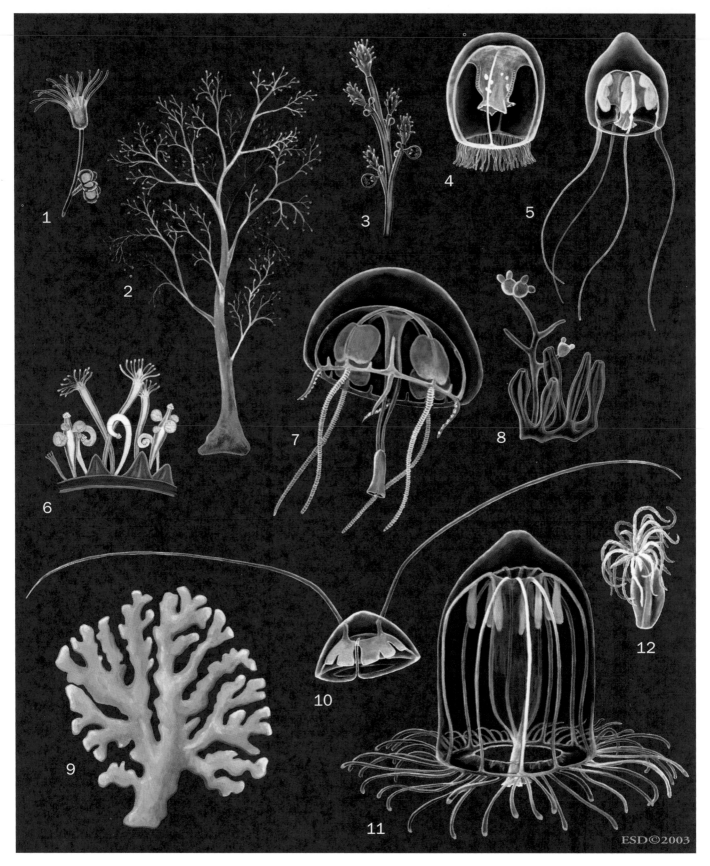

1. *Eudendrium glomeratum* polyp; 2. *E. glomeratum* medusa; 3. Immortal jellyfish (*Turritopsis nutricula*) polyp; 4. *T. nutricula* medusa; 5. *Hydrichthys mirus* medusa; 6. *Hydractinia echinata* polyp; 7. *Liriope tetraphylla* medusa; 8. *H. mirus* polyp; 9. *Distichopora violacea* polyp; 10. *Solmundella bitentaculata* medusa; 11. *Aglantha digitale*; 12. *Halammohydra schulzei* polyp. (Illustration by Emily Damstra)

1. *Aequorea victoria* medusa; 2. *Obelia dichotoma* medusa; 3. *A. victoria* polyp; 4. *O. dichotoma* polyp; 5. Zappa's jellyfish (*Phialella zappai*) polyp; 6. *P. zappai* medusa; 7. *Laingia jaumotti* medusa; 8. *Aglaophenia pluma* polyp. (Illustration by Emily Damstra)

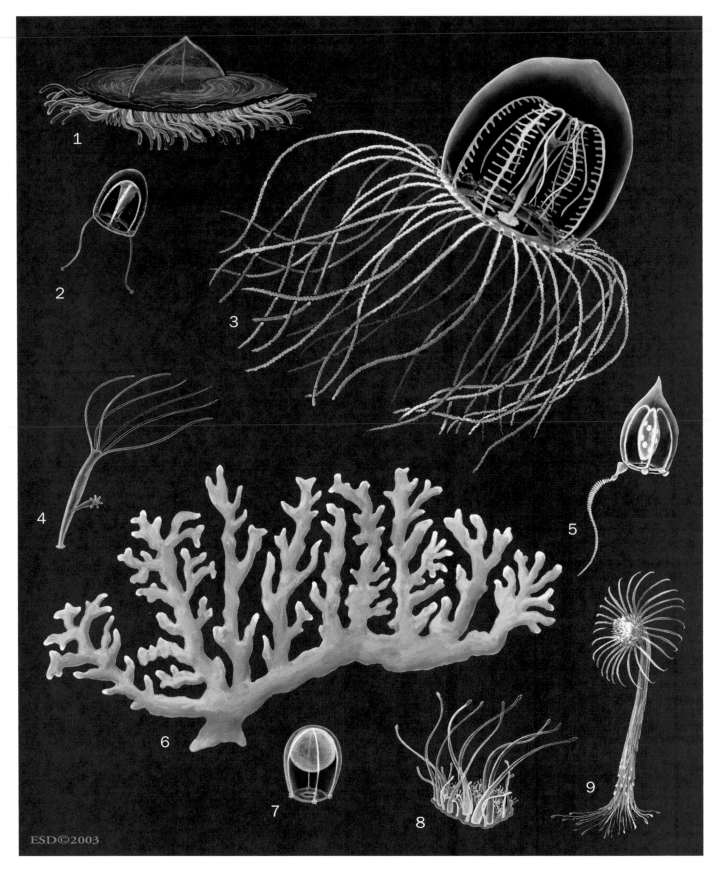

1. By the wind sailor (*Velella velella*) polyp; 2. *V. velella* medusa; 3. *Polyorchis penicillatus* medusa; 4. *Hydra vulgaris* polyp; 5. *Corymorpha nutans* medusa; 6. Fire coral (*Millepora alcicornis*) polyp; 7. *M. alcicornis* medusa; 8. *Paracoryne huvei* polyp; 9. *C. nutans* polyp. (Illustration by Emily Damstra)

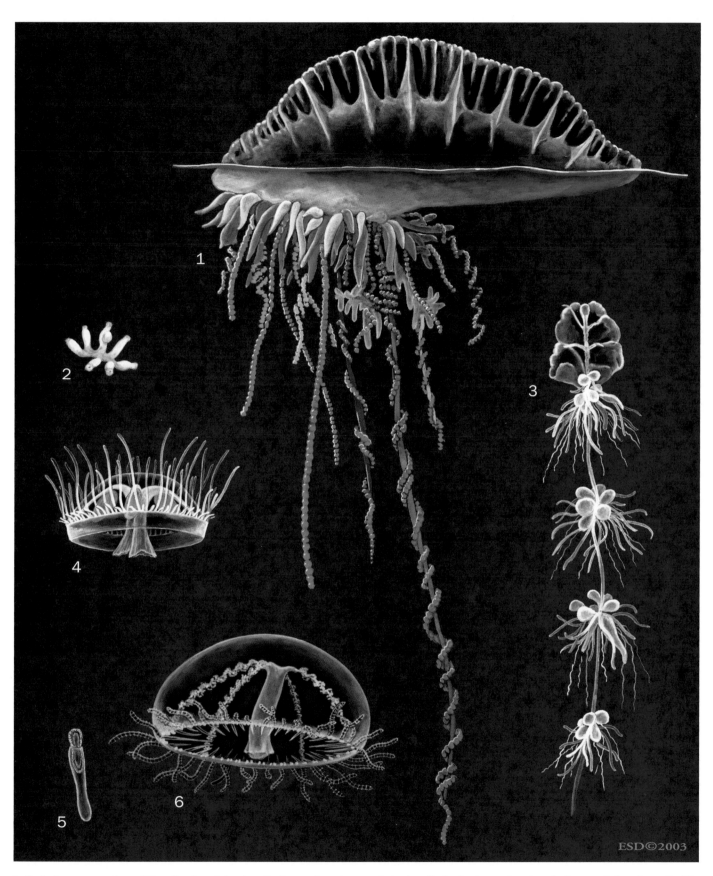

1. Portuguese man of war (*Physalia physalis*) polyp; 2. *Craspedacusta sowerbyi* polyp; 3. *Apolemia uvaria* polyp; 4. *C. sowerbyi* medusa; 5. *Olindias phosphorica* polyp; 6. *O. phosphorica* medusa. (Illustration by Emily Damstra)

Species accounts

No common name
Halammohydra schulzei

ORDER
Actinulida

FAMILY
Halammohydridae

TAXONOMY
Halammohydra schulzei Remane, 1927, North Sea.

OTHER COMMON NAMES
None known.

PHYSICAL CHARACTERISTICS
Body as a long gastric tube (manubrium) with a terminal mouth, with a small aboral cone, separated from manubrium by a neck, bearing an adhesive organ; aboral nerve ring; one aboral whorl of 28 (sometimes up to 32) amphicoronate solid tentacles with a large basal thickening, alternating with 12 ecto-endodermic statocysts; gonochoric; generally two opposite gonads; without brood pouch. Body 0.027–0.031 in (0.7–0.8 mm) high.

DISTRIBUTION
Cosmopolitan. (Specific distribution map not available.)

HABITAT
Interstices of marine sand.

BEHAVIOR
Eretant in the marine sand.

FEEDING ECOLOGY AND DIET
Nothing is known.

REPRODUCTIVE BIOLOGY
Dioecious, sex cells released in the sea, reproduction through a special actinulid larvae or halhydrula.

CONSERVATION STATUS
Not listed by the IUCN.

SIGNIFICANCE TO HUMANS
None known. ◆

No common name
Corymorpha nutans

ORDER
Capitata

FAMILY
Corymorphidae

TAXONOMY
Corymorpha nutans M. Sars, 1835, Norway.

OTHER COMMON NAMES
None known.

PHYSICAL CHARACTERISTICS
Hydroid: large, solitary, hydrocaulus subcylindrical, with short sensory papillae at base and numerous long, anchoring filaments; hydrocaulus with parenchymatic endoderm with numerous longitudinal anastomizing peripheral canals; hydrocaulus protected by a transparent membranous tube; hydranth flask-shaped to vasiform with about 20–80 oral filiform tentacles in several irregular whorls, and 20–32 aboral considerably longer filiform tentacles; hydranth and distal part of hydrocaulus bend downward; medusa buds in dense clusters on about 15–20 branching peduncles just above aboral tentacles.

Medusa: umbrella up to 0.23 in (6 mm) high (including apical process), cylindrical, with a high, pointed apical process and a long, narrow umbilical canal; mesoglea thick; velum wide; manubrium large, cylindrical, on short gastric peduncle, about two-thirds of the length of subumbrella, in full extension reaching slightly beyond exumbrellar margin; mouth simple, tube like, armed with cnidocysts; four radial canals and circular canal fairly broad; gonads completely surrounding manubrium, but living peduncle and mouth free; one single, long, marginal tentacle, moniliform; three perradial non-tentacular bulbs smaller than the tentacular one.

DISTRIBUTION
Atlantic, the Mediterranean, and Black Sea. (Specific distribution map not available.)

HABITAT
The hydroid lives at moderate depth with its basal part embedded in soft bottoms. The medusae are planktonic in coastal waters.

BEHAVIOR
Nothing is known.

FEEDING ECOLOGY AND DIET
The hydroid, with its typical posture, is oriented with the mouth near to the bottom and it probably captures epibenthic vagile animals. The medusae feed on planktonic animals.

REPRODUCTIVE BIOLOGY
Both hydroids and medusae are sharply seasonal, occurring in correspondence with spring plankton blooms. Hydroids are not produced immediately, fertilization and the sexual reproduction originate an encysted embryo that spends the unfavorable season in the sediments, to become a polyp at the onset of the following favorable season.

CONSERVATION STATUS
Not listed by the IUCN.

SIGNIFICANCE TO HUMANS
None known. ◆

Fire coral
Millepora alcicornis

ORDER
Capitata

FAMILY
Milleporidae

TAXONOMY
Millepora alcicornis Linnaeus, 1758, West Indies.

OTHER COMMON NAMES
English: Sea ginger; French: Corail de feu; German: Feuerko-rallen.

PHYSICAL CHARACTERISTICS
Hydroid: colony forms massive, calcareous exoskeletons (coenosteum), forming horn-shaped, upright branches or plates; coenosteum with an internal complex network of coenosarcal tubes, covered externally by a thin ectodermal layer, coenosteal surface perforated by pores; margins of pores not protruding from surface of coenosteum; large gastropores surrounded by smaller dactylopores, forming indistinct cyclosystems; no gastrostyles or dactylostyles; polyps polymorphic; gastrozooids relatively short and stout, with an oral whorl of 4–6 short capitate tentacles, arising from gastropores; dactylozooids long, slender, mouth less, with scattered capitate tentacles, arising from dactylopores; cnidome with macrobasic mastigophores; gonophores arising from coenosarc within ampullae embedded in the coenosteum. Color variable, usually yellow-brown because of algal symbionts.

Medusa: short-lived, free-swimming eumedusoids with exumbrellar cnidocyst patches, narrow velum, four radial and circular canals, gonads occupying the place of an indistinct manubrium; without tentacles, without sense organs (this description applies to the eumedusoid of *M. complanata*, the only *Millepora* eumedusoid known in detail).

DISTRIBUTION
Tropical Pacific, Indian Ocean, Red Sea, and Caribbean. (Specific distribution map not available.)

HABITAT
Common to abundant in shallow waters, or in inner ledges and outer reefs, usually not exceeding 98.4 ft (30 m), correlated with dependence of the colonies on their symbiotic unicellular algae or zooxanthellae that need light for their processes of assimilation.

BEHAVIOR
Nothing is known.

FEEDING ECOLOGY AND DIET
Feed upon planktonic animals, mostly crustaceans.

REPRODUCTIVE BIOLOGY
Sexual reproduction by free-swimming eumedusoids, usually ripe from April–July.

CONSERVATION STATUS
Not listed by the IUCN.

SIGNIFICANCE TO HUMANS
Millepora are severe stingers and can cause violent, uncomfortable burning, even in dried state. ◆

No common name
Paracoryne huvei

ORDER
Capitata

FAMILY
Paracorynidae

TAXONOMY
Paracoryne huvei Picard, 1957, Mediterranean coast of France.

OTHER COMMON NAMES
None known.

PHYSICAL CHARACTERISTICS
Stolonal colonies with encrusting hydrorhizae composed of naked coenosarc, with gastrozooids, dactylozooids, and gonozooids; gastrozooids with 12–26 tentacles irregularly distributed in four verticils at distal half; dactylozooids large, without tentacles or mouth; gonozooids without tentacles or mouth; gonophores as fixed sporosacs (cryptomedusoids), male and female in different colonies; male gonophores in number of 2–10 per gonozooid, ovoids, with a characteristic apical prolongation; female gonophores 2–4 per gonozooid, bigger and more spherical than males; young gonophores with an apical prolongation (as in the males), which is absent and replaced by an orifice at maturity.

DISTRIBUTION
Western Mediterranean. (Specific distribution map not available.)

HABITAT
The colonies live on rocks and mussels, from the intertidal to the very shallow subtidal. They form very evident pink patches on bare rocks.

BEHAVIOR
Nothing is known.

FEEDING ECOLOGY AND DIET
Nothing is known.

REPRODUCTIVE BIOLOGY
The dioecious colonies are present in the winter and disappear in the summer. After sexual reproduction, the planulae encyst and remain dormant until the onset of the following favorable season.

CONSERVATION STATUS
In spite of the great tradition of biodiversity exploration in the Mediterranean, this very distinctive species has been discovered in relatively recent times. It became very frequent and abundant in the 1970s and 1980s, but is now extremely rare again. The resting stages surely play an important part in these patterns of abundance.

SIGNIFICANCE TO HUMANS
None known. ◆

No common name
Polyorchis penicillatus

ORDER
Capitata

FAMILY
Polyorchidae

TAXONOMY
Melicertum penicillatum Eschscholtz, 1829, California, United States.

OTHER COMMON NAMES
None known.

PHYSICAL CHARACTERISTICS
Hydroid: unknown.

Medusa: bell up to 2.3 in (60 mm) high, with a gastric peduncle, up to 160 marginal hollow tentacles in a simple row along exumbrellar margin; tentacular bulbs tubular, adnate, with bright red ocellus on short spur; four radial canals with numerous, short, blind lateral diverticula; ring canal with centripetal diverticula; manubrium prismatic, pendulous, sausage shaped; four crenulated oral lips with distinct cnidocyst row; gonads pendulous and long, hanging from peduncular manubrium pouches.

DISTRIBUTION
Pacific Ocean, from the coasts of the North American continent to Hawaii. (Specific distribution map not available.)

HABITAT
Bays and gulfs.

BEHAVIOR
It is a benthic medusa, but it can periodically come to the surface and sink back to the bottom.

FEEDING ECOLOGY AND DIET
Foraging takes place mostly on the bottom of bays, where the medusae feed on crustaceans, mainly copepods.

REPRODUCTIVE BIOLOGY
The medusae are present all year round and are mostly ripe. In spite of this widespread occurrence of the medusae, the polyp stage remains unknown.

CONSERVATION STATUS
Concern has been expressed about the impact of coastal development on this animal. Because of its preference for bays, it is very sensitive to coastal pollution.

SIGNIFICANCE TO HUMANS
It is used in aquarium displays for educational purposes. ◆

By the wind sailor
Velella velella

ORDER
Capitata

FAMILY
Porpitidae

TAXONOMY
Medusa velella Linnaeus, 1758, Mediterranean Sea.

OTHER COMMON NAMES
None known.

PHYSICAL CHARACTERISTICS
Hydroid: the colony is a flattened oval to slightly S-shaped float, with a triangular sail and concentric air chambers; up to 1.5 in (40 mm) long and 0.78 in (20 mm) wide, higher in the center than at the edges. Float and sail are supported by chitin covered by mantle tissue; margin of float is soft and flexible. The center of colony underside is a single large gastrozooid encircled by a ring of medusa-producing gastro-gonozooids and a peripheral band of dactylozooids. Central-feeding zooid oval, with an elongated hypostome, without tentacles or medusa buds. Gastro-gonozooids are spindle shaped, with a swollen mouth region, lacking tentacles, but with warts of cnidocyst clusters concentrated in distal half; on proximal half of hydranth, numerous medusa buds growing in groups from short blastostyles. Dactylozooids are mouth less, long, and tapering, oval in cross section, with cnidocysts concentrated in two lateral bands on the narrow sides. The float is deeply blue when alive, medusa buds yellow-olive from symbiotic algae.

Medusa: with four exumbrellar cnidocyst rows, four radial canals; two pairs of opposite, perradial tentacles, a short adaxial one and a long abaxial one, each with a large terminal cnidocyst cluster; two perradial marginal bulbs without tentacles; manubrium conical with quadrate base; mouth tubular; gonads irregularly arranged perradially and interradially.

DISTRIBUTION
Atlantic and Indo-Pacific. (Specific distribution map not available.)

HABITAT
The hydroid floats, with the sail out of the water and the zooids hanging down in the hyponeuston. The medusae are planktonic, in shallow water and possibly in deep water as well, where the young colonies have been recorded, before climbing to the surface.

BEHAVIOR
Nothing is known.

FEEDING ECOLOGY AND DIET
The hydroid colony feeds on surface plankton, from crustaceans to appendicularians, and especially on fish eggs.

REPRODUCTIVE BIOLOGY
Floating hydroids can occur in immense offshore swarms, sometimes stranding along shorelines. The colonies liberate rarely observed young medusae that sink toward the bottom. Development of new colonies occurs in deep waters.

CONSERVATION STATUS
Not listed by the IUCN.

SIGNIFICANCE TO HUMANS
The massive strandings can entrap sand and protect the shore from erosion. Feeding on fish eggs makes *Velella* a potential competitor for man in the use of fish resources. ◆

No common name
Aequorea victoria

ORDER
Conica

FAMILY
Aequoreidae

TAXONOMY
Mesonema victoria Murbach and Shearer, 1902, British Columbia, Canada.

OTHER COMMON NAMES
None known.

PHYSICAL CHARACTERISTICS
Hydroid: small colonies, unbranched, with creeping hydrorhiza; pedicels twice as long as hydrotheca. Hydrotheca thin, with an operculum made by many flaps not articulated to hydrothecal rim that converge and close the theca when the hydranth is contracted. Hydranth has 20 tentacles, with an intertentacular membrane at base; gonotheca of thin perisarc, originating right below the theca; a single medusa bud.

Medusae are small at liberation, with two tentacles and four radial canals. They can grow up to 4.7 in (120 mm) wide and 1.5 in (40 mm) high. Umbrella saucer shaped to hemispherical, a thick hemispherical projection of mesoglea protrudes into the manubrium; lips of manubrium are fringed and can close the manubrium completely on the sides of the mesogleal projection. Tentacles can be up to 150, lip lobes are usually half the number of tentacles. Each tentacular bulb has an abaxial pore connecting the circular canal to the outside.

DISTRIBUTION
Northeastern Pacific. (Specific distribution map not available.)

HABITAT
The hydroid grows on mussel shells. The medusae are planktonic, in coastal and open waters.

BEHAVIOR
The function of the hemispherical mesogleal projection into the manubrium is to allow the closed mouth to rotate around it, thus pushing the contents of the stomach into the radial canals. The pores in the circular canal are used to eject undigested materials and are more anal pores than excretory pores (as they are commonly called).

FEEDING ECOLOGY AND DIET
Unknown for the hydroid, whereas the medusae have been reported to feed on fish larvae and on gelatinous plankton.

REPRODUCTIVE BIOLOGY
The hydroid is tiny and rarely observed, as are the newly released medusae. Medusa production, however, is very intense, since the species can be present in swarms. The swarms produce great quantities of planulae to support future blooms through the hydroid generation.

CONSERVATION STATUS
The massive use of this species to extract the bioluminescent enzyme aequorein might have some influence on population viability.

SIGNIFICANCE TO HUMANS
By feeding on larvae, it might have a negative impact on fish recruitment; but it has a great role in experimental biology because of its bioluminescent enzyme. ◆

No common name
Aglaophenia pluma

ORDER
Conica

FAMILY
Aglaopheniidae

TAXONOMY
Sertularia pluma Linnaeus, 1758, North Sea, United Kingdom.

OTHER COMMON NAMES
None known.

PHYSICAL CHARACTERISTICS
The colonies resemble a feather, with a straight stem (hydrocaulus) and two series of alternate branches (hydrocladia). Hydrocaulus monosiphonic up to 5.9 in (150 mm), brown, unbranched or dichotomously branched; basal part athecate, followed by one or two prosegments; remainder internodes, each with three nematothecae and a pseudonematotheca; nodes oblique; hydrocladia alternate, with whitish cormidia separated by transverse nodes. Hydrotheca deep, rim with nine cusps of varied length; intrathecal adcauline septum usually well developed, median nematothecae two-thirds adnate, not reaching the margin of the hydrothecae; lateral nematothecae reaching the rim of the hydrothecae; aperture of nematothecae gutter shaped. Male and female corbulae white, with free costa; male close, with slit-like openings between the costae, female with fused costae and smaller slits. Hydranth small, transparent, cylindrical; hypostome held at level of hydrothecal rim, rounded and low; 10 tentacles emerging at hydrothecal rim.

DISTRIBUTION
Usually considered cosmopolitan, but many other species of the same genus have probably been identified as this nominal species. (Specific distribution map not available.)

HABITAT
It grows on algae, in shallow rocky bottoms.

BEHAVIOR
Nothing is known.

FEEDING ECOLOGY AND DIET
The shape of the colonies is suited to intercept food particles suspended in the water column. The feeding hydranths are small and in great number; they have been observed to beat their tentacles to create a current towards the mouth. This might be an adaptation for active filter feeding of small food items such as phytoplankton.

REPRODUCTIVE BIOLOGY
Colonies dioecious, almost continuously fertile.

CONSERVATION STATUS
Not listed by the IUCN.

SIGNIFICANCE TO HUMANS
None known. ◆

Zappa's jellyfish
Phialella zappai

ORDER
Conica

FAMILY
Phialellidae

TAXONOMY
Phialella zappai Boero, 1987, Bodega Bay, California, United States.

OTHER COMMON NAMES
None known.

PHYSICAL CHARACTERISTICS
Hydroid: colony simple, unbranched; hydranth very extensile, with about 14 tentacles, alternately held upward and downward. Oral part of the hydranth is globular, separated from the rest of the body. Hydrotheca cylindrical, elongated, with an operculum of about seven cusps separated from the hydrothecal wall by a thin line, not always evident. Diaphragm present. Pedicel is as long as the hydrotheca or a little shorter, annulated throughout. Gonothecae on the stolon, arising from short, annulated pedicels, with wavy or smooth walls, tapering below and truncate above, with one or two medusa buds.

Medusa: newly released medusae sub-spherical, about 0.002 in (0.6 mm) in diameter, with four tentacles, four interradial tentacular bulbs deprived of tentacles, and four radial canals with medial darker areas. Eight statocysts, with 1–3 statoliths, are on the inner edge of the ring canal, supported by a cushion of cells. Manubrium is short (one third of the bell cavity), with four short lips. Tentacular nematocysts are in clusters, giving a moniliform appearance to the tentacles. Perradial tentacular bulbs are almost round or triangular, interradial tentacular bulbs much smaller, but evident. The medusae grow rapidly, reaching 0.11 in (3 mm) in diameter in 10 days, dome shaped with four well-developed inter-radial tentacles, and eight developing adradial tentacles. At this age, the eggs are clearly visible. Manubrium cruciform, lips more evident and starting to bend upward. Tentacular bulbs still round, tending to elongate. Tentacles moniliform. Development continues with an increase in size and number of tentacles (36 the highest number observed). Adult specimens dome shaped, with gonads almost in the middle of the radial canals. Manubrium cruciform, with folded lips bending upward, with four gastric pouches; four black spots may be present at its base. Tentacular bulbs triangular, but still rounded. Tentacles evidently moniliform.

DISTRIBUTION
Endemic to the west coast of the United States (California). (Specific distribution map not available.)

HABITAT
The hydroid grows on mussel shells. The medusa is in coastal plankton.

BEHAVIOR
Nothing is known.

FEEDING ECOLOGY AND DIET
Under laboratory conditions, both polyps and medusae feed on *Artemia nauplii*; in the wild they probably feed on small crustaceans.

REPRODUCTIVE BIOLOGY
Each gonotheca produces one or two medusae that, under laboratory conditions, continue to develop also after reaching sexual maturity. Some specimens become mature 10 days after liberation. They can develop a new gonad after the first spawning.

CONSERVATION STATUS
Not listed by the IUCN.

SIGNIFICANCE TO HUMANS
This species was named after the modern music composer Frank Zappa (1940–1993), who, in exchange, wrote a song about it ("Lonesome Cowboy Nando"). ◆

Portuguese man of war
Physalia physalis

ORDER
Cystonectae

FAMILY
Physaliidae

TAXONOMY
Holothuria physalis Linné, 1758, Atlantic Ocean.

OTHER COMMON NAMES
English: Blue bottle; French: Galère, frégate, vaisseau de guerre hollandais, vaisseau de guerre portugais.

PHYSICAL CHARACTERISTICS
The only pleustonic siphonophore. Colonies consist of a large, purplish blue horizontal pneumatophore that floats on the sea surface, reaching 11.8 in (30 cm) in length in the largest specimens; pneumatophore carries the polyps, which form cormidia at the oral end; at its top, an erectile, longitudinal crest or sail that may be left- or right-handed, all drifting at the mercy of the winds. Pneumatophore asymmetrical; two forms, each the mirror image of the other. Each cormidium consisting of a gastrozooid associated with a tentacle and a gonodendron; however, unlike other siphonophores, tentacle separating from the basigaster during the later stages of development. The tentacles may attain several meters in length; a continuous formation of new cormidia. As the cormidia mature, new gastrozooids gradually lose their tentacles and becoming palpons. Small medusoid gonophores develop at the bases of the terminal palpons. Functional gonophores of a given colony are of a single sex only.

DISTRIBUTION
Widely distributed in tropical and subtropical regions in the three great oceans and in the Mediterranean. (Specific distribution map not available.)

HABITAT
Only pleustonic siphonophore can occur in great numbers (navies); frequently blown ashore by strong winds.

BEHAVIOR
The effect of the wind on the sail is to move the left-handed specimens to the right of the wind direction and vice versa for the right-handed ones.
feeding ecology and diet
Feed on small planktonic organisms or small fishes.

REPRODUCTIVE BIOLOGY

Dioecious, release their sexual cells in the sea, development through siphonula larvae.

CONSERVATION STATUS

Not listed by the IUCN.

SIGNIFICANCE TO HUMANS

The tentacles can stretch out to many feet (meters) below the float and can inflict powerful and very painful stings on swimmers who become entangled in their tentacles. ◆

Immortal jellyfish

Turritopsis nutricula

ORDER
Filifera

FAMILY
Clavidae

TAXONOMY
Turritopsis nutricula McCrady, 1859, South Carolina, United States.

OTHER COMMON NAMES
None known.

PHYSICAL CHARACTERISTICS
Hydroid: small colonies stolonal, monosiphonic (simple stem), larger ones erect, irregularly branched and increasing in diameter from base to apex, polysiphonic (compound stem); branches basally adnate to hydrocaulus or other branches, then curving away at an acute angle and becoming free; hydrocaulus and hydrocladia covered by a firm bi-layered perisarc, mostly encrusted with detritus and algae, terminating below hydranth base; hydranths terminal, naked, elongated, fusiform, with 12–38 filiform tentacles scattered over distal three quarters of column, proximal ones shorter than distal; hypostome elongated conical; medusa buds arising mostly one by one from short pedicels below hydranths, pear shaped, enclosed in perisarc.

Medusa: umbrella 0.15–0.43 in (4–11 mm) high, bell shaped to piriform, higher than wide, mesoglea thicker at apex; manubrium large, cross-shaped in transverse section, red; four radial canals passing through the four compact vacuolated endodermal masses situated above digestive part of manubrium; four-lipped mouth with a continuous row of sessile cnidocyst clusters along margin; 80–120 closely spaced marginal tentacles; gonads interradial, mature females often with developing embryos and planulae; with adaxial ocelli.

DISTRIBUTION
Atlantic, Indo-Pacific, and the Mediterranean. (Specific distribution map not available.)

HABITAT
The hydroids live in shallow, highly oxygenated water, often under overhangs. The medusae are members of the coastal plankton.

BEHAVIOR
Nothing is known.

FEEDING ECOLOGY AND DIET

In the laboratory, both hydroid and medusae can survive with a diet of *Artemia nauplii*; the diet in the wild is unknown.

REPRODUCTIVE BIOLOGY

Medusae are produced by the hydroid in summer months and can survive in the laboratory for one month. A peculiarity of this species is the possibility of ontogeny reversal under laboratory conditions: the medusae, in fact, can rearrange their tissues and go back to the polyp stage if subjected to sub-lethal stress and also at the end of their lifespan, after spawning. For this reason, this medusa has been called "the immortal jellyfish" by the media.

CONSERVATION STATUS

Not listed by the IUCN.

SIGNIFICANCE TO HUMANS

The possibility for this species to perform ontogeny reversal in the laboratory offers a unique opportunity to study the genetic control of aging and the mechanisms of rejuvenation. ◆

No common name

Eudendrium glomeratum

ORDER
Filifera

FAMILY
Eudendriidae

TAXONOMY
Eudendrium glomeratum Picard, 1951, Banjuls-sur-Mer, Mediterranean Sea.

OTHER COMMON NAMES
None known.

PHYSICAL CHARACTERISTICS
Colonies up to 11.8 in (30 cm) high, composed of compound and branched hydrocauli; older perisarc brown and thick; thin and yellowish to transparent in younger regions. Polyp is urn shaped, hypostome peduncled, tentacles filiform, 24–28 in one whorl. Gonophores are fixed. Female mature gonophore with unbranched spadix; male gonophore provided with one or two chambers; mature balstostyle with either a normal number of tentacles or a reduced number of partly atrophied tentacles. Diagnostic feature is holotrichous macrobasic euryteles (24 × 10–28 × 11μm), concentrated in several groups characteristically conspicuous at the basal half of the polyp, with long butt (four times length of capsule), spirally coiled around the main axis of the cnidocyst.

DISTRIBUTION
Atlantic, Indo-Pacific; mostly in temperate waters. (Specific distribution map not available.)

HABITAT
Rocky cliffs, from the surface to 328 ft (100 m) and below. It can be very abundant under favorable conditions, forming a facies.

BEHAVIOR
The hydranths use the peduncled hypostome as a sphincter dividing the coelenteron from the outside. They stretch their tentacles in the water to catch their prey.

FEEDING ECOLOGY AND DIET

Some species of the genus *Eudendrium* have been reported to feed on the plankton surrounding their colonies. They can also ingest the mucus that they produce and possibly digest organic particles and organisms trapped into it, acting as mucous filter feeders.

REPRODUCTIVE BIOLOGY

Fertilization occurs while eggs are still on the gonophores; the planulae can remain attached to the mother colony by mucous threads to be able to settle in its vicinities. Planulae can aggregate and form chimerical hydroid colonies.

CONSERVATION STATUS

Not listed by the IUCN.

SIGNIFICANCE TO HUMANS

None known. ◆

No common name
Hydractinia echinata

ORDER
Filifera

FAMILY
Hydractiniidae

TAXONOMY
Alcyonium echinatum Fleming, 1828, (Scotland, United Kingdom).

OTHER COMMON NAMES
None known.

PHYSICAL CHARACTERISTICS

Hydrorhiza about 0.11 in (3 mm) thick, with numerous blunt conical chitinous spines with jagged edges, encrusting on gastropod shells but also other solid substrata; white to pale pink, giving rise to different kinds of polyps: gastrozooids slender, widening upwards, about eight tentacles in one whorl, hypostome conical; gonozooids shorter than gastrozooids, with few tentacles and a ring of gonophores; dactylozooids long, slender.

Medusae reduced to fixed gonophores, sexes generally on different gonozooids; male gonophore yellow to white, ovoid; female gonophore pink and spherical.

DISTRIBUTION

Atlantic, Arctic, and the Mediterranean. (Specific distribution map not available.)

HABITAT

Hydroids grow on gastropod shells inhabited by hermit crabs, on soft bottoms.

BEHAVIOR

Under laboratory conditions, the colonies can be aggressive with the formation of hyperplasic stolons that sting other organisms while competing for the substrate.

FEEDING ECOLOGY AND DIET

Gastrozooids feed on benthic organisms (nematodes, crustaceans, etc.) while being transported by the hermit crab. They also can feed on the larvae of the crab.

REPRODUCTIVE BIOLOGY

Eggs and sperms are released in the water, where fertilization occurs and planulae are formed. Gamete production is almost continuous, but at a slower pace in winter time.

CONSERVATION STATUS

Not listed by the IUCN.

SIGNIFICANCE TO HUMANS

This species, and in general many Hydractiniids, is easy to rear under laboratory conditions and is widely used for experimental biology as a laboratory animal. ◆

No common name
Hydrichthys mirus

ORDER
Filifera

FAMILY
Pandeidae

TAXONOMY
Hydrichthys mirus Fewkes, 1887, New England.

OTHER COMMON NAMES
None known.

PHYSICAL CHARACTERISTICS

Hydroid: hydrorhiza as a thin encrusting plate on the skin of fish host. Polyps of two types: gastrozooids are elongate, tubular, with distal mouth, and no tentacles; nearly all with a small bud at base; mouth with an armature of some microbasic euryteles and desmonemes; no nematocysts in other regions of colony, apart from medusa buds; gonozooids very contractile, with up to three lateral branches bearing clusters of medusa buds at different stages of development; no sign of perisarc. Medusa buds on gonozooids, hydrorhiza, and possibly also on hydranths; exumbrellar nematocysts present; well-developed buds stiff, with two big tentacular bulbs and no apparent sign of tentacles.

Medusa: umbrella height 0.15 in (4 mm), dome shaped, with apical projection; jelly soft. Manubrium with well-developed folds in stomach wall, bearing the gametes; manubrial folds nearest to mouth much developed, almost covering lips. Above these main folds, two or three additional folds sometimes developed. Gametes almost completely covering interradial portions of stomach.

DISTRIBUTION

Indo-Pacific; very rare. (Specific distribution map not available.)

HABITAT

The hydroid lives symbiotically with fish; the medusa is planktonic, possibly with a tendency to forage near the bottom.

BEHAVIOR

The gastrozooids can bend to touch the surface of the fish. They have a distal annular contractile ring resembling a sucker, but feeding on fish was not observed. Under laboratory conditions, the hydroid do not take *Artemia nauplii*. After the death of the fish, colonies can detach and become free living. It is therefore possible that they can survive in the plankton if the host dies. In the laboratory, the medusae tend to stay on the bottom of rearing jars, with their long tentacles completely extended, attached to the bottom of the finger bowl, stretching them for a long distance while swimming parallel with the bottom. Tentacles are then contracted and the bell is drawn quickly backwards. Prey is captured in lips during this fast backward movement. Medusae possibly have a benthic feeding habit. After feeding, they swim vigorously for some hours.

FEEDING ECOLOGY AND DIET
The hydroid most probably feeds on the supporting fish, using the mouth as a sucker. The medusa feeds on crustaceans.

REPRODUCTIVE BIOLOGY
Nothing is known.

CONSERVATION STATUS
Not listed by the IUCN.

SIGNIFICANCE TO HUMANS
None known. ◆

No common name
Distichopora violacea

ORDER
Filifera

FAMILY
Stylasteridae

TAXONOMY
Millepora violacea Pallas, 1766, Pacific Ocean.

OTHER COMMON NAMES
None known.

PHYSICAL CHARACTERISTICS
Hydroid: colony violet-blue, small, and flabellate; branches compressed and granular, with low longitudinal ridges, ending with two flattened lobes, usually whitish, growing in opposite directions; gastro- and dactylopores very long, extending for a great distance down the center of the lobes; pore rows on lateral edges of lobes and main stem; dactylopores in equal number on both sides of pore rows; gastrostyles ridges bearing tall, slender, often fused, spines; no ring palisade; ampullae in groups, opening to surface by irregularly shaped pores.

DISTRIBUTION
Indo-Pacific. (Specific distribution map not available.)

HABITAT
Coral formations in shallow water.

BEHAVIOR
Nothing is known.

FEEDING ECOLOGY AND DIET
Nothing is known.

REPRODUCTIVE BIOLOGY
Nothing is known.

CONSERVATION STATUS
Like almost all stylasterids, it is protected by international treaties.

SIGNIFICANCE TO HUMANS
None known. ◆

No common name
Laingia jaumotti

ORDER
Laingiomedusae

FAMILY
Laingiidae

TAXONOMY
Laingia jaumotti Bouillon, 1978, Laing Island, Bismarck Sea.

OTHER COMMON NAMES
None known.

PHYSICAL CHARACTERISTICS
Hydroid: unknown.

Medusa: umbrella lobed, divided by peronial grooves or similar structures, with no exumbrellar cnidocyst tracks; four radial canals; no typical circular canal, but a solid core of endodermal cells around umbrellar margin; tentacles solid, bent shortly after their point of origin, inserted on the exumbrellar surface above bell margin; marginal bulbs largely displaced towards exumbrella, forming peronial-like structures; manubrium simple, quadrangular, tubular, or conical; mouth opening quadrangular to circular; gonads on manubrium in four interradial pouches; no sense organs.

DISTRIBUTION
Bismarck Sea, Papua New Guinea. (Specific distribution map not available.)

HABITAT
Superficial layers of the sea.

BEHAVIOR
Nothing is known.

FEEDING ECOLOGY AND DIET
Nothing is known.

REPRODUCTIVE BIOLOGY
Nothing is known.

CONSERVATION STATUS
Not listed by the IUCN.

SIGNIFICANCE TO HUMANS
None known. ◆

No common name
Craspedacusta sowerbyi

ORDER
Limnomedusa

FAMILY
Olindiidae

TAXONOMY
Craspedacusta sowerbyi Lankester, 1880, found on a water-lily in a tank at Regent's Park, London.

OTHER COMMON NAMES
None known.

PHYSICAL CHARACTERISTICS

Hydroid: freshwater, solitary or forming small colonies of 2–4, rarely 7 polyps; hydranths without tentacles, cylindrical, with apical mouth (hypostome) surrounded by cnidocysts forming a spherical capitulum under which the polyp is slightly tapering, forming a distinct neck; basal portion of hydranths with periderm covering, attaching colony to substrate; medusa buds lateral, on the middle or lower part of body column, often becoming terminal by hydranth reduction; asexual reproduction by frustules, transversal division and resting stages (cysts).

Medusa: umbrella 0.39–0.78 in (10–20 mm) wide, slightly flatter than a hemisphere; mesoglea fairly thick; with well-developed, marginal cnidocyst ring; velum broad and well developed; manubrium large, upper portion conical with broad square base, tapering downwards to cross shaped distal region; mouth with four simple or slightly folded lips, extending beyond umbrella margin; four straight radial canals and circular canal broad and massive; four large, smooth, triangular pouch-like gonads, with rounded corners, hanging down into subumbrellar cavity from points of junction of radial canals with manubrium; with 200–400 or more hollow marginal tentacles, in several series situated at different levels on umbrella margin; oldest four perradial marginal tentacles being largest and highest; bases of marginal tentacles adherent to exumbrella; surface of marginal tentacles covered with evenly distributed papillae, each with 3–10 cnidocysts; 100–200 or more statocysts, usually about half number of marginal tentacles; statocysts situated in velum, forming centripetal tubes with basal enlargements near umbrella margin.

Numerous species of *Craspedacusta* have been described, mainly from China; it is not excluded that they represent anything more than variations of a single species.

DISTRIBUTION

Cosmopolitan in freshwaters and sometimes in brackish waters of temperate and tropical areas. (Specific distribution map not available.)

HABITAT

Freshwater surfaces and calm rivers, often found in tanks and aquaria.

BEHAVIOR

Hydroids live on water plants; the medusae are active swimmers, living usually near surface.

FEEDING ECOLOGY AND DIET

Feed on small freshwater planktonic organisms, mainly protozoa, rotifers, crustaceans, worms, and fish larvae.

REPRODUCTIVE BIOLOGY

Dioecious, sex cells released in freshwater; the planula gives rise to polyp colonies.

CONSERVATION STATUS

Not listed by the IUCN.

SIGNIFICANCE TO HUMANS

The medusae may damage and injure fishes in fish farms. ◆

No common name
Olindias phosphorica

ORDER
Limnomedusa

FAMILY
Olindiidae

TAXONOMY
Oceania phosphorica Delle Chiaje, 1841, Mediterranean Sea.

OTHER COMMON NAMES
None known.

PHYSICAL CHARACTERISTICS

Hydroid: the polyps have not been yet found in field. Weill (1936) described from laboratory observations a small solitary hydranth without tentacles enclosed in a cylindrical or irregularly curved hydrothecae covering more than half its length, and much longer than the polyp itself; mouth distal surrounded by large cnidocysts.

Medusa: umbrella 0.78–1.5 in (40–60 mm) wide, almost hemispherical, mesoglea fairly thick; 11–19 centripetal canals per quadrant; 30–60 primary tentacles; usually two ecto-endodermal statocysts at base of each primary tentacle; 100–120 secondary tentacles; 100–170 marginal clubs.

DISTRIBUTION

Atlantic and the Mediterranean. (Specific distribution map not available.)

HABITAT

Lives near the bottom or at the surface near shores.

BEHAVIOR

Lives usually fixed to algae or *Posidonia*.

FEEDING ECOLOGY AND DIET

Feeds on small planktonic organisms.

REPRODUCTIVE BIOLOGY

Dioecious, the sex cells are released in the sea; the planula gives rise to a reduced polyp stage.

CONSERVATION STATUS

Not listed by the IUCN.

SIGNIFICANCE TO HUMANS

The sting of the medusae may cause skin irritations. ◆

No common name
Hydra vulgaris

ORDER
Moerisiida

FAMILY
Hydridae

TAXONOMY
Hydra vulgaris Pallas, 1766, European freshwaters.

OTHER COMMON NAMES
French: Hydre d'eau douce.

PHYSICAL CHARACTERISTICS

Solitary freshwater hydroids; 0.47 in (12 mm) in height; with 7–12 hollow filiform tentacles, but often moniliform distally, in one whorl under hypostome; hermaphrodic species, eggs and sperm developed directly in ectoderm of polyps in wart-like protuberances, "testis" developing on upper part of hydranth, "ovaries" on lower part, with up to eight eggs enveloped; in a chitinous embryotheca when fecundated, embryotheca with long, thin spines; asexual reproduction by lateral buds, leading only to temporary colonies; lower part of hydranth with simple pedal disc and with central pore, no perisarc except on encysted embryos.

DISTRIBUTION

Cosmopolitan. (Specific distribution map not available.)

HABITAT

Freshwater surfaces, rivers, and ponds.

BEHAVIOR

Fixed on freshwater plants, stones, empty freshwater shells, and worm larval tubes (*Trichoptera*); able to move with the help of the tentacles.

FEEDING ECOLOGY AND DIET

Feed on small freshwater planktonic organisms: primarily protozoa, rotifers, crustaceans, and worms.

REPRODUCTIVE BIOLOGY

Protanderic hermaphoditic; female sex cells fertilized when fixed on mother and enveloped in a chitinous spiny embryotheca.

CONSERVATION STATUS

Not listed by the IUCN.

SIGNIFICANCE TO HUMANS

Hydra is one of the most popular laboratory animals and has greatly contributed to the development of experimental biology. ◆

No common name
Solmundella bitentaculata

ORDER
Narcomedusa

FAMILY
Aeginidae

TAXONOMY
Charybdea bitentaculata Quoy and Gaimard, 1833.

OTHER COMMON NAMES
None known.

PHYSICAL CHARACTERISTICS
Umbrella up to 0.4 in (12 mm) wide, usually much smaller, rounded apex, keel-shaped along the axis leading to tentacles, apical mesoglea very thick, lateral walls thin; velum well developed; manubrium short, lenticular, with eight rectangular manubrial pouches with rounded edges; mouth circular, simple; two long, tapering, opposite tentacles issuing from umbrella above manubrium, near apex; gonads in subumbrellar wall, under manubrial pouches; four peronia in deep grooves; no peripheral system or otoporpae; 8–32 statocysts. Monotypic genus.

DISTRIBUTION

Widely distributed in all oceans and in the Mediterranean. (Specific distribution map not available.)

HABITAT

From surface to bathypelagic zone.

BEHAVIOR

Often parasitized by larvae of another Narcomedusae, *Cunina peregrine*.

FEEDING ECOLOGY AND DIET

Small planktonic animals.

REPRODUCTIVE BIOLOGY

Dioecious, sex cells released in sea water, direct development.

CONSERVATION STATUS

Not listed by the IUCN.

SIGNIFICANCE TO HUMANS

None known. ◆

No common name
Obelia dichotoma

ORDER
Proboscoida

FAMILY
Campanulariidae

TAXONOMY
Sertularia dichotoma Linnaeus, 1759, coast of southwest England.

OTHER COMMON NAMES
None known.

PHYSICAL CHARACTERISTICS
Hydroid: colonies extremely varied in size and shape. Stems erect, up to 13.7 in (35 cm) high, monosiphonic and branched, flexuous to straight, thickened in old colonies. Internodes have several annuli at their base. Hydrothecae alternate, lateral, borne on completely annulated pedicels at the upper part of the internodes. Hydrotheca bell shaped, usually not very deep, thin walled; rim to crenate, slightly flared; diaphragm transverse to oblique. Gonothecae usually inverted, conical on annulated pedicels, truncated at the distal end, with a short distal neck when mature, where the aperture is.

Medusa: umbrella 0.09–0.23 in (2.5–6 mm) wide, circular, flat; mesoglea very thin; without gastric peduncle; mouth with four simple lips; four radial canals; gonads sac like, hanging from middle to end of the radial canals; numerous short, stiff, solid, not extensible marginal tentacles; tentacles with short endodermal roots extending into bell mesoglea; eight statocysts situated on underside of basal bulbs of some marginal tentacles. The velum is absent.

DISTRIBUTION
Almost cosmopolitan, even though the identity of the many nominal species of *Obelia* is still matter of debate, as is their distribution. (Specific distribution map not available.)

HABITAT
The hydroid is very common on rocks and algae on shallow hard bottoms; the medusa is very common in coastal waters.

BEHAVIOR
Both polyps and medusae can move their tentacles in a rhythmic way to create currents that bring food particles toward the mouth.

FEEDING ECOLOGY AND DIET
Both hydroids and medusae feed on plankton, usually crustaceans, but also possibly phytoplankton.

REPRODUCTIVE BIOLOGY
Both medusae and polyps are quite constant in presence in their respective environments and reproduction occurs continuously.

CONSERVATION STATUS
Not listed by the IUCN.

SIGNIFICANCE TO HUMANS
Obelia is traditionally depicted in all zoology textbooks as a paradigmatic hydrozoan, so that many generations of students have learned its name. Unfortunately, the medusae of *Obelia*, being deprived of velum, are far from being the archetype of the medusae of the Hydrozoa. Furthermore, the figures of *Obelia* medusae present in many textbooks depict a campanulate medusa, and not a flat one, as is the medusa of this genus. ◆

No common name
Apolemia uvaria

ORDER
Physonectae

FAMILY
Apolemiidae

TAXONOMY
Stephanomia uviformis Lesueur, 1811, Nice (France).

OTHER COMMON NAMES
French: Stéphanomie pamproide, stèphanomie à, grains de raisins.

PHYSICAL CHARACTERISTICS
Pneumatophore bulb shaped, widening near the apex. Extended colonies are up to 32.8–98.4 ft (10–30 m). Nectosome with up to 12 nectophores in two parallel rows on the stem; largest is 0.14 in (3.7 mm) high by 0.13 in (3.4 mm) wide by 0.16 in (4.2 mm) deep. Nectophore consisting of two wings looking like those of a butterfly, with a deep ventral furrow. Nectosac large. Lateral radial canals form S-shaped bends with short branches on the upper bend. Groups of five or six nectosomal tentacles issuing from the base of the nectophores near the pedicular canal, at the base of the muscular lamellae. Siphosome measuring up to several feet (meters) in length, composed of several cormidia. Each cormidium consists of a gastrozooid and about 50 palpons, both with thin filiform tentacles of a single type issuing from their bases. Palpons are very long and delicate. Opaque spots bearing cnidocysts on the outer surface cover bracts, like the nectophores. Only physonect siphonophore whose nectophores are separated from each other by a cluster of 5–6 nectosomal tentacles.

DISTRIBUTION
The Mediterranean and Atlantic Ocean. (Specific distribution map not available.)

HABITAT
Occurs in the top 328 ft (100 m) of water.

BEHAVIOR
Epiplanktonic.

FEEDING ECOLOGY AND DIET
Feeds on other planktonic organisms such as crustaceans, polychaetes, mollusks, tunicates, and even small fishes.

REPRODUCTIVE BIOLOGY
Dioecious, release their sexual cells in the sea, development through siphonula larvae.

CONSERVATION STATUS
Not listed by the IUCN.

SIGNIFICANCE TO HUMANS
None known. ◆

No common name
Polypodium hydriforme

ORDER
Polypodiozoa

FAMILY
Polypodiidae

TAXONOMY
Polypodium hydriforme Ussov, 1885, Black Sea.

OTHER COMMON NAMES
None known.

PHYSICAL CHARACTERISTICS
Polypodium lifecycle as a succession of a free-living stage and of a stage parasitizing the eggs of some Acipenseridae and Polyodontidae fishes. The earliest known stage is a binucleate cell, parasitizing previtellogenetic fish oocytes. Further development may last several years, leading to a convoluted didermic stolonal structure, with inverted germ layers, forming numerous inverted buds. Before fish spawning, eversion takes place and the germ layers take their normal position (ectoderm outside, endoderm inside). The stolon exits the egg and becomes fragmented into individual buds, each giving rise to a free-creeping globular stage that multiplies by longitudinal fission. Globular stages can move and feed, having an oral mouth-cone and either 24, 12, or six tentacles, according to season. Germ cells are endodermal. So-called females have two kinds of gonads, each with a gonoduct opening in the gastral cavity; so-called males deprived of gonoducts, their gonads forming gametophores carrying cnidocysts. It is not known how the parasites get into young previtellogenic fish oocytes. The free-living stages are presumably homologous to sexual medusae, the parasitic stages being considered as polypoid. By their stolonal parasitic budding stage and their cnidome, the Polypodiozoa seem to present some affinities with the Narcomedusae, to which they were previously associated. *P. hydriforme* was, until recently, the only known metazoan adapted to an intracellular parasitic life.

DISTRIBUTION
Freshwater basins of Russia, Romania, Iran, and North America. (Specific distribution map not available.)

HABITAT
Fish gonads.

BEHAVIOR
Parasitic.

FEEDING ECOLOGY AND DIET
Parasitic.

REPRODUCTIVE BIOLOGY
Early developmental stages intracellular parasites of the eggs of Acipenseridae and Polyodontidae fishes; free-living stages a small medusae.

CONSERVATION STATUS
Not listed by the IUCN.

SIGNIFICANCE TO HUMANS
Polypodium parasitize and destroy the eggs of the fishes producing caviar and, consequently, has a great economical impact. ◆

No common name
Liriope tetraphylla

ORDER
Trachymedusa

FAMILY
Geryoniidae

TAXONOMY
Geryonia teraphylla Chamisso and Eysenhardt, 1821, Indian Ocean.

OTHER COMMON NAMES
None known.

PHYSICAL CHARACTERISTICS
Umbrella 0.39–1.1 in (10–30 mm) wide, hemispherical, apex somewhat flattened; mesoglea thick, rigid; velum broad; manubrium small, on long, cylindrical gastric peduncle, longer than umbrellar diameter; mouth with four simple lips lined with cnidocysts; with normally four radial canals (sometimes more); 1–7 centripetal canals in each quadrant; with marginal cnidocyst ring; typically four long hollow perradial tentacles with cnidocyst rings and four small solid interradial tentacles with adaxial cnidocyst clusters; with gonads variable in shape and size, generally heart shaped, on either side of the middle of radial canals; eight statocysts. Extremely variable species.

DISTRIBUTION
Worldwide distribution, common in temperate regions of all oceans and the Mediterranean. (Specific distribution map not available.)

HABITAT
Chiefly near surface, sometimes in great shoals.

BEHAVIOR
Bold and rapid medusa.

FEEDING ECOLOGY AND DIET
Very rapacious, feeds on other small planktonic animals and fishes sometimes three times its size.

REPRODUCTIVE BIOLOGY
Dioecious, sex cells released in sea water, direct development.

CONSERVATION STATUS
Not listed by the IUCN.

SIGNIFICANCE TO HUMANS
None known. ◆

No common name
Aglantha digitale

ORDER
Trachymedusa

FAMILY
Rhopalonematidae

TAXONOMY
Medusa digitale O. F. Müller, 1776, North Atlantic.

OTHER COMMON NAMES
French: Méduse doigtier.

PHYSICAL CHARACTERISTICS
Umbrella cylindrical, 0.39–1.5 in (10–40 mm) high, about twice high than wide, with a small conical projection; lateral mesogleal walls thin, subumbrellar muscles strong; peduncle slender, long, conical, almost as long as subumbrellar cavity; manubrium small; mouth with four simple lips; eight long, sausage-shaped gonads, arising from the radial canals near the apex of the subumbrella and hanging freely in subumbrellar cavity; 80 or more solid marginal tentacles with a core of single endodermal chordal cells; eight free statocysts.

DISTRIBUTION
Worldwide distribution in all oceans and entering the Mediterranean. (Specific distribution map not available.)

HABITAT
From surface down to considerable depths, 1,970 ft (600 m).

BEHAVIOR
Prolate medusae using the entire bell for jet propulsion; extremely active medusae.

FEEDING ECOLOGY AND DIET
Other small planktonic animals.

REPRODUCTIVE BIOLOGY
Dioecious, sex cells released in sea water, direct development.

CONSERVATION STATUS
Not listed by the IUCN.

SIGNIFICANCE TO HUMANS
None known. ◆

Resources

Books

Boero, F., J. Bouillon, S. Piraino, and V. Schmid. "Asexual Reproduction in the Hydrozoa (Cnidaria)." In *Reproductive Biology of Invertebrates*. Volume XI, *Progress in Asexual Reproduction*, edited by R. N. Hughes. New Delhi: Oxford and IBH Publishing Co., 2002.

Bouillon, J., F. Boero, F. Cicogna, and P. F. S. Cornelius, eds. *Modern Trends in the Systematics, Ecology and Evolution of Hydroids and Hydromedusae*. Oxford: Clarendon Press, 1987.

Bouillon, J., C. Gravili, F. Pages, J. M. Gili, and F. Boero. *An Introduction to Hydrozoa*. In press.

Cornelius, P. F. S. "North-west European Thecate Hydroids and Their Medusae. Part 1. Laodiceidae to Haleciidae." In *Synopses of the British Fauna (New Series)*, edited by R. S. K. Barnes and J. H. Crothers. The Linn. Soc. London and Est. Coas. Sci. Assoc., Field Studies Council. 50: i–vii, 1995.

———. "North-west European Thecate Hydroids and Their Medusae. Part 2. Sertulariidae to Campanulariidae." In *Synopses of the British Fauna (New Series)*, edited by R. S. K. Barnes and J. H. Crothers. The Linn. Soc. London and Est. Coas. Sci. Assoc., Field Studies Council. 50: i–vii, 1995.

Periodicals

Boero, F. "The Ecology of Marine Hydroids and Effects of Environmental Factors: A Review." *Marine Ecology—Pubblicazioni della Stazione Zoologica di Napoli I* 5 (1984): 93–118.

Boero, F., J. Bouillon, and S. Piraino. "Classification and Phylogeny in the Hydroidomedusae (Hydrozoa, Cnidaria)." *Scientia Marina* 60 (1996): 17–33.

Boero, F., C. Gravili, P. Pagliara, S. Piraino, J. Bouillon, and V. Schmid. "The Cnidarian Premises of Metazoan Evolution: From Triploblasty, to Coelom Formation, to Metamery." *Italian Journal of Zoology* 65 (1998): 5–9.

Bouillon, J. "Classe des Hydrozoaires." *Traité de Zoologie* 3, no. 2 (1995): 29–416.

Bouillon, J., and F. Boero. "Phylogeny and Classification of Hydroidomedusae. The Hydrozoa: A New Classification in the Light of Old Knowledge." *Thalassia Salentina* 24 (2000): 1–46.

———. "Phylogeny and Classification of Hydroidomedusae. Synopsis of the Families and Genera of the Hydromedusae of the World, with a List of the Worldwide Species." *Thalassia Salentina* 24 (2000): 47–296.

Bouillon, J., F. Boero, F. Cicogna, J. M. Gili, and R. G. Hughes, eds. "Aspects of Hydrozoan Biology." *Scientia Marina* 56, no. 2–3 (1992): 99–284.

Bouillon, J., D. Medel, F. Pages, J. M. Gili, F. Boero, and C. Gravili. "Fauna of the Mediterranean Hydrozoa." *Scientia Marina* In press.

Carré, C., and D. Carré. "Ordre des Siphonophores." In *Traité de Zoologie*, edited by P. P. Grassé, and D. Doumenc, vol. 3, no. 2 (1995): 523–596.

Gili, J. M., and R. G. Hughes. "The Ecology of Marine Benthic Hydroids." *Oceanography and Marine Biology: An Annual Review* 33 (1995): 351–426.

Kirkpatrick, P. A., and P. R. Pugh. "Siphonophores and Velellids." *Synopsis of the British Fauna (New Series)* no. 29 (1984): 154.

Miglietta, M. P., et al. "Approaches to the Ethology of Hydroids and Medusae (Cnidaria, Hydrozoa)." *Scientia Marina* 64, Suppl. 1 (2000): 63–71.

Mills, C. E., F. Boero, A. Migotto, and J. M. Gili. "Trends in Hydrozoan Biology—IV." *Scientia Marina* 64, Supl. 1 (2000): 1–284.

Piraino, S., F. Boero, B. Aeschbach, and V. Schmid. "Reversing the Life Cycle: Medusae Transforming into Polyps and Cell Transdifferentiation in *Turritopsis nutricula* (Cnidaria, Hydrozoa)." *Biological Bulletin* 190 (1996): 302–312.

Piraino, S., F. Boero, J. Bouillon, P. F. S. Cornelius, and J. M. Gili, eds. "Advances in Hydrozoan Biology." *Scientia Marina* 60, no. 1 (1996): 1–243.

Other

"Cheating Death: The Immortal Life Cycle of *Turritopsis*." [July 10, 2003]. <http://zygote.swarthmore.edu/intro6.html>.

"Hydrozoa." Cnidaria Home Page, University of California, Irvine. [July 10, 2003]. <http://www.ucihs.uci.edu/biochem/steele/hydrozoa.html>.

"Hydrozoa Nematocysts." September 15, 1995 [July 10, 2003]. <http://www.pitt.edu/AFShome/s/s/sshostak/public/html/cnidocyst_database/hydr.html>.

"Hydrozoan Society: Dedicated to the Study of Hydrozoan Biology." [July 10, 2003]. <http://www.ucmp.berkeley.edu/agc/HS/>.

"Hydrozoan Taxonomy." [July 10, 2003]. <http://www.biology.duke.edu/hydrodb/databases.html>.

Mills, C. E. "Hydromedusae." June 10, 2003 [July 10, 2003]. <http://faculty.washington.edu/cemills/Hydromedusae.html>.

Peard, Terry. "Freshwater Jellyfish." January 13, 2003 [July 10, 2003]. <http://nsm1.nsm.iup.edu/tpeard/jellyfish.html>.

Raskoff, Kevin A. "Midwater Medusae." January 2003 [July 10, 2003]. <http://www.mbari.org/kraskoff/medusae2.htm>.

Schuchert, Peter. "Hydrozoa." June 2003 [July 10, 2003]. <http://www.geocities.com/peterschuchert/Hydrozoa.htm>.

Ferdinando Boero
Jean Bouillon

Cubozoa
(Box jellies)

Phylum Cnidaria
Class Cubozoa
Number of families 2

Thumbnail description
Gelatinous animals that possess image-forming eyes and complex behavior, including internal fertilization in at least one species; often known as "killer box jellies," although only one of the seven genera (*Chironex*, in the family Chirodropidae) is potentially lethal to humans

Photo: The southern sea wasp (*Carybdea rastoni*) can give painful stings with its tentacles. (Photo by Animals Animals ©Gowlett-Holmes, OSF. Reproduced by permission.)

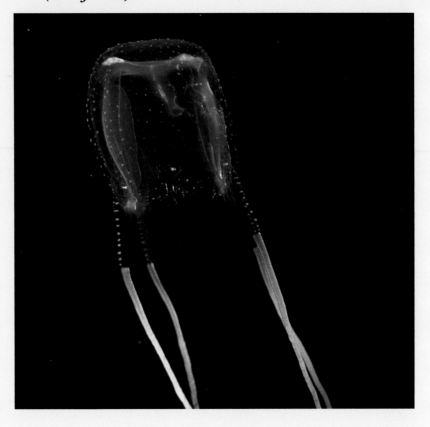

Evolution and systematics

At one time cubozoans were placed within the class Scyphozoa (true jellyfishes), but recent research supports the Cubozoa at the class level. There is one order, Cubomedusae, with two families, Chirodropidae and Carybdeidae, and 19 described species. There is only one recognized fossil species, *Anthracomedusa turnbulli*, which was found near Chicago, Illinois, United States. This fossil dates from the Pennsylvanian age (323–290 million years ago), and with its square shape and clusters of tentacles, was likely within the family Chirodropidae.

Physical characteristics

The most obvious characteristic of cubozoans is the box-like shape of their bell. The manubrium and mouth are located inside the bell, and the velarium is along the edge of the bell. Pedalia are muscular extensions of the bell and the tentacles are attached to the pedalium. Chirodropids have multiple tentacles attached to each pedalium; carybdeids have only have one tentacle per pedalium. Between the pedalia are the unique sensory structures known as rhopalia. Each rhopalia has a statocyst and six eyes, four simple eyes, and two relatively complex eyes composed of an epidermal cornea, spheroidal cellular lens, and an upright retina. Antisera testing has revealed that cubozoans possess blue-, green-, and ultraviolet-sensing opsins in the both the small and large complex eyes. The cubomedusan nervous system is complex compared to that of other cnidarians. A diffuse synaptic nerve net throughout the bell region is connected to a subumbrellar nerve ring, and nerve processes extend from this ring into the rhopaliar stalks. The processing of visual images probably occurs on this nerve ring.

Distribution

Cubozoans can be found in most tropical and subtropical waters *Carybdea marsupialis* and *Carybdea rastoni* have been found in temperate waters as well. It is likely that the distribution records are incomplete as the transparent nature of these medusae make them very difficult to locate despite their shallow and coastal distribution.

Habitat

The preferred habitat of cubozoans appears to be over sandy substrate, with box jellies located just above the bottom during the day and moving up toward the surface at night. Field observation is extremely difficult, as the jellies react to the presence of divers by rapidly moving away. *Carybdea*

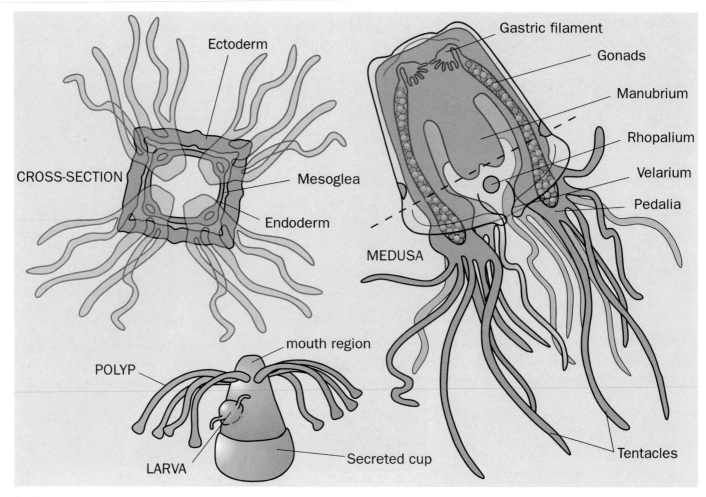

Box jelly anatomy. (Illustration by Christina St. Clair)

sivickisi possesses adhesive pads on the exumbrellar surface which enables it to attach to various substrates in the field.

Behavior

Cubozoans have the most complex behavior of any cnidarian. They are active swimmers capable of moving 9.8–19.7 ft (3–6 m) per minute. They are positively phototactic (move toward light) and are active during the day and night, although they may feed only during the night or predawn hours. Vision clearly plays a role in both feeding and reproduction. One characteristic that makes them interesting, their image forming eyes, also makes them difficult to study in the field or the laboratory because they react to the presence of their human observers by swimming away.

Feeding ecology and diet

Cubozoans are important predators in the nearshore ecosystem, feeding mainly on fishes and crustaceans. Additional prey items include polychaetes, crab megalopae (post-larvae), isopods, amphipods, stomatopod ("mantis shrimps") larvae, and chaetognaths (arrow worms). As active predators,

cubozoans chase, catch, and eat fishes and other organisms. Feeding behavior varies slightly between species. The prey are caught on the tentacles and brought up towards the pedalia by the tentacle contracting. The medusa then either remains upright in the water or turns upside down. The pedalium with the prey item than bends inward toward the manubrium, and the prey are then engulfed.

Reproductive biology

Little is known about the reproductive biology of cubozoans. The life history includes a benthic stage, the polyp, which can reproduce asexually by budding, and a pelagic stage, the medusa. Eggs and sperm combine to form a ciliated larva (planula), which settles on the bottom and becomes a polyp. Unlike scyphozoans, cubozoan polyps do not undergo strobilation (asexual reproduction by division into body segments); rather, the entire polyp becomes the juvenile medusa. *C. sivickisi* uses spermatophores that can be stored by the female, and it has been hypothesized that female *C. rastoni* may collect sperm strands produced by the males. Other cubozoans may have internal fertilization as well, but most broadcast their gametes.

Conservation status

No species of cubozoans is listed by the IUCN.

Significance to humans

The presence of "stinger-resistant enclosures" on beaches in northern Queensland, Australia, is indicative of the impact that cubozoans can have on humans. Although deaths are reported almost every year from encounters with *C. fleckeri*, the overall impact of cubozoans is much greater as stings are not always reported. *Carukia barnesi* is now recognized as the cause of "Irukandji syndrome," which results in severe backache, muscle pains, chest and abdominal pains, headache, localized sweating, and piloerection, as well as nausea and reduced urine output. There is a box jelly antivenom that binds to both *C. fleckeri* and *C. barnesi*, and vinegar can inhibit unfired nematocysts from firing (although it stimulates nematocyst firing with other cnidarians).

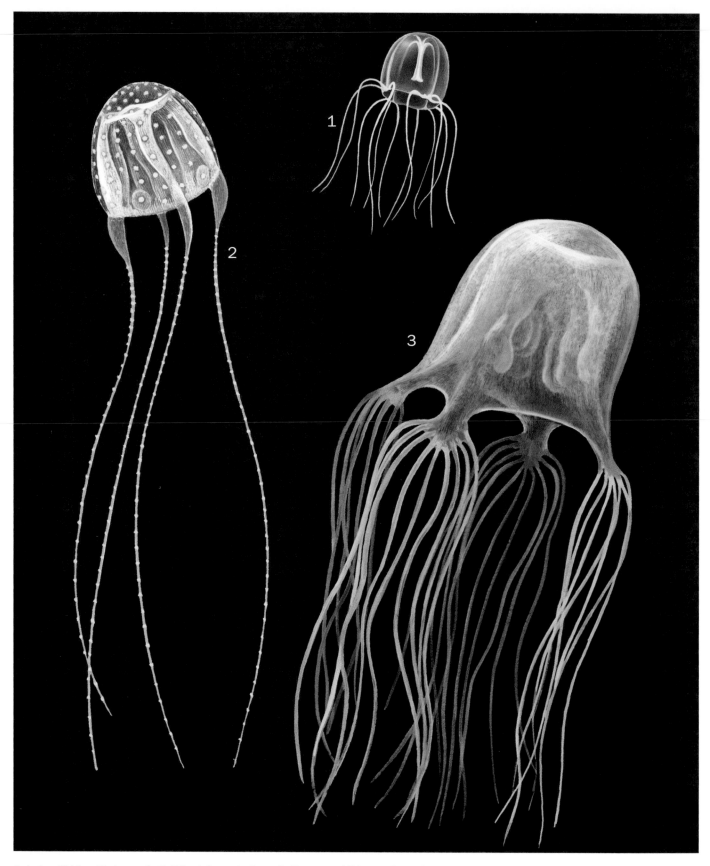

1. Irukandji (*Carukia barnesi*); 2. *Tripedalia cystophora*; 3. Sea wasp (*Chironex fleckeri*). (Illustration by John Megahan)

Grzimek's Animal Life Encyclopedia

Species accounts

Irukandji
Carukia barnesi

FAMILY
Carybdeidae

TAXONOMY
Carukia barnesi Southcutt, 1967, Cairns, Australia.

OTHER COMMON NAMES
English: Box jelly.

PHYSICAL CHARACTERISTICS
The bell is transparent, reaching to 0.8 in (20 mm) in diameter and 1 in (25 mm) in depth; each of the four tentacles may extend to 25.6 in (650 mm).

DISTRIBUTION
Tropical Australian waters from central Queensland (Agnes Water) to Broome, Western Australia.

HABITAT
Occurs in deeper waters of the reef, although often swept inshore by currents.

BEHAVIOR
Nothing is known.

FEEDING ECOLOGY AND DIET
Nothing is known.

REPRODUCTIVE BIOLOGY
Nothing is known about the life cycle. The only time juveniles were collected was in 1997, when four small jellies were collected 3.1 mi (5 km) north of Mackay Harbor in Queensland. This led researchers to postulate that the early part of the life cycle may be in creeks or rivers as in *C. fleckeri.*

CONSERVATION STATUS
Not listed by the IUCN.

SIGNIFICANCE TO HUMANS
Stingings were named the Irukandji syndrome (after a local aboriginal tribe) as early as 1943, but the species was not collected and identified until 1964 to 1966. The venom is extremely potent, and may be responsible for some fatalities. Research in 2002 focused on the use of magnesium infusions to counter the toxic sting. ◆

Sea wasp
Chironex fleckeri

FAMILY
Chirodropidae

TAXONOMY
Chironex fleckeri Southcutt, 1956, Australia.

OTHER COMMON NAMES
English: Box jelly.

PHYSICAL CHARACTERISTICS
Reaches up to 11.8 in (30 cm) in diameter, but is difficult to see despite its large size. The as many as 15 tentacles in each corner can reach up to 98.4 ft (30 m) distance from the bell.

DISTRIBUTION
Tropical waters around Australia, from Exmouth, Western Australia, to Bustard Heads, Queensland, as well as around the Indo-west Pacific Ocean near Papua New Guinea, the Philippines, and Vietnam. Full extent of the distribution has not yet been determined.

HABITAT
Usually shallow waters around creek or mangrove outlets; often reported swimming around pier pilings in search of food.

BEHAVIOR
Swims around pier pilings.

FEEDING ECOLOGY AND DIET
Primary diet items are fishes and prawns.

REPRODUCTIVE BIOLOGY
The search for the polyp of *Chironex fleckeri* took years before polyps were found attached under some rocks in a northern Australian estuary. Polyps have been found in mangrove swamps and river outlets, but not much is known about how the planulae find their way to these locations. Polyps start to metamorphose into medusae in the Australian spring (September)

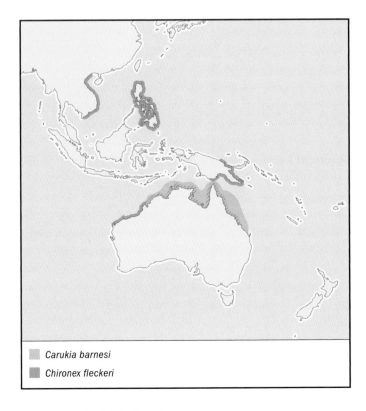

▨ *Carukia barnesi*
▨ *Chironex fleckeri*

and continue until the first large summer rains (January), when they are flushed out into the ocean.

CONSERVATION STATUS
Not listed by the IUCN.

SIGNIFICANCE TO HUMANS
The venom is neurotoxic, cardiotoxic, and dermonecrotic (causes skin tissue to be damaged). Death can occur extremely rapidly while the victim is still in the water or on the beach. Antivenom is available but must be administered quickly. Vinegar can be used to remove undischarged nematocysts. ◆

No common name
Tripedalia cystophora

FAMILY
Carybdeidae

TAXONOMY
Tripedalia cystophora Conant, 1898, Kingston Harbor, Puerto Rico.

OTHER COMMON NAMES
None known.

PHYSICAL CHARACTERISTICS
Small, mature jellies 0.02–0.4 in (0.5 mm–1.0 cm) in diameter. Bell slightly yellow brown, blends well with the mangrove habitat.

DISTRIBUTION
Puerto Rico and the Caribbean.

HABITAT
Shallow waters among mangrove roots.

BEHAVIOR
As *D. oculata* aggregate in vertical lift shafts generated by the sun shining through the mangrove roots, *T. cystophora* change their swimming speed and turning rate in the light shafts, re-

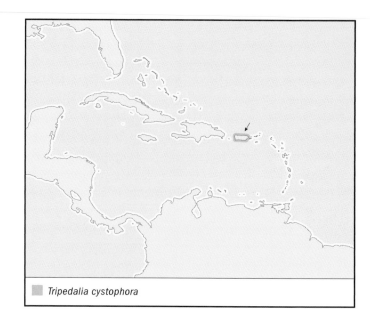

Tripedalia cystophora

sulting in the jellies spending more time in the light shafts where they can feed on the copepods.

FEEDING ECOLOGY AND DIET
Preys on dense swarms of the copepod *Dioithona oculata* in the mangrove prop-root habitat of Puerto Rico, where it has been intensely studied.

REPRODUCTIVE BIOLOGY
Unique in cubozoans, broods fertilize eggs until they develop into planulae within gastral pockets. Planulae are released upon maturity. Show a preference for settling on substrates in the dark where they develop directly into a medusa.

CONSERVATION STATUS
Not listed by the IUCN.

SIGNIFICANCE TO HUMANS
Not harmful to humans, but plays an important role in marine education as it is exhibited in many aquaria. ◆

Resources

Periodicals

Collins, A. G. "Phylogeny of Medusozoa and the Evolution of Cnidarian Life Cycles." *Journal of Evolutionary Biology* 15, no. 3 (2002): 418–432.

Hartwick, R. F. "Distributional Ecology and Behaviour of the Early Life Stages of the Box-Jellyfish *Chironex fleckeri*." *Hydrobiologia* 216/217 (1991): 181–188.

Martin, V. J. "Photoreceptors of Cnidarians." *Canadian Journal of Zoology.* 80 (2002): 1703–1722.

Matsumoto, G. I. "Observations on the Anatomy and Behaviour of the Cubozoan *Carybdea rastonii*, Haacke." *Marine and Freshwater Behaviour and Physiology* 26 (1995): 139–148.

Wiltshire, C. J., S. K. Sutherland, K. D. Winkel, and P. J. Fenner. "Comparative Studies on Venom Extracts from Three Jellyfish: the Irukankji (*Carukia barnesi*), the Box Jellyfish (*Chironex fleckeri*, Southcott) and the Blubber (*Catostylus mosaicus*)." *Toxicon* 36 (1998): 1239.

George I. Matsumoto, PhD

Scyphozoa
(Jellyfish)

Phylum Cnidaria

Class Scyphozoa

Number of families 20

Thumbnail description
Large, soft-bodied, gelatinous marine invertebrates that swim by contracting their umbrella-shaped swimming bell and catch small prey by means of stinging tentacles

Photo: *Chrysaora dactylometre.* (Photo by William H. Amos. Bruce Coleman, Inc. Reproduced by permission.)

Evolution and systematics

The class Scyphozoa includes four orders, 20 families, 66 genera, and about 200 species. The four orders are Stauromedusae, the stalked jellyfish; Coronatae, the crown or grooved jellyfish; Semaeostomeae; and Rhizostomeae.

Animals in the phylum Cnidaria may have one or both of two body forms, the benthic polyp and the pelagic medusa. The four orders within the class Scyphozoa emphasize these two forms to different degrees. Specifically, in the order Stauromedusae, there is only a benthic stage, which is considered the medusa. In the orders Coronatae, Semaeostomeae, and Rhizostomeae both stages occur in most species, with the medusa stage being the largest and most conspicuous.

The phylum Cnidaria is considered to be of early evolutionary origin, but the position of the Scyphozoa relative to other cnidarian classes (Anthozoa [corals and anemones], Cubozoa [box jellyfish], and Hydrozoa [hydroids, hydromedusae, fire corals, and siphonophores]) is uncertain. It is debated whether the polyp or the medusa form is most primitive. The scyphozoans are related most closely to cubozoans, which were placed in the same class until recently. They have similar body forms and life cycles. The scyphozoans are related least to the Anthozoa. Molecular evidence suggests that An-

thozoa represents the most primitive class in the phylum Cnidaria.

The fossil record of Scyphozoa is poor. Radially symmetrical impressions have been interpreted to be casts of primitive scyphomedusae. Recently, a large number of scyphomedusae that apparently were stranded and buried on a beach was discovered in central Wisconsin in the United States. Other fossil groups that may be ancient scyphozoan polyps are the conulariids, which were similar to modern coronate polyps and were present from the Ordovician to the late Triassic, and *Bryonia* from the Upper Cambrian and Ordovician, which is from the extinct order Bryoniida of the Scyphozoa.

Physical characteristics

Most species in the class Scyphozoa, excluding the order Stauromedusae, have two life stages, the predominant medusa stage (up to 80 in, or 2 m, in diameter) and the small, inconspicuous polyp stage (less than 0.13 in, or 4 mm, long). The medusa, or jellyfish, stage has a saucer- to umbrella-shaped body with two epithelial layers (epidermis and gastrodermis) separated by a thick layer of mesenchyme, a gelatinous connective tissue containing cells. Near the edge of the bell in the orders Coronatae and Semaeostomeae are tentacles used in

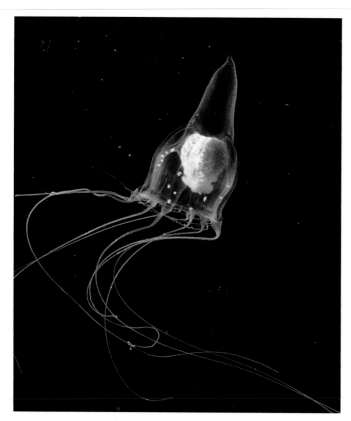

The medusa of *Leuckatiara octona* has a jelly filled projection at the tip of the bell-like body which is a distinguishing feature of this genus. (Photo by ©Sinclair Stammers/Science Photo Library/ Photo Researchers, Inc. Reproduced by permission.)

feeding. The tentacles have millions of microscopic intracellular organelles called nematocysts that evert a hollow thread from a capsule and may inject toxin into or entangle their small prey (zooplankton, fish eggs and larvae, or other gelatinous species). In the Semaeostomeae and Rhizostomeae, there are four oral or mouth arms on the underside (concave) of the bell, which also have stinging nematocysts for feeding. The polyps, called scyphistomae, can form colonies of individuals by budding or, in the case of coronate polyps, true colonies that have a chitinous sheath. Polyps are cup-shaped, attached to the substrate by a "foot," and with the central mouth surrounded by a single ring of tentacles with nematocysts.

Stalked jellyfish in the order Stauromedusae attach to seaweed or sea grasses by an aboral stalk. The main body (calyx) is funnel- or goblet-shaped and grows to 1.2 in (3 cm) wide, with eight arms, each bearing a cluster of as many as 100 short, clubbed tentacles. In the common genus *Haliclystus*, between each arm there is an adhesive disk, by which the animal can attach to move about. The gonads extend down the arms. The mouth is located at the inside center of the funnel-shaped body. Coloration varies; it may be shades of green, brown, yellow, or maroon and often matches the color of the substrate, making these jellyfish difficult to see. This form is considered the medusa stage, and there is no polyp or swimming stage.

Jellyfish in the order Coronatae have a deep groove around the aboral surface that separates the swimming bell into a cen-

tral disk and a peripheral zone, which has lappets. One thick tentacle emerges between lappets on the upper surface of the bell; depending on the species, there are between eight and 36 tentacles. The mouth opens into a large, pouchlike stomach on the underside of the bell. Most of the species are deep living, and thus the central disk is colored dark red to maroon, making it invisible at depths and presumably concealing the bioluminescence emanating from consumed prey. Most of the medusae are small, less than 2 in (5 cm) in diameter or bell height, but some species may attain 6 in (15 cm) in diameter. The known polyps are colonial and covered with a chitinous sheath.

Adult jellyfish in the order Semaeostomeae generally are large, up to 80 in (2 m) in diameter, but usually less than 12 in (30 cm). The swimming bell is flat to hemispherical in shape. The bell edge may have lappets, or it may be smooth. From eight to hundreds of tentacles are present at the bell margin or beneath the bell. In the center of the concave side of the bell are four diaphanous or frilled oral arms that lead to the central mouth. The bell ranges from translucent to opaque and from white to dark orange in color, and it may have radiating stripes in some species. The polyp stage is small and may form groups of individuals by budding.

The rhizostome medusae also are large, up to 80 in (2 m) in diameter. The swimming bell is hemispherical and very firm in texture and lacks tentacles at the edge. The bell margin has eight or 16 lappets. The four oral arms of rhizostome medusae are fused and usually very elaborate, with many tiny tentacles and small mouths for feeding. There may be clublike projections from the oral arms. The medusae are translucent to opaque, with colors ranging from white to dark red; some have patterns that include stripes and spots. The polyp stage is small and may form clusters of individuals by budding.

Distribution

Scyphozoans are found in all marine waters. Most Stauromedusae are found in cool waters along temperate to subpolar shorelines spring through autumn. Coronate medusae generally occur at great depths, where temperatures are a cool 40–46°F (5–8°C), but a few species occur in subtropical to tropical waters. Semaeostome species are the predominant large medusae in polar to temperate oceans, and they also inhabit subtropical and tropical waters. The scyphistomae of those species are active only in warm months; however, they can become dormant and survive through winter months. Rhizostome species live mostly in tropical waters, with only a few species found in subtropical or temperate regions. The medusae of shallow-living scyphozoans are seen during late spring to early autumn in surface waters of temperate to polar seas. In tropical waters and among deep-dwelling species, medusae may be present all year.

Habitat

Scyphozoan medusae are found from surface waters to abyssal depths, and the polyps are attached to hard surfaces, such as pilings, shells, and rocks, at various depths, depending on the species. Most Stauromedusae are seen at intertidal

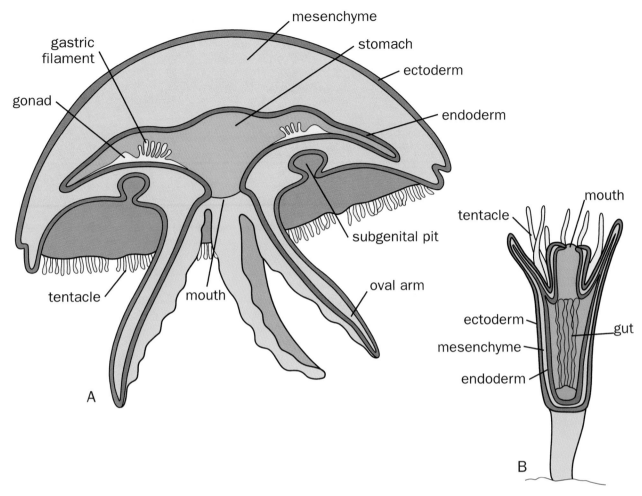

Jellyfish anatomy. A. Medusa; B. Polyp. (Illustration by Patricia Ferrer)

down to shallow subtidal depths, usually attached to benthic plants (algae or sea grasses). One species is known from deep hydrothermal vent communities. Coronate medusae typically are found at mesopelagic depths (1,625–4,875 ft, or 500–1,500 m), but a few species occur near the surface. Deep-living species may have polyps at abyssal depths, but the polyps of shallow-living species are on shallow substrates. Semaeostome and rhizostome medusae occur most abundantly near shore in surface waters above 165 ft (150 m), where food supplies are greatest. Their polyps also are found at shallow depths, often on the underside of structures away from direct light. Some semaeostome species are deep living, and their polyps generally are not known. There are no known deep-dwelling rhizostome species.

Behavior

Jellyfish behavior generally is simple, owing to their simple nervous system. Stauromedusae move around on the substrate by somersaulting, which they accomplish by alternating adhesion of the basal disc with that of tentacles or adhesive pads located between the tentacles of some genera. The most noticeable behavior of jellyfish is the rhythmic pulsation of the swimming bell, which moves them through the water for feed-

ing and respiration. The swimming pulsations are coordinated by nerve centers around the edge of the bell. At the bell margin there also are sensory clubs (rhopalia), each consisting of a light-sensing organ (ocellus) and a gravity-sensing organ (statocyst); thus, medusae can sense light and dark and can determine their orientation in the water column. Semaeostome and rhizostome jellyfish swim continuously. This is important for oxygen exchange, which occurs over the entire body surface, and for feeding. The swimming of several species is known to be against flow in the water column; the result is that they all swim in the same direction and may become concentrated in convergences, like bales of hay stacked up to dry. Some species move up in the water column at night and down in the day ("diel vertical migration"). The scyphistomae (polyps) are able to move by the so-called foot and its extensions. They feed when prey makes contact with their tentacles, which have nematocysts; the jellyfish contract the tentacles and bring the prey to their mouths.

Feeding ecology and diet

All scyphozoans feed with tentacles or tentacle-like projections that have millions of microscopic intracellular organelles called "nematocysts." Some nematocysts act to

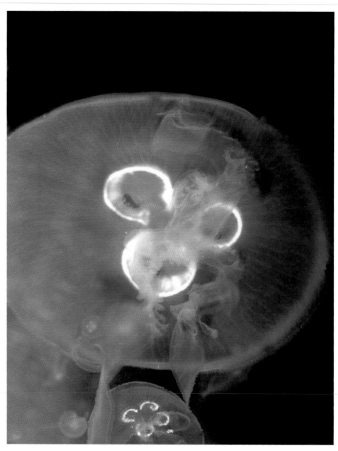

Moon jellies (*Aurelia aurita*). ©Shedd Aquarium. Photo by Edward G. Lines, Jr. Reproduced by permission.)

paralyze or kill the prey, whereas others entangle them. Stauromedusae catch prey by the tentacles and fold the arm inward to bring the prey to the mouth. Many coronate medusae do not swim actively while feeding but instead remain nearly motionless with their tentacles extended above the bell. For semaeostome and rhizostome medusae, the pulsations of the swimming bell force water through the tentacles and create vortices that may bring prey into contact with the tentacles and oral arms. For semaeostome medusae, when a prey item is immobilized on a tentacle, the tentacle contracts and transfers the prey to an oral arm. The prey is moved by cilia up the inside of the folded oral arm to the mouth and into one of the four gastric (stomach) pouches, where short, fingerlike projections wrap around the prey and secrete digestive enzymes. For rhizostome medusae, prey capture is by the small tentacles on the oral arms, which transfer the prey to one of the many small mouths nearby.

Most species feed on small crustaceans that predominate in most habitats. Stauromedusae consume epibenthic crustaceans, including gammarid amphipods and harpacticoid copepods. Medusae in the other orders primarily eat abundant calanoid copepods but also eat other small zooplankton, such as cladocerans, larvaceans (= appendicularians), and chaetognaths. Many semaeostome species also feed on other gelatinous species, including scyphomedusae, hydromedusae,

siphonophores, and ctenophores. It is of particular interest that several species are known to consume the eggs and larvae of fish. Thus, scyphomedusae may be detrimental to fish populations, both by consuming the zooplankton foods needed by fish larvae and zooplanktivorous fish species, like herring, and by feeding on the young fish directly.

Reproductive biology

Scyphozoans generally reproduce both asexually and sexually. The benthic forms, Stauromedusae and scyphistomae (polyps) of the other orders, reproduce asexually by budding new polyps or cysts from the body or foot. Scyphistomae, which are present in most species of all orders except Stauromedusae, also produce the medusa stage by an asexual budding process called strobilation. Strobilation typically takes place at a certain time of year and is triggered by environmental factors, which differ by species; these factors include rising (spring) or falling (autumn) temperatures or changes in light levels. During strobilation, the polyp undergoes transverse segmentation, forming one to several small medusae, called "ephyrae." The process requires days to weeks, depending on temperature.

The fully formed ephyrae break free by swimming pulsations. The ephyrae grow into sexually mature medusae over the course of a month or longer. The medusae of most species have separate sexes, but a few species are sequential hermaphrodites. The males and females are indistinguishable except by examination of the gonads. No mating occurs. Sperm strands are released into the water by males and are taken up by the females during feeding. The gonads surround the gastrovascular cavity, and eggs may be fertilized in the ovary or after they are released into the gastrovascular cavity. In most species the fertilized eggs develop into small ciliated larvae (planulae) that swim to a suitable substrate, attach, and develop into polyps. In some species, the planulae are retained (brooded) by the female before settlement. Some species lack a polyp stage.

Conservation status

No species of Scyphozoa is listed by the IUCN.

Significance to humans

Scyphozoan jellyfish have direct and indirect effects on humans, many of which are negative. Swimmers fear them for their painful stings. All jellyfish sting, but the stings of small specimens and those with short tentacles often are not painful to humans. The genera *Chrysaora* and *Cyanea* are known for painful stings. Scyphozoan stings are painful but not deadly. More painful and dangerous stingers are in the class Cubozoa (box jellyfish) and the class Hydrozoa (specifically, the Portuguese man of war, *Physalia physalis*).

Fish populations and commercial fisheries may be affected detrimentally by jellyfish. Jellyfish may occur in great abundance, and, if they are caught in fishing nets, their great weight may cause the nets to rip or the fish catch to be damaged. Jellyfish eat the pelagic eggs and larvae of fish as well as the small zooplankton prey of fish larvae and zooplanktivorous fish species. Therefore, jellyfish both eat fish and compete with them for food. Jellyfish also appear to be intermediate hosts for some parasites of fish. Jellyfish have been a nuisance to fish farms, where they break up on the fish impoundments and sting and kill the fish, and to power plants, where they may clog the cooling water intake screens, sometimes causing the plants to suspend operations. On the positive side for fish and fisheries, the juveniles of at least 80 species of fish, many of which are commercially important, associate with large jellyfish. While the relative advantages of such associations are not known, they are thought to benefit the fish partners most.

Jellyfish also have a place of value in human enterprise. In Japan and China jellyfish are an important food and have been exploited for more than 1,700 years. In China they are considered a culinary delicacy and are thought to have medicinal value. A multimillion-dollar commercial fishery exists for at least 10 species of rhizostome medusae throughout Southeast Asia, and a fishery for *Stomolophus meleagris* has been started in the Gulf of Mexico. The swimming bell of the jellyfish is processed in a mixture of salt and alum and packaged for distribution. The semidried jellyfish is rehydrated, desalted, blanched, and served in a variety of dishes. The prepared jellyfish has a special crunchy texture.

Owing to their great beauty and the relaxing effect of their swimming pulsations, jellyfish have been a great success as specimens in public aquariums and even as household pets. Over the past decade, considerable advances have been made in jellyfish husbandry, and several species are on display at aquariums worldwide. In Japan jellyfish are kept as pets in special aquariums.

1. Sea nettle (*Chrysaora quinquecirrha*); 2. Lion's mane jellyfish (*Cyanea capillata*); 3. Cannonball jellyfish (*Stomolophus meleagris*); 4. Moon jelly (*Aurelia aurita*). (Illustration by Joseph E. Trumpey)

1. Nightlight jellyfish (*Pelagia noctiluca*); 2. Thimble jelly (*Linuche unguiculata*); 3. Stalked jellyfish (*Haliclystus auricula*); 4. Crown jellyfish (*Periphylla periphylla*); 5. Golden jellyfish (*Mastigias papua*); 6. Upside-down jellyfish (*Cassiopea xamachana*). (Illustration by Joseph E. Trumpey)

Species accounts

Thimble jelly
Linuche unguiculata

ORDER
Coronatae

FAMILY
Linuchidae

TAXONOMY
Linuche unguiculata Schwartz, 1788, American Tropical Atlantic.

OTHER COMMON NAMES
None known.

PHYSICAL CHARACTERISTICS
The medusae grow only to 1 in (2.5 cm) in height. As the name implies, they are thimble-shaped, with a shallow groove near the top of the bell. They have eight very short tentacles and eight rhopalia alternating between the 16 lappets at the bell margin. The outside of the bell is transparent, with numerous warts of stinging cells. The inner part of the bell is white with greenish brown spots. The polyps form colonies and are covered by a thin, chitinous sheath.

DISTRIBUTION
This species lives in tropical and subtropical waters worldwide.

HABITAT
The medusae occur near the surface in near shore waters in spring and summer, which is unusual for coronate species. Polyps occur on coral rubble.

BEHAVIOR
The thimble jelly usually is found in large groups, up to 0.6 mi^2 (1 km^2) in area, just beneath the surface. They are very active swimmers, moving horizontally in circles. Surface convection cells cause them to become concentrated, and their swimming behavior helps them remain in an aggregation.

FEEDING ECOLOGY AND DIET
The medusae catch a variety of zooplankton prey on the lappets. The colored spots in the bell are filled with intracellular algae (zooxanthellae) that transfer some of their photosynthesized carbon to the medusa, contributing to the animal's nutrition.

REPRODUCTIVE BIOLOGY
The fertilized eggs of the thimble jelly form large larvae that remain planktonic for three to four weeks. They settle and form an unbranched colony of polyps. Each polyp can produce an unusually large number (up to 40) of ephyrae that grow into sexually mature medusae.

CONSERVATION STATUS
Not listed by the IUCN.

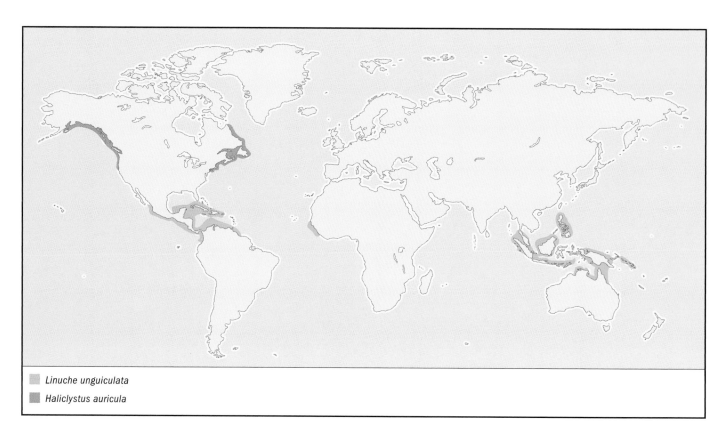

Linuche unguiculata

Haliclystus auricula

SIGNIFICANCE TO HUMANS

Stings from the planktonic larvae or medusae of the thimble jelly cause the syndrome called "seabathers' eruption." The problem is aggravated when they become trapped underneath a swimsuit. This syndrome, characterized by a prickling sensation and red bumps persisting for 7–12 days, is an irritating but not a dangerous condition. ◆

Crown jellyfish
Periphylla periphylla

ORDER
Coronatae

FAMILY
Periphyllidae

TAXONOMY
Periphylla periphylla Peron and Lesueur, 1809, equatorial Atlantic Ocean.

OTHER COMMON NAMES
None known.

PHYSICAL CHARACTERISTICS

These medusae have a conical swimming bell that is up to 8 in (20 cm) tall and 6.5 in (17 cm) in diameter, but specimens from oceanic waters usually are less than 2 in (5 cm) in size. The bell of small specimens is transparent and reveals the reddish brown stomach. In large specimens, the bell is opaque and maroon in color. The bell has a pronounced groove with 16 deeply notched lappets beneath it. Twelve thick tentacles

emerge from the bell surface above the clefts, in a repeating pattern of three tentacles and one rhopalium. The stomach is baglike. Oral arms and the polyp stage are lacking.

DISTRIBUTION
This species is found at mesopelagic depths in all oceans worldwide. Populations that are several orders of magnitude greater than they are in the open ocean have been found in Norwegian fjords, especially Lurefjorden.

HABITAT
The crown jellyfish generally is found at depths below 3,000 ft (900 m), where water temperatures remain a cool 45°F (7°C) or less all year. At high latitudes they inhabit shallower depths, 650–1,300 ft (200–400 m) in the daytime and from the surface to 650 ft (200 m) at night.

BEHAVIOR
These medusae undergo vertical migration from deepwater in the daytime to shallower depths at night, presumably following their prey. Exposure to white light causes rapid downward swimming. As with other deep-dwelling coronate medusae, when they are disturbed, the bell and ovaries are brilliantly bioluminescent. They also produce copious amounts of luminescent mucus that contains stinging cells.

FEEDING ECOLOGY AND DIET
The natural behavior of medusae was observed with red light and video cameras on a Remotely Operated Vehicle (ROV) in Lurefjorden, Norway. The medusae hold their tentacles up alongside the bell or at right angles to the bell. They swim downward for about 30 ft (10 m) and then drift upward. The tentacles sometimes quickly arch toward the mouth, coil, and enter the stomach. Few prey (copepods, ostracods, and chaetognaths) were found in ROV-collected specimens. These

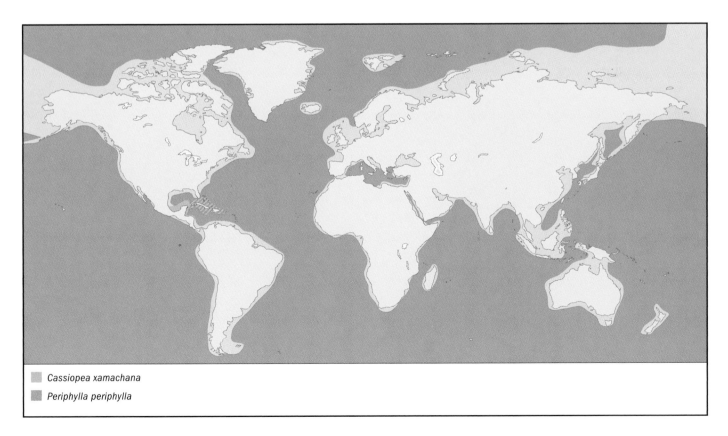

Cassiopea xamachana
Periphylla periphylla

medusae have low metabolic rates and apparently survive on few (less than 35) prey items daily.

REPRODUCTIVE BIOLOGY
Fertilized eggs are released in deepwater, where they drift, not feeding for several months. Eggs and larvae are present all year in Lurefjorden, suggesting lack of seasonality in the relatively constant environment of the deep ocean. The larvae develop directly into medusae without polyp or ephyra stages.

CONSERVATION STATUS
Not listed by the IUCN.

SIGNIFICANCE TO HUMANS
In Lurefjorden the crown jellyfish may have nearly excluded small fish, which are common in other fjords. There are possible medical applications for the bioluminescent proteins. Deep oceanic populations have no apparent significance for humans. ◆

Upside-down jellyfish
Cassiopea xamachana

ORDER
Rhizostomeae

FAMILY
Cassiopeidae

TAXONOMY
Cassiopea xamachana R. P. Bigelow, 1892, Jamaica, West Indies.

OTHER COMMON NAMES
None known.

PHYSICAL CHARACTERISTICS
Medusae in the genus *Cassiopea* are unique in resting bell side down on the ocean bottom. The swimming bell may reach 6 in (15 cm) in diameter. It is flattened, with 40 lappets. The oral arms are about 1.5 times the bell radius. They branch laterally and have numerous tiny tentacle-like projections. The upside-down medusae appear to be clumps of algae because of the bushy greenish brown oral arms that cover the bell. The topside of the bell can be marked boldly with regular spots and stripes in brown and blue. Tentacles are lacking.

DISTRIBUTION
This species is found in the eastern tropical Atlantic, including Florida, the Caribbean, and Mexico. Other members of the genus occur in tropical waters worldwide.

HABITAT
These jellyfish are found year-round in sunny, shallow tropical lagoons and mangroves and in coral back-reef areas.

BEHAVIOR
Although the medusae rest bell side down on the bottom, regular swimming pulsations are important in terms of feeding and water exchange for respiration. The medusae do not swim in the water column unless they are disturbed.

FEEDING ECOLOGY AND DIET
Nutrition of the upside-down jellyfish depends on a combination of zooplankton feeding and symbiosis with intracellular al-gae (zooxanthellae). Pulsations of the bell sweep small epibenthic crustaceans and zooplankton into the oral arms, where they are caught by the numerous tiny tentacle-like projections and passed to the many small mouths along the oral arms. The greenish brown color of the oral arms comes from the zooxanthellae in the tissues. These algae contribute photosynthetic products to the medusa's nutrition. The symbiosis represents a mutual exchange of nutrients that, together with zooplankton, supports algae and medusa metabolism, growth, and reproduction. The polyps also contain zooxanthellae and feed on small crustaceans.

REPRODUCTIVE BIOLOGY
The life cycle is typical of rhizostome scyphozoans. Metamorphosis of the larvae into polyps requires a peptide derived from the cell walls of decomposing plants, such as mangroves. Polyps can reproduce asexually by budding off chains of special individuals that settle to form polyps. Strobilation produces only one ephyra per polyp.

CONSERVATION STATUS
Not listed by the IUCN.

SIGNIFICANCE TO HUMANS
The creation of protected lagoons for resorts or aquaculture and associated eutrophication (nutrient pollution) of those areas have resulted in increased medusa populations in the Florida Keys and Mexico. The medusae are harmless to humans but can be irritating if they are handled. ◆

Golden jellyfish
Mastigias papua

ORDER
Rhizostomeae

FAMILY
Mastigiidae

TAXONOMY
Mastigias papua Lesson, 1830, Japan.

OTHER COMMON NAMES
None known.

PHYSICAL CHARACTERISTICS
The firm bell is hemispherical and up to 3.5 in (9 cm) in diameter. The bell is translucent with white spots and a granular appearance. The eight oral arms are frilled and flare near the bell, tapering to smooth clublike structures at the end. Tissue along the bell margin and in the frilly oral arms is golden in color, from intracellular algae (zooxanthellae). Tentacles are lacking.

DISTRIBUTION
This species is found in the tropical southern and central Pacific Ocean and into the Indian Ocean, Malaysia, Japan, Fiji, the Philippines, and Palau.

HABITAT
Medusae occur in tropical surface waters near shore. They have tremendous populations in the marine lakes of Palau.

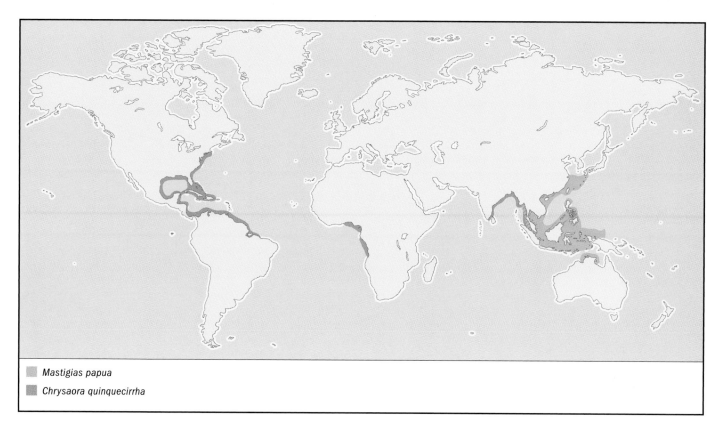

■ *Mastigias papua*
■ *Chrysaora quinquecirrha*

BEHAVIOR

These medusae undergo daily horizontal migrations across the lakes of Palau, staying in the sun or avoiding the shade, which maximizes photosynthesis by their symbiotic algae. The medusae also migrate vertically at night down to nutrient-rich deeper layers in the lakes, which provide nutrients for algal photosynthesis.

FEEDING ECOLOGY AND DIET

The medusae have symbiotic algae (zooxanthellae) in the tissues that supply much of their nutrition. They also eat zooplankton, catching them with small tentacle-like projections on the oral arms. The polyps also have zooxanthellae and eat zooplankton.

REPRODUCTIVE BIOLOGY

The medusa (sexual) and polyp (asexual) generations alternate, as is typical of rhizostome medusae. In the marine lakes in Palau, asexual reproduction can occur all year, but medusa production (strobilation) is inhibited at high temperatures that kill the zooxanthellae.

CONSERVATION STATUS

Not listed by the IUCN.

SIGNIFICANCE TO HUMANS

The very large populations of medusae (1.5 million) in marine lakes have become an important tourist attraction in Palau. The jellyfish in the lakes also have been featured in National Geographic and an IMAX film. ◆

Cannonball jellyfish
Stomolophus meleagris

ORDER
Rhizostomeae

FAMILY
Stomolophidae

TAXONOMY
Stomolophus meleagris L. Agassiz, 1862, Atlantic Ocean, East Coast of United States.

OTHER COMMON NAMES
English: Cabbagehead jellyfish.

PHYSICAL CHARACTERISTICS
The firm, almost spherical swimming bell grows up to 7 in (18 cm) in diameter and lacks tentacles. The oral arms are fused to form a rigid, short mound below the bell. The bell color ranges from nearly white to bluish, and it may darken to reddish orange with blue speckles toward the bell margin; the oral arms are white.

DISTRIBUTION
The cannonball jellyfish occurs in subtropical and tropical Atlantic and Pacific North American waters. It is abundant from North Carolina through the Gulf of Mexico. It also is found in

Cyanea capillata

Stomolophus meleagris

the Caribbean, in South America to northern Argentina, and in the Pacific from southern California to Ecuador.

HABITAT
The medusae are found in coastal waters in summer.

BEHAVIOR
These medusae are strong swimmers. Within a group, individuals swim horizontally in the same direction, oriented with or against the wind and surface-wave or current direction.

FEEDING ECOLOGY AND DIET
The medusae lack obvious long feeding tentacles and oral arms, but they pump water containing zooplankton through the oral arms and elaborate filtering structures within the bell. There, minute tentacle-like projections with nematocysts catch prey. The medusae eat a variety of zooplankton prey, including mollusk veligers, copepods, tintinnids, larvaceans (= appendicularians), and fish eggs.

REPRODUCTIVE BIOLOGY
This species exhibits the typical life cycle of rhizostome scyphozoans, having both medusa (sexual) and polyp (asexual) stages. Only one or two ephyrae are produced per polyp.

CONSERVATION STATUS
Not listed by the IUCN.

SIGNIFICANCE TO HUMANS
This is one of the 11 species of rhizostome medusae that are fished commercially for human consumption. This species has no stinging tentacles but produces mucus that is shed in strings laden with nematocysts that irritate skin on contact. They also act as a host for juvenile butterfish *Peprilis* species that depend on the jellyfish for protection. ◆

Lion's mane jellyfish
Cyanea capillata

ORDER
Semaeostomeae

FAMILY
Cyaneidae

TAXONOMY
Cyanea capillata Linnaeus, 1758, North Sea.

OTHER COMMON NAMES
None known.

PHYSICAL CHARACTERISTICS
Medusae are reported to grow to 80 in (2 m) in diameter but are usually less than an eighth that size. The edge of the swimming bell is divided into eight lobes, giving it the appearance of a flower viewed from above. As many as 150 long tentacles are present in groups beneath each of the eight lobes. The oral arms form a diaphanous mass beneath the bell that is a little longer than it is wide. The bell often is brownish orange, and the oral arms are maroon, but color varies from pale yellow to dark red.

DISTRIBUTION
Medusae are present during summertime in boreal latitudes, but in wintertime they occur in temperate zones. Reported from Arctic, northern European, North American Atlantic and Pacific, southern Australian, and Antarctic waters.

HABITAT
The medusae prefer cool temperatures below 70°F (20°C). They often are found in surface waters of estuaries or coastal bays.

BEHAVIOR

The swimming bell of the medusa orients upward toward the water surface. The swimming beat is slow, just maintaining the position of the medusa in the water column. The tentacles may spread out several meters around the bell. They do not form aggregations.

FEEDING ECOLOGY AND DIET

The diet includes planktonic crustaceans, such as copepods and cladocerans, and fish eggs and larvae. It also contains large proportions of pelagic tunicates (larvaceans = appendicularians), ctenophores (comb jellies), hydromedusae, and scyphomedusae. The long tentacles snare such relatively large prey and bring them to the oral arms, where they are enveloped and digested. Predation effects have been studied in Australia, Norway, and Alaska, where populations of medusae remove only a small percentage of the zooplankton daily.

REPRODUCTIVE BIOLOGY

The life cycle is typical of semaeostome scyphomedusae. Sperm are shed into the water and fertilize eggs in the female. Larvae are brooded along the edges of the oral arms. Sexual reproduction takes place near the end of the life of the medusa, in early spring in warm climates and during the summer at high latitudes. Larvae attach to the undersides of hard surfaces and form polyps, which asexually produce medusae and other polyps.

CONSERVATION STATUS

Not listed by the IUCN.

SIGNIFICANCE TO HUMANS

Medusae have a painful sting that annoys swimmers and fishermen retrieving fishing nets. ◆

Sea nettle
Chrysaora quinquecirrha

ORDER
Semaeostomeae

FAMILY
Pelagiidae

TAXONOMY
Chrysaora quinquecirrha Desor, 1848, Nantucket Bay, Massachusetts, United States.

OTHER COMMON NAMES
None known.

PHYSICAL CHARACTERISTICS
The swimming bell may reach 10 in (25 cm) in diameter, but medusae generally are much smaller. The edges of the swimming bell appear scalloped, with 16 or more lappets. One large tentacle emerges from between every other lappet, and twice as many small tentacles arise from beneath the lappets. Eight rhopalia are present in alternate clefts between lappets. The narrow oral arms are long and diaphanous. Medusa color ranges from milky white to white with radiating purplish red stripes on the bell.

DISTRIBUTION
This species is found near shore in temperate to subtropical Atlantic Ocean waters of North, Central, and South America

above the equator and in the Gulf of Guinea and Angola, Africa. They also are reported from the western Pacific Ocean in the Philippines, southern China, Malaysia, and the Bay of Bengal.

HABITAT
Medusae are most abundant during the summer in estuaries, where they thrive at salinity levels as low as 7 ppt. In Chesapeake Bay unusually low salinity levels in spring result in fewer medusae, with the distribution shifted to waters of higher salinity.

BEHAVIOR
The medusae swim constantly in slow circles, owing to the drag of their oral arms and tentacles. As seems to be true for other scyphomedusa species, the sea nettle feeds continuously.

FEEDING ECOLOGY AND DIET
This species has been studied extensively in Chesapeake Bay and its tributaries, where it occurs in great numbers. When they are abundant, medusae may reduce copepod populations. This species also feeds on comb jellies (*Mnemiopsis leidyi*) and can eliminate them from tributaries. Bay anchovy (*Anchoa mitchilli*) spawns during peak medusa abundance, and medusae may eat 50% of the fish eggs and larvae daily, on average. Surprisingly, the polyps eat and digest oyster larvae (veligers), but the medusae do not digest them.

REPRODUCTIVE BIOLOGY
The life cycle is typical of semaeostome scyphomedusae, having both a polyp and a medusa stage. In temperate Chesapeake Bay, the polyps become dormant during the cold winter months. They excyst and undergo strobilation in spring when water temperatures exceed 60°F (17°C). More ephyrae are produced at salinity levels between 10 and 25 ppt than at lower or higher salinity levels. Spawning takes place around dawn. Larvae are not brooded by the females.

CONSERVATION STATUS
Not listed by the IUCN.

SIGNIFICANCE TO HUMANS
This species was so abundant in Chesapeake Bay during the 1960s that a legislative bill was passed to provide money for research on it. The medusae have an irritating sting, which deters swimming, especially in the shallow tributaries where they are most abundant. The species may be of overall benefit in the food web, by controlling populations of comb jellies, which consume oyster veligers and much more zooplankton than do the medusae. Thus, more zooplankton may be available for zooplanktivorous fishes, such as bay anchovy, which are prey for favorite sport fish, such as striped bass and bluefish. ◆

Nightlight jellyfish
Pelagia noctiluca

ORDER
Semaeostomeae

FAMILY
Pelagiidae

TAXONOMY
Pelagia noctiluca Forskål, 1775, Mediterranean Sea.

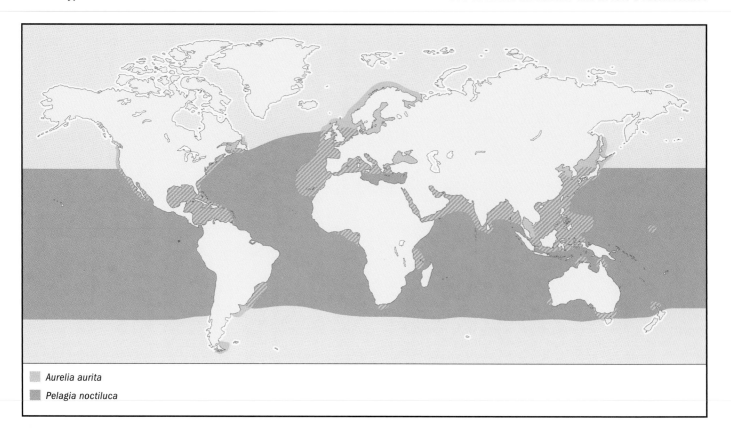

Aurelia aurita

Pelagia noctiluca

OTHER COMMON NAMES
English: Mauve baubler.

PHYSICAL CHARACTERISTICS
The swimming bell usually is less than 3.5 in (9 cm) in diameter. It has a bumpy surface from clusters of stinging cells, which impart a purple or yellowish color to the translucent bell. There are four elongate oral arms and eight long tentacles that alternate with eight rhopalia in the clefts between lappets. There is no polyp stage.

DISTRIBUTION
The medusa is found in tropical to warm temperate Atlantic and Pacific waters and in the Mediterranean Sea. They are present throughout the year in warm waters.

HABITAT
They are found in surface waters of the open ocean, but sometimes can be advected into shallow coastal waters.

BEHAVIOR
This species is unusual among semaeostome scyphomedusae in being bioluminescent. Medusae emit a blue-green light when they are touched or injured. Mucus released from the damaged area continues to glow. The bioluminescence is believed to serve a protective function. Medusae look like glowing balls at night in a boat's wake.

FEEDING ECOLOGY AND DIET
This species has a diverse diet that includes all types of zooplankton, such as crustaceans, mollusk larvae, chaetognaths, pelagic tunicates (larvaceans, salps, and doliolids), siphonophores, hydromedusae, ctenophores, and fish eggs and larvae. It feeds in the same manner as other semaeostome scyphomedusae. Populations of medusae may require a 5–8% diet of zooplankton daily to balance their metabolism.

REPRODUCTIVE BIOLOGY
This species is unusual among semaeostome scyphomedusae in lacking a polyp stage. The larvae develop directly into ephyrae, without ever settling on the bottom. Medusae reproduce throughout the year in tropical waters.

CONSERVATION STATUS
Not listed by the IUCN.

SIGNIFICANCE TO HUMANS
The nightlight jellyfish periodically develops large populations in the Mediterranean Sea. The medusae can become concentrated near the shore in crowded tourist areas. They have a painful sting, which can cause a severe reaction and has led to two international scientific conferences on jellyfish blooms in the Mediterranean Sea. Dramatic fluctuations in their abundance in the Mediterranean is linked to 12-year climate cycles. ◆

Moon jelly
Aurelia aurita

ORDER
Semaeostomeae

FAMILY
Ulmaridae

TAXONOMY
Aurelia aurita Linnaeus, 1758, Baltic Sea.

OTHER COMMON NAMES
None known.

PHYSICAL CHARACTERISTICS
The diameter of the medusa swimming bell may reach 20 in (50 cm). The eight lobes of the bell are marked by shallow indentations, each with a rhopalium. The bell is translucent, usually with a pink tinge. The gonads resemble a pink four-leaf clover, as seen inside the semitransparent bell. Hundreds of short, fine tentacles hang in a single circle from the bell margin. The oral arms extend only to about the edge of the bell and may have bright reddish orange larvae brooded in pockets at the edges. Polyps are white and about 0.1–0.2 in (2–4 mm) long, with a single ring of tentacles. They often occur in large aggregations. Developing ephyrae are orange.

DISTRIBUTION
The moon jelly is reported from all oceans, from tropical to temperate waters between 70°N and 55°S latitude. They are common in coastal European, North American, and Japanese waters. They also are reported from some locations in Asia, Australia, Pacific Islands, South America, and Africa. This species may be endemic to Europe and introduced elsewhere. It closely resembles other species in the genus. Recent molecular studies indicate perhaps six species in the genus that may be easily confused.

HABITAT
This species occurs in a wide range of conditions, in waters with near-freezing winter temperatures, to 90°F (32°C), and with salinity levels of 14–38 ppt. Medusae generally are present only in warm months in temperate locations, but they occur throughout the year in tropical and some temperate locations. They often are found in estuaries, fjords, and bays, usually at the surface to 100 ft (30 m). Large populations may form in semi-enclosed bays with restricted tidal exchange. The polyps are found at depths above 65 ft (20 m) on the undersides of hard structures.

BEHAVIOR
The moon jelly is remarkable in forming large aggregations of medusae. Aggregations may be a mile or more (2 km) in length and can contain millions of individuals. These groups may be seen from low-flying airplanes and detected by "fish finders" on fishing boats. The aggregations are formed because of the tendency of the medusae to swim either up or down against directional water flow. They also are reported to swim horizontally by orienting to a specific compass direction in sunlight, causing them to gather in certain locations.

FEEDING ECOLOGY AND DIET
The fine tentacles of the medusae catch mainly small crustacean zooplankton, such as copepods and cladocerans. They also feed on fish and mollusk larvae, small hydromedusae, and even microzooplankton, such as ciliates. They may catch food in the mucus on the outer surface of the swimming bell.

REPRODUCTIVE BIOLOGY
The life cycle is typical of semaeostome scyphomedusae, having both a medusa (sexual) stage and a polyp (asexual) stage. Moon jelly aggregations are believed to increase fertilization success by bringing females and males into proximity. The males release sperm strands into the water, which are taken up by the females during feeding. The fertilized eggs form larvae that are brooded in pockets on the oral arms. The larvae attach to hard surfaces and become polyps, which then bud medusae.

CONSERVATION STATUS
Not listed by the IUCN.

SIGNIFICANCE TO HUMANS
Aggregations sometimes have clogged the seawater intakes of power plants in Asia. The medusae also may reduce commercial herring populations by feeding on larvae in Kiel Bight, Germany. Their sting is not painful to humans. They are raised easily for aquarium exhibits and sometimes are kept in special home aquariums. ◆

Stalked jellyfish
Haliclystus auricula

ORDER
Stauromeduasae

FAMILY
Lucernariidae

TAXONOMY
Haliclystus auricula Rathke, 1806, Norway.

OTHER COMMON NAMES
None known.

PHYSICAL CHARACTERISTICS
In the Stauromedusae the medusa is stalked and remains attached to the substrate. The "bell" (calyx) is funnel-shaped and grows to 1.25 in (3 cm) in diameter, with the mouth at the bottom of the funnel. The calyx has eight arms, with clusters of 30–80 knobbed tentacles at the tips. Between the arms are eight bean-shaped sticky pads. The stalk is as long as the calyx is wide. The color of this jellyfish often matches the substrate—green, brown, yellow, or reddish-purple.

DISTRIBUTION
This species is found on the northern Atlantic and Pacific coasts above 40°N latitude.

HABITAT
These stauromedusae live attached to rock, algae, or eelgrass in shallow intertidal and subtidal areas late spring through autumn.

BEHAVIOR
Stauromedusae can move about by somersaulting. The stalk bends toward the substrate, where the calyx can attach temporarily by a sticky pad between the arms or by the tentacles. The stalk releases its hold, arches over, and reattaches to the substrate in a different location as the calyx releases its attachment.

FEEDING ECOLOGY AND DIET
Stauromedusae feed on epibenthic animals, including amphipods, copepods, and gastropods, that bump into their stinging tentacles. The arms bend toward the center to bring captured prey to the central mouth.

REPRODUCTIVE BIOLOGY
The stalked jellyfish spawns into the water at first daylight. Fertilized eggs form creeping, nonswimming larvae. When the larvae attach to the substrate, they form a minute polyp, which grows directly into an adult.

CONSERVATION STATUS
Not listed by the IUCN.

SIGNIFICANCE TO HUMANS
None known. ◆

Resources

Books

Arai, Mary N. *A Functional Biology of Scyphozoa*. London: Chapman and Hall, 1997.

Cornelius, Paul F. S. "Keys to the Genera of Cubomedusae and Scyphomedusae (Cnidaria)." In *Proceedings of the Sixth International Conference on Coelenterate Biology*, edited. by J. C. Den Hartog. Leiden, The Netherlands: National Natuurhistorich Museum, 1997.

Franc, André. "Classe des Scyphozoaires." In *Traité de Zoologie: Anatomie, Systématique, Biologie*. Vol. 3, *Cnidaires, Cténaires*, edited by Pierre-P. Grassé. Paris: Masson, 1993.

Mayer, A. G. *Medusae of the World*. Vol. 3, *The Scyphomedusae*. Washington, DC: Carnegie Institution, 1910.

Mianzan, Hermes W., and Paul F. S. Cornelius. "Cubomedusae and Scyphomedusae." In *South Atlantic Zooplankton*, edited by Demetrio Boltovskoy. Leiden, The Netherlands: Backhuys, 1999.

Purcell, Jennifer E., W. Monty Graham, and Henri J. Dumont. *Jellyfish Blooms: Ecological and Societal Importance*. Developments in Hydrobiology, no. 155. Dordrecht, The Netherlands: Kluwer Academic, 2001.

Purcell, Jennifer E., Alenka Malej, and Adam Benovic. "Potential Links of Jellyfish to Eutrophication and Fisheries." In *Ecosystems at the Land-Sea Margin: Drainage Basin to Coastal Sea*, edited by Thomas C. Malone, Alenka Malej, Larry W. Harding Jr., Nenad Smodlaka, and R. Eugene Turner. Washington, DC: American Geophysical Union, 1999.

Wrobel, David, and Claudia Mills. *Pacific Coast Pelagic Invertebrates: A Guide to the Common Gelatinous Animals*. Monterey, CA: Monterey Bay Aquarium, 1998.

Periodicals

Brodeur, Richard D., Claudia E. Mills, James E. Overland, Gary E. Walters, and James D. Schumacher. "Evidence for a Substantial Increase in Gelatinous Zooplankton in the Bering Sea, with Possible Links to Climate Change." *Fisheries Oceanography* 8, no. 4 (1999): 296–306.

Kramp, P. L. "A Synopsis of the Medusae of the World." *Journal of the Marine Biological Association of the United Kingdom* 40 (1961): 1–469.

Larson, Ronald J. "Feeding in Coronate Medusae (Class Scyphozoa, Order Coronatae)." *Marine Behavior and Physiology* 6 (1979): 123–129.

———. "Scyphomedusae and Cubomedusae from the Eastern Pacific." *Bulletin of Marine Science* 47 (1990): 546–556.

———. "Diet, Prey Selection and Daily Ration of *Stomolophus meleagris*, a Filter-Feeding Scyphomedusa from the NE Gulf of Mexico." *Estuarine Coastal and Shelf Science* 32 (1991): 511–525.

———. "Riding Langmuir Circulations and Swimming in Circles: A Novel Clustering Behavior by the Scyphomedusa *Linuche unguiculata*." *Marine Biology* 112 (1992): 229–235.

Mills, Claudia E. "Jellyfish Blooms: Are Populations Increasing Globally in Response to Changing Ocean Conditions?" *Hydrobiologia* 451 (2001): 55–68.

Purcell, Jennifer E. "Predation on Zooplankton by Large Jellyfish (*Aurelia labiata, Cyanea capillata, Aequorea aequorea*) in Prince William Sound, Alaska, USA." *Marine Ecology Progress Series* 246 (2003): 137–152.

Purcell, Jennifer E., and Mary N. Arai. "Interactions of Pelagic Cnidarians and Ctenophores with Fishes: A Review." *Hydrobiologia* 451 (2001): 27–44.

Other

Gershwin, Lisa. "Medusozoa Home Page." [June 13, 2003]. <http://www.medusozoa.com>.

Jellies and Other Ocean Drifters. Video. Monterey Bay Aquarium, 1996.

"The Jellies Zone." [June 13, 2003]. <http://jellieszone.com>.

Ocean Drifters. Video. National Geographic Explorer, 1993.

Mills, Claudia E. Home Page. June 10, 2003 [June 13, 2003]. <http://faculty.washington.edu/cemills>.

Monterey Bay Aquarium. [June 13, 2003]. <http://mbayaq.org>.

The Shape of Life: Life on the Move: Cnidarians. Video. Sea Studios, Monterey, CA, 2001.

Jennifer E. Purcell, PhD

Ctenophora
(Comb jellies)

Phylum Ctenophora

Number of families 20

Thumbnail description
Primarily pelagic animals, and the largest organisms that use cilary propulsion for their main locomotory mode

Photo: Ctenophores "bloom" in surface waters of Apo Reef, Mindoro Strait, Philippine Islands. Lobate ctenophore (top), cestid ctenophore (*Cestum veneris*), "Venus girdle," about 20 in (50 cm) long (bottom). (Photo by ©Gregory G. Dimijian, M. D./Photo Researchers, Inc. Reproduced by permission.)

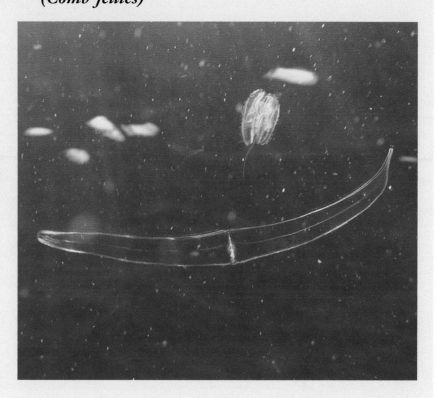

Evolution and systematics

Determination of phylogenetic relationships within the phylum Ctenophora and with other phyla is a difficult task. Although the Ctenophora have been closely aligned with Cnidaria, numerous alternative schemes exist. Ctenophores have been linked with several phyla (Cnidaria, Platyhelminthes, Porifera, and Echinodermata), but molecular evidence supports them as a unique phylum linked most closely to the Cnidarians (based on ribosomal RNA analysis). Ctenophores were once classified as coelenterates based on the presence of nematocysts in one species (*Haeckelia rubra*), which were later found to be kleptocnidae. There are currently two classes (Cydippida and Nuda) in this phylum, seven orders, including Lobata, Cydippida, Cestida, Thalassocalycida, Ganeshida, Platyctenida, and Beroida, 20 families, and 100–150 described species. There have been several new species described since 1995 and there are many more awaiting descriptions. There are two recognized fossil species found dating from the Devonian age (417–354 million years ago) that are recognizable as cydippids.

Physical characteristics

Ctenophores, or "comb bearers," are named for the characteristic eight rows of macrociliary plates that they all possess at some point during their life. The body consists of two tissue layers, the endodermis and the ectodermis, which en-

close a poorly differentiated gelatinous layer of acellular mesoglea. Ctenophores have a statocyst that consists of a pit of modified ciliated epithelium containing a calcareous statolith. This statocyst is mechanically coupled to the ctene rows by four pairs of ciliary tracts and four sets of balancer cilia. The subectodermal nerve net is associated with this statocyst. Ctenophores lack the epithelial-muscle cells characteristic of cnidarians, but do have both smooth and striated muscle. The gastro-vascular system does have two openings, however secretions of material out of the anal pores is rare, with defecation of undigested material occurring primarily through the mouth. There is a complex network of circulatory canals that often form the basis for ctenophore taxonomy. The phylum Ctenophora is unique in the possession of specialized adhesive structures called colloblasts. These organelles are utilized to capture prey in an analogous fashion to the nematocysts in the Cnidaria, but the colloblast and nematocyst morphologies are very different. The atentaculate ctenophores lack colloblasts.

Distribution

Ctenophores are exclusively marine and can be found in all of the world's oceans, from the poles to the equator and from the surface to as deep as 23,950 ft (7,300 m) in the ocean. It is likely that they occur even deeper, but researchers have not been able to spend much time below this depth.

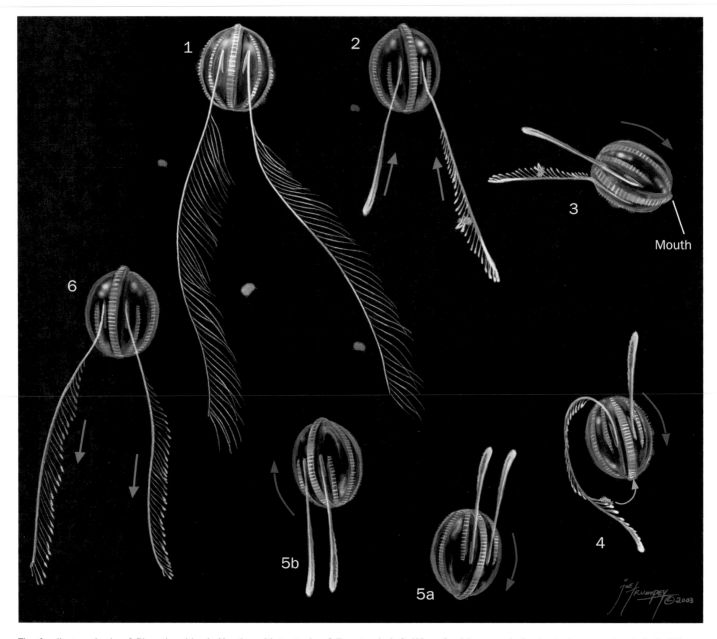

The feeding methods of *Pleurobrachia*. 1. Hunting with tentacles fully extended; 2. When food is snared, the tentacles are retracted; 3. Cilia on ctene rows pulsate, allowing the animal to rotate; 4. Tentacle with prey is relaxed and scraped over mouth; 5. The animal reorients itself in the water column; 6. Tentacles are extended to hunt again. (Illustration by Joseph E. Trumpey)

Habitat

The majority of ctenophores are pelagic species, free swimming in the water column. They can be found in patches or as individuals. One order (Platyctenida) is benthic in nature. Species in this order lose their ctene rows as they settle onto the bottom, but retain their tentacles for feeding. At least one genus is parasitic (*Lampea*), feeding on salps by locating them with tentacles and then settling down and feeding on the salp with the oral region pressed against them.

Behavior

Ctenophores are predators either actively foraging or serving as an ambush predator. Ctenophores are capable of a wide suite of complex behavioral patterns that belie their apparent structural simplicity. Water flow patterns around a number of ctenophore species have been observed in the field and in the laboratory, with the findings that ctenophores manipulate water flow in a manner that can enhance locomotion (jet propulsion and "flying"), feeding ability (capture and handling of prey), and the ability to escape from predators.

Feeding ecology and diet

Ctenophores are predators feeding on a wide variety of prey items, including but not limited to jellies, copepods, ctenophores, mollusks, fish larvae, and salps. Some ctenophores exhibit prey selectivity, feeding along thermoclines or along the ocean bottom. Metabolic rates vary among ctenophores, with respiratory and excretory rates a direct linear function of animal weight (ash-free dry weight) and these rates are very temperature sensitive. The differences in metabolic rates between genera are exemplified by the basic metabolic requirement of 1.5 mg carbon (C) per day for *Beroe*, 260 mg C per day for *Bolinopsis*, and 2,600 mg C per day for *Mnemiopsis*.

Reproductive biology

Ctenophores do not possess an alternation in morphologies (as seen in most scyphozoans and many hydrozoans) nor has the formation of a colonial structure ever been documented. Reproduction is primarily external, with the majority of ctenophores being simultaneous hermaphrodites. Pelagic ctenophores seem to be able to self-fertilize with close to 100% normal development achieved in the laboratory, although the sequential release of sperm and eggs and the mass spawning behavior that is often noted may reduce the percentage of self- fertilization in the field. Protandry has been documented for some platyctenids. Platyctene ctenophores are also unusual in that internal fertilization, brooding of embryos, and asexual reproduction by laceration is common.

Hermaphroditism is postulated as the ancestral state for the phylum and one genus is known to be dioecious, *Ocyropsis*. Breeding season for most ctenophores is year-round, with spawning peaks in the spring and the summer. Ctenophores may be utilizing photoreceptor cells for spawning cues, with different species spawning in response to different light/dark regimes. Other factors affecting gametogenesis are nutrients and possibly endocrine secretions. Diisogamy or paedogenesis has been noted for *Pleurobrachia*, *Bolinopsis*, *Dryodora*, and *Mnemiopsis*.

Conservation status

No species of ctenophore is listed by the IUCN.

Significance to humans

Most ctenophores have very little effect on humans, but the introduction of *Mnemiopsis leidyi* into the Black Sea during the 1970s (presumably from ballast water discharge) has been implicated in the collapse of fisheries in that area. Since then, *Beroe ovata* has also been introduced into the area and appears to be controlling *Mnemiopsis leidyi* populations.

1. Venus's girdle (*Cestum veneris*); 2. Bloody belly (*Lampocteis cruentiventer*); 3. Sea gooseberry (*Pleurobrachia bachei*); 4. *Beroe forskalii*; 5. Sea walnut (*Mnemiopsis leidyi*); 6. *Vallicula multiformis*; 7. *Thalassocalyce inconstans*. (Illustration by Joseph E. Trumpey)

Species accounts

No common name
Beroe forskalii

ORDER
Beroida

FAMILY
Beroidae

TAXONOMY
Beroe forskalii Milne Edwards, 1841.

OTHER COMMON NAMES
None known.

PHYSICAL CHARACTERISTICS
Flat and triangular in shape with the aboral end at the apex of the triangle and the oral region being the base. There are eight rows of cilia running the length of the body and numerous canals under and between the ctene rows.

DISTRIBUTION
Worldwide distribution. A related species (*Beroe ovata*) was recently introduced into the Black Sea and Sea of Azov.

HABITAT
Open ocean and near shore, from the surface waters to depths of 1,640 ft (500 m).

BEHAVIOR
An active predator, foraging often includes a spiral swimming pattern. Once food is located, members of this family use specially modified macrocilia to manipulate and cut prey items. These macrocilia are very distinct in morphology and, for *Beroe forskalii*, these macrocilia are arranged in stripes inside the oral region.

FEEDING ECOLOGY AND DIET
Primary food is other ctenophores.

REPRODUCTIVE BIOLOGY
Hermaphroditic.

CONSERVATION STATUS
Not listed by the IUCN.

SIGNIFICANCE TO HUMANS
None known. ◆

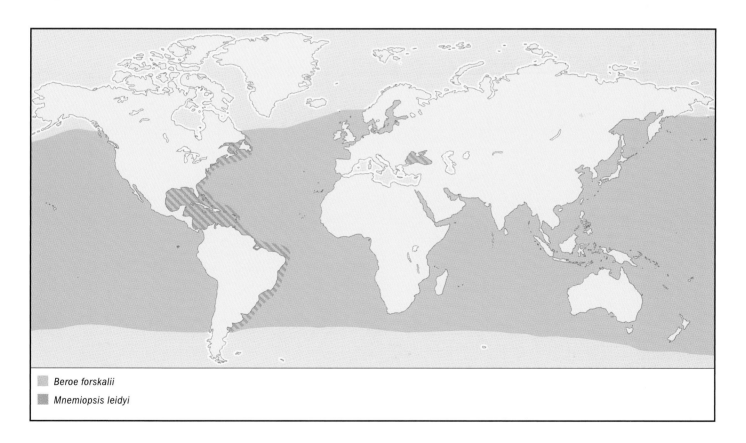

▨ *Beroe forskalii*

▨ *Mnemiopsis leidyi*

Sea walnut
Mnemiopsis leidyi

ORDER
Cydippida

FAMILY
Bolinopsidae

TAXONOMY
Mnemiopsis leidyi A. Agassiz, 1865.

OTHER COMMON NAMES
None known.

PHYSICAL CHARACTERISTICS
Lobate ctenophore reach 3.9 in (10 cm) in length and with four deep lateral furrows from the edges of the oral lobes to the statocyst. The oral lobes are smooth and elongate, as are the auricles.

DISTRIBUTION
West coast of the Atlantic Ocean and, more recently, the Sea of Azov and the Black Sea.

HABITAT
Shallow waters, near shore, and in bays and estuaries.

BEHAVIOR
There are two distinct feeding behaviors: either swimming (oral end first) or being stationary (with the oral end down). The majority of time is spent actively swimming vertically, with the latter behavior being seen primarily in the lab and likely being an energy-conserving behavior.

FEEDING ECOLOGY AND DIET
Feeds on barnacle nauplii, small zooplankters, copepods, and both fish eggs and larvae.

REPRODUCTIVE BIOLOGY
Nothing is known.

CONSERVATION STATUS
Not listed by the IUCN.

SIGNIFICANCE TO HUMANS
None known. ◆

Venus's girdle
Cestum veneris

ORDER
Cydippida

FAMILY
Cestidae

TAXONOMY
Cestum veneris Lesueur, 1813 (= *Cestus veneris* Chun, 1879 and 1880).

OTHER COMMON NAMES
None known.

☐ *Cestum veneris*
☐ *Vallicula multiformis*

PHYSICAL CHARACTERISTICS
Ribbon-shaped, reaching lengths of 4.9 ft (1.5 m) but only 3.1 in (8 cm) in width. The ctene rows are all on one side of this ribbon, with the mouth on the other.

DISTRIBUTION
Atlantic, Pacific, Antarctic, and Mediterranean waters.

HABITAT
Surface waters.

BEHAVIOR
Has an escape behavior that consists of a snakelike undulation of the long body enabling the ctenophore to move several body lengths in seconds.

FEEDING ECOLOGY AND DIET
Swims horizontally in the water column, moving 3.2–6.5 ft (1–2 m) before moving vertically (up or down 1.9–3.9 in [5–10 cm]) and reversing direction. This behavior results in the cestid retracing its original path offset by 1.9–3.9 in (5-10 cm) above or below. Prey capture is on the tentacles lying over the body and the ctenes generate small vortices that may enhance prey movement and capture as the cestid moves back and forth through the water. Prey includes copepods and small mollusks.

REPRODUCTIVE BIOLOGY
Eggs and sperm develop in the meridional canals in a similar fashion as most other ctenophores.

CONSERVATION STATUS
Not listed by the IUCN.

SIGNIFICANCE TO HUMANS
None known. ◆

No common name
Vallicula multiformis

ORDER
Cydippida

FAMILY
Coeloplanidae

TAXONOMY
Vallicula multiformis Rankin, 1956.

OTHER COMMON NAMES
None known.

PHYSICAL CHARACTERISTICS
Nothing is known.

DISTRIBUTION
First described from Jamaica; has been found in warm shallow waters around the Pacific.

HABITAT
Living on mangroves or tunicates in Jamaica; often found on manmade objects such as piers. Also found on sea grasses (and often exported to aquaria accidentally).

BEHAVIOR
Glides over the substrate, stopping and forming small peaks through which the tentacles rise into the waters to forage for food. Most benthic ctenophores are highly substrate selective, but *Vallicula* is more of a generalist.

FEEDING ECOLOGY AND DIET
Copepods and larval decapods have been observed being caught and ingested. Diatoms have also been found within the stomach pouches. Unwanted food is wrapped up in mucus and ejected.

REPRODUCTIVE BIOLOGY
Viviparous, shedding embryos that are almost competent to settle.

CONSERVATION STATUS
Not listed by the IUCN.

SIGNIFICANCE TO HUMANS
None known. ◆

Bloody belly
Lampocteis cruentiventer

ORDER
Cydippida

FAMILY
Lampocteidae

TAXONOMY
Lampocteis cruentiventer Harbison, Matsumoto, and Robison, 2001. Also known as *Lampoctena sanguineventer*.

OTHER COMMON NAMES
None known.

PHYSICAL CHARACTERISTICS
In the order Lobata, but differs from other families by the presence of a deep notch between adjacent subtentacular ctene rows and by the blind aboral ending for all meridional canals. The gut is always a deep red color while the body coloration varies from clear to deep purple (and all shades in between).

DISTRIBUTION
Currently known only from the east coast of the Pacific, however it is believed to have a much wider distribution in the Pacific and in other oceans.

HABITAT
Inhabits the deep sea from 984–3,320 ft (300–1,012 m).

BEHAVIOR
Nothing is known.

FEEDING ECOLOGY AND DIET
Nothing is known.

REPRODUCTIVE BIOLOGY
Nothing is known.

CONSERVATION STATUS
Not listed by the IUCN.

SIGNIFICANCE TO HUMANS
None known. ◆

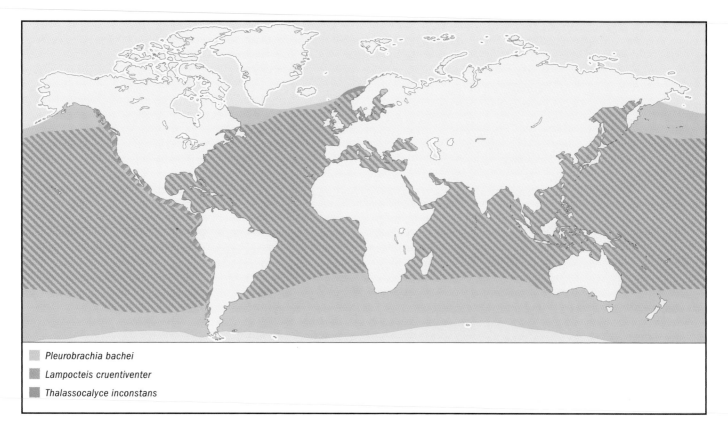

Pleurobrachia bachei
Lampocteis cruentiventer
Thalassocalyce inconstans

Sea gooseberry
Pleurobrachia bachei

ORDER
Cydippida

FAMILY
Pleurobrachiidae

TAXONOMY
Pleurobrachia bachei A. Agassiz, 1860.

OTHER COMMON NAMES
None known.

PHYSICAL CHARACTERISTICS
Globular in shape with two tentacles (each of which have secondary tentacles). The tentacle sheaths are found at a distance from the infundibulum rather than next to it. Of particular note is that species in this family do not appear to have bioluminescent capabilities (unlike other ctenophores).

DISTRIBUTION
Worldwide distribution.

HABITAT
Surface waters both near shore and open ocean.

BEHAVIOR
Has the oral end at the opposite end to where the tentacles originate. Due to this, it exhibits a stereotypical feeding behavior consisting of contracting the tentacle with prey and rapidly spinning around, wrapping the tentacle around the body, and eventually swiping the tentacle across the mouth region and ingesting the prey.

FEEDING ECOLOGY AND DIET
Predators on small crustaceans, it exhibits some prey selectivity. Patchy in distribution and have the potential to strip a water column of small zooplankton.

REPRODUCTIVE BIOLOGY
Hermaphroditic, with direct development.

CONSERVATION STATUS
Not listed by the IUCN.

SIGNIFICANCE TO HUMANS
None known. ◆

No common name
Thalassocalyce inconstans

ORDER
Cydippida

FAMILY
Thalassocalycidae

TAXONOMY
Thalassocalyce inconstans Madin and Harbison, 1978.

OTHER COMMON NAMES
None known.

PHYSICAL CHARACTERISTICS
Ctenophore with tentacles that have lateral filaments, a medusa-like body, short ctene rows, and a mouth at the apex of a central conical peduncle. The order Thalassocalycida has one family, one genus, and one species currently described. It

is morphologically distinct from the Cydippida and the Lobata, appearing to be an intermediate step between these two orders.

DISTRIBUTION
Originally described from the slope waters off New England; found in other oceans as well. It is more of an open-ocean species, its relative fragility making it too delicate to survive near shore.

HABITAT
Found from the surface waters down to 9,070 ft (2,765 m).

BEHAVIOR
Likely an ambush predator, waiting for and capturing prey within its bell. When startled, it rapidly squeezes the bell, forc-ing water out and propelling the animal backwards. This escape behavior is limited to that single clap and distances traveled are 1–2 body lengths.

FEEDING ECOLOGY AND DIET
Copepods.

REPRODUCTIVE BIOLOGY
Presumably hermaphroditic.

CONSERVATION STATUS
Not listed by the IUCN.

SIGNIFICANCE TO HUMANS
None known. ◆

Resources

Books

Harbison, G. R. "On the Classification and Evolution of the Ctenophora." In *The Origins and Relationships of Lower Invertebrates*, edited by S. C. Morris, J. D. George, R. Gibson, and H. M. Platt. Oxford: Oxford University Press, 1985.

Matsumoto, G. I. *Phylum Ctenophora (Orders Lobata, Cestida, Beroida, Cydippida, and Thalassocalycida): Functional Morphology, Locomotion, and Natural History.* Los Angeles: University of California, Los Angeles, PhD Dissertation, 1990.

Periodicals

Harbison, G. R., G. I. Matsumoto, and B. H. Robison. "*Lampocteis cruentiventer* gen. nov., sp. nov.: A New Mesopelagic Lobate Ctenophore, Representing the Type of a New Family (Class Tentaculata, Order Lobata, Family Lampoctenidae, fam. nov.)." *Bulletin of Marine Science*, 68(2) (2001): 299–311.

Madin, L. P., and G. R. Harbison. "*Thalassocalyce inconstans*, New Genus and Species, an Enigmatic Ctenophore Representing a New Family and Order." *Bulletin of Marine Science* 28(4) (1978): 680–687.

Rankin, J. J. "The Structure and Biology of *Vallicula multiformis*, gen. et sp. nov., a Platyctenid Ctenophore." *Journal of the Linnean Society*, 43 (1956): 55–71.

Stanley, G. D., and W. Sturmer. "A New Fossil Ctenophore Discovered by X-rays." *Nature* Vol. 327, No. 6125 (1987): 61–63.

Organizations

Monterey Bay Aquarium Research Institute. 7700 Sandholdt Road, Moss Landing, CA 95039 Phone: (831) 775-1700. Fax: (831) 775-1620. E-mail: mage@mbari.org Web site: <http://www.mbari.org>

Other

Mills, C. E. "Phylum Ctenophora: List of All Valid Species Names." March 1998 [June 10, 2003]. <http://faculty.washington.edu/cemills/Ctenolist.html>.

George I. Matsumoto, PhD

Acoela
(Acoels)

Phylum Platyhelminthes

Class Acoela

Number of families 20

Thumbnail description
Tiny marine wormlike metazoans; the most primitive living animals with bilateral symmetry

Photo: A close-up of acoels (*Waminoa* sp.) on soft corals, Madang, Papua New Guinea. (Photo by A. Flowers & L. Newman. Reproduced by permission.)

Evolution and systematics

The name Acoela comes from two Greek words that mean "without a body cavity"; it refers to a distinguishing feature of this order (or phylum) of tiny wormlike multicellular marine invertebrates. Species in this group have no true body cavity or *coelom*. A true coelom is a fluid-filled body cavity formed from mesodermal tissue. It lies between the outer body wall of epidermal tissue and the gut or digestive tract.

The Acoela have long been included as an order within the phylum Platyhelminthes, the flatworms, and the class Turbellaria, the marine flatworms. On the other hand, very recent morphological, developmental, and molecular studies, including comparisons of 18S ribosomal DNA from 61 species representing 25 animal phyla, support the hypothesis that the Acoela are a separate phylum representing the most primitive living animals showing bilateral symmetry, or regularity of body form on the right and left sides (Ruiz-Trillo et al.). Bilateral symmetry is a basic characteristic of all *triploblastic* animals, those with three tissue layers in their embryonic stages: the outer ectoderm, middle mesoderm or mesenchyme, and inner endoderm. In contrast to this pattern, *diploblastic* animals have only two layers of tissue as embryos, lacking the mesodermal layer.

The Acoela may be direct descendants of the earliest line of animals to diverge from diploblastic organisms with the beginnings of triploblastic features: a middle tissue layer and bilateral symmetry. The DNA studies suggest that the acoels or their direct ancestors diverged from the diploblasts much earlier than did the main line of triploblastic animals (including all living triploblasts other than Acoela), in pre-Cambrian times, well before the so-called Cambrian Explosion era of

540–500 million years ago, when most if not all modern animal phyla appeared.

The work of Ruiz-Trillo and his colleagues has been challenged by Tyler and his working group, who maintain that the order Acoela should be considered the earliest or most primitive group within phylum Platyhelminthes. Lundin considers the Acoela to have likely evolved from ancestors within or related to the Nemertodermatida.

Nevertheless, other features of the Acoela that support their being a phylum unto themselves include their simple nervous system, as compared with platyhelminth species, and their manner of embryonic development. According to a study by Raikova et al., which compared the brains of acoels with those of other platyhelminths, the acoel brain and nervous system are simpler and much different in structure from those of other platyhelminths, suggesting major differences between the two groups and supporting the classification of Acoela as a separate phylum.

As of 2003, Acoela is included as an order together with Nemertodermatida in the species complex Acoelomorpha within the phylum Platyhelminthes on the basis of similarities between Acoela and Nemertodermatida. According to the European Register of Marine Species, there are 20 families within the order Acoela.

Physical characteristics

Acoels are tiny; the members of most species are no longer than 0.078 in (2 mm), although *Convolutriloba retrogemma* can reach lengths of 0.23–0.28 in (6–7 mm) and *Convoluta roscoffensis* can grow up to 0.59 in (15 mm) long. The bodies

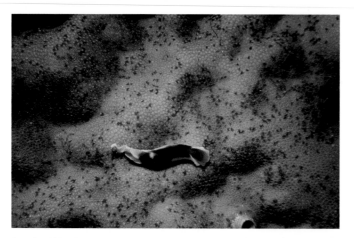

Acoels (*Waminoa* sp.) on hard coral, being sucked up by the sea slug (*Chelidonura electra*), Coral Bay, WA. (Photo by A. Flowers & L. Newman. Reproduced by permission.)

of acoels may be either oval or cylindrical in shape, and flattened dorsoventrally.

Acoels have either a simple pharynx or none at all; the pharynx or mouth is situated on the ventral (lower) surface. They have no digestive tract (gut), no protonephridia (primitive kidneys for excretion and osmotic balance), and no distinct gonads. The brain is quite simple, unlike the more complex bilobed brain found in most platyhelminth species. The acoel nervous system is a loose net of nerve fibers strung throughout the body. Most species also have simple eyes known as ocelli. Individuals of nearly all species carry a statocyst, a tiny, spherical organ for balance and orientation.

Distribution

Acoela are found worldwide in mostly shallow waters in all oceans from tropical to polar regions.

Habitat

Acoela live in marine or brackish water. Some species drift or swim in the open sea, or live unobtrusively among sand grains on sea bottoms in shallow coastal waters. Some even live in cold brine channels within Antarctic ice floes.

Behavior

Acoela are free-living, either planktonic (swimming or drifting) or interstitial (living between sand grains on the sea bottom). A few are commensals on other invertebrates. They move by means of cilia, or tiny hairlike projections, covering the entire outer side of their epidermis. The species that live among sand grains reportedly behave much as do ciliated protozoans, particularly *Paramecium* and *Opalina*, which share space with the Acoela.

Feeding ecology and diet

Acoels have no true gut, but have instead a *digestive syncytium*, a simple, not formed inner cavity. The digestive syncytium is not formed from mesodermal tissue as is a true coelom. Various acoel species feed on algae, microorganisms and detritus, ingesting food through a simple pharynx or even simpler mouth located on the ventral surface. Some carry endosymbiotic algae within their epidermis and absorb nourishment manufactured by the algae.

The mouth or pharynx of an acoel leads to a packed or loose mass of endodermal cells that serves as the digestive organ. Food particles ingested by the animal are absorbed and digested by individual cells in the endodermal mass. This mode of digestion is known as phagocytosis.

Reproductive biology

Acoels reproduce sexually, although some species also reproduce asexually by fission (asexual reproduction). Acoela have no distinct gonads; gametes are formed directly from the mesenchyme, or middle tissue layer. Acoel spermatozoa bear two flagella on each sperm cell, another primitive feature.

Acoels differ from other bilaterally symmetrical animals in the way in which their embryonic cells divide during development. A fertilized acoel egg divides once; then the two resulting cells subdivide further into many smaller cells. This pattern contrasts with the eggs of all other bilateral animals, each of which divides first into two and then into four cells that go on to divide into many smaller cells. According to Henry et al., the acoel pattern of embryonic development supports the hypothesis that the acoels branched off from the ancestral line of bilateral animals very early, and may thus represent an earlier evolutionary experiment in body structure.

Some adult acoela reproduce asexually by fission in one of three possible ways: *architomy*, in which smaller pieces separate from the maternal animal prior to organ differentiation; *paratomy*, or transverse fission, in which organ differentiation takes place within the fragments before separation; and *budding*, or local tissue reorganization in which a small bud or outgrowth detaches itself from the parent organism and lives independently.

Conservation status

Acoels are quite obscure yet so widespread and numerous that no species are considered threatened.

Significance to humans

Acoels have no known direct significance to humans as of 2003. Their main value to humanity is scientific, in that studies of them may shed light on the earliest evolution of bilateral triploblastic animals.

Species accounts

No common name
Convoluta roscoffensis

FAMILY
Convolutidae

TAXONOMY
Convoluta roscoffensis
Graff, 1891.

OTHER COMMON NAMES
None known.

PHYSICAL CHARACTERISTICS
Individuals of *Convoluta roscoffensis* are oval wormlike creatures that may be as much as 0.59 in (15 mm) long and have a characteristic green color from the inclusion of numerous individuals of the photosynthetic alga Tetraselmis convolutae in their tissues. There may be as

Convoluta roscoffensis

many as 25,000 individual algae living within an individual of *C. roscoffensis*.

DISTRIBUTION
Southern coast of the English Channel.

HABITAT
Shallow marine sediments and tidal pools.

BEHAVIOR
Adult individuals of *C. roscoffensis* congregate in temporary tidal pools, swarming near the water surface to allow maximal photosynthesis to take place in their onboard alga. These acoels may be present in the summer in such numbers that they turn the water green. When the tide returns, the acoels are swept back out to sea.

FEEDING ECOLOGY AND DIET
Juvenile individuals of *C. roscoffensis* take in but do not digest individuals of the alga *Tetraselmis convolutae*. At maturity, the pharynx and mouth disappear while the alga distributes itself throughout the worm's body. The worm is then entirely dependent upon the alga for its nourishment. *Tetraselmis* supplies the acoel with sugars and oxygen while the acoel donates nitrogen wastes to the alga.

REPRODUCTIVE BIOLOGY
Reproduces sexually by mating between male and female individuals (dermal impregnation). The eggs are fertilized internally and then released.

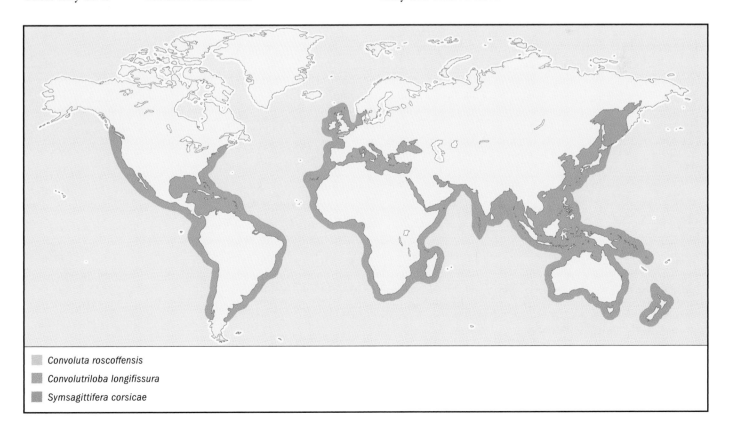

Convoluta roscoffensis
Convolutriloba longifissura
Symsagittifera corsicae

CONSERVATION STATUS
Not threatened.

SIGNIFICANCE TO HUMANS
None known. ◆

No common name
Convolutriloba longifissura

FAMILY
Sagittiferidae

TAXONOMY
Convolutriloba longifissura Bartolomaeus and Balzer, 1997.

OTHER COMMON NAMES
None known.

PHYSICAL CHARACTERISTICS
Oval wormlike creatures, colored green by the presence of symbiotic algae in their tissues. *Convolutriloba longifissura*, like other members of Sagittiferidae, bears sagittocysts, which are tiny hooklike projections that form in and arise from the epidermis. Sagittocysts are used for defense and capturing prey. *Convolutriloba longifissura* carries its sagittocysts on its dorsal (upper) surface.

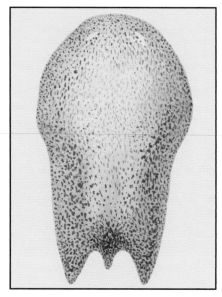
Convolutriloba longifissura

DISTRIBUTION
Widespread.

HABITAT
Shallow marine sand beds.

BEHAVIOR
Nothing is known.

FEEDING ECOLOGY AND DIET
Ingests smaller organisms, while carrying symbiotic *Tetraselmis* algae within its epidermis as well as internally. Unlike *Convoluta roscoffensis*, however, *Convolutriloba longifissura* regularly digests individual members of its onboard complement of algae.

REPRODUCTIVE BIOLOGY
Reproduces sexually by mating between male and female individuals (dermal impregnation). The eggs are fertilized internally and then released. It also reproduces asexually by fission. In fission, the hindmost fourth of the maternal animal separates itself in a transverse direction and drops away. The fragment divides longitudinally and the new individuals form eyes and mouths over a period of 2–3 days. Meanwhile, the maternal animal regrows the lost section and repeats the fissioning process, thus launching a new group of offspring every four days.

CONSERVATION STATUS
Not threatened.

SIGNIFICANCE TO HUMANS
None known. ◆

No common name
Symsagittifera corsicae

FAMILY
Sagittiferidae

TAXONOMY
Symsagittifera corsicae Gschwentner, 2000, Corsica, Mediterranean Sea.

OTHER COMMON NAMES
None known.

PHYSICAL CHARACTERISTICS
Like other members of the Sagittiferidae family, *S. corsicae* bears sagittocysts, which form in and arise from the epidermis, and are used for defense and capturing prey. The needle-shaped sagittocysts are produced in specialized gland cells called sagittocytes, whose roots are surrounded by muscles. The worm contracts the muscles to eject the sagittocysts. *Symsagittifera corsicae* carries its sagittocysts on its rear ventral surface.

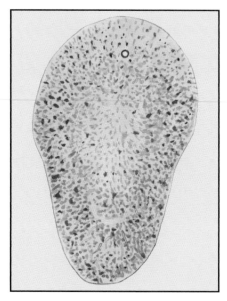
Symsagittifera corsicae

DISTRIBUTION
Marine; off the coastline of Corsica in the Mediterranean Sea.

HABITAT
Shallow marine sediments.

BEHAVIOR
Lives and feeds in shallow marine waters.

FEEDING ECOLOGY AND DIET
Ingests small prey animals, while also carrying symbiotic *Tetraselmis* algae.

REPRODUCTIVE BIOLOGY
Reproduces sexually, via male-female mating and internal fertilization of eggs, which are then released.

CONSERVATION STATUS
Not threatened.

SIGNIFICANCE TO HUMANS
None known. ◆

Resources

Books

Balzer, I. "Symbiotic Association Between the Acoel *Convolutriloba longifissura* and the Alga *Tetraselmis* sp." In *Endocytobiology VII*, edited by E. Wagner, J. Normann, H. Greppin, J. H. P. Hackstein, R. G. Hermann, K. V. Kowallik, H. E. A. Schenk, and J. Seckbach. Geneva: Geneva University Press, 1999.

Caira, Janine N., and D. Timothy J. Littlewood. "Worms, Platyhelminthes." In *Encyclopedia of Biodiversity*, vol. 5, edited by S. Levin. San Diego: Academic Press, 2001.

Littlewood, D. T. J., and R. A. Bray, eds. *Interrelationships of the Platyhelminthes*. London: Taylor and Francis, 2001.

Periodicals

Akesson, B., R. Gschwentner, J. Hendelberg, P. Ladurner, J. Müller, and R. Rieger. "Fission in *Convolutriloba longifissura*: Asexual Reproduction in Acoelous Turbellarians Revisited." *Acta Zoologica* 82, no. 3 (2001): 231–239.

Balzer, I. "Symbiotic Association Between the Plathelminth *Convolutrilova longifissura* and the Alga *Tetraselmis* sp." *Nova Hedwigia* 112 (1996): 461–475.

Bartolomäus, Thomas, and Ivonne Balzer. "*Convolutriloba longifissura* nov. spec. (Acoela)— The First Case of Longitudinal Fission in Plathelminthes." *Microfauna Marina* 11 (1997): 7–118.

Gschwentner, Robert, Sanja Baric, and Reinhard Rieger. "New Model for the Formation and Function of Sagittocysts: *Symsagittifera corsicae* n. sp. (Acoela)." *Invertebrate Biology* 121, no. 2 (1999): 95–103.

Gschwentner, Robert, Peter Ladurner, Willi Salvenmoser, Reinhard Rieger, and Seth Tyler. "Fine Structure and Evolutionary Significance of Sagittocysts of *Convolutriloba longifissura* (Acoela, Platyhelminthes)." *Invertebrate Biology* 118, no. 4 (1999): 332–345.

Henry, J. Q., M. Q. Martindale, and B. C. Boyer. "The Unique Developmental Program of the Acoel Flatworm, *Neochildia fusca*." *Developmental Biology* 220 (2000): 285–295.

Janssen, Hans Heinrich, and Rolf Gradinger. "Turbellaria (Archoophora: Acoela) from Antarctic Sea Ice Endofauna— Examination of their Micromorphology." *Polar Biology* 21, no. 6 (1999): 410–416.

Raikova, Olga, M. Reuter, Elena Kotikova, and Margaretha K. S. Gustafsson. "A Commissural Brain! The Pattern of 5-HT Immunoreactivity in Acoela (Plathelminthes)." *Zoomorphology* 118, no. 2 (1998): 69–77.

Ruiz-Trillo, I., M. Riutort, T. J. Littlewood, E. A. Herniou, and J. Baguna. "Acoel Flatworms: Earliest Extant Bilaterian Metazoans, Not Members of Platyhelminthes." *Science* 283 (1999): 1919–1923.

Other

"Acoel Flatworms Misrepresented?" University of Maine, Department of Biological Sciences. 1999 (July 15, 2003). <www.umesci.maine.edu/Science/letter.htm>.

"Order Acoela." *Alien Travel Guide* (July 15, 2003). <www.alientravelguide.com/science/biology/life/animals/platyhel/turbella/acoela/>.

"Project: Morphology and Phylogeny of the Acoelomorpha (Platyhelminthes)." Göteborgs Universitets Marina Forskningscentrum. <www.gmf.gu.se/old_english_gmf/Researchers/KennetLundin.html>.

Kevin F. Fitzgerald, BS

Turbellaria
(Free-living flatworms)

Phylum Platyhelminthes

Class Turbellaria

Number of families 102

Thumbnail description
Mostly free-living flatworms with a cellular epidermis that is usually ciliated; the mouth leads to a stomodeal pharynx and incomplete gut

Photo: A flatworm (*Pseudoceros bimarginatus*) showing warning coloration. (Photo by ©A. Flowers & L. Newman/Photo Researchers, Inc. Reproduced by permission.)

Evolution and systematics

The phylogeny of the platyhelminth classes is not clear. Recent morphological and molecular studies have generated numerous hypotheses as to their relation to each other and to other phyla. Traditionally, the class Turbellaria was thought to be the basal ancestor of the parasitic classes (Trematoda, Cestoda, Monogenea) within the phylum. However, some researchers believe that the parasitic classes should be separated into a separate phylum (Neodermata) based on their unique tegument, the neodermis that may be adaptive to a parasitic existence. Both morphological and molecular studies also suggest that the Turbellaria are paraphyletic and that the orders Acoela and Nemertodermatida should be placed into a separate phylum. The Acoela have a primitive nerve net (no brain as in other flatworms), a simple pharynx when present, and a syncytial cellular gut without a cavity, entolecithal ova, and a lack of protonephridia. They may be the closest relatives to the acoeloid ancestor that gave rise to bilateral metazoa. It also has been postulated that the acoelomate condition (no body cavity) of the other platyhelminths may be secondarily derived from more advanced protostomes. The Acoela and Nemertodermatida are considered a separate, distinct taxon.

The ten recognized orders constitute the remaining members of the class Turbellaria. There are more than 4,500 described species within the Acoela and Turbellaria combined; however, many species have yet to be discovered and described. The characteristics of each order is the following:

- Order Catenulida has a simple pharynx and sac-like gut; the mesenchyme is poorly differentiated; ova are enotlecithal. They are elongate forms that occur in freshwater and marine habitats.

- Order Haplopharyngida are small worms with a simple pharynx, the proboscis is simple and ventral to the anterior tip of body (reminiscent of nemereans), anal pore is weakly developed; the brain is encapsulated with two ventral-lateral nerve cords; the ovary is simple without accessory organs; the male pore has a circle of hard straight stylets, anterior to female pore; the are free living and marine. They were once considered to be macrostomids. They contain two species.

- Order Lecithoepitheliata's ordinal status is questionable; the pharynx is complex and somewhat variable in the anterior of the body, the gut is simple; the ova are not entolecithal and surrounded by vitelline cells. There are about 30 species.

- Order Macrostomida have a simple pharynx, gut is a simple sac; the posterior end of the body may be broadened into an adhesive disc; there is no asexual zooid formation; the ovaries are often paired, the

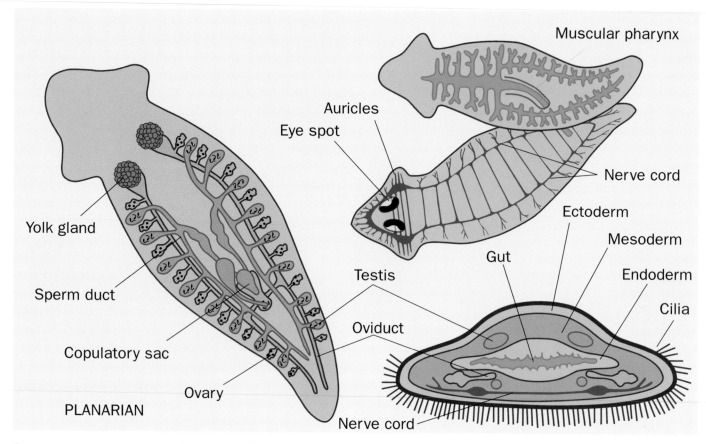

Flatworm anatomy. (Illustration by Christina St. Clair)

eggs entolecithal; the male pore is usually separate. They are mostly small interstitial marine and freshwater forms.

- Order Polycladida have ruffled plicate pharynx, the gut is multibranched with diverticula that may be anastomosing; the ovaries are scattered, with entolecithal ova; testes are scattered follicular, the male pore is usually anterior to the female pore. They are mostly large free-living marine flatworms that may be brightly colored.

- Order Prolecithophora's ordinal status is questionable. They have a plicate or bulbous pharynx, the gut is simple; the ovaries, and testes are follicular or compact, vitellaria is diffuse, eggs are ectolecithal. They are small, free-living or commensal, freshwater and marine forms.

- Order Proplicastomata is similar to the Acoela based on a few specimens; they have an elongate plicate pharynx; no statocysts; entolecithal ova. They are free-living marine forms.

- Order Proseriata is closely related to Tricladida. They have a cylindrical plicate pharynx, a simple gut, small, compact, paired ovaries at the end of the vitelline duct, the vitellaria is arranged along duct, ectolecithal ova. They are free-living marine forms.

- Order Rhabdocoela is a large diverse group with four suborders (Dalyellioida, Typhloplanoida, Kalptorhynchia, and Temnocephalida). They have a bulbous pharynx, a simple sac-like gut; the mesenchyme is fairly open; protonephridea are paired when present; the anterior brain and ventral nerve trunks are usually with cross connections, no statocyst; testes are compact; ovaries separate or joined with vitellaria, ova ectolecithal, a uterus is sometimes present. They have marine, freshwater, and terrestrial forms, many of which are symbiotic.

- Order Tricladida have a cylindrical plicate pharynx posteriorly directed, gut has one anterior and two posterior branches and numerous diverticula; mesenchyme is thick; no statocyst; male and female copulatory structures are complex, posterior to pharynx; follicular testes; one pair of small ovaries is usually anterior, vitellaria is extensive over most of lateral body, ectolecithal ova. They are usually large, flattened, and sometimes elongate worms with marine, freshwater, and terrestrial forms.

Physical characteristics

The class Turbellaria share the following characteristics with other classes within the Platyhelminthes:

- triblobastic (three tissue layers)

- acoelomate (no fluid filled body cavity or coelom)

- bilaterally symmetrical

- dorsoventrally flattened

- spiral cleavage and mesoderm derived from the 4d cell

- complex, incomplete gut (no anus)

- cephalized, with cerebral ganglion (brain) and longitudinal nerve cords that form a ladder-like nervous system

- numerous sense organs at the anterior end of the body, and tactile receptors distributed over the body, especially around the pharynx

- protonephridia that function in excretion and osmoregulation

- no circulatory system, which restricts the size and shape of these animals

- hermaphroditic with complex reproductive system

Turbellarians also are free-living or commensal (not usually parasitic), usually aquatic, and have a stomodeal pharynx. Their cellular epidermis is usually ciliated and contains mucous secreting cells and structures called rhabdoids that can produce copious mucus to prevent desiccation. Most turbellarians also have pigment-cup occelli for detecting light; some have an anterior pair where larger species may posses numerous pairs along the body.

Distribution

Turbellarian species are distributed worldwide, mostly in freshwater and marine environments with a few taxa occurring on land.

Habitat

Many of the minute species occur interstitially between grains of sand in aquatic habitats. Larger species are pelagic (marine) or live among submerged substrates such as rocks, coral, and algae. Many species, especially of the order Rhabodocoela, are symbiotic with various invertebrates and fishes. A few genera of the order Tricladida are terrestrial, living in damp leaf litter and soil.

Behavior

Turbellarians display a number of behaviors that prevent them from straying beyond their normal habitats and allows to them to maintain orientation within those habitats. For instance, most turbellarians are positively thigmotactic (touch) ventrally and negatively thigmotactic dorsally. This allows them to maintain their ventral side against the substrate in benthic forms. In other species where touch may not be the best way to orient to a substrate (such as interstitial and pelagic

A colorful terrestrial flatworm (phylum Platyhelminthes) 8 in (20 cm) long, crawling slowly over leaf litter on the forest floor in lowland Amazon rainforest of northeasten Peru. (Photo by ©Gregory G. Dimijian, M. D./Photo Researchers, Inc. Reproduced by permission.)

forms), they have statocysts so that they can orient to gravity (geotaxis). Most species are also negatively phototactic, which prevents worms from coming out in the daylight where they may get eaten or dry out in the case of terrestrial forms.

All turbellarians have a strong sense of smell that can be used to find food or mates (chemotaxis). Chemosensors are concentrated on each side of the head to help them determine the direction that the chemical trail is coming from. The heads of freshwater species are often expanded into auricles that have sensors. Some species have tentacles and ciliated pits to assist in chemotaxis. *Dugesia* swings its head back and forth to help determine the proper direction of the food source. Other species use trial and error to determine the proper direction to find food. They move in one direction until the signal gets weaker, and then continue switching direction until the signal is strongest. Some species also have been shown to orient to currents in order to find food (rheotaxis).

Feeding ecology and diet

Most turbellarians are carnivorous predators or scavengers. Carnivores feed on organisms that they can fit into their mouths, such as protozoans, copepods, small worms, and minute mollusks. Some species use mucus that may have poisonous or narcotic chemicals to slow or entangle prey. Some have specific diets and feed on sponges, ectoprocts, barnacles, and tunicates. Several species have commensal relationships with various invertebrates and few actually border on being parasitic because they graze on their live hosts. Terrestrial species feed on earthworms and land snails. A few species feed on microalgae that may be incorporated into the body, forming a symbiotic relationship in which the algae supply the worm with carbohydrates and fats and the worm supplies the algae with nitrogen waste products and a safe haven.

The pharynx and gut cells produce digestive enzymes that breakdown food extracellularly. Nutritive cells in the gastrodermis then phagotize partially digested material that is distributed throughout the body. Because these worm lack a circulatory system, larger species have extensive anastomosing guts to aid in distribution. Since these worms have incomplete guts, all waste must pass back out of the mouth.

Reproductive biology

Asexual reproduction is a common method of reproduction in freshwater and terrestrial turbellarians. Many of the triclads divide by transverse fission: the body splits transversely behind the pharynx and each part generates the rest of the body. The posterior portion attaches to the substrate and the anterior portion crawls away until it tears in two. In species such as *Dugesia*, the cells tend to vary in their ability to regenerate. The cells in the middle portion of the body have the strongest ability to regenerate. Experiments have shown that if just the tail is cut off, it will not grow a new body, whereas the main portion of the body will regenerate a new tail. The ability of the fissioned portions of these worms to regenerate the proper half has interested scientists for years in investigating why the head portion grows a tail and why the tail portion regenerates a new head. In the genera *Catenula, Microstomum,* and *Stenostomum* (orders Catenulida and Macrostomida), multiple transverse planes develop that lead to a train of individuals called zooids that do not detach until they reach a certain stage. Other species (e.g., *Phagocata* and some terrestrial species) detach fragments that become encysted and eventually develop into new individuals.

Turbellarians are hermaphroditic and their sexual reproductive systems are quite complicated. The male system may have one, two, or multiple testes that drain via sperm ducts that may lead to a storage area called the seminal vesicle. Prostate glands may be present that produce seminal fluid that mixes with the sperm in the seminal vesicle. The sperm then exits the worm via the protrusible penis or eversible cirrus with help from a muscular ejaculatory duct. The female system is more variable among the Turbellaria, depending on whether they produce entolecithal (Macrostomida and Poly-

cladida) or ectolecithal (Rhabdocoela, Prolecithophora, and Tricladida) ova. A germovitellarium, which may be single or paired, produces entolecithal ova (yolk reserves within ova). A germarium or ovary, which is separate from the vitellaria, produces yolk-free ova that eventually are surrounded by separate yolk cells in a tanned protein capsule to form the ectolecithal egg. Sperm also are included within the egg capsule to insure fertilization. Eggs pass through the oviduct that may be differentiated into a seminal receptacle or uterus before deposition.

Cross fertilization (mating) usually occurs when worms align themselves with each other, and the cirrus or penis of each worm is inserted into the female gonopore or atrium of the other and deposits sperm. The worms then go their own way with the sperm stored in their seminal receptacles. In some species, mating occurs by hypodermic impregnation in which the male copulatory organ penetrates the body wall of the mate and deposits sperm in the mesenchyme. The sperm then make their way to the ova.

Turbellarians have either direct development or produce a pelagic larva. Polyclads often produce a pelagic Muller's larva that settles to the bottom and goes through metamorphosis in a few days. This larva has eight ventrally directed ciliated lobes, which it uses to swim. *Stylochus*, a parasitic polyclad, produces the Gotte's larva, which has only four ciliated lobes.

Conservation status

No turbellarians are considered threatened by the IUCN Red Book.

Significance to humans

The regenerative abilities of *Dugesia* have been studied extensively by scientists to better understand the healing and cell regeneration processes in humans. Several species parasitize commercially important species such as oysters and a few species cause pathological problems in marine ornamental fishes kept in aquaria.

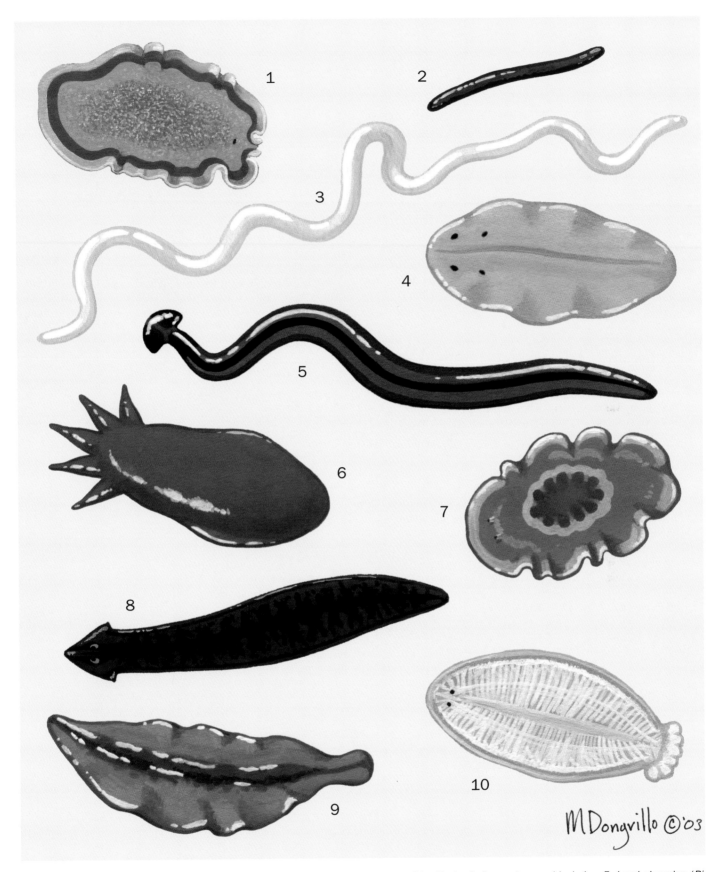

1. *Pseudoceros ferrugineus*; 2. Male *Kronborgia amphipodicola*; 3. Female *K. amphipodicola*; 4. *Paravortex scrobiculariae*; 5. Land planarian (*Bipalium pennsylvanicum*); 6. *Temnocephala chilensis*; 7. Oyster leech (*Stylochus inimicus*); 8. Freshwater planarian (*Dugesia tigrina*); 9. *Notoplana acticola*; 10. *Bdelloura candida*. (Illustration by Marguette Dongvillo)

Species accounts

No common name
Notoplana acticola

ORDER
Polycladida

FAMILY
Leptoplanidae

TAXONOMY
Notoplana acticola (Boone, 1929), Monterey Bay, California, United States.

OTHER COMMON NAMES
None known.

PHYSICAL CHARACTERISTICS
Adults are 1–2.4 in (25–60 mm) long when extended, body usually widest anteriorly, tapering posteriorly; tentacular eyes in rounded clusters with scattered eyes lying anterior, posterior, and sometimes lateral to them; about 25 cerebral eyes occur in an elongate band; eyes consist of single, cup-shaped pigment cell covering 6–10 retinal cells; color tan or pale gray with darker markings along midline.

DISTRIBUTION
Pacific coast of United States.

HABITAT
Rocky intertidal and subtidal zones.

Notoplana acticola
Bipalium pennsylvanicum

BEHAVIOR
Experimental severing of main nerve pathways resulted in rapid repair of nerves and pathways.

FEEDING ECOLOGY AND DIET
Can ingest prey up to half its size. Feeds on limpets (*Collisella digitalis*), small acorn barnacles, and captive worms, and have been observed eating the red nudibranch (*Rostanga pulchra*) in captivity.

REPRODUCTIVE BIOLOGY
Functional hermaphrodites throughout the year with mature sperm and eggs present; a small number of worms were found with only ovaries in the spring and 10–50% of the population only had testes throughout the year; two-thirds of all worms had sperm in their seminal receptacles, indicating that mating had taken place; egg deposition occurs from late spring to early fall.

CONSERVATION STATUS
Not threatened.

SIGNIFICANCE TO HUMANS
None known. ◆

No common name
Pseudoceros ferrugineus

ORDER
Polycladida

FAMILY
Pseudoceridae

TAXONOMY
Pseudoceros ferrugineus Hyman 1959 Palau, Indonesia.

OTHER COMMON NAMES
None known.

PHYSICAL CHARACTERISTICS
Elongate worm 0.7 in long by 0.4 in wide (18 mm long by 11 mm wide); may get larger. Two slightly developed nuchal tentacles. Body is deep red with white flecks or dots centrally located; yellowish band around the margin with deeper shades of purple between margin and central area with flecks. This genus is speciose on coral reefs and colors vary considerably between species ranging from bright pink and orange to blue and green.

DISTRIBUTION
Tropical coral reefs in South Africa, Red Sea, Indo-Pacific, and Hawaii.

HABITAT
Coral reefs often under rubble.

BEHAVIOR
Displays aposematic coloration to prevent predation.

FEEDING ECOLOGY AND DIET
Many species of the genus *Pseudoceros*, including *P. ferrugineus*, feed on colonial tunicates.

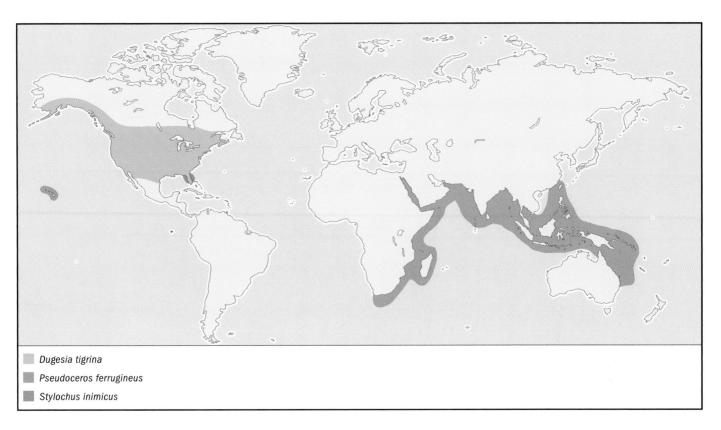

■ *Dugesia tigrina*
■ *Pseudoceros ferrugineus*
■ *Stylochus inimicus*

REPRODUCTIVE BIOLOGY
Hermaphroditic, eggs hatch as Muller's larva that will settle out in a few days and metamorphose into a juvenile resembling the adult.

CONSERVATION STATUS
Not threatened.

SIGNIFICANCE TO HUMANS
Aesthetic value for scuba divers. ◆

Oyster leech
Stylochus inimicus

ORDER
Polycladida

FAMILY
Stylochidae

TAXONOMY
Stylochus inimicus Polombi, 1931, Florida, United States.

OTHER COMMON NAMES
None known.

PHYSICAL CHARACTERISTICS
Body is oval or discoid with retractile nuchal tentacles; marginal eyes in band, cerebral and tentacular eyes present; pharynx long in middle of body, intestinal branches not anastomosing; male and female gonopores close to each other in posterior of body.

DISTRIBUTION
Florida, United States.

HABITAT
Under rocks and algae and in oyster shells and other invertebrates.

BEHAVIOR
Tends to hide under debris or in shells of oysters and barnacles.

FEEDING ECOLOGY AND DIET
Carnivorous feeding on animal matter, including oyster tissue. *Hypsoblennius*, a small blenny, has been shown to feed on it.

REPRODUCTIVE BIOLOGY
Following mating, which may last nine hours and involve more than four partners, an individual can deposit 7,000–21,000 eggs (0.007 in [0.18 mm]) during a summer month. Egg masses usually attached to clean oyster shell and irregular in shape and only one layer thick. Worms observed covering eggs, protecting them. Ciliated larvae with six eyes hatch from eggs. Positively phototactic, causing them to swim up and away from the bottom helping them to disperse.

CONSERVATION STATUS
Not threatened.

SIGNIFICANCE TO HUMANS
Observed entering, devouring, and killing oysters. It has been blamed for the large oyster kill in Florida in 1932–1933. However, it is likely that other environmental factors were also compromising the oysters. The eradication of the worms from oyster beds is difficult; it has been suggested that flooding with freshwater may control worms. ◆

No common name
Kronborgia amphipodicola

ORDER
Rhabdocoela

FAMILY
Fecampiidae

TAXONOMY
Kronborgia amphipodicola Christensen and Kanneworff, 1964, Øresund, Denmark.

OTHER COMMON NAMES
None known.

PHYSICAL CHARACTERISTICS
Parasitic in *Ampelisca macrocephala*; have separate sexes; males are cylindrical red worms, 0.12–0.20 in (3–5 mm) long, lack eyes, mouth, pharynx, and intestine, but possess distinctive gonopore at one end, active and able to swim using cilia; females white with no gonopore, also lack eyes and gut, 1.5 in (40 mm) long, fragile, sluggish and unable to swim.

DISTRIBUTION
Host occurs through out the North Atlantic Ocean. It is not known whether it infects *A. macrocephala* throughout its range. It has only been identified from the northeast Atlantic area.

HABITAT
Males occur in hemocoel and females wrap themselves around the intestine of host.

BEHAVIOR
Each host only harbors either male or female worms.

FEEDING ECOLOGY AND DIET
Most likely feed on host blood and gut contents.

REPRODUCTIVE BIOLOGY
Dioecious, separate sexes. Males and females leave host at the site of the anus by dislodgement of the intestine. Females immediately secrete cocoon longer than its body using epidermal glands; cocoon attaches to host amphipod tube. One or more males enter cocoon through trumpet-shaped opening and inseminate females most likely by hypodermal impregnation. After laying numerous capsules, weakened female crawls out of cocoon and dies. Cocoon morphology is quite elaborate and host specific. Ciliated larvae hatch in 50–60 days and are able to continue swimming searching for a host for two to three days. Larvae have two rhabdomeric eyes for detecting light. After contacting a host's carapace, larvae encyst in five minutes. They produce concentrated enzymes that create a small hole in carapace where larva enters host and migrates to the hemocoel.

CONSERVATION STATUS
Not threatened.

SIGNIFICANCE TO HUMANS
None known. ◆

No common name
Paravortex scrobiculariae

ORDER
Rhabdocoela

FAMILY
Graffillidae

TAXONOMY
Paravortex scrobiculariae (Graff, 1903), Great Britain.

OTHER COMMON NAMES
None known.

PHYSICAL CHARACTERISTICS
Elongate worm approximately 0.08 in (2 mm) long, lacks adhesive organs; mouth anterior with subterminal opening that passes through small ciliated buccal cavity into muscular pharynx and then into simple saccate gut; testis sacciform, ovary single.

DISTRIBUTION
Its host, the intertidal bivalve *Scrobicularia plana*, occurs in northeast Atlantic, Mediterranean, and West Africa. It is unknown if it occurs throughout its host's range.

HABITAT
Lives inside gut and digestive gland of host.

BEHAVIOR
Adult worms migrate within host based on the tidal cycle that relates to feeding cycle of host. Worms use cilia to swim forward from intestine through stomach to digestive gland during the ebb tide and return to intestine during the flood tide.

FEEDING ECOLOGY AND DIET
Feeds on semi-digested components of host food and residual cells released by digestive glands, which it ingests through muscular pharynx, as it makes its migrations from intestine to digestive gland. It appears that it is not capable of producing all the enzymes necessary to digest food and relies on ability to

■ *Kronborgia amphipodicola*
■ *Paravortex scrobiculariae*

acquire enzymes along with food during its migrations to digestive gland. This strategy reduces the need to expend energy on producing these enzymes.

REPRODUCTIVE BIOLOGY
Hermaphroditic and viviparous (releases juvenile worms); up to 40 capsules, each containing two embryos and numerous vitelline cells, which enter parenchyma from female atrium, fully developed embryos leave the capsules and move freely in parenchyma until they escape parent through mouth.

CONSERVATION STATUS
Not threatened.

SIGNIFICANCE TO HUMANS
One species of *Paravortex* commonly called tang turbellaria or black ich are pathogenic on skin of several species of marine ornamental fishes commonly kept in aquaria. ◆

No common name
Temnocephala chilensis

ORDER
Rhabdocoela

FAMILY
Temnocephalidae

TAXONOMY
Temnocephala chilensis (Moquin-Tandon, 1846), Santiago, Chile.

OTHER COMMON NAMES
None known.

PHYSICAL CHARACTERISTICS
Body ovoid, extensible with five anterior prehensile tentacles, posterior attachment disc present, body color is uniform orange-brown; epidermis composed of four syncytial plates, epidermal cilia are absent; eyespots with distinctive red pigment; gonopore is mid-ventral in posterior third of body, two testes occur on each side in posterior of body, cirrus on left side; ovary lies on right side at level of genital pore, vitellaria are extensive.

DISTRIBUTION
Follows distribution of its hosts in Chile and Argentina.

HABITAT
Ectosymbiotes of freshwater decapod and isopod crustaceans. Some live on gills, while others occur on carapace and limbs. It associates with species of the anomuran crabs of genus *Aegla* and species of the crayfish *Samastacus*.

BEHAVIOR
Can move in leech-like movement using posterior disc and tentacles.

FEEDING ECOLOGY AND DIET
Uses its prehensile tentacles to capture free-living and co-symbiotic invertebrates of suitable size such as protozoans, rotifers, oligochaetes, insect larvae, copepods, or ostrcods.

REPRODUCTIVE BIOLOGY
After mating, it attaches eggs to exoskeleton of host. After eggs are deposited, detritus tends to collect around eggs, which at-

■ *Bdelloura candida*
■ *Temnocephala chilensis*

tracts protozoans and other organisms that provide food for newly hatched worms. Development is direct without larval stage.

CONSERVATION STATUS
Not threatened.

SIGNIFICANCE TO HUMANS
None known. ◆

No common name
Bdelloura candida

ORDER
Tricladida

FAMILY
Bdellouridae

TAXONOMY
Bdelloura candida (Girard, 1850), Massachusetts, United States.

OTHER COMMON NAMES
None known.

PHYSICAL CHARACTERISTICS
Body lanceolate or oval shaped, lateral margins undulated, measure 0.6 x 0.2 in (15 x 4 mm) ranging up to 0.9 in (25 mm). Caudal adhesive disk wide as body and set off from the rest of the body; pharynx large about one-third of body; eye lenses absent; numerous testes distributed throughout body; whitish in color.

DISTRIBUTION
Same distribution as host *Limulus polyphemus* along eastern seaboard of United States from Maine south to the Gulf of Mexico.

HABITAT
Ectocommensal on the last pair of cleaning legs and gills of host.

BEHAVIOR
Obligate commensal in that it does not occur off its host.

FEEDING ECOLOGY AND DIET
Most likely feeds on debris and food particles stirred up by its host.

REPRODUCTIVE BIOLOGY
Lays cocoons from May until mid-August on inner surface of gill lamellae of host. Cocoons are 0.15 in (4 mm) in length by 0.07 in (2 mm) in diameter and located on a pedicel 0.39 in (1 mm) high. Development is direct.

CONSERVATION STATUS
Not threatened.

SIGNIFICANCE TO HUMANS
None known. ◆

Land planarian
Bipalium pennsylvanicum

ORDER
Tricladida

FAMILY
Bipaliidae

TAXONOMY
Bipalium pennsylvanicum Ogren, 1987, Pennsylvania, United States.

OTHER COMMON NAMES
None known.

PHYSICAL CHARACTERISTICS
Long, brownish yellow in color with three dorsal stripes; head is half-rounded or lunate, body retracted and coiled on self at rest, during locomotion over a flat dry surface, the body is greatly extended, undulates, and head is raised above surface moving from side to side; head bordered by numerous, small eyes that extend posteriorly along the body; mouth with eversible pharynx located in mid-region of body; gonopore is just posterior to mouth.

DISTRIBUTION
Considered to be introduced into United States, it has only been found in Pennsylvania in outdoor habitats, though other species occur throughout eastern United States.

HABITAT
Lives in damp areas under stones and pieces of wood and are capable of over-wintering in soil where air temperatures reach freezing.

BEHAVIOR
During day, it remains in damp, dark areas under rocks and wood, and in soil. Under dry conditions, it may move further into soil to find favorable conditions.

FEEDING ECOLOGY AND DIET
It captures earthworms by attaching ventral side of head and tail. Thrashing by prey results in planarian getting a better grasp on prey. Arthropods and mollusks are not preferred foods; slugs were only eaten if they were torn open first.

REPRODUCTIVE BIOLOGY
No evidence for fragmentation, but it can regenerate if damaged. After mating, a cream-colored swelling is apparent in mid-region of body where cocoon is developing. Cocoons are yellowish when first deposited and change to light red and eventually shiny black and measure 0.13 in (3.3 mm). One to three juveniles hatch from each cocoon. Juveniles are grayish and lack stripes. It may take 100 days to reach adult size.

CONSERVATION STATUS
Not threatened.

SIGNIFICANCE TO HUMANS
None known. ◆

Freshwater planarian
Dugesia tigrina

ORDER
Tricladida

FAMILY
Dugesiidae

TAXONOMY
Dugesia tigrina (Girard, 1847), New Jersey, United States.

OTHER COMMON NAMES
None known.

PHYSICAL CHARACTERISTICS
Body lanceolate with auricles laterally on head, light to dark brown in color with some forms having a stripe down its midline, light spots on a dark background or dark spots on a light background; pharynx large in middle of body, gut with two posterior and one anterior directed branch as others in order.

DISTRIBUTION
Widespread in North America and has scattered distribution in Europe, where it may have been introduced with aquatic plants.

HABITAT
Under rocks, plants, and debris in clear freshwater ponds, streams, and springs.

BEHAVIOR
Tends to hide under rocks during day. When hunting, will swing head side to side to better sense sources of chemicals coming from food or prey.

FEEDING ECOLOGY AND DIET
Carnivorous, feeds on various invertebrates, including mosquito larvae.

REPRODUCTIVE BIOLOGY
Asexual reproduction by transverse fission. Transverse fission may be more common method of reproducing since sexual organs have not been observed in some populations. It has strong capacity to regenerate. Following mating, several egg capsules are deposited per worm. Development is direct and juveniles hatch from eggs.

CONSERVATION STATUS
Not threatened.

SIGNIFICANCE TO HUMANS
Studied intensively as a model to better understand regeneration of cells in humans and other animals. ◆

Resources

Books

Brusca R. C., and G. J. Brusca. *Invertebrates.* 2nd ed. Sunderland, MA: Sinauer Associates, Inc., 2003.

Cannon, L. R. G. *Turbellaria of the World—A Guide to Families and Genera.* Queensland, Australia: Queensland Museum, 1986.

Kearn, G. C. *Parasitism and the Platyhelminths.* New York: Oxford University Press, 1998.

Prudhoe, S. *A Monograph on Polyclad Turbellaria.* New York: Pemberley books, 1985.

Sluys, R. *A Monograph of the Marine Triclads.* Brookfield, MA: A. A. Balkema, 1989.

Dennis A. Thoney, PhD

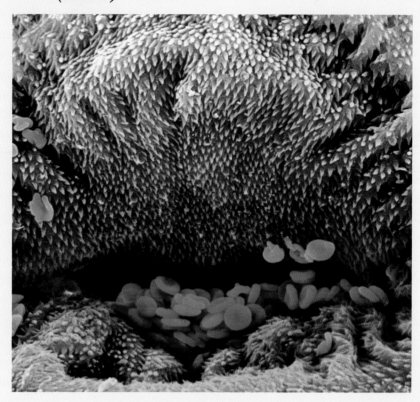

Trematoda
(Flukes)

Phylum Platyhelminthes

Class Trematoda

Number of families Approximately 176

Thumbnail description
Parasitic flatworms ranging from less than 0.04 in (1 mm) to 23 ft (7 m) long, but usually 0.2–2 in (0.5–5 cm)

Photo: Color scanning electron micrograph of red blood cells in the mouth of a *Schistosoma mansoni* fluke, a parasite causing the disease bilharzia. It lives in the veins of the intestines and bladder of humans and feeds on blood cells. (Photo by © Oliver Meckes/Eye of Science/Photo Researchers, Inc. Reproduced by permission.)

Evolution and systematics

The class Trematoda has about 6,000 species, although the number given varies considerably among different researchers. The class is commonly divided into two subclasses, Aspidogastrea and Digenea. Both are sometimes elevated to class level or dropped to order status; however, they will be treated in this text as subclasses of the class Trematoda. At one time, the class Monogenea was considered a subclass or order of the class Trematoda.

Recent studies of the evolution of trematodes indicate that the class may be paraphyletic—that is, it may represent only some of the descendants of a common ancestor; and that most of the species actually should be grouped together with the mollusks and annelids in the taxon Lophotrochozoa. Part of the difficulty in defining the placement of trematodes is that their fossil record is so sparse. Other than some fossil eggs dating from the Pleistocene epoch (1.8 million–8,000 years ago), historical evidence of trematodes consists mostly of trace fossils. Trace fossils are signs thought to have been left by an organism. For example, a trace fossil from a trematode might be a slight indentation on a fossil snail shell.

Despite the poor fossil record, scientists believe that Aspidogastrea is an ancient group, because many of its species, including those in the families Multicalycidae, Rugogastridae,

and Stichocotylidae, use cartilaginous fishes as hosts. These fishes, which are in the class Chondrichthyes, evolved at an earlier point in time than the mammals and teleosts (bony fishes) generally used as hosts by digenetic trematodes. Species within the digenetic genera *Nagmia* and *Probolitrema* are exceptions to the rule, and will invade cartilaginous fishes.

A particularly noticeable synapomorphy of the two subclasses is the presence of a posterior sucker, which is manifested as the large, ventral disk in the subclass Aspidogastrea. In addition, both subclasses have life cycles that involve mollusks and vertebrates. Scientists still disagree about whether the former or the latter were the original hosts evolutionarily.

The subclass Aspidogastrea has four families with about 80 species and is often split into four orders. The subclass Digenea has about 6,000 species that are generally split into 10 major orders and numerous smaller orders. Frequently, taxa are switched from family to order status, so a total number of families and orders in this large subclass is difficult to ascertain.

Physical characteristics

Trematodes are parasitic flatworms, usually leaflike in appearance, with holdfast organs that they use to adhere to their hosts. Digenetic species typically use muscular, oral and/or

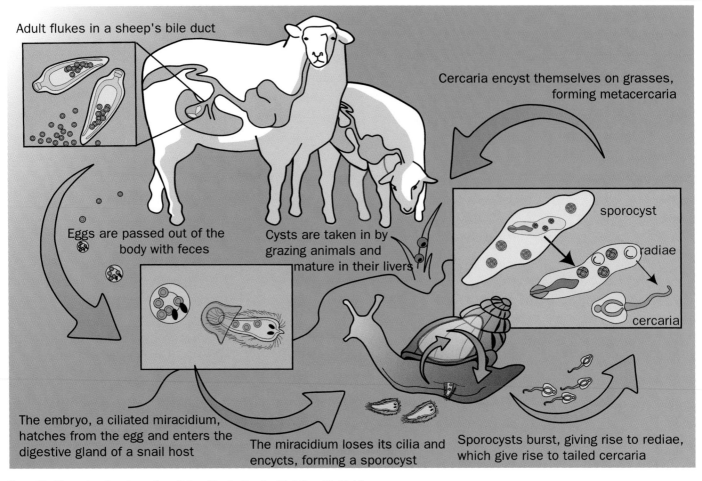

Adult flukes in a sheep's bile duct

Cercaria encyst themselves on grasses, forming metacercaria

sporocyst

radiae

cercaria

Eggs are passed out of the body with feces

Cysts are taken in by grazing animals and mature in their livers

The embryo, a ciliated miracidium, hatches from the egg and enters the digestive gland of a snail host

The miracidium loses its cilia and encycts, forming a sporocyst

Sporocysts burst, giving rise to rediae, which give rise to tailed cercaria

Parasitic life cycle of a sheep liver fluke. (Illustration by Christina St. Clair)

ventral suckers to attach to their hosts. Members of the subclass Aspidogastrea have a large ventral disk that they use for adhesion. This disk is covered with many small suckers known as alveoli.

Ventral suckers are generally about the same size as the oral suckers, and located centrally on the anterior, ventral surface. Often, the ventral suckers are quite close behind the oral sucker. Trematodes generally have one or two large, sometimes branched, testes, a comparatively small ovary, an often long and looping uterus, and a single common genital pore. They typically have a bifurcated intestine and blind caeca that exit the body through an anus or via an excretory vesicle. The integument lacks a cuticle.

Distribution

Trematodes are found around the world. The presence of their host species is often the limiting factor in their geographic distribution. For example, a flatworm that uses a mollusk as its host occurs only where that mollusk lives. Nonetheless, the distribution of a particular trematode can span large areas, particularly if their host species live in a broad range of habitats; or if members of the host species, like birds, are able to run or fly over vast regions.

In many cases, the species' distributions follow their hosts. For example, hosts of *Uvulifer ambloplitis* include snails, fishes, and birds. With such a complex life cycle, the trematode's distribution is a combination of the ranges of all three animals. As another example, the aspidogastrids infect mollusks, teleost fishes, and turtles, presenting another complex distribution pattern.

Habitat

The habitat of trematodes frequently varies over the course of their lives. Members of the subclass Digenea are endoparasites with indirect life cycles, which means that they infect different hosts during the various stages of their life cycles. Digenetic trematodes typically have two larval stages and at least two hosts. Several, including *Halipegus occidualis*, have four hosts, ending up at the base of the tongue of a green frog (*Rana clamitans*), which is the definitive host of *H. occidualis*. The life cycles of members of the subclass Aspidogastrea, which sometimes take a shorter path with only one host and no asexually reproducing larvae, are described as direct.

A typical digenetic individual begins its life as an egg in its so-called definitive host, and then passes with the host's feces either into water or onto land. After hatching, the larva takes up residence in a first intermediate host, which is often a mollusk. It then exits and moves to a second intermediate host, which is frequently either another mollusk, a fish, or an amphibian. The characteristic digenetic life cycle continues when the definitive host eats the secondary host, at which point the trematode infects the definitive host. Definitive hosts often include predatory mammals or birds.

Among species of Aspidogastrea, the life cycle is a bit simpler with usually only one host. After birth, the ectoparasitic species generally latch immediately onto the outside of a host organism, usually the skin or gills of bivalves or fishes. Endoparasitic species of Aspidogastrea infect such taxa as mollusks and fishes, but also the shark-like holocephalans and elasmobranchs.

Behavior

As described, the life cycle pattern of the digenetic trematodes is quite complex, involving first- and second-stage larvae as well as adults that are dependent on a host species for survival. Upon hatching, the first-stage larvae, known as miracidia, infect the first intermediate host, usually a mollusk. Many miracidia take advantage of light and gravitational cues to reach an area suitable to their hosts, then hone in on an individual host by following its chemical signature, which may be fatty acids or amino acids specific to the organism. Other miracidia, however, seem to stumble upon rather than track their host. In still other trematode species, the egg does not develop until it is eaten by the host species.

Once in the first intermediate host, the miracidium sheds the ciliated epidermal cells that the aquatic forms typically use to navigate the water. In the first intermediate host, the miracidia travel to a specific site, depending on the species of the trematode and of the host. Once there, the miracidium may develop into a saclike sporocyst and/or redia before generating cercariae. Cercariae are the larvae that finally exit from the mollusk and begin actively seeking the second intermediate host. Cercariae are generally propelled by a variety of different types of tails, although a few species, like *Maritrema arenaria*, have small or no tails and move by crawling rather than swimming. Cercariae use environmental cues, like light or water turbulence (in aquatic species), to seek out their secondary host. A few species, like *Schistosoma mansoni* and *Trichobilharzia ocellata*, also follow a chemical trail laid by the secondary host. Because cercariae are nonfeeding organisms, they must find a host quickly. The strong-swimming species generally need to locate a host more rapidly than slow-crawling species, with swimmers surviving only about 24 hours without a host, while crawlers can continue living without a host for several days.

Among digenetic trematodes, the species of several families of blood flukes, including the Schistosomatidae, Spirorchiidae, and Sanguinicolidae, skip the second intermediate host; their cercariae invade the definitive hosts directly. A few, like those of the families Azygiidae and Faciolidae, encyst on vegetation rather than in a secondary host. The definitive host then becomes infected by eating the trematode-infested vegetation. Other trematodes, such as *Alaria* species, add a larval stage that

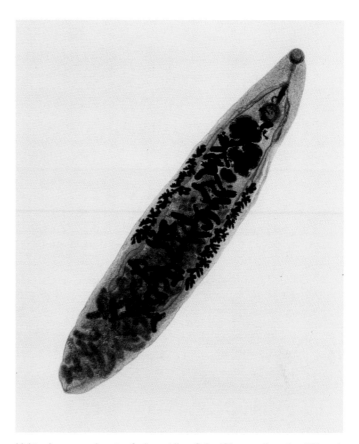

Light microscope image of a lancet liver fluke (*Dicrocoelium dendriticum*), a parasite of sheep. The eggs are picked up and carried by snails and the larva are transferred to ants through the snail's excretions. The maturing larvae cause cramping in the ant's mandible musculature which causes the ant to clamp down on blades of grass and get stuck. The larva are then ingested by herbivores such as sheep and cattle where they develop to the adult stage in the bile duct of the liver. Eggs are excreted by the sheep, starting the life cycle over again. (Photo by Oliver Meckes/Photo Researchers, Inc. Reproduced by permission.)

forms after the cercaria and develops into the metacercaria. This stage is called the mesocercaria. *Plagioporus sinitsini* is an atypical trematode that can apparently take any of three paths from egg to adult. In the most intriguing of the three, these trematodes are able to bypass their usual second intermediate and definitive hosts, and develop into adults while still in the sporocysts of the first intermediate host. As a result, they can produce and hatch eggs, and release already-developed miracidia through the feces of their first intermediate hosts.

Cercariae enter the secondary host, drop their tails, and develop into metacercariae. Metacercaria look like the adults, except they lack reproductive organs. Often, the host can survive, ostensibly ignoring even heavy infestations with metacercariae. Young bluegills (*Lepomis macrochirus*), on the other hand, die if they are infected with too many metacercaria of the species *Uvulifer ambloplitis*. In some cases, metacercariae can cause behavioral or morphological changes in the second intermediate host species that have the effect of making it more susceptible to predation by the definitive host. Trematodes in the genus *Ribeiroia*, for example, can trigger a frog to produce additional or fewer hind limbs. Such deformities effectively cripple the host, making it an easier target for birds, the definitive host.

Transmission to the definitive host usually results from the host's inadvertent ingestion of the metacercariae. From the digestive system, the metacercariae migrate to the target site, which varies by species, and reach their sexual maturity in that location.

Studies of trematode communities in the first intermediate host indicate that a single host carries no more than four different species of trematodes. The reason for this low number is under some debate. One hypothesis suggests that interspecific predation by redia keeps the number in check. Other researchers believe that temporal and/or spatial separation of trematode species accounts for the limit of four trematode species per host. There may also be limitations on space or other resources.

Members of the subclass Aspidogastrea have a simpler lifestyle than the digenetic trematodes. It usually involves just one host, often freshwater mussels or snails, in which development from egg to adult occurs. The larvae (cotylocidia) generally lack cilia, but some, like the larvae of *Lophotaspis vallei*, have cilia and are good swimmers. Nonciliated larvae, in contrast, creep rather than swim.

Feeding ecology and diet

Trematodes are obligate parasites, which means that they require nourishment from a host organism. Adult digenetic trematodes die soon after removal from the host, while members of Aspidogastrea may survive independently for a month or more.

Digenetic trematode first-stage larvae, or miracidia, do not feed. For this reason, they must find a first intermediate host very quickly, usually within one or two days of hatching. The eggs of some species don't even hatch until they are eaten by the first intermediate host. Miracidia develop into sporocysts, which also do not feed. The redia stage that often develops from the mother sporocyst in some taxa use their mouths to rip away and eat bits of host tissue. In fact, rediae will eat just about anything, including sporocysts of other species.

Adult trematodes attach to the host organism using suckers. They will eat blood cells, mucus, and loose cells; in some cases, they secrete enzymes that begin to digest tissue before consumption. Some genera, such as *Schistosoma* and *Faciola* species, adhere to the host in the blood vessels, liver, and other sensitive areas, inflicting damage to the host by releasing toxic materials through their excretory pores. In some cases, high numbers of trematodes can even plug small body passages or openings, like the host's ureter or bile duct; or interfere with digestion or respiration. Occasionally, mollusk hosts are so irritated by these tiny invaders that they begin coating them with layers of nacreous material, turning the trematodes into pearls.

Reproductive biology

Digenetic flukes engage in both asexual and sexual reproduction. When the first-stage larvae reach their destination within the first intermediate host, a polyembryonic asexual reproductive phase begins. After the larvae develop into sporocysts, the sporocysts of some species produce a second generation of sporocysts, while others produce rediae. *Trichobilharzia physellae* and *Diplostomum flexicaudum* take the two-generations-of sporocysts route. *Metorchis conjunctus* and *Proterometra dickermani* develop from miracidium to sporocyst to redia. A few species, including *Stichorchis subtriquetres*, go directly from miracidium to redia, bypassing the sporocyst stage altogether. The rediae are also capable of asexual reproduction to produce a second generation of rediae. Via sporocysts and rediae, the number of invading trematodes can multiply very quickly within the first intermediate host.

Both sporocysts and rediae produce the next stage, the cercariae and metacercariae. Cercariae are free-living and swim to the second intermediate host, which is typically a prey for the definitive host. Once on or in the second intermediate host, the cercariae transform themselves into metacercariae. Neither the cercariae nor metacercariae reproduce. The metacercariae finally enter the definitive host and become adults, which participate in sexual reproduction either by mating with a second fluke or by self-fertilization. Almost all trematodes are hermaphrodites. Some of these species, including members of the family Hemiuridae, even have fused male and female reproductive ducts. Members of the family Schistosomatidae are the exception. The schistosomatids have separate sexes with the adult females usually found coupled with adult males.

The eggs of both subclasses of trematodes are typically light-colored oval structures. The eggs of some species are embryonated (*Heterophyes heterophyes*) while others are not (*Paragonimus westermani*). Several species, including *Metagonimus yokogawai* and *Opisthorchis felineus*, have eggs that contain mature miracidia in addition to embryos. In digenetic trematodes, the eggs develop in a large loop-shaped uterus. The number of eggs varies by species, but production of several hundred per day is not unusual. The common digenetic marine trematode *Cryptocotyle lingua* can produce thousands of eggs over a 24-hour period. Trematode eggs exit the host through the feces. In some species, miracidia form in the egg before it is shed; more typically, however, the miracidia develop two or three weeks after oviposition. In a few digenetic species, the miracidium forms only after the egg is taken into the first intermediate host.

The life cycle of the subclass Aspidogastrea is simpler; it has no asexual phase of reproduction. The entire life cycle typically occurs in one host, usually a mollusk. Predators of the host species have been known to eat infected mollusks and temporarily harbor the flatworms. Flatworms can survive in the predator's digestive tract for a short period, but cannot reproduce or develop there. There are other exceptions. *Mulicalyx cristata* is thought to use marine crustaceans as intermediate hosts. After the crustaceans are eaten by sharks or rays, *M. cristata* matures in the gall bladder of the shark or ray.

Conservation status

No species of trematodes are listed by the IUCN as threatened.

Significance to humans

Trematodes pose a significant health threat to humans, particularly those living in developing countries. A common illness in developing countries is schistosomiasis. This condition, caused by three species of *Schistosoma*, affects more than 40 million individuals who live in tropical and subtropical countries, causing weakness, diarrhea, hemorrhage, fever, enlargement of the spleen, and other severe symptoms.

Other trematodes also infect humans, including the trematode *Opisthorchis*, which is transmitted to humans through eating infected fishes. Prevalent in parts of Russia, the fluke currently infects 1.2 million people, which is more than 4 percent of the region's population.

The cost for treatment of human fluke infections ranges into billions of dollars, in part because these conditions are frequently misdiagnosed. Treatment for infection by such organisms as *Paragonimus* species costs about $1, but the patient's illness is often misinterpreted as tuberculosis, which calls for years of expensive treatment.

Trematodes that infect such household pets as rabbits, dogs, and cats may cause gastrointestinal symptoms requiring veterinary treatment. In the case of dogs, the trematode *Nanophyetus salmincola* or so-called salmon-poisoning fluke, may cause a fatal disease resembling distemper because it carries a rickettsia (a type of bacterium) to which dogs are susceptible. The rickettsia, however, does not produce clinical disease in either humans or cats.

Trematodes can also infect livestock, sport and commercial fishes, and game mammals, which can have negative economic impacts on agriculture, sport fishing, and commercial fishing.

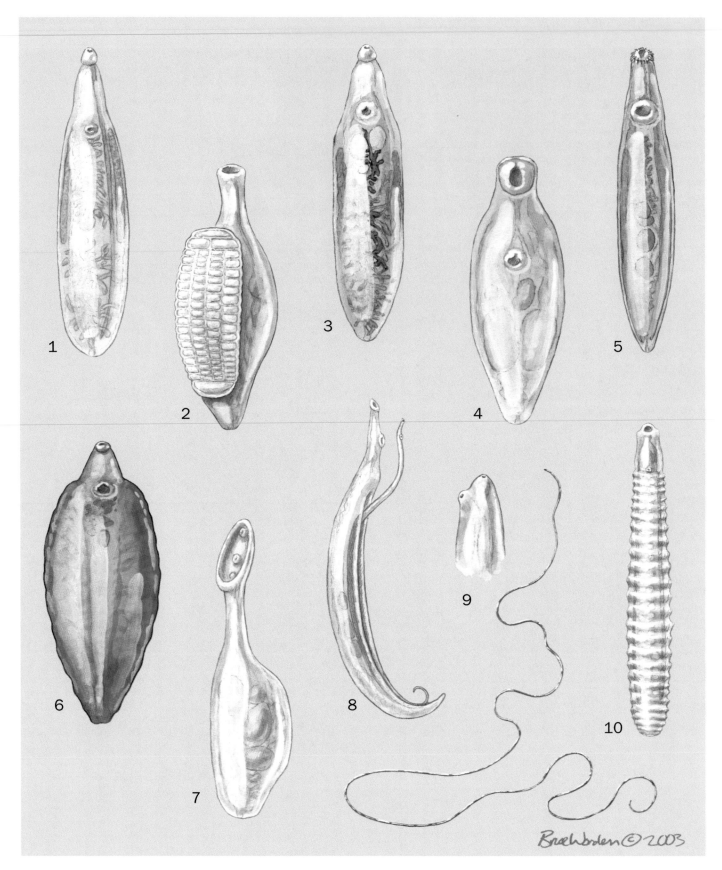

1. Oriental liver fluke (*Clonorchis sinensis*); 2. *Aspidogaster conchicola*; 3. Human blood fluke (*Dicrocoelium dendriticum*); 4. Salmon-poisoning fluke (*Nanophyetus salmincola*); 5. Echinostome (*Echinostoma revolutum*); 6. *Fasciola hepatica*; 7. Black-spot flatworm (*Uvulifer ambloplitis*); 8. *Schistosoma mansoni*; 9. *Nematobothrium texomensis*; 10. *Rugogaster hydrolagi*. (Illustration by Bruce Worden)

Species accounts

No common name
Rugogaster hydrolagi

SUBCLASS
Aspidogastrea

FAMILY
Rugogastridae

TAXONOMY
Rugogaster hydrolagi Schell, 1973, vicinity of San Juan Island, Washington, United States.

OTHER COMMON NAMES
None known.

PHYSICAL CHARACTERISTICS
Adults have an elongated body 0.27–0.59 in (7–15 mm) long and 0.039–0.078 in (1–2 mm) wide without spines, but with a row of 17–25 transverse ridges or rugae on a small ventral sucker. It also has numerous testes and two caeca (large intestine) rather than the one caecum and one or two testes characteristic of other members of its subclass.

DISTRIBUTION
Southeastern coast of Australia; northwestern coast of United States and British Columbia.

HABITAT
Hosts include the ratfish (*Hydrolagus colliei*). The parasite inhabits the coecal glands, sometimes protruding into the rectum.

BEHAVIOR
Adult flatworms infect the rectal glands of holocephalan fishes. (chimaeras).

FEEDING ECOLOGY AND DIET
These parasites rely for their nutritional needs on their host fishes, but little else is known about their feeding ecology.

REPRODUCTIVE BIOLOGY
A large uterus contains numerous amber-colored eggs, which are oval and operculate, ranging from 0.006–0.0066 in (154–168 μm) long and 0.0036–0.004 in (92–102 μm) wide. The larvae, which do not reproduce, are 0.0104–0.011 in (265–288 μm) by 0.0027–0.0032 in (70–82 μm), and leaf-shaped with a rounded anterior end and a pointed posterior end. They have no cilia, but do have a posterior sucker.

CONSERVATION STATUS
Not listed by IUCN

SIGNIFICANCE TO HUMANS
None known. ◆

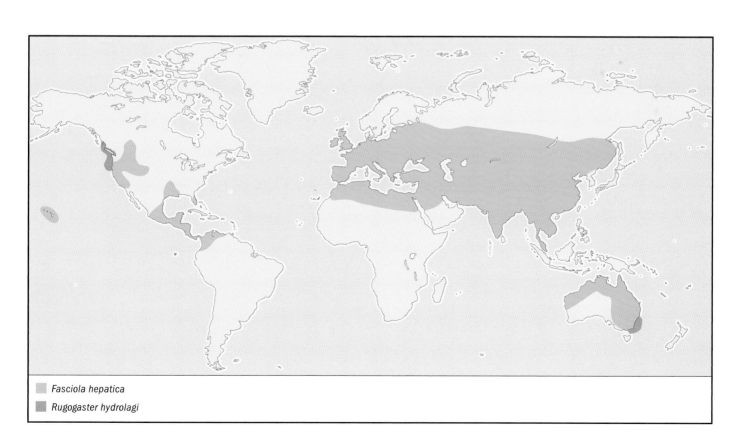

▨ *Fasciola hepatica*
▨ *Rugogaster hydrolagi*

No common name
Aspidogaster conchicola

ORDER
Aspidogastrida

FAMILY
Aspidogastridae

TAXONOMY
Aspidogaster conchicola von Baer, 1826, Anodonta, Unio Prussia.

OTHER COMMON NAMES
None known.

PHYSICAL CHARACTERISTICS
Adults are about 0.08 in (2.5–2.7 mm) long and 0.04 in (1.1–1.2 mm) wide, with a large sucker that covers almost the entire ventral surface of the organism. The body narrows to a neck-like region with a mouth at the front tip. The large ventral sucker has a windowpane pattern of sucking grooves. A lone testis is located in the middle of the posterior half of the body, and the ovary sits in the center and to one side of the animal. A common genital pore opens anteriorly about a quarter of the body length behind the mouth sucker.

DISTRIBUTION
Europe, North America, China, and Egypt.

HABITAT
Hosts are snails, including *Viviparus malleatus* and *Goniobasis livescens*; and such mussels as *Anadonta grandis*. *Aspidogaster*

species use only one host. Sometimes, other species, like turtles, will eat infected mussels. The flukes can survive temporarily in the turtle's stomach.

BEHAVIOR
Aspidogaster larvae use snails and mollusks as their hosts. Larvae apparently infect the host mollusk by entering its siphon. Various reports indicate that the flatworms enter the host either as larvae or as eggs containing larvae. *Aspidogaster* eggs hatch after they are eaten by freshwater snails or mussels. Eggs that are not eaten can survive for about two weeks. In snails, the newly hatched worms lack cilia and move about by creeping. The larvae migrate to the snail's intestines and then to the hepatopancreas, where they mature and lay eggs.

FEEDING ECOLOGY AND DIET
Parasites feed on host epithelium and mucus using the anteriroly located mouth.

REPRODUCTIVE BIOLOGY
The entire life cycle of *Aspidogaster* from egg to adult to production of the next generation of eggs takes about 270 days at 68°F (20°C). Unlike digenetic trematodes, the life cycle of this species and other members of the subclass Aspidogastrea occurs in just one host and does not involve asexual reproduction. The eggs are oblong, operculate (having a small covering structure), and about 0.005 in (128–130 µm) long and 0.0019 in (48–50 µm) wide. Juveniles, which have a shape similar to the adults, grow from about 0.02 in (0.5 mm) in length to 0.05–0.06 in (1.25–1.5 mm).

CONSERVATION STATUS
Not listed by IUCN.

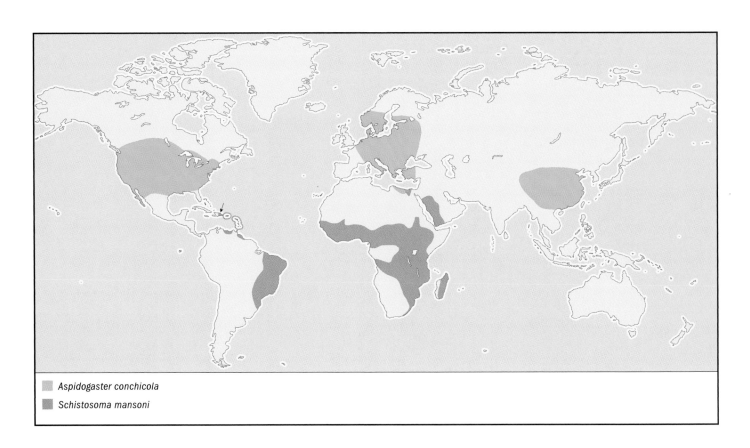

▢ *Aspidogaster conchicola*
◼ *Schistosoma mansoni*

SIGNIFICANCE TO HUMANS
None known. ◆

No common name
Nematobothrium texomensis

ORDER
Azygiida

FAMILY
Didymozoidae

TAXONOMY
Nematobothrium texomensis McIntosh and Self, 1955, entangled among buffalo fish ovaries, Lake Texoma, Willis, Oklahoma, United States.

OTHER COMMON NAMES
None known.

PHYSICAL CHARACTERISTICS
One of a group of worms that are generally long and thin, the adults of this species can reach 8.2 ft (2.5 m) in length. A ventral sucker is not evident and may be lacking in most specimens. It has little musculature, even in its oral sucker, which is completely enclosed in its integument (covering). It is transparent to slightly opaque.

DISTRIBUTION
Arizona, Arkansas, Kentucky, and Oklahoma.

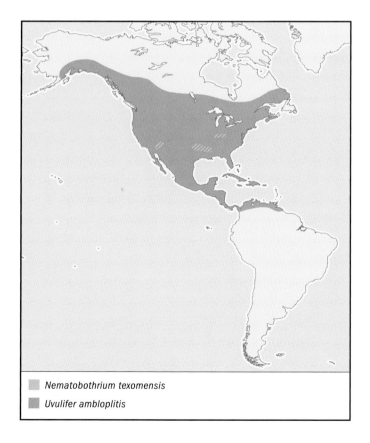

☐ *Nematobothrium texomensis*
☐ *Uvulifer ambloplitis*

HABITAT
This species is generally found in the gravid (pregnant) ovaries and occasionally in mature testes of the buffalo fish (*Ictiobus* species). In some cases, the worm can extend several inches outside the host organism through the fish's genital opening. Including this flatworm species, only a few other didymozoids infect freshwater fishes. Its snail host is unknown as of 2003.

BEHAVIOR
The hermaphroditic adult worms live only as long as the fish remains gravid, and the worms probably shed their eggs as the adults perish and disintegrate into fragments.

FEEDING ECOLOGY AND DIET
Like other parasitic flatworms, this species is dependent on its host, the buffalo fish, to fulfill its nutritional needs.

REPRODUCTIVE BIOLOGY
N. texomensis reproduces in the spring, shedding its eggs according to the reproductive cycle of its host. Eggs of this species are round and thin-shelled structures ranging from 0.0005–0.001 in (13–30 µm) in diameter. Usually within a few days of being shed, miracidia develop. The miracidia resemble amebas in shape, lack cilia, and move slowly. They are armed with small spines.

CONSERVATION STATUS
Not listed by the IUCN

SIGNIFICANCE TO HUMANS
None known. ◆

Echinostome
Echinostoma revolutum

ORDER
Echinostomida

FAMILY
Echinostomidae

TAXONOMY
Echinostoma revolutum Froelich, 1802.

OTHER COMMON NAMES
French: Échinostome.

PHYSICAL CHARACTERISTICS
Adult echinostomes range from 0.24–1.3 in (6–30 mm) long and 0.02–0.06 in (0.6–1.6 mm) wide. Echinostomes have a tegument (outer surface) carrying spines or papillae (small rounded projections) and an anterior collar consisting of 37 spines arranged in a characteristic pattern. The ventral sucker is slightly behind the oral sucker. The uterus takes up much of the front half of the body and two testes follow, one after another, in the posterior half. The comparatively small ovary is located in the center of the animal.

DISTRIBUTION
Europe and Asia.

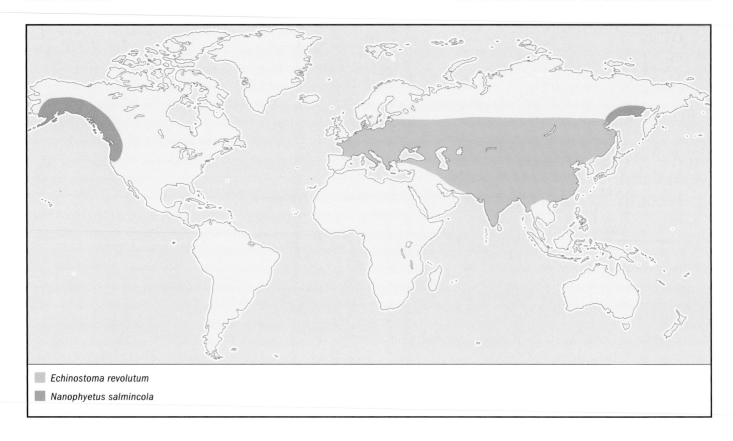

Echinostoma revolutum

Nanophyetus salmincola

HABITAT

Echinostome eggs hatch in freshwater. The first intermediate hosts are lymnaeid snails. Common snail hosts in the United States and Europe belong to the genus *Lymnaea*. Second intermediate hosts include frogs, mussels, turtles, and snails; definitive hosts include various waterfowl, chickens, and other birds. Flukes attach to the mucosa or the ileum.

BEHAVIOR

Echinostome eggs hatch in nine to 12 days in freshwater habitats. The miracidia can survive for only six to eight hours, so these fast swimmers must find snail hosts quickly. Penetration typically occurs along the mantle edge or foot of the snail. After developing into sporocysts, the eggs then undergo two or three generations of asexual reproduction as rediae, finally resulting in the development and release of cercariae. The progress from miracidia to cercariae takes about a month. The cercariae then swim and/or crawl to a second snail, a fingernail clam, a tadpole, or a silurid fish. The larval forms of this fluke use chemotaxis (orientation toward or away from a chemical stimulus) to detect snail slime trails and find their first intermediate hosts. They also engage in a searching method that uses such environmental stimuli as light and gravity. The cercariae burrow into snails and clams, where they encyst in soft tissue; but enter tadpoles and fish through the cloaca, eventually encysting in the kidneys.

Birds then devour the infected snails, clams, tadpoles, or fishes, and become the definitive hosts of echinostomes. Humans who eat raw or undercooked frogs also become definitive hosts. The metacercariae migrate to the cecum, small intestine, or rectum, and attach themselves to these organs

with oral and ventral suckers. The hermaphrodites mature in the host's digestive tract, engage in self-fertilization, and lay eggs. In the laboratory, adult echinostomes live four to eight weeks. The eggs are transmitted to the environment via the feces.

FEEDING ECOLOGY AND DIET

As with other trematodes, these parasitic flukes depend on host species to supply their nutrition. These species include *Lymnea* or *Planorbis* snails as first intermediate hosts; snails or tadpoles as second intermediate hosts; and ducks, geese, chickens, partridge, pigeons, or humans as definitive hosts.

REPRODUCTIVE BIOLOGY

Echinostomes reproduce asexually as larvae, and sexually as adults. The yellowish eggs range from 0.0034–0.0044 in (88–113 μm) in length and about 0.024 in (61 μm) in width; the flukes may produce as many as 3000 per day. The miracidia have four rows of epidermal plates and papillae on the body. The cercariae have a 37-spined anterior collar like that of the adults, and a robust unforked tail. Mature flukes are hermaphrodites and self-fertilize.

CONSERVATION STATUS

Not listed by IUCN.

SIGNIFICANCE TO HUMANS

People who eat raw snails or frogs may become infected with this parasite. Symptoms appear about two to three weeks later. Hemorrhagic enteritis (inflammation of the intestines) may result from severe infections. Milder infections typically cause weakness and weight loss. ◆

Liver fluke
Fasciola hepatica

ORDER
Echinostomida

FAMILY
Fasciolidae

TAXONOMY
Fasciola hepatica Linnaeus, 1758, "in aquis dulcibus ad radices lapidum, inque hepate pecorum. Diss. de Ovibus;" Europe.

OTHER COMMON NAMES
English: Sheep liver fluke; French: Grande douve du foie, douve du foie de mouton; German: Großer Leberegel.

PHYSICAL CHARACTERISTICS
Adult liver flukes may reach 1.7–2.2 in (4–5 cm) in length and 0.6 in (1.5 cm) wide. They are typically about 1.3 in (3 cm) long, 0.4 in (1 cm) wide, and have a spiny tegument. They taper toward the rear. The front end bears an oral sucker and a cone-shaped tip. The sucker on the fluke's ventral (lower) surface is larger than the oral sucker. The ventral sucker is about a third of the body length behind the oral sucker. The branched ovary is situated behind and to the side of the ventral sucker about a third of the way back in the body. The testes are also branched and extend throughout the body behind the ovary.

DISTRIBUTION
Worldwide, but found most often in Europe and Latin America in habitats congenial to their freshwater snail and definitive hosts.

HABITAT
Liver flukes are found in swampy, generally wet freshwater areas inhabited by snails, especially of the species *Lymnaea truncatula*, *Stagnicola bulimoides*, and *Fossaria modicella*. Snails are their sole intermediate host. The definitive hosts of this fluke include grazing herbivores (in the bile ducts) primarily, including sheep and cattle, but also dogs, cats, rabbits, and humans.

BEHAVIOR
The eggs, deposited in the environment in the definitive host's feces, hatch in freshwater areas, usually within about 10 days, longer if temperatures are cool. They have been known to survive in particularly cold water for several years. The embryos develop into miracidia, which quickly swim to and penetrate the soft tissue of snails. Miracidia can survive only 24 hours in the free-living state. Sporocysts form and produce first-generation rediae, which in turn produce second-generation rediae and eventually numerous cercariae. The cercariae live in the snail for 4–8 weeks, then exit and swim to vegetation lying just below the water line. There, they drop their tails and encyst. Passing herbivores become infected when they eat the vegetation, often grass. Humans typically become infected by drinking water containing flukes or by eating vegetation such as watercress. The flukes travel to the abdominal cavity in the first 24 hours, then to the liver over the next few days. Research indicates that immature flukes are able to orient during their migration from the duodenum (the first section of the small intestine) toward the liver. Within six to eight weeks, they reach the bile ducts, sometimes spreading to the lungs, where they mature and lay eggs. The eggs are then carried to the duodenum and pass into the feces.

FEEDING ECOLOGY AND DIET
This parasitic digenetic fluke has two hosts: *Lymnaea* species as the intermediate host, and wild or domesticated ruminants as the definitive host. Humans may become secondary hosts. The fluke feeds on bile duct lining, causing calcification of the duct.

REPRODUCTIVE BIOLOGY
The tan or yellow eggs are about 0.0048–0.006 in (120–150 µm) long and 0.0025–0.0035 in (65–90 µm) wide. In warm water (78.8°F or 26°C), they develop into miracidia in less than two weeks. The miracidia are ciliated, and somewhat triangular in shape with the front end being broader than the rear. The front end also has a noticeable slender outgrowth with two eyespots behind it. The cercariae, which range from 0.0098–0.013 in (250–350 µm) long, resemble tadpoles in shape with a bulbous anterior end and long tail making up about two-thirds of the overall length. In artificial laboratory conditions, adult flukes have been known to survive as long as 11 years.

CONSERVATION STATUS
Not listed by IUCN.

SIGNIFICANCE TO HUMANS
Infected humans may develop symptoms ranging from skin inflammation to pneumonia. Fluke infection can result in massive hemorrhages in horses, a reduction of milk production in dairy cattle, and mortality in sheep. Sheep mortality is often caused by bacterium *Clostridium novyi*, which thrives on the infected livers of sheep. ◆

Oriental liver fluke
Clonorchis sinensis

ORDER
Opisthorchiida

FAMILY
Opisthorchiidae

TAXONOMY
Clonorchis sinensis Cobbold, 1875. Some scientists now use the genus designation *Opisthorchis* for this fluke.

OTHER COMMON NAMES
English: Chinese liver fluke; French: Douve du foie chinoise, douve du foie orientale; German: Chinesischer Leberegel.

PHYSICAL CHARACTERISTICS
The adults are flattened cigar-shaped flatworms 0.4–1.0 in (10–25 mm) long and 0.1–0.2 in (3–5 mm) wide. The pointed front end has an oral sucker at its tip. Much of the anterior half of the animal is filled with a looping uterus. A small, slightly lobed ovary follows with two large branching testes located in the posterior half of the body. The genital pore opens about a fifth of the body length behind the oral sucker. The ventral sucker is circular and located slightly behind the oral sucker.

DISTRIBUTION
East Asia, including Japan, much of China, and Korea.

HABITAT
This organism begins its life as an egg passed in the feces of a human or other mammal. It then infects snails, followed by one of more than 100 species of fish, and finally humans or other fish-eating mammals.

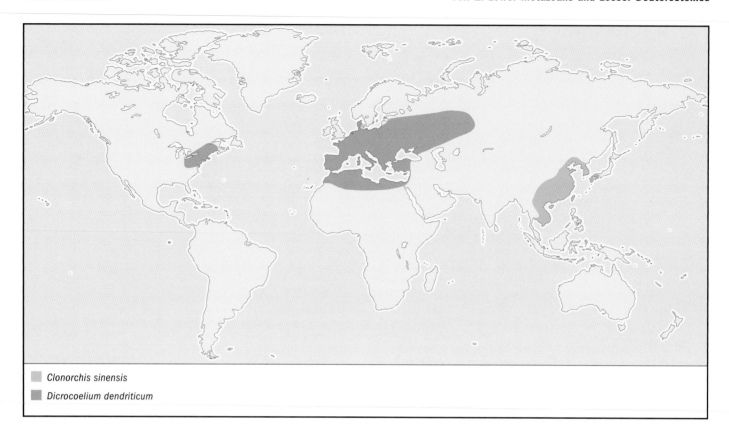

Clonorchis sinensis

Dicrocoelium dendriticum

BEHAVIOR
After the eggs are passed into the water, they are eaten by the first intermediate host, which is a snail (*Bulimus* species or *Parafossarulus manchouricus*). The miracidia hatch in the intestine, form a sporocyte, generate rediae, and finally cercariae. Individual rediae can produce as many as 50 cercariae, which exit the snail into the water and begin the search for a second intermediate host. *Clonorchis sinensis* has an unusual method of searching; its cercariae sink until they hit either the bottom or some other structure, then swim to the surface to sink again. While bouncing back and forth between the surface and the bottom, cercariae that encounter a fish burrow through the skin and encyst in its muscle tissue. These second intermediate hosts are usually freshwater fishes of the Cyprinidae family, in addition to at least nine species of fishes from eight other families and possibly a few freshwater shrimp. The cercariae remain in the second intermediate host until the definitive host, which may be a human or other mammal, eats the fish. The flukes, now metacercariae, encyst in the host's small intestine and transform themselves into immature flukes, which then migrate to the bile ducts in the host's liver. There, the fluke matures in about three weeks and lays eggs that exit the host in the feces.

FEEDING ECOLOGY AND DIET
As a parasitic organism, *Clonorchis sinensis* relies on hosts to meet its dietary needs. Snails serve as hosts for its miracidia, fishes for its cercariae and metacercariae, and humans or other fish-eating predators for the adult flukes. Sporocysts form in snails. Metacercarial encyst in fishes. Adults live in the small bile ducts of the liver of definitive hosts and dine on their blood.

REPRODUCTIVE BIOLOGY
Adults are hermaphroditic. Adult liver flukes can lay as many as 4000 eggs per day. The operculate eggs are shaped like

light bulbs, yellowish brown in color, and range from 0.0009–0.001 in (25–35 µm) long and 0.0004–0.0007 in (11–20 µm) wide. The miracidia are ciliated and ovate with a short anterior protrusion. The cercariae resemble tadpoles with long tails. The life span of *Clonorchis sinensis* is at least 10 years, reportedly reaching 30–40 years.

CONSERVATION STATUS
Not listed by IUCN.

SIGNIFICANCE TO HUMANS
Clonorchiasis is a disease that commonly affects people in parts of Asia, where people may eat uncooked infected fishes. Infections in other parts of the world result from eating imported infected fishes. Transmission of *Clonorchis sinensis* via infected dried, salted, smoked, or pickled fish has also been reported. ◆

Lancet fluke
Dicrocoelium dendriticum

ORDER
Plagiorchiida

FAMILY
Dicrocoeliidae

TAXONOMY
Dicrocoelium dendriticum Rudolphi, 1819, intestine of *Xiphias gladius*, a swordfish (probably in error).

OTHER COMMON NAMES
English: Lancet liver fluke; French: Petite douve du foie; German: Kleiner Leberegel, Lanzettegel.

PHYSICAL CHARACTERISTICS

Adult lancet flukes have translucent bodies shaped like long, thin leaves. Both oral and ventral suckers are located toward the front of the body, with the foremost oral sucker a bit smaller than the others. This species averages about 0.02–0.06 in (5–15 mm) long and 0.04–0.08 in (1–2.5 mm) wide.

DISTRIBUTION

Northeastern United States; Australia; northern and central Europe; Asia; and Africa.

HABITAT

Dicrocoelium dendriticum prefers dry habitats. It begins its life as an egg in the feces of its definitive hosts. Its first intermediate hosts are terrestrial snails, including *Helicella* species and *Cionella lubrica*; its second intermediate hosts are ants (*Formica fusca*); and its definitive hosts include various mammals, including sheep, cattle, and pigs, as well as cottontail rabbits, deer, and woodchucks. In snails the miracidia hatch in the intestine, then migrate through the intestinal wall. In ants, metacercariae encyst in the gaster. In the definitive hosts, immature flukes leave the metacercaria cysts and migrate to the common bile duct.

BEHAVIOR

The eggs of lancet flukes are found on dry land, where they are eaten by snails, including *Cionella lubrica* in the United States. Hatched miracidia develop into sporocysts that in turn produce a second generation of sporocysts. The sporocysts transform into cercariae and migrate to the snail's lung cavity. The snail encases as many as 400 cercariae in mucus, then ejects the slime balls thus formed through its respiratory pore. Ants (*Formica fusca*) become the second intermediate host by ingesting the slime balls. While most of the infection in ants occurs in the hemocoel (spaces between the cells and tissues), some still-immature flukes make their way into the subesophageal ganglion (cluster of nerve cells underneath the eosphagus), which causes the ants' behavior to change. The ants are impelled to climb to the tips of grass blades. When the temperature drops at night, the ants' jaws clamp onto the grass, and keep them attached to the grass until temperatures rise the next morning. Herbivores grazing early in the morning inadvertently eat the ants with the foliage and become infected with the flukes. The metacercariae travel to the new host's bile duct, gall bladder, and pancreatic ducts, where they mature and lay their eggs. The mammals eliminate the eggs in their feces, and the life cycle repeats.

FEEDING ECOLOGY AND DIET

Lancet flukes meet their nutritional needs through various hosts, including snails, ants, and plant-eating mammals. In snails, mother sporocysts form, giving rise to daughter sporocysts and finally cercariae.

REPRODUCTIVE BIOLOGY

Adults are hermaphroditic. The dark brown, ovate eggs are operculate, about 0.0014–0.0019 in (36–48 μm) long and 0.0008–0.0012 in (22–30 μm) wide. The cercariae are long with nipped-in "waists." The fluke's anterior portion is ovate and the posterior portion tapered.

CONSERVATION STATUS

Not listed by the IUCN.

SIGNIFICANCE TO HUMANS

When left untreated, infected livestock may develop progressive hepatic cirrhosis. In sheep, the condition can reduce reproductive capacity and wool production. ◆

Salmon-poisoning fluke
Nanophyetus salmincola

ORDER
Plagiorchiida

FAMILY
Troglotrematidae

TAXONOMY
Nanophyetus salmincola Chapin, 1927, dogs, *Canis lestes*, *Procyon psora pacifica*, *Lynx fasciatu*.

OTHER COMMON NAMES
None known.

PHYSICAL CHARACTERISTICS

The ovate adult form of *N. salmincola* is about 0.03–0.08 in (0.8–2.5 mm) long and about 0.011–0.3 in (0.3–0.8 mm) wide. Two similar-sized suckers are present, one located at the anterior edge, and the other about a third of the way back on the body. The tegument is spiny. The pair of testes are large oval structures that extend from the central to the posterior body. In contrast, a small round ovary is located near the ventral sucker. A genital pore opens just behind the ventral sucker.

DISTRIBUTION

North Pacific basin, stretching from the Columbia River basin through British Columbia to Alaska.

HABITAT

N. salmincola is a fluke of freshwater streams when it is not infesting other animals. The eggs hatch in fast-moving streams, the miracidia infect snails (commonly *Oxytrema silicula*), and the cercariae penetrate the skin of frogs and fish, especially salmon and trout. The definitive hosts include birds and mammals, particularly skunks and raccoons, and occasionally humans. Immature and adult flukes inhabit the wall of the small intestine. Cercariae live free (not embedded) in the snail's tissues and enter the muscular tissue of fishes.

BEHAVIOR

The eggs of this parasitic flatworm incubate in freshwater streams for a period of about three to seven months. The newly hatched miracidia randomly find and then burrow into snails, where they develop into cercariae. The cercariae exit the snails and move in random patterns through the water. They can survive in this free-living form for as long as two days. When they encounter a frog or fish, they quickly penetrate its tissue—a process that takes 30–120 seconds—and migrate to the muscles, kidneys and fins. The host sometimes dies from heavy infestations. Migratory fishes like salmon may carry *N. salmincola* many miles. Predatory birds and mammals ingest infected fish and frogs, and the flatworms encyst in the small intestine, where they mature within seven days and lay eggs. The eggs pass from the definitive host in the feces, and the life cycle begins again.

Salmon poisoning in dogs and other canids is often blamed on *N. salmincola*. The actual poisoning agent is a rickettsia bacteria known as *Neorickettsia helminthoeca*, which infects all life stages of *N. salmincola*. If untreated, infection by *N. helminthoeca* causes distemper-like symptoms and may be fatal in dogs. It does not, however, affect humans or cats. Dogs should never be fed raw salmonids.

FEEDING ECOLOGY AND DIET

N. salmincola is dependent on hosts to meet its nutritional needs. Its hosts include the snail *Oxytrema silicula* as the first intermediate host, salmonid fish as the second intermediate host, and a variety of fish-eating mammals, including dogs, cats, and humans, as definitive hosts. Metacercarial encyst in fishes. Immature flukes in dogs attach to the intestinal wall.

REPRODUCTIVE BIOLOGY

The unembryonated eggs of *N. salmincola* are yellowish brown in color, and about 0.003–0.0039 in (80–100 μm) long and 0.0015–0.0019 in (40–50 μm) wide. Incubation may take from 87–200 days at room temperature. The rediae are elongated with birth pores located toward the front. The cercariae are also elongated, but have a ventral sucker and short tail. Thousands of cercariae may be deposited at a time in long strings of mucus.

CONSERVATION STATUS

Not listed by IUCN.

SIGNIFICANCE TO HUMANS

Humans who eat raw or underprocessed infected fish are susceptible to the disease called nanophyetiasis or "fish flu," but such cases are rare. The symptoms of fish flu, which arise 5–8 days after eating infected fish, may include diarrhea or abdominal discomfort. Humans are unaffected by *N. ricksettia*, which is carried by *N. salmincola*, but such an infection can be fatal to dogs if left untreated. Heavy infestations in salmon and trout may cause some losses in populations of these sport fishes. ◆

Black-spot flatworm

Uvulifer ambloplitis

ORDER

Strigeatida

FAMILY

Diplostomatidae

TAXONOMY

Uvulifer ambloplitis (Hughes 1927) *Ceryle alcyon, Helisoma trivolvis, H. campanulatum, Ambloplitis rupestris, Micropterus dolomieu, Aplites salmoides, Eupomotis gibbosus, Apomotis cyanellus, Enneacanthus obesus.*

OTHER COMMON NAMES

English: Black grub.

PHYSICAL CHARACTERISTICS

Adults are spoon-shaped with a broad body, thinner "neck," and a slightly broader cocked "head." The oral sucker is small and foremost on the fluke, with a small ventral sucker about halfway back on the head, and a holdfast organ just behind the central sucker. Adults range from 0.08–0.09 in (1.8–2.3 mm) long. Two testes are present, one in front of the other in the posterior half of the animal. The comparatively small ovary is located in front of the anterior testis. A common genital pore opens at the posterior tip of the animal.

DISTRIBUTION

Northern tip of South America, Central America, and much of North America.

HABITAT

The first intermediate hosts are snails, including *Helisoma trivolvis*. Second intermediate hosts are commonly green sunfish

(*Lepomis cyanellus*) and bluegill (*L. macrochirus*). The most common definitive host is the bird known as the belted kingfisher (*Megaceryle alcyon*). Sporocysts migrate to the digestive gland and liver of snails. Cercarial bore through the skin of fish, and cysts appear.

BEHAVIOR

Eggs of this species hatch into miracidia within a month. The ciliated miricidia burrow into their snail intermediate hosts, and develop into first-generation sporocysts. Second-generation sporocysts spread to the snail's digestive gland and liver, where they produce cercariae about six weeks after the initial infection. The cercariae leave the first intermediate host, swim to, and penetrate the skin of sunfish, bass, perch, and other fishes. They drop their tails and develop into metacercariae that encyst and are surrounded by host melanocytes that appear as hard black spots in the fish's skin. (Other species of Diplostomatidae and Strigeidae also produce black spots in fishes.) When kingfishers eat the infected fish, the worms mature within the host's intestine in about a month. Eggs deposited are passed by the bird through its feces into the snail's environment.

FEEDING ECOLOGY AND DIET

The black-spot flatworm depends for its nutrition on host species, including the ram's horn snail (*Helisoma* species); percid fishes; and birds, especially belted kingfishers.

REPRODUCTIVE BIOLOGY

Eggs are about 0.00004 in (1 μm) long and 0.00002 in (0.5 μm) wide; the miracidia are ovate and broader at the anterior end. The life cycle of the black-spot flatworm follows the prevalence and activity of its two host species. Cercariae begin appearing in the fish hosts in late spring following a 21-day development period. The snails continue to shed cercariae throughout the summer. The cercariae resemble tadpoles with a forked tail.

CONSERVATION STATUS

Not listed by IUCN.

SIGNIFICANCE TO HUMANS

Infections with metacercariae appear as black spots in several species of game fish. Their appearance sometimes discourages humans from eating them, even though there is no danger, especially when cooked. "Blackspot disease" is sometimes fatal to fish fry. ◆

Human blood fluke

Schistosoma mansoni

ORDER

Strigeatida

FAMILY

Schistosomatidae

TAXONOMY

Schistosoma mansoni Sambon, 1907, Africa.

OTHER COMMON NAMES

French: Schistosome intestinal; German: Pärchenegel

PHYSICAL CHARACTERISTICS

A sexually dimorphic species, the female is thin and cylindrical, reaching between 0.5–1 in (1.2–2.5 cm) in length. The male is elongated but thicker, and reaches 0.39–0.78 in (1–2 cm) in length. He also has small spiny oral and ventral suckers, and a wrinkled dorsal surface dotted with tubercles (small nodules). A female typically exists in the male's gynecophoric canal, absorbing nutrients by diffusion from the male.

DISTRIBUTION
The human blood fluke is found in warm regions around the world, particularly in developing countries in South America, Africa, the Caribbean, and the Middle East.

HABITAT
Larvae of this digenetic parasitic flatworm infect the hepatic pancreas of freshwater snails of the genus *Biomphilaria*. Adult worms infect mesenteric veins of mammals ranging from rodents, dogs, cattle, and baboons to humans.

BEHAVIOR
Eggs, which have spines, lodge in the intestinal mucosa and cause ulceration as well as the formation of granulomata (lesions that result from a chronic inflammation). The eggs hatch in freshwater areas and develop into miracidia, which follow chemical, light and gravitational cues to find and then penetrate the soft tissues of the snails. A sporocyst then forms. Cercariae develop within the sporocyst. The cercariae leave the snail by swimming and actively seek out their next host, apparently targeting certain fatty acids of the skin. They then penetrate the skin of a secondary host, which may be a human being or other mammal. Once in the host, they become immature flukes called schistosomules, migrate to the circulatory system, and travel to a site (specifically the superior and inferior mesenteric veins and related smaller veins) near the large intestine. Once in the mesenteric veins, they mature, mate, and

lay eggs, many of which leave the host's body with the feces. The cycle begins again when the eggs make their way into the freshwater habitat of the snail.

FEEDING ECOLOGY AND DIET
The hosts of the human blood fluke include *Biomphilaria* snails and mammals, including humans. They enter both snails and mammals by burrowing into the skin and other soft tissue. They feed on blood from the hepatic and mesenteric veins.

REPRODUCTIVE BIOLOGY
Adult males are typically found conjoined with the females, with the female remaining in the male's spine-covered gynecophoric canal, a groove that runs along the lower surface of the body. Larger spines cover the lateral margins of the canal, which suggests that they may play a role in mating. The eggs, which bear a lateral spine, range from 0.0045–0.0068 in (115–175 μm) long and 0.0017–0.0027 in (45–70 μm) wide. The free-swimming miracidia are tadpole-like in appearance with a tail slightly smaller than the body. The cercariae are about 0.019–0.039 in (0.5–1 mm) long, with constricted waists and forked tails.

CONSERVATION STATUS
Not listed by IUCN.

SIGNIFICANCE TO HUMANS
Infection with this species results in schistosomiasis, also known as bilharziasis, in humans. The condition causes abdominal pain, dysentery, lethargy, and anemia, leaving the victim weak and susceptible to other diseases. ◆

Resources

Books

Doss, Mildred A. *Index Catalogue of Medical and Veterinary Zoology: Trematoda*, Parts 1–8. Washington, DC: U. S. Government Printing Office, 1966.

Olsen, O. Wilford. *Animal Parasites: Their Biology and Life Cycles*. Minneapolis: Burgess Publishing Co., 1967.

Schell, Stewart C. *How to Know the Trematodes*. Dubuque, IA: William C. Brown Co., Publishers, 1970.

Periodicals

Combes, C., A. Fournier, H. Moné, and A. Théron. "Behaviours in Trematode Cercariae That Enhance Parasite Transmission: Patterns and Processes." *Parasitology* 109 (1994): S3–S13.

Esch, G. W., M. A. Barger, and K. Joel Fellis. The Transmission of Digenetic Trematodes: Style, Elegance, Complexity." *Integrative and Comparative Biology* 42, no. 2 (2002): 304–312.

Esch, G. W., E. J. Wetzel, D. A. Zelmer, and A. M. Schotthoeffer. "Long-Term Changes in Parasite Population and Community Structures: A Case History." *American Midland Naturalist* 137 (1997): 369–387.

Kanev, I. "Life-Cycle, Delimitation, and Redescription of *Echinostoma revolutum* (Froelich, 1802) (Trematoda: Echinostomatidae)." *Systemic Parasitology* 28 (1994): 125–144.

Machado-Silva, J. R., R. M. Lanfredi, and D. C. Gomes. "Morphological Study of Adult Male Worms of *Schistosoma*

mansoni Sambon, 1907 by Scanning Electron Microscopy." *Memorias do Instituto Oswaldo Cruz Online* 92, no. 5 (1997): 647–653.

Maurice, J. "Beware the Flukes of Nature." *World Press Review* 41 (June 1994): 41.

McCarthy, H. O., S. Fitzpatrick, and S. W. B. Irwin. "Life History and Life Cycles: Production and Behavior of Trematode Cercariae in Relation to Host Exploitation and Next-Host Characteristics." *Journal of Parasitology* 88, no. 5 (2002): 910–918.

Self, J. T., L. E. Peters, and E. D. Davis. "The Egg, Miracidium and Adult of *Nematobothrium texomensis*." *Journal of Parasitology* 49, no. 5 (1963): 731–736.

Yoshimura, H. "The Life Cycle of *Clonorchis sinensis*: A Comment on the Presentation in the Seventh Edition of Craig and Faust's *Clinical Parasitology*." *Journal of Parasitology* 51, no. 6 (1965): 961–966.

Organizations

American Society of Parasitologists. Web site: <http://asp.unl.edu>

Helminthological Society of Washington. c/o Allen Richards, Ricksettsial Disease Department, Naval Medical Research Center, 503 Army Grant Ave., Silver Spring, MD 20910-7500 United States.

Leslie Ann Mertz, PhD

Monogenea
(Monogeneans)

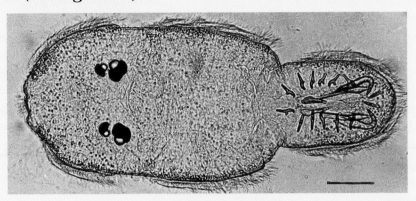

Phylum Platyhelminthes

Class Monogenea

Number of families 53

Thumbnail description

Flatworm parasites that live mainly on fish skin and gills, with a haptor (posterior attachment organ) containing hooks, and a direct (single host) life cycle

Photo: Electronic flash micrograph of a living cilated larva of *Entobdella soleae*. (Photo by G. C. Kearn. Reproduced by permission.)

Evolution and systematics

The origins of the monogeneans are obscure, but it is assumed that they evolved from free-living ancestors similar to modern turbellarians (aquatic flatworms). Their nearest relatives are the cestodes (tapeworms). The naming of their upper-level taxa (biological categories) is controversial. "Monogenea" is widely accepted as the name of the class itself, as is the subdivision of the class into two orders, namely the Monopisthocotylea and the Polyopisthocotylea. More recently, the name "Monogenoidea" has been proposed for the class, with the class being subdivided in turn into three subclasses. The first-mentioned, older, and more widely supported scheme is followed here. Some believe that the Monopisthocotylea and the Polyopisthocotylea have separate origins; that is, they think that Monogenea is not a monophyletic (descended from a common ancestor) class. As of 2003, 53 families of monogeneans are recognized.

Physical characteristics

Like other platyhelminths, monogeneans are acoelomate animals; that is, they do not have a body cavity lying between the body wall and the digestive tract. Their bodies are covered by a living syncytial tegument. The digestive system consists of a pharynx, which is a muscular tube used to suck in food, and a saclike or branched intestine with no anus. The pharynx may or may not be glandular. Monogeneans range in length from about 0.04–0.08 in (1–2 mm), as in some gyrodactylids, to more than 0.75 in (2 cm), as in some capsalids. Large monogeneans tend to be flat and leaf-shaped, but the smaller parasites are usually cylindrical. In general, these flatworms are colorless and semitransparent. When on fish skin some may be virtually invisible to the human eye, either by virtue of their glass-like transparency or because they contain scattered pigment that matches pigment in the host's skin. These features may protect the parasites from being eaten by predatory fishes or crustaceans ("cleaner" organisms). The

brown/black coloration of the polyopisthocotyleans is associated with their digestive system and derived from their blood meals.

The most significant anatomical feature of monogeneans is their possession of a posterior attachment organ or haptor armed with hooks. The hooks usually fall into two groups: small hooklets, which are often called marginal hooklets, and larger hooks called hamuli or anchors. The hooklets are essentially found in larvae although they often persist, usually without further growth, in adult monogeneans. The hooklets are specialized for attachment to the upper layer of cells (epidermal cells or Malpighian cells) in the host's skin; they fasten themselves in the web of filaments made of keratin known as the terminal web, which lies beneath the apimal membrane of the host epidermal cell. A basic and maximum number of 16 hooklets occurs in monopisthocotyleans and in polyopisthocotyleans, although they may be reduced to 14, 10, or lost altogether. In many but not in all monogeneans, the hooklets are supplemented as the parasite grows by one or two pairs of hamuli, which are usually large enough to penetrate through the epidermis (the outer layer of the host's skin) into the dermis, which is the thicker layer of skin just below the epidermis. Some monopisthocotyleans have haptors that become more elaborate during their development, either by acquiring glands or by subdividing into separate small compartments that function as suckers. As typical polyopisthocotyleans grow, they develop three or four pairs of muscular suckers or clamps; each one is at the site of a hooklet. Many monopisthyocotyleans are able to move like leeches since they have suckers or glands that secrete sticky material on the lateral borders of the head.

The ovary consists of a germarium, which produces egg cells, and an extensive vitellarium that produces vitelline cells. The vitelline cells do not contain genetic material; they secret substances that form a chemically and physically resistant eggshell of tanned protein (sclerotin) and provide food for the growing embryo. Gyrodactylids are exceptional viviparous. Monogeneans are hermaphrodites; many have hard structures (sclerites) supporting the penis, while others have an eversible cirrus. There may be one or more vaginae, often with supporting sclerites, but hypodermic impregnation (through the skin) also takes place.

Distribution

Monogeneans are found worldwide in freshwater and marine environments.

Habitat

Many monogeneans are strictly host-specific; that is, they are limited to a single or a few closely related hosts. Skin parasites may be widespread on the surface of the host's skin or concentrated in specific areas. Many monopisthocotyleans use two pairs of counter-rotating hamuli, one pair on their ventral surface and one pair on the dorsal surface, to attach themselves between two adjacent secondary gill lamellae on one of the fish's primary lamellae. The monocotylid monopisthocotyleans are parasites of elasmobranch fishes, and must occupy internal sites such as their nasal fossae (cavities), body cavity, and cloaca. Another monopisthocotylean, *Amphibdella*, spends its early life inside the heart of an electric ray, moving to the gills when it is an adult and releasing its eggs into the ray's gill cavity. Many polyopisthocotyleans use their clamps to grip one or two secondary gill lamellae. Other polyopisthocotyleans (polystomatids) use suckers to attach themselves inside the bladders of frogs and toads, or the bladders or mouths of freshwater turtles. *Oculotrema hippopotami* is the only monogenean that infests a mammal; it is a polystomatid that lives beneath the eyelids of the hippopotamus. The gyrodactylid *Isancistrum* lives on the skin of squids and is the only monogenean parasite that infests an invertebrate.

Behavior

Many monogeneans migrate, moving like leeches from their site of initial attachment to the host to the site where they mate and lay their eggs. How they find their way within the host is mostly unknown as of 2003. Many monopisthocotyleans retain the ability to change their location on the host throughout their lives. Most adult polyopisthocotyleans, however, are sedentary. Some skin parasites ventilate their bodies by undulating while some juvenile and adult parasites are able to swim.

Feeding ecology and diet

Most monopisthocotyleans feed on the epidermis of their host, which is eroded by a protrusible glandular pharynx. Polyopisthocotyleans are blood feeders. These parasites accumulate indigestible residues of brown/black hematin (a pigment found in blood) from their blood meals in their digestive tract, which they eject at intervals through their mouths.

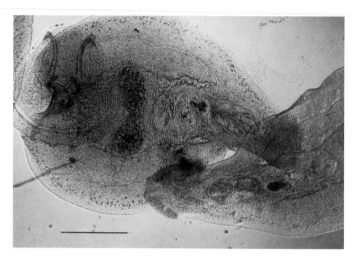

Two *Entobdella soleae* exchanging spermatophores. (Photo by G. C. Kearn/Cambridge University Press. Reproduced by permission.)

Reproductive biology

Monogeneans are hermaphrodites, with the male reproductive system usually first to mature. This characteristic is known as protandry. Mutual or unilateral insemination may occur, although self-insemination also takes place. The tanned eggs are assembled in an egg mold or ootype; the vitelline cells provide the raw material for the shell and resources for the developing embryo. With the exception of the viviparous gyrodactylids, the eggs are released into the environment and produce infective larvae (oncomiracidia) which are able to swim freely with the help of cilia. The larvae of many monogeneans hatch spontaneously at a particular time of day, which often coincides with times when the host is particularly vulnerable to invasion. Hatching may also be triggered by such host-derived cues as chemicals, mechanical disturbance, or shadows. The oncomiracidia do not feed until they reach the host, which means that their survival as free-living organisms and their potential for host infection are limited, usually to a period of several hours. The oncomiracidia throw aside their ciliated cells when they reach the host. *Entobdella soleae* can infect new hosts by direct transfer of adults or juveniles, as well as by eggs and oncomiracidia. The juveniles of a related parasite can swim and may reach new hosts in this way.

The gyrodactylids are unique among monogeneans. They have abandoned freely deposited eggs and free-swimming oncomiracidia; they are viviparous, producing offspring that are usually full size at birth. They increase their reproductive rate by telescoping generations; their offspring already contain a partly developed embryo at birth and sometimes a second smaller embryo within the first. The appearance of the male reproductive system is delayed until after the female reproductive system is operational. This characteristic, which is known as protogyny, helps to concentrate the organism's resources on embryo development. The first two offspring are probably produced asexually. Gyrodactylids usually spread to new hosts by direct transfer when infected hosts make contact with one another. In addition, however, the hosts may be infected by gyrodactylids drifting freely in the water column or by making contact with parasites attached to the substrate. There is also one record of a gyrodactylid that can swim.

Conservation status

No species of monogeneans are listed by the IUCN.

Significance to humans

In the wild, the number of monogeneans living on an individual host is generally low, and infestations of these parasites do not usually cause disease. In crowded fish farms, however, parasite populations often increase uncontrollably and the hosts may be damaged or killed.

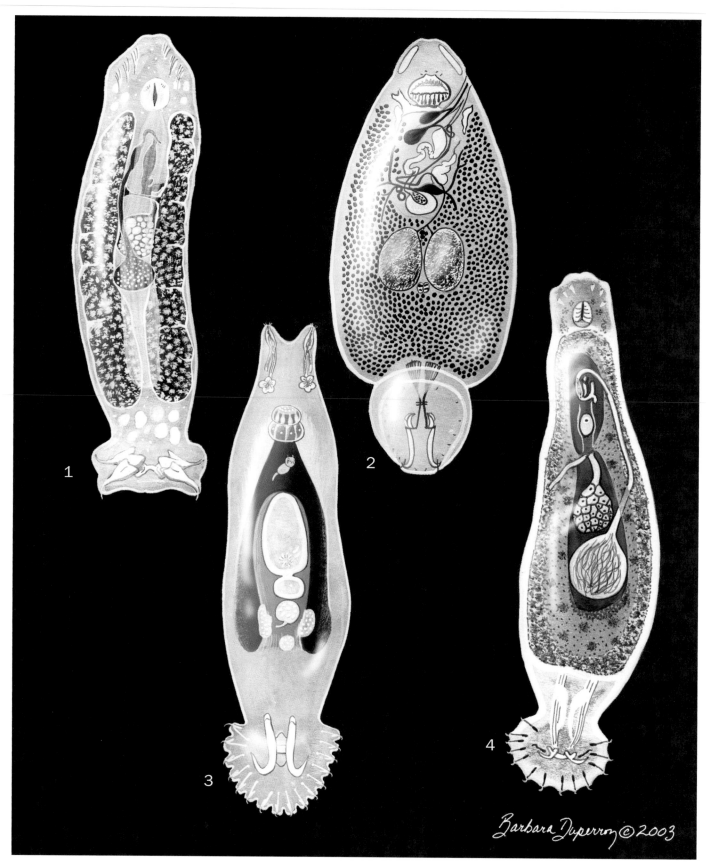

1. *Tetraonchus monenteron*; 2. *Entobdella soleae*; 3. *Gyrodactylus pungitii*; 4. *Dactylogyrus vastator*. (Illustration by Barbara Duperron)

1. *Gastrocotyle trachuri*; 2. *Polystoma integerrimum*; 3. *Pseudodiplorchis americanus*; 4. *Diclidophora merlangi*. (Illustration by Barbara Duperron)

Species accounts

No common name
Entobdella soleae

ORDER
Monopisthocotylea

FAMILY
Capsalidae

TAXONOMY
Entobdella soleae van Beneden & Hesse, 1864; Johnston, 1929.

OTHER COMMON NAMES
None known.

PHYSICAL CHARACTERISTICS
Adults 0.07–0.23 in (2 –6 mm) long. Body white or yellowish, flattened with disc-shaped haptor. Embedded in haptor: two pairs of hamuli; one pair of central accessory sclerites that may represent a modified central pair of hooklets; and 14 tiny peripherally located hooklets. Mouth on ventral surface. Conspicuous glandular pharynx. Four eyespots. Adhesive pad on each side of head. Two testes and a large muscular unarmed "penis."

DISTRIBUTION
Not systematically mapped. Parasite of common sole, *Solea solea*, sand sole, *Pegusa lascaris* and Senegalese sole, *Solea senegalensis* on eastern Atlantic seaboard of Europe.

HABITAT
Adult parasites on lower surface of sole. Fifty percent of common soles off English coast carry 1–6 adult parasites.

BEHAVIOR
Body muscles generate haptor suction by lifting anterior pair of hamuli relative to accessory sclerites. Edge of haptor sealed by valve. Soles partly bury themselves in sediments with low levels of oxygen; parasites respond to this condition by undulating their bodies. Their flat bodies also spread and become thinner. These responses increase the availability of oxygen to the organism and enhance its uptake. Adults and juveniles orientate themselves with respect to the host's scales, most probably using their sense of touch. May use scales as clues when moving forward along the upper surface of the host. Movement achieved by alternately attaching the haptor and the sticky pads.

FEEDING ECOLOGY AND DIET
Feed on host's epidermis, which is eroded by the parasite's protrusible pharynx.

REPRODUCTIVE BIOLOGY
Mutual exchange of spermatophores occurs. Parasite lays tetrahedral eggs, each with long stalk. Two eggs laid per hour at 53°F (12°C). Eggs attach to sand grains— not to fish— by sticky droplets on stalks. After incubation for 4 weeks at 53°F (12°C), eggs hatch soon after dawn in absence of host. Host is

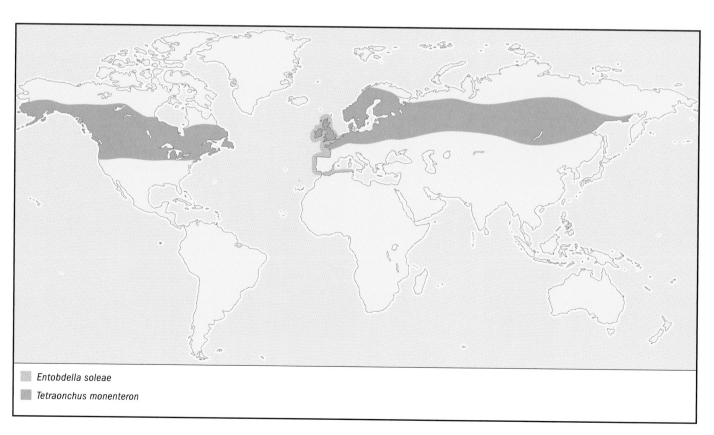

■ *Entobdella soleae*
■ *Tetraonchus monenteron*

nocturnal and rests during daylight; hence freshly hatched larvae have a stationary target. There is, however, an unknown ingredient in the mucus of the sole's skin that stimulates hatching at any time of day or night. Oncomiracidia usually attach to sole's upper surface and migrate forward, moving from the head to the host's lower surface where they reach sexual maturity. When two fishes make contact, adult and juvenile parasites may move from the lower surface of one fish to upper surface of the other. Parasites arriving on the recipient host find their way forward and move onto its lower surface.

CONSERVATION STATUS
Not threatened.

SIGNIFICANCE TO HUMANS
None known as of 2003. *Entobdella*, however, has the potential to kill soles by superinfection in captive situations. Its close relative *Neobenedenia melleni* poses an especially serious threat to fish farms and aquaria, because it combines the potential for massive superinfections with a remarkably low specificity. *Neobenedenia melleni* is capable of infecting over a hundred wild and captive fish species belonging to more than 30 families from 5 different orders. ◆

No common name
Dactylogyrus vastator

ORDER
Monopisthocotylea

FAMILY
Dactylogyridae

TAXONOMY
Dactylogyrus vastator Nybelin, 1924. *Dactylogyrus* is the largest helminth genus with more than 900 species. They live mostly on the gills of cypriniform fishes.

OTHER COMMON NAMES
None known.

PHYSICAL CHARACTERISTICS
About .045 in (1.25 mm) long. Two pairs of hamuli, with ventral hamuli reduced to spicules. One or two supporting bars. Fourteen hooklets, each with an enlarged handle. Three pairs of eversible adhesive sacs. Four eyespots. Male copulatory organ has hardened penis tube with accessory sclerite. Single testis.

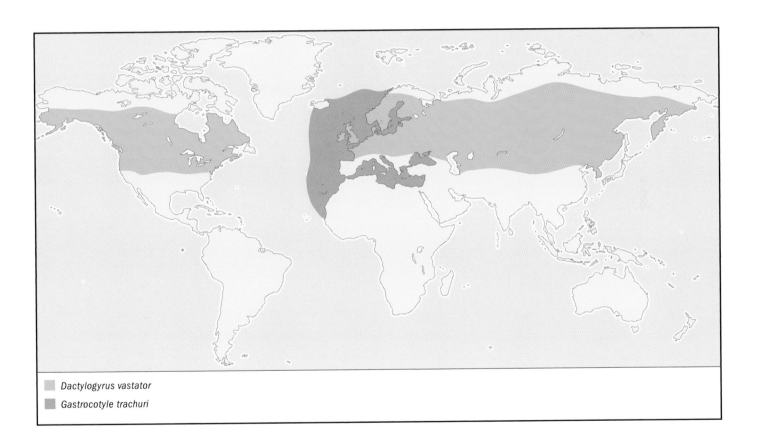

☐ *Dactylogyrus vastator*

☐ *Gastrocotyle trachuri*

DISTRIBUTION
Widespread in Palearctic region on common carp (*Cyprinus carpio*), Crucian carp (*Carassius carassius*), and goldfish (*Carassius auratus*).

HABITAT
Haptor lodges between two secondary gill lamellae.

BEHAVIOR
Little is known about behavior. Parasite secures itself to host principally by dorsal hamuli.

FEEDING ECOLOGY AND DIET
Feeds on gill epithelium.

REPRODUCTIVE BIOLOGY
Ovoid eggs are washed out of host's gill cavity and sink to bottom. Ciliated oncomiracidium emerges in 3–5 days, depending on water temperature. Larvae drawn into gill cavity by current attach themselves to host's gills. Some larvae may first attach to host's skin and then migrate to gills. Eggs are thought to spend the winter in a state of diapause, or period of inactivity.

CONSERVATION STATUS
Not threatened.

SIGNIFICANCE TO HUMANS
Causes mass mortality among fingerling carp in fish-rearing ponds. Abnormal multiplication of cells in the gill epithelium interferes with carp's respiratory function. Parasite is especially significant in the former Soviet Union and eastern and northern Europe where carp are bred for food. ◆

No common name
Gyrodactylus pungitii

ORDER
Monopisthocotylea

FAMILY
Gyrodactylidae

TAXONOMY
Gyrodactylus pungitii Malmberg, 1964. About 400 species attributed to *Gyrodactylus*. Morphologial and anatomical differences between the many species of *Gyrodactylus* are relatively small, making their taxonomy difficult.

OTHER COMMON NAMES
None known.

PHYSICAL CHARACTERISTICS
Length 0.023–0.03 in (0.6–0.9 mm). May contain developing embryo. Haptor fan-shaped with one pair of hamuli linked by two bars; sixteen hooklets arranged around periphery. Single pair of eversible adhesive sacs on head. No eyespots. Vitellarium absent or possibly reduced. Germarium reduced. Single testis; penis armed with hooks. Penis absent in young specimens.

DISTRIBUTION
Not systematically mapped, but likely to occur wherever *Pungitius pungitius*, its specific host, occurs.

HABITAT
Skin, fins (pharynx) of stickleback, *Pungitius pungitius*.

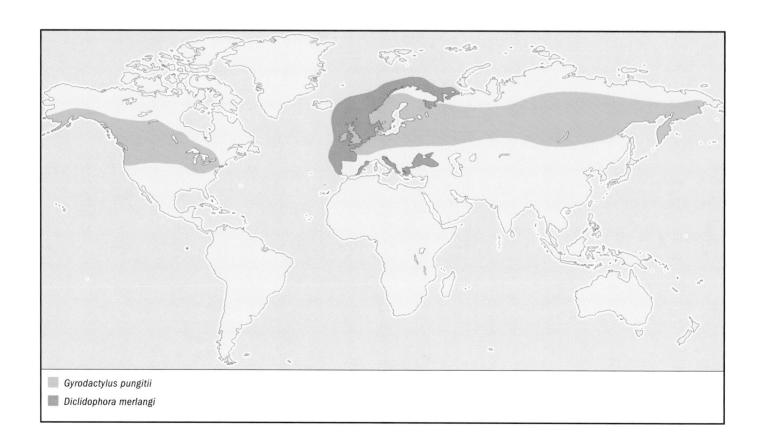

Gyrodactylus pungitii
Diclidophora merlangi

BEHAVIOR
Gyrodactylids rely on hamuli and hooklets for attachment. They move like leeches.

FEEDING ECOLOGY AND DIET
Feeds on fishes' epidermis.

REPRODUCTIVE BIOLOGY
Unique among monogeneans. New hosts usually infected by transfer of parasites when hosts make contact with one another. Population growth of many gyrodactylids is limited by host reaction. Mechanism of limitation largely unknown as of 2003.

CONSERVATION STATUS
Not threatened.

SIGNIFICANCE TO HUMANS
None known for *G. pungitius*. There are, however, strains of Norwegian salmon (*Salmo salar*) unable to control population growth of *G. salaris*. Young salmon (salmon parr), rarely exceeding 6 in (15 cm) in length, may support more than 10,000 parasites. The parasites' heavy grazing on the fishes' outer layer of skin, both in farms and in the wild, leads to loss of osmotic integrity, invasion by secondary disease agents, and death. Parasite probably introduced to Norway by Baltic salmon that have some resistance to it. ◆

No common name
Tetraonchus monenteron

ORDER
Monopisthocotylea

FAMILY
Tetraonchidae

TAXONOMY
Tetraonchus monenteron Wagener, 1857; Diesing, 1858.

OTHER COMMON NAMES
None known.

PHYSICAL CHARACTERISTICS
Body elongated, about 0.04 in (1 mm) in length. Haptor with four similar hamuli, arranged in lateral pairs, each pair comprising one dorsally orientated and one ventrally orientated hamulus. Transverse supporting bar between hamulus pairs. Sixteen hooklets present; one pair central on ventral surface, two pairs dorsally orientated. Haptor glands present. Head region with four eyespots and three pairs of adhesive sacs that can be turned inside out. Male copulatory organ is a hardened tube with associated accessory sclerite. Single testis.

DISTRIBUTION
Recorded in Europe, North America and Russia on its specific host, the freshwater pike, *Esox lucius*. Not systematically mapped.

HABITAT
Abundant on gills of *Esox lucius*. Haptor lodges between two secondary gill lamellae.

BEHAVIOR
The parasite secures itself to the host by counter-rotating its ventral and dorsal hamuli until they push through the sec-

ondary gill lamellae. Each laterally situated pair of hamuli is operated by a single muscle. This muscle gives rise to a long tendon, which is threaded through loops attached to the hamuli. The arrangement resembles a pulley system and is likely to confer mechanical advantage. The hamuli are assisted by hooklets and possibly by glands. Sticky eversible sacs on head permit leech-like movement along the gill.

FEEDING ECOLOGY AND DIET
Feeds on gill epithelium; may take blood when the secondary gill lamella is ruptured.

REPRODUCTIVE BIOLOGY
Mating not recorded. Ovoid eggs presumably leave the gill chamber and settle to the bottom. Ciliated oncomiracidium hatches after 3–4 days. Immature parasites have been found on host skin.

CONSERVATION STATUS
Not threatened.

SIGNIFICANCE TO HUMANS
None known. ◆

No common name
Diclidophora merlangi

ORDER
Polyopisthocotylea

FAMILY
Diclidophoridae

TAXONOMY
Diclidophora merlangi Nordmann, 1832; Kroyer, 1838.

OTHER COMMON NAMES
None known.

PHYSICAL CHARACTERISTICS
Length may exceed 0.27 in (7 mm). Haptor has four pairs of clamps on stalks (peduncles); no hooks in adult. Mouth terminal (anterior) located at one end; buccal cavity contains two buccal suckers. No sticky pads or eversible sacs. No eyes. Small pharynx. Numerous testes. Penis bulb armed with spines.

DISTRIBUTION
Likely to be present wherever specific host whiting, *Merlangius merlangus*, is found, but not systematically mapped.

HABITAT
Gill parasite, strictly host-specific to *Merlangius merlangus*; 1–12 parasites per fish. Adults prefer first gill arch.

BEHAVIOR
Parasite attaches itself to secondary gill lamellae using clamps. No evidence that adult is able to change location on gill. Each clamp has two jaws supported by sclerites. Jaws drawn together by suction generated by lifting a diaphragm near hinge line of clamp. Clamp peduncles enable parasite to span as many as five primary gill lamellae. Parasite's body lies between two hemibranchs of a gill. Gill-ventilating current washes both sides of parasite equally; consequently parasite has symmetrical shape.

FEEDING ECOLOGY AND DIET
Feeds on blood.

REPRODUCTIVE BIOLOGY
Juveniles probably migrate mainly to first gill, where adults settle near one another. Insemination is unilateral in detached adults. Sperms enter recipient via breaches in tegument made by genital spines.

CONSERVATION STATUS
Not threatened.

SIGNIFICANCE TO HUMANS
None known. ◆

No common name
Gastrocotyle trachuri

ORDER
Polyopisthocotylea

FAMILY
Gastrocotylidae

TAXONOMY
Gastrocotyle trachuri van Beneden and Hesse, 1863.

OTHER COMMON NAMES
None known.

PHYSICAL CHARACTERISTICS
Oncomiracidia and early post-larvae are symmetrical; however, 25–35 clamps develop on one side of body, producing a strongly asymmetrical adult that may exceed 0.15 in (4 mm) in length. Small terminal (posterior) flap bearing hamuli and hooklets. Mouth terminal located at one end; buccal cavity encloses two buccal suckers. No sticky pads or eversible sacs. No eyes. Small pharynx. Testis follicular. Muscular penis bulb with penis tube and ring of spines.

DISTRIBUTION
Likely to be present wherever its specific host the scad *Trachurus trachurus* occurs, but not systematically mapped.

HABITAT
Attaches to secondary gill lamellae of *Trachurus trachurus* by clamps. Single row of clamps attaches to the upstream, outer (efferent) region of secondary lamellae; body then drifts downstream between hemibranchs. Gill ventilating current washes only one side of parasite, promoting asymmetrical development. Asymmetry may be on either the right or left side, depending on which side of primary lamella the parasite attaches itself.

BEHAVIOR
Clamps close, not by suction, but by relatively simple mechanical arrangement. Tendon from single muscle threads through hole in supporting sclerite of one jaw and attaches to other jaw. When muscle contracts, jaws are drawn together. Adult parasite probably unable to change its site on the gill.

FEEDING ECOLOGY AND DIET
Feeds on blood.

REPRODUCTIVE BIOLOGY
Male copulatory organ probably serves for hypodermic (through the skin) impregnation. Infection of scad takes place on sea bottom; in summer at Plymouth, U.K., however, scad

become planktonic. Parasites appear to anticipate this change either by ceasing to lay eggs or by laying eggs that enter diapause. Reproductive behavior of parasites may be controlled by hormonal changes in the host. The parasite has access to these hormonal changes via its blood meals.

CONSERVATION STATUS
Not threatened.

SIGNIFICANCE TO HUMANS
None known. ◆

No common name
Polystoma integerrimum

ORDER
Polyopisthocotylea

FAMILY
Polystomatidae

TAXONOMY
Polystoma integerrimum Froelich, 1791; Rudolphi, 1808.

OTHER COMMON NAMES
None known.

PHYSICAL CHARACTERISTICS
Adult about 0.39 in (10 mm) in length. Haptor has six muscular suckers, one pair of hamuli and 16 hooklets, each sucker containing one hooklet. Oral sucker around mouth. Testis follicular.

DISTRIBUTION
Likely to be present wherever its specific host, the brown frog, *Rana temporaria*, occurs, but not systematically mapped.

HABITAT
Bladder of *Rana temporaria*.

BEHAVIOR
Nothing is known for adult.

FEEDING ECOLOGY AND DIET
Feeds on blood.

REPRODUCTIVE BIOLOGY
Adult parasite accumulates reserves, but assembles no eggs while frog is living on land during most of year. When frogs enter water to spawn in spring, the parasite assembles and lays its eggs. Hormonal changes in the host may control parasite reproduction; parasites feed on blood and would have access to hormones circulating in the blood. Ciliated oncomiracidia invade frog tadpole gills. If attached to young tadpole, parasites become precociously sexually mature and lay a few eggs. Oncomiracidia infecting older tadpoles remain immature; when the host undergoes metamorphosis, they migrate via the digestive tract and possibly the skin to the host's bladder where they mature. Single egg retained in uterus of adult develops and hatches; larva remains in host of its parent, increasing the host's parasite burden by autoinfection. In *Polystoma nearcticum*, which infests North American tree frogs, the parasite's egg assembly switches on and off abruptly as sexual activity of host switches on and off during nocturnal spawning episodes on successive nights.

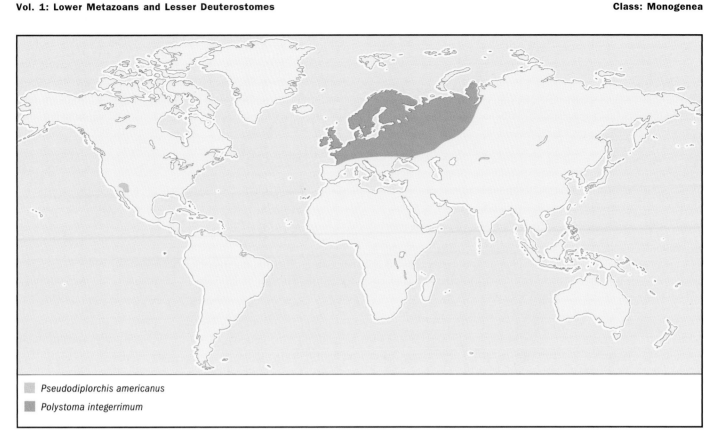

Pseudodiplorchis americanus
Polystoma integerrimum

CONSERVATION STATUS
Not threatened.

SIGNIFICANCE TO HUMANS
None known. ◆

No common name
Pseudodiplorchis americanus

ORDER
Polyopisthocotylea

FAMILY
Polystomatidae

TAXONOMY
Pseudodiplorchis americanus Rodgers and Kuntz, 1940; Yamaguti, 1963.

OTHER COMMON NAMES
None known.

PHYSICAL CHARACTERISTICS
Length of adult about 0.31–0.39 in (8–10 mm). Haptor with six muscular suckers, each with an embedded hooklet. Sixteen hooklets. Fully developed, encapsulated oncomiracidia accumulate in huge uterus. Vitellarium greatly reduced. Pair of compact testes, positioned laterally in the anterior region of the body. Oral sucker present.

DISTRIBUTION
Deserts of Arizona, United States.

HABITAT
Bladder of spadefoot toad, *Scaphiopus couchii*.

BEHAVIOR
No information available for adult.

FEEDING ECOLOGY AND DIET
Feeds on blood.

REPRODUCTIVE BIOLOGY
Desert toad spends most of year buried about 3 ft (1 m) below surface of sand. Parasite in toad's bladder accumulates large numbers of eggs in uterus. Toads emerge following torrential annual rains; they spawn in temporary rain pools over a period of two or three nights. As toad enters pool, eggs of parasite are released and hatch immediately. Eggshell is reduced to thin membranous sac. Ciliated oncomiracidia are exceptionally large: 0.0236 in (600 μm) in length, compared to 0.00984 in (250 μm) for the oncomiracidia of *Entobdella soleae*. Larvae survive for 48 hours at 77–81°F (25–27°C). They attach to the toad, migrating over its skin to the nostrils. Larvae likely to be subjected to dehydration and high temperatures, but survive drying for up to one hour at 90°F (32°C) and 45% relative humidity. Period of development spent in lungs; then juveniles migrate through gut to bladder. Tegumental secretion probably protects migrating juvenile.

CONSERVATION STATUS
Not threatened.

SIGNIFICANCE TO HUMANS
None known. ◆

Resources

Books

Boeger, W. A., and D. C. Kritsky. "Phylogenetic Relationships of the Monogenoidea." In *Interrelationships of the Platyhelminths*, edited by D. T. J. Littlewood and R. A. Bray. London: Taylor & Francis, 2001.

Kearn, G. C. *Parasitism and the Platyhelminths.* London: Chapman & Hall, 1998.

Periodicals

Bakke, T. A., P. D. Harris, and J. Cable. "Host Specificity Dynamics: Observations on Gyrodactylid Monogeneans." *International Journal for Parasitology* 32 (2002): 281–308.

Kearn, G. C. "Evolutionary Expansion of the Monogenea." *International Journal for Parasitology* 24 (1994): 1227–1271.

———. "The Survival of Monogenean (Platyhelminth) Parasites on Fish Skin." *Parasitology* 119 (1999): S57–S88.

Tinsley, R. C. "Parasite Adaptation to Extreme Conditions in a Desert Environment." *Parasitology* 119 (1999): S31–S56.

Whittington, I. D., L. A. Chisholm, and K. Rohde. "The Larvae of Monogenea (Platyhelminthes)." *Advances in Parasitology* 44 (2000): 139–232.

Graham Clive Kearn, DSc

Cestoda
(Tapeworms)

Phylum Platyhelminthes
Class Cestoda
Number of families 72

Thumbnail description
Internal parasitic flatworms lacking a gut during all stages of their development; adults parasitic in vertebrates, larvae in invertebrate and vertebrate hosts

Photo: Light micrograph of the scolex of an adult pork tapeworm (*Taenia solium*). At the top are the hooks, at lower left and right are the rounded suckers. These are used by the tapeworm to attach itself to the intestinal walls of its host. Humans are the main host of this parasite, which infects some four million people worldwide. (Photo by ©Alfred Pasieka/Science Photo Library/Photo Researchers, Inc. Reproduced by permission.)

Evolution and systematics

The class Cestoda encompasses about 5,100–5,200 species, 680 genera, and 72 families. During 1992–2002, annually 30–40 newly discovered species have been added, mostly recorded from tropical habitats (terrestrial and freshwater) or from marine fishes (especially from sharks and rays).

There is no generally accepted concept on the classification of the tapeworms. As of 1999–2002, the validity of the following 15 orders is widely recognized:

1. Gyrocotylidea. In holocephalan fishes (Chimaeriformes). Intermediate hosts unknown (life cycle without intermediate hosts was hypothesized). One family (Gyrocotylidae), one to two genera, and about 10 species.

2. Amphilinidea. In freshwater and marine fishes and freshwater turtles. Intermediate hosts: crustaceans. Two families (Amphilinidae and Schizochoeridae), six genera, and about eight species.

3. Caryophyllidea. In siluriform and cypriniform freshwater fishes. Intermediate hosts: tubificid annelids. Four families (Balanotaeniidae, Lytocestidae, Caryophyllaeidae, and Capingentidae), about 45 genera, and 140 species.

4. Pseudophyllidea. Mostly in freshwater and marine teleost fishes, also in amphibians, reptiles, birds, and mammals. Intermediate hosts: crustaceans (first or only), usually fishes, rarely other vertebrates (second). Six families (Bothriocephalidae, Philobythiidae, Echinophallidae, Triaenophoridae, Diphyllobothriidae, and Cephalochlamydidae), about 60 genera, and 280 species.

5. Spathebothriidea. In chondrostean and teleost fishes; marine and fresh water. Intermediate hosts: amphipod crustaceans. Two families (Spathebothriidae and Acrobothriidae), five genera, and six to seven species.

6. Haplobothriidea. In relict freshwater fishes (Amiiformes). Intermediate hosts: copepod crustaceans (first) and fishes (second). One family (Haplobothriidae), one genus, and two species.

7. Diphyllidea. In elasmobranch fishes. Intermediate hosts unknown. Three families (Echinobothriidae, Macrobothriidae, and Ditrachybothriidae), three genera, and about 35 species.

8. Trypanorhyncha. In elasmobranch fishes. Intermediate hosts: copepod crustaceans, possibly also

other marine invertebrates (first), marine teleost fishes (second). Nineteen families (Tentaculariidae, Paranybeliniidae, Hepatoxylidae, Sphyriocephalidae, Tetrarhynchobothriidae, Eutetrarhynchidae, Gilquiniidae, Shirleyrhynchidae, Otobothriidae, Rhinoptericolidae, Pterobothriidae, Grillotiidae, Molicolidae, Lacistorhynchidae, Dasyrhynchidae, Hornelliellidae, Mustelicolidae, Gymnorhynchidae, and Mixodigmatidae), about 50 genera, and 300–350 species.

9. Tetraphyllidea. In elasmobranch and holocephalan fishes. First intermediate hosts unknown, larvae found in marine teleost fishes (possible second intermediate or paratenic hosts). Seven families (Cathetocephalidae, Disculicipitidae, Prosobothriidae, Dioecotaeniidae, Onchobothriidae, Phyllobothriidae, and Chimaerocestidae), about 60 genera, and 800 species.

10. Litobothriidea. In lamniform sharks. Intermediate hosts unknown. One family (Litobothriidae), one genus, and eight species.

11. Lecanicephalidea. In elasmobranch fishes. Intermediate hosts unknown. Four families (Polypocephalidae, Anteroporidae, Tetragonocephalidae, and Lecanicephalidae), about 12 genera, and 45 species.

12. Proteocephalidea. Mostly in freshwater fishes, also in amphibians and reptiles connected with freshwater habitats. *Thaumasioscolex didelphidis*, described in 2001 from opossums in Mexico, is the only species of this order known from a mammalian host. Intermediate hosts: copepods (only or first), fishes and amphibians (second). Two families (Proteocephalidae and Monticelliidae), about 50 genera, and 320 species.

13. Nippotaeniidea. In freshwater teleost fishes. Intermediate hosts: copepod crustaceans. One family (Nippotaeniidae), two genera, and about six species.

14. Tetrabothriidea. In marine birds and mammals. Intermediate hosts unknown. One family (Tetrabothriidae), six genera, and about 50 species.

15. Cyclophyllidea. In tetrapods: mostly in birds and mammals, some species in reptiles and amphibians. Intermediate hosts: arthropods, annelids, mollusks or mammals (only or first), fishes, amphibians, reptiles, birds or mammals (second). Eighteen families (Mesocestoididae, Anoplocephalidae, Linstowiidae, Inermicapsiferidae, Thysanosomatidae, Catenotaeniidae, Nematotaeniidae, Progynotaeniidae, Acoleidae, Dioecocestidae, Amabiliidae, Davaineidae, Dilepididae, Dipylidiidae, Hymenolepididae, Paruterinidae, Metadilepididae, and Taeniidae), about 380 genera, and 3,100 species.

The phylogenetic relationships and the classification of the tapeworms are often disputed. According to the traditional views, the tapeworms have been considered the most primitive group among the parasitic flatworms. Two subclasses have been recognized within this class: Cestodaria, including the monozoic orders Gyrocotylidea and Amphilinidea, and Eucestoda, comprising remaining orders (mostly polyzoic but also monozoic). Some authorities consider Amphilinidea and Gyrocotylidea as distinct classes within the phylum Platyhelminthes (as Amphilinida and Gyrocotylida, respectively). Sometimes the order Caryophyllidea (encompassing monozoic worms) is placed out of the Eucestoda and believed to be close to the amphilinideans and gyrocotylideans.

However, as of 1999–2002, mostly as a result of extensive phylogenetic studies based on morphology (including ultrastructure) and molecular data, a wide consensus has been achieved on several points. The Cestoda, comprising Gyrocotylidea, Amphilinidea, and Eucestoda, are believed to form a monophyletic and highly derived flatworm group. One of the major characters supporting their monophyly is the lack of an intestine in all stages of their development. Some further characters, mostly connected with ultrastructural peculiarities of the osmoregulatory system and the tissue covering the body, also confirm their origin from a common ancestor.

The tapeworms, together with the monogeneans and the trematodes, belong to a monophyletic taxon named Neodermata (i.e., "having new skin"). This name reflects the fact that the ciliated epidermis of the larvae is replaced during the metamorphosis by a peculiar syncytial tissue (tegument or neodermis) occurring in adult worms. Main functions of the tegument are protective (against host's immune reactions and enzymes) and digestive (as a major site of absorption, metabolic transformations, and transport of nutrients). The tegument consists of a surface syncytial layer (distal cytoplasm) connected by cytoplasmic bridges with cell bodies (cytons). The cytons, containing nuclei and possessing powerful secretory apparatus, are situated deeply beneath the superficial muscle layers; therefore, they are well protected against host's reactions. Their secretions permanently renovate the distal cytoplasm, which acts as a contact zone between the parasite body and the host's tissues and fluids. Thus, the tegument is an important adaptation for parasitic life (all the adult neodermatans are parasitic). In addition, a range of further characters also is important for defining the Neodermata as a monophyletic group. These are some peculiarities in the structure of the locomotory cilia and the sensory receptors and common patterns of some processes in the course of the spermatogenesis and the formation of the excretory organs.

The monogeneans are believed to be the closest relatives of the tapeworms. The two groups are included in the superior taxon Cercomeromorphae. Their phylogenetic relationships are mostly supported by the presence of set of hooks in the posterior end of the body. In the Eucestoda, these hooks occur in larvae (six embryonic hooks), rarely in both larvae and adults (10 hooks in Gyrocotylidea and Amphilinidea). In eucestode larvae, the embryonic hooks are often situated in a distinct portion of the body (cercomer), which is usually delimited by a constriction from the anterior part of the body. As a rule, the cercomer together with embryonic hooks is detached in the course of the development of the eucestode larvae before its transmission to the final host. In contrast, the caudal hooks of the monogenean larvae are persistent in adult worms as important elements of their attachment apparatus (haptor). The belief that the eucestode cercomer is homologous to the monogenean haptor (not explicitly supported by recent studies) gave the name of the Cercomeromorphae.

Within the Cestoda, the Gyrocotylidea have a basal position to the branch containing the remaining taxa (Amphilinidea plus eucestode orders). Among the Eucestoda, the monozoic order Caryophyllidea is considered basal to the remaining orders. Among the polyzoic orders, these having as a rule four suckers or bothridia on the scolex (known as tetrafossate, e.g., Tetraphyllidea, Proteocephalidea, and Cyclophyllidea) are considered more derived than those having two bothria or bothridia (difossate, e.g., Pseudophyllidea and Diphyllidea).

The monophyly of the Neodermata and the Cercomeromorphae as well as the phylogenetic interrelations within the Cestoda also are well supported by comparisons of their gene sequences.

There are no fossil records of cestodes. However, the phylogenetic studies and the analyses of the evolutionary associations with hosts suggest a long period of the eucestode-vertebrate coevolution, perhaps since the Devonian (before 350–420 million years).

Physical characteristics

The body of the tapeworms is usually dorso-ventrally flattened, narrow, and highly elongate. It resembles a tape, which may explain the etymology of both common and scientific names of the class ("cestus" in Latin means belt, girdle, or ribbon). The size range varies very much: from 0.02 in (0.6 mm) length of the cyclophyllidean *Mathevolepis petrotschenkoi* (parasite of shrews) to 98 ft (30 m) length of the pseudophyllidean *Hexagonoporus physeteris* (from the sperm whale). As a rule, tapeworms are whitish because as internal parasites living in darkness they do not possess any pigments.

Typically, the body of tapeworms consists of three distinct regions: scolex (plural scoleces), neck, and strobila (plural strobila).

The scolex (sometimes referred to as "head") is the anterior end of the body. Its major function is the attachment of the parasite to the wall of the intestine. By this reason, it may bear spines, hooks, glands releasing adhesive secretions, grooves, suckers, tentacles, etc., or various combinations of these depending on the ordinal or family affiliation of the worm. The suckers are the most widespread attachment organs (e.g., in Cyclophyllidea, Proteocephalidea, Lecanicephalidea, and in some Tetraphyllidea). They are usually cup shaped, with powerful muscular walls, four in number, two dorsal and two ventral. The bothridium (plural bothridia) is an ear-shaped muscular outgrowth projecting sharply from the scolex and often possessing leaflike mobile margins. The bothridia are usually four in number; they might be sessile or situated at the end of elongate stems connecting them with the scolex. Bothridia occur in several orders (Trypanorhyn-

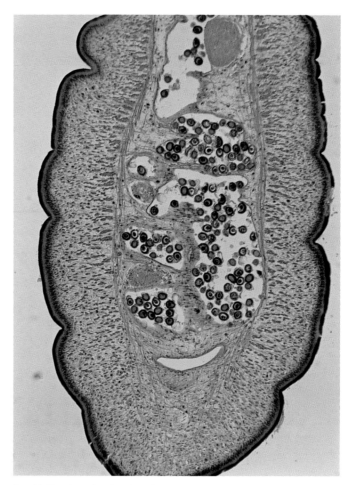

Gravid *Taenia*, a beef and pork tapeworm. Shows branched uterus full of eggs. Tapeworms occur in the digestive tract of every vertebrate species. They have no head, mouth, or digestive system, but absorb food from the host's gut through their body surface. (Photo by ©Biophoto Associates/Photo Researchers, Inc. Reproduced by permission.)

cha, Diphyllidea, Tetraphyllidea, and Tetrabothriidea). The bothria (singular bothrium) are simple longitudinal grooves on the scolex, two in number (e.g., in Pseudophyllidea).

The scoleces of the worms of many orders are characterized by the presence of an apical organ consisting mostly of muscular and (or) glandular tissue. In some tetraphyllideans, this organ is represented by a well-developed gland at the apex of the scolex. In the majority of the proteocephalideans, the apical organ has more or less a structure identical to that of the suckers (often referred to as "apical sucker"); however, there are species of this order with an apical organ transformed into a sac filled up of glandular tissue. An immense variability of the structure of the apical organ can be seen in the Cyclophyllidea, where it is usually marked as a rostellum. It is protrusible, often dome-shaped, and in the most common case provided by one or two rows of hooks. In some families, the rostellum can be withdrawn in a special muscular pouch (rostellar sac). The protruded rostellum penetrates deeply into the intestinal wall of the host, anchoring there by the crown of hooks situated on its top. In addition, some cyclophyllideans may have accessory circles of spines or strongly

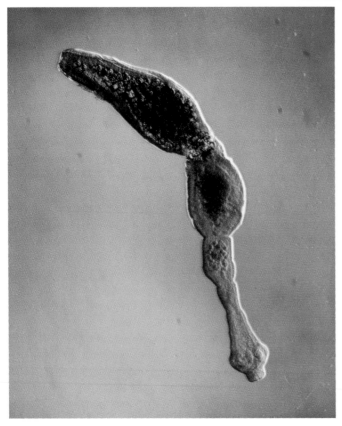

Light micrograph image of a fox tapeworm (*Echinococcus multilocularis*). The most common host is the fox, with intermediary hosts of mice and humans. (Photo by ©Nicole Ottawa/Eye of Science/Photo Researchers, Inc. Reproduced by permission.)

- Premature proglottides, with primordia of genital organs only.

- Mature proglottides, with developed and functioning male and female genital systems (almost all tapeworms are hermaphroditic).

- Postmature proglottides, in which the uteri are filled of developing eggs and gonads gradually degenerate.

- Gravid proglottides, containing uteri with ripe eggs.

As a rule, the gravid proglottides having terminal position in the strobila detach after completing their development. They pass to the environment with the host's feces or disintegrate along their route and only eggs are released.

Formation of proglottides occurs in 11 of the orders listed above. However, the members of the orders Gyrocotylidea, Amphilinidea, and Caryophyllidea have only a single set of genital organs per body. Consequently, they have no proglottides. The Spathebothriidea exhibits an intermediate pattern of body organization: an internal multiplication of reproductive organs down the strobila occurs but no distinct proglottides are formed.

In the past, an interpretation of the body organization of the tapeworms was proposed, considering them as colonial organisms, i.e., each worm was believed to represent a linear colony of numerous zoids (individuals). In other words, each proglottis was recognized as an individual. This view is known as the "polyzoic" concept or theory. The majority of recent workers do not support it. However, the terms arising from this concept, "monozoic" (for cestodes with a single set of genital organs per body) and "polyzoic" (for cestodes with proglottides), are widely used for describing the organization of the body of the tapeworms. Thus, Gyrocotylidea, Amphilinidea, and Caryophyllidea are monozoic, and the remaining orders (excluding Spathebothriidea exhibiting intermediate features) are polyzoic.

Tapeworms lack a gut during all the stages of their development. They feed through the tegument covering the body. Under the tegument, there are several layers of superficial musculature (usually three). Inside, most of the body is made up of parenchyma. Powerful longitudinal muscular bundles, responsible for the movements of the body, pass along the entire strobila. They separate the parenchyma in the center of each proglottis from that in the periphery (the relevant parts of the parenchyma are named medullar and cortical). The nervous system is represented by paired ganglia situated in the scolex and arising from them major anterior and posterior longitudinal nerves, the latter running through the strobila. There are also numerous transverse commissures connecting longitudinal nerves and smaller nerves emanating from them and reaching to the musculature and the receptors.

The excretory system includes flame cells scattered in the parenchyma. Small ducts connect these cells with the major longitudinal canals of the system passing along the strobila. Usually, the longitudinal canals are two dorso-lateral and two ventro-lateral. In addition to the excretion of metabolic prod-

developed glands associated with the rostellum, also facilitating the reliable attachment to the intestinal wall.

The neck is the region of the body just posterior to the scolex. It is usually short. This is a zone of proliferation, containing numerous stem cells. The latter are responsible for giving rise of the strobila.

The strobila is posterior to the neck. It consists of proglottides arranged in a linear series. The proglottis is a distinct portion of the body containing a set of reproductive organs. The stobila may contain from few (two in *Mathevolepis petrotschenkoi*), several dozens (in the majority of tapeworms), or numerous (more than thousand in *Taenia saginata*) proglottides. Each proglottis starts its development at the neck, as a result of the division of the stem cells. Typically, the formation of proglottids one by one at the neck is a permanent process lasting the whole life of the tapeworm in the final host. Just posterior to the neck, the proglottides are short and narrow, containing undifferentiated cells (juvenile proglottides). With the appearance of a new proglottis at the neck, already formed proglottides are pushed in posterior direction, which coincides with their growth and the gradual development of the reproductive organs in them. After the juvenile proglottides, each strobila typically contains the following types of proglottides (from anterior to posterior direction):

ucts, this system also eliminates the excess water from the body of the worm. By this reason, it also is known as osmoregulatory system.

As a rule, each mature proglottis (or each body of a monozoic cestode) contains one male reproductive system and one female reproductive system.

The male reproductive system includes testes, from one in the genus *Aploparaksis* (Hymenolepididae) to several hundreds in the genus *Taenia* (Taeniidae). Each testis is provided with a narrow outgoing duct (vas efferens). These ducts unite into a common wider duct (vas deferens), which transports the sperm to the male copulatory organ. The vas deferens leads into a muscular pouch (cirrus pouch) containing the copulatory organ (cirrus). Along its course, vas deferens may form seminal vesicles before entering the cirrus pouch (external seminal vesicle) and (or) within it (internal seminal vesicle), or to be highly convoluted, in order to have greater sperm storage capacity. The cirrus is a muscular organ, often with spines on its surface. It is able to invaginate (to be withdrawn) in the cirrus pouch or to evaginate (project) through the pore of the cirrus sac.

The female reproductive system consists of ovary, vitellarium, ootype, uterus, vagina, seminal receptacle, and ducts connecting them. The sperm enters the female reproductive system through the vagina during the copulation and is stored in the seminal receptacle. The ovary is variable in location, shape, and size. As oocytes mature, they pass from the ovary into the oviduct. A narrow duct coming from the seminal receptacle joins to the oviduct, which is the place of the fertilization. Vitellarium may be a compact organ (in Cyclophyllidea, Tetrabothriidea, and Nippotaeniidea) or may consist of numerous vitelline follicles scattered in the parenchyma and possessing outgoing ducts uniting into a common vitelline duct (in the majority of orders). The vitelline duct also is connected with the oviduct, and one or more vitelline cells join to each zygote. Together they pass into the ootype. It is usually surrounded by glandular tissue (known as Mehlis's gland) producing a secretion, forming a thin envelope encompassing the zygote and associated vitelline cells. The young eggs pass from the ootype through the uterine duct into the uterus where they complete their development. In the majority of the tapeworms, the eggs leave the final host together with the proglottis in which they have developed. However, some cestodes have uterine pores and eggs can be released one by one.

Distribution

The tapeworms are widespread throughout the world. They occur in almost all terrestrial, marine, brackish, and freshwater habitats where vertebrate animals live. Their diversity appears to be great but poorly explored at the tropical latitudes (most of the newly described species during the last decade originate from tropical habitats). They also are abundant at the temperate latitudes (e.g., the number of species found in terrestrial and freshwater habitats in Europe exceeds 900, and the number of the species recorded from a small territory such as Bulgaria (southeastern Europe) is 310).

Light micrograph of the scolex (head) and neck of the adult beef tapeworm (*Taenia saginata*). At upper left are the two rounded structures which are the suckers used to attach the tapeworm to the intestinal wall of its host. The head narrows into a segmented neck. (Photo by ©Alfred Pasieka/Science Photo Library/Photo Researchers, Inc. Reproduced by permission.)

More than hundred species were reported from marine birds and mammals in Arctic and Antarctic habitats.

About one third of the cestode species are parasites of marine fishes, mostly of sharks and rays. Several pseudophyllideans were described from deep-sea teleost fishes (e.g., *Probothriocephalus alaini* was collected at a depth of 2,590–3,350 ft (790–1,020 m), *Probothriocephalus muelleri* and *Phylobythoides stunkardi* at a depth between 5,550 and 7,520 ft (1,690 and 2,290 m), all from the North Atlantic).

The cestode orders have cosmopolitan distributions, with three exceptions only. The litobothriideans are known from the Pacific Ocean only, off California, Mexico, and Australia. The nippotaeniideans were recorded from freshwater fishes in Japan, China, Russian Far East, and New Zealand. The haplobothriideans have the most restricted geographical range. The two species of this order occur in North America, in the "living fossil" bowfin (*Amia calva*).

Habitat

When considering habitats of parasitic organisms, parasitologists make difference between microhabitats and macrohabitats. The microhabitat of a parasite species is an organ or a tissue in the host inhabited by parasite individuals. In broader sense, the whole host individual also could be recognized as a microhabitat. The macrohabitat of a parasite species is that place or environment where its hosts (final, intermediate, and paratenic) live.

The usual microhabitat of the adult tapeworms is the intestine of vertebrate animals. Only the species of the order Amphilinidea inhabit the body cavity of their final hosts (fishes and turtles). Some species of the order Cyclophyllidea occur in other parts of the digestive system (e.g., *Cloacotaenia megalops* [Hymenolepididae] lives in the cloaca of ducks, *Thysanosoma actinioides* [Thysanosomatidae] inhabits the bile ducts of ruminants, and *Gastrotaenia cygni* [Hymenolepididae] occurs under the horny lining of the gizzard of swans).

The intestine of vertebrate animals is not a homogeneous habitat. Its portions differ from each other by their content of nutrients, range of enzymes, pH, and structure of the intestinal wall. It is known for many cestode species that they can be found only in a certain portion of the intestine, e.g., in the duodenum or in the most posterior part (at the ileocaecal junction). A detailed study on the distribution of the parasitic worms along the small intestine of grebes in Canada showed that each of four grebe species was parasitized by 9–17 tapeworm species. Some of these occurred in a high proportion of the intestine that could be interpreted as a broad tolerance for conditions along it. A quarter of cestode species, however, occupied portions no longer than 20% of the intestine length.

Tapeworm larvae have diverse locations in intermediate hosts. Species occurring in crustaceans or insects are usually in the body cavity. The members of the family Taeniidae are parasitic in carnivore mammals and humans and their larvae develop also in mammals. Depending on the cestode species, the larvae of taeniids can be situated in the liver, lungs, musculature, body cavity, brain, and mesentery, sometimes even in the eyes. The most typical locations of cestode larvae in fishes are the musculature, body cavity, and gall bladder.

Concerning the range of the final hosts of cestodes, it is not exaggerated to say that they are not recorded only in vertebrate species, which have not been examined. Only the Cyclostomata (lampreys and hagfishes) have no cestode parasites.

Each tapeworm species is characterized by its host spectrum. This phenomenon is usually marked as "parasite specificity to the host," or simply as "host specificity." Some cestode species are found in one vertebrate species only (oioxenous parasite species). A good example is *Dollfusilepis hoploporus* (Hymenolepididae). This worm lives in the great crested grebe (*Podiceps cristatus*) only, even in lakes where this host co-occurs with other grebe species. Other cestode species are found in a small group of related host species belonging to one genus or one family (stenoxenous parasite species) (e.g., *Tatria biremis* [Amabiliidae] occurs in Eurasia and America in three species of grebes). The majority of tapeworm species are stenoxenous. A few cestodes are euryxenous parasites, i.e., live in unrelated (phylogenetically distant) hosts. In the latter case, a prerequisite for the formation of such parasite-hosts associations is the convergence of the ecology (in terms of similar habitats and diets) of the hosts. An example for this can be *Ligula intestinalis* (Diphyllobothriidae) occurring as adult in birds of various orders: gulls of Charadriiformes, herons of Ciconiiformes, grebes of Podicipediformes, cormorants of Pelecaniformes, and mergansers of Anseriformes. However, the infective larvae of this parasite live in fishes of the family Cyprinidae only. Therefore, *L. intestinalis* is euryxenous relative to the final hosts but stenoxenous relative to the second intermediate host.

A few tapeworm species are known that mature in invertebrate animals. Thus, *Archigetes sieboldi*, *Archigetes limnodrili*, and *Archigetes iowensis* (Caryophyllidea) can mature in the body cavity of freshwater annelids, and *Diplocotyle olriki* and *Cyathocephalus truncatus* (Spathebothriidea) in amphipod crustaceans. The same invertebrate groups are intermediate hosts of the two cestode orders. Alternatively, all these forms can mature in the intestine of fishes (final hosts). The life cycles with the participation of only one invertebrate host are considered secondarily simplified, allowing shortening time needed for maturation of worms in the final host. However, there are data showing that at least some of these species can live in water basins where appropriate fish final hosts are lacking.

Macrohabitats of tapeworms are diverse, from deep sea to high mountains and from tropical forests to tundra. There are even species living in extreme conditions (e.g., several species parasitizing flamingos, avocets, and plovers live as larvae in brine shrimps [*Artemia* spp.] in salinas. Such species is *Flamingolepis liguloides* [Hymenolepididae]).

Behavior

Little is known about the behavior of tapeworms in the intestine of the host. It seems that most of the tapeworms are permanently attached during their entire life at a certain site of the intestinal wall. However, there are well-documented observations on circadian migrations of *Hymenolepis diminuta* (Hymenolepididae) from one microhabitat to another in the intestine of rats. This migration depends on the host feeding and digestion. When the gut of the rat is empty, the worms of this species are situated in the posterior region of the small intestine. However, as the content of the stomach passes into the intestine, they rapidly migrate towards the duodenum.

Feeding ecology and diet

All the cestodes feed through the body surface. The nutrient molecules (carbohydrates as glucose and galactose, amino acids, purines and pyrimidines, fatty acids, monoglycerides, sterols, and some vitamins) are absorbed through the tegument.

Reproductive biology

The majority of cestodes are hermaphroditic (only the members of the tetraphyllidean family Dioecotaeniidae and the cyclophyllidean family Dioecocestidae are dioecious). As a rule, each proglottis contains one set of male reproductive organs and one set of female reproductive organs. Often the maturation of the male organs and female organs do not coincide in time. In the majority of the families, the male organs mature first and proglottides initially act as male. This type of strobilar development is known as protandry. In some families (e.g., in the Progynotaeniidae), female reproductive organs develop first. In the latter case, the sperm production in testes coincides with the development of the eggs in the uterus. This type of development is known as proterogyny. Both types of strobilar development should be considered adaptations allowing more complete utilization of resources—though hermaphroditic, each proglottis is initially functionally male and after that functionally female (or visa versa), thus being able to produce larger number of gametes. It also prevents self-fertilization.

The life cycle of each cestode species includes at least two hosts, final and intermediate. The final or definitive host is that harboring adult (reproducing sexually) worms. The intermediate host is that where larvae (known also as metacestodes) develop. The two hosts are in close ecological associations, facilitating the transmission of the parasite. The intermediate host lives in habitats where the final host feeds and defecates. The intermediate host is also a common component of the diet of the final host. The transmission of the cestode from the intermediate host to the final host is along the food chains only. Often the transmission is facilitated by a parasite-induced modification of the intermediate host's behavior, color, or health condition, in order to make it easier prey for the final host.

The general scheme of the life cycle is as follows. The cestode eggs pass with host's feces into the environment. Each egg encompasses an embryo named oncosphere. The latter possesses six embryonic hooks and several glandular cells and is surrounded by several envelopes. The egg is eaten by the intermediate host. In the gut of this host, the oncosphere hatches and, using its hooks and glands, penetrates through the intestinal wall and locates in the body cavity or in any internal organ. There, it metamorphoses into an infective larval stage (metacestode). In the most common case, the metacestode has a fully developed scolex identical to that of the adult worm. The final host is infested by eating infected intermediate hosts. The scolex of the metacestode attaches to the intestinal wall of the final host. The neck of the worms starts the production of proglottides and thus the strobila is formed. With the further development of the proglottides, the worm starts producing eggs, which are released with feces into the environment.

There are numerous cestode species exhibiting details in the life cycle differing from that described in the general scheme. Some species have two intermediate hosts or mobile embryos able to swim (see the account for *Diphyllobothrium latum*). The embryos of Amphilinidea and Gyrocotylidea are not oncospheres but lycophoras (see the account for *Amphilina foliacea*). The range of the intermediate hosts used is also impressive.

Metacestodes of various orders and families exhibit an immense morphological variability. Procercoid is the metacestode in the first intermediate host that has an elongate body and cercomer (e.g., in Pseudophyllidea). Entering into the second intermediate host, it develops into the plerocercoid. The latter possesses a differentiated scolex and is able to infect the final host. Embryos of other cestodes directly develop into plerocercoids (e.g., Proteocephalidea and Nippotaeniidea).

The most widespread type of metacestodes in the order Cyclophyllidea is the cysticercoid. It is a solid-bodied organism with fully developed scolex retracted into the body. Among the Taeniidae, the most widespread metacestode is the cysticercus. Its scolex is introverted.

As in 2001, life cycles of about 200 cestode species are known (out of 5,100–5,200 species described). Obviously, the discovery of their enormous diversity, in terms of the range of hosts utilized and the morphological specialization of metacestodes, is a matter for the future.

Conservation status
No cestode species is listed by the IUCN.

Significance to humans
A total of 57 cestode species were reported from humans. Some of these are not "true" parasites of humans (the infections with them are accidental). However, six species are considered of great public health significance because they are agents of serious and widespread diseases. In 1999, when the human population was almost 6 billion persons, the estimate of the numbers of infected humans (in millions) was as follows: *Taenia saginata*—77.0, *Hymenolepis nana*—75.0, *Taenia solium* —10.0, *Diphyllobothrium latum*—9.0, and *Echinococcus granulosus* and *Echinococcus multilocularis* (considered together)—2.7.

Numerous cestode species are of primary importance for veterinary medicine. Several species of Taeniidae are major parasites of domestic ruminants (*Echinococcus granulosus*, *Taenia multiceps*, *T. hydatigena*, and *T. ovis*). Members of the family Anoplocephalidae are important parasites of horses, ruminants, and rabbits. Some taeniids, mesocestoidids, and dipylidiids are frequent parasites of dogs and cats. Among the parasites of domestic birds, the most common are members of the families Hymenolepididae and Davaineidae.

Cestodes of the order Caryophyllidea, Pseudophyllidea, and Proteocephalidea may cause significantly reduced production in fish farming operations.

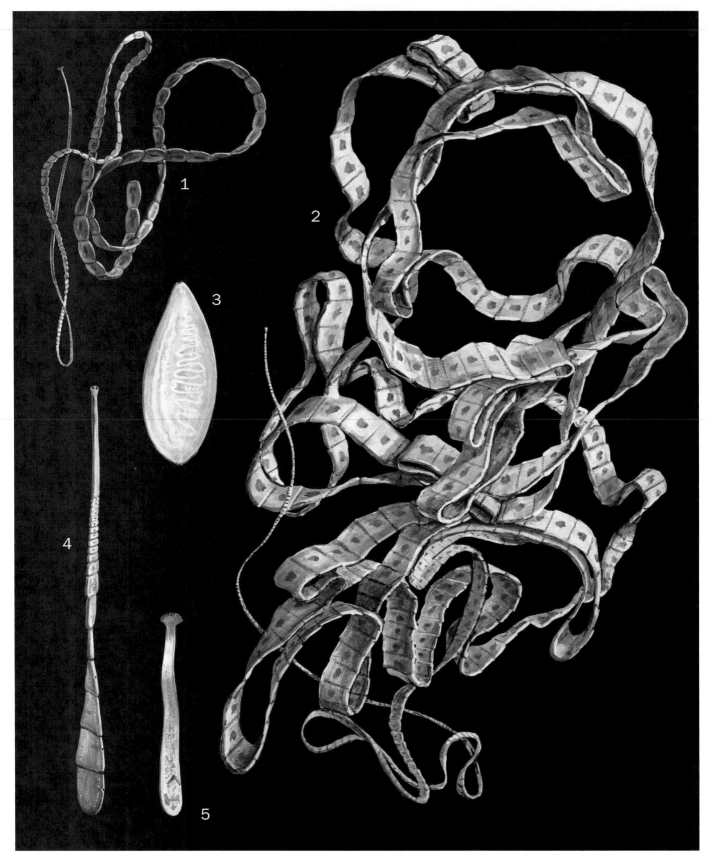

1. *Phyllobothrium squali*; 2. Broad fish tapeworm (*Diphyllobothrium latum*); 3. *Amphilina foliacea*; 4. *Proteocephalus longicollis*; 5. *Caryophyllaeus laticeps*. (Illustration by Brian Cressman)

1. Dog tapeworm (*Echinococcus granulosus*); 2. Beef tapeworm (*Taenia saginata*); 3. *Davainea proglottina*; 4. *Moniezia benedeni*; 5. *Tatria biremis*; 6. *Progynotaenia odhneri*. (Illustration by Brian Cressman)

Species accounts

No common name
Amphilina foliacea

ORDER
Amphilinidea

FAMILY
Amphilinidae

TAXONOMY
Monostomum foliaceum Rudolphi, 1819, Italy.

OTHER COMMON NAMES
None known.

PHYSICAL CHARACTERISTICS
Body monozoic, dorso-ventrally flattened, oval or leaf-shaped in outline, 1–2.6 in (28–65 mm) long and 0.67–1.2 in (17–30 mm) wide. Anterior end pointed, with slightly expressed apical invagination. Uterine orifice situated in anterior end. Orifice of ejaculatory duct on posterior end. Vaginal pore postero-lateral, at some distance from male pore.

DISTRIBUTION
Europe and Siberia.

HABITAT
Adults are parasitic in the body cavity of sturgeons (*Acipenser sturio, A. nudiventris, A. ruthenus, A. stellatus, Huso huso,* etc.).

Larvae develop in freshwater amphipod crustaceans. The macrohabitats of *Amphilina foliacea* are large rivers in Eurasia. Though it cannot develop in marine amphipods, it can be found also in marine sturgeons (which are anadromous).

FEEDING ECOLOGY AND DIET
Not studied. Apparently absorb nutrients through the tegument.

BEHAVIOR
Nothing is known.

REPRODUCTIVE BIOLOGY
A. foliacea is hermaphroditic. There are no data how often cross-fertilization or self-fertilization may occur. Eggs develop in uterus and are released through its orifice. Each egg contains larva named lycophora. It is not known how eggs pass from the body cavity of the sturgeon into the water. The eggs are swallowed by amphipods (intermediate hosts). The lycophora leaves the egg envelope in the intestine of the intermediate host and passes through its wall into the body cavity. There, it develops into a larva (about 0.16 in [4 mm] long) for some six weeks. Feeding on crustaceans containing fully developed larvae infects sturgeons. Larvae pass through the wall of the stomach into the body cavity. They become mature after six to seven months and live several years.

CONSERVATION STATUS
Not listed by the IUCN.

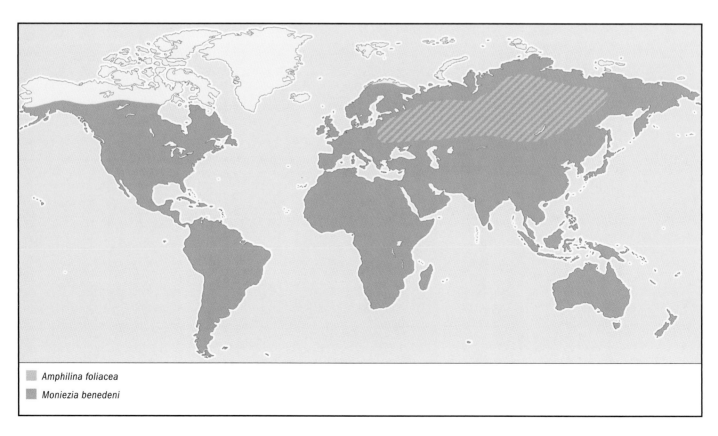

Amphilina foliacea
Moniezia benedeni

SIGNIFICANCE TO HUMANS
None known. ◆

No common name
Caryophyllaeus laticeps

ORDER
Caryophyllidea

FAMILY
Caryphyllaeidae

TAXONOMY
Taenia laticeps Pallas, 1781, Russia.

OTHER COMMON NAMES
None known.

PHYSICAL CHARACTERISTICS
Body monozoic, longitudinally elongate, 0.79–1.6 in (20–40 mm) long and 0.04–0.08 in (1–2 mm) wide. Anterior end wider, forming some folds. Male genital pore and female genital pore situated on the ventral surface of the body at some distance from its posterior end.

DISTRIBUTION
North Eurasia, in Europe, Siberia, Central Asia, and Russian Far East.

HABITAT
Originally described as a parasite of the common bream (*Abramis brama*). Known also from about 30 species of fresh-water fishes, mostly of the family Cyprinidae. Recorded also in some predatory fishes belonging to other families (e.g., pike and perch). Larvae recorded in tubificid annelids. The macrohabitats are slow rivers, lakes, ponds, marches, reservoirs, etc.

FEEDING ECOLOGY AND DIET
Internal parasite absorbing nutrients through the tegument.

BEHAVIOR
Nothing is known.

REPRODUCTIVE BIOLOGY
Caryophyllaeus laticeps is hermaphroditic. Eggs released with feces of the final hosts (fishes) need to stay in water for about three months in order to become infective. Intermediate hosts (aquatic tubificids) become infected by eating them. The oncosphere hatches in the intestine and penetrates into the body cavity. The larva develops for about six months. When fully developed, it is about 0.08 in (2 mm) long, with primordia of reproductive organs. Eating tubificids containing larvae infects fishes.

CONSERVATION STATUS
Not listed by the IUCN.

SIGNIFICANCE TO HUMANS
Together with another parasite of the same genus (*C. fimbriceps*), *C. laticeps* may cause a disease of farmed carps known as caryophyllaeasis. If parasites are few, they are not pathogenic for the host. It is believed that about 20 parasites per fish may cause some disorders of the digestive system and anemia. ◆

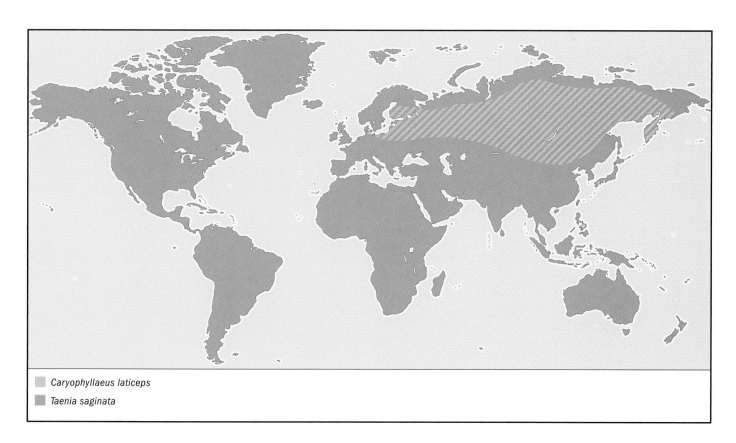

■ *Caryophyllaeus laticeps*
■ *Taenia saginata*

No common name
Tatria biremis

ORDER
Cyclophyllidea

FAMILY
Amabiliidae

TAXONOMY
Tatria biremis Kowalewski, 1904, Ukraine.

OTHER COMMON NAMES
None known.

PHYSICAL CHARACTERISTICS
Body 0.08–0.16 in (2–4 mm) long and 0.02–0.03 in (0.6–0.8 mm) wide. Strobila consists of 20–30 proglottides. Rostellum, bearing 10 hooks, can protrude very much in order to penetrate deeply into the intestine wall.

DISTRIBUTION
Eurasia and North America.

HABITAT
Parasite of grebes, mostly of *Podiceps auritus*, *P. nigricollis*, and *P. grisegena*. Larvae recorded in aquatic insects. The macrohabitats include freshwater lakes and slow rivers.

FEEDING ECOLOGY AND DIET
Internal parasite absorbing nutrients through the tegument.

BEHAVIOR
Nothing is known.

REPRODUCTIVE BIOLOGY
Tatria biremis is hermaphroditic. Gravid proglottides are released into the environment with feces. They are eaten by aquatic insects which are intermediate hosts of this parasite. The only record of larvae is from *Sigara concinna* (Heteroptera, Corixidae) from Kazakhstan.

CONSERVATION STATUS
Not listed by the IUCN.

SIGNIFICANCE TO HUMANS
None known. ◆

No common name
Moniezia benedeni

ORDER
Cyclophyllidea

FAMILY
Anoplocephalidae

TAXONOMY
Taenia benedeni Moniez, 1879, France.

OTHER COMMON NAMES
None known.

PHYSICAL CHARACTERISTICS
Body 8.2–13 ft (2.5–4 m) long and 0.99–1.0 in (25–26 mm) wide. Scolex with four suckers, without rostellum. Proglottides

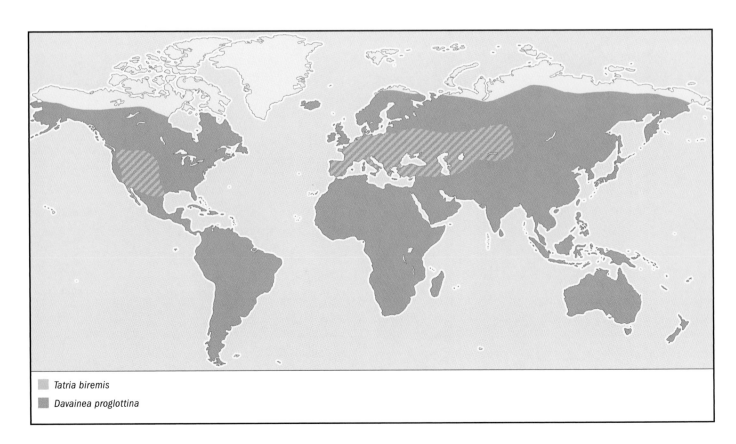

■ *Tatria biremis*
■ *Davainea proglottina*

transversely elongate. Each proglottid has two sets of reproductive organs, including two genital pores situated on both lateral margins.

DISTRIBUTION
Cosmopolitan.

HABITAT
Microhabitats of adult worms are intestines of domestic (cattle, sheep, goats) and some wild ruminants (moose, antelopes, deer, etc.). Larvae develop in oribatid mites. Macrohabitats include grasslands, forests, and pastures.

FEEDING ECOLOGY AND DIET
Internal parasite absorbing nutrients through the tegument.

BEHAVIOR
Nothing is known.

REPRODUCTIVE BIOLOGY
This species is hermaphroditic. Oribatid mites are infected by eating eggs of the parasite. Depending on the temperature, larvae (cysticercoids) develop in them for four to seven months. Ruminants eat infected mites while grazing on grass. Worms become mature after about 50 days.

CONSERVATION STATUS
Not listed by the IUCN.

SIGNIFICANCE TO HUMANS
This and another parasite of the same genus, *M. expansa*, are agents of a disease (monieziasis) of sheep, goats, and cattle. It is more dangerous for young animals than for adults. Histological changes of the intestinal walls and intoxication of the infected animals have been described. ◆

No common name
Davainea proglottina

ORDER
Cyclophyllidea

FAMILY
Davaineidae

TAXONOMY
Taenia proglottina Davaine, 1860, France.

OTHER COMMON NAMES
None known.

PHYSICAL CHARACTERISTICS
Body 0.02–0.04 in (0.5–1.0 mm) (sometimes up to 0.12 in [3 mm]) long. Scolex with four suckers armed with spines and rostellum armed with 60–90 hammer-shaped rostellar hooks. Proglottides, five to nine, usually six in number.

DISTRIBUTION
Cosmopolitan.

HABITAT
Adults in intestines of poultry. Larvae in the body cavity of slugs. This parasite is very common in farms in humid areas where slugs are abundant.

FEEDING ECOLOGY AND DIET
Internal parasite absorbing nutrients through the tegument.

BEHAVIOR
Nothing is known.

REPRODUCTIVE BIOLOGY
Each adult parasite produces one gravid proglottis per day, which is released with feces of the final host. Slugs, which are often coprophagous, eat gravid proglottides. The development of the larva (named cysticercoid) continues in the body cavity of the mollusk for 15–22 days (depending on temperature regime). Hens are infested as a result of eating infected slugs. Worms mature in the intestine of the final host for 12–16 days.

CONSERVATION STATUS
Not listed by the IUCN.

SIGNIFICANCE TO HUMANS
Davainea proglottina causes a parasitic disease of poultry. Its pathogenesis is mostly connected with an inflammation of the duodenum. Its acute phase continues three to five days. The mortality can reach up to 60%. ◆

No common name
Progynotaenia odhneri

ORDER
Cyclophyllidea

FAMILY
Progynotaeniidae

TAXONOMY
Progynotaenia odhneri Nybelin, 1914, Sweden.

OTHER COMMON NAMES
None known.

PHYSICAL CHARACTERISTICS
Body minute, 0.08–0.12 in (2–3 mm) long and 0.02–0.03 in (0.6–0.8 mm) wide, wedge-shaped. Strobila consisting of 8–12 proglottides. Scolex provided with rostellum bearing 12 rostellar hooks. Genital pores on the lateral margins of the proglottides, regularly alternating on the left and the right sides.

DISTRIBUTION
Eurasia and Africa.

HABITAT
Parasitic in plovers, mostly in ringed plover (*Charadrius hiaticula*) and snowy plover (*Charadrius alexandrinus*). Most of the records are from seashores or salt lakes, which seem to be the macrohabitats of this parasite.

FEEDING ECOLOGY AND DIET
Intestinal parasite absorbing nutrients through the tegument.

BEHAVIOR
Nothing is known.

REPRODUCTIVE BIOLOGY
Progynotaenia odhneri is hermaphroditic. However, it does not possess a vagina. The copulation is traumatic: the heavily armed cirrus penetrates through the tegument and the parenchyma and directly inseminates the seminal receptacle. The intermediate hosts of this species are not known.

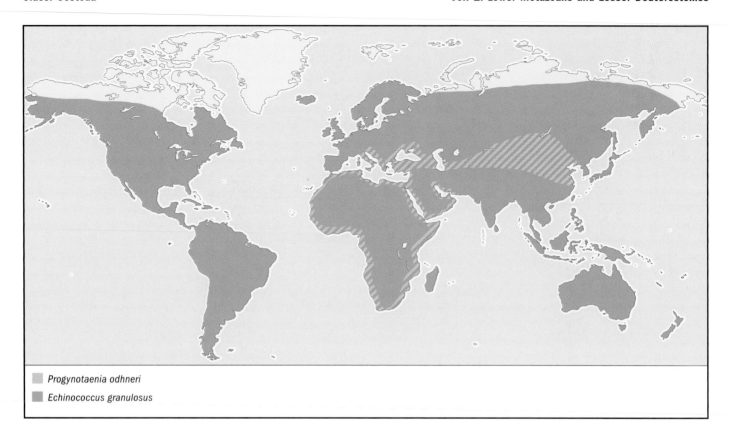

■ *Progynotaenia odhneri*

■ *Echinococcus granulosus*

CONSERVATION STATUS
Not listed by the IUCN.

SIGNIFICANCE TO HUMANS
None. ◆

Dog tapeworm
Echinococcus granulosus

ORDER
Cyclophyllidea

FAMILY
Taeniidae

TAXONOMY
Hydatigena granulosa Batsch, 1786, Germany.

OTHER COMMON NAMES
French: Ténia échinocoque; German: Hülsenwurm.

PHYSICAL CHARACTERISTICS
Adult 0.12–0.24 in (3–6 mm) long, consisting of scolex, short neck, and three to five proglottides. Scolex bearing 30–36 (rarely more) rostellar hooks. Gravid proglottis highly elongate.

DISTRIBUTION
Cosmopolitan.

HABITAT
Microhabitats are intestines of carnivore mammals, mostly of the family Canidae (dogs, wolves, jackals, etc). Larvae occur in internal organs (liver, lungs, musculature) of herbivorous mammals. Macrohabitats include natural ecosystems where parasites circulate along the food chain wild herbivores—wild carnivores or in habitats associated with humans (pastures, farms, villages) where major final hosts are dogs and intermediate hosts are domestic animals (sheep, cattle, camels, pigs, goats, horses, etc.).

FEEDING ECOLOGY AND DIET
Internal parasite absorbing nutrients through the tegument.

BEHAVIOR
The released gravid proglottis is able to crawl. Some data indicate that such proglottides may climb up grasses, where they are able to disperse their eggs more efficiently and contaminate a larger area of grassland. Some gravid proglottides may stay around the anus of the dog, contaminating its fur with eggs.

REPRODUCTIVE BIOLOGY
Oncospheres hatching from eggs in the intestine of the intermediate host migrate to the liver or the lungs, sometimes to the musculature or even to the eyes. They grow very slowly and transform into a cyst named "unilocullar hydatid." Its wall consists of two layers. The inner layer is able to produce numerous scoleces (i.e., asexual reproduction occurs during the larval development). The inner layer also is able to produce daughter cysts, situated within the mother cyst, which also can produce numerous scoleces. The development of the hydatid may continue for 20–30 years. When a carnivorous mammal eats a liver or another organ containing a hydatid, it becomes infected. In its intestine, each scolex produces an adult tapeworm.

CONSERVATION STATUS
Not listed by the IUCN.

SIGNIFICANCE TO HUMANS

Echinococcosis, or hydatid disease, is one of the most serious parasitic diseases of human in Asia, Africa, South America, and Europe. Humans are infected as intermediate hosts, i.e., hydatids develop in the internal organs. Some recent attempts for drug treatment are very promising. However, in 2000–2002, surgery remains the only routine method of treatment.

Echinococcus granulosus is also of primary veterinary importance because the hydatid disease is dangerous for many domestic herbivores (sheep, pigs, goats, cattle, camels, horses, etc.).

Beef tapeworm
Taenia saginata

ORDER
Cyclophyllidea

FAMILY
Taeniidae

TAXONOMY
Taenia cucurbitina grandis saginata Goeze, 1782, Germany. *Taeniarhynchus saginatus* (Goeze, 1782) Weinland, 1858.

OTHER COMMON NAMES
French: Ténia inerme; German: Rinderbandwurm.

PHYSICAL CHARACTERISTICS
Body usually 9.8–16 ft (3–5 m) long (exceptionally, some specimens reach a length of 66 ft [20 m]), with maximum width (0.20–0.28 in [5–7 mm]) at gravid proglottides. Scolex lacks rostellum and hooks. Gravid proglottides (found in feces of humans) can be distinguished from those of the other common human *Taenia*, *T. solium*, by the uterus having 15–20 (or more) lateral branches (versus 7–13 lateral branches in *T. solium*).

DISTRIBUTION
Cosmopolitan.

HABITAT
The microhabitats of this species are the intestines of humans (for adult worms) and the body musculature of cattle (for larvae). The macrohabitats can be all places where uninspected raw or undercooked beef is eaten, or where cattle graze on grass contaminated by eggs released with human feces.

FEEDING ECOLOGY AND DIET
Internal parasite absorbing nutrients through the tegument.

BEHAVIOR
The gravid proglottis can migrate out of the anus of the infected human and then can be found in the bed or in the underpants. However, it is usually released with feces. It actively crawls and can go at some distance from feces. With drying up, a rupture appears on the ventral surface of the proglottis, allowing eggs to disperse in the environment.

REPRODUCTIVE BIOLOGY
Russian researchers carried out a long-term study on the egg production of *T. saginata* in infected humans, including by provoking experimental self-infections during the period 1930–1940. They found that the adult parasite could live in the human intestine for more than 10 years. The daily output can reach up to 28 proglottides. Each proglottis may contain up to 175,000 eggs. Thus, an infected person may release up to 5

million eggs per day. Eggs are eaten by cattle with contaminated grass. They hatch in the duodenum. Oncospheres penetrate through the intestinal wall and enter blood vessels. They reach the musculature and turn into infective larvae (cysticerci) in about 8–10 weeks. Humans become infected by eating beef containing alive cysticerci (raw or undercooked). The adult worms start to produce gravid proglottids about one to three months after entering the final host.

CONSERVATION STATUS
Not listed by the IUCN.

SIGNIFICANCE TO HUMANS
Among tapeworms, *T. saginata* is the most widespread agent of parasitic diseases of humans in the world. The parasitic disease caused by it is known as taeniiasis or taeniarhynchiasis. The latter originates from the name of the genus *Taeniarhynchus* where this parasite was placed before 1994. The validity of this genus is not supported by most authors publishing since then.

Most of the infected people have no symptoms (except releasing gravid proglottides). Sometimes symptoms such as abdominal pain, diarrhea, headache, loss of appetite, and allergic reactions may occur. Several anthelmintic drugs are very efficient against this worm. The prevention is not difficult: beef is not dangerous when fully cooked and no longer pink inside. Cysticerci are killed at 131–140°F (55–60°C).

An alternative viewpoint of the significance of *T. saginata* (and the other two species using humans as final hosts, *T. solium* and *T. asiatica*, both with larvae developing in pigs) to humankind may be relevant. The traditional opinion among scientists is that with the domestication of intermediate hosts (cattle and swine) the species of Taenia have become associated with humans (some 10,000 years ago). However, phylogenetic studies during 2000–2001 based on comparative morphology and gene sequences suggest that hominids obtained taeniids before the origin of the modern humans. Probably some 2 million years ago, large African hominids preyed on antelopes and other bovids in the savanna, where large cats and hyenas also lived. They were parasitized by taeniid tapeworms because of their similar diet with that of the carnivore mammals. Once hominids acquired taeniid tapeworms, they also contributed to the rate of the infestation of herbivore mammals by providing additional tapeworm eggs in the grasslands. Wild bovids, with their musculature heavily infected by cysticerci, were probably easier prey than uninfected animals. Thus, *T. saginata* (or its ancestor living in our ancestors) helped early hominids have meat on their menu more frequently. At least part of the evolutionary success of our species may be due to tapeworms!

No common name
Proteocephalus longicollis

ORDER
Proteocephalidea

FAMILY
Proteocephalidae

TAXONOMY
Alyselmintus longicollis Zeder, 1800, type locality unknown.

OTHER COMMON NAMES
None known.

PHYSICAL CHARACTERISTICS
Body up to 8.7 in (220 mm) long and about 0.06–0.08 in (1.5–2 mm) wide. Scolex small (diameter up to 0.016 in [0.4 mm]), provided with small suckers and small sucker-like apical organ. Gravid proglottides up to 0.06–0.08 in (1.5–2 mm) long. Genital pores on the lateral margins of proglottides, irregularly alternating.

DISTRIBUTION
Eurasia and North America.

HABITAT
The microhabitats are the intestines of freshwater fishes (Salmonidae, Coregonidae, and Osmeridae, sometimes in others). This species inhabits lakes and rivers in the Northern Hemisphere.

FEEDING ECOLOGY AND DIET
Internal parasite absorbing nutrients through the tegument.

BEHAVIOR
Nothing is known.

REPRODUCTIVE BIOLOGY
This species is hermaphroditic. Eggs released in water are eaten by copepods, which are intermediate hosts. The infective larva develops in the body cavity of the crustacean. The fishes become infected by eating copepods containing cestode larvae.

CONSERVATION STATUS
Not listed by the IUCN.

SIGNIFICANCE TO HUMANS
Parasitic in fishes, including in those in fish farms. No data of economic significance. ◆

Broad fish tapeworm
Diphyllobothrium latum

ORDER
Pseudophyllidea

FAMILY
Diphyllobothriidae

TAXONOMY
Taenia lata Linnaeus, 1758, type locality unknown.

OTHER COMMON NAMES
German: Fischbandwurm.

PHYSICAL CHARACTERISTICS
Polyzoic. Strobila with length about 30 ft (9 m) (there are data about specimens reaching up to 66 ft [20 m]), consisting of 3,000–4,000 proglottides. Scolex finger-shaped, with two bothria. Uterus rosette-shaped in gravid proglottides.

DISTRIBUTION
Scandinavia, Baltic States, Russia, United States and Canada (Great Lakes area, Pacific Coast, Arctic), Ireland, Japan, around some lakes and large rivers in Africa, and South America.

HABITAT
Adults are common intestinal parasites of fish-eating mammals (dogs, cats, bears, seals, humans). Larvae develop in crustaceans (first intermediate hosts) and fishes (second intermediate hosts). The macrohabitats include rivers and freshwater lakes.

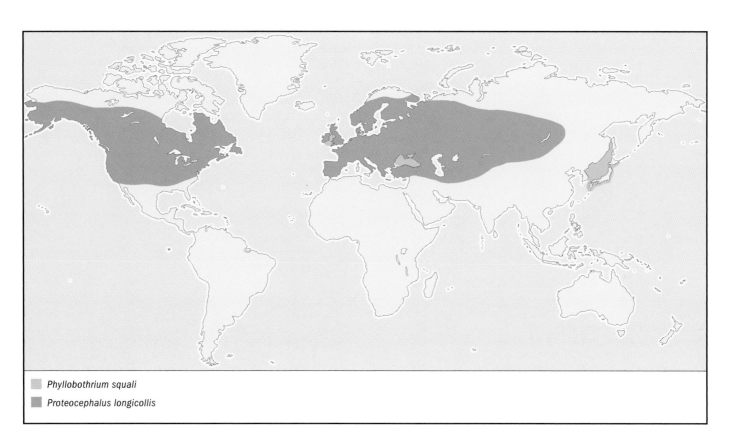

Phyllobothrium squali
Proteocephalus longicollis

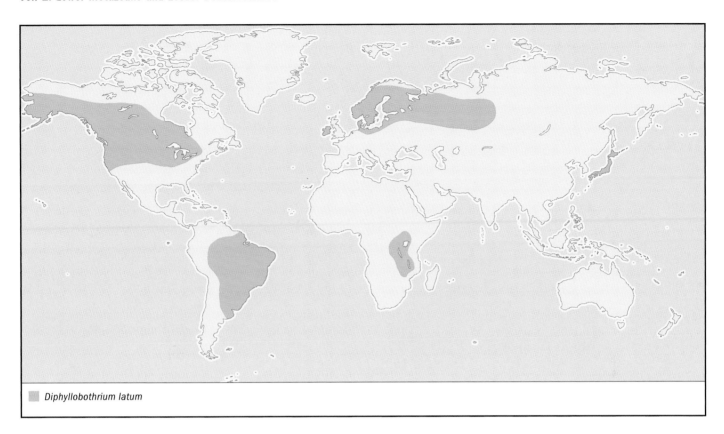

Diphyllobothrium latum

FEEDING ECOLOGY AND DIET
Internal parasite absorbing nutrients through the tegument.

BEHAVIOR
Nothing is known.

REPRODUCTIVE BIOLOGY
As almost all cestodes, this species is hermaphroditic. Eggs are released through the uterine pore and pass into the environment with feces of the host. The embryo encompassed in the egg needs one or several weeks (depending on the temperature) to become infective. The fully developed embryo (known as coracidium) hatches from the egg in water. It is covered by ciliated epidermis and can swim several hours until being eaten by a copepod crustacean (first intermediate host). In the intestine of the copepod, the coracidium loses its ciliated cover and penetrates into the body cavity. There, the embryo feeds on the nutrients contained in the hemolymph. For about 20–25 days, it turns into an elongate larva (procercoid, up to 0.0197 in [500 μm] long) possessing a cercomer at its posterior end. The procercoid is pathogenic for the copepod in terms of interfering with its motility and thus turning it into an easy prey for fishes. When the infected copepod is eaten by a fish (second intermediate host), the procercoid migrates from the intestine into the body musculature and turns into the next larval stage (plerocercoid). The predation of infected fishes, or even eating undercooked fish dishes in a restaurant, is the way of the transmission of the plerocercoid (from several millimeters to a few centimeters long) to the final host. In the host's intestine, the worm grows and becomes mature for some two weeks.

CONSERVATION STATUS
Not listed by the IUCN.

SIGNIFICANCE TO HUMANS
This species is of great medical importance. The disease caused by it is named diphyllobothriasis. Among the most widespread parasitic diseases caused by tapeworms, it is ranked fourth. Its symptoms are diarrhea, abdominal discomfort, and weakness, in some cases anemia. The contemporary drug treatment is very efficient. ◆

No common name
Phyllobothrium squali

ORDER
Tetraphyllidea

FAMILY
Phyllobothriidae

TAXONOMY
Phyllobothrium squali Yamaguti, 1952, off Onahama, Japan.

OTHER COMMON NAMES
None known.

PHYSICAL CHARACTERISTICS
Body 5.5–24 in (14–60 cm) long, 0.08–0.16 in (2–4 mm) wide. Scolex 0.12–0.18 in (3–4.5 mm) wide, with four foliose bothridia with folded margins. Anterior end of each bothridium provided with small sucker. Apex of the scolex provided with glandular organ. Genital pores on lateral margins of proglottides, irregularly alternating.

DISTRIBUTION
Japanese waters, Black Sea, and Irish Sea.

HABITAT
In the spiral intestine of piked dogfish, *Squalus acanthias* (Squaliformes, Squalidae).

FEEDING ECOLOGY AND DIET
Internal parasite absorbing nutrients through the tegument.

BEHAVIOR
Nothing is known.

REPRODUCTIVE BIOLOGY
Hermaphroditic. The life cycle is unknown.

CONSERVATION STATUS
Not listed by the IUCN.

SIGNIFICANCE TO HUMANS
None known. ◆

Resources

Books

Brooks, D. R., and D. A. McLennan. *Parascript: Parasites and the Language of Evolution*. Washington and London: Smithsonian Institution Press, 1993.

Combes, C. *Parasitism: The Ecology and Evolution of Intimate Interactions*. Chicago and London: University of Chicago Press, 2001.

Gibson, D. I., R. A. Bray, and C. B. Powell. "Aspects of the Life History and Origins of *Nesolecithus africanus* (Cestoda: Amphilinidea)." Journal of Natural History 21 (1987): 785–794.

Hoberg, E. P., N. L. Alkire, A. de Queiroz, and A. Jones. "Out of Africa: Origins of the *Taenia* Tapeworms in Humans." *Proceeding of the Royal Society, London. Series B* 268 (2000): 781–787.

Kearn, G. C. *Parasitism and the Platyhelminths*. London: Chapman and Hall, 1998.

Khalil, L. F., A. Jones, and R. A. Bray. *Keys to the Cestode Parasites of Vertebrates*. Wallingford: CAB International, 1994.

Littlewood, D. T. J., and R. A. Bray, eds. *Interrelations of the Platyhelminthes*. London and New York: Taylor and Francis, 2001.

Roberts, L. S., and J. Janovy. *Gerald D. Schmidt and Larry S. Roberts' Foundations of Parasitology*. 6th ed. Boston: McGraw-Hill Co., 2000.

Ruppert, E. E., and R. D. Barnes. *Invertebrate Zoology*. 6th ed. Forth Worth, TX: Saunders College Publishing, 1994.

Schmidt, G. D. *CRC Handbook of Tapeworm Identification*. Boca Raton, FL: CRC Press, Inc., 1986.

Scholz, T., and V. Hanzelová. *Tapeworms of the Genus* Proteocephalus *Weinland, 1858 (Cestoda: Proteocephalidae), Parasites of Fishes in Europe*. Prague: Academia, 1998.

Periodicals

Bandoni, S. M., and D. R. Brooks. "Revision and Phylogenetic Analysis of the Amphilinidea Poche, 1922 (Platyhelminthes: Cercomeria: Cercomeromorpha)." *Canadian Journal of Zoology* 65 (1987): 1110–1128.

Beveridge, I., and R. A. Campbell. "Redescription of *Diesingium lomentaceum* (Diesing, 1850) (Cestoda: Trypanorhyncha)." *Systematic Parasitology* 27 (1994): 149–157.

Caira, J. N., K. Jensen, and C. J. Healy. "On the Phylogenetic Relationships Among Tetraphyllidean, Lecanicephalidean and Diphyllidean Tapeworm Genera." *Systematic Parasitology* 42, no. 2 (February 1999): 77–151.

Cañeda–Guzmán, I. C., A. Chambrier, and T. Scholz. "*Thaumasioscolex didelphidis* n.gen., n.sp. (Eucestoda: Proteocephalidae) from the Black-eared Opossum *Didelphis marsupialis* from Mexico, the First Proteocephalidean Tapeworm from a Mammal." *Journal of Parasitology* 87, no. 3 (June 2001): 639–646.

Crompton, D. W. T. "How Much Human Helminthiasis Is There in the World?" *Journal of Parasitology* 85, no. 3 (June 1999): 397–403.

Faliex, E., G. Tyler, and L. Euzet. "A New Species of *Ditrachybothridium* (Cestoda: Diphyllidea) from *Galeus* sp. (Selachii, Scyliorhynidae) from the South Pacific Ocean, with a Revision of the Diagnosis of the Order, Family, and Genus and Notes on Descriptive Terminology of Microtriches." *Journal of Parasitology* 86, no. 5 (October 2000): 1078–1084.

Hanzelová, V., and T. Scholz. "Species of *Proteocephalus* Weinland, 1858 (Cestoda: Proteocephalidae), Parasites of Coregonid and Salmonid Fishes from North America: Taxonomic Reappraisal." *Journal of Parasitology* 85, no. 1 (February 1999): 94–101.

Hoberg, E. P. "Phylogenetic Relationships Among Genera of the Tetrabothriidae (Eucestoda)." *Journal of Parasitology* 75, no. 4 (August 1989): 617–626.

Hoberg, E. P., S. L. Gardner, and R. A. Campbell. "Systematics of the Eucestoda: Advances Toward a New Phylogenetic Paradigm, and Observations on the Early Diversification of Tapeworms and Vertebrates." *Systematic Parasitology* 42, no. 1 (January 1999): 1–12

Hoberg, E. P., J. Mariaux, J.-L. Justine, D. R. Brooks, and P. J. Weekes. "Phylogeny of the Orders of the Eucestoda (Cercomeromorphae) Based on Comparative Morphology: Historical Perspectives and a New Working Hypothesis." *Journal of Parasitology* 83, no. 6 (December 1997): 1128–1147.

Jensen, K. "Four New Genera and Five New Species of Lecanicephalideans (Cestoda: Lecanicephalidea) from Elasmobranchs in the Gulf of California, Mexico." *Journal of Parasitology* 87, no. 4 (August 2001): 845–861.

Justine, J.-L. "Spermatozoa as Phylogenetic Characters for the Eucestoda." *Journal of Parasitology* 84, no. 2 (April 1998): 385–408.

Littlewood, D. T. J., K. Rohde, and K.A. Clough. "The Interrelations of All Major Groups of Platyhelminthes: Phylogenetic Evidence from Morphology and Molecules." *Biological Journal of the Linnean Society* 66 (1999): 75–114.

Resources

Neifar, L., G. A. Tayler, and L. Euzet. "Two New Species of *Macrobothridium* (Cestoda: Diphyllidea) from Rhinobatid Elasmobranchs in the Gulf of Gabès, Tunisia, with Notes on the Status of the Genus." *Journal of Parasitology* 87, no. 3 (June 2001): 673–680.

Nikolov, P. N., and B. B. Georgiev. "The Morphology and New Records of Two Progynotaeniid Cestode Species." *Acta Parasitologica* 47, no. 2 (June 2002): 121–130.

Olson, P. D., and J. N. Caira. "Two New Species of *Litobothrium* Dailey, 1969 (Cestoda: Litobothriidea) from Thresher Sharks in the Gulf of California, Mexico, with Redescriptions of Two Species in the Genus." *Systematic Parasitology* 48, no. 3 (March 2001): 159–177.

Olson, P. D., D. T. J. Littlewood, R. A. Bray, and J. Mariaux. "Interrelations and Evolution of the Tapeworms (Platyhelminthes: Cestoda)." *Molecular Phylogenetics and Evolution* 19, no. 3 (June 2001): 443–467.

Rego, A. A., A. Chambrier, V. Hanzelová, E. Hoberg, T. Scholz, P. Weekes, and M. Zehnder. "Preliminary Phylogenetic Analysis of Subfamilies of the Proteocephalidea (Eucestoda)." *Systematic Parasitology* 40, no. 1 (May 1998): 1–19.

Scholz, T., and R. A. Bray. "*Probothriocephallus alaini* n. sp. (Cestoda: Triaenophoridae) from the Deep-sea Fish *Xenodermichthys copei* in the North Atlantic Ocean." *Systematic Parasitology* 50, no. 3 (November 2001): 231–235.

Stock, T. M., and J. Holmes. "Functional Relationships and Microhabitat Distributions of Enteric Helminths of *Grebes* (Podicipedidae): The Evidence for Interactive Communities." *Journal of Parasitology* 74, no. 2 (April 1988): 214–227

Tyler, G. A., and J. N. Caira. "Two New Species of *Echinobothrium* (Cestoidea: Diphyllidea) from Myliobatiform Elasmobranchs in the Gulf of California, Mexico." *Journal of Parasitology* 85, no. 2 (April 1999): 327–335.

Tyler, G. A. "Diphyllidean Cestodes of the Gulf of California, Mexico with Descriptions of Two New Species of *Echinobothrium* (Cestoda: Diphyllidea)." *Journal of Parasitology* 87, no. 1 (February 2001): 173–184.

Vasileva, G. P., G. I. Dimitrov, and B. B. Georgiev. "*Phyllobothrium squali* Yamaguti, 1952 (Tetraphyllidea, Phyllobothriidae): Redescription and First Record in the Black Sea." *Systematic Parasitology* 53, no. 1 (September 2002): 49–59.

Vasileva, G. P., D. I. Gibson, and R. A. Bray. "Taxonomic Revision of *Tatria* Kowalewski, 1904 (Cestoda: Amabiliidae): Redescriptions of *T. biremis* Kowalewski, 1904 and *T. minor* Kowalewski, 1904, and the Description of *T. gulyaevi* n. sp. from Palaearctic *Grebes*. "*Systematic Parasitology* 54, no. 3 (March 2003): 177–198.

Organizations

Academy of Sciences of the Czech Republic, Institute of Parasitology. Ceské Budejovice, Czech Republic.<http://www.paru.cas.cz/structure/Lab_of_par_flat worms/index.html>

Agricultural Research Service. Web site: <http://www.ars.usda.gov/is/AR/archive/may01/worms0501.htm>

Bulgarian Academy of Sciences, Central Laboratory of General Ecology. Sofia, Bulgaria. <http://www.diplectanum.dsl.pipex.com/bas/>

Natural History Museum, Department of Invertebrates. Geneva, Switzerland. <http://www.ville-ge.ch/musinfo/mhng/index.htm>

The Natural History Museum, Department of Zoology, Parasitic Worms Division. London, UK. <http://www.nhm.ac.uk/zoology/home/home.htm>

National Museum of Natural History, Laboratory of Biology of Parasites, Protistology and Helminthology. Paris, France. <http://www.mnhn.fr/mnhn/bpph/>

Russian Academy of Sciences, Zoological Institute, Department of Parasitic Worms. St. Petersburg, Russia. <http://www.zin.ru/labs/worms/eng/main_eng2.htm>

U. S. Department of Agriculture, Agricultural Research Service, U. S. National Parasite Collection,.Beltsville, MD USA. <http://www.lpsi.barc.usda.gov/bnpcu/>

University of Connecticut, Department of Ecology and Evolutionary Biology. Storrs, CT USA. <http://collections2.eeb.uconn.edu/tapewormsdotorg/home.htm>

University of Nebraska State Museum, The Harold W. Manter Laboratory of Parasitology. Lincoln, NE USA. <http://www-museum.unl.edu/research/parasitology/>

University of Toronto, Department of Zoology. Toronto, Canada. <http://brooksweb.zoo.utoronto.ca/index.html>

Boyko B. Georgiev, PhD

Anopla
(Anoplans)

Phylum Nemertea

Class Anopla

Number of families 11

Thumbnail description
Worms that have an unarmed proboscis; mouth situated below or posterior to the cerebral ganglia; and central nervous system situated within the body wall, or between the body musculature and epidermis

Photo: Ribbon worms such as *Parborlasia* sp. are roving predators, intent on hunting other invertebrates with their good eyesight and ability to get into small hiding places, Heron Island, the southern Great Barrier Reef, Australia. (Photo by A. Flowers & L. Newman. Reproduced by permission.)

Evolution and systematics

The fossil record of nemertines is extremely sparse, as would be expected from a soft-bodied animal. There is a trace fossil (genus *Archisymplectes*) from the Pennsylvania-age from central Illinois that may represent an anoplan nemertine. The anoplan palaeonemerteans have been regarded as the phylogenetically basal nemerteans based on a simpler nervous system and cerebral organs, but recent molecular studies are inconclusive.

Class Anopla used to be divided in three subclasses, Archinemertea, Palaeonemertea, and Heteronemertea, but Archinemertea has been shown to be paraphyletic and the name is not used by most authors. Recent studies based on nucleotide sequences established that the paleonemertea is a non-monophyletic group but did not formally reclassify the order. The term palaeonemerteans and Paleonemertea thus refer to a presumed paraphyletic assemblage and are used for convenience and tradition without reflecting a monophyletic group. The Palaeonemertea includes those species with two (outer circular and inner longitudinal) or three (outer circular, middle longitudinal, and inner circular) body muscle layers. The subclass Heteronemertea includes those species with primarily three body wall muscles, although there exist species with an additional inner circular layer. The central nervous system is situated between the outer longitudinal and the middle circular muscular layers in heteronemerteans, and in the inner longitudinal muscle layer, or external to body wall muscles, in the palaeonemerteans.

The class is currently divided into 11 families and 93 genera comprising approximately 500 described species. The largest genus, *Cerebratulus* contains 116 species, and the second largest, *Lineus*, has 80 species.

Physical characteristics

Anoplans are differentiated from nemerteans in the class Enopla by the proboscis, which is armed (i.e., there is a stylet attached to it) in enoplans but unarmed in anoplans. Other differences between the two groups include the position of the lateral nervous system and the relationship between the mouth and proboscis; enoplans have a common opening for the mouth and proboscis, while anoplans have separate openings for the two structures. Anopla contains the largest nemerteans. The longest species, *Lineus longissimus*, is described as being up to 98 ft (30 m) long, and several other hetero- and palaeonemerteans are big and solid worms. *Parborlasia corrugatus*, for example, is abundant and widely distributed in the Antarctic and can reach lengths up to 6.5 ft (2 m). While most nemerteans are rather drab in color, the most colorful and strikingly pigmented are found among the anoplans. There have not been many experiments with nemerteans, but a few studies indicate that the coloration and pigmentation are cases of warning coloration. Most nemerteans have toxic and noxious substances in their body—undoubtedly to deter predators. When these substances are accompanied with bright colors and a lifestyle (day active, no hiding, etc.) that does not encourage them to hide from predators, the interpretation of aposematisms seems well founded. In the field, a nemertean is generally recognized by the way it moves. The normal movement is gliding over the surface by help of cilia on the ventral side in combination with mucus produced by the worm. Some species may, under certain circumstances, swim with undulating movements, but only for a short period of time.

Distribution

Anoplan nemerteans are known from all continents and all seas.

Parborlasia corrugatus feeding on eggs of the Antarctic dragon fish (*Gymnodraco acuticeps*). (Photo by Dirk Schories. Reproduced by permission.)

Habitat

Anoplan nemerteans are benthic and found in the littoral as well as in deeper waters. In general, heteronemerteans are more common in the littoral, while palaeonemerteans seems to be more abundant at deeper soft bottoms. However, the ecology of nemerteans is, in general, poorly known, as is their actual spatial distribution. While larger species may be found simply by turning over boulders, smaller species are not found unless special techniques are utilized. An easy way of collecting nemerteans is to place seaweed and smaller algae in a bucket of seawater and let it stand for a few hours, and up to a couple of days, depending on weather and temperature. The worms will crawl to the sides of the bucket, where they are easily observed and collected, as the oxygen concentration decreases in the water. Although nemerteans are abundant, especially in temperate waters, their presence is often overlooked because they are not easily observed.

Behavior

Most nemerteans are solitary, free-living animals. No anoplan nemerteans are known to be parasitic or commensal.

Feeding ecology and diet

Nemerteans are common predators in a variety of habitats, but there is little published information on food and feeding of anoplans. Palaeonemerteans appear to feed on annelids that are compatible with their size. What is known about heteronemerteans indicates that their food is living or dead polychaetes, but they also feed on mollusks, crustaceans, and other nemerteans. Scavenging may be a way of life for many species, but clear-cut evidence exists only for the Antarctic *Parborlasia corrugatus*. There are no parasitic or commensal known hetero- or palaeonemerteans.

Reproductive biology

Most nemerteans are dioecious, although there are a few hermaphroditic hoplonemerteans (class Enopla). Palaeonemerteans are traditionally regarded as direct developers, whereas heteronemerteans all produce pilidium larvae (free-swimming, planktotropic larvae), or larvae that seem to be derived from a pilidium. These heteronemerteans are broadcast spawners, but there are also heteronemerteans (*Lineus rubber* and *L. viridis*) known to reproduce by a kind of pseudocopulation in which the

Unsegmented ribbon worm (*Lineus* sp.) seen near Panama. (Photo by Kjell B. Sandved/Photo Researchers, Inc. Reproduced by permission.)

fertilized eggs are deposited in a gelatinous cocoon producing pilidium-like larvae known as Schmidt's and Desor's larvae, respectively. Only a few species are known to reproduce vegetatively by fission (a few species in genus *Myoisophagus*).

Conservation status

Very little is known about nemertean ecology, distribution, or abundance. It is clear, however, that certain species are the most abundant invertebrate group in some habitats/some localities. Whether other species are threatened is almost impossible to say. The most recent IUCN Red List of Threatened Species includes only six nemerteans, all belonging to class Enopla.

Significance to humans

It is documented that the heteronemertean, *Cerebratulus lacteus*, (tapeworm) was frequently collected and used as bait by sport fishermen in the United States, at least into the middle of the twentieth century. Another large heteronemertean used as bait in South Africa and Mozambique is *Polybrachiorhynchus dayi*, also referred to as tapeworm by fishermen. Other uses of anoplan nemerteans are not known.

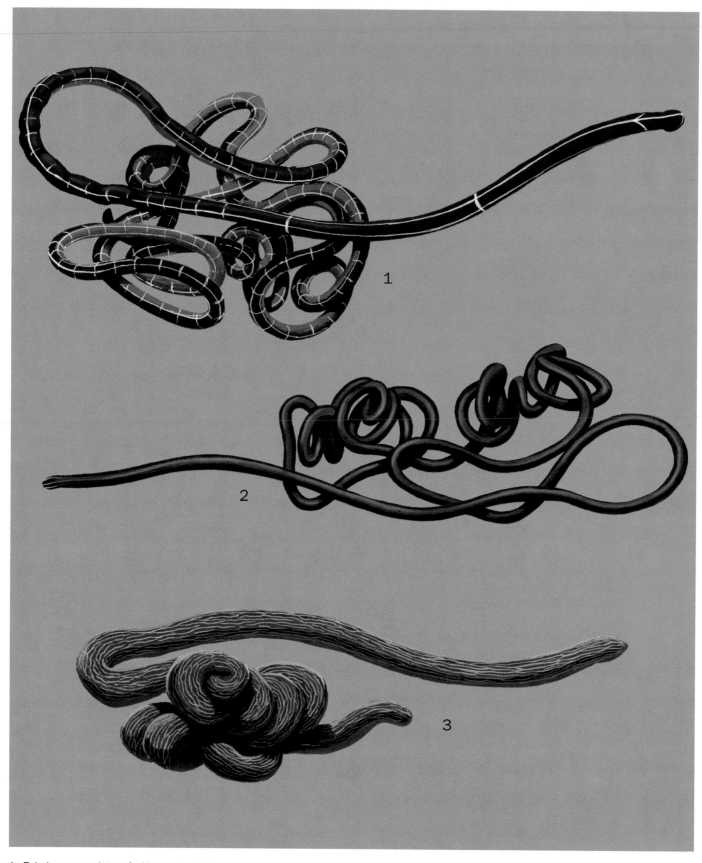

1. *Tubulanus annulatus*; 2. *Lineus longissimus*; 3. *Baseodiscus delineatus*. (Illustration by John Megahan)

Species accounts

No common name
Baseodiscus delineatus

ORDER
Heteronemertea

FAMILY
Baseodiscidae

TAXONOMY
Baseodiscus delineatus (Delle Chiaje, 1825), Italy.

OTHER COMMON NAMES
None known.

PHYSICAL CHARACTERISTICS
May attain lengths of 3.3 ft (1 m) or more, but rarely more than 0.07–0.11 in (2–3 mm) wide. Head slightly bi-lobed, demarcated from rest of body, with numerous black or dark brown eyes, mainly distributed along the cephalic margins, and form two large dorsolateral groups near back of head. Background color uniform dull yellowish fawn to light brown, marked by reddish brown, interrupted longitudinal stripes extending full body length.

DISTRIBUTION
From Iceland eastwards to the Atlantic and North Sea coasts of Europe.

HABITAT
Lower shore beneath boulders or sublittorally on coarse shell or gravel substrata containing some mud or sand.

BEHAVIOR
Nothing is known.

FEEDING ECOLOGY AND DIET
Nothing is known.

REPRODUCTIVE BIOLOGY
Separate sexes, external fertilization with pelagic larvae.

CONSERVATION STATUS
Not threatened.

SIGNIFICANCE TO HUMANS
None known. ◆

No common name
Lineus longissimus

ORDER
Heteronemertea

FAMILY
Lineidae

TAXONOMY
Lineus longissimus (Gunnerus, 1770), Norway.

OTHER COMMON NAMES
None known.

Baseodiscus delineatus

PHYSICAL CHARACTERISTICS
The longest nemertean known, individuals of 16.4–32.8 ft (5–10 m) length are common and specimens of up to 98.4 ft (30 m) are reported. Flaccid body contracts and extends in series of irregular muscular waves when disturbed. Colors range from dark olive-brown or rich chocolate brown in smaller specimens to blackish brown to black in larger animals. Flickering purplish iridescence often evident, resulting from activity of epidermal cilia. Ventral color may be paler than dorsal. Tip of head pale or whitish and usually appearing bi-lobed. There are 10–40 reddish brown or black eyes in a row on each side of the snout.

DISTRIBUTION
From Iceland eastwards to the Atlantic and North Sea coasts of Europe.

HABITAT
Typically found on the lower shore beneath boulders on muddy sands, but also in rockpools or in deeper sublittoral locations on muddy, sandy, stony, or shelly bottoms.

BEHAVIOR
Epidermis contains toxic/noxious substances; probably a defense mechanism.

FEEDING ECOLOGY AND DIET
Nothing is known.

REPRODUCTIVE BIOLOGY
Separate sexes, external fertilization with pelagic larvae.

CONSERVATION STATUS
Not threatened.

SIGNIFICANCE TO HUMANS
None known. ◆

No common name
Tubulanus annulatus

ORDER
Palaeonemertea

FAMILY
Tubulanidae

TAXONOMY
Tubulanus annulatus (Montagu, 1804), England.

OTHER COMMON NAMES
None known.

PHYSICAL CHARACTERISTICS
Strikingly colored, vivid brick-red, orange-red, garnet-red, or brownish red marked with white longitudinal stripes and rings. Mid-dorsal stripe extends on to the cephalic lobe and terminates at a transverse white band on head. The two lateral stripes do not reach head. Ventral surface is paler than dorsal. Length 29.5 in (75 cm) or more, but rarely exceeds 0.11–0.15 in (3–4 mm).

DISTRIBUTION
In Northern Hemisphere from Pacific coast of North America eastwards to the Atlantic, North Sea, and Mediterranean coasts of Europe.

HABITAT
Found intertidally beneath stones or on sand or mud near low water level, but more common sublittorally on a wide variety of substrata at depths down to 130 ft (40 m) or more.

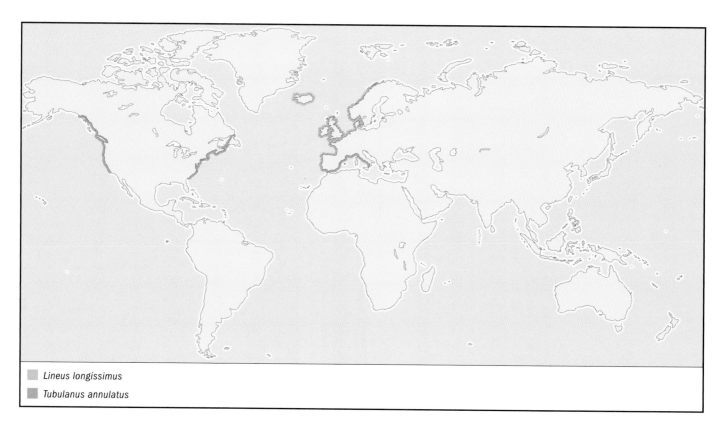

Lineus longissimus

Tubulanus annulatus

BEHAVIOR
Epidermis contains toxic/noxious substances and the worm is avoided by fishes. In view of the conspicuous appearance in shallow waters, this species may be a case of aposematic coloration.

FEEDING ECOLOGY AND DIET
Nothing is known.

REPRODUCTIVE BIOLOGY
Separate sexes.

CONSERVATION STATUS
Not threatened.

SIGNIFICANCE TO HUMANS
None known. ◆

Resources

Books
Gibson, Ray. *Nemerteans.* London: Hutchinson University Library, 1972.

———. *Nemerteans. Synopses of the British Fauna.* Dorchester: The Linnean Society of London, 1994.

Gibson, Ray, Janet Moore, and Per Sundberg. *Advances in Nemertean Biology.* Dordrecht: Kluwer Academic Publishers, 1993.

Sundberg, Per, Ray Gibson, and Gunnar Berg. *Recent Advances in Nemertean Biology.* Dordrecht: Dr. W. Junk Publishers, 1988.

Periodicals
Gibson, Ray. "The Invertebrate Fauna of New Zealand: Nemertea (Ribbon Worms)." *NIWA Biodiversity Memoir* 118 (2002): 1–87.

Sundberg, Per, Turbeville J. McClintock, and Susanne Lindh. "Phylogenetic Relationships among Higher Nemertean (Nemertea) Taxa Inferred from 18S rDNA Sequences." *Molecular Phylogenetics and Evolution.* 20 (2001): 327–334.

Thollesson, Mikael, and Jon L. Norenburg. "Ribbon Worm Relationships: A Phylogeny of the Phylum Nemertea." *Proceedings of the Royal Society of London Series B—Biological Sciences.* 270 (2003): 407–415.

Per A. Sundberg, PhD

Enopla
(Enoplans)

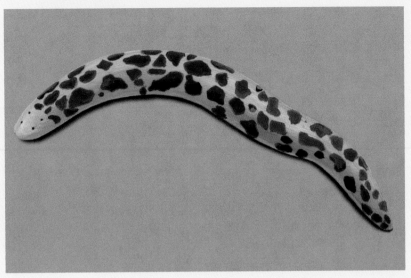

Phylum Nemertea

Class Enopla

Number of families 30

Thumbnail description
Worms with an armed proboscis; mouth situated anterior to the cerebral ganglia; and the central nervous system internal to body wall, which consists of three muscle layers

Illustration: *Oerstedia dorsalia.* (Illustration by Emily Damstra)

Evolution and systematics

The fossil record of nemertines is extremely sparse, as would be expected from a soft-bodied animal. The Cambrian fossil, Amiskwia, has been interpreted as a nemertean based on its resemblance to some pelagic ribbon worms; however, this interpretation is disputed by many paleontologists. The enoplan nemerteans have been regarded as highly derived based on a more complicated muscle arrangement in the body wall and a more complex nervous system. However, whether this is a plesiomorphic or apomorphic character is not clear, and recent molecular studies are inconclusive in this respect.

Class Enopla used to be divided in two subclasses, Hoplonemertea and Bdellonemertea, but recent phylogenetic analyses based on nucleotide sequences show that Bdellonemertea should be included in Hoplonemertea. Hoplonemertea (in the old sense) contains two suborders, Monostilifera and Polystilifera. The former encompasses those animals with a proboscis armature consisting of a single central stylet on a large cylindrical basis. The Polystilifera are armed with a pad, or shield, bearing numerous small stylets. The Polystilifera are further divided in two taxa, one (Pelagica) containing the pelagic species, and the other (Reptantia) with crawling or burrowing forms.

The class is currently divided into 30 families and 155 genera with approximately 650 described species. The two largest genera, *Amphiporus* and *Tetrastemma* contain 230 species, i.e., one third of all named species in the class. However, it must be made very clear that the systematics and classification of nemerteans are not based on a phylogenetic approach, and recent studies question the classification.

Physical characteristics

Enoplan nemerteans are generally small, from less than 0.4 in (1 cm) up to 4 in (10 cm), although larger species exist. While most nemerteans are rather drab in color, others are more conspicuous with striking pigment patterns and coloration. However, the more brightly colored forms are more common in the class Anopla. A nemertean is generally recognized in the field by the way it moves. Its normal movement is gliding over the surface by help of cilia on the ventral side in combination with mucus produced by the worm. Some species may, under certain circumstances, swim with undulating movements, but only for a short period of time. Enoplans are differentiated by the proboscis, which is armed (i.e., there is a stylet attached to it) in enoplans but unarmed in anoplans. Enoplans have a common opening for the proboscis and mouth, whereas anoplans have separate openings for the two structures.

Distribution

Enoplan nemerteans are known from all continents and all seas. Terrestrial nemerteans are mainly known from islands in the tropical and subtropical regions, although there are few more widespread species. Freshwater species are also reported from all continents, except the Antarctic.

Habitat

Enoplan nemerteans are typically found in the sea, in the littoral among algae. While larger species may be found simply by turning over boulders, smaller species are not found unless special techniques are utilized. An easy way of collecting nemerteans is to place seaweed and smaller algae in a bucket of sea water and let it stand for a few hours, and up

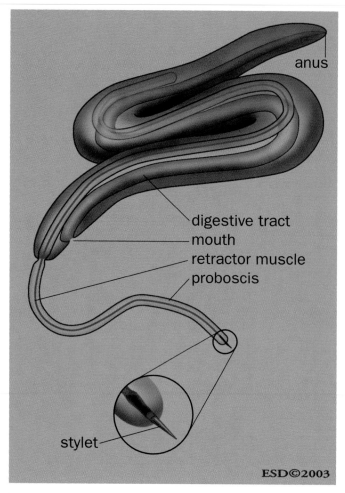

Diagram showing the position of the stylet at the end of the proboscis. (Illustration by Emily Damstra)

to a particular island. These species live in damp places under stones and among rotting woods.

Behavior

Most nemerteans are solitary, free-living animals.

Feeding ecology and diet

Nemerteans are common predators in a variety of habitats. Benthic marine enoplans are suctorial feeders and prey

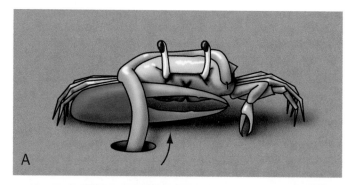

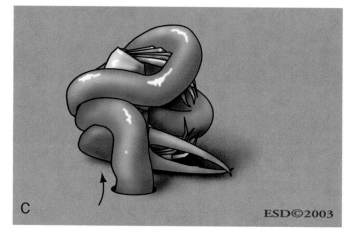

A hoplonemertean attacking the crab *Uca musica*. A. The everted proboscis rapidly entwines the prey and the stylet injects toxins to immobilize the prey, and probably injects digestive enzymes at the same time. B. The struggle is over and the worm inverts its proboscis. C. The worm emerges and searches for a place to enter the crab, whose tissues will be drained in about an hour. (Illustration by Emily Damstra)

to a couple of days, depending on weather and temperature. The worms will crawl to the sides of the bucket, where they are easily observed and collected, as the oxygen concentration decreases in the water. Although nemerteans are abundant, especially in temperate waters, their presence is often overlooked because they are not easily observed. Enoplan nemerteans do not appear to be equally common sublittorally, but this may be a result of biased sampling (less accessible environments). The majority of enoplan ribbon worms are marine and benthic, but there are approximately 100 named and described species of pelagic nemerteans. These creatures inhabit the water column of the world oceans, commonly found at depths of between a few hundred feet (meters) and several thousand feet (meters), and they are most abundant at 2,130–8,200 ft (625–2,500 m). There are a few freshwater species recorded, of which most are placed in the genus *Prostoma*. This genus is also by far the most widespread, especially the two species, *P. eilhardi* and *P. graecense*. The latter has been recorded from Europe, Africa, Japan, and Australia. The spreading of these animals is probably a result of the exportation and importation of freshwater vegetation. There are 13 known species of terrestrial nemerteans; a typical feature of these species is that their distribution tends to be restricted mainly on crustaceans. The proboscis is everted and the central armature (the stylet) is used to pierce and immobilize the prey. After inversion of the proboscis, the worm uses its head to probe among the crustacean appendages, seeking a place where it can penetrate the prey; eventually, the head is wedged past the opening and the anterior gut is everted into the opening. It is uncertain whether proteolytic enzymes are inserted through the stylet-produced hole in the exoskeloton—histology of central armature suggests this—but at some stage, enzymes are injected to dissolve the prey's body tissue. Free-living marine suctorial nemerteans appear to be food specialists feeding primarily on amphipods. There are some enoplan species known to feed upon barnacles, limpets, and polychaetes. There are also examples of macrophagus hoplonemerteans that engulf the entire prey after paralyzing it with a blow by the stylet. Freshwater hoplonemerteans are known to feed on oligochaetes, unicellular organisms, insect larvae, and other crustaceans. Very little is known about the ecology of pelagic nemerteans, including diet and feeding behavior.

There is one group of parasitic enoplan nemerteans (family Carcinonemertidae) found among the egg masses of certain crab species that feed on the host's embryos. There are also commensal enoplans (in family Bdellonemertidae) that live in the mantle cavities of bivalves where they feed on plankton from the mantle cavity. Obviously, the proboscis is not used to capture prey and has been (perhaps secondarily) reduced in these species.

Reproductive biology

Most nemerteans are dioecious, although there are a few hermaphroditic hoplonemerteans. Most species are oviparous, i.e., produce eggs that are laid and hatched externally. Mode of spawning is unknown for most species, but where known, it ranges from widespread release of gametes into surrounding waters, to pseudocopulation with eggs attached in a gelatinous matrix to a benthic substratum. A few species bear living young.

Conservation status

Very little is known about nemertean ecology, distribution, or abundance. It is clear, however, that certain species are the most abundant invertebrate group in some habitats/some localities. Whether other species are threatened is almost impossible to say, but the 1996 IUCN Red List of Threatened Species includes six terrestrial nemerteans. Two species (*Antiponemertes allisonae* and *Katechonemertes nightingaleensis*) are considered Threatened, and *Argonemertes hillii* as Near Threatened.

Significance to humans

There is no direct significance to humans. However, many nemerteans produce toxins of which some are nicotinic agonists. Some of these toxins, originally found in a nemertean, have been synthesized and tested in pre-clinical trials as a possible memory enhancer in the treatment of Alzheimer's disease.

Species accounts

No common name
Oerstedia dorsalis

ORDER
Hoplonemertea

FAMILY
Prosorhochmidae

TAXONOMY
Planaria dorsalis Abildgaard, 1806, Denmark.

OTHER COMMON NAMES
None known.

PHYSICAL CHARACTERISTICS
Small, rather stout, up to 1.2 in (30 mm); most specimens
0.4–0.6 in (10–15 mm) long and 0.04–0.07 in (1–2 mm) wide.
Head bluntly rounded, not demarcated from body, with four eyes. Extremely variable in color; more or less uniform brown to reddish brown with mid-dorsal white stripe, or with irregular light or dark brown speckles varying in size and shape between specimens. Some specimens have yellowish dots, some specimens are cream without pigmentation, others with regular dark bands on light background. Ventral surface usually paler than dorsal.

Oerstedia dorsalis

DISTRIBUTION
Widely distributed in the northern hemisphere; found in the Baltic Sea, the North Sea, the Mediterranean, on eastern Atlantic coasts from northern Europe to Madeira, and both Atlantic and Pacific coasts of North America down to Mexico.

HABITAT
Marine, littoral and sublittorally, down to 260 ft (80 m). Generally associated with small algae, especially species of *Ceramium* and *Corallina*.

BEHAVIOR
Lives among algae. No other details known.

FEEDING ECOLOGY AND DIET
Not known in detail, but probably small crustaceans and other worms.

REPRODUCTIVE BIOLOGY
Dioecious, external fertilization.

CONSERVATION STATUS
Not listed by the IUCN. Common.

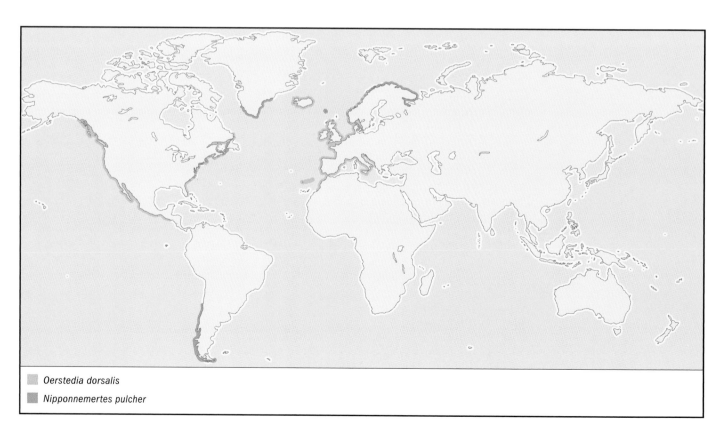

- *Oerstedia dorsalis*
- *Nipponnemertes pulcher*

SIGNIFICANCE TO HUMANS
None. ◆

No common name
Nipponnemertes pulcher

ORDER
Hoplonemertea

FAMILY
Cratenemertidae

TAXONOMY
Nemertes pulchra Johnston, 1837, England.

OTHER COMMON NAMES
None known.

PHYSICAL CHARACTERISTICS
Up to 4 in (10 cm) long and 0.04–0.2 in (1–5 mm) broad; rather stout, somewhat dorso-ventrally compressed, body gradually tapering posteriorly to the bluntly pointed tail. Distinct head like a rounded triangle or shield, with two pairs of cephalic furrows and numerous eyes irregularly distributed near the lateral cephalic margins; number of eyes

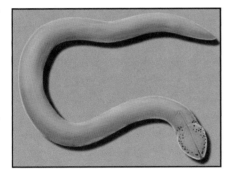

Nipponnemertes pulcher

increase with age, up to 70–80 in specimens 1.5–1.9 in (4–5 cm). Dorsally brown, red, or pink, but lateral and ventral surfaces always much lighter. Color affected by age, degree of sexual maturation, and habitat.

DISTRIBUTION
Reported from east coast of North America, Greenland, Faroe Islands, the White Sea, and northern Europe from the Atlantic coast of France to Scandinavia, and from Chile and many parts of the Antarctic and sub-Antarctic areas.

HABITAT
Marine, sublittoral among corallines or on coarser sediments such as sand, gravel, or shelly debris at depths down to 787 ft (240 m); also on muddy or stony bottoms down to 1,870 ft (570 m).

BEHAVIOR
Nothing is known.

FEEDING ECOLOGY AND DIET
Known to feed on crustaceans, *Haploops* spp. and *Corophium volutator*. Feeding rate measured to one prey per every fifth day (in laboratory conditions).

REPRODUCTIVE BIOLOGY
Dioecious, external fertilization. Swim above bottom surface by undulating movements when releasing their gametes.

CONSERVATION STATUS
Not threatened.

SIGNIFICANCE TO HUMANS
None known. ◆

Resources

Books

Gibson, Ray. *Nemerteans.* London: Hutchinson University Library, 1972.

———. *Nemerteans: Synopses of the British Fauna.* Dorchester: The Linnean Society of London, 1994.

Gibson, Ray, Janet Moore, and Per Sundberg. *Advances in Nemertean Biology.* Dordrecht: Kluwer Academic Publishers, 1993.

Sundberg, Per, Ray Gibson, and Gunnar Berg. *Recent Advances in Nemertean Biology.* Dordrecht: Dr. W. Junk Publishers, 1988.

Periodicals

Gibson, Ray. "The Invertebrate Fauna of New Zealand: Nemertea (Ribbon Worms)." *NIWA Biodiversity Memoir* 118 (2002): 1–87.

Sundberg, Per, Turbeville J. McClintock, and Susanne Lindh. "Phylogenetic Relationships among Higher Nemertean (Nemertea) Taxa Inferred from 18S rDNA Sequences." *Molecular Phylogenetics and Evolution* 20 (2001): 327–334.

Thollesson, Mikael and Jon L. Norenburg. "Ribbon Worm Relationships: A Phylogeny of the Phylum Nemertea." *Proceedings of the Royal Society of London Series B—Biological Sciences* 270 (2003): 407–415.

Per A. Sundberg, PhD

Rotifera
(Rotifers)

Phylum Rotifera
Number of families 34

Thumbnail description
Group of microscopic animals characterized by the presence of a complex jaw apparatus and a ciliary wheel organ used for locomotion and feeding

Photo: Light micrograph of a common pond dwelling rotifer *Philodina* sp. seen feeding among organic matter and filamentous algae. The head of this rotifer is funnel-shaped, terminating in a mouth that is fringed with cilia or fine hairs. These hairs beat rapidly against the water, creating currents that draw particles in food into the mouth. (Photo by ©John Walsh/Science Photo Library/Photo Researchers, Inc. Reproduced by permission.)

Evolution and systematics

The Rotifera traditionally have been considered part of a group called Aschelminthes or Pseudocoelomata that comprised most of the microscopic animal groups without a true body cavity. Modern phylogenetic analyses have rejected this group, however, and today most studies support a close relationship between Rotifera, Acanthocephala (thorny-headed worms), Gnathostomulida (jaw worms), and the recently described Micrognathozoa (jaw animals) and unite them in a superphylum named Gnathifera (meaning, "those that possess jaws"). The acanthocephalans usually are considered a sister group to Rotifera, but this has been questioned by molecular data, which imply that acanthocephalans are highly advanced rotifers. This hypothesis still needs support from morphological data. The phylogenetic position of Gnathifera remains uncertain, but the most recent phylogenetic analyses suggest that they either are a basal group in Spiralia or form a monophyletic group with Gastrotricha and Platyhelminthes. In 2003 Rotifera included about 1,817 species, distributed among five orders and 34 families and divided into three classes: Seisonidea, Bdelloidea, and Monogononta. The latter class contains approximately 80% of the known species and displays the greatest morphological diversity.

Physical characteristics

Rotifers may range in size from less than 0.00394 in (100 µm) to 0.098 in (2,500 µm), but most species measure between 0.00591 and 0.0197 in (150–500 µm). The body generally is divided into a head, a trunk, and a foot region, but this basic pattern may vary greatly. The most conspicuous organ in the head is the wheel organ, also called the corona, which is composed of metachronously beating cilia that are arranged in different, distinct bands. Generally, the corona comprises a large buccal field that surrounds the mouth and a circumapical band that encircles the aciliate apical head region. Numerous modifications from this basic plan have made the corona morphological highly variable. In the class Bdelloidea, for example, the circumapical band is divided medially so that it forms two trochal discs. In live swimming or feeding animals, the ciliary beat of the disc gives the illusion that the animal carries two small, rotating wheels, and the whole phylum is named after this feature (*rota*, meaning "wheel," and *fero*, meaning "to bear"). The corona is used both for locomotion and feeding. The ciliary beat leads food particles toward the mouth opening, which always is located more or less ventrally.

The trunk may vary in shape from a very elongated form, sometimes divided into telescopic, retractable pseudo-segments,

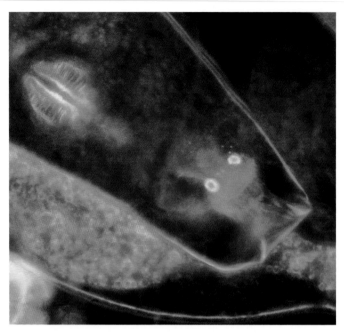

Rotifer embryo with parent. (Photo by Animals Animals ©P. Parks, OSF. Reproduced by permission.)

developed head and are responsible mainly for the movement of the unci. The morphological features of the four principal elements differ greatly, and the rotifer trophi have been divided into nine different types, depending on the size and shape of the principal sclerites. These types are highly significant for rotiferan taxonomy; several families and genera can be recognized solely on the basis of jaw type.

From the pharynx a short esophagus leads to the stomach, where the food is digested. Digestive enzymes are produced in syncytial gastric glands that empty into the stomach. After the stomach comes the gut, which terminates in a dorsal cloaca. Pairs of protonephridia control excretion and maintenance of osmotic balance. Each protonephridium is composed of one or more multiciliated terminal cells and multiciliated canal cells that lead to the collecting tubules, which guide the wastes on to the urinary bladder. The female reproductive system comprises one or two syncytial germovitellaria that each consists of a germinal region and a yolk-producing vitellarium, surrounded by a follicular layer. The classes Seisonidea and Bdelloidea have paired gonads, whereas the Monogononta have only a single germovitellarium. An oviduct formed by the follicular layer may connect the germovitellarium with the cloaca. The male organs comprise an unpaired testis and a penis. (See Reproductive biology for a more detailed description of the complex rotifer reproduction.)

Distribution

Rotifers have been recorded from all parts of the world, and the distribution pattern for the different species may vary from truly cosmopolitan to endemic. In particular, the bdelloid genera *Philodina* and *Rotaria* and various monogonont genera, such as *Brachionus*, *Keratella*, *Lecane*, and *Lepadella*, contain some extremely abundant species that have been recorded from most places in the world. Other species also are rather common but have a more limited distribution, being restricted, for example, to the Eurasian continent or the Holarctic or pantropical regions. Furthermore, some species appear to be endemic. For instance, no less than 11 species from the genus *Notholca* are endemic to Lake Baikal.

Among the key factors of rotifer success in terms of distribution are their cryptobiotic capabilities and their ability to produce resting eggs (see Habitat and Reproductive biology). Both resting eggs and dormant bdelloids may disperse over large distances with the aid of wind or water. Furthermore, resting eggs from many species have sculptured shells with tiny spines and hooks that enable them to attach to other animals (for example, birds) and spread by epizoic dispersal.

Habitat

Rotifers are found in all aquatic and semiaquatic habitats, but they reach the greatest diversity and largest population sizes in freshwater. They may inhabit the sediment, live in association with submerged plants (live as well as dead and partly decayed), or be restricted to plankton. They also may be adapted for more special habitats, such as the ice of the Arctic Ocean, terrestrial mosses and lichens, or meltwater ponds on glaciers. Furthermore, some species are specialized para-

to a much more globular or sacciform form. The foot is composed of one to several pseudo-segments, and it often has two terminal toes with adhesive gland openings at their tips. Both the foot and the toes may be reduced in several genera and species. The rotifer integument is syncytial, which means that the epidermal cells are not separated by cell membranes. An outer cuticle, which is present in most other invertebrates, is lacking in rotifers, and instead they have an intracellular filamentous lamina. The thickness of the lamina varies, and in some taxa parts of it are so thick that it forms a heavy body armor. Such taxa are called loricate, whereas those with a thinner lamina are referred to as illoricate.

The mouth is located ventrally or apically in the corona. It leads to the pharynx, which contains a complex masticatory apparatus referred to as the mastax. The mastax is made up of hard jaw parts, called trophi, and minute muscles that connect the jaw elements. The trophi comprise four principal elements: paired rami, paired unci, paired manubria and an unpaired fulcrum. There also may be different associated elements that together are referred to as epipharyngeal elements. The rami are the central elements in the trophi and often are equipped with teeth or denticles. They are joined caudally with a flexible ligament such that they are able to open and close and in that way crush food items. The only unpaired principal element, the fulcrum, extends caudally from the articulation point of the rami. It primarily serves as a muscle attachment point for different muscle fibers, for example, the large abductor muscles that run to the rami. The unci are located rostral or ventral to the rami. They may be rod-shaped with a single sharp tooth or plate-shaped with several strong teeth, and they may be used to grab and manipulate food particles or, in some predatory rotifers, to penetrate the integument of the prey. Proximally, the unci join the manubria. The manubria often are rod-shaped with a well-sites and live in the intestines or gills of various invertebrates. The optimal environment for many species is warm, nutrient-rich, slightly alkaline freshwater, but several species also are capable of surviving in more demanding habitats, such as temporarily dry or frozen ponds. Many bdelloid rotifers have cryptobiotic capabilities, which means that they are able to stop their metabolism, dehydrate their cells, and enter a state of dormancy. When entering cryptobiosis, the animal is capable of surviving under conditions that normally would be hazardous, such as complete dehydration, freezing, or oxygen deficiency. Many monogonont rotifers may survive under similar conditions by producing thick-shelled resting eggs. When the circumstances become more optimal, the populations may grow rapidly, because of their ability to reproduce asexually. The population and species richness is generally lower in marine habitats. The greatest diversity is found in the periphyton, but many species also may inhabit the interstices of sand grains on beaches. Marine rotifers are found mostly in the plankton or in the littoral zone and are extremely rare in deeper waters.

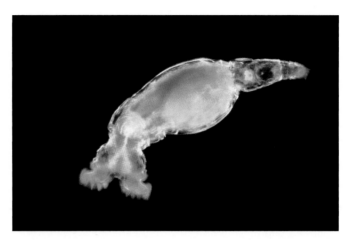

Light micrograph of the Antarctic species *Philodina gregaria*. This red colored rotifer survives the winter in cyst form and hatches out when pools and lakes of meltwater form during the austral summer. It reproduces in vast numbers, sometimes coloring the floor of the lake red. (Photo by ©John Walsh/Science Photo Library/Photo Researchers, Inc. Reproduced by permission.)

Behavior

Studies on rotifer behavior are extremely limited. With their minute size and simple nervous system, there is no basis for a complex type of behavioral biology. Various observations can be made concerning their locomotive patterns, escape behavior, and communication in relation to reproduction. When rotifers move, they typically swim or, alternatively, crawl on the substratum. Swimming specimens often move gently through the water in a characteristic helical motion that makes them easy to distinguish from most other microinvertebrates. Their sensory structures enable them to avoid obstacles, which distinguishes them from rotifer-like protozoa that often bump into obstacles and subsequently perform a rapid jump backward. Bdelloid rotifers either crawl or swim with their two ciliary discs. When crawling, they use the adhesive glands in the foot and in the rostrum. First, they adhere to the substratum with their pedal glands, and then they extend the body and attach to the substratum with the rostrum. Subsequently, the foot is detached, moved forward, and attached again. This inchworm-like way of crawling is very characteristic, and it makes it easy to distinguish crawling bdelloids from other animals.

When a rotifer is touched by another animal or by a thin dissecting needle, it often reacts by retracting the corona or else changes its swimming direction. Some species have special features that are used to "escape." For example, species in the genus *Polyarthra* have bundles of leaflike fins; when the animals are disturbed, they can flick with the fins and rapidly move as far as twelve body lengths. In the genus *Scaridium* the foot is equipped with strong muscles that allow it to act as a spring, which enables the animal to make a quick jump away from the source of disturbance.

Feeding ecology and diet

Rotifers typically feed on suspended organic particles, microalgae, ciliates, or bacteria. Most rotifers are filter feeders or suspension feeders and collect food with a water current created by the ciliary beat of the corona. There also are several forms of specialization. Instead of filtering, some rotifers creep along plants or sediment particles and graze on the bacterial layer. Others are predatory and may feed on algae, flagellates, or even other rotifers. In the latter case, it often is possible to identify the ingested prey by analyzing its indigestible trophi in the predator's stomach contents. Species that feed on algae often employ their unci to penetrate the filament, consequently using the large hypopharyngeal muscles to create a vacuum and in this way suck out the cytoplasm of the prey.

Reproductive biology

Information on rotifer mating behavior is scarce. It has been shown, however, that females of *Brachionus plicatilis* carry a glycoprotein on the surfaces of the body that acts as a sex pheromone. This protein can bind to chemoreceptors in the corona of conspecific males. Hence, the pheromone probably serves as a mating recognition signal that helps rotifers avoid mating with nonspecific specimens.

The rotifer reproductive cycle differs between the three classes. Whereas seisonids reproduce sexually and bdelloids are solely asexual, the monogonont rotifers have a complex cycle that includes both a sexual and an asexual phase. In seisonids males and females are the same size, and both probably are diploid (contrary to the haploid monogonont dwarf males, discussed later). The male stores sperm in a spermatophore that is transferred to the female. Fertilization and initial cell divisions occur inside the female's germarium; later, the female attaches the eggs to the host, the crustacean *Nebalia*, where they stay until the juveniles hatch. In bdelloids there are only females, and they reproduce exclusively by asexual parthenogenesis. This means that the maternal individual produces diploid eggs via mitosis and that these eggs can develop into new embryos without initial fertilization. As a consequence, the daughters are cloned individuals that always are genetically identical to the mother.

The monogonont reproductive cycle is divided into an asexual (the amictic) and a sexual (the mictic) phase. The amictic phase resembles the bdelloid cycle, with parthenogenetically reproducing amictic females and complete absence of males; during this phase the population is capable of growing very quickly. Certain physical stimuli may induce the production of another kind of female, named a mictic female. Mictic females are morphologically similar to amictic females, but they produce eggs by meiotic cell division, which means that the eggs become haploid. The haploid mictic egg either waits to be fertilized by a male or, if it is not fertilized, starts to develop into a haploid male. Males are much smaller than females, and internal organs, such as the alimentary canal, often are reduced. The short-lived males seek mictic females immediately after hatching and fertilize eggs by hypodermic impregnation. Fertilization results in a thick-shelled resting egg that can survive extreme conditions, such as freezing and dehydration; after a period of dormancy an amictic female hatches from the egg and enters the amictic phase again. Most rotifers have direct development, and mitosis never occurs after hatching. Larvalike stages are found among permanently attached rotifers, so that the newly hatched animals can move to a suitable place before they settle permanently.

Conservation status

No species is listed by the IUCN.

Significance to humans

Rotifers in the wild have little significance to humans. They may have some economic significance, however, because many species are cultured as a food source for aquariums and cultured filter-feeding invertebrates and fish fry. They also may be used as biological pollution indicators.

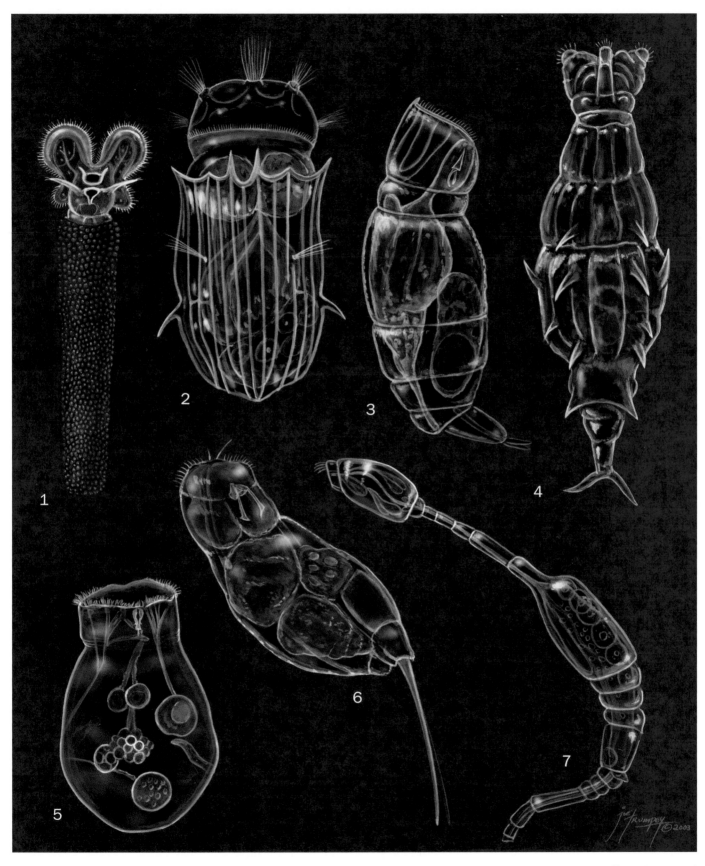

1. *Floscularia ringens*; 2. Ikaite rotifer (*Notholca ikaitophila*); 3. Astrid's rotifer (*Encentrum astridae*); 4. *Dissotrocha aculeata*; 5. *Asplanchna priodonta*; 6. *Cephalodella gibba*; 7. *Seison nebaliae*. (Illustration by Joseph E. Trumpey)

Species accounts

No common name
Asplanchna priodonta

FAMILY
Asplanchnidae

TAXONOMY
Asplanchna priodonta Gosse, 1850, Hyde Park, London.

OTHER COMMON NAMES
None known.

PHYSICAL CHARACTERISTICS
Females measure 0.00984–0.0591 in (250–1,500 µm) and males 0.00787–0.0197 in (200–500 µm). The species is illoricate, and the shape varies from sacciform to elongate. Feet and toes are reduced. Trophi belong to the incudate type, which is characterized by very large, forceps-like rami and partly reduced manubria and unci. The species is distinguished by the shape of the germovitellarium combined with details of the trophi.

DISTRIBUTION
Cosmopolitan.

HABITAT
The species is planktonic and lives in freshwater lakes and occasionally in brackish water.

BEHAVIOR
The species displays seasonal cyclomorphosis, which means that its body shape changes with the seasons, so that the sum-mer form is much more elongate than the sacciform spring and autumn form.

FEEDING ECOLOGY AND DIET
Algae, ciliates, and other rotifers.

REPRODUCTIVE BIOLOGY
The species is viviparous and has heterogamic reproduction. It can produce mictic and amictic females. In addition, it may produce a third type of female, called the amphoteric female, which is capable of generating both diploid eggs that hatch into parthenogenetic females and haploid eggs that yield dwarf male offspring.

CONSERVATION STATUS
Not listed by the IUCN.

SIGNIFICANCE TO HUMANS
None known. ◆

Ikaite rotifer
Notholca ikaitophila

FAMILY
Brachionidae

TAXONOMY
Notholca ikaitophila Sørensen and Kristensen, 2000, Ikka Fjord, Greenland.

Seison nebaliae
Asplanchna priodonta
Notholca ikaitophila

OTHER COMMON NAMES
None known.

PHYSICAL CHARACTERISTICS
Measures 0.00744–0.00925 in (189–235 μm). The species has a well-developed lorica, but the feet and toes are completely reduced. The lorica is rounded posteriorly and has six spines on the dorsal edge of the anterior opening. The dorsal plate is ornamented with a distinct longitudinal striation and a pair of lateral movable spines. Trophi belong to the malleate type, with thick rami and plate-shaped unci having several teeth. *Notholca ikaitophila* is distinguished from other closely related species by the dimensions of the lorica structures, the shape of the dorsal antennae, and details of the trophi.

DISTRIBUTION
Ikka Fjord, southwestern Greenland.

HABITAT
The species lives in association with the unique tufa columns on the bottom of the Ikka Fjord. The columns are made up of the unique mineral ikaite (calcium carbonate hexahydrate), which gives rise to submarine springs. The mineral is dissolved in the brackish seep water but precipitates and forms columns up to 6 ft (20 m) high when it meets cold and calcium-rich marine water. *Notholca ikaitophila* lives in the brackish water inside these columns.

BEHAVIOR
Nothing is known.

FEEDING ECOLOGY AND DIET
Diatoms and microalgae.

REPRODUCTIVE BIOLOGY
Obligate parthenogenetic or heterogamic. Males never have been recorded.

CONSERVATION STATUS
Not listed by the IUCN.

SIGNIFICANCE TO HUMANS
None known. ◆

Astrid's rotifer
Encentrum astridae

FAMILY
Dicranophoridae

TAXONOMY
Encentrum astridae Sørensen, 2001, Bermuda.

OTHER COMMON NAMES
None known.

PHYSICAL CHARACTERISTICS
Measures 0.0121–0.0153 in (308–388 μm). Body is illoricate, elongate, and fusiform. Foot is relatively long and composed of one pseudo-segment, with two closely set parallel-sided toes. Trophi belong to the forceps-like forcipate type that always has relatively slender elements. The species is recognized easily by its very long unci and supramanubria.

DISTRIBUTION
The species has been recorded from Bermuda and Denmark and probably has a North Atlantic to mid-Atlantic distribution.

HABITAT
It lives in the interstices of sand grains in the tidal and subtidal zones of sandy beaches.

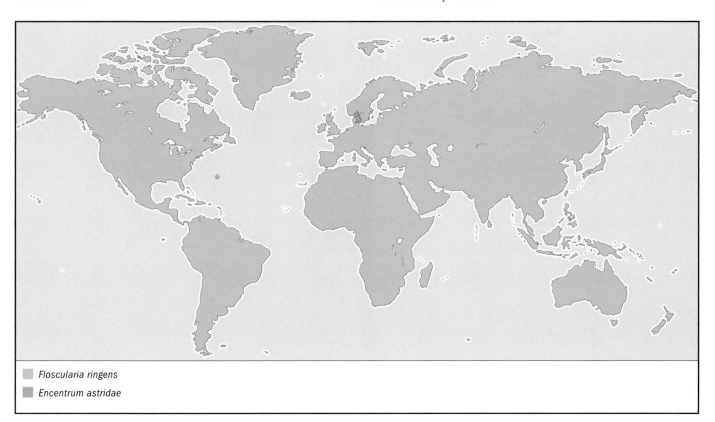

Floscularia ringens

Encentrum astridae

BEHAVIOR
Nothing is known.

FEEDING ECOLOGY AND DIET
Nothing is known.

REPRODUCTIVE BIOLOGY
Obligate parthenogenetic or heterogamic. Males never have been recorded.

CONSERVATION STATUS
Not listed by the IUCN.

SIGNIFICANCE TO HUMANS
None known. ◆

No common name
Floscularia ringens

FAMILY
Flosculariidae

TAXONOMY
Serpula ringens Linnaeus, 1758, Europe.

OTHER COMMON NAMES
None known.

PHYSICAL CHARACTERISTICS
Females measure about 0.0748 in (1,900 µm) and males 0.0157–0.0236 in (400–600 µm). Corona is large, with four lobes. Trunk and foot are elongated, and the foot terminates in an adhesive disc. Trophi belong to the malleoramate type, which resembles the ramate type but has a fulcrum and more devel-

oped manubria. *Floscularia ringens* is sessile and lives in a tube made of detritus and fecal pellets. The species is recognized most easily by the appearance of the tube, which is dark yellow to brownish in color and composed of relatively small pellets.

DISTRIBUTION
Cosmopolitan.

HABITAT
In freshwater attached to stalks or leaves of submerged plants.

BEHAVIOR
Nothing is known.

FEEDING ECOLOGY AND DIET
Filters microalgae and bacteria from the water.

REPRODUCTIVE BIOLOGY
Heterogamic cycle with mictic and amictic phases.

CONSERVATION STATUS
Not listed by the IUCN.

SIGNIFICANCE TO HUMANS
None known. ◆

No common name
Cephalodella gibba

FAMILY
Notommatidae

TAXONOMY
Furcularia gibba Ehrenberg, 1832, Germany.

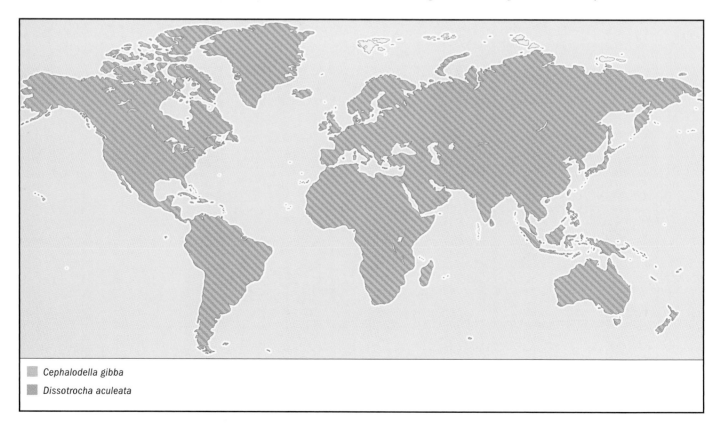

■ *Cephalodella gibba*
■ *Dissotrocha aculeata*

OTHER COMMON NAMES
None known.

PHYSICAL CHARACTERISTICS
Females measure 0.00984–0.0177 in (250–450 μm), males 0.00539–0.00906 in (137–230 μm). Body is gibbous and laterally compressed with a thin lorica. Foot is short and has two relatively long and slender toes that are straight or bent slightly dorsally. Trophi belong to the virgate type. The general appearance of this species is characteristic for *Cephalodella*, and *C. gibba* is easily distinguished from other *Cephalodella* by its relatively large size, long toes, and large trophi.

DISTRIBUTION
Cosmopolitan.

HABITAT
Lives among the vegetation in freshwater ponds, lakes, and streams. Occasionally also present in brackish water.

BEHAVIOR
Nothing is known.

FEEDING ECOLOGY AND DIET
Feeds on algae, flagellates and other microinvertebrates.

REPRODUCTIVE BIOLOGY
Heterogamic cycle with mictic and amictic phase.

CONSERVATION STATUS
Not listed by the IUCN.

SIGNIFICANCE TO HUMANS
None known. ◆

No common name
Dissotrocha aculeata

FAMILY
Philodinidae

TAXONOMY
Philodina aculeata Ehrenberg, 1832, Germany.

OTHER COMMON NAMES
None known.

PHYSICAL CHARACTERISTICS
Measures 0.0138–0.0197 in (350–500 μm). Color may vary from grayish to red or brown. Head has long rostrum and two eyespots. Dorsal side of the trunk has several characteristic large, thornlike spines. The foot has four telescopic retractable pseudo-segments and terminates in four toes. Trophi belong to the ramate type that is characteristic of all bdelloids. A fulcrum is lacking, and the manubria are thin bands that flank the lateral rims of the unci. The unci have numerous arrow-like teeth and are used to grind small food objects.

DISTRIBUTION
Cosmopolitan.

HABITAT
Lives in freshwater among plants and mosses.

BEHAVIOR
Nothing is known.

FEEDING ECOLOGY AND DIET
Bacteria and small algae.

REPRODUCTIVE BIOLOGY
Obligate parthenogenesis. Viviparous or ovoviviparous. Males do not exist. Juveniles are brooded inside their mother and hatch when fully developed. Their cells may increase slightly in size, but cell division never occurs after hatching.

CONSERVATION STATUS
Not listed by the IUCN.

SIGNIFICANCE TO HUMANS
None known. ◆

No common name
Seison nebaliae

FAMILY
Seisonidae

TAXONOMY
Seison nebaliae Grube, 1861, Adriatic Sea near Trieste.

OTHER COMMON NAMES
None known.

PHYSICAL CHARACTERISTICS
Measures 0.0315–0.0984 in (800–2,500 μm). Males and females are of similar size. The head is egg-shaped and has a long neck with telescopic, retractable segments. The trunk is oval, and the foot is long and segmented. Toes are missing; instead, there is an adhesive disc that allows attachment to the host, *Nebalia*. Trophi belong to the special fulcrate type.

DISTRIBUTION
Found on *Nebalia* from the Mediterranean Sea and the western European Atlantic coast, but there also is one confirmed record from the Sakhalin Islands in the Sea of Okhotsk.

HABITAT
Lives as a commensal on the pleopods of the leptostracan crustacean *Nebalia*. Commensalism refers to the condition in which a parasite neither harms nor provides any benefits to the host.

BEHAVIOR
Nothing is known.

FEEDING ECOLOGY AND DIET
Feeds exclusively on bacteria.

REPRODUCTIVE BIOLOGY
Obligate sexual reproduction.

CONSERVATION STATUS
Not listed by the IUCN.

SIGNIFICANCE TO HUMANS
None known. ◆

Resources

Books

Donner, J. *Ordnung Bdelloidea.* Berlin: Akademie Verlag, 1965.

Koste, Walter. *Rotatoria: Die Rädertiere Mitteleuropas.* Stuttgart: Gebrüder Borntraeger, 1978.

Nogrady, T., R. L. Wallace, and T. Snell. *Rotifera.* Vol. 1, *Biology, Ecology, and Systematics.* Guides to the Identification of the Microinvertebrates of the Continental Waters 4, edited by H. J. F. Dumont. Amsterdam: SPB Academic Publishing, 1993.

Wallace, R. L., and C. Ricci. "Rotifera." In *Freshwater Meiofauna: Biology and Ecology,* edited by S. D. Rundle, A. L. Robertson, and J. M. Schmid-Araya. Leiden, The Netherlands: Backhuys Publishers, 2002.

Periodicals

Ricci, C., and G. Melone. "Key to the Identification of the Genera of Bdelloid Rotifers." *Hydrobiologia* 418 (2000): 73–80.

Martin Vinther Sørensen, PhD

Gastrotricha
(Gastrotrichs)

Phylum Gastrotricha

Number of families 13

Thumbnail description
Microscopic, aquatic, strap-shaped, and tenpin-shaped ciliated worms with cuticular adhesive tubes

Photo: *Paraturbanella* sp. (Macrodasyida, Turbanellidae). Species of this genus have been reported from the world over; the image is an adult specimen of a new species found along the sea shore of the State of Saõ Paolo, Brazil, during 2002. Like other gastrotrichs this species feeds preferentially on bacteria and microscopic algae. (Photo by M. Antonio Todaro. Reproduced by permission.)

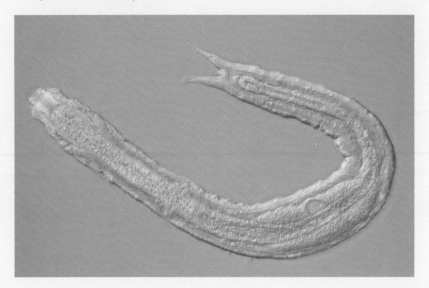

Evolution and systematics

The phylum Gastrotricha is divided into two orders, Macrodasyida and Chaetonotida. The order Macrodasyida contains six families, 31 genera, and approximately 210 marine species. One species occurs in freshwater. The six families are Dactylopodolidae, Lepidodasyidae, Macrodasyidae, Planodasyidae, Turbanellidae, Thaumastodermatidae. The order Chaetonotida contains two suborders, seven families, 29 genera, and approximately 400 marine and freshwater species. The seven families are Neodasyidae, Chaetonotidae, Dasydytidae, Dichaeteuridae, Neogosseidae, Proichthydidae, Xenotrichulidae.

Evolutionary relationships within the phylum are not well known. There is no fossil record. Within the Macrodasyida, the Dactylopodolidae is the most primitive family. Relationships among the five remaining families are unknown; Lepidodasyidae is probably a polyphyletic taxon. The Chaetonotida is divided into two suborders, the Multitubulatina and Paucitubulatina. The Multitubulatina contains a single family, Neodasyidae, and is basal within the Chaetonotida. Species of Neodasyidae are superficially similar to macrodasyidans but possess a chaetonotidan-type pharynx. The remaining six families of Chaetonotida make up the suborder Paucitubulatina. Most members of Paucitubulatina have tenpin-shaped bodies, sculptured cuticles, and a combination of hermaphroditic and parthenogenetic reproduction. The largest family, Chaetonotidae, may be an unnatural taxon.

Physical characteristics

Gastrotrichs are aquatic, strap-shaped to tenpin-shaped worms, 0.002–0.14 in (0.05–3.5 mm) long. The body is flat ventrally and arched dorsally. A multilayered, translucent cuticle covers the entire body. The ventral epidermis is ciliated; the cilia are covered with a thin layer of epicuticle. Epidermal cells may be monociliated or multiciliated. The body generally is divided into head and trunk regions. The head bears a terminal mouth, anterior myoepithelial pharynx, and sometimes eyes or tentacles or both. The trunk contains a straight tubular intestine, at least one pair of protonephridia, reproductive organs, and a ventral anus. There is no body cavity. Cuticular duo-gland adhesive tubes may occur on the head or trunk. Muscles are present in circular, longitudinal, and helical orientations; they may be cross-striated, obliquely striated, or, rarely, smooth.

The order Macrodasyida contains strap-shaped animals, 0.006–0.14 in (0.15–3.5 mm) long. The pharynx has an inverted Y-shaped lumen and pores connecting it to the outside. Pharyngeal pores are absent in *Lepidodasys*. The ventral epidermal cells may be monociliated or multiciliated. Epidermal glands generally are present. Adhesive tubes often are numerous and occur anteriorly behind the mouth and posteriorly; adhesive tubes also may be present in lateral, dorsolateral, and ventral positions. The cuticle is smooth in most species, except for the species of Thaumastodermatidae and a few others, where the cuticle forms scales, spines, or hooks. Macrodasyida are simultaneous or sequential hermaphrodites with complex male and female reproductive organs.

The order Chaetonotida contains vermiform and tenpin-shaped animals, 0.002–0.04 in (0.05–0.9 mm) long. The pharynx has a Y-shaped lumen and no pharyngeal pores. There is a pharyngeal plug at the junction between the pharynx and the intestine. Adhesive tubes typically are present only on the posterior caudal furca. Some species lack adhesive tubes (e.g., *Dasydytes*), while species of *Neodasys* (suborder Multitubulatina) possess papilla-like lateral adhesive tubes. The cuticle often

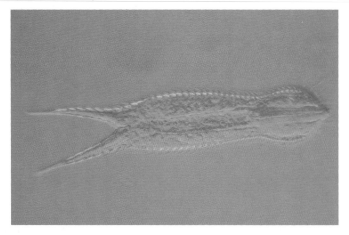

Musellifer delamarei is a rare Mediterranean species inhabiting coarse, organogenous sediment. In contrast with the vast majority of chaetonotidans, which reproduce by parthenogenesis, *M. delamarei* is a hermaphroditic species and reproduces by internal fertilization. (Photo by M. Antonio Todaro. Reproduced by permission.)

bears scales or spines or both, except in species of *Neodasys* and the Proichthydidae. The epidermis is monociliated in Multitubulatina and multiciliated in Paucitubulatina. Cross-striated muscles occur in *Neodasys*, and obliquely striated muscles are seen in all other species. Chaetonotida are hermaphroditic; several species of Paucitubulatina also are parthenogenetic. An anomalous reproductive organ, the X-organ, usually is present in Paucitubulatina.

Distribution

Gastrotrichs are found in all tropical, subtropical, and temperate waters worldwide. Families and genera are cosmopolitan. Several species are transoceanic; *Dactylopodola baltica*, for example, occurs on both sides of the Atlantic. Other species have a more restricted distribution, but this may be a result of inadequate sampling. Several freshwater gastrotrichs are cosmopolitan.

Habitat

All gastrotrichs are aquatic. Approximately half the known species are marine, found living in the voids between sand grains (interstitial spaces) on coastal beaches and continental shelves. Some species are known from the deep sea. Marine gastrotrichs generally prefer well-oxygenated sediments, although some species are present in low-oxygen and even dysoxic (oxygen-free) sediments. Grain size and the consolidation of the sediment also may be important factors. Freshwater gastrotrichs typically live on submerged or floating vegetation; some species are semiplanktonic and others may be interstitial.

Behavior

Little is known about gastrotrich behavior. Locomotion relies entirely on the ventral cilia, and muscles are used to change direction during episodes of ciliary gliding. Marine species are thigmotactic (i.e., move along/toward solid objects such as sand or ground) and adhere to the substratum with their adhesive tubes. Creeping, inchworm-like movements are known for some species. Most gastrotrichs show some form of negative phototaxis (orientation away from light). Chemotaxis (orientation toward the source of a chemical stimulus) may play a significant role in mating and in the general distribution of gastrotrichs. Copulation often involves active flexion and contact between partners.

Feeding ecology and diet

Marine gastrotrichs generally feed on diatoms, foraminiferans, bacteria, and minute protists. Freshwater gastrotrichs probably are bacteriovores but also may consume microalgae and organic detritus. It is thought that gastrotrichs are preyed upon by larger macrofauna.

Reproductive biology

Gastrotrichs are primitively hermaphroditic, with frequent protandry (male organs develop first). Most species possess paired testes and ovaries. Loss or reduction of the testes is common in several lineages. (Some Thaumastodermatidae have lost the left testis, for example, and many Chaenotonotida have reduced testes). The ovary is single in species of Lepidodasyidae. Fertilization is via indirect transfer of sperm or spermatophores. Complex reproductive organs may facilitate transfer of sperm in some species. Development is direct with no larval stage. Freshwater chaetonotidans often are parthenogenetic, with a later hermaphroditic phase.

Conservation status

No species of Gastrotricha is listed by the IUCN. One species, *Hemidasys agaso*, a facultative ectocommensal (does not require its host for survival) on the annelid *Nereilepas caudata*, is thought to be extinct.

Significance to humans

The importance of gastrotrichs remains undetermined. As bacteriovores and detritrovores, gastrotrichs may contribute to the aesthetics of coastal beaches by consuming washed-up debris, preventing its decay and associated odor. The study of gastrotrichs also may be used to augment our knowledge of animal origins, evolution, and relationships. *Lepidodermella squamata* is a commercially available freshwater gastrotrich.

1. *Dactylopodola baltica*; 2. *Lepidodermella squamata*. (Illustration by John Megahan)

Species accounts

No common name
Lepidodermella squamata

ORDER
Chaetonotida

FAMILY
Chaetonotidae

TAXONOMY
Lepidodermella squamata (Dujardin, 1841), River Seine, Paris, France.

OTHER COMMON NAMES
None known.

PHYSICAL CHARACTERISTICS
A short, tenpin-shaped gastrotrich that grows to 0.007 in (0.19 mm) in length. Distinct, five-lobed head separated from the body by a short neck. Trunk has posterior caudal furca and two adhesive tubes. Cuticle consists of scales without ridges or spines. Cilia present on the lateral margin of the head and ventrally in two rows.

DISTRIBUTION
Freshwater bodies across the United States (Arkansas, Ohio, Michigan, New Hampshire, and North Carolina), Brazil, Uruguay, Japan, and much of Europe. Probably distributed worldwide.

HABITAT
Found on aquatic vegetation in lakes, ponds, swamps, and streams. Also may occur interstitially in sandy sediments.

BEHAVIOR
Slow ciliary glider, with spectral sensitivity to blue light.

FEEDING ECOLOGY AND DIET
Diet consists of microalgae, bacteria, and organic detritus.

REPRODUCTIVE BIOLOGY
The life cycle begins with parthenogenetic reproduction and the deposition of up to four eggs. Eggs usually are opsiblastic (slow developing) and can survive desiccation and freezing; some eggs are tachyblastic (fast developing). The parthenogenetic phase is complete within a few days, after which the animal becomes a simultaneous hermaphrodite.

CONSERVATION STATUS
Not listed by the IUCN.

SIGNIFICANCE TO HUMANS
Commercially available for laboratory study. ◆

■ *Dactylopodola baltica*
■ *Lepidodermella squamata*

No common name
Dactylopodola baltica

ORDER
Macrodasyida

FAMILY
Dactylopodolidae

TAXONOMY
Dactylopodola baltica (Remane, 1926), Kiel, Germany.

OTHER COMMON NAMES
None known.

PHYSICAL CHARACTERISTICS
Body reaches a length of 0.01 in (0.3 mm), with a well-defined head, paired eyespots, and a bifid posterior. Adhesive tubes are present anteriorly, laterally, and posteriorly. Epidermal cells are monociliated, and the muscles are cross-striated.

DISTRIBUTION
Atlantic coast of the United States. Also known from northern Ireland, Wales, the Scilly Isles, and the Baltic Sea to the west coast of France and the Mediterranean Sea.

HABITAT
Marine coastal beaches in a variety of sediment types; middle to low intertidal zone.

BEHAVIOR
A slow ciliary glider; common but generally not found in abundance.

FEEDING ECOLOGY AND DIET
Feeds on diatoms.

REPRODUCTIVE BIOLOGY
Sequential protandric hermaphrodite with paired ovaries and testes. Spermatophores are passed indirectly to partners. Reproductive activity is greatest during the summer.

CONSERVATION STATUS
Not listed by the IUCN.

SIGNIFICANCE TO HUMANS
None known.

Resources

Books

Hummon, William D. "Gastrotricha." In *Synopsis and Classification of Living Organisms,* edited by S. P. Parker. New York: McGraw-Hill, 1982.

Ruppert, Edward E. "Gastrotricha." In *Introduction to the Study of Meiofauna,* edited by Robert P. Higgins and Hjalmar Thiel. Washington, DC: Smithsonian Institution Press, 1988.

———. "Gastrotricha." In *Microscopic Anatomy of Invertebrates.* Vol. 4, *Aschelminthes,* edited by Fredrick W. Harrison and Edward E. Ruppert. New York: Wiley-Liss, 1991.

Periodicals

Hummon, W. D. "The Marine and Brackish-water Gastrotricha in Perspective." *Contributions to Zoology* 76 (1971): 21–23.

Rick Hochberg, PhD

Kinorhyncha

(Kinorhynchs)

Phylum Kinorhyncha

Number of families 10

Thumbnail description
Meiobenthic, superficially segmented, and spined marine free-living cephalorhynch worms

Photo: A species of *Echinoderes* from Australia. (Photo by Rick Hochberg. Reproduced by permission.)

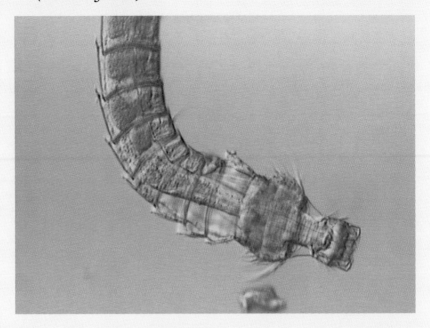

Evolution and systematics

The name "Kinorhyncha" comes from the Greek words *kinema* (motion) and *rhynchos* (proboscis or snout). Kinorhyncha is considered either a class within the phylum Aschelminthes or a separate phylum with close relationships to aschelminth worms. However, the Kinorhyncha was later included as a class in the phylum Cephalorhyncha, established for four classes of Aschelminthes: Priapulida, Kinorhyncha, Loricifera, and Nematomorpha.

Treated here as a phylum, Kinorhyncha encompasses two orders, five suborders, 10 families, 15 genera, and about 150 species. The two orders are: Cyclorhagida (with families Zelinkaderidae, Antigomonidae, Cateridae, Semnoderidae, Centroderidae, Echinoderidae, Dracoderidae, and Cephalorhynchidae), and Homalorhagida (with families Pycnophyidae and Neocentrophyidae). There is no fossil record for the Kinorhyncha.

Physical characteristics

Adult kinorhynchs range in length from 0.008 in (0.2 mm) in certain *Echinoderes*, to 0.05 in (1.2 mm) in the arctic *Pycnophyes*. Most are transparent; but there are yellowish and reddish species that live on macroalgae. A few have red or brown cup-shaped eyes located in the head near the brain.

The body is subdivided into two main regions: an eversible head, or introvert, and trunk, metamerically segmented into 11 cuticularized trunk segments. The spherical head is joined to the trunk by a short eversible neck with closing plates, called placids. The head and neck are not serially go-

mologous to trunk segments but are traditionally considered as the first and second segments; thus, there is a total of 13 body segments.

The head terminates with a protrusible mouth cone surrounded by nine oral styles. Internally, the terminal mouth is followed by 20 pentamerously arranged buccal styles, which are partly eversible when feeding. The spherical head bears 5–7 rings of as many as 89–91 posteriorly directed spines called scalids. Scalids are sensory and are also used for forward locomotion. The neck is composed of a series of closing plates, which retract over the head when it is withdrawn into the trunk. Cyclorhagids typically have 14–16 radially arranged placids. Homalorhagids have 2–4 dorsal and 2–4 ventral trapezium-like placids.

Kinorhynchs are characterized by metamerical trunk armor (exoskeleton). The trunk segments are variously subdivided longitudinally into a series of cuticular plates (dorsal tergites and ventral sternal plates). In most Cyclorhagida, segment 3 (the first trunk segment), is entire and composed of a complete ring of cuticle. In the Semnoderidae, segment 3 is divided into bivalved plates forming a clamshell-like closing apparatus, which acts with placids to close off the inverted head. In the Pycnophyidae, segment 3 is composed of one arched dorsal plate, or tergite, and three mobile ventral, or sternal, plates. Mobile ventral plates close the anterior region of the trunk when the head is introverted. Segment 4 is entire in the Echinoderidae or subdivided into one tergal and two sternal plates by midventral and lateral articulations extending posteriorly from segment 4 through segment 13 in most other kinorhynchs. Articulations, the flexible junctions of the arthrocorial cuticle allow movement between plates and segments, as well as inversion and eversion of the head.

Trunk spines are usually located middorsally (DS), laterally (LS), midterminally (MTS), or lateroterminally (LTS). In some species, there are accessory lateral (LAS) and accessory lateroterminal (LTAS) spines, middorsal processes (MP), and modified spinelike appendages. Some spines are adhesive tubes (AT). The spines are sensorial and related to locomotion.

The internal anatomy of kinorhynchs is related to the outer segmentation, nervous system, muscles, and glandular system, which are all distinctively segmented. Simultaneous contraction of segmental dorsoventral muscles increases the pressure of the body-cavity fluid in the trunk, displaces it forward, and everts the head. Scalids move forward, plow backward through the interstices around and propel the kynorhynch forward. Special head retractor muscles retract the introvert back into the trunk in synchrony with relaxation of the dorsoventral muscles.

Distribution

Kinorhynchs occur worldwide, from polar to tropical seas.

Habitat

Kinorhynchs are eurybathic (from zero to 17,390 ft [0 to 5,300 m]), euryhaline (from 7 ppt in estuaries to 60 ppt in tide pools), and eurythermic (from 29.3 to 104°F [-1.5 to 40°C]).

Mesobenthic or interstitial species of kinorhynchs live in the interstices between large sediment particles; endobenthic species burrow by displacing small sediment particles in muddy sediments; and epibenthic species live at the water-sediment interface or in the suspended flocculent material on the surface of marine algae and invertebrates. They usually are found within the few uppermost inches of muddy sediments, depending on the oxygen gradient. In sand or shell-gravel of high-energy beaches, kinorhynchs are found at depths of 3.3 ft (1 m) or more. Most prefer mud, or mud mixed with sand, with a high organic content.

Behavior

Kinorhynchs are relatively abundant representatives of the permanent meiofauna, occasionally ranking third or fourth in number within a meiobenthic sample. The mean annual population density is about 10,000–15,000, but may reach 50,000 specimens per square meter. Sex ratio in the populations is about 1:1.

Being placed into sedimented particles, kinorhynchs can agglutinate the detritus and amass the particles into concretions glued together by mucus. As also noted for nematodes, these activities result in soft bottoms acquiring a particular framework.

Parasitic Suctoria and sessile Peritricha infusorians (Ciliophora) are often found on the trunk surface of kinorhynchs.

Feeding ecology and diet

Kinorhynchs are herbivorous and detrivorous. Most cyclorhagids are diatom-feeders. They collect pennate diatoms using their rigid articulated oral styles as multiple pincers to locate one end of the shell and manipulate the diatom to their terminal mouth. Kinorhynchs bring diatoms into their buccal cavity with an action of their mobile buccal styles and with pumping movements of their sucking pharynx. The sharp buccal styles then move against the diatom, damaging the girdle and causing the two valves to separate. The diatom cell is emptied by the sucking of the pharynx, and the emptied frustule is rejected in the substrate. *Echinoderes* also can strip off terminal spines and collect diatoms between their head scalids. The diatoms are then collected by the oral styles using a repetitive withdrawing of the scalids and a protrusion of the mouth cone.

Most homalorhagids are selective deposit-feeders. They have flexible nonarticulated oral styles, which they use to collect detritus and bacterial clots. The sharp buccal teeth of the eversible anterior part of the buccal cavity also act as scrapers to graze on bacterial film and fungi.

Numerous glands secrete mucus, which is released onto the trunk surface. This secretion is also used to entrap detritus, bacteria, diatoms, and fungi, which are subsequently browsed on together with the mucus. Flexible and spiny cyclorhagids also can strip off terminal spines and collect attached diatoms between their head scalids. The diatoms are then collected by the oral styles using a repetitive withdrawing of the scalids and a protrusion of the mouth cone.

The parasitic protozoan *Kinorhynchospora japonica* (phylum Microspora) infects all midgut cells of *Kinorhynchus yushini*. Sulfur gram-negative bacteria are found in a few special midgut cells (bacteriocytes) of *Pycnophyes kielensis* from the intertidal mud with sulfuretted hydrogen and are thought to be symbiotic.

Reproductive biology

Kinorhynchs are dioecious, with external sexual dimorphic characters (AT and penile spines in males, and LTAS in females). Sexes copulate with spermatophore transfer; fertilization is internal. Mature ovaries commonly develop a single large oocyte. Fertilized eggs (60–80 microns) are attached to sand grains or detritus. Cleavage has not been observed. The early embryo in the egg shows no segmentation. Just before hatching, the juvenile kynorhynch has 11 segments: the head with scalids, the neck with placids, and nine trunk segments, which are not divided into sternal and tergal plates. The time required for development from oviposition to hatching is about 10 days. At the time of hatching, the juvenile straightens and simultaneously everts its head, thus tearing open the egg envelope. Kinorhynchs have a direct development through five to six juvenile stages, each derived from a molt.

Just after hatching, the first juvenile can feed on diatoms and detritus. The first molt establishes a 12-segmented juvenile. All trunk segments are discernible in the third stage, but the complete separation of the most posterior segments becomes evident in the fourth or fifth stages.

To distinguish various juvenile stages of kinorhynchs, terms previously proposed for some invalid genera are used. In Cyclorhagida, two morphological states of juveniles are designated: "*Centropsis*," with MTS and "*Habroderes*," having LTS and without MTS. In Homalorhagida, there are three morphological states of juveniles: "*Centrophyes*," with MTS; "*Hyalophyes*," with LTS and without MTS; and "*Leptodemus*," without any terminal spines. Cyclorhagids with MTS usually have series of only "*Centropsis*" stages. In Echinoderidae, first three "*Centropsis*" stages are usually followed by three "*Habroderes*" stages. In Pycnophyidae, three "*Centrophyes*" or "*Leptodemus*" stages are usually followed by three "*Hyalophyes*" stages or there are only "*Leptodemus*" stages.

Conservation status

No species of Kinorhyncha is listed by the IUCN.

Significance to humans

None known.

1. *Echinoderes sensibilis*; 2. *Kinorhynchus yushini*; 3. *Centroderes eisigii*; 4. *Cephalorhyncha asiatica*. (Illustration by Amanda Humphrey)

Species accounts

No common name
Centroderes eisigii

ORDER
Cyclorhagida

FAMILY
Centroderidae

TAXONOMY
Centroderes eisigii Zelinka, 1928, Mediterranean Sea, Bay of Naples.

OTHER COMMON NAMES
None known.

PHYSICAL CHARACTERISTICS
Trunk length (TL) 0.0118–0.0138 in (300–350 μm). Trunk transparent; neck placids alternating in widths; MTS at least twice as long as LTS; LS of segment 3 (AT) longer than two first trunk segments together; DS on segments 3–13; LS on segments 4, 7, 10–11; MTS about 60% of TL.

DISTRIBUTION
Eastern Atlantic boreal species; also Mediterranean and Black Seas.

HABITAT
Abundant in phaseoline mud; subtidal, at 98–656 ft (30–200 m) deep.

BEHAVIOR
Nothing is known.

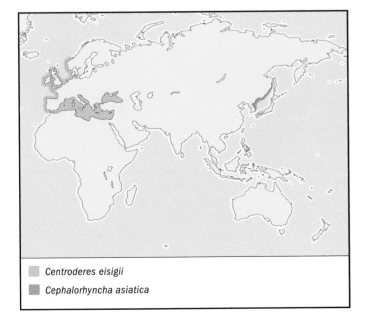

Centroderes eisigii

Cephalorhyncha asiatica

FEEDING ECOLOGY AND DIET
Herbivorous and selective deposit feeder.

REPRODUCTIVE BIOLOGY
Six juvenile stages of type *"Centropsis."*

CONSERVATION STATUS
Not listed by the IUCN.

SIGNIFICANCE TO HUMANS
None known. ◆

No common name
Cephalorhyncha asiatica

ORDER
Cyclorhagida

FAMILY
Cephalorhynchidae

TAXONOMY
Cephalorhyncha asiatica Adrianov, 1999, Northwest Pacific Ocean, Sea of Japan, Peter the Great Bay.

OTHER COMMON NAMES
None known.

PHYSICAL CHARACTERISTICS
Trunk length (TL) 0.0138–0.0157 in 350–400 μm. Trunk transparent; only segment 3 composed of complete ring of cuticle, segment 4 composed of arched tergite and sternal plate incompletely subdivided by midventral articulation into 2 substernites; DS on segments 6–10; LS on segments 4, 7–11; segment 12 with a pair of subdorsal spines; LTS 65% of TL.

DISTRIBUTION
Western coast of the Sea of Japan.

HABITAT
Mud and muddy sand with a high organic content near estuaries; subtidal, 10–33 ft (3–10 m) deep.

BEHAVIOR
Nothing is known.

FEEDING ECOLOGY AND DIET
Herbivorous, diatom feeder.

REPRODUCTIVE BIOLOGY
Six juvenile stages (*"Centropsis"* and *"Habroderes"*).

CONSERVATION STATUS
Not listed by the IUCN.

SIGNIFICANCE TO HUMANS
None known. ◆

No common name
Echinoderes sensibilis

ORDER
Cyclorhagida

FAMILY
Echinoderidae

TAXONOMY
Echinoderes sensibilis Adrianov Murakami, et Shirayama, 2002; Northwest Pacific, Honshu Island, Japan.

OTHER COMMON NAMES
None known.

PHYSICAL CHARACTERISTICS
Trunk length (TL) 0.0126–0.0138 in (320–350 µm); trunk yellowish, head with 91 scalids arranged in 7 circlets (10, 10, 20, 10, 20, 6, 15(6+9); DS on segments 6–10; LS on segments 4, 7–12; subventral fields of minute cuticular hairs on segments 5–12; LTS 43–47% of TL.

DISTRIBUTION
Off Pacific coast of Japan.

HABITAT
Intertidal pools, abundant on red algae *Corallina pilulifera*.

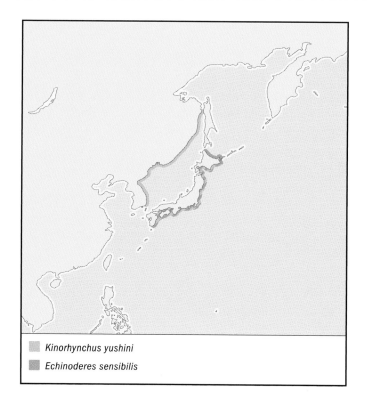

Kinorhynchus yushini

Echinoderes sensibilis

BEHAVIOR
Nothing is known.

FEEDING ECOLOGY AND DIET
Herbivorous, diatom feeder.

REPRODUCTIVE BIOLOGY
Six juvenile stages ("*Centropsis*" and "*Habroderes*").

CONSERVATION STATUS
Not listed by the IUCN.

SIGNIFICANCE TO HUMANS
None known. ◆

No common name
Kinorhynchus yushini

ORDER
Homalorhagida

FAMILY
Pycnophyidae

TAXONOMY
Kinorhynchus yushini Adrianov, 1989, Northwest Pacific, Sea of Japan, Peter the Great Bay.

OTHER COMMON NAMES
None known.

PHYSICAL CHARACTERISTICS
Trunk length (TL) 0.0177–0.0220 in (450–560 µm); trunk hemitransparent; 4 dorsal and 2 ventral neck placids; MP on segments 4–11, bifurcated on segments 4–10, single on segment 11; segment 11 with very long lateral daggerlike elements.

DISTRIBUTION
Russian, Korean, and Japanese coasts of the Sea of Japan; off Pacific coast of Japan.

HABITAT
Intertidal and subtidal muddy sediments with a high organic content, at zero to 98 ft (zero to 30 m) deep.

BEHAVIOR
Very abundant in muddy sediments in shallow waters.

FEEDING ECOLOGY AND DIET
Selective deposit feeder on detritus and bacteria.

REPRODUCTIVE BIOLOGY
Six juvenile stages of "*Leptodemus*" type.

CONSERVATION STATUS
Not listed by the IUCN.

SIGNIFICANCE TO HUMANS
None known. ◆

Resources

Books

Adrianov A. V., and V. V. Malakhov. *Kinorhyncha: Structure, Development, Phylogeny and Classification.* Moscow: Nauka Publishing, 1994.

————. *Cephalorhyncha of the World Oceans.* Moscow: KMK Scientific Press Ltd., 1999.

————. "Kinorhyncha." In *Reproductive Biology of Invertebrates.* Vol. 9, edited by K. Adiyodi, R. Adiyodi, and B. Jamieson. [n.p.], 1999.

Higgins, R. P. "Kinorhyncha." In *Reproduction of Marine Invertebrates.* Vol. 1, edited by A. Giese and J. Pearse. [n.p.], 1974.

Kristensen, R. M., and R. P. Higgins. "Kinorhyncha." In *Microscopic Anatomy of Invertebrates,* Vol. 4, Aschelminthes, edited by F. W. Harrison and E. E. Ruppert. New York: Wiley-Liss Inc., 1991.

Zelinka, K. *Monographie der Echinoders.* Leipzig: Engelman, 1928.

Periodicals

Adrianov A. V., and V. V. Malakhov. "Phylogeny and Classification of the Class Kinorhyncha." *Zoosystematica Rossica* 4 (1995): 23–44.

Higgins, R. P. "A Historical Overview of Kinorhynch Research." *Smithsonian Contributions to Zoology* 76 (1971): 25–31.

————. "The Atlantic Barrier Reef Ecosystem at Carrie Bow Cay, Belize, II Kinorhyncha." *Smithsonian Contributions to the Marine Sciences* 18 (1983): 1–131.

Kozloff, E. N. "Some Aspects of Development in *Echinoderes* (Kinorhyncha")* Transactions of the American Microscopical Society* 91 (1972): 119–130.

Neuhaus, B., and R. P. Higgins. "Ultrastructure, Biology, and Phylogenetic Relationships of Kinorhyncha." *Integrative and Comparative Biology* 42 (2002): 619–632.

Andrey Adrianov, PhD

Adenophorea
(Roundworms)

Phylum Nematoda

Class Adenophorea

Number of families 96

Thumbnail description
Primarily free-living marine, freshwater, and terrestrial nematodes; considered to be the most primitive form of nematodes

Photo: Trichina worm (*Trichinella spiralis*) in muscle tissue section. (Photo by Bob Gossington. Bruce Coleman, Inc. Reproduced by permission.)

Evolution and systematics

Nematoda, the phylum above the class Adenophorea, have left very few fossil remains. The earliest fossils that contained nematode remnants were found in Eocene strata (the era from about 55–38 million years ago). More authenticated fossils are of nematodes preserved in amber, such as those from fossilized shark muscles and mammals frozen in permafrost. The fossil record is too fragmented to explain much about nematode origins, and conclusions about nematode phylogeny have been mostly based on observations of living species. It is hypothesized that nematodes originated during the Precambrian era, what was the Proterozoic period (about one billion years ago).

As of 1994, about 20,000 species of nematodes have been described, with an estimate by various researchers of the total number of nematode species living on the planet at 80,000–1,000,000. This phylum is considered to have the lowest number of yet-to-be-described species of any animal. Hyman divided nematodes into 17 orders, whereas Chitwood separated them into two main classes, Aphasmidia (now Adenophorea) and Phasmidia (now Secernentea). Controversies still exist, but for the most part, many scientists such as A. R. Maggenti, who helped to develop the classifications, treat nematodes as a separate phylum with two classes, Adenophorea and Secernentea, which were divided based on molecular and morphological characteristics. These two classes are primarily separated (along with other important criteria) with respect to whether they do not possess phasmids (as in Adenophorea) or do possess phasmids (as in Secernentea). Two subclasses are recognized: Enoplia and Chromadoria. In addition, there are

11 orders and approximately 96 families. The total number of species of adenophoreans is estimated at about 12,000 worldwide. Scientific surveys of seabed mud, along with other reliable evidence, suggest that a great number of species are yet-to-be discovered.

Physical characteristics

The majority of adenophoreans are free-living, microbotrophic, and aquatic nematodes. Only a few species are plant parasitic, invertebrate parasites, or vertebrate parasites. They range in size from microscopic to as long as 3.25 ft(1 m) in exceptional cases. Adenophoreans are considered non-segmented pseudocoelomates; that is, creatures possessing a three-tissue-layered body that has a fluid-filled body cavity (pseudocoelom) between the endoderm and the mesoderm (the innermost and middle tissue layers).

A flexible but durable collagenous cuticle protects the body with a series of grooves across the body from head to tail. The non-cellular cuticle, which generally has a smooth surface, can sometimes contain transverse or longitudinal striations and has four layers: endocuticle, epicuticle, exocuticle, and mesocuticle. The cellular hypodermis is the subcuticular layer that secretes the cuticle. Phasmids—which are minute pore-like chemoreceptors that (when present) are usually paired—are generally absent from adenophoreans. Their sensory system contains well-developed amphid apertures, which are post-labial (past the lips) in position, with some species having apertures that are labial (on the lips). The apertures are variable

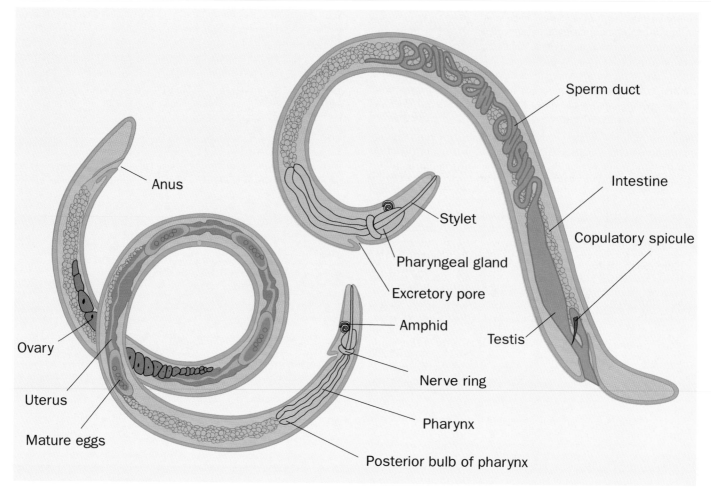

Roundworm anatomy. (Illustration by Christina St. Clair)

in external shape, being sometimes circular, pocketlike, pore-like, or spiral.

Somatic and cephalic setae, which are elongated structures joined with the cuticle, are common. These tactile sensory organs are usually located around the oral openings. The cephalic sensory organs, which number 16, are setiform to papilloid, and post-labial or labial in position about the head. Deirids, which are paired, porelike organs located in the lateral fields near the nerve ring, are present in some species.

Usually present, hypodermal glands are a thin tissue layer beneath the cuticle that thickens to make the dorsal, lateral, and ventral chords and extend the body's length. In general, they are tactile sensory organs usually located around the oral openings. They contain uninucleate hypodermal cells. A layer of longitudinal muscles underlies the hypodermis.

Bursae (or caudal alae) are rarely found. The ventrally located excretory system, when present, is usually single-celled, usually with non-cuticularized terminal ducts, and lacking collecting tubules. A rectal gland is usually absent. When present, there are three caudal glands located near the posterior region. The muscular esophagus or pharynx (the tube that moves food from the mouth/head to the stomach/intestine) varies in configuration, but the majority of adenophoreans

have three esophageal glands: two that are subventral and one that is dorsal. The subventral glands open into the posterior metacarpus. The dorsal gland opens anteriorly into the procorpus or the anterior metacarpus. Its basic structure is corpus (the anterior part is cylindrical) with the basal (bottom) region sometimes swollen in the shape of a bulb. The glands empty their contents into the esophagus to aid in digestion. The tail is the region between the anus and the posterior tip. The male tail is smooth, and lateral cuticular caudal extensions of the tail rarely occur.

Distribution

Adenophoreans are found worldwide, especially in water-filled areas around vegetation and in soils, on plants, and in animals.

Habitat

Adenophoreans are commonly found in most kinds of habitat (except for desert regions) throughout the world, particularly in marine sediments. They inhabit a wide variety of habitats, including both water and soil.

Behavior

Their behavior is classified as being usually free living, but with some species behaving as parasites.

Feeding ecology and diet

The free-living species are predators, carnivores, herbivores, and omnivores, feeding mostly on bacteria, fungi, and soil organisms. The parasitic species feed on nutrients (such as blood, body fluid, intestinal contents, mucus) found within their hosts. Adenophorean worms generally have some form of stylet, a hard, sharp spear, which is used for feeding. Muscles move the stylet in and out, allowing the worms to puncture cells. Once opened, the worm will empty the contents of the cell.

Reproductive biology

Male adenophoreans sometimes have thin cuticle extensions on both sides of the anus called bursae, which are fluid-filled body sacs. When present, bursae are used to hold females during copulation. Male testes, when present, are two in number and are shaped similar to the female ovary. The spicules, the male copulatory organ, are paired. Males have a single, ventral series of papilloid or tabloid pre-anal supplements. Sperm accumulate in the seminal vesicle and exit through the anus. During mating, the rigid spicules insert into the vagina and form a passageway for the sperm. The female ovary contains germ cells that produce eggs. Fertilization, mostly initiated by male sperm, takes place in the uterus, with eggs released through the vagina. (Most species produce males and females, but some species only produce hermaphrodites, in which both male and female structures are contained in the same individual.) Most eggs are about the same size and shape.

Conservation status

No adenophoreans are listed on the 2002 IUCN Red List. They are found worldwide and frequently occur in very great densities.

Female pinworm (*Enterobius vermicularis*). Pinworms are the most common nematode parasite of humans. (Photo by E. R. Degginger. Bruce Coleman, Inc. Reproduced by permission.)

Significance to humans

Living free and as parasites of animals, insects, and plants, adenophoreans constitute an important part of nature and to the activities of humans. As crops are cultivated to feed the world's population, adenophoreans become more numerous as they feed on agricultural plants. Parasitic adenophoreans cause yield losses by themselves or they may join with other organisms such as viruses, fungi, and bacteria to advance disease development in plants. They also cause loss of nutrients and water in the plant, thus increasing the plant's susceptibility to other dangers. Adenophoreans, when infecting a human, can cause various diseases, and in some circumstances, death to the human host. On the other hand, studies have shown that the appearance of adenophoreans is a good indicator of biodiversity, which is important to the health and survival of humans. Adenophoreans help to cycle carbon and nitrogen and to breakdown organic matter in the soil environment.

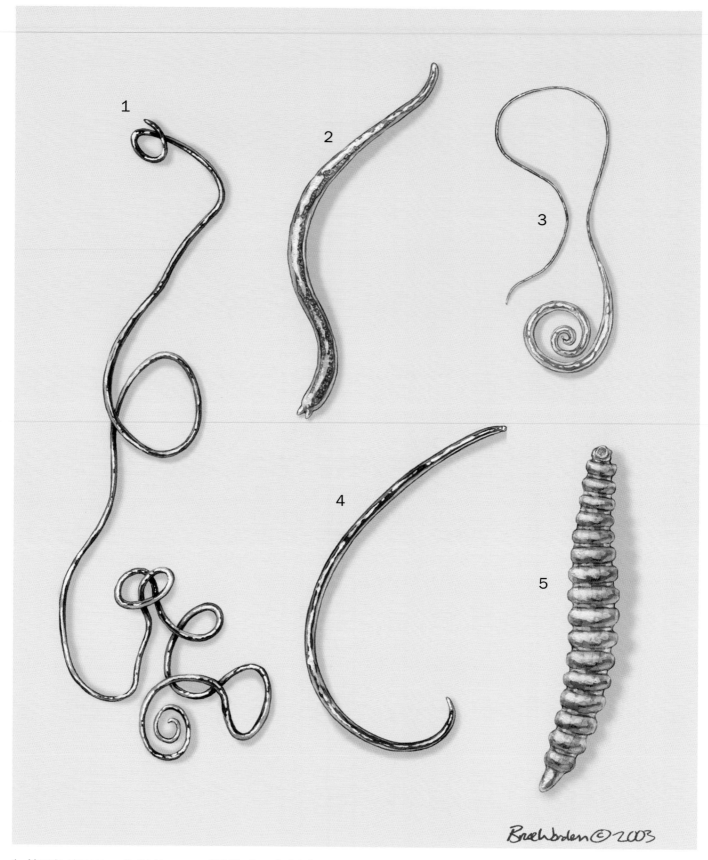

1. *Mermis nigrescens*; 2. Trichina worm (*Trichinella spiralis*); 3. Human whipworm (*Trichuris trichiura*); 4. *Nygolaimus parvus*; 5. *Desmoscolex squamosus*. (Illustration by Bruce Worden)

Species accounts

No common name
Desmoscolex squamosus

ORDER
Desmoscolecida

FAMILY
Desmoscolecidae

TAXONOMY
Protricomoides squamosus Timm, 1970.

OTHER COMMON NAMES
None known.

PHYSICAL CHARACTERISTICS
Characterized by the annulation and ornamentation of the body cuticles, the shape of the head, the shape and arrangement of the somatic setae, and the copulatory apparatus. They have a small body that is tapered at both ends; the female body is longer than the male. The cuticle is annulated, with only the two anterior and the three posterior main rings being considered complete; that is, they are covered with a continuous narrow or wide layer of secretion and concretion material. The body cuticle has 70–71 annules (when counting complete main rings as single annules) in the holotype male and 73 annules in the paratype female. The annules contain a transverse row of minute pores with a small accumulation of secretion and concretion particles around them. at the level of the insertion of the somatic setae, the layer of concretion is larger.

The somatic setae are arranged in pairs, inserted on short peduncles or almost directly on the body cuticle, but they differ sub-dorsally and sub-ventrally. The sub-dorsal setae taper distally to a shorter spatulate tip, which is 0.0000787–0.000118 in (2–3 µm) in length. The sub-ventral setae are shorter than the sub-dorsal ones, and are all about equally long, except for the longer elongated terminal pair of sub-dorsal setae. The head is generally wider than long, with a wide truncated anterior end. The cuticle is thickened and sclerotized, except in the labial region and at the level of the amphids. From the insertion of the cephalis setae to about halfway along the head length, the cuticle is covered with concretion material. The labial zone has six papillae that are very small. the cephalic setae is jointed, with a length of 0.000118–0.000158 in (3–4 µm) at the base and a fine distal part.

Amphids are rounded and vesicular, mostly covering the head. The amphidial opening is large, and situated in the posterior region of the head. The digestive system is typical for its genus. the stoma is very small. The pharyngo-intestinal junction is opposite the posterior end of the second main ring. The intestine overlaps the rectum by a large blind sac, extending to halfway to the end ring of the paratype female. There are very small ochrous pigment spots situated at the level of annule 13. The cloacal tube protrudes from the ventral body wall in annule 64. Spicules are about 0.000945 in (24 µm) in length, and gradually taper to a tip. the gubernaculum is not observed. The tail has two main rings, with females also having two partial rings. The phasmata is not observed.

DISTRIBUTION
Not known; although the family Demoscolecida is found primarily in marine waters and occasionally in freshwater and soil. (Specific distribution map not available.)

HABITAT
Nothing is known.

BEHAVIOR
Nothing is known.

FEEDING ECOLOGY AND DIET
Nothing is known.

REPRODUCTIVE BIOLOGY
The reproductive system is typical for the genus. It is obscured because of the very enlarged intestine that contains many large globules. The female vulva is situated within annule 48. Males have one testis.

CONSERVATION STATUS
Not listed by the IUCN.

SIGNIFICANCE TO HUMANS
None known. ◆

No common name
Greeffiella minutum

ORDER
Desmoscolecida

FAMILY
Greeffiellidae

TAXONOMY
Greeffiella minutum Steiner, 1916.

OTHER COMMON NAMES
None known.

PHYSICAL CHARACTERISTICS
Adults are less than 0.00315 in (80 µm) in length, and considered by many scientists as the smallest of all the nematode species. The prominent annulation is homogeneous and, as a result, not mixed with smaller annules. Each annule bears a ring of elongated spines that do not pass through the cuticle. In addition, there are large subdorsal and subventral tubular setae along the body.

DISTRIBUTION
Live in the seas. (Specific distribution map not available.)

HABITAT
Found in marine habitat or, rarely, in brackish estuarine waters in sediment.

BEHAVIOR
Free-living worms.

FEEDING ECOLOGY AND DIET
Nothing is known.

REPRODUCTIVE BIOLOGY
Nothing is known.

CONSERVATION STATUS
Not listed by the IUCN.

SIGNIFICANCE TO HUMANS
None known. ◆

No common name
Nygolaimus parvus

ORDER
Dorylaimida

FAMILY
Nygolaimidae

TAXONOMY
Nygolaimus parvus Thorne, 1974.

OTHER COMMON NAMES
None known.

PHYSICAL CHARACTERISTICS
Classified as predatory and free-living, with a length of 0.051–0.055 in (1.3–1.4 mm). They are moderate-sized nematodes with relatively thin cuticles. The body is twisted so that when observed from an upper and lower viewpoint, only the head and tail can be seen. The neck takes up about two-thirds of the body length. The lip region is apparently set off by a modest constriction, with papillae low and rounded. The lateral field is one-fourth of the body width. The mural tooth is deltoid in shape, set in the left sub-ventral wall of the cheilostomal region, of length 0.000866 in (22 µm), a length that is somewhat less than the width of the lip region. The complete stoma is divided into three parts: a thin-walled cheilostomal vestibule, the cheilostome with the mural tooth, and an elongated thick-walled esophastome. The esophagus begins as a slender tube, and then slowly expands two-thirds of the way to the posterior, which then nearly fills the body cavity. A sheath encloses the posterior portion (postcorpus) of the esophagus. The three glands at the junction of the esophagus and stomach are well developed and discrete. At this site, the opening (cardia) is an elongated hemispheroid, which is one-fourth the body width and is usually obscured by the cardiac glands. The intestinal granules are colored a light brown, and sometimes form a slightly tessellated pattern. The rectum and pre-rectum both are about as long as the anal body diameter. Both sexes have similarly shaped tails, generally conoid and bluntly rounded.

DISTRIBUTION
Midwestern to western United States; especially in South Dakota, North Dakota, Minnesota, and Montana. (Specific distribution map not available.)

HABITAT
Live in native soils.

BEHAVIOR
Classified as predatory and free-living.

FEEDING ECOLOGY AND DIET
Predacious on other small, soil-inhabiting organisms.

REPRODUCTIVE BIOLOGY
The female vulva is transverse, and is smaller and of a different form than those of other species within the genus. The vagina extends about 40% of the body. The reflexed ovaries extend about two-thirds the distance back to the vulva. The eggs are 3–4 times as long as the body width. Males have lateral guiding pieces and poorly developed ventromedial supplements.

CONSERVATION STATUS
Not listed by the IUCN.

SIGNIFICANCE TO HUMANS
None known. ◆

No common name
Mermis nigrescens

ORDER
Stichosomida

FAMILY
Mermithidae

TAXONOMY
Mermis nigrescens Dujardin, 1842.

OTHER COMMON NAMES
None known.

PHYSICAL CHARACTERISTICS
A free-living adult and parasitic larvae species that infects the body cavity of grasshoppers. They are an unusually long and slender nematode, with a length of 1.97–7.87 in (5–20 cm), with females longer than males. Adult females are colored reddish brown at the anterior extremity, supposedly as an aid for light sensitivity. The terminal mouth has two closely associated lip papillae and four cephalic papillae placed further back. The crisscross fibers are distinct and the lateral chords are wide. The thick-walled brownish-colored eggs are small, divided in half by a distinct equatorial groove, and possessing prominent byssi (filamentous branches). The size of the eggs (larvae) is about 0.00276 in (70 µm) in length by 0.00118 in (30 µm) in width. When the grasshopper host ingests the eggs found on vegetation, which is usually grasses and plants, the juveniles will break open the egg at the equatorial groove and grow to a length of 3.9 in (10 cm) in a period lasting 1–3 months.

DISTRIBUTION
Commonly occur in the British Isles, Europe, and North America. (Specific distribution map not available.)

HABITAT
During their stage as developing larvae, they live primarily within locusts and grasshoppers, but may also infect other insect species. They may be located anywhere within the host's hemocoel (the body cavity where blood circulates). Once juveniles burrow out of a host, they will dig 6–8 in (15–20 cm)

into the soil where they molt into adults. They are presumed to live in temperate forests and grasslands, rainforests, and mountains.

BEHAVIOR
Adults are very agile and readily climb plants, especially during rainy seasons. During their adult stage while in the soil, there is no social interaction, except possibly to mate. Most nematodes will move away from light, but this is usually not the case for this species. Females may remain in the soil for several years before emerging pregnant into a lighted, moist environment in order to find a site on grasses and plants to lay eggs. They usually stay in the soil during the cold winter, sheltering under debris, and then emerging in late spring, during periods of overcast, humid weather. It is believed that they are able to discriminate between light levels with the use of melanin, which is a binary system (either off or on). Even though they sometimes will move toward light, they will die from continued exposure of direct sunlight.

The action of females crawling on plants in order to deposit their eggs on vegetation above ground is considered an important behavior modification for insect parasitism (most nematodes do not infect their hosts by depositing eggs on plant foliage that will be eaten). Extending from the sides of the eggs are long, filamentous brances, called byssi, which become entangled with the plant, thus holding them in place. After infecting a host, juveniles penetrate through the gut wall into the hemocoel. Once located within the hemocoel, the parasites consume the hemolymph. Many worms may infect the same host. By late summer, the abdomen of many grasshoppers will be packed with these parasites. Such infections seriously stress and sterilize infected grasshoppers. After this period of infestation, the grasshopper will usually die, upon which the species exits the body of its dead host. The remainder of life is largely spent in the soil, except when adult females emerge for egg deposition aboveground.

FEEDING ECOLOGY AND DIET
Developing juveniles consume large amounts of amino acids, lipids, and carbohydrates from the hemolymph of the host. The free-living adults do not eat, so they must gain all of their nutrients while in the insect host.

REPRODUCTIVE BIOLOGY
During their post-parasitic stage, they live in the soil where they reach sexual maturity. Females may mate in the soil, but males are not necessarily needed for egg production. During or just after spring, females climb out of the soil onto vegetation where they deposit their eggs to await ingestion. Eggs can survive on foliage throughout the summer. They have an unusually large free-living stage when compared to other nematodes. A complete life cycle usually comprises of 2–3 years.

CONSERVATION STATUS
Not listed by the IUCN. The parasitic species is common and widespread.

SIGNIFICANCE TO HUMANS
Since the larvae kill the host grasshopper upon emerging, they have the potential to be used as a biological control agent if their rate of survival within the soil should be extended and if their numbers could be supplemented when released into the soil. However, nematode ecology is poorly understood, especially in nature, and such control practices are still far off into the future. ◆

Trichina worm
Trichinella spiralis

ORDER
Stichosomida

FAMILY
Trichinellidae

TAXONOMY
Trichinella spiralis Owens, 1835.

OTHER COMMON NAMES
English: Pork worm.

PHYSICAL CHARACTERISTICS
Classified as animal (mammal) parasites and the causal organism of the disease trichinosis, they are small roundworms that live mainly in rats and other small mammals such as pigs that pick up the worm while rooting for food. Adults have a length of 0.055–0.158 in (1.4–4.0 mm), with males measuring 0.055–0.063 in (1.4–1.6 mm) in length and females 0.118–0.158 in (3.0–4.0 mm) in length. Males and females have distinct features. Females possess a uterus and vulva. The vulva is located near the middle of the esophagus, which is about one-third the length of the body. Males have a single gonad, but no copulatory spicule, and have an ejaculatory duct. Structures identifiable on both sexes include the muscular esophagus, stichosome, and intestine. Stichosomes are formed by a single short row of stichocytes, following a short muscular esophagus. The color of the external surface of the adult is translucent and white. Both sexes are more slender at the anterior than at the posterior, but do not have two distinct sections. The anus is nearly terminal and has a large papilla on each side of it.

DISTRIBUTION
They are found worldwide, especially in South America and Africa. (Specific distribution map not available.)

HABITAT
Found more commonly in temperate rather than tropical regions. Unlike many parasites that show a high degree of host specificity, they can be found in many species of carnivores and omnivores, and in virtually all mammals. They are found attached to or buried in the mucosa (mucous membrane) of the duodenum of their host.

BEHAVIOR
Have no stages outside a host, which is unusual for helminth parasites. Their lifecycle begins when viable encysted larvae are eaten with the flesh of any meat-eating animal. When these infectious cysts are ingested, the juveniles are freed from the hard cysts by the dissolving action of stomach acids. The liberated juveniles pass to the duodenum where they go through four rapid molting periods within 27–29 hours and then mature within 1–2 days. Adults live in the gastrointestinal tract of their host. Mature males fertilize the females and then die. Mature females will penetrate the mucosa of the intestine and, within a few days, will begin laying pre-juveniles into lymph vessels. Females die after producing the juveniles. The new pre-juveniles are then carried to the lymphatic and mesenteric veins, and are found throughout the arterial circulation 7–25 days after the initial infection. The pre-juveniles move in the hepatoportal system through the liver, and then to the heart, lungs, and the arterial system, which distributes them throughout the body. They are transported to striated muscles, where

they penetrate individual fibers. Within the muscles, the immature worms curl into a ball and encyst. Juvenile cysts can grow to 0.12 in (3 mm) in diameter. These newly infective juveniles are now ready to be eaten by another host (which eats the infected muscle of the previous host). If another host does not eat the previous host, then juveniles may remain viable for many years, for example, up to 25 years in humans and 11 years in pigs. Their lifecycle is considered to be complete when a definitive host ingests the intermediate host. There is general agreement among scientists that there are four juvenile stages of the worm, but there is distinct disagreement as to whether or not nematode development occurs within the cyst.

FEEDING ECOLOGY AND DIET
As parasites, they feed off of their hosts.

REPRODUCTIVE BIOLOGY
The single female uterus is filled with developing eggs in its posterior portion, while the anterior region contains fully developed hatching juveniles. The size of the larvae is about 0.00394 in (100 μm) in length by 0.000394 in (10 μm) in width. After impregnating the female, the male dies. The females subsequently increase to their maximum size and burrow deeper into the mucosa. Females are ovoviviparous (very thinly shelled or shell-less eggs are developed within the female and juveniles hatch before leaving the uterus). A female produces about 1,500 young over a period of 4–6 weeks, after which the female dies. Both males and females live for only a short time in the intestinal epithelium of a wide variety of mammals. The juveniles then migrate into the intestinal lymphatic and mesenteric venules, on to the heart and lungs, and eventually into the arterial circulatory system. Concurrently, they molt three times in order to reach adulthood. When reaching striated muscle, they encyst, remaining viable for more than five years within the cysts. The termination of their lifecycle is reached at this time, and the larvae must wait for their host to be devoured in order to continue other stages.

CONSERVATION STATUS
Not listed by the IUCN.

SIGNIFICANCE TO HUMANS
Trichinosis is a zoonotic infection associated with the colonization of worms in muscles. It is often found in humans because of the consumption of uncooked or insufficiently cooked pork products (though other animals are also potential sources). Trichinella is the third most common worm that infects humans. They cause nausea, dysentery, puffy eyes, and colic. They also cause pain and more severe problems such as edema, cardiac and pulmonary problems, deafness, delirium, muscle pain, muted reflexes, nervous disorders, and pneumonia. Their natural hosts are flesh-eating animals, especially humans, pigs, rats, and other mammals. Humans are considered accidental hosts because, under normal conditions, the parasite ends its cycle; that is, no other animals eat humans in order to transfer their larvae to other hosts. But concern for trichinosis is not as great today with improved pork production practices. Still, an estimated 5–6 million human infections are present at any one time in North America (about 2% of the population). ◆

Human whipworm
Trichuris trichiura

ORDER
Stichosomida

FAMILY
Trichuridae

TAXONOMY
Trichuris trichiusa Linnaes, 1771.

OTHER COMMON NAMES
English: Whipworm, threadworm.

PHYSICAL CHARACTERISTICS
Exceptionally thin for nematodes. They derive their name from the characteristic "whip-like" shape. Adults are 1.2–2.0 in (30–50 mm) in length, with a thread-like anterior end that becomes thicker at the posterior end. Both sexes have two distinct body regions. Males are 1.2–1.8 in (30–45 mm) in length, while adult females are 1.4–2.0 in (35–50 mm) in length. Females are very attenuated on the anterior three-fifths of the body, and become greatly expanded in the posterior two-fifths. Males are similarly shaped, but the swollen posterior is less pronounced. For most of the body's length, there is an area designated as the bacillary band, which is a combination of hypodermal and glandular tissues. The glandular tissue opens up to the exterior through cuticular pores. They have a mouth with a simple opening, and do not have lips. The buccal cavity is tiny and is provided with a minute spear. The esophagus is very long, occupying about two-thirds of the body length and consists of a thin-walled tube surrounded by large, unicellular glands, the stitchocytes. The entire structure is called a stichosome. The anterior end of the esophagus is somewhat muscular. The transition from the anterior, filiform portion of the esophagus and the posterior, stout portion is sudden. Both sexes have a single gonad, and the anus is near the tip of the tale. Males have a single spicule that is surrounded by a spiny spicule sheath. The ejaculatory duct joins the intestine anterior to the cloaca. In the female, the vulva is near the junction of the esophagus and the intestine. The uterus contains many unembryonated eggs. The excretory system is absent. The ventral surface of the esophageal region bears a wide band of minute pores, leading to underlying glandular and non-glandular cells.

DISTRIBUTION
Found worldwide, but are most commonly in Europe, followed by Asia, Africa, and South America. (Specific distribution map not available.)

HABITAT
Most often found in temperate and tropical areas, especially in moist areas. The final hosts are humans, where they primarily inhabit the colon.

BEHAVIOR
After 10–14 days in soil, their eggs become infective. For the host to become infective, the eggs must be swallowed. Upon entering the host's body, larvae hatch in the small intestine where, over a short period of time, they grow and molt at various points along the intestines, including the lumen and ileocecal area. Adults eventually bury their thin, thread-like anterior half into the mucosa of the large intestine, and then feed on tissue secretions, but not blood. They complete maturation at this time with their large posterior end breaking out of the mucosa and protruding. Adults are characterized by a lack of a

tissue migration phase. They can reach the lungs by way of the lymph and blood systems. In the lungs, the larvae break out of the pulmonary capillaries into the air sacs, ascend into the throat, and descend to the small intestine again, where they grow. From the time of ingestion of the eggs to the development of mature worms is approximately three months. The prepatent period in the final host is 4–12 weeks.

FEEDING ECOLOGY AND DIET
As parasites, they feed off of their human hosts, primarily receiving nutrients from tissue secretions mostly within the intestines; it does not feed on blood.

REPRODUCTIVE BIOLOGY
The reproductive system is found at the esophago-intestine region. Both males and females have single reflexed gonads. The male has only one spicule. The eggs are operculate, and the females are oviparous. In early larval development, there are two rows of cells. When they infect a host, the eggs are "sticky." Eggs have, smooth outer shells. Adult females can lay eggs for up to five years. Egg production is estimated at 1,000–7,000 per day following copulation, and may contain up to 46,000 eggs at any one time. The eggs can be expelled with the feces of the host. Embryonation is completed in about 21 days in soil at about 86°F (30°C), where it is moist and shady. Adults can live for several years, so large numbers can accumulate in humans.

CONSERVATION STATUS
Not listed by the IUCN.

SIGNIFICANCE TO HUMANS
Occasionally cause the condition eosinophilia, with the cecum and colon being the most commonly infected sites in the host. The particular disease is trichuriasis, or "whip worm" infection. In the United States, whipworm infection is rare overall, but is less rare in the rural southeast, where more than 2.2 million people are infected. Internationally, human whipworm infection is common in less-developed countries, with about 300–500 million people infected worldwide, but with some estimates as high as one billion people. Human whipworm infection is rarely fatal, but rectal prolapse may occur in heavily infected hosts. Fewer than 100 worms rarely cause clinical symptoms. Only hosts with very heavy infections become symptomatic. For those with heavy infection, common symptoms include abdominal discomfort, anemia, abdominal tenderness, secondary bacterial infections, and diarrhea; for more severe cases, symptoms include rectal prolapse, insomnia, nervousness, loss of appetite, vomiting, urticaria (skin rash), prolonged diarrhea, constipation, and flatulence. The infection is usually diagnosed by observing eggs in the patient's feces, and on occasion the presence of larval or adult worms in the feces. ◆

Resources

Books
Farrand, John, ed. *The Audubon Society Encyclopedia of Animal Life*. New York: Clarkson N. Potter, 1982.

Bird, Alan F. *The Structure of Nematodes*. New York: Academic Press, 1971.

Chitwood, B. G., and M. B. Chitwood. *Introduction to Nematology*. Baltimore: University Park Press, 1950.

Croll, Neil A., and Bernard E. Matthews. *Biology of Nematodes*. Glasgow and London: Blackie and Son Limited, 1977.

Levin, Simon Asher, ed. *Encyclopedia of Biodiversity*. San Diego: Academic Press, 2001.

Maggenti, Armand. *General Nematology*. New York: Springer-Verlag, 1981.

Malakhov, V. V. (translated by George V. Bentz, edited by W. Duane Hope). *Nematodes: Structure, Development, Classification, and Phylogeny*. Washington, DC, and London, U.K.: Smithsonian Institution Press, 1994.

Mehlhorn, Heinz, ed. *Encyclopedic Reference of Parasitology: Diseases, Treatment, Therapy*, 2nd ed. New York: Springer, 2001.

The New Larousse Encyclopedia of Animal Life. New York: Bonanza Books, 1981.

Parker, Sybil P., ed. *Synopsis and Classification of Living Organisms*. New York: McGraw-Hill Book Company, 1982.

Poinar, George O. Jr. *The Natural History of Nematodes*. Englewood Cliffs, NJ: Prentice-Hall, Inc., 1983.

Stone, A. R., H. M. Platt, and L. F. Khalil, eds. *Concepts in Nematode Systematics, Special Volume No. 22*. London: Academic Press, 1983.

Wharton, David A. *A Functional Biology of Nematodes*. London: Croom Helm, 1986.

William Arthur Atkins

Secernentea

(Secernenteans)

Phylum Nematoda

Class Secernentea

Number of families 60+

Thumbnail description
Almost exclusively terrestrial nematodes, generally parasitic of plants and invertebrate and vertebrate animals; they are bilaterally symmetrical and non-segmented

Illustration: *Criconema* sp. (Illustration by John Megahan)

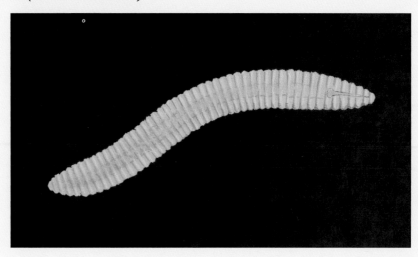

Evolution and systematics

Nematoda, the phylum above the class Secernentea, has left very few fossil remains. The earliest fossils that contained nematode structures were found in Eocene strata (the era from about 55–38 million years ago). More authenticated fossils are of nematodes preserved in amber within such items as fossilized shark muscles and mammals frozen in permafrost. The fossil record is too fragmented to explain much about nematode origins, so most conclusions about phylogeny are based on observations of living species. It is hypothesized that nematodes originated during the Precambrian era, what was the Proterozoic period (about one billion years ago).

Earlier in the classification process Chitwood separated hematodes into two main classes, the Phasmidia (now Secernentea) and Aphasmidia (now Adenophorea). Controversies still exist, but for the most part, scientists, such as A. R. Maggenti, who helped to develop the classifications under this system, treat nematodes as a separate phylum with two classes, Adenophorea and Secernentea, which have been divided based on molecular and morphological characteristics. These two classes are primarily separated (along with other important criteria) with respect to whether they possess phasmids (as in Secernentea) or do not possess phasmids (as in Adenophorea). The total number of species of secernenteans is estimated at about 8,000 worldwide, with scientific surveys suggesting that an enormous number of species has yet-to-be discovered. There are six orders and the number of families ranges from 60 to 89.

Physical characteristics

Secernenteans vary greatly in size from microscopic to several feet long. The largest known secernentean, which is up to 30 ft (9 m) in length, lives in the placentas of female sperm whales. The body of secernenteans consists of a flexible cylinder that tapers at both ends, with a pointed tail and a blunt head. They are considered non-segmented pseudocoelomates; that is, creatures possessing a three-layered body that has a fluid-filled body cavity (pseudocoelom) between the endoderm and the mesoderm (the innermost and middle tissue layers).

A flexible but tough collagenous cuticle surrounds the body with a system of grooves across the body from head to tail, which protects them internally. The non-cellular cuticle varies from four to two layers and is almost always transversely striated. Laterally for most of the body length, the cuticle is generally modified into a wing area that is marked by longitudinal ridges; generally, this region is only slightly above the normal body contour. However, in some parasitic forms, it may extend out a distance equal to the body's diameter. The cellular hypodermis is the subcuticular layer that secretes the cuticle.

The sensory system contains phasmids, which are a pair of bilateral cuticular, glandular organs, situated laterally in the caudal (posterior to the anus) region and opening to the surface by a slit or pore. Also known as precaudal glands, phasmids are unique to the secernenteans, in which their function is believed to be sensory. At the other end are pore-like amphid apertures, which are a pair of glandular chemosensory organs situated dorso-laterally in the cephalic (head or anterior) region and opening through the cuticle. Although usually pore-like, in isolated instances the aperture can be an oval or a cleft. The apertures show little variation throughout the secernenteans. The amphids are always labial (located on the lips). The external amphidial aperture is usually less well developed than in adenophorean worms.

Somatic and cephalic setae, which are elongated structures jointed with the cuticle, are rare. When present, the cephalic sensilla are located on the labial region, and they are pore-like or papilliform. In males, there may be caudal setae. In females, somatic setae are absent. Generally, sixteen sensilla are present in the shape of two circles (an inner circle of six, and an outer circle of 10). In some parasitic groups, the number of cephalic sensilla may be reduced. Deirids, pairs of pore-like sensilla that usually protrude above the surface of the cuticle, are usually present on the cervical region near the level of the nerve ring.

Secernenteans contain no hypodermal glands, but the hypodermal cells of the hypodermis (a thin tissue layer beneath the cuticle that thickens to form the dorsal, ventral, and two lateral hypodermal chords, and extends the length of the body) are usually multinucleate (syncytial: more than two nuclei), but may also be uninucleate (cellular: one nuclei). These divide the muscle cells into four fields. A layer of longitudinal muscles underlies the hypodermis.

Bursae, or caudal alae, are sublateral projections generated by a longitudinal widening of the cuticle. It is common for them to be present within male secernenteans, each looking like a fluid-filled body sac. The thin cuticle extensions are located on both sides of the anus, specifically on either side of (or sometimes surrounding) the cloaca (the urogenital opening) of males. The well-developed excretory system is in the shape of an H or U. It is a simple tubular system that is located in one or both of the lateral hypodermal chords, and embedded between the three cell bodies in the hypodermal chord. The system opens ventromedially by way of a cuticularized duct. The rectal part of the gland system is usually present. They have no caudal glands.

The muscular esophagus or pharynx varies in configuration, but the majority of secernenteans have three esophageal glands: two that are subventral and one that is dorsal. The subventral glands open into the posterior metacarpus. The dorsal gland opens anteriorly into the procorpus or the anterior metacarpus. Its basic structure is corpus (the anterior part is cylindrical), with the basal (bottom) region sometimes swollen in the shape of a bulb. The glands empty their contents into the esophagus to aid in digestion. The tail is the region between the anus and the back tip of the body.

Distribution

Secernenteans are distributed worldwide in terrestrial environments, and only rarely in regions containing water habitats.

Habitat

Secernenteans show considerable diversity in their habitats, including free-living microbotrophs, plant parasites, invertebrate parasites, and vertebrate parasites. They are rarely found in marine and freshwater habitats, being almost exclusively terrestrial. They are mostly parasites, living on or within their plant or animal hosts. When free living, they often live in soil.

Behavior

This class encompasses most of the plant, invertebrate, and vertebrate parasites in the Nematoda phylum. Within the class, there are more than 3,000 animal parasites and 2,000 plant parasites. Their lifestyle is classified primarily as parasitic, but there are some free-living species. When parasitic or a combination of parasitic and free-living, secernentean behavior primarily revolves around their hosts. Each species has developed particular ways to infect their hosts, adapting their behaviors to best suit their needs while living off their hosts.

Feeding ecology and diet

They feed mostly on bacteria, fungi, and soil organisms, as well as on the nutrients such as blood, body fluid, intestinal contents, and mucus found within their hosts. Secernentean worms generally have some form of stylet, a hard, sharp spear that is used for feeding. Muscles move the stylet in and out, allowing them to puncture cells. Once opened, the worm will empty the contents of the cell.

Reproductive biology

The female ovary contains germ cells that give rise to eggs. Fertilization, most of the time by male sperm, takes place in the uterus and eggs are released through the vagina. Males have one testis (monarchic), when testes are present. Most species produce males and females (that is, dioecious), but some species only produce females, in which both male and female structures are contained in the same individual. In those few cases the species are hermaphroditic. Most eggs are about the same size and shape. Males produce sperm in the testes, which are shaped similar to female ovaries. Sperm accumulate in the seminal vesicle and exit through the anus. Males have accessory copulatory organs, called spicules. During mating, a spicule becomes rigid so it can be inserted into the vagina in order to form a passageway for the sperm. Male bursae, usually numbering four, are used to clasp the female during copulation. Males possess paired preanal supplement glands (or genital papillae) in two sublateral rows on the ventral side of the body. The glands are used for secretion and attachment, and function during copulation.

The life cycle is generally a straight cycle going from fertilized egg, through four juvenile (often called larval) stages involving a set number of cell divisions, and into adulthood. Secernenteans generally produce offspring through internal sexual reproduction. Fertilization occurs in the female's uterus, where the zygote is then placed inside a tough shell. Development from egg to adult is generally similar for all species, although many differences exist to the norm (often times brought about by environmental conditions and special types of life, as two examples). Usually a young juvenile hatches from the egg. It usually resembles the adult except in size and in the maturity of its sex organs. Each of the four larval stages (usually referred to as L1, L2, L3, and L4) is separated from one another by the complete shedding of the outer layer (what is called ecdysis, or molting in other animals), including the lining of the mouth and rectum. There is no increase in the number of cells after hatching, with growth coming exclusively from an increase in the size of the original cells. Development in successive stages is gradual over all, however organs themselves can develop at rapid rates.

Conservation status

No secernenteans are listed on the 2002 IUCN Red List of Threatened Species. They are found worldwide and frequently occur in very great densities.

Significance to humans

As crops are cultivated to feed the world's population, secernenteans become more numerous as they feed on agricultural plants. Parasitic secernenteans cause yield losses and they may join with other soil-living organisms such as bacteria, fungi, and viruses to advance disease development in plants. They also can cause loss of nutrients and water into the plant, thus increasing the plant's susceptibility to other dangers. Secernenteans, when infecting humans, can cause various diseases and, in some circumstances, death to the human host. When infected animals are used as food or kept as pets for humans, these parasites can cause and transmit various diseases and, often, death to the hosts. On the other hand, free-living forms of secernenteans can be good indication of biodiversity, and important to the health and survival of humans. They help to cycle carbon and nitrogen and to breakdown organic matter in the soil environment.

1. Cod worm (*Phocanema decipiens*); 2. Barber's pole worm (*Haemonchus contortus*); 3. African river blindness nematode (*Onchocerca volvulus*); 4. Dog hookworm (*Ancylostoma caninum*); 5. Rat lungworm (*Angiostrongylus cantonensis*); 6. Threadworm (*Strongyloides stercoralis*); 7. Canine heartworm (*Dirofilaria immitis*); 8. Citrus spine nematode (*Criconema civellae*); 9. Maw-worm (*Ascaris lumbricoides*). (Illustration by John Megahan)

Species accounts

Maw-worm
Ascaris lumbricoides

ORDER
Ascaridida

FAMILY
Ascarididae

TAXONOMY
Ascaris's lumbricoides Linnaeus, 1758, *Homo sapiens*, Europe.

OTHER COMMON NAMES
English: Large roundworm.

PHYSICAL CHARACTERISTICS
One of the largest and most common parasites, it measures 8–18 in (20–46 cm) in length by 0.12–0.24 in (3–6 mm) in width for females; males generally more slender and shorter at 6–12 in (15–30 cm) in length by 0.08–0.16 in (2–4 mm) in width. Bodies are cylindrically shaped, with a head that is slightly narrower. Males have a curved tail with two spicules, but no copulatory bursa; females have a vulva approximately one-third of the length of the body down from the head, and a blunt tail. Characterized with smooth, finely striated cuticle; lip region is well developed and separated from the cervical region. The mouth has three lips, each equipped with small papillae. Internally, they have a cylindrical esophagus opening into a flattened ribbon-like intestine. Female ovaries can be 3.3 ft (1 m) in length). The eggs have thick shells consisting of a transparent inner shell covered in a warty, albuminous coat, and are sticky to the touch. The size of the eggs are 0.00197–0.00295 in (50–75 µm) in length by 0.00158–0.00197 in (40–50 µm) in width. The excretory system empties every three minutes or so.

DISTRIBUTION
Cosmopolitan. In 1986 they caused over 64 million infections worldwide. (Specific distribution map not available.)

HABITAT
Most common in warm, moist climates or in regions with temperate or tropical climates. They live in the small intestines of humans and pigs.

BEHAVIOR
Have a direct and simple lifestyle, with no intermediate hosts, unlike many parasites. Adults live in the lumen of the small intestines and eggs are passed in the feces. The eggs hatch in the small intestine. Once they grow into an adult worm and mate, they become as large as 12 in (31 cm) in length by 2 in (4 cm) in width.

FEEDING ECOLOGY AND DIET
Feeds on the semi-digested contents of the gut, although there is some evidence that it can bite the intestinal mucous membrane and feed on blood and tissue fluids.

REPRODUCTIVE BIOLOGY
Females can produce 200,000 to 2 million eggs per day, and 70 million in a year with very developed ovaries. The ovaries can contain 27 million eggs at one time. The fertilized eggs are broadly oval, and the shell is thickened, tuberculate and measure about 0.0098 in (250 µm) in length by 0.00059 in (15 µm) in width. About 2–3 weeks after passage in the feces and with ideal environmental conditions, the eggs contain an infective juvenile, and humans are infected when they ingest such infective eggs. Adults can live in the small intestine for six months or longer. In the intestine, eggs are only embryonated mass of cells, with further differentiation occurring outside the host. Eggs can stay alive in the soil for many years if conditions are adequate. The cycle from egg ingestion to new egg production takes approximately two months.

CONSERVATION STATUS
Not listed by the IUCN.

SIGNIFICANCE TO HUMANS
About a sixth of the world's population suffers from ascariasis, a maw-worm infection often called "large roundworm infection." An estimated 25% of the people in developing countries are believed to be infected. In the United States, occurrence is uncommon with only four million people infected, mostly in the rural southeast. Infection occurs from ingestion of raw food such as fruit or vegetables. Infection with the species is rarely fatal, but death may occur because of mechanical intestinal obstruction. ◆

Threadworm
Strongyloides stercoralis

ORDER
Rhabditida

FAMILY
Strongyloididae

TAXONOMY
Anguilla stercoralis (Bavay in Normond, 1876), In Toulon, France, on host *Homo sapiens*, but host from Cochin China.

OTHER COMMON NAMES
English: Dwarf threadworm.

PHYSICAL CHARACTERISTICS
An important parasite of humans, primates, dogs, and other animals, being one of the smallest to inhabit the human body. Females measure from 0.079–0.098 in (2.0–2.5 mm) in length; males slightly smaller. Size of the eggs is about 0.00158 in (40 µm) in length by 0.00118 in (30 µm) in width. Both sexes have small buccal cavities and a long esophagus. Females have a cylindrical pharynx with no posterior bulb swelling. Both free-living male and female worms have a prominent rhabdiform pharynx.

DISTRIBUTION
Occur all over the world, but exist in large numbers in Southeast Asia and South America, and in Eastern Europe and in the Mediterranean region. In the United States, they are rare, al-

though are more prevalent in the rural southeast. (Specific distribution map not available.)

HABITAT
Found primarily in temperate climates, and to a lesser degree in tropical climates. Female worms are found in the superficial tissues of the small intestines of vertebrates, mainly humans and dogs. Males are believed not to be parasitic, but free living in the soil.

BEHAVIOR
Heterogonic lifecycle, i.e., a parasitic generation is interspersed with a free-living one. There can be multiple cycles of each phase. Females are more invasive than males. This parasite generally replicates within its definitive hosts. Females lay partially embryonated eggs that are released into the host's submocosa or the intestinal lumen. Once in the external environment (after passing through the host's feces), the larvae either remain as infective stages or develop to free-living adults.

FEEDING ECOLOGY AND DIET
As parasites, they feed off of their hosts, taking nutrients, mostly mucosa from the intestine, from various locations within the body as they travel through the circulatory system, lungs, stomach, and small intestines. As free living, they feed on organic matter, bacteria, and other nutrients in the soil.

REPRODUCTIVE BIOLOGY
Females parthenogenetic; males exist, but are only rarely found, and then only in feces. Free-living males and females mate, produce more larvae, and some of these larvae will develop into free-living larvae, while others will develop into parasitic juveniles.

CONSERVATION STATUS
Not listed by the IUCN.

SIGNIFICANCE TO HUMANS
Septicemia is caused by the infection of this worm into the human body. More than 70 million people are infected, mostly in the temperate regions of the world. ◆

Canine heartworm
Dirofilaria immitis

ORDER
Spirurida

FAMILY
Onchocercidae

TAXONOMY
Filaria immitis (Leidy, 1856), *Canis familiaris*, United States.

OTHER COMMON NAMES
English: Dog heartworm.

PHYSICAL CHARACTERISTICS
Females are 9-12 in (25-30 cm) in length by only about 0.13 in (5 mm) in width, while males are about half the size of females with a length of 5-6 in (12-16 cm). They have somatic (coelomyarian) muscles. (Somatic muscles provide the shape to individual muscle cells thus, supplying a portion of the hydrostatic skeleton in nematodes). Specifically, coelomyarian muscles (a type of muscle cell arrangement) make a protoplasmic zone that bulges into the pseudocel and fibrillar zone, which extends up the sides of the cell.) The average diameter of the filaments are 0.007-0.017 µm. The cuticle, which is the elastic covering of the body and all body openings, consists of a thin polysaccharide-rich layer. It does not possess longitudinal ridges.

DISTRIBUTION
Widespread throughout the world and exists anywhere mosquitoes live. (Specific distribution map not available.)

HABITAT
Primarily live in the tropics, subtropics, and some temperate areas. They are found in dogs, cats, foxes, wolves, and other wild carnivores as well as in sea lions and humans. Within the host, adults live in the right ventricle and the adjacent blood vessels from the posterior vena cava, hepatic vein, and anterior vena cava to the pulmonary artery.

BEHAVIOR
Usually infest the heart of its hosts through about 30 species of mosquitoes. Adults live in the peripheral branches of the pulmonary arteries and produce large numbers of microfilaria that circulate throughout the bloodstream. They usually infect dogs, but also cats, ferrets, and seals. These hosts are the definitive hosts, while mosquitoes are the intermediate hosts. Lifecycle begins when a dog with circulating microfilaria is bitten by a mosquito. They are passed into the bloodstream where they remain active for up to one year or more, but are incapable of further development until ingested by a mosquito. Microfilaria matures into infective larvae inside the mosquito within about 14 days. When the infective mosquito bites a host, larvae are injected into the host's skin and begin to mature into adults in the subcutaneous tissues, muscles, and fatty tissues; they develop to 0.98–4.33 in (25–110 mm) in length. They arrive in the right lower chamber of the heart at 2–4 months. After infection, an additional four months are required for the worms to reach maturity; microfilariae first appear in the peripheral blood circulation about eight months after infection. Adults may live and continue to produce microfilariae for several years. Adults live in the right ventricle and the adjacent blood vessels from the posterior vena cava, hepatic vein, and anterior vena cava to the pulmonary artery.

FEEDING ECOLOGY AND DIET
Live off nutrients of their hosts, primarily through the blood (mostly pulmonary arteries) in and around the heart and lungs.

REPRODUCTIVE BIOLOGY
Females produce large numbers of microfilaria after about six months, which circulate throughout the bloodstream. Up to 5,000 microfilariae are shed into the host's bloodstream each day, and can remain alive and infective in a host's bloodstream for up to three years.

CONSERVATION STATUS
Not listed by the IUCN.

SIGNIFICANCE TO HUMANS
One dog may be infected with 25–50 canine heartworms. In heavy infestations, with 50–100 worms, a dilation of the heart is apparent, as well as pathological problems with lungs, liver, and kidney. Pharmacueticals or surgery are used to remove the worms from the infected host. ◆

African river blindness nematode
Onchocerca volvulus

ORDER
Spirurida

FAMILY
Onchocercidae

TAXONOMY
Filiria volvulus (Lenckart in Mason, 1893), originally *Filaria Homo sapiens*, West Africa.

OTHER COMMON NAMES
English: River blindness nematode.

PHYSICAL CHARACTERISTICS
Filarial parasites of primates, primarily humans. They have an adult length of 0.8–27.6 in (2–70 cm) with females measuring 13.0–27.6 in (33–70 cm) in length by 0.011–0.016 in (270–400 µm) in width; males measure 0.8–1.6 in (2–4 cm) in length by 0.00512–0.00827 in (130–210 µm) in width. Microfilariae measure 0.00866–0.0142 in (220–360 µm) in length by 0.000197–0.000354 in (5–9 µm) in width, are unsheathed, and have a lifespan of almost two years. The epicuticle is folded separately from the underlying cuticle, with the adult male epicuticle showing a honeycomb-like pattern. The intestinal cells of adult females are very thick. The body is slender and blunt at both ends. Lips and a buccal capsule are absent, and two circles of four papillae each surround the mouth. The esophagus is not conspicuously divided. The female vulva is behind the posterior end of the esophagus. The male tail is curled and lacks alae; it bears four pairs of adanal and six or eight pairs of postanal papillae. The microfilariae are unsheathed.

DISTRIBUTION
Distributed throughout the world, including central Africa, Central America, northern South America, and Mexico. (Specific distribution map not available.)

HABITAT
Live primarily in the tropics and subtropics near fast flowing rivers where the *Simuliam* black fly breeds. (They are normally transmitted by the flies' bites.) They accumulate in raised nodules found under the skin and in the lymphatic system of connective tissues of the human host. Also found occasionally in peripheral blood, urine, and sputum. They can also enter the eye, leading to the formation of lesions and cataracts.

BEHAVIOR
Complex lifecycle begins when an infected female takes a blood-meal from a human host. The larvae enter the host's subcutaneous tissues and slowly mature into adults in one year. Adults can live for 15 years, and many males and females can live together. The microfilariae reach about 11.8 in (300 mm) in length by 0.03 in (0.8 mm) in diameter, and are sheath-less with sharply pointed, curved tails. Many thousands of microfilariae migrate in the subcutaneous tissue. When the infected host is bitten by another female fly, microfilariae are transferred from the host to the black fly where they develop into infective larvae, and the lifecycle continues.

FEEDING ECOLOGY AND DIET
Live off the nutrients located in such places as subcutaneous connective tissues, peripheral blood, urine, sputum, and skin found within the host.

REPRODUCTIVE BIOLOGY
Females may produce 1,000 microfilariae per day, which are already hatched when they are born (unsheathed). After mating eggs inside the female develop into microfilariae ovivipariously, which leave the worm one by one.

CONSERVATION STATUS
Not listed by the IUCN.

SIGNIFICANCE TO HUMANS
Causes onchocersiasis, which has infected an estimated 18 million people worldwide (mostly in Central and South America and sub-Saharan Africa). It has also caused more than 270,000 cases of bilateral blindness and more than one million cases of visual impairment. It rarely causes death, and is the second most common cause of infectious blindness. The severity of this disease has far reaching economic consequences but, fortunately, advancements have been recently made in reducing the disease. Controlling black flies is the prime way to control the disease. As a result, this species has almost been eliminated from some locations. ◆

Dog hookworm
Ancylostoma caninum

ORDER
Strongylida

FAMILY
Ancylostomatidae

TAXONOMY
Sclerostoma caninum (Ercolani, 1859), originally *Sclerostoma Canis familiaris*, Europe.

OTHER COMMON NAMES
English: Creeping eruption (when found in humans).

PHYSICAL CHARACTERISTICS
The most widespread of the hookworm species, they target dogs and other canids. Male adults are 0.43–0.55 in (1.1–1.4 cm) in length, with a copulatory bursa, two large lateral lobes, and two equal filiform spicules; females 0.5–0.8 in (1.3–1.9 cm) in length, with no vulvular flap. Adults have an anterior end that is bent dorsally, a buccal capsule that is deep and supported by thick cuticle, three pairs of teeth (three on each side of the ventral margin) that are located at the anterior stoma, and a strongyliform esophagus. They are often red in color because of the blood in their gut; otherwise, they are gray in color. Their heads "hook" into the small intestines of the host, where they begin to eat away at the tissue and suck blood. Eggs are 0.00209–0.00272 in (53–69 µm) by 0.00142–0.00209 in (36–53 µm), ovoid-shaped, and thin-shelled embryos. The first-stage larvae have a mouth tube, bulbed rhabditiform esophagus, and straight tail. No genital rudiment is visible.

DISTRIBUTION
Found in eastern Asia, Central and South America, and Australia. (Specific distribution map not available.)

HABITAT
Generally live in temperate forests and rainforests, temperate grasslands, tropical deciduous forest, tropical rainforests, and

tropical scrub forests. Within the host, they live within the small intestines of dogs and other canids, and can also live in foxes and cats. They can also live under the skin of humans.

BEHAVIOR
Intermediate hosts are generally not present, however, paratenic hosts (hosts that act as transfer hosts where the parasite does not develop) are normally encountered. The first eggs appear 60–75 days after the initial exposure. On reaching the small intestine, they proceed to develop to adult males and females. They suck the blood and tissues; both the plasma and corpuscles undergo at least partial digestion. In some cases, some larvae may go dormant, but can be reactivated later by unknown means.

FEEDING ECOLOGY AND DIET
As parasites, feeds off its hosts. In the free-living larvae stage, they feed on organic matter.

REPRODUCTIVE BIOLOGY
Dioecious; following copulation, females lay eggs in the intestines of the host. Females lay 7,000–30,000 eggs per day. The embryonated eggs are carried out in the host's feces. Eggs develop in the soil/feces under favorable environmental conditions of moisture, oxygen, and temperature. About three weeks later, the juveniles are unsheathed, non-feeding, and infective. When they reach the small intestine of the host, they molt a final time and develop to maturity in about five weeks.

CONSERVATION STATUS
Not listed by the IUCN.

SIGNIFICANCE TO HUMANS
In humans, the infection of the species is called "creeping eruption," and causes a severe rash. This species causes disease primarily in puppies; in its severest form, the disease is life threatening to puppies and sometimes to adult dogs. ◆

Rat lungworm
Angiostrongylus cantonensis

ORDER
Strongylida

FAMILY
Angiostrongylidae

TAXONOMY
Pulmonema catonesis (Chen, 1935), Dominican Republic, originally *Pulmonema raltucnorvegicus*, Canton, China.

OTHER COMMON NAMES
None known.

PHYSICAL CHARACTERISTICS
Moderate sized, adult males measure 0.79–0.87 in (20–22 mm) in length by 0.0126–0.0165 in (320–420 µm) in width, and adult females measure 0.87–1.34 in (22–34 mm) in length by 0.0134–0.0221 in (340–560 µm) in width.

DISTRIBUTION
Distribution can only be identified by the disease caused by this species, which has been reported in Asia (Philippines, Indonesia, Malaisia, Thailand, Vietnam, Taiwan, Hong-Kong, Japan); Oceania (Pacific Islands [Tahiti, New Caledonia],

Papua New Guinea, Australia); Cuba, Puerto Rico, Hawaii; United States; and Africa (Madagascar). (Specific distribution map not available.)

HABITAT
They utilize a wide variety of invertebrate intermediate hosts where adults live primarily in the blood vessels of the lungs of the host.

BEHAVIOR
Adults live within the blood vessels of the lungs of the rodent host. Females produce eggs that hatch in the lungs and then attach to the terminal branches of the pulmonary arteries. First-stage juveniles enter the respiratory tract and, from there, the juveniles move up the trachea, where they are then swallowed and later passed in the host's feces; juveniles can be detected in feces 40–60 days after infection. Requires an intermediate host (usually snails, but it can be found in almost any invertebrate such as oysters, slugs, and crabs) to complete its lifecycle. The rodent or human host is infected when it ingests an intermediate host (such as snails and slugs) containing infective juveniles. The infective juveniles develop to adults through two stages in 2–3 weeks. Adults enter the pulmonary arteries and the lungs where they become mature; they eventually enter the rodent host's brain. Adults then migrate back to the host's lungs via the venous circulation. In human hosts, the parasites enter the brain, but do not develop further and die.

FEEDING ECOLOGY AND DIET
Live off nutrients of their hosts, specifically around the lungs (and pulmonary arteries) and brains of rodent hosts, and usually only around the lungs of human hosts, but also in the blood vessels between the lungs and brain.

REPRODUCTIVE BIOLOGY
After mating, oviposition takes place in the lungs. Eggs are then coughed up, swallowed, and are subsequently passed in the host's feces. Hatched larvae develop in about two weeks to the infective third stage.

CONSERVATION STATUS
Not listed by the IUCN.

SIGNIFICANCE TO HUMANS
Especially dangerous to humans because it ruptures vessels in the brain. The presence of juveniles in the blood vessels, meninges, or tissue of the human brain can result in symptoms such as headache, fever, paralysis, neurological disorders, and even coma and death. The appearance of worms is often associated with eosinophilia, and is the primary cause of eosinophilic meningoenciphalitis, a disease that occurs when humans eat infected, raw snails. ◆

Barber's pole worm
Haemonchus contortus

ORDER
Strongylida

FAMILY
Trichostrongylidae

TAXONOMY
Strongylus contours (Rudolphi, 1803), originally *Strongylus ovisaries* (?), Europe.

OTHER COMMON NAMES
English: Barber pole worm, sheep stomach worm, wire worm.

PHYSICAL CHARACTERISTICS
A stomach parasitic roundworm found inside ruminants such as sheep, goats, cattle, and wild ruminants. Males have a length of 0.7–0.8 in (18–21 mm); females 0.7–1.2 in (18–30 mm). Females possess white uteri and ovaries that spiral around their red blood-filled intestine, which gives a twisted ("barber pole") appearance. The small buccal capsule contains a curved dorsal tooth. There are two distinctive lateral spike-like cervical papillae near the connection of the first and second quarters of the esophagus. The male bursa has long lateral lobes and slender rays with a flap-like dorsal lobe, which is located asymmetrically near the bases of the left lateral lobe. Spicules are 0.018–0.020 in (450–500 μm) in length, each with a terminal barb, and the gubernaculums is navicular (that is, its structure is considered to be shaped like a small boat). Usually, an anterior thumb-like flap covers the vulva, and may be reduced to a mere knob in some worms. The oval eggs are 0.00276–0.00335 in (70–85 μm) in length by 0.0016–0.00173 in (41–44 μm) in width.

DISTRIBUTION
Distributed in the Arctic and immediately adjacent temperate regions of Europe, North Africa, and Asia, north of the tropics. (Specific distribution map not available.)

HABITAT
Found in coastal and high rainfall areas, especially in areas where hosts such as goats and sheep are plentiful. They are found in many terrestrial habitats, including tundra, taiga, temperate forest, and rainforest, temperate grassland, chaparral, tropical deciduous and scrub forests, tropical savanna and grasslands, and mountains. They live inside the abomasums (fourth stomach) of ruminants. Egg development is limited to areas and seasons where pastures are moist during warm months. (They are more prevalent in warm, moist regions than in cold, dry ones.) However, juveniles can survive for some time, particularly during cool conditions, and can infect hosts even without favorable periods of development. Pastures that remain green over the summer, perennial pastures (especially with kikuyu grass), irrigated pastures, and areas near creeks, troughs, and seepage points are preferred.

BEHAVIOR
Do not require an intermediate host. The first three juvenile stages are free-living. After the infective stage is reached, the organisms move to an area that will optimize its chances of being eaten by a host. The organisms have a short lifecycle (90 days) and must find a host quickly after it has completed its first stages of growth.

FEEDING ECOLOGY AND DIET
As a parasitic species, they feed on their hosts. They parasitize the abdomen or stomach of its host, using a single dorsal tooth to make cutting movements in the host tissues. A secreted anticoagulant allows them to feed on blood, but cell contents and other fluids are also consumed. Adults in the fourth stage of life are able to form a clot and feed from it.

REPRODUCTIVE BIOLOGY
Following internal fertilization, females lay eggs that pass out in the feces of the host. Each female can deposit 5,000–10,000 eggs per day. The first microscopic larvae hatch 14–17 hours after being passed through the feces; in 3–5 days the organism will have the ability to infect a host. First- and second-stage juveniles feed on bacteria. Third stage juveniles retain the second stage cuticle as a sheath; they do not feed and are infective for the vertebrate host. In a sheep's gut, larvae develop to adults in about three weeks. Mating of adults occurs and egg production commences. The eggs hatch in soil or water. Infections by third stage juveniles may also occur through the skin. Enormous numbers of juveniles may accumulate on heavily grazed pastures. However, many die during low temperatures.

CONSERVATION STATUS
Not listed by the IUCN.

SIGNIFICANCE TO HUMANS
Cause haemonchosis, which is related to the degree of blood loss. Large numbers of worms can accumulate very rapidly, causing host deaths often without warning, especially in young animals. Vaccination of sheep and goats (de-worming) is used for control. When the worms move through the soil, they help to aerate the soil, which helps to control erosion and keeps the soil from clumping and hardening. ◆

Citrus spine nematode
Criconema civellae

ORDER
Tylenchida

FAMILY
Criconematidae

TAXONOMY
Criconema civellae (Steiner, 1949), Greenhouse, Beltsville, Maryland, United States. Latest name: *Crossonema civellae*, Menta and Raski, 1971.

OTHER COMMON NAMES
None known.

PHYSICAL CHARACTERISTICS
Females are of length 0.0114–0.0339 in (0.29–0.86 mm), with 40–93 annules. Contains spear muscles that are well developed and large compared to other species. The muscles are 0.000591–0.00394 in (15–100 μm) in length. The labial cap is not easily distinguished from the overall body contour, except for a narrowing that leads to the oral disk. The esophagus is distinctive, the corpus is typical for the family, and the postcorpus is very distinctive. The isthmus and glandular regions merge into an almost cylindrical form that is slightly expanded at the end. This general region is narrower than the corpus, and equal to or shorter than the corpus in length as measured from the spear knobs. The external cuticle is ornamented with annulations. The annules are round—not lobed—and with a continuous edge of fine, blunt spines on most of the body. The spines are generally simple in shape, except in the posterior region where they are bifurcate, knobbed, or clubbed. The lip area is elevated (with the lip generally pointed), and the submedian lobes are absent. The stylet is 0.00244–0.00453 in (0.06–0.12 mm) in length. The vulva is 3–16 annules from the terminus.

DISTRIBUTION
Not known. (Specific distribution map not available.)

HABITAT
Migratory.

FEEDING ECOLOGY AND DIET
As parasites, they live off the nutrients found on their hosts.

BEHAVIOR
Plant parasites, specifically, ectoparasites, living on the outside of the plant host.

REPRODUCTIVE BIOLOGY
Males are degenerate and incapable of breeding.

CONSERVATION STATUS
Not listed by the IUCN.

SIGNIFICANCE TO HUMANS
None known. ◆

Cod worm
Phocanema decipiens

ORDER
Ascaridida

FAMILY
Anisakidae

TAXONOMY
Ascaris's decipens (Krabbe, 1878), originally *Cristophora cristata*, Greenland coast. Latest name: *Pseudoterranova decipens*.

OTHER COMMON NAMES
English: Seal worm.

PHYSICAL CHARACTERISTICS
Often found in cod, but are also found in many other species of fish. In the larval stage, they are 0.20–2.28 in (5–58 mm) in length by 0.012–0.047 in (0.3–1.2 mm) in width, and yellowish, reddish, or brownish in color. They have well-developed and distinct lips. The excretory system is elongated and cord-like, while the adult esophagus is cylindrical in shape.

DISTRIBUTION
Located in the Atlantic Ocean. (Specific distribution map not available.)

HABITAT
Found in the guts or flesh of fish such as cod. In their final hosts, they are found mostly in gray seals and other similar animals.

BEHAVIOR
When inside fish, they are usually found tightly coiled in the flesh and guts of fish. They are often found in considerable numbers, particularly in the belly flaps of fish, where they can remain for extended periods of time encased in a sack-like membrane produced by the fish tissue. Adults also live in the stomach of gray seals and other similar creatures. Eggs of the parasite pass into the waters with the mammal's excreta, and when the eggs hatch, the microscopic larvae must invade a new host in order to develop. Small shrimp-like crustaceans, euphausiids (often called krill), and other parasitic crustaceans eat the larval worms. When a fish eats these infested crustaceans, the larval worms are released into its stomach. They then bore through the stomach wall and eventually become encased in the guts or in the flesh of the host fish. The lifecycle is completed when a suitable marine mammal eats an infested fish. The incidence of infection in fish generally increases with length, weight, and age of the fish host.

FEEDING ECOLOGY AND DIET
As a parasitic species, they live off of nutrients of their hosts, primarily from the guts and tissues of fishes.

REPRODUCTIVE BIOLOGY
Females have ovaries and uteri, while males have copulatory spicules.

CONSERVATION STATUS
Not listed by the IUCN.

SIGNIFICANCE TO HUMANS
The cause of human illness in countries where there is ingestion of raw or lightly cured fish. The disease is called anisakiasis and can be easily prevented because larvae are killed in only one minute at a temperature of 140°F (60°C) or higher. In the wrong host the worms get the wrong signal and migrate through the tissues of its host, causing hemmoraging and bacterial infections. This kind of migration is known as "visceral larval migranes." ◆

Resources

Books

The Audubon Society Encyclopedia of Animal Life. New York: Clarkson N. Potter, 1982.

Bird, Alan F. *The Structure of Nematodes.* New York: Academic Press, 1971.

Chitwood, B. G., and M. B. Chitwood. *Introduction to Nematology.* Baltimore: University Park Press, 1950.

Croll, Neil A., and Bernard E. Matthews. *Biology of Nematodes.* Glasgow and London: Blackie and Son Limited, 1977.

Levin, Simon Asher, ed. *Encyclopedia of Biodiversity.* San Diego: Academic Press, 2001.

Mehlhorn, Heinz, ed. *Encyclopedic Reference of Parasitology: Diseases, Treatment, Therapy.* 2nd edition. New York: Springer, 2001.

Maggenti, Armand. *General Nematology.* New York: Springer-Verlag, 1981.

Malakhov, V. V. (translated by George V. Bentz, edited by W. Duane Hope). *Nematodes: Structure, Development, Classification, and Phylogeny.* Washington, DC, and London: Smithsonian Institution Press, 1994.

The New Larousse Encyclopedia of Animal Life. New York: Bonanza Books, 1981.

Parker, Sybil P., ed. *Synopsis and Classification of Living Organisms.* New York: McGraw-Hill Book Company, 1982.

Poinar, George O., Jr. *The Natural History of Nematodes.* Englewood Cliffs, NJ: Prentice-Hall, Inc., 1983.

Stone, A. R., H. M. Platt, and L. F. Khalil, eds. *Concepts in Nematode Systematics.* Special volume no. 22. London: Academic Press, 1983.

Wharton, David A. *A Functional Biology of Nematodes.* London: Croom Helm, 1986.

William Arthur Atkins

Nematomorpha
(Hair worms)

Phylum Nematomorpha
Number of families 2

Thumbnail description
Parasitic worms as juveniles in marine or terrestrial arthropods, free-living as adults

Photo: *Paragordius varius* worms emerging from *Gryllus firmus* cricket. The worms are undergoing the transition from a parasitic life cycle within the host, to a free-living one. (Photo by Ben Hanelt. Reproduced by permission.)

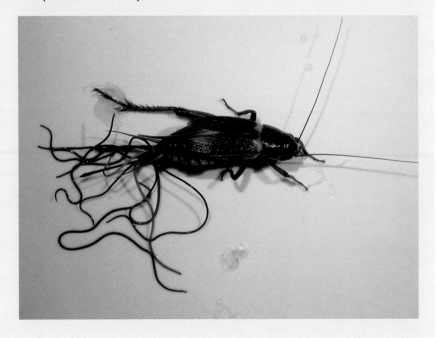

Evolution and systematics

The Nematomorpha consists of two clades: the class Nectonematoida and the class Gordiida. The nectonematids are parasites of marine crustaceans such as crabs and shrimp. The gordiids are usually parasites of terrestrial arthropods such as crickets, grasshoppers, beetles, mantids, and cockroaches. Based on molecular evidence, the phylum Nematomorpha has been shown to be the sister group to nematodes. The phylum contains two orders, two families, and two genera—all corresponding to the two clades. Overall, the phylum contains approximately 230 species.

Fossil gordiids have been found emerging from a cockroach trapped in fossilized amber dated at 15–45 million years ago. However, it has been suggested that this group might date back to the Carboniferous.

Physical characteristics

Nematomorphs are long, thin, cylindrical worms. Their shape leads many to refer to these worms as hair worms. Gordiid adults can be from 2–118 in (5–300 cm) long and 0.02–0.40 in (0.5–10 mm) thick. The color of gordiids ranges from black and brown to yellow and white depending on sex, and species. Most have a white anterior tip immediately followed by a thin dark band or collar. Nectonematid adults can be up to 11.8 in (300 mm) long and 0.07 in (1.7 mm) in diameter. Most gordiids contain surface ornamentations, areoles, which are made up of raised bumps. The color of nectonematids ranges from grayish white to yellow. Most nematomorph males can be distinguished from females by having a slight inward curving posture of their posterior ends.

Distribution

Gordiids have been recorded from every continent except Antarctica. Nectonematids have been reported from oceans around the globe, including the shores of North America, South America, Europe, Japan, New Zealand, and Indonesia.

Habitat

Nematomorphs are parasites as juveniles but free-living as adults. Adult gordiids are usually found in slow-moving freshwater streams or ponds. In streams, worms are either attached to vegetation hanging over the banks or in between rocks on the bottom. Worms attach by winding their muscular bodies into tight coils. Gordiids have also been recorded from larger rivers such as the Mississippi River and lakes such as the Great Lakes. The habitat of adult nectonematids is largely unknown. Adults have often been found at the surface in the littoral zone, but have also been dredged from the seafloor several hundred feet (meters) deep.

Behavior

Gordiids tend to entangle in large knots during mating. Often, hundreds of individuals can be found in a seemingly

Horsehair worm. (Photo by A. Captain/R. Kulkarni/S. Thakur. Reproduced by permission.)

undoable tangle. This behavior has led the gordiids to be called Gordian worms from the Greek myth of Gordius. The behavior of nectonematids has not been studied.

Feeding ecology and diet

Adult nematomorphs do not feed. The gut is greatly reduced and in adults is non-functional. In some adults, the mouth may be closed by skin, while in others, parts of the gut are missing. Worms get all of their energy from their hosts. Although the exact mechanism has not yet been resolved, juvenile worms are able to move host nutrients through their cuticle and into their gut.

Reproductive biology

Adult gordiids found in temperate climates appear in late spring or summer. In some species, mating is immediate, while in others mating can be delayed for as long as several months after emergence. Males wrap around the female with their posterior end and glides along the female surface until he reaches her cloaca. The male then deposits a drop of sperm onto the posterior end of the female, which usually covers the tip of the female. Within a month, females lay up to 6 million eggs, soon after which they die. Larvae hatch from eggs

within 20 days, and penetrate and encyst in aquatic organisms. Cysts within aquatic insects are carried to land when the insect metamorphoses into a fly. Arthropod hosts are infected upon eating a fly containing cysts. The reproductive biology of tropical gordiids or nectonematids is unknown.

Conservation status

No species of nematomorphs are listed by the IUCN. Some evidence indicates that gordiid populations are not affected by human modifications of the landscape.

Significance to humans

Although several hundred reports exist of humans supposedly infected with gordiids, humans do not serve as hosts for these worms. Most, if not all, of these cases are due to incidental associations. Worms have been noted from the human digestive tract by being spit up or passed through the intestine. These worms are likely to have been swallowed as adults. No evidence exists that these worms are able to live within a human for an extended period. It is also likely that worms discovered in toilets or chamber pots were present in these vessels before use, carried in by insect hosts. Nectonematids do not appear to have immediate importance to humans.

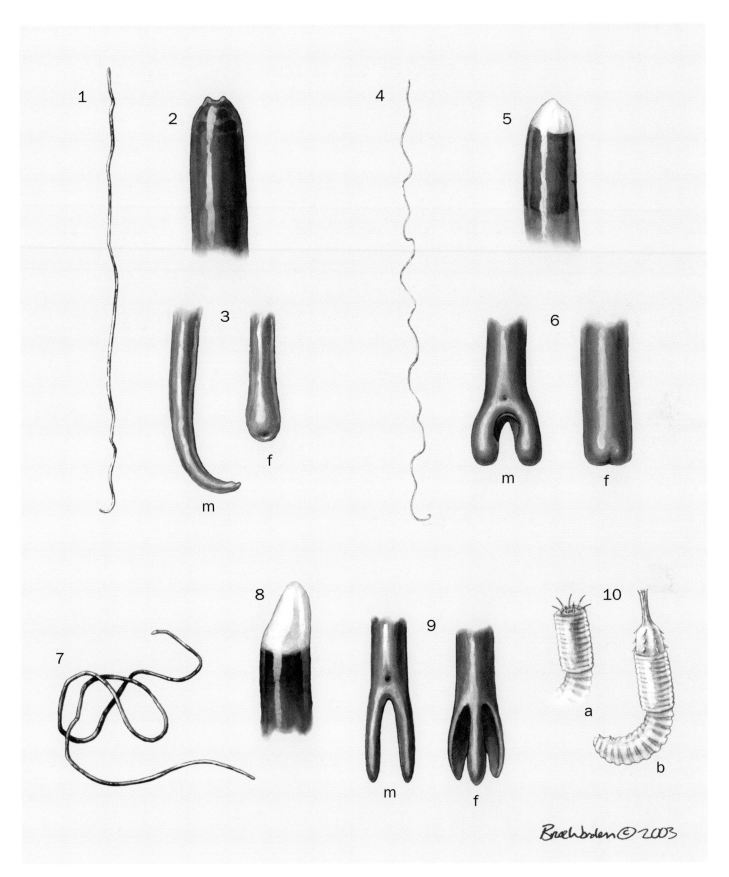

1. Adult *Nectonema agile*; 2. *N. agile* anterior end; *3. N. agile* male and female posterior ends; 4. Adult *Gordius aquaticus*; 5. *G. aquaticus* anterior end; 6. *G. aquaticus* male and female posterior ends; 7. Adult *Paragordius varius*; 8. *P. varius* anterior end; 9. *P. varius* male and female posterior ends; 10. *P. varius* larva; a. proboscis contracted inside the body; b. proboscis extended outward. (Illustration by Bruce Worden)

Species accounts

No common name
Gordius aquaticus

ORDER
Gordioidea

FAMILY
Gordiidae

TAXONOMY
Gordius aquaticus Linnaeus, 1758, Europe.

OTHER COMMON NAMES
None known.

PHYSICAL CHARACTERISTICS
Body ranges from light to dark brown. Body length ranges from 4.7–19.3 in (120–490 mm) with a maximum diameter of 0.02 in (0.6 mm). Male posterior is bifurcating, female posterior entire. Male has postcloacal crescent, which is a parabolic fold of the cuticle just past the cloaca on the posterior end. Areoles are completely lacking.

DISTRIBUTION
Found throughout Europe. In the south, they have been reported from southern France to Turkey. In the north, they have been reported from Belgium to Finland.

HABITAT
Adults are free-living in freshwater environments. This species is often collected from ponds and from slower streams and temporary waters such as wheel ruts or puddles filled with rainwater. Juveniles develop within coleopteran hosts such as ground beetles.

BEHAVIOR
The behavior of this group has not been intensively studied.

FEEDING ECOLOGY AND DIET
Parasitic during larval stage, non-feeding as adult.

REPRODUCTIVE BIOLOGY
These worms reproduce once per year. Worms either over winter within coleopteran hosts or as free-living adults in sediments or leaf litter. Adult worms emerge from hosts during late summer or early fall, and begin mating. Egg production begins about a month after copulation, and can last as long as four weeks. Larvae hatch from eggs and enter aquatic insect larvae such as midges or mayflies. Within these insects, parasites form cysts able to survive insect metamorphosis to adult flies. Flying insects, carrying cysts, are eaten by beetle hosts, completing the life cycle. Development to adult worms in beetle hosts may take up to three months.

CONSERVATION STATUS
Not threatened.

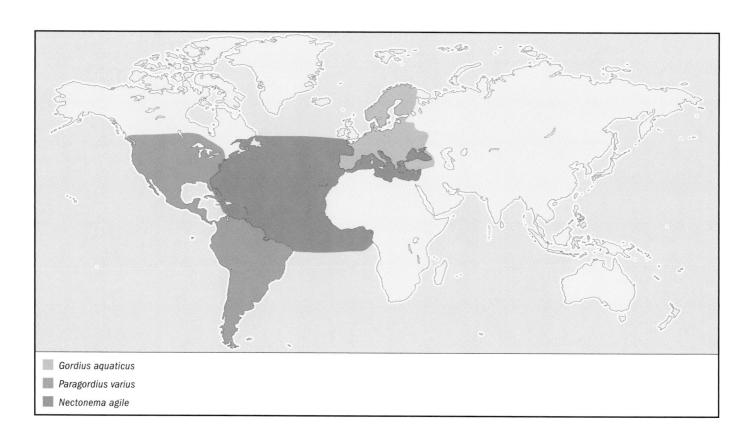

Gordius aquaticus
Paragordius varius
Nectonema agile

SIGNIFICANCE TO HUMANS

As with all gordiids, this species does not infect humans. However, the accidental expulsion of this worm from its insect host in domestic water supplies, toilets, pools, and livestock watering tanks has caused much unwarranted distress. ◆

No common name
Paragordius varius

ORDER
Gordioidea

FAMILY
Gordiidae

TAXONOMY
Paragordius varius Leidy, 1851, type locality unknown, although it was likely on the East Coast of the United States.

OTHER COMMON NAMES
None known.

PHYSICAL CHARACTERISTICS
Body ranges from light yellow to nearly black. Body length ranges from 3.9–13.8 in (100–350 mm) with a maximum diameter of 0.03 in (0.7 mm). Male posterior is bifurcating, female posterior trifurcating. Male lacks postcloacal crescent. Only one kind of areoles are present, which are small and flat.

DISTRIBUTION
Found throughout the Americas. They have been reported in North America from Canada and the United States; in Central America, they have been reported from Costa Rica and Guatemala; and in South America from countries including Columbia, Brazil, and Argentina. Within the United States, they have been reported from 26 states, including Hawaii. They have also been reported from Cuba.

HABITAT
Adults are free-living in freshwater environments. This species is often collected in slower streams and has often been encountered in temporary waters such as rain puddles and many places where rainwater collects. Juveniles develop within orthopteran hosts such as crickets and grasshoppers.

BEHAVIOR
The behavior of this group has not been intensively studied.

FEEDING ECOLOGY AND DIET
Parasitic during larval stage, non-feeding as adult.

REPRODUCTIVE BIOLOGY
These worms produce several generations annually. Worms most likely overwinter as cysts in aquatic insects. During the spring, aquatic insects, carrying cysts, metamorphose into flying adults. Crickets and grasshoppers are infected when they scavenge dead insects harboring cysts. Development to adult worms takes up to one month. The fast development of this species allows up to three generations to be produced during a single year in temperate climates. The reproductive biology of this group in more tropical regions is unknown.

CONSERVATION STATUS
Not threatened.

SIGNIFICANCE TO HUMANS
As with all gordiids, this species does not infect humans. However, this species is often encountered by humans in the United States. Such findings of adult worms in a pet's water dish, puddles, toilets, hot tubs, or even near childrens' playgrounds result in numerous calls to local health authorities annually. ◆

No common name
Nectonema agile

ORDER
Nectonematoidea

FAMILY
Nectonematidae

TAXONOMY
Nectonema agile Verill, 1873, Woods Hole, Massachusetts, United States.

OTHER COMMON NAMES
None known.

PHYSICAL CHARACTERISTICS
Body ranges from grayish white to pale cream. Semi-transparent anterior tips. Body length ranges from 1.2–7.9 in (32–200 mm) with a maximum diameter of 0.03 in (0.75 mm). Both male and females ends entire. Sensory bristles in two rows cover the length of the adult body.

DISTRIBUTION
Found sporadically throughout the world. The locations include the eastern seaboard of North America, Brazil, the Mediterranean, and the Black Sea.

HABITAT
Adults are free-living in marine environments. This species is often collected in intertidal areas, especially at night in the glare of artificial light. Juveniles are parasites of decapod crustaceans such as crabs and shrimp.

BEHAVIOR
Nothing is known.

FEEDING ECOLOGY AND DIET
Parasitic during larval stage, non-feeding as adult.

REPRODUCTIVE BIOLOGY
Nothing is known.

CONSERVATION STATUS
Not threatened.

SIGNIFICANCE TO HUMANS
None known. ◆

Resources

Books

Poinar, George O. "Nematoda and Nematomorpha." In *Ecology and Classification of North American Freshwater Invertebrates*, edited by James H. Thorp and Alan P. Covich. San Diego: Academic Press, 2001.

Schmidt-Rhaesa, Andreas. *Süsswasserfauna von Mitteleuropa: Nematomorpha*. Stuttgart, Germany: Gustav Fisher Verlag, 1997.

Periodicals

Hanelt, B., and J. Janovy Jr. "Untying a Gordian Knot: The Domestication and Laboratory Maintenance of a Gordian Worm. *Paragordius varius* (Nematomorpha: Gordiida)." *Journal of Natural History* In press, 2003.

Poinar, G., Jr., and A. M. Brockerhoff. "*Nectonema zealandica* n. sp. (Nematomorpha: Nectonematoidea) Parasitising the Purple Rock Crab *Hemigrapsus edwardsi* (Brachyura: Decapoda) in New Zealand, with Notes on the Prevalence of Infection and Host Defense Reactions." *Systematic Parasitology* 50, no. 2 (2001): 149–157.

Schmidt-Rhaesa, A. "Phylogenetic Relationships of the Nematomorpha: A discussion of Current Hypothesis." *Zoologischer Anzeiger* 236 (1998): 203–216.

Schmidt-Rhaesa, A. "The Life Cycle of Horsehair Worms (Nematomorpha)." *Acta Parasitologica* 46, no. 3 (2001): 151–158.

Ben Hanelt, PhD

Acanthocephala

(Thorny headed worms)

Phylum Acanthocephala
Number of families 22

Thumbnail description
Parasitic thorny headed worms with complex life cycles; sexes separated; adults found in intestines of vertebrates (definitive host), larvae found in hemocoel (body cavity) of arthropods (intermediate hosts) and sometimes in body cavities of vertebrates (paratenic or transport hosts)

Photo: Scanning electron micrograph of the proboscis of *Hypoechinorhynchus thermaceri*, taken from the eelpout fish *Thermarces andersoni*, found near deep sea hydrothermal vents. (Photo by Isaure de Buron. Reproduced by permission.)

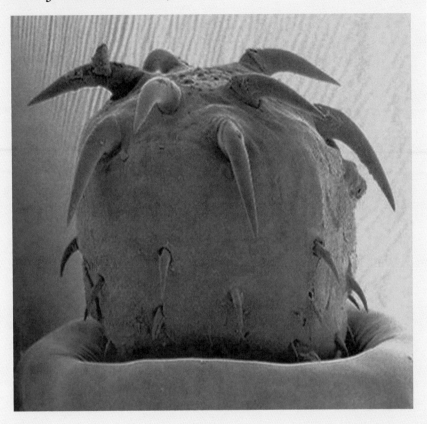

Evolution and systematics

As with most soft-bodied parasites, no fossil record of acanthocephalans is known. However, prehistoric human coprolites at archeological sites in the United States and Brazil have revealed infections by acanthocephalans. Additionally, some 9,000-year-old animal coprolites also from Brazil have been found to contain acanthocephalan eggs. The phylum Acanthocephala is divided into 3 major taxonomic groups: the Archiacanthocephala, Eoacanthocephala, and Palaeacanthocephala, which are considered by some to be classes and others to be orders. A fourth group, the Polyacanthocephala, has been proposed but its status is controversial. Taxonomic groups are based on morphological characters of the worms as well as their hosts' taxonomy and ecology, and such division is supported by molecular data. Molecular, morphological, and ultrastructural analysis of 18S ribosomal DNA sequences has revealed that acanthocephalans and their closest living relatives, members of the phylum Retifera, should be in one clade—referred to as the Syndermata. Acanthocephalans include 22 families and about 1,000 species.

Physical characteristics

Adults are cylindrical or slightly flattened worms that are usually white or colorless; however, some species may be yellow, brown, red, or orange. Acanthocephalans are never segmented, although some species exhibit superficial pseudosegmentation. As adults acanthocephalans measure from less than an inch (a few millimeters) to more than 24 in (60 cm) in length, with the archiacanthocephalans being the largest ones. These worms are sexually dimorphic with females usually larger than males. Structurally, the worms may be divided into three body regions: a proboscis, a neck, and a trunk. The proboscis harbors hooks that may be arranged either in rows or longitudinal lines depending upon the species. Some species harbor an apical organ at the tip of the proboscis that is presumably sensory. The proboscis invaginates into a proboscis receptacle hanging into the anterior part of the trunk. The neck is unarmed but may show lateral organs that may be involved in sensory perception. The trunk may or may not be armed with spines whose distribution is an important criterion for species identification. Acanthocephalans are pseudocoelomate with a syncitial tegument within which runs a lacunar system (an interconnected fluid-filled network of cavities). These worms also have hollow tubular muscles and lemnisci (sac-like structures hanging from the base of the neck into the pseudocoel) that are connected to the proboscis lacunar system. Unique features of acanthocephalan genitalia are a uterine bell in females and cement glands, the organ of Saefftigen, and the copulatory bursa in males. Both males and females have a cerebral ganglion in the proboscis receptacle, and males have genital ganglia and a bursal ganglion. All internal organs are derived from a ligament running down the center of the pseudocoel.

Distribution

Acanthocephalans are found throughout the world, including in fish at deep-sea hydrothermal vents.

Habitat

Adults are intestinal parasites of mammals, birds, fishes, amphibians, and reptiles. Larvae develop in the hemocoel of mandibulate arthropods (crustaceans, myriapods, and insects).

Behavior

Acanthocephalans usually occupy precise niches within the intestines of their definitive hosts. However, some species have been shown to migrate along the intestinal tract during the term of infection. Such migration is correlated with both host diet and sexual maturity of the worms. Nothing is known of acanthocephalan behavior and communication, and there is no evidence of chemical attractants being released to assist in mates finding each other within the hosts' digestive tract. However, acanthocephalans are known to modify the behavior of both their definitive host (e.g., to induce giddiness) and intermediate hosts (e.g., to induce positive phototropism or decrease/alter intermediate host evasive responsiveness). Further, some species also selectively alter the coloration of their intermediate hosts. Such alterations are known to favor transmission of acanthocephalan infective stages to their definitive hosts via increasing intermediate host susceptibility to predation. Disruption of definitive host behavior is speculated to affect transmission by altering host distribution and selection of habitat. Because acanthocephalans attach themselves to the intestinal wall of their host, they may induce pathology, such as inflammation of the surrounding tissues, perforation of the intestinal wall, peritonitis, enlargement of the intestine, and edema, in their host. Results of infection may be fatal while other cases may appear to be mild; oftentimes, more numerous the worms, the more serious the infection. Further, some acanthocephalans have been shown to disrupt host digestion and energy metabolism, which has been shown to have serious detrimental effects during periods of host stress.

Feeding ecology and diet

Acanthocephalans have no digestive tract but selectively absorb nutrients from the host's intestine across their tegument. However, knowledge in this area is limited and is based on only a few species whose feeding and metabolism have been studied (see below *Moniliformis moniliformis*). The major substrate for acanthocephalan metabolism is carbohydrate, with ethanol being the main end product. Uptake of monosaccharides appears to involve active transport mechanisms. Some species may store glucose in the proboscis re-

ceptacle muscle, and intense labeling of the cytoplasmic core of the hollow muscles following uptake of radiolabled glucose occurs. However, it has not been reported whether this latter area acts as a storage site, or whether it plays a role in distributing nutrients throughout the body. The plasma membrane at the surface of the tegument shows hydrolytic activity and tegumental surface crypts within which various enzymatic activities have been localized. These crypts increase the absorptive surface area and are considered to be extra-cytoplasmic digestive organelles. Various amino acids are known to be absorbed through the tegumental surface, but their role in metabolism is not clear. Routes of absorption also vary according to the amino acid studied. Lipids are absorbed and then stored in the lemnisci, although controversy exists as to whether the primary site of lipid absorption is the body wall or the neck/proboscis region. While large amounts of lipids may be deposited in acanthocephalans, they are not thought to be used in metabolism.

Reproductive biology

Female and male acanthocephalans copulate in the intestine of their vertebrate definitive hosts. Fertilization is internal and it is thought that males initiate copulation. The male bursa, the spines that both males and females may harbor at the very posterior end of their trunks, as well as the "cement" (mucilaginous and proteinaceous material) discharged from the male's cement gland(s), all appear to play a role in strengthening the copulatory union. Cement plugs/caps are often observed at the posterior extremities of females, although the role of these plugs is still under discussion. The favored hypothesis is that the cap, which lasts a few days, prevents the loss of injected sperm and further prevents sperm from competing males to enter the female. Cement plugs are also often observed on males, which may serve to prevent male competitors from copulating.

Not all acanthocephalans show seasonality in their life cycle, but there are many instances in which they do. When seasonal life cycles exist there is often a direct link to change in host diet, water temperature, and/or the presence of intermediate hosts in the environment. Maturation of acanthocephalans is sometimes correlated to the maturation of their definitive hosts. Females have ovaries that break up into ovarian balls, which in turn produce oocytes. Once fertilized oocytes become eggs, they float freely in the pseudocoel while the larvae within them develops into an acanthor. During larval maturation eggs are sorted via the uterine bell, which returns immature eggs to the pseudocoel while directing mature eggs containing acanthors to the uterus where they are stored until release. Mature eggs are released via a gonopore into the lumen of the host's intestine and are excreted with the feces. Once outside the host, the eggs must be consumed by the intermediate host to assure transmission. The eggs of acanthocephalans are the only free-living stages of the parasites. Eggs have four envelopes (exceptions exist) that are separated by interstices. Although all analyzed acanthocephalan eggs contain keratin in their second shell, other chemical structures exist and differences among palae-, archi-, and eoacanthocephalan shell structure probably reflect differences in the nature of the intermediate hosts among the three groups. Eggs contain the acanthor, which is the infective stage of the parasite for the intermediate hosts. Acanthors harbor at their anterior end a boring structure called the aclid organ, which consists of a pair of "blades." Acanthors of most species are covered with spines that decrease in size posteriorly. Acanthors of some species also exhibit hooks on their anterior part. Once the intermediate host ingests the egg, the acanthor hatches and uses the aclid organ to penetrate the intestinal wall and then develops, often first beneath the intestinal serosa and then in the hemocoel. Development generally lasts several weeks as the acanthor first transforms into the acanthella and then into a cystacanth, which is the infective stage for the definitive host.

With the exception of size and sexual maturity, cystacanths show all the morphological features of an adult. A thin membrane whose origin is still under discussion surrounds the cystacanths of most species and may serve to protect against the invertebrate host's immuno-defense system. Once intermediate hosts carrying cystacanths are ingested by the correct definitive host, the life cycle is completed. Some cycles include paratenic (transport) hosts, which are vertebrates that ingest infected intermediate hosts and within which the larvae do not develop further. Paratenic hosts oftentimes accumulate large numbers of infective cystacanths and may be required to assure that the parasites reach a definitive host that is higher in the food web. In paratenic hosts the cystacanths most often penetrate the intestinal wall and stay in the body cavity. Within these hosts, accumulated cystacanths are usually attached to the intestinal mesenteries, awaiting paratenic host ingestion by an appropriate definitive host. Postcyclic para-sitism may also occur when predators of a definitive host ingest adult acanthocephalans and in turn become parasitized themselves. As such, acanthocephalans may be transferred via cannibalism within a definitive host population.

Conservation status

Little is known about the status of most acanthocephalans. Declines, occurrences, or commonality would all be linked directly to the status of the life cycle, i.e., to the presence of all hosts, and can thus be adversely affected by habitat loss/disruption.

Significance to humans

Very few species of acanthocephalans are known to induce acanthocephaliasis in humans. Symptoms such as giddiness, acute abdominal pain, tinnitus, edema, constipation, diarrhea, undernutrition, and underdevelopment have been reported. The disease is fairly rare (only several hundred cases reported) but may be fatal. Humans obtain the parasite by ingesting infected intermediate hosts either accidentally, as part of their regular diet, or for medicinal purposes. In the former case children are most often affected. Eating sashimi (or raw fish in general) may also be a way for humans to become infected with acanthocephalans. Acanthocephalans, particularly those parasitizing fish, are known to selectively accumulate toxic heavy metals, such as lead and cadmium, in extremely high proportion relative to their surrounding host tissues and host environment. Consequently, their potential use in monitoring polluted environments is an active avenue of research.

Species accounts

No common name
Moniliformis moniliformis

ORDER
Moniliformida

FAMILY
Moniliformidae

TAXONOMY
Moniliformis moniliformis (Bremser, 1811) Travassos, 1915.

OTHER COMMON NAMES
None known.

PHYSICAL CHARACTERISTICS
Worm filiform, often coiled, with distinct pseudosegmentation and bead-like appearance when mature. Females: 4–11 in (10–27 cm) long and 0.08 in (2 mm) maximum in width; males: 1.6–2 in (4–5 cm) long. Trunk unarmed. Proboscis nearly cylindrical. Twelve longitudinal rows of 7–8 hooks. Eight cement glands.

DISTRIBUTION
Cosmopolitan.

HABITAT
Definitive hosts: numerous wild rodents, particularly rats, dogs, and cats. Intermediate hosts: beetles and cockroaches. Paratenic hosts: toads and lizards.

BEHAVIOR
A few weeks post infection: migration of worms from posterior part to anterior half of intestine of definitive host. Optimal attachment site influenced by sugar gradient in host intestine. Several, but not all, species of roaches containing larvae exhibit altered behavior (e.g., decreased evasive responsivness). Altered intermediate host behavior likely increases probability of parasite transfer to definitive host.

FEEDING ECOLOGY AND DIET
Major sources of energy are host dietary carbohydrates. Mannose, glucose, fructose, galactose, and starch all known to influence worm growth, longevity, and reproduction. Absorption across tegumental surface. Primary sites of absorption are extracytoplasmic crypts opening to the outside via narrow necks and pores. Pinocytotic and enzymatic activity observed within crypts. Cystacanths store large amounts of glycogen that likely act as an energy source during activation and establishment in the definitive host. Little known about lipid metabolism: available evidence indicates larval stages accumulate lipids. Adult dispersion

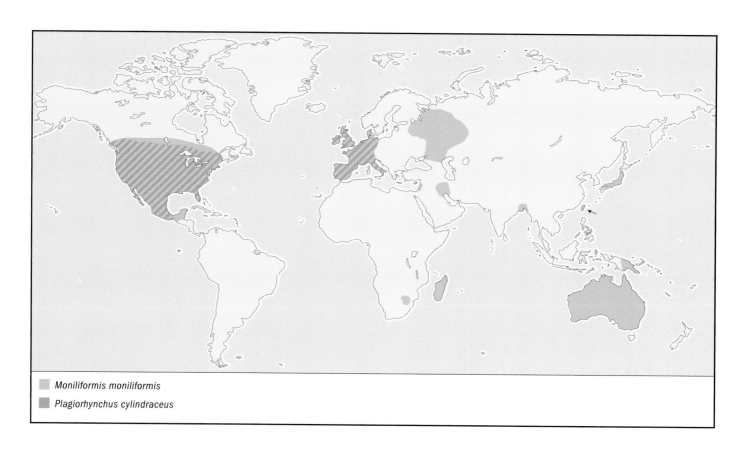

■ *Moniliformis moniliformis*
■ *Plagiorhynchus cylindraceus*

in host intestine influenced by nature of lipids in host diet. Specific amino acids readily taken up and catabolized; role in energy metabolism still unclear. Site of absorption of amino acids unknown. Both uridine and thymine transported from host lumen but nucleoside transport mechanism not known.

REPRODUCTIVE BIOLOGY

Adults mature in 5–6 weeks in intestine of definitive host. Hatching of acanthor occurs between 15 minutes and 48 hours post ingestion by intermediate host. Larvae develop into cystacanth in adult roach in about two months at 81°F (27°C).

CONSERVATION STATUS

Not threatened. Most likely flourishing because of widespread distribution and abundance of hosts.

SIGNIFICANCE TO HUMANS

Human pathogen. Symptoms include fatigue, tinnitus, and diarrhea. ◆

Giant thorny-headed worm

Macracanthorhynchus hirudinaceus

ORDER

Oligacanthorhynchida

FAMILY

Oligacanthorhynchidae

TAXONOMY

Macracanthorhynchus hirudinaceus (Pallas, 1781) Travassos, 1917.

OTHER COMMON NAMES

None known.

PHYSICAL CHARACTERISTICS

Very large worms. Females up to 26 in (65 cm) long, 0.32–0.36 in (8–9 mm) at their largest width and ventrally curved. Males up to 4 in (10 cm) long. Body unarmed, grayish brown, with deep grooves on surface. Globular proboscis with six spiral rows of six hooks each. Eight cement glands.

DISTRIBUTION

Cosmopolitan.

HABITAT

Adults in swine *Sus scrofa* and other mammals (e.g., fox squirrel [*Sciurus niger*], eastern mole [*Scalopus aquaticus*], hyena [*Hyaena hyaena*], and dog [*Canis familiaris*]). Thirty-three species of intermediate hosts reported (e.g., the cockroach *Periplaneta americana* and scarab [*Polyphylla rugosa*]).

BEHAVIOR

Causes serious pathology in pigs.

FEEDING ECOLOGY AND DIET

Little known. Metabolism is likely to be carbohydrate based. Uptake of amino acids via undefined transport mechanisms.

REPRODUCTIVE BIOLOGY

Immense number of eggs released by each female. Eggs remain viable for up to 3.5 years. Cold temperatures improve egg survival. Larval development in 4–5 months in the intermediate

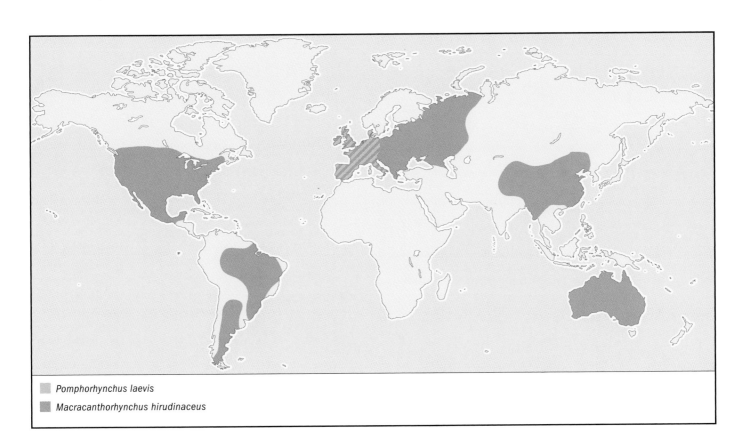

Pomphorhynchus laevis
Macracanthorhynchus hirudinaceus

host. Adult maturity reached in definitive hosts in 70–110 days. Life span: 10–23 months.

CONSERVATION STATUS
Not listed by the IUCN.

SIGNIFICANCE TO HUMANS
Most frequent agent of human acanthocephaliasis (macracanthorhynchosis or macracanthorhynchiasis). Symptoms: constipation, abdominal pain, fever, perforation of the intestinal wall. Also economical effect in countries where macracanthorhynchosis causes heavy losses in pig farms. ◆

No common name
Plagiorhynchus cylindraceus

ORDER
Polymorphida

FAMILY
Plagiorhynchidae

TAXONOMY
Plagiorhynchus (Prosthorhynchus) cylindraceus (Goeze, 1782) Schmidt and Kuntz, 1966.

OTHER COMMON NAMES
None known.

PHYSICAL CHARACTERISTICS
Small worms with elliptical, unarmed, and milky-white body. Females: 0.35–0.60 in (9–15 mm) long. Males 0.32–0.51 in (8–13 mm) long. Cylindrical proboscis with 15–18 longitudinal rows of 11–15 hooks each. Six cement glands.

DISTRIBUTION
Cosmopolitan.

HABITAT
Definitive hosts: passerine birds such as robins (*Turdus migratorius*) and starlings (*Sturnus vulgaris*), although virtually any bird can be infected. Intermediate hosts: terrestrial isopods *Armadillidium vulgare* and *Porcellio scaber*. Paratenic hosts: crested anole (*Anolis cristatellus*) and short-tailed shrew (*Blarina brevicauda*).

BEHAVIOR
Behavior of adult worms: not known. Detrimentally affects definitive host metabolism and digestive abilities. Behavior change of infected isopods: found more on white surfaces and low humidity areas than uninfected isopods. Evidence indicates such altered behavior makes them easier prey to bird definitive hosts.

FEEDING ECOLOGY AND DIET
Nothing is known.

REPRODUCTIVE BIOLOGY
Development in isopods takes about 60–65 days. Maturity reached within several weeks following ingestion of cystacanths by definitive host.

CONSERVATION STATUS
Not threatened. Based upon great diversity of definitive hosts, an unlikely candidate for extinction.

SIGNIFICANCE TO HUMANS
None known. Indirect effect by negatively affecting populations of passerine birds such as mountain bluebirds. ◆

No common name
Pomphorhynchus laevis

ORDER
Echinorhynchida

FAMILY
Pomphorhynchidae

TAXONOMY
Pomphorhynchus laevis (Zoega in O. F. Muller, 1776) Van Cleave, 1924 (*nec laeve*).

OTHER COMMON NAMES
None known.

PHYSICAL CHARACTERISTICS
Average-sized worms with a long and cylindrical neck. Neck dilated in its anterior part into the shape of a bulb. Females are 0.5–1.1 in (13–28 mm) long. Males: 0.24–0.63 in (6–16 mm) long. Body unarmed and most often orange. Cylindrical proboscis with 18–20 longitudinal rows of 12–13 hooks each. Short lemnisci. Two testes in tandem. Six cement glands.

DISTRIBUTION
Palaearctic.

HABITAT
Definitive hosts: numerous freshwater fishes (e.g., sharp-nosed eel (*Anguilla vulgaris*), common bream (*Abramis brama*), chub (*Leuciscus cephalus*), barbel (*Barbus barbus*), goldfish (*Carassius auratus*), etc. Intermediate hosts: amphipods: *Corophium volutator, Gammarus bergi, G. fossarum, G. lacustris, G. pulex,* and *Pontagammarus robustoides*. Fish for paratenic hosts (e.g., *Phoxinus phoxinus*).

BEHAVIOR
Not known. Host dietary carbohydrates likely the major energy source. Adults perforate all layers of intestinal wall with their proboscis and thus never change position in intestine. Infected intermediate hosts exhibit photophilic behavior. Cystacanths bright orange, making infected amphipod intermediate hosts more visible to fish predators.

FEEDING ECOLOGY AND DIET
Little known. Lipid analysis indicates neck and lemnisci function in lipid absorption and storage.

REPRODUCTIVE BIOLOGY
Larvae mature in intermediate hosts within several weeks. Gravid females in fish intestine carry immense numbers of eggs. Once released in water, spindle-shaped eggs appear to be diatom-like.

CONSERVATION STATUS
Not listed by the IUCN.

SIGNIFICANCE TO HUMANS
Not pathological to humans. Possible economic effect by affecting fingerling development in aquaculture conditions. ◆

Resources

Books

Crompton, D. W. T., and Brent B. Nickol. *Biology of the Acanthocephala*. Cambridge: Cambridge University Press, 1985.

Moore, Janice. *Parasites and the Behavior of Animals*. New York and Oxford: Oxford University Press, 2002

Muller, Ralph. *Worms and Human Diseases*. Cambridge, MA: CABI Publishing, 2002.

Neafie, Ronald C., and Aileen M. Marty. "Acanthocephaliasis." In *Pathology of Infectious Diseases*, vol.1, *Helminthiases*, edited by W. M. Meyers. Armed Forces Institute of Pathology, American Registry of Pathology, 2000.

Taraschewski, Horst. "Host-Parasite Interactions in Acanthocephala: A Morphological Approach." In *Advances in Parasitology*, vol. 46, edited by J. R. Baker, R. Muller, and D. Rollinson. San Diego: Academic Press, 2000.

Periodicals

Garcia-Varela, M., M. P. Cummings, G. Perez-Ponce de Leon, S. L. Gardner, and J. P. Laclette. "Phylogenetic Analysis Based on 18S Ribosomal RNA Gene Sequences Supports the Existence of Class Polyacanthocephala (Acanthocephala)." *Molecular Phylogenetics and Evolution* 23 (2002): 288–292.

Garcia-Varela, M., G. Perez-Ponce de Leon, P. de la Torre, M. P. Cummings, S. S. S. Sarma, and J. P. Laclette. "Phylogenetic Relationships of Acanthocephala Based on Analysis of 18S Ribosomal RNA Gene Sequences." *Journal of Molecular Evolution* 50: 532–540.

Golvan, Y. J. "Nomenclature of the Acanthocephala." *Research and Reviews in Parasitology* 54, no.3 (1994): 135–205.

Goncalves, M. L. C., A. Araujo, and L. F. Ferreira. "Human Intestinal Parasites in the Past: New Findings and a Review." *Memorias do Instituto Oswaldo Cruz* 98, suppl.1 (2003): 103–118.

Herlyn, H., O. Piskurek, J. Schmitz, U. Ehlers, and H. Zischler. "The Sundermatan Phylogeny and the Evolution of Acanthocephalan Endoparasitism as Inferred from 18S rDNA Sequences." *Molecular Phylogenetics and Evolution* 26 (2003): 155–164.

Isaure de Buron, PhD
Vincent A. Connors, PhD

Entoprocta
(Entoprocts)

Phylum Entoprocta (Kamptozoa)
Number of families 4

Thumbnail description
Colonial or solitary tiny benthic animals with a tentacular crown on top, and a slender stalk that attaches basally to the substratum

Photo: *Loxosomella* sp. inhabiting parapodia of a polynoid polychaete found at Noto Peninsula, Japan. (Photo by Tohru Iseto. Reproduced by permission.)

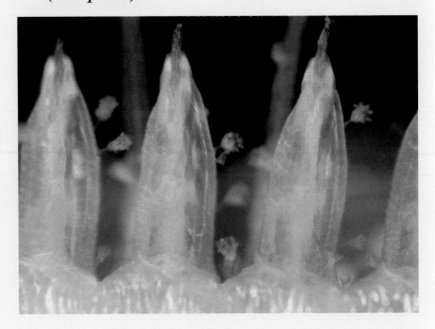

Evolution and systematics

There has been only one fossil record of Entoprocta, belonging to the extant genus *Barentsia*, which was collected from the Upper Jurassic of England. The phylogenetic relationships of Entoprocta to other invertebrate phyla are still obscure, but Entoprocta may have affinities to spiralians (animal groups that show spiral cleavage patterns). The phylum encompasses two orders, four families, sixteen genera, and approximately 170 species. The four families are: Barentsiidae, colonial species with a muscular swelling at the base of the stalk; Pedicellinidae, colonial species without basal muscular swelling, each zooid of a colony interrupted by stolon; Loxokalypodidae, colonial species without basal muscular swelling, component zooids of a colony erect from a common basal plate, not interrupted by stolons; and Loxosomatidae, which encompasses all solitary species.

The phylum name Entoprocta means "inside anus;" the phylum has this name because of its unique plan. The animal's anus opens inside its tentacular crown. Kamptozoa is another scientific name for this phylum; the name means "bending animal," and comes from the very active movement of these animals.

Physical characteristics

The calyx, or main body, of an entoproct contains a U-shaped gut, a ganglion, a pair of gonads, a pair of protonephridia, and has a tentacular crown on top. Both the mouth and anus open inside the tentacular crown. The calyx is supported by a slender stalk that attaches basally to the substratum. In colonial species, zooids of a colony are generally connected by a highly branched stolon that creeps over the substratum. The solitary species have an attaching organ at the base of the stalk. In some species, however, the attaching organ degenerates in the adult stage and the adult animals are cemented onto the substratum.

Distribution

Entoprocts have been reported from tropical, temperate, and polar marine waters, and from shallow seashore to deep seas of more than 1,640 ft (500 m). One colonial species, *Urnatella gracilis*, occurs worldwide in inland waters.

Habitat

Colonial species live on a wide variety of substrata, including rocks, stones, shell remains, human-made objects, and occasionally on other animals. Most solitary species have been known to live on the bodies of specific host animals, such as polychaetes, bryozoans, sponges, and sipunculans, and on the inner side of the tube of polychaetes.

Behavior

In response to irritation, entoprocts contract their tentacles and bend at the stalk. Some solitary species can glide over the substratum as slugs do. One solitary species (*Loxosoma agile*) somersaults across the substratum, and another species (*Loxosomella bifida*) can walk on the substratum similar to the way humans do, using a unique foot with two elongated, leg-like extensions. Newly liberated buds of solitary species often swim using ciliary tentacles, contributing to the dispersal of the species.

Loxomitra kefersteinii crawling. (Illustration by Emily Damstra)

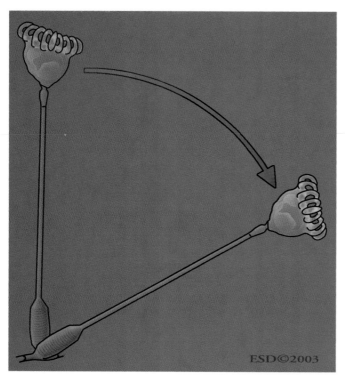

Barentsia discreta bending. (Illustration by Emily Damstra)

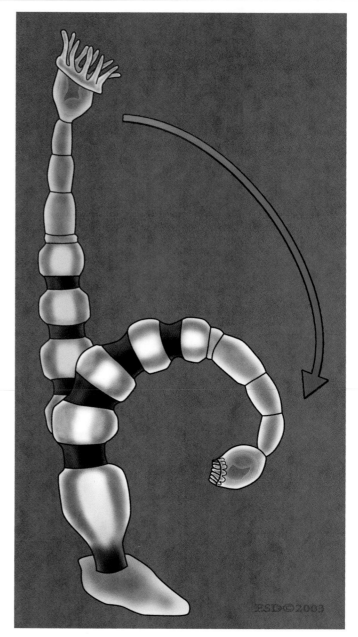

Urnatella gracilis bending. (Illustration by Emily Damstra)

Feeding ecology and diet

All entoprocts are suspension feeders, feeding on phytoplankton or other organic particles in a water current they create using the cilia along their tentacles.

Reproductive biology

Each zooid of a colonial species is generally dioecious, male or female, but both sexes occur in a single colony. Solitary species are generally protandrous hermaphrodites, namely, animals are males in the early stage but later convert into females. Eggs are fertilized in the ovary and transferred to a brood pouch, a deep depression between the mouth and the anus, where embryos develop to trochophorelike larvae. Asexual reproduction (budding) is vigorous in all entoprocts. Buds occur from the tips of developing stolon, or from a basal disc and stalks in colonial species. In solitary species budding usually occurs at two latero-frontal areas of the calyx.

Conservation status

Entoprocts may be common in worldwide seas. However, their distribution and abundance are still poorly documented, and their responses to human activities have not been monitored. No species is listed by the IUCN.

Significance to humans

Entoprocts have no significance to humans.

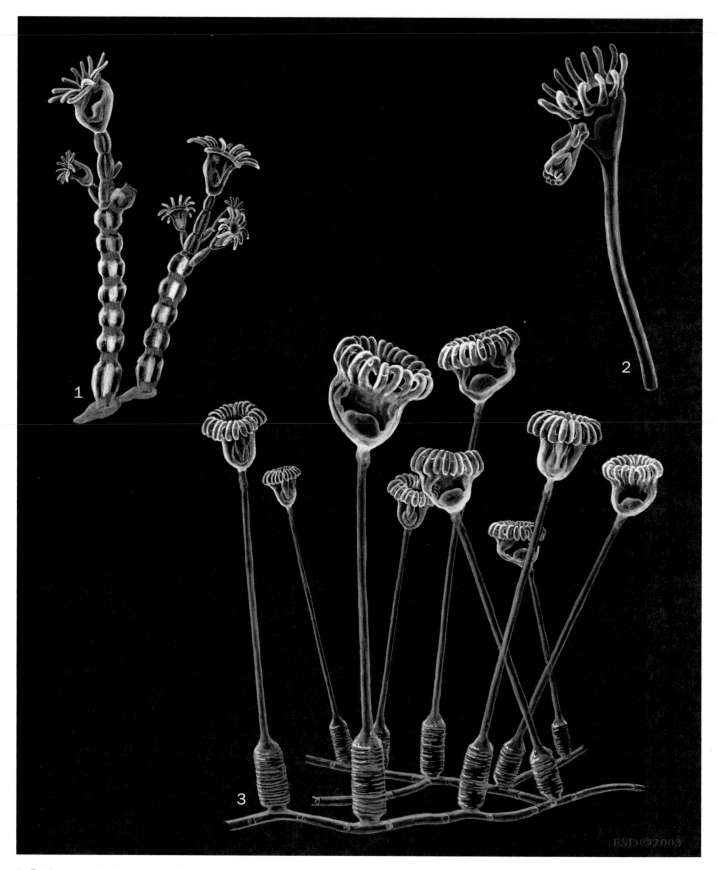

1. Freshwater colonial entoproct (*Urnatella gracilis*); 2. Solitary entoproct (*Loxomitra kefersteinii*); 3. Marine colonial entoproct (*Barentsia discreta*). (Illustration by Emily Damstra)

Species accounts

Marine colonial entoproct
Barentsia discreta

ORDER
Coloniales

FAMILY
Barentsiidae

TAXONOMY
Ascopodaria discreta Busk, 1886, Nightingale Island, Tristan de Cunha.

OTHER COMMON NAMES
None known.

PHYSICAL CHARACTERISTICS
Total length of each zooid up to 0.4 in (9.5 mm), usually 0.1–0.2 in (3–6 mm). Tentacles number about 20. Stalk is thin and long, about three to eight times longer than calyx, with many tiny pores, and a muscular swelling at base of stalk. Each zooid of a colony is interconnected by a stolon.

DISTRIBUTION
Cosmopolitan, but not in northern Europe. Ranges from shallow coastal zone to deep seas of more than 1,640 ft (500 m).

HABITAT
On any nonliving substrata, including rocks, stones, and dock pilings, as well as on living substrata, such as worm tubes.

BEHAVIOR
In response to a disturbance, it bends from basal muscular swelling, but the stalk itself does not curve. Action of one individual leads actions of surrounding zooids.

FEEDING ECOLOGY AND DIET
Suspension feeder. Feeds on phytoplankton and organic particles.

REPRODUCTIVE BIOLOGY
Buds at tips of developing stolon. Single colony contains both male and female zooids. Embryos brooded at brood pouch. Larva trochophorelike.

CONSERVATION STATUS
Not listed by the IUCN.

SIGNIFICANCE TO HUMANS
None known. ◆

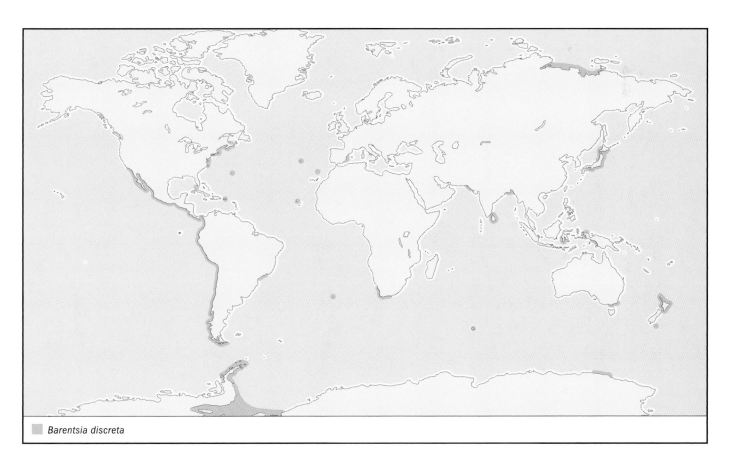

Barentsia discreta

Freshwater colonial entoproct
Urnatella gracilis

ORDER
Coloniales

FAMILY
Barentsiidae

TAXONOMY
Urnatella gracilis Leidy, 1951, Schuylkill River, Pennsylvania, United States.

OTHER COMMON NAMES
None known.

PHYSICAL CHARACTERISTICS
Height of the colony is 0.08–0.12 in (2–3 mm), and each zooid has 12–16 tentacles. In addition to a basal muscular swelling common to Barentsiidae species, numerous muscular swellings are present throughout the stalk. One or two original stalks erect from a basal plate, successive stalks branch from older stalks. Calyx at tip of each branch of stalk. No interrupting stolons.

DISTRIBUTION
Belgium, Germany, Hungary, Romania, Russia (Don River), India, Japan, Africa (Nile River, Congo River, and Lake Tanganyika), United States, South America (Parana and Uruguay Rivers).

HABITAT
In freshwater on stones, twigs, and remains of shells.

BEHAVIOR
In response to a disturbance, it bends from muscular swellings of the stalks. Fragments of colony with two to three young zooids detach from the colony, drift in water current, or creep on substratum. Eventually, they fix on a favorable habitat and generate a new colony.

FEEDING ECOLOGY AND DIET
Suspension feeder. Feeds on phytoplankton and organic particles.

REPRODUCTIVE BIOLOGY
Buds at basal plate and stalks. Larva trochophorelike. Sexual reproduction is very rare. Calyx often degenerates in low temperature or in any insufficient condition, but regenerates in favorable conditions.

CONSERVATION STATUS
Originally found in the United States. Current worldwide distribution may be due to human activities but the transfer mechanism is still unknown. Not listed by the IUCN.

SIGNIFICANCE TO HUMANS
None known. ◆

☐ *Urnatella gracilis*

■ *Loxomitra kefersteinii*

Solitary entoproct
Loxomitra kefersteinii

ORDER
Solitaria

FAMILY
Loxosomatidae

TAXONOMY
Loxosoma kefersteinii Claparède, 1867, Naples, Italy.

OTHER COMMON NAMES
None known.

PHYSICAL CHARACTERISTICS
Total length up to 0.06 in (1.5 mm). Tentacles number 10–14. Stalk slender, about three times longer than calyx. Bud has tiny attaching organ at base of stalk, but adult loses it and fixes onto substratum.

DISTRIBUTION
South Wales, Great Britain; Naples, Italy; Red Sea; Florida, United States.

HABITAT
Solitary species. Lives on substrata such as body surface of other animals, settlement panels, and water pipes.

BEHAVIOR
In response to a disturbance, it bends along slender, flexible stalk. Newly liberated bud lays its body horizontally and crawls on substratum, twisting the whole body in search of a favorable point to attach to.

FEEDING ECOLOGY AND DIET
Suspension feeder. Feeds on phytoplankton and organic particles.

REPRODUCTIVE BIOLOGY
Budding occurs at latero-frontal areas of calyx. Up to 12 buds on a single animal. Sexual reproduction has not been observed.

CONSERVATION STATUS
Not listed by the IUCN.

SIGNIFICANCE TO HUMANS
None known. ◆

Resources

Books

Hyman, L. H. *The Invertebrates: Acanthocephala, Aschelminthes, and Entoprocta*. Vol. 3, *The Pseudocoelomate Bilateria*. New York: McGraw-Hill, 1951.

Nielsen, C. "Entoprocts." In *Synopses of the British Fauna (New Series)* 41, edited by Doris M. Kermack and R. S. K. Barnes. Leiden, The Netherlands: E. J. Brill., 1989.

Periodicals

Wasson, K. "A Review of the Invertebrate Phylum Kamptozoa (Entoprocta) and Synopses of Kamptozoan Diversity in Australia and New Zealand." *Transactions of the Royal Society of South Australia* 126 (2002): 1–20.

Tohru Iseto, PhD

Micrognathozoa
(Jaw animals)

Phylum Micrognathozoa
Number of families 1

Thumbnail description
Microscopic animal group with one described species, *Limnognathia maerski,* characterized by the presence of intracellular plates in the dorsal and lateral epidermis; the ventral epidermis is covered with frontal and ventral ciliation (ciliophores); sensory structures consist of serially arranged tactile bristles; the digestive system has a highly complex jaw apparatus, a simple midgut, and a dorsal, periodically functioning anus

Illustration: *Limnognathia maerski.* (Illustration by Emily Damstra)

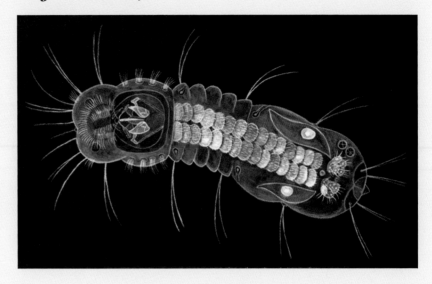

Evolution and systematics

Micrognathozoa contains only a single species, *Limnognathia maerski,* which was discovered in 1994 but first described in 2000 by R. M. Kristensen and P. Funch. It was cited as a new class, but owing to its numerous unique character traits, it could not be assigned to any known phylum. The special morphological features of its integument and the presence of a complex jaw apparatus, however, pointed toward a close relationship with the Rotifera, and Micrognathozoa therefore were assigned to the superphylum Gnathifera (that is, "those that possess jaws"), which, besides the Rotifera, contains the phyla Gnathostomulida (jaw worms) and Acanthocephala (thorny-headed worms). The phylum Micrognathozoa contains one class, Micrognathozoa; one order, Limnognathida; and one family, Limnognathiidae.

The taxonomy for this species is *Limnognathia maerski* Kristensen and Funch, 2000, Disko Island, Greenland.

Physical characteristics

Limnognathia maerski is a microscopic, acoelomate animal that ranges in size from 0.004 to 0.006 in (100–150 µm). Its body is divided into a head, an accordion-like thorax, and an abdomen. The dorsal and lateral parts of the integument have no ciliation, except scattered sensory bristles that are made up of joined cilia. The epidermis consists of several epidermal plates, and each plate is made up of two to four epidermal cells. It lacks a true cuticle, and the only cover is a very thin glycocalyx. Instead of an extracellular cuticle, it has a filamentous intracellular lamina that helps make the integument stiffer. Most of the ventral part of the integument lacks both cuticle and the intracellular lamina; instead, it has a much thicker extracellular glycocalyx. It does have one large cuticular area, called the oral plate, posterior to the mouth open-

ing. Most of the ventral side is covered with cilia that are arranged in four distinct groups.

The frontal part of the head has bands of cilia that beat metachronously (using coordinated waves) and resemble those found in Rotifera and various protostome larvae. On each side of the oral plate there are four groups of cilia. The cilia in each group beat in unison, and all cilia arise from a single cell, a so-called ciliophore. The ventral side of the thorax and abdomen is covered by a dense mat of cilia that arise from a double row of ciliophores. The fourth group of cilia is located on the caudal part of the abdomen and is formed by cilia that arise from an adhesive pad that consists of ten cells.

The mouth is located ventrally, anterior to the oral plate, and leads to the pharynx, which contains a highly complex jaw apparatus. It consist of several independent elements that are interconnected by ligaments and fine, cross-striated muscles. The central part of the jaws consists of the large main jaws that anteriorly form a pair of forceps with long teeth. A pair of large droplet-shaped elements, the fibularia, are located lateral to the main jaws, and two cuticular bridges, named the reinforced web, connect the anterior parts of the fibularium with the forceps of the main jaw. A pair of rod-shaped elements with distal teeth is located on the dorsal side of the fibularium. These elements are called the dorsal jaws. A cuticular membrane covers the ventral side of the main jaw and fibularium, and ventrally on this membrane other elements are present, including the oral lamellae, the ventral jaws and the basal plates. If the jaw elements are investigated with scanning or transmission electron microscopy, they can be seen to be composed of extremely fine, rodlike fibrillae, and this is identical to the ultrastructure in the jaws of Rotifera (wheel animalcules) and Gnathostomulida (jaw worms), hence supporting a close relationship between these groups.

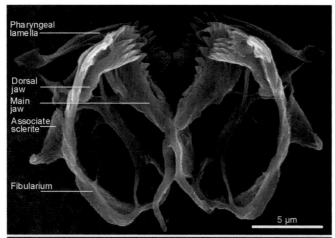

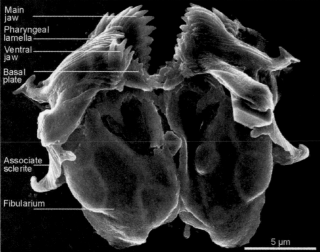

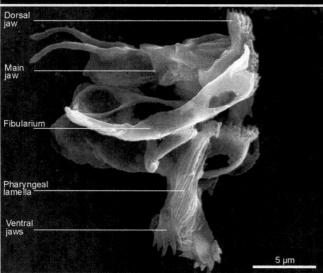

Scanning micrographs of the jaws of the species *Limnognathia maerski*. Upper: dorsal view; center: ventral view; lower: lateral view. (Scanning micrograph by Martin V. Sørensen. Reproduced by permission.)

A short esophagus leads to the gut and then to the rectum. A normal, permanent anus apparently is lacking. Instead, the rectum gradually narrows and at its terminus, near the in-

tegument, is blocked by an anal plate. The anal plate has a pair of small muscles, which suggests that it can be moved so that the anus becomes periodically functional. However, defecation has never been observed. Two pairs of protonephridia control excretion and maintenance of osmotic balance. The protonephridial terminal cells are monociliated, contrary to all other ciliated cells in *L. maerski*.

Distribution

Micrognathozoans have a bipolar distribution. *Limnognathia maerski* was found in the cold Isunngua spring at the bay of Aqajarua (Muddy Bay) on the northeastern part of Disko Island, western Greenland, in the summer of 1994. Since its discovery, several hundred specimens have been collected from this spring, but recordings from nearby springs are very few. Three specimens from a collection made in 1979 from a spring close to Lymnaea Lake in the valley of Sullorsuaq/Kvandalen (Angelica Valley) were rediscovered in 1995. Lymnaea Lake is located about 10.5 mi (17 km) from the Isunngua spring. In 1997 the species was collected from different freshwater bodies on the Crozet Islands, located between Africa and Australia, 1,500 mi (2,400 km) north of Antarctica. It has not been determined how the species has this peculiar distribution, but one explanation could be that *L. maerski* has very narrowly defined demands in terms of habitat and climate and that such localities still are poorly investigated. The search for *L. maerski* has been intense in all kinds of homothermic springs (those with a constant temperature throughout the year) as well as heterothermic springs (those that may freeze in the winter) throughout the world, without any further findings.

Habitat

Limnognathia maerski lives in mosses in running or stagnant freshwater, but it also may inhabit sandy bottom sediments. Knowledge of its ecology is extremely limited, but the distribution suggests that it prefers low-temperature waters. On Disko Island it has been found only in cold springs that often are frozen for seven to eight months of the year. The adult animals cannot survive freezing, but it is likely that their thick-shelled resting eggs (see Reproductive biology) can tolerate such extreme conditions. Interestingly, Disko Island has more than a thousand homothermic springs that maintain a constant temperature throughout the year and therefore also run during the winter, but *L. maerski* clearly avoids this kind of spring, which suggests that the deep-frozen period is important for its life cycle.

Behavior

When *L. maerski* moves, it either crawls or swims. While swimming, it moves slowly in a characteristic spiral, which makes it easy to distinguish from other microinvertebrates. The swimming motion is created by the ventral trunk ciliophores; hence, the ciliary bands in the head are not involved in locomotion. When crawling, it also uses ventral ciliophores and glides slowly on the substratum. This slow gliding very much resembles the movement of chaetonotid gastrotrichs

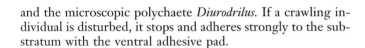

Limnognathia maerski

and the microscopic polychaete *Diurodrilus*. If a crawling individual is disturbed, it stops and adheres strongly to the substratum with the ventral adhesive pad.

Feeding ecology and diet

Limnognathia maerski feeds on bacteria, blue-green algae, and diatoms. During the search for food the head moves slowly from side to side while the ciliary bands on the head beat food particles toward the mouth. When a food item reaches the mouth, it is quickly grabbed by the ventral jaws, dragged into the pharynx, and processed by the main jaws. The species appears to be highly selective in its food choice; if it by accident grabs and swallows an unwanted item, it quickly rejects it. Following its characteristic "vomit behavior," it lifts the forehead to a vertical position that pushes the whole jaw apparatus out of the mouth; while the animal stands in this position, the swallowed item is ejected.

Reproductive biology

Even though *L. maerski* has been sampled frequently throughout the short Arctic summer over the course of several years, male individuals have never been seen, and it is certain that reproduction is based on parthenogenesis, at least in part. The female reproductive system is rather simple and comprises a pair of ovaries. The ovaries consist mostly of oocytes, and yolk appears to be produced inside each developing egg from small molecules that diffuse through the thin shell. Oviposition can be provoked by pricking the animal gently with a thin needle or by increasing the temperature by

a few degrees. The egg is laid through a small pore posterior to the ventral adhesive pad.

Limnognathia maerski is capable of producing two kinds of eggs: a thin-shelled type and a type with a much thicker, sculptured shell. The latter egg resembles the resting eggs that are produced during the sexual phase of the reproductive cycle of monogonont rotifers. If the micrognathozoan cycle can be compared to the rotiferan one, it would suggest that the thin-shelled eggs are products of asexual parthenogenetic reproduction, whereas sexual reproduction is involved in the formation of the sculptured type. Hence, males could be present, at least periodically. One possibility is that *L. maerski*, like rotifers, produces dwarf males that live only for a very short period and therefore have not yet been found. Another possibility is that Micrognathozoa are protandric hermaphrodites, which means that the animals hatch as males and then quickly develop into females. Both solutions are purely hypothetical.

Conservation status

The type locality of *L. maerski* is protected by the international wetlands convention (RAMSAR site). Both the bay of Aqajarua and the valley of Sullorsuaq, with the many eutrophic lakes, are important as feeding sites for numerous species of Arctic birds.

Significance to humans

None known.

Resources

Books

Funch, Peter, and Reinhardt M. Kristensen. "Coda: The Micrognathozoa—a New Class or Phylum of Freshwater Meiofauna?" In *Freshwater Meiofauna: Biology and Ecology*, edited by S. D. Rundle, A. L. Robertson, and J. M. Schmid-Araya. Leiden, The Netherlands: Backhuys Publishers, 2002.

Periodicals

Ahlrichs, W. H. "Epidermal Ultrastructure of *Seison nebaliae* and *Seison annulatus*, and a Comparison of Epidermal Structures within the Gnathifera." *Zoomorphology* 117 (1997): 41–48.

De Smet, Willem H. "A New Record of *Limnognathia maerski* Kristensen & Funch, 2000 (Micrognathozoa) from the Subantarctic Crozet Islands, with Redescription of the Trophi." *Journal of Zoology (London)* 258 (2002): 381–393.

Kristensen, Reinhardt M. "An Introduction to Loricifera, Cycliophora, and Micrognathozoa." *Integrative and Comparative Biology* 42 no. 3 (June 2002): 641–651.

Kristensen, Reinhardt M., and Peter Funch. "Micrognathozoa: A New Class with Complicated Jaws like Those of Rotifera and Gnathostomulida." *Journal of Morphology* 246, no. 1 (October 2000): 1–49.

Sørensen, M. V. "Further Structures in the Jaw Apparatus of *Limnognathia maerski* (Micrognathozoa), with Notes on the Phylogeny of the Gnathifera." *Journal of Morphology* 255, no. 2 (February 2003): 131–145.

Martin Vinther Sørensen, PhD
Reinhardt Møbjerg Kristensen, PhD

Gnathostomulida
(Gnathostomulids)

Phylum Gnathostomulida
Number of families 12

Thumbnail description
Marine group of microscopic free-living worms
characterized by an entirely monociliated
epidermis and complex cuticular mouthparts

Photo: A few species of Filospermoidea, such as
this *Haplognathia ruberrima*, are bright red, while
all other gnathostomulids are colorless-opaque.
(Photo by Wolfgang Sterrer. Reproduced by permission.)

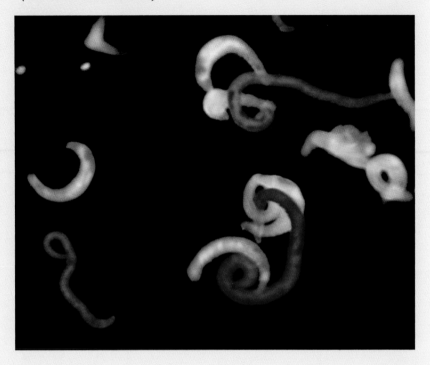

Evolution and systematics

There is no fossil record for this group. Described in 1956
as aberrant Turbellaria (flatworms), the gnathostomulids are
now considered related to the rotifers (Rotifera) and Micrognathozoa. The phylum (and single class) encompasses two
orders, Filospermoidea and Bursovaginoidea, the latter with
two suborders, Conophoralia and Scleroperalia. The 12 families contain 25 genera, with fewer than 100 valid species.

Physical characteristics

These thread-shaped worms range from 0.01 in (0.3 mm)
to more than 0.1 in (3 mm) in length. Most species are colorless or transparent, but a few are bright red. The anterior end
of Filospermoidea is pointed, while that of Bursovaginoidea
appears as a rounded head. The posterior end is rounded or
extends into a tail. The epidermis is completely monociliated,
i.e., each cell carries a single, long locomotory cilium. Some
cilia, singly or in paired groups, may have sensory functions.
The nervous system, which is largely situated at the basis of
the epidermis, consists of an unpaired frontal ganglion (brain)
and an unpaired buccal ganglion from which paired nerves
originate. The musculature is simple and rather weak, except
for a complex pharynx. Circulatory and respiratory organs are
lacking. The digestive tract provides the greatest number of
distinguishing characters. The mouth is located ventrally, behind the anterior end, and there is no permanent anus. In the
majority of species the complex, muscular, bilaterally symmet-

ric pharynx contains cuticular, hard mouthparts consisting of
an unpaired basal plate in the lower lip, and paired jaws. The
basal plate may be flat and dorsally set with ridges or teeth, as
in the family Haplognathiidae; transverse rod shaped, as in
Pterognathiidae; or consist of wings and set with rows of teeth,
as in the Gnathostomulidae. The jaws may be solid and for-
cepslike, as in the order Filospermoidea, or hollow like a pair
of forward-pointing funnels, with muscles inserting from behind, as in most Bursovaginoidea. In most species, the inner,
anterior parts of the jaw are set with groups or rows of teeth.

Distribution

Gnathostomulids are distributed worldwide, with most
species known from the North Atlantic and South Pacific
Oceans.

Habitat

Gnathostomulids occur exclusively, sometimes in large
numbers, in detritus-rich marine sand as typically found on
sheltered beaches, near sea grasses and mangroves, and between coral reefs. Such habitats are often characterized by low
oxygen but high hydrogen sulfide concentrations (that create
a rotten egg smell), which gnathostomulids seem to tolerate.
Most species have been found in the intertidal and shallow
subtidal zones, with occasional finds at 1,310 ft (400 m).

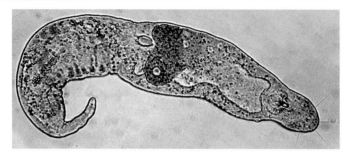

Light micrograph of *Gnathostomula armata*. (Photo by Matthew Hooge. Reproduced by permission.)

Behavior

Gnathostomulids glide through the interstices between sand grains, propelled slowly by their sparse ciliation, and contracting when disturbed. Some species spin a mucous cocoon which may let them survive a deteriorating environment.

Feeding ecology and diet

Belying their fearsome jaws, gnathostomulids are not predators but seem to graze on the microflora (bacteria and fungal threads) attached to sand grains. Scientists do not yet know what role the basal plate and jaws play in feeding.

Reproductive biology

All gnathostomulids are hermaphrodites. The male organs are located in the posterior part of the body. They consist of either an unpaired or paired testes, and a copulatory organ (penis) which, in Scleroperalia, contains a tubelike penis stylet. The sperm is diverse. In Filospermoidea, it is threadlike, with a single ciliary tail of 9+2 microtubules. In Bursovaginoidea, it is aflagellate and droplet shaped, in Conophoralia, aflagellate and cone shaped. Sperm is transferred by copulation, and stored either freely between gut and epidermis, or in a storage pouch (the bursa copulatrix). In all species, the single, unpaired ovary is located dorsally, behind the mouth. Only one large egg matures at a time, and is presumably fertilized by sperm from the bursa before being laid via rupture of the dorsal body wall. Development is direct, and cleavage probably follows the spiral pattern.

Conservation status

No species of gnathostomulid is listed by the IUCN.

Significance to humans

None known.

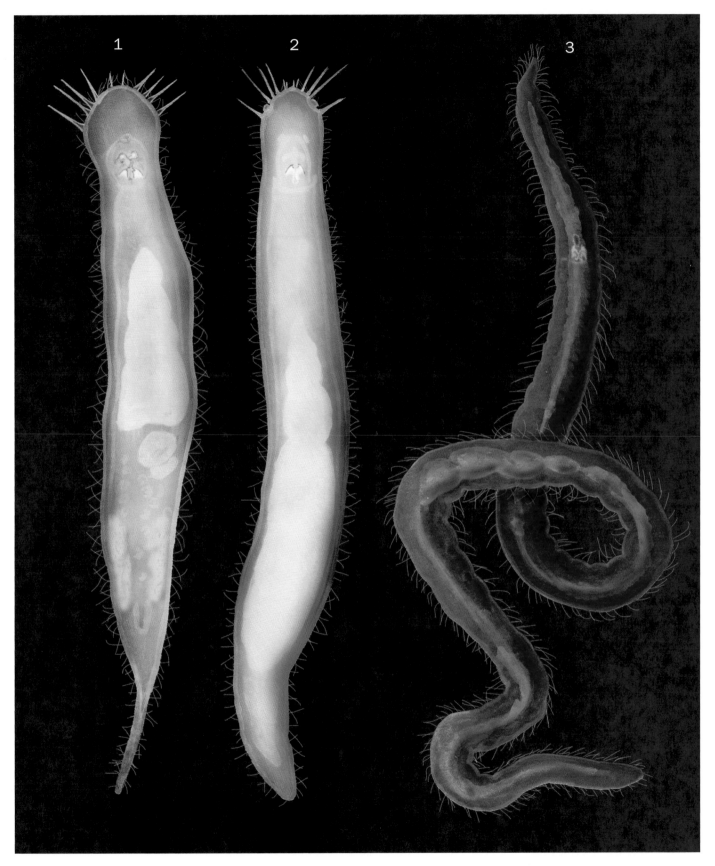

1. *Gnathostomula paradoxa*; 2. *Austrognathia australiensis*; 3. Red haplognathia (*Haplognathia ruberrima*). (Illustration by Dan Erickson)

Species accounts

Red haplognathia
Haplognathia ruberrima

FAMILY
Haplognathiidae

TAXONOMY
Haplognathia ruberrima Sterrer, 1966, Swedish west coast.

OTHER COMMON NAMES
None known.

PHYSICAL CHARACTERISTICS
Length to 0.138 in (3,500 µm); diameter 0.0051 in (140 µm); one of the largest gnathostomulids. Most specimens are uniformly brick red, reddish brown, or pink, owing to pigment granules in the epidermis. Rostrum (head) slender and pointed, without paired sensory bristles; posterior end uniformly rounded. Jaws solid, with large winglike apophyses and many sharp denticles; basal plate shieldlike, set with dorsal thorns. Sperm threadlike, with corkscrewlike head.

DISTRIBUTION
The most globally distributed species, found in Australia, Fiji, and Hawaii, as well as on both sides of the North Atlantic and the Mediterranean.

HABITAT
Like most gnathostomulids, prefers detritus-rich sand in shallow sublittoral areas.

BEHAVIOR
An animal that has been isolated from the sediments usually coils up by muscular action, then uncoils again by means of its cilia, often from both ends simultaneously, with the rostrum pulling forward and the posterior end pulling backward.

FEEDING ECOLOGY AND DIET
Seems to graze on fungal hyphae and bacteria adhering to sand grains.

REPRODUCTIVE BIOLOGY
Single egg is laid by rupture of the dorsal body wall. Egg then sticks to a sand grain until a fully ciliated hatchling 330 µm long emerges.

CONSERVATION STATUS
Not listed by the IUCN.

SIGNIFICANCE TO HUMANS
None known. ◆

Haplognathia ruberrima
Gnathostomula paradoxa
Austrognathia australiensis

No common name
Gnathostomula paradoxa

FAMILY
Gnathostomulidae

TAXONOMY
Gnathostomula paradoxa Ax, 1956, Kieler Bucht (Kiel Bay), North Sea.

OTHER COMMON NAMES
None known.

PHYSICAL CHARACTERISTICS
Length to 0.0394 in (1,000 µm), diameter to 0.00295 in (75 µm). Body colorless; rostrum rounded, headlike, set with paired sensory bristles; posterior drawn out into a short tail. Jaws hollow, with three rows of teeth; basal plate with paired lateral and rostral wings. Reproductive system with a cuticular bursa and a penis stylet. Sperm small, aflagellate, droplet shaped.

DISTRIBUTION
North Sea.

HABITAT
Occurs in detritus-rich sand, straying into clean sand.

BEHAVIOR
Nothing is known.

FEEDING ECOLOGY AND DIET
Grazes on attached microflora.

REPRODUCTIVE BIOLOGY
Nothing is known.

CONSERVATION STATUS
Not listed by the IUCN.

SIGNIFICANCE TO HUMANS
None known. ◆

No common name
Austrognathia australiensis

FAMILY
Austrognathiidae

TAXONOMY
Austrognathia australiensis Sterrer, 2001, Lizard Island, Queensland, Australia.

OTHER COMMON NAMES
None known.

PHYSICAL CHARACTERISTICS
Length to 0.0315 in (800 µm), diameter to 0.00315 in (80 µm). Body colorless, slender; head rounded, set with paired sensory bristles; posterior end pointed. Jaws hollow, with two rows of teeth; basal plate with pronounced median and a pair of lateral lobes. Bursa and penis without hard structures; sperm large, cone shaped, aflagellate.

DISTRIBUTION
Queensland, Australia.

HABITAT
Occurs in detritus-rich sand near sea grass beds and patch reefs.

BEHAVIOR
Nothing is known.

FEEDING ECOLOGY AND DIET
Nothing is known.

REPRODUCTIVE BIOLOGY
Nothing is known.

CONSERVATION STATUS
Not listed by the IUCN.

SIGNIFICANCE TO HUMANS
None known. ◆

Resources

Periodicals
Sterrer, W. "Gnathostomulida from the (Sub)tropical Northwestern Atlantic." *Studies on the Natural History of the Caribbean Region* 74 (1998): 1–178.

———. "Gnathostomulida from Australia and Papua New Guinea." *Cahiers de Biologie Marine* 42 (2001): 363–395.

Sørensen, M. V., and Sterrer, W. "New Characters in the Gnathostomulid Mouth Parts Revealed by Scanning Electron Microscopy." *Journal of Morphology* 253 (2002): 310–334.

Wolfgang Sterrer, PhD

Priapulida
(Priapulans)

Phylum Priapulida
Number of families 3

Thumbnail description
Cylindrical body, the anterior part of which—the introvert—is covered with chitinous teeth and can be rolled inward; at the posterior end are one or two caudal appendages

Illustration: *Tubilucus corallicola.* (Illustration by John Megahan)

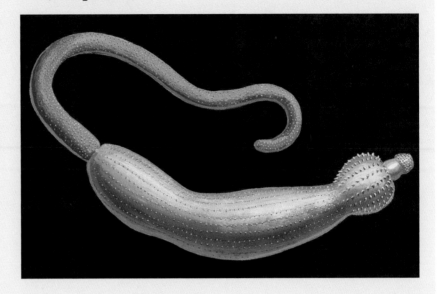

Evolution and systematics

Priapulans once were grouped among a diverse collection of invertebrates known as the Aschelminthes. Modern comparative studies place the priapulans with kinorhynchs and loriciferans in a group called Cephalorhyncha or Scalidophora. These three phyla have chitinous cuticles and rings of scalids on the introvert. Sixteen living and approximately 14 fossil priapulans are known. The living species are generally classified into three families.

Physical characteristics

The body of a priapulan is divided into three parts—introvert, trunk, and caudal appendage. The introvert can be pulled completely into the trunk by a pair of retractor muscles. Chitinous scalids of various sizes and shapes cover the entire surface of the introvert. Sometimes the scalids at the anterior end of the introvert are larger than those nearer the trunk. Within the introvert is a muscular pharynx armed with cuticular teeth. The trunk houses the internal body organs, in particular the digestive system and reproductive organs. The body of priapulans is filled with fluid that acts a hydrostatic skeleton as the body wall muscles contract. During movement, the fluid moves around in the body cavity and serves the functions of circulation, excretion, and respiration. The caudal appendage is continuous with the body cavity of the trunk. The function of the caudal appendage has not been established; it may serve a respiratory function.

Distribution

All oceans from shallow water to the deepest parts of the sea. In some areas only one or two records exist, most likely because of infrequent collecting. Larger priapulans have been found in colder ocean waters, especially in the Northern Hemisphere. Small and interstitial priapulans are most common in the shallow tropics.

Habitat

Priapulans inhabit soft sediments of all kinds but are uncommon in areas with rocks. Larger species, such as *Priapulus caudatus*, are generally found in very soupy muddy bottoms. Some of the smaller species live in sandier sediments, where there is considerable interstitial space for movement. In the tropics, priapulans commonly are found in poorly sorted coral sand.

Behavior

Having no legs or other appendages, priapulans depend on the hydrostatic skeleton for movement. When extended, the introvert acts as an anchor in the sediment, as does the anterior part of the trunk when the circular body wall muscles of the central trunk are contracted. Once anchored, the priapulan can pull itself through the sediment by contraction of the longitudinal body wall muscles. Peristaltic contractions of the body wall muscles move the body through the sediment.

Feeding ecology and diet

The food source of the larger priapulans has not been ascertained. These animals apparently are capable of capturing and ingesting larger, slow-moving polychaetes. With the introvert fully extended, priapulans can grasp prey with the teeth of the pharynx and rapidly roll it inward. In aquaria *P. caudatus* has been seen eating a variety of marine worms. Smaller individuals, however, may be mud eaters, as are smaller species. Examination of stomach contents has proved inconclusive about priapulan feeding habits.

Reproductive biology

The sexes are separate in all priapulans that have been studied. Gonadal products are released freely into the water, where fertilization occurs. In larger species, a loricate larva forms and lives in the bottom mud. As the larva grows, it sheds the cuticular covering and gradually grows into a juvenile priapulan. In at least one meiofaunal priapulan, the female broods the embryos, which hatch as juveniles.

Conservation status

Priapulans are not protected in any region, although they are designated as animals of special significance in Maine in the United States. No species is listed by the IUCN.

Significance to humans

This phylum was apparently very abundant in the Cambrian period. Species alive today are in some ways living fossils and should be conserved.

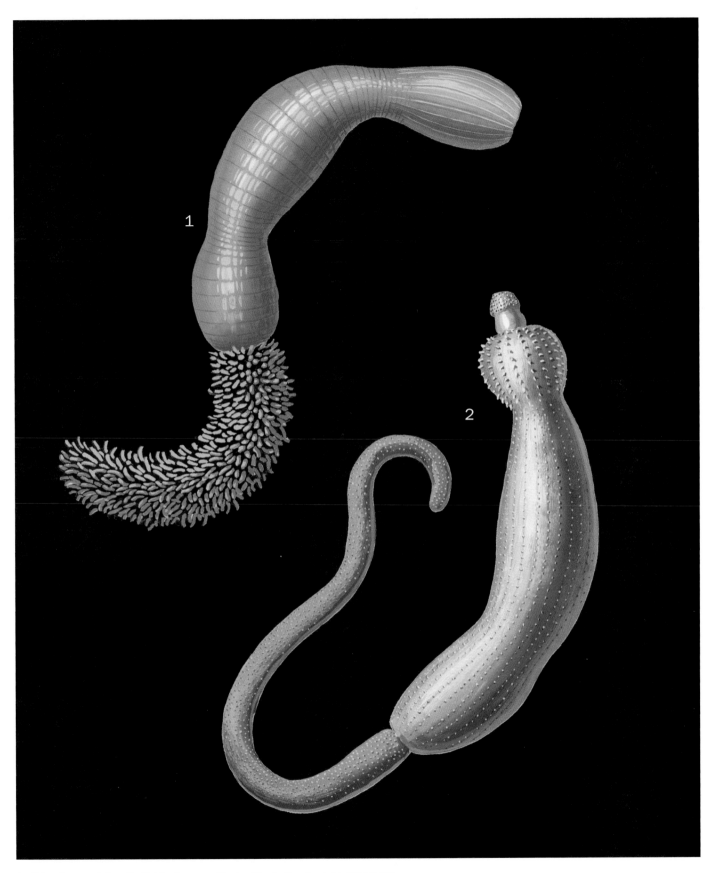

1. *Priapulus caudatus*; 2. *Tubiluchus corallicola*. (Illustration by John Megahan)

Species accounts

No common name
Priapulus caudatus

ORDER
Priapulimorphida

FAMILY
Priapulidae

TAXONOMY
Priapulus caudatus Lamarck, 1816.

OTHER COMMON NAMES
None known.

PHYSICAL CHARACTERISTICS
The body is large, up to 8 in (200 mm). This species is the typical priapulan. Its body is strongly tubular and ringed with many annulations. The introvert can be quite long when extended, reaching as much as a third of the length of the trunk. At the posterior end are a pair of caudal appendages.

DISTRIBUTION
Circumpolar Northern Hemisphere to Mediterranean Sea in eastern Atlantic Ocean and California in eastern Pacific Ocean.

HABITAT
Soft muddy bottoms.

BEHAVIOR
The hydrostatic skeleton is used for movement. The introvert and anterior part of the trunk act as an anchor in the sediment. Once anchored, the animal pulls itself through the sediment by peristaltic contraction of body wall muscles.

FEEDING ECOLOGY AND DIET
Most probably a detritus feeder when young and a predator as an adult.

REPRODUCTIVE BIOLOGY
Sexes are separate. Gonadal products are released freely into the water, where fertilization occurs. A loricate larva lives in the bottom mud. As the larva grows, it sheds the cuticular covering and gradually grows into a juvenile.

CONSERVATION STATUS
Not listed by the IUCN.

SIGNIFICANCE TO HUMANS
None known. ◆

■ *Priapulus caudatus*
■ *Tubiluchus corallicola*

No common name

Tubiluchus corallicola

ORDER
Priapulimorphida

FAMILY
Tubiluchidae

TAXONOMY
Tubiluchus corallicola van der Land, 1968.

OTHER COMMON NAMES
None known.

PHYSICAL CHARACTERISTICS
Generally less than 0.1 in (5 mm) long. The introvert is moderately short, making up less than one fourth of the total body length in extended individuals. The trunk is not annulated and is covered with small spines. The single caudal appendage is very long, usually much longer than the trunk.

DISTRIBUTION
Bermuda, Curacao, Bonaire, and Barbados.

HABITAT
Poorly sorted coral sands in shallow tropical waters.

BEHAVIOR
In laboratory dishes, the animals always use their muscular caudal appendage to maintain contact with the substrate.

FEEDING ECOLOGY AND DIET
Most likely ingests small organic particles or may eat other small meiofaunal organisms.

REPRODUCTIVE BIOLOGY
Copulation is probable with internal fertilization. Females produce only up to 20 eggs. Release of embryos has not been observed, but larvae can be collected from the sediment and reared in the laboratory.

CONSERVATION STATUS
Not listed by the IUCN.

SIGNIFICANCE TO HUMANS
None known. ◆

Resources

Periodicals

Kirsteuer, E., and J. van der Land. "Some Notes on *Tubiluchus corallicola* (Priapulida) from Barbados, West Indies." *Marine Biology* 7 (1970): 230–8.

van der Land, J. "Systematics, Zoogeography, and Ecology of the Priapulida." *Zoologische Verhandelingen* 112 (1970): 1–118.

Les Watling, PhD

Loricifera

(Girdle wearers)

Phylum Loricifera

Number of families 2

Thumbnail description
Group of microscopic, bilateral symmetrical animals characterized by a body with five sections: a protrusive mouth cone; a head (introvert) with up to nine rows of scalids (in the adults); a neck with trichoscalids; a thorax; and an abdomen with a lorica consisting of plates or plicae (folds)

Illustration: Bucket-tailed loriciferan (*Rugiloricus cauliculus*). (Illustration by Barbara Duperron)

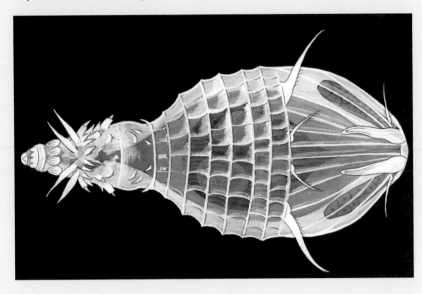

Evolution and systematics

Loricifera is traditionally considered a phylum in the group Introverta. The closest relatives of the loriciferans are the Kinorhyncha (kinorhynchs) and Priapulida (priapulans). Loricifera includes 11 described species as of 2003. All species have been assigned to one order named Nanaloricida that includes two families, the Nanaloricidae and the Pliciloricidae. As of 2003, more than 100 species have been collected that have not yet been described.

Physical characteristics

Adult loriciferans range in size from 0.008 to 0.0157 in (200–400 µm). The largest specimen ever recorded was a giant larva from the deep sea that measured about 0.027 in (700 µm). The body is divided into five sections: a mouth cone; a head that can be introverted or drawn into the body; a neck; a thorax; and an abdomen. The mouth cone consists of 6–16 oral ridges and a cuticulanized, extrudable buccal tube. Furthermore, some nanaloricids may have six oral stylets.

The head, or introvert, consists of nine rows of sensory or locomotory structures called scalids. The scalids in the first row are called clavoscalids whereas the scalids in the remaining eight rows are called spinoscalids. There are usually eight clavoscalids; however, males of the family Nanaloricidae always have more clavoscalids than the females. The second and third rows of scalids vary greatly in the number of scalids in each row (from 7–15) and types. There are different kinds of leg-shaped scalids and a ventral double organ. The number of scalids in the fourth through the eighth rows is always 30. In the fourth row, the scalids may alternate between the typical spinoscalids and a noticeable claw-shaped variation on the typical form. The spinoscalids in the fifth through the sev-

enth rows always belong to the common, simple type. The spinoscalids in the eighth row resemble the trichoscalids of the neck region; sometimes they have alternating plates. The scalids in the ninth row in pliciloricids have beak-like structures. These scalids are not present in nanaloricids.

The neck of loriciferans consists of three rows of plates with 15 plates in each row and 15 appendages known as trichoscalids. The trichoscalids are flattened with serrated margins; they either alternate between eight single and seven double scalids or have 15 single trichoscalids. The thorax has no appendages. The abdomen consists of a lorica with either 6–10 strong cuticularized plates or 22–40 plicae. A varying number of sense organs are present toward the rear of the abdomen. These specialized receptors are called flosculi and can be divided into two types; *Nanaloricus*-flosculi and *Pliciloricus*-flosculi.

Internally the body consists of a pharynx bulb, a digestive system with a short esophagus, and a straight midgut. The reproductive system consists of a pair of sack-shaped gonads with a pair of protonephridia, or primitive excretory organs, inside the gonads. The protonephridia are divided into anterior and posterior parts. Their presence inside the gonads is unique to Loricifera. In addition to these organ systems, the body of loriciferans contains a complex muscular system as well as a nervous system with a large brain and ventral nerve cord with ganglia (groups of nerve cells located outside the brain).

The postlarvae in the family Nanaloricidae look like adults although they lack one row of scalids and a reproductive system. The postlarvae in the family Pliciloricidae, however, differ significantly from the adults because their scalids are reduced in size and much simpler in appearance.

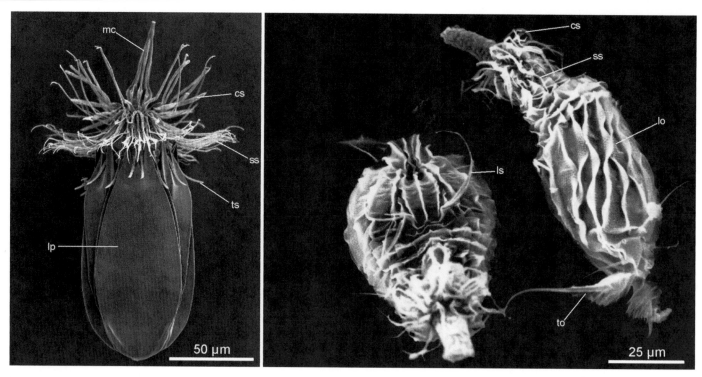

Scanning micrograph of *Nanaloricus mysticus*. Left: Adult male. Right: The Higgins larvae. Abbreviations: cs: clavoscalid; lo: lorica; lp: lorica plate; ls: locomotory seta; mc: mouth cone; ss: spinoscalid; to: toe; ts: trichoscalid. (Scanning micrograph by Reinhardt M. Kristensen. Reproduced by permission.)

The Higgins larva ranges in size from 0.003–0.0197 to 0.027 in (80–500 to 700 μm) and is easily recognized by the presence of two conspicuous toes. It is named for Professor Robert P. Higgins in honor of his longtime study of the Introverta. The body of the Higgins larva is divided into the same sections as the adult. The mouth cone consists of either 6–12 double oral stylets (thin rods) and/or six oral ridges. The head, or introvert, consists of seven rows of scalids with eight clavoscalids in the first row. The second through the seventh row have spinoscalids as in the adult. The bilateral symmetry is pronounced in the arrangement of the larval scalids. The neck sometimes has a collar-like area that can close over the introvert when it is retracted into the body. The thorax resembles an accordion, having 5–6 ventral longitudinal folds. The abdomen also has longitudinal folds, but only on its sides and dorsal surface. Two or three pairs of sensory or locomotory setae (bristles) are located on the larva's ventral surface in the area between the thorax and the abdomen, with two or three pairs of setae and a pair of toes located toward the rear of the lorica. The toes are used for movement and have adhesive glands. There are several flosculi of the Nanaloricus-type on the abdomen of the Higgins larva.

Distribution

As of 2003, information is limited regarding the distribution of Loricifera. Representative organisms from this group have been recorded in relatively few locations. It is very likely, however, that these records reflect the scarcity of sampling rather than the actual distribution of loriciferans. The group is probably present in the marine meiofauna throughout the world, from shallow to deep coastal and abyssal waters. Meiofauna refers to a category of microorganisms that are large enough to see with a basic microscope but too small to see with the naked eye. Loriciferans have been sampled in the coastal waters of France, Denmark, and North Carolina (USA); the Faroe Islands of the North Atlantic; the Mediterranean; the Coral Sea off Queensland, Australia; the Angola basin in the southeastern Atlantic; and from the Great Meteor Seamount in the North Atlantic. They have been sampled in abyssal waters in the Gulf of Mexico at a depth of 9,708 ft (2,959 m) and the Izu-Ogasawara trench at a depth of 27,100 ft (8,260 m) off Japan. Loriciferans are common in polar waters, especially in the deep sea.

Habitat

Loriciferans are found exclusively in marine habitats, and live in the spaces between sand grains or in the mud at the bottom of the deep sea. The nanaloricids prefer sand with low levels of detritus (material derived from the decomposition of once-living organisms) or clean shell gravel, whereas the pliciloricids are often found in such deep-sea sediments as the white abyssal *Globigerina* ooze and the red deep-sea clay from the hadal zone (below 20,000 ft; 6,100 m). Both families of Loricifera have been recorded in Australian marine caves, and a new nanaloricid genus was discovered in 1998 in sediments influenced by hydrothermal vents in the deep waters of the Kilinailau Trench off Papua New Guinea.

Behavior

Studies of loriciferan behavior are extremely scarce since most of the animals die before they can be examined. They have been studied live a few times, however, and researchers have observed their movement patterns and mating behavior.

The adults and the larvae both adhere strongly to the sediment with a kind of glue made by the adhesive glands located toward the rear of the adults and on the toes of the larvae. The adults usually crawl by using their many scalids and their mouth cones. Observers have noticed that the mouth cone telescopes out to its full length, fastens itself to a sand grain, and then draws in again so that the loriciferan's body is pulled forward.

The larvae may use their scalids and setae to crawl between grains of sand. They can also swim by using their toes. Larvae of the family Nanaloricidae, whose members have large and flipper-shaped toes, have sometimes have been found swimming free among the plankton. The larvae of the family Pliciloricidae have thin spine-shaped toes and are probably unable to swim; their movement is apparently restricted to crawling.

Little is known about the mating behavior of loriciferans. It is assumed, however, that the branched clavoscalids in the males of the family Nanaloricidae are used as chemoreceptors to locate female sex pheromones. The males in this family also have one ventral modified hook-shaped pair of trichoscalids that hold the female during copulation.

Feeding ecology and diet

Loriciferans probably feed on suspended organic particles, microalgae, and bacteria. They eat bacteria or algae by piercing them with their oral stylets and then sucking out the contents. The spinoscalids of nanaloricids are often completely covered with bacteria. It is still uncertain why the bacteria are present in such numbers. Some researchers suggest that the animals collect the bacteria from the interstitial water, whereas others maintain that the animals cultivate the bacteria on the mucus that they secrete. What is certain is that the loriciferans feed on these bacteria; this is confirmed by the fact that their digestive tracts are filled in some cases with partially digested bacteria.

Reproductive biology

Loricifera have very complicated life cycles with both sexual and asexual forms of reproduction. Members of the family Nanaloricidae reproduce only sexually and have distinct sexual dimorphism; that is, different body forms related to gender. The dimorphism is illustrated in the males by some of the clavo- and trichoscalids that are modified pheromone receptors as well as hooks to hold the female during copulation. Fertilization may be either internal or external. The primary larva, the Higgins larva, hatches from the fertilized egg and grows by molting. After at least 2–5 larval instars (stages between molts), the larva metamorphoses into a postlarva. The postlarva is a dormant stage and never feeds. A male or female with fully developed gonads emerges from the postlarva and the life cycle repeats itself.

The members of the family Pliciloricidae follow two different life cycles, named the *Rugiloricus cauliculus* and *Rugiloricus carolinensis* cycles respectively. The *Rugiloricus cauliculus* cycle includes a cycle of sexual reproduction like the one in the family Nanaloricidae. The only difference is that the postlarva looks different from the adult. The scalids of the postlarva are reduced in size and much simpler in form. The *Rugiloricus cauliculus* cycle, however, also comprises an asexual cycle in which the Higgins larva develops a mature ovary. The maturation of larval ovary is usually called neoteny, hence the matured larval stage is named the neotenous larva. Several (4–8) Higgins larvae form inside the neotenous larva. The larvae are formed by parthenogenesis, which means that they develop from female gametes without fertilization and are therefore genetically identical to the neotenous larva. The parthenogenetic larvae are completely identical to the Higgins larva and may develop either into new neotenous larvae or into a postlarva that will molt later into an adult male or female.

In the *Rugiloricus carolinensis* cycle, the postlarva is reduced so that only the cuticle of the postlarva is left inside the larva. The adult then emerges directly from the larva. Several reductions occur in the life cycle of an undescribed order from the Faroe Islands in the North Atlantic. The cuticles of the postlarva and the adult are found inside a cyst-like larva, and a mature ovary with several eggs develops inside these cuticles. The eggs mature into normal Higgins larvae that eat the maternal individual. The new larvae emerge from an opening in the rear of the empty exuvium (cast-off covering) of the cyst-like larva. This mode of reproduction is called pedogenesis and is usually seen only in nematodes and insects.

Conservation status

No species of loriciferans are listed by the IUCN.

Significance to humans

Loriciferans may have some significance to humans. They are found in large numbers in areas where methane seeps from the sediment. It may be possible in the future to use loriciferans or other meiofauna as indicators for the presence of methane or other gases.

1. Japanese deep-sea girdle wearer (*Pliciloricus hadalis*); 2. Bucket-tailed loriciferan (*Rugiloricus cauliculus*); 3. French girdle wearer (*Nanaloricus mysticus*); 4. American diamond girdle wearer (*Rugiloricus carolinensis*). (Illustration by Barbara Duperron)

Species accounts

French girdle wearer
Nanaloricus mysticus

ORDER
Nanaloricida

FAMILY
Nanaloricidae

TAXONOMY
Nanaloricus mysticus Kristensen, 1983, off the coast of Roscoff, France.

OTHER COMMON NAMES
None known.

PHYSICAL CHARACTERISTICS
The adults measure 0.00894–0.00925 in (227–235 μm). The mouth cone has eight oral ridges. The introvert has eight rows of true scalids; however, the first row of tricoscalid plates is fused with a row of scalids, giving the impression of a ninth row. Both sexes have eight clavoscalids. The male's clavoscalids are all branched except for the midventral pair, which resembles the female's clavoscalids. The last two rows on the introvert are beak- or tooth-like scalids. The lorica consists of six plates with a fine honeycomb structure. Flosculi are present in two pairs on the plates along the back and sides, and as a single flosculum on the anal plate.

The Higgins larvae measure 0.0047–0.0072 in (120–185 μm). The mouth cone lacks an armature (protective structure). The introvert has seven rows of scalids. The accordion-like thorax consists of 5–6 folds. Three ventral pairs of setae are located between the thorax and the abdomen. The lorica has coarse honeycomb ornamentation. Three pairs of anterior setae are modified to form a locomotory organ with grasping function. Three pairs of posterior setae are located on the lorica together with a pair of toes. The toes are flattened by a leaf-like structure called the mucro.

DISTRIBUTION
Roscoff, France. Two previously published records from Florida and the Azores Islands respectively turned out to refer to two undescribed species.

HABITAT
Found in the upper surface layer of *Dentalium*-sand (a specific type of coarse sand with high concentrations of detritus) at a depth of 65.6–82 ft (20–25 m).

BEHAVIOR
The Higgins larva may be able to swim by using its large leaf-like toes as a propeller.

FEEDING ECOLOGY AND DIET
In the first description of the Loricifera, it was postulated that this species is ectoparasitic because one specimen was attached

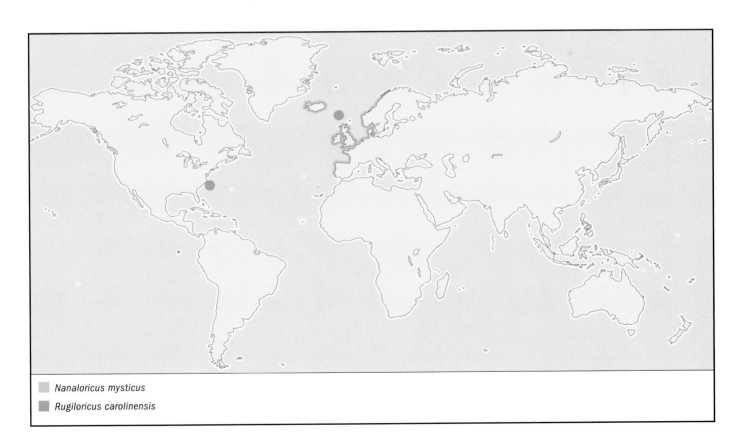

Nanaloricus mysticus

Rugiloricus carolinensis

to a copepod. This finding, however, turned out to be an artifact, or misleading result arising from the way the specimen was prepared. Nanaloricids feed exclusively on bacteria.

REPRODUCTIVE BIOLOGY
Sexual reproduction.

CONSERVATION STATUS
Not listed by the IUCN.

SIGNIFICANCE TO HUMANS
None known. ◆

Japanese deep-sea girdle wearer
Pliciloricus hadalis

ORDER
Nanaloricida

FAMILY
Pliciloricidae

TAXONOMY
Pliciloricus hadalis Kristensen & Shirayama, 1988, Izu-Ogasawara Trench, Western Pacific.

OTHER COMMON NAMES
None known.

PHYSICAL CHARACTERISTICS
Adults measure 0.0058–0.0086 in (149–219 µm). The males are smaller than the females. Mouth cone is small and without

mouth tube. The leg-shaped scalids in the second row are very large and robust. Two of the scalids in the second row are modified into a short and thick double organ. The fourth row has 15 claw-shaped scalids and 15 simple spinoscalids. The trichoscalids (eight single and seven double) are very long, about 0.0039 in (100 µm). The abdomen consists of 20 plicae. A single pair of *Pliciloricus*-flosculi is located caudally. Higgins larvae are long and slender, measuring 0.0103 in (262 µm). There are seven rows of scalids. The spinoscalids are long, over 0.00197 in (50 µm). Two pairs of setae are situated between the thorax and the abdomen. The toes are very long, about 0.006 in (153 µm), straight and pointed. Three pairs of setae are located toward the rear of the abdomen.

DISTRIBUTION
The hadal bathymetric zone of the deep sea (below 20,000 ft; 6,100 m). Found in the Izu-Ogasawara Trench off Japan at a depth of 27,100 ft (8,260 m).

HABITAT
Red deep-sea clay.

BEHAVIOR
One specimen was found 1.57 in (4 cm) deep in the clay, which indicates that the animal is capable of burrowing or living in association with a tube-dwelling macrofauna animal.

FEEDING ECOLOGY AND DIET
Probably bacteria.

REPRODUCTIVE BIOLOGY
The mode of reproduction of this species is not understood. The lack of seminal receptacles in the female suggests that this species has external fertilization.

▢ *Pliciloricus hadalis*
◾ *Rugiloricus cauliculus*

CONSERVATION STATUS
Not listed by the IUCN.

SIGNIFICANCE TO HUMANS
None known. ◆

Bucket-tailed loriciferan
Rugiloricus cauliculus

ORDER
Nanaloricida

FAMILY
Pliciloricidae

TAXONOMY
Rugiloricus cauliculus Higgins & Kristensen, 1986, off the coast of South Carolina, United States.

OTHER COMMON NAMES
None known.

PHYSICAL CHARACTERISTICS
Adults measure 0.0071–0.0103 in (180–264 µm). Mouth cone is very small. Introvert with nine rows of scalids. Appearance of the clavoscalids displays sexual dimorphism. Two types of clavoscalids (four large dorsal and four small ventral) are present in the male, but not in the female. The fourth row has 15 claw-shaped scalids and 15 simple spinoscalids. Lorica has 60 plicae. Anal cone pointed. The larvae measure 0.0108 in (275 µm), but are insufficiently described.

DISTRIBUTION
Found on the continental shelf off North and South Carolina, USA; in the Mediterranean Sea; and near the Faroe Islands. The identity of the Faroe Islands specimens is slightly uncertain, however; they may represent a new species.

HABITAT
Coarse phosphorite or oolytic sand at a depth of 656–1,640 ft (200–500 m).

BEHAVIOR
Nothing is known.

FEEDING ECOLOGY AND DIET
Probably feeds exclusively on methano-bacteria.

REPRODUCTIVE BIOLOGY
Sexual and asexual life cycles, including parthenogenesis with neotenous larvae (see Reproductive behavior in main chapter).

CONSERVATION STATUS
Not listed by the IUCN.

SIGNIFICANCE TO HUMANS
None known. ◆

American diamond girdle wearer
Rugiloricus carolinensis

ORDER
Nanaloricida

FAMILY
Pliciloricidae

TAXONOMY
Rugiloricus carolinensis Higgins & Kristensen, 1986, South Carolina, United States.

OTHER COMMON NAMES
None known.

PHYSICAL CHARACTERISTICS
Adults measure 0.008 in (205 µm). Introvert has nine rows of scalids. The clavoscalids in the first row are all uniform. The second row has between 9–15 leg-shaped spinoscalids. The fourth row has 30 uniform spinoscalids. The neck region consists of 15 single trichoscalids. The lorica consists of 30 plicae with a pair of Pliciloricus-flosculi and a large anal field.

DISTRIBUTION
Found on the continental shelf off North and South Carolina, USA, and near the Faroe Islands in the North Atlantic. The specimens from the Faroe Islands may be a new species.

HABITAT
Coarse phosphorite and carbonate sand.

BEHAVIOR
The Higgins larvae may have a very patchy distribution. More than 200 larvae were collected in one sample in phosphorite sand at a depth of 964 ft (294 m) east of Cape Romain, South Carolina.

FEEDING ECOLOGY AND DIET
Feeds on microalgae and bacteria.

REPRODUCTIVE BIOLOGY
Sexual reproduction with reduced postlarvae. The postlarval stage is very short and the cuticle consists only of a thin membrane without scalids. Therefore, it looks as if the Higgins larva molts directly into a mature adult. This species may also have an asexual life cycle with parthenogenetic neotenous larvae.

CONSERVATION STATUS
Not listed by the IUCN.

SIGNIFICANCE TO HUMANS
None known. ◆

Resources

Books

Kristensen, Reinhardt M. "Loricifera." In *Microscopic Anatomy of Invertebrates*. Vol. 4, *Aschelminthes*, edited by F. W. Harrison and E. E. Ruppert. New York: Wiley-Liss, 1991.

Periodicals

Higgins, R. P., and R. M. Kristensen. "New Loricifera from Southeastern United States Costal Waters." *Smithsonian Contributions to Zoology* 438 (1986): 1–70.

Kristensen, R. M. "Loricifera, A New Phylum with Aschelminthes Characters from the Meiobenthos." *Zeitschrift für Zoologische Systematik und Evolutionsforschung* 21 (1983): 163–180.

———. "An Introduction to Loricifera, Cyclophora, and Micrognathozoa." *Integrative and Comparative Biology* 42 (2002): 641–651.

Kristensen, R. M., and Y. Shirayama. "*Pliciloricus hadalis* (Pliciloricidae), A New Loriciferan Species Collected from the Izu-Ogasawara Trench, Western Pacific." *Zoological Science* 5 (1988): 875–881.

Todaro, M. A., and R. M. Kristensen. "A New Species and First Report of the Genus *Nanaloricus* (Loricifera, Nanaloricida, Nanaloricidae) from the Mediterranean Sea." *Italian Journal of Zoology* 65 (1998): 219–226.

Iben Heiner, MSc
Martin Vinther Sørensen, PhD
Reinhardt Møbjerg Kristensen, PhD

■
Cycliophora
(Wheel wearers)

Phylum Cycliophora
Number of families 1

Thumbnail description
Microscopic acoelomate animal group living as commensals on the mouthparts of the Norway lobster (*Nephrops norvegicus*); characterized by a highly complex life cycle with six life stages; the dispersal stage is the chordoid larva

Illustration: *Symbion pandora*. (Illustration by Bruce Worden)

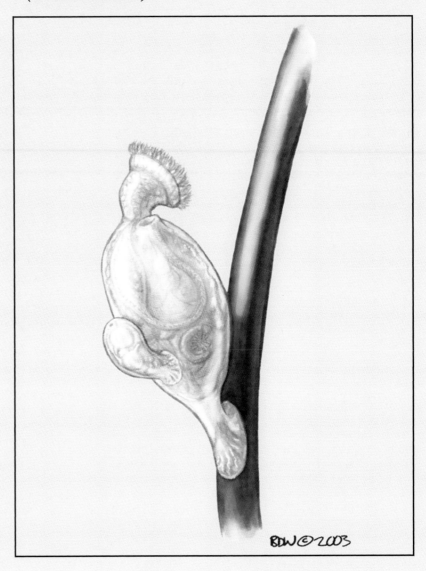

Evolution and systematics

The Cycliophora contains one described species, *Symbion pandora*, but two other undescribed species are known from the European lobster (*Homarus gammarus*) and the American lobster (*H. americanus*). Cycliophorans were first described by P. Funch and R. M. Kristensen in 1995, but they have been known to several other scientists since 1969. The morphology of the cycliophoran life stages suggests various phylogenetic positions. The ultrastructure of the cuticle in the feeding stages is very similar to the cuticle in some nematodes; its formation of new individuals by internal budding resembles the internal budding found in some bryozoans (Ectoprocta); and the food-collecting organ in the feeding stages and the presence of dwarf males resemble the conditions found in rotifers. Ultrastructural investigations of the epidermis have revealed structures in its basal lamina that are very similar to those on the basal lamina in Entoprocta, and based on this single character, the Cycliophora and Entoprocta are usually considered as sister groups. However, the complicated life cycle of the cyclio-phorans, with several attached and several free-living stages, occurs in neither the Rotifera nor the Ectoprocta/Entoprocta, and the chordoid larva is unique for a protostomian animal, hence the phylogenetic position of Cycliophora is still uncertain. The phylum Cycliophora contains one class, Eucyclio-phora; one order, Symbiida; and one family, Symbiidae.

Physical characteristics

The most prominent stage in the cycliophoran life cycle is that of the sessile feeding individuals that are attached to the setae of the lobster's maxillae and maxillipeds. The feeding-stage individuals measure approximately 0.138 in (350 μm) and are composed of a bell-shaped buccal funnel, a trunk, and a stalk with an adhesive disc. The buccal funnel is covered with cilia on its internal surface, but lacks external ciliation. The rim of the buccal funnel is composed of about 50 alternating ciliated and myoepithelial (muscular) cells. The ciliated cells form a circular ciliary band that is used for food-collection, and the myoepithelial cells enable the animal to close the mouth rim when it stops feeding.

The buccal funnel tapers proximally and connects to the trunk through a narrow, movable neck. The trunk is elongate or slightly egg shaped, and contains a U-shaped alimentary channel that leads from the mouth, through the stomach, to the anus, which is situated close to the neck. The stomach has no real lumen, but is filled with cells that produce digestive enzymes. The feeding individuals have no excretory system or sensory structures and the nervous system is simple, comprising a nerve concentration in the buccal funnel and pair of nerves that run through the trunk. A true brain has not been observed in the sessile individual.

The trunk narrows posteriorly to a stalk that leads to the adhesive disc. Both the stalk and the disc are purely cuticular, and therefore lack living tissue. (Illustration shown in chapter introduction.)

Distribution

The distribution of *S. pandora* mostly follows that of the Norway lobster (*Nephrops norvegicus*), which is the Northwest Atlantic, including the North Sea and Danish waters. Lobsters infected with *S. pandora* have been recorded from the Faroe Islands, Denmark, and the west coast of Sweden. *Symbion pandora* has also been found on the mouth appendages of Norway lobsters collected at about 1,640 ft (500 m) in the Mediterranean Sea, however investigated Norway lobsters from Iceland and the Caribbean were not infected. The two new *Symbion* species on the American and European lobsters have been found on lobsters from waters off the American northeast coast (Massachusetts and Maine) and the western coast of Europe, respectively.

Habitat

Symbion pandora lives exclusively in the nutrient-rich environment on the setae of the lobster's maxillae and maxillipeds, where the feeding stage individuals sit in dense populations. The number of individuals depends on the age of the lobster, and varies from some hundred on young lobsters to thousands on older specimens. A young lobster molts more frequently than an older one, and since the sessile population dies when the cuticle is shed, populations are always largest on the oldest specimens.

Behavior

Information about the behavior of the cycliophoran life stages is extremely scarce. The sessile feeding stage has been recorded on videotape, and the recordings show that they are capable of slowly moving the buccal funnel, but the funnel was never observed to retract into the trunk. Most free-living stages, except for the dwarf male and the chordoid larva (the dispersal stage), are poor swimmers.

Feeding ecology and diet

The sessile feeding stage is the only stage in the life cycle during which individuals feed. When the lobster seizes its food, food particles and nutrients are suspended in the water around its mouthparts. When the cilia on the cyclophoran buccal funnel beat, they create a current that generates an inflow of particles into the buccal funnel, where they are grabbed by the internal ciliation of the buccal funnel and transported toward the stomach.

Reproductive biology

Symbion pandora has an extremely complicated life cycle, with a sexual phase and two asexual phases.

The feeding stage individuals do not have a specific sex, but are capable of producing other sexual or asexual individuals, including males, females, and asexual larvae called Pandora larvae. They are also capable of performing "self-renewal," where nearly all the living tissue is replaced by a new set of organs. This self-renewal is initiated by the formation of an internal bud inside the feeding individual. Inside this bud, a new set of organs, including the alimentary system, nervous system, and buccal funnel, start to develop. When all organs are fully developed, they slowly move forward and push out all the old organs, until the new buccal funnel can emerge through the neck and replace the old one. The only parts of the old individual that are reused are the trunk and the adhesive disc. A feeding stage may repeat this self-renewal process several times, and scientists are still uncertain why it is necessary. One explanation could be that

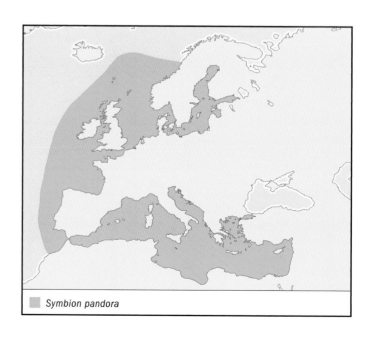

Symbion pandora

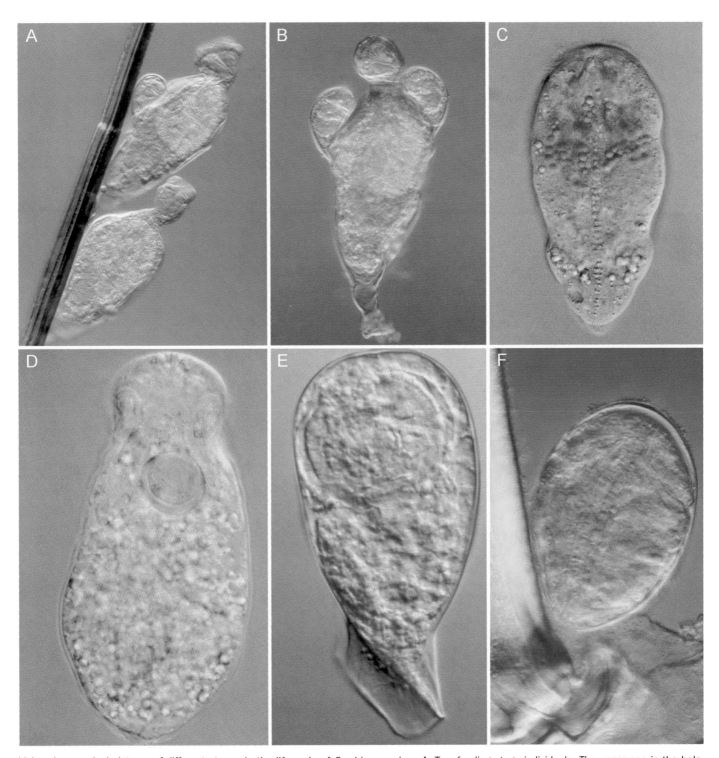

Light microscopical pictures of different stages in the life cycle of *Symbion pandora*. A. Two feeding stage individuals. The upper one is the holotype for the entire phylum, Cycliophora. B. Feeding stage individual with two attached Prometheus larvae. C. The chordoid larva. D. The female. E. Cyst with Pandora larva. The developing mouth ring for the feeding stages is already visible in the upper part of the animal. F. Cyst with chordoid larva. (Scanning micrograph by Reinhardt M. Kristensen. Reproduced by permission.)

because the feeding stages lack an excretory system, wastes are accumulated, and that self-renewal enables the individual to eliminate these wastes before they reach toxic concentrations.

Other than the internal, self-renewing buds, the feeding individuals can also produce Pandora larvae. These larvae also develop from internal buds, and always develop simultaneously with the formation of new organs of the feeding stage.

The fully developed Pandora larva is released from the feeding stage through the mother's anus. It is a poor swimmer, but may crawl using a ciliated field on the ventral part of its head. When it has escaped from its mother, it creeps slowly on the lobster's setae, trying to locate a suitable settling spot with its cephalic sensory structures. After the Pandora larva settles, it adheres to a seta, the larval traits immediately degenerate, and a cystlike structure is formed.

When the Pandora larva is still inside its mother, a new feeding stage individual is already visible inside the larva, and this new feeding stage contains an internal bud with clear traces of a developing mouth ring that will be ready for self-renewal as soon as the new feeding stage is fully developed. When the settled Pandora larva turns into a cyst, the new feeding stage starts to grow, and after a short time, the buccal funnel emerges through the cyst and the new individual starts to feed. The formation of Pandora larvae mostly happens in young colonies, typically after the lobster has molted, when the population has to grow quickly.

Under certain circumstances, reproduction may shift from an asexual to a sexual phase. In the sexual phase, the feeding stages may produce either females or the so-called Prometheus larvae that give rise to males. Both are, like the Pandora larva, produced by internal budding. The Prometheus larva can only move over very short distances, and immediately after it is released from its maternal feeding individual it seeks the closest feeding individual with a developing female inside and attaches to this individual close to its anal opening. When the Prometheus larva is attached, one or two dwarf males begin to develop inside it. The dwarf males are good swimmers and are easily recognized by their large cuticular penises. When the female escapes through the anus of the feeding stage, the Prometheus larva releases the dwarf males, which quickly find and fertilize the female. The female looks very similar to the Pandora larva, but contains a single very large oocyte (egg). Usually she settles on the flagellum of the lobster's mouth limbs, and not in the colonies of the sessile feeding individuals. When she has attached, she starts to degenerate and form a cyst, as did the Pandora larva. Inside the cyst a new larva, the chordoid larva, starts to develop. The chordoid larva has a dense ventral ciliation, and is a much better swimmer than any of the other stages in the life cycle. It is therefore able to swim to a new lobster or to stay in the free water while its host is molting. When the chordoid larva has settled, it turns into a cyst and a new feeding stage starts to develop.

Conservation status

The Cycliophora are not listed by the IUCN.

Significance to humans

The Cycliophora have no significance for humans, and as they apparently do not harm their lobster host, have no economic significance.

Resources

Books

Funch, Peter, and Reinhardt M. Kristensen. "Cycliophora." In *Microscopic Anatomy of Invertebrates*. Vol. 13, *Lophophorates, Entoprocta, and Cycliophora*, edited by F. W. Harrison and R. M. Woollacott. New York: Wiley-Liss, 1995.

———. "Cycliophora." In *Encyclopedia of Reproduction*, Vol. 1. New York: Academic Press, 1999.

Kristensen, R. M. "Cycliophora." In *Encyclopedia of Life Sciences*. Vol. 5. London: Macmillan Reference, 2002.

Kristensen, R. M., and Peter Funch. "Phylum Cycliophora." In *Atlas of Marine Invertebrate Larvae*, edited by C. M. Young, M. A. Sewell and M. E. Rice. London: Academic Press, 2002.

Periodicals

Funch, P. "The Chordoid Larva of *Symbion pandora* (Cycliophora) Is a Modified Trochophore." *Journal of Morphology* 230 (1996): 231–263.

Funch, P., and R. M. Kristensen. "Cycliophora Is a New Phylum with Affinities to Entoprocta and Ectoprocta." *Nature* 378 (1995): 711–714.

———. "An Introduction to Loricifera, Cycliophora, and Micrognathozoa." *Integrative and Comparative Biology* 42, no. 3 (June 2002): 641–651.

Obst, M., and P. Funch. "Dwarf Male of *Symbion pandora* (Cycliophora)." *Journal of Morphology* 255, no. 3 (March 2003): 261–278.

Martin Vinther Sørensen, PhD
Reinhardt Møbjerg Kristensen, PhD

Crinoidea

(Sea lilies and feather stars)

Phylum Echinodermata

Class Crinoidea

Number of families 25

Thumbnail description

Stalked or stalkless organisms with a crown composed of a calyx, five or multiple arms, an anal cone, and a mouth pointing upward

Photo: Feather star arms trapping plankton near Sipadan Island, Borneo, in the South China Sea. (Photo by ©Jeff Rotman/Photo Researchers, Inc. Reproduced by permission.)

Evolution and systematics

Crinoids are a living lineage of echinoderms more than 500 million years old. The first crinoids were stalked forms (the sea lilies), whose probable ancestors are the extinct rhombiferans or the extinct edrioasteroid echinoderms. The first fossil record dates from the Lower Ordovician (510 million years ago[mya]). During the Paleozoic era (550–245 mya), there were at least two major expansions and declines in crinoid diversity. In the early Carboniferous (360 mya) crinoid diversity reached its zenith, exceeding the total diversity of all other echinoderm taxa. During the Permo-Triassic extinction (240 mya), the crinoids suffered a catastrophic decline and only one lineage survived, which gave rise to the earliest subclass, Articulata. Throughout the Mesozoic era, this lineage had begun to diversify and, about the time of the early Jurassic (210 mya), the order Comatulida (stalkless crinoids, the feather stars) appeared. The disappearance of stalked crinoids from shallow waters and their restriction to deeper sites coincides with the Mesozoic radiation of predatory bony fishes. About 6,000 species of crinoids have lived and died out in past geological ages.

There are about 600 feather star species distributed among 150 genera and 17 families in one order, and 95 extant sea lily species distributed among 25 genera (50% of them are monospecific), 8 families, and 4 orders. The living crinoids orders are: Millericrinida, Cyrtocrinida, Bourgueticrinida, and Isocrinida (all sea lilies); and Comatulida (feather stars).

The class Crinoidea is the ancestor group of all other echinoderm classes. The relationships among extant orders are still obscure, but some attempts have been made to elucidate them. Among the orders, Millericrinida and Isocrinida are the most ancient. The comatulids diverged from a group

of Isocrinida, and the bourgueticrinids, due to the retention of larval stem, diverged from the comatulids. Cyrtocrinids possibly diverged from the millericrinids.

Physical characteristics

Crinoids are pentamerous organisms that differ from other echinoderm classes because of the upward position of their mouth. Numerous calcareous plates, more or less firmly joined together, form their endoskeleton. The main body part is the crown, which is made up of the calyx, the tegmen, and the arms. The calyx, a rigid cup formed by the calcareous plates, carries the digestive tract, the mouth, esophagus, gut, rectum, and anus. An upper membrane, called the tegmen, bears the openings of mouth and anus and is perforated by numerous small pores that connect the interior of the crinoid with the external environment. The mouth is usually located near or at the center of the tegmen, although it is displaced peripherally in the family Comasteridae. The anus, displaced from the center, is elevated at the tip of a cone or tube. In all but adult comatulide there is a cylindrical or polygonal stalk (stem, column) below the crown, which elevates the crown above the substratum. In comatulids, a cluster of appendages, called cirri, takes the place of the column. The cirri may or may not be present along columns of stalked crinoids. All crinoids have five arms, developing from the calyx, that usually branch one or more times, giving rise to up to 200 arms. Small branches called pinnules border each arm. Those nearest to the month are called oral pinnules. Gonads usually occur in the next group of pinnules, called the gonadal pinnules, although they may also occur in the arm axis, but almost never in the central mass of the body. After the gonadal pinnules are the so-called distal pinnules. Ambulacral grooves (food

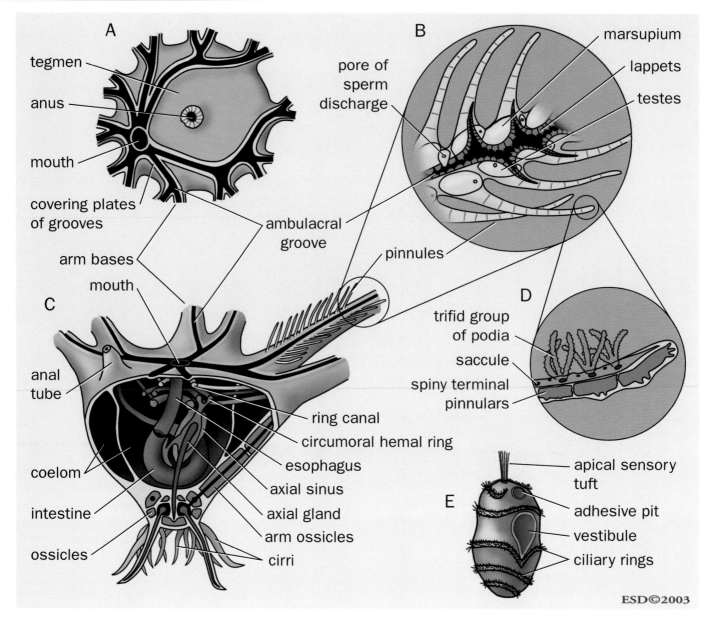

A. Oral view of a comasterid; B. Close-up of a male's arm; C. Feather star anatomy; D. Close-up of a pinnule; E. Doliolaria larva. (Illustration by Emily Damstra)

grooves) occur in the oral surface of the calyx and reach the distal extremity of each arm and pinnule. Ambulacral podia (tube feet) line each groove, but the pinnular podia are organized in groups of three podia of different sizes (each podia with different functions during feeding).

In living crinoids, the arms range in size from 0.39 to 13.8 in (1 to 35 cm), depending on the species. The stem of living sea lilies reaches about 3.3 ft (1 m) long, but was much longer in some fossil species, up to more than 65.6 ft (20 m). Comatulids may be of almost any color, white through black, purple, red, green, brown, or violet. The species may be uniform in color or have a combination of colors. Usually, the deeper the organism, the paler the color.

Distribution

Crinoids are found from substidal fringe zones to great depths in tropical, temperate, and polar waters, although they are more diversified in coral reefs of the tropical Indo-Pacific and Caribbean (although fewer species are present in the Caribbean). Stalked crinoids are restricted to the deep sea, with just a few species living at depths of 200–490 ft (60–150 m). There are three major areas of sea-lily biodiversity: the tropical West Pacific, where three members of the order Isocrinida predominate, the pentacrinids at 660–1,970 ft (200–600 m) and the bathycrinids and hyocrinids at 4,920–9,840 ft (1,500–3,000 m); the tropical western Atlantic, where more diversity occurs at upper water levels; and the Northeastern Atlantic, where more diversity occurs at a deeper water level.

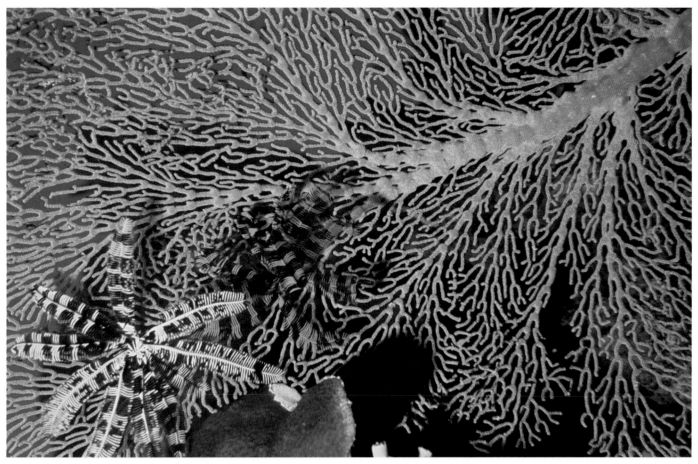

Feather stars (order Comatulida) on a gorgonia sea fan. (Photo by Bill Wood. Bruce Coleman, Inc. Reproduced by permission.)

Habitat

Crinoids frequently live on hard substratum. Some live in areas of high current flow, usually use the vertical filtration fan posture (described in the next section). Others avoid high streams and use the radial feeding posture. Nevertheless, crinoid community is probably determined by substratum complexity, independent of water flow. A highly complex substratum may trigger a high diversity crinoid community, and a homogenous substratum carry a low diversity crinoid community.

Behavior

Feather stars usually live in clumps, preferring to attach to crevices, lateral surfaces, or in other places in which they can hide their central mass. This behavior prevents and avoids injuries to vital body parts caused by predators, and also optimizes filtration by enhancing the baffle effect, which improves the chance of food particles touching the feeding structure. They frequently emerge at night, exposing part, or all of, the arm, or even the entire body, although some species emerge during daylight, and others are exposed both during the day and at night.

Stalked crinoids also occur in dense clusters, but do not have a diel pattern of emergence because of the lack of light in deep water. Most can also be found attached to a hard sub-

stratum. The depth distribution of stalked-crinoid diversity seems to be controlled by variations of both crinoid hydrodynamic vulnerability and abundance of food particles reaching the sea floor.

Crinoids can also regenerate lost body parts. Feather stars can regenerate their arms as long as at least one arm and an intact dorsal nerve center remain. Sea lilies can regenerate an entire crown.

Feather stars are able to crawl over the substratum utilizing their arms. Some comatulids have been observed swimming. They swim by alternating their arms up and down, and descend through the water by extending their arms out like parachutes. Only a few sea lilies are able to crawl over the substratum, and none have been observed swimming.

Feeding ecology and diet

Feather stars assume a vertical filtration fan posture in areas of high current flow. In this posture, the arms are deployed in a planar fan, with pinnules held in the same plane and the food grooves usually directed downstream. The vertical filtration posture serves to present the maximum cross-sectional area of food-collecting surfaces to the incoming water flow, and also acts to baffle through-flowing water, pos-

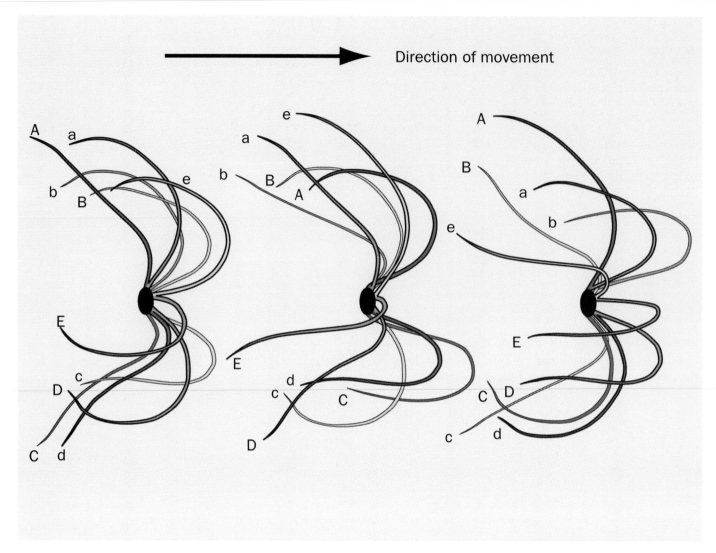

A feather star swimming. Three groups of arms (red, yellow, blue) move separately to achieve forward motion. (Illustration by Emily Damstra)

sibly facilitating the capture of food particles by the tube feet. Sea lilies assume a similar feeding posture, although they recurve their arms almost 270° upstream to form a parabolic filtration fan. The mouth may be oriented laterally downstream, with the food grooves also turned downstream, or it may be oriented upward in slack currents. Feather stars living in low-current areas use a radial feeding posture, orienting their arms in many directions with the pinnules extended radially in four rows. The radial feeding posture serves to maximize the surface area of the feeding structures so that more particles will settle on them.

The crinoid diet consists of phyto- and zooplankton and detritus, and varies with habitat and seasonal availability. The size of the particles captured depends on the width of the food groove. The primary podium (the largest in the group of three) collects particles in the water column and folds them back into the groove. Relatively large particles are captured by podia partly curling over them; small particles adhere to the mucous layer. The podia transfer the par-

ticles to the food groove by brushing them away with the ciliary tract or the tertiary podia. The secondary podia behave as the primary and secondary podia do, collecting particles in the water column and folding them back into the food groove.

Reproductive biology

All crinoid species are gonochoric (although some individuals may present hermaphroditism), and they probably do not reproduce asexually. Depending on the species, the ova vary in size from 0.004–0.012 in (100 to 300 μm). The maturing oocyte enters the ovarian lumen through a temporary opening in the layer of nongerminal cells in the inner epithelium, a process called ovulation. Inside the ovarian lumen, the oocytes undergo two maturation divisions and become ova. Crinoids take 12 to 18 months to reach maturity. The gametogenic cycle usually takes one year, although in some species it takes several months and in others takes almost three years. The spawning season, the period of the year during which gametes are released, varies among species and populations of each species and can last from one hour to many months.

Sperm are released directly from the testes into sea water. Females of most species also spawn freely into sea water, but in some feather stars, the ova are retained on the outer surface of the mother's genital pinnule. In these species, the ova may be kept for days and then released, or may enter into brood pouches (where such pouches exist). Almost all crinoids develop by lecititrophic larvae (short-lived, nonfeeding, planktonic larvae called doliolaria larvae) followed by a benthic, nonfeeding, stalked stage that metamorphoses to a benthic stalked juvenile. Most crinoids have only doliolaria larvae, which are ovoid with four or five transverse bands of cilia and a tuft of apical cilia. Only one species is known to have internally brooded vitellaria larvae, which lack the ciliated bands.

Conservation status

No species are listed by the IUCN.

Significance to humans

None known.

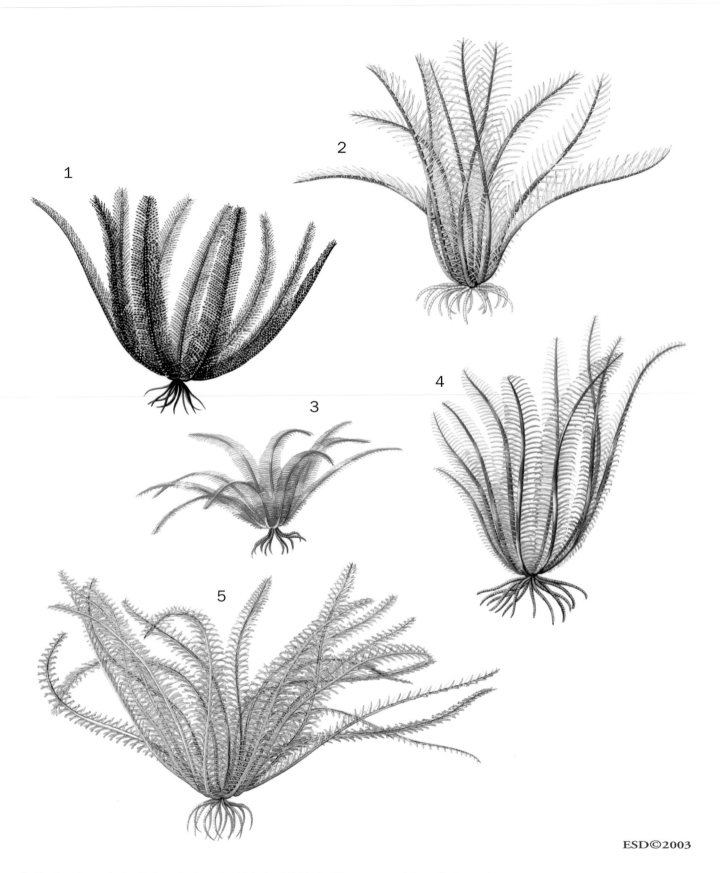

1. *Tropiometra carinata*; 2. Rosy feather star (*Antedon bifida*); 3. *Oligometra serripinna*; 4. *Comactinia echinoptera*; 5. Orange sea lily (*Nemaster rubiginosa*). (Illustration by Emily Damstra)

ESD©2003

1. West Atlantic stalked crinoid (*Endoxocrinus parrae*); 2. Great West Indian sea lily (*Cenocrinus asterius*). (Illustration by Emily Damstra)

ESD©2003

Species accounts

Rosy feather star
Antedon bifida

ORDER
Comatulida

FAMILY
Antedonidae

TAXONOMY
Antedon bifida Pennant, 1777, western coast of Scotland.

OTHER COMMON NAMES
None known.

PHYSICAL CHARACTERISTICS
Feather star with 10 arms, 2–4 in (50–100 mm) long; pink, red
yellow, or orange, frequently banded, usually with white pinnules.

DISTRIBUTION
Northeastern Atlantic, from Shetland Isles to Liberia and west
to the Azores. Intertidal zone to 1,476.4 ft (450 m).

HABITAT
Lives in current-agitated shallow water attached to hard substrata, such as cliff faces and boulders.

BEHAVIOR
Broods offspring by bringing the arms near the body and folding the pinnules against the arm axis.

FEEDING ECOLOGY AND DIET
Feeds through the vertical filtration fan posture.

REPRODUCTIVE BIOLOGY
On the external wall of the genital pinnules, females brood
their eggs in a mucus net until they hatch as early larvae.
Spawning occurs in late spring, although mature gametes may
occur for many months throughout the year.

CONSERVATION STATUS
Not listed by the IUCN.

SIGNIFICANCE TO HUMANS
None known. ◆

No common name
Oligometra serripinna

ORDER
Comatulida

FAMILY
Colobometridae

TAXONOMY
Oligometra serripinna Carpenter, 1881, Andai, New Guinea.

OTHER COMMON NAMES
None known.

PHYSICAL CHARACTERISTICS
Feather star with 10 arms, 0.4–1.8 in (10–45 mm) long. Proximal pinnule segments longer than they are broad; yellow to orange.

DISTRIBUTION
Western Indian Ocean to the South Pacific, through the Great
Barrier Reef. From at least 20–80 ft (6–25 m) in depth. (Spe-
cific distribution map not available.)

HABITAT
Usually fully exposed on unsheltered perches and attached on
hard substratum.

FEEDING ECOLOGY AND DIET
Feeds through the radial filtration fan posture.

BEHAVIOR
No diel pattern.

REPRODUCTIVE BIOLOGY
Probably reproduces twice a year, around February and June.
Spawns freely on sea water and develops through lecititrophic
larvae. By the spawning time, a large range of gamete sizes exists, but presumably only the larger gametes are released by repeated trickle spawning.

CONSERVATION STATUS
Not listed by the IUCN.

SIGNIFICANCE TO HUMANS
Nothing known. ◆

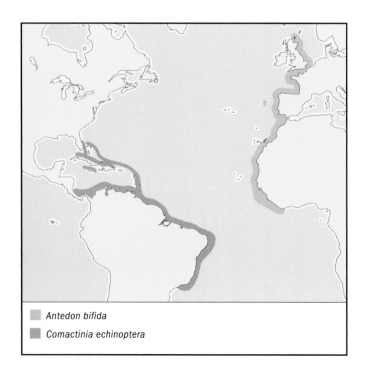

■ *Antedon bifida*
■ *Comactinia echinoptera*

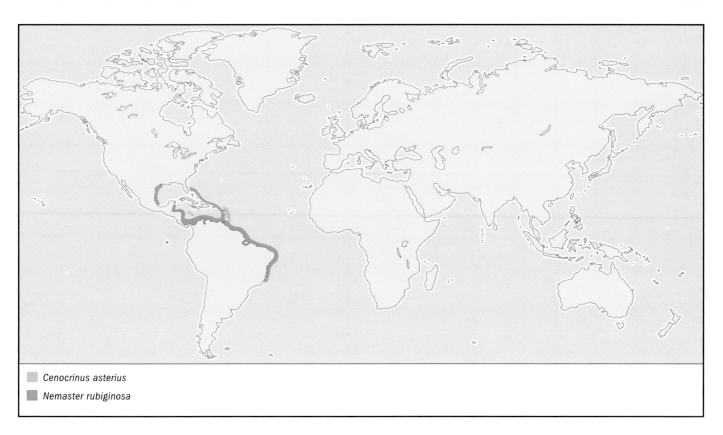

■ Cenocrinus asterius
■ Nemaster rubiginosa

No common name
Comactinia echinoptera

ORDER
Comatulida

FAMILY
Comasteridae

TAXONOMY
Comactinia echinoptera Müller, 1840, type locality unknown.

OTHER COMMON NAMES
None known.

PHYSICAL CHARACTERISTICS
Feather star with 10 arms, 1.9–6.5 in (30–165 mm) long, slightly broader in middle than at base, and with comb-bearing proximal pinnules. Arms most distal from mouth are less than half the length of those nearest to mouth. Has central anal cone and marginal mouth. Color is extremely variable, but arms have a reddish background color of varying shades and may present white, yellow, or brown spots. Pinnules are red to orange or yellow.

DISTRIBUTION
From southeastern Florida to Cabo Frio (and perhaps Alcatrazes Island), Brazil; through the Bahamas, Turks, and Caicos Islands, and the Antillean Arc, and Caribbean coasts of Central and South America. Intertidal to 295 ft (90 m), possibly to 590 ft (180 m).

HABITAT
Areas of current in shallow water, attached to hard substratum.

BEHAVIOR
Occurs in high densities; nocturnal; extends the longest arms above the substratum, usually with central mass hidden, while attached in crevices.

FEEDING ECOLOGY AND DIET
Feeds through the vertical filtration fan posture.

REPRODUCTIVE BIOLOGY
Nothing is known.

CONSERVATION STATUS
Not listed by the IUCN.

SIGNIFICANCE TO HUMANS
None known. ◆

Orange sea lily
Nemaster rubiginosa

ORDER
Comatulida

FAMILY
Comasteridae

TAXONOMY
Nemaster rubiginosa Portalès, 1869, off Orange Key, Bahama Bank and off Tortugas.

OTHER COMMON NAMES
None known.

PHYSICAL CHARACTERISTICS
Feather star with 20 (up to 35) arms 3.9–7.9 in (100–200 mm) long, bright orange with a black stripe running along dorsal side. Mouth and anal cone about equal distance from center of tegmen.

DISTRIBUTION
Western Gulf of Mexico, southeastern Florida, Bahamas, Barbados; Caribbean coast of Central and South America from Belize to Bahia, Brazil. At 3.3–1,100 ft (1–334 m) deep.

HABITAT
Lives in shallow water, frequently sheltered from current, attached on hard substratum, favors the fore-edges of reef escarpments.

BEHAVIOR
Central mass is hidden while attached to undersurfaces of hard substratum. Only arms are visible by day, but entire body may be exposed at night.

FEEDING ECOLOGY AND DIET
Feeds through the radial feeding posture.

REPRODUCTIVE BIOLOGY
Produces gametes throughout the reproductive cycle (usually spring). By spawning time, a complete range of gamete development stages exists, from recently produced oocytes to fully mature ova.

CONSERVATION STATUS
Not listed by the IUCN.

SIGNIFICANCE TO HUMANS
None known. ◆

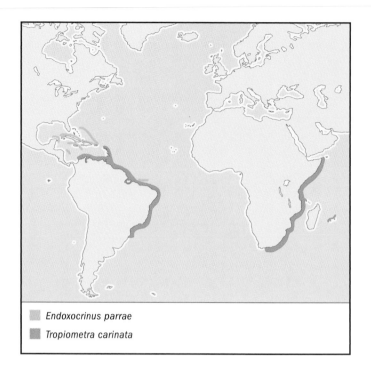

<table>
<tr><td>▨</td><td>Endoxocrinus parrae</td></tr>
<tr><td>▨</td><td>Tropiometra carinata</td></tr>
</table>

No common name
Tropiometra carinata

ORDER
Comatulida

FAMILY
Tropiometridae

TAXONOMY
Tropiometra carianta Lamarck, 1816, Mauritius.

OTHER COMMON NAMES
None known.

PHYSICAL CHARACTERISTICS
Feather star with 10 arms, 3.9–7.1 in (100–180 mm) long. Arms and pinnules usually dark brown with yellow bands, although individuals may be yellow, yellow with brown bands, or orange.

DISTRIBUTION
Antillean Arc from Guadeloupe south; South America from Cartagena, Colombia, through Trinidad Tobago, to Santa Catarina, Brazil. Santa Helena. East Africa from Cape of Good Hope to Northern Somalia; Madagascar Islands from Comoros to the Seychelles, Mascarene Islands, and Cargados Carajos Shoals. Intertidal fringe to 276 ft (84 m).

HABITAT
Lives on current agitated shallow water, attached to hard substratum.

BEHAVIOR
Specimens usually found in clumps, mainly of small individuals that tend to live near larger ones. Individuals may or may not have the central mass hidden while attached on lateral surfaces or in crevices or nooks (where more frequently found). No diel pattern of emergence.

FEEDING ECOLOGY AND DIET
Feeds through the vertical filtration fan posture.

REPRODUCTIVE BIOLOGY
Free spawning in sea water and have lecititrophic larvae. Differs from other species because it spawns immature ova.

CONSERVATION STATUS
Not listed by the IUCN.

SIGNIFICANCE TO HUMANS
None known. ◆

Great West Indian sea lily
Cenocrinus asterius

ORDER
Isocrinida

FAMILY
Isocrinidae

TAXONOMY
Cenocrinus asterius Linnaeus, 1775.

OTHER COMMON NAMES
None known.

PHYSICAL CHARACTERISTICS
Sea lily with up to 50 arms and a stalk, to 3.3 ft (1 m) long.

DISTRIBUTION
Northwest Providence Channel; Antillean Arc from Saba to Barbados, including Jamaica. At 600–1,920 ft (183–585 m) deep.

HABITAT
Groups of cirri along a curved stalk anchor individuals on hard substratum.

BEHAVIOR
Individual arms wave rapidly up and down to prevent the settlement of other organisms or undesired particles on the crown. Can also move from place to place by crawling over substratum using their arms. Relocation may be stimulated by the presence of other individuals or other organisms as possible predators.

FEEDING ECOLOGY AND DIET
Feeds through the parabolic filtration fan posture.

REPRODUCTIVE BIOLOGY
Nothing is known.

CONSERVATION STATUS
Not listed by the IUCN.

SIGNIFICANCE TO HUMANS
None known. ◆

West Atlantic stalked crinoid
Endoxocrinus parrae

ORDER
Isocrinida

FAMILY
Isocrinidae

TAXONOMY
Endoxocrinus parrae Gervais, 1835, off Cuba.

OTHER COMMON NAMES
None known.

PHYSICAL CHARACTERISTICS
Sea lily similar to *Cenocrinus asterius*, but with a shorter stalk and a denser filtration fan. The arm-branching pattern is also different because its arms divide until there are a total of eight arms.

DISTRIBUTION
Off Cape Canaveral; Bahama Island; Antillean Arc; Yucatán Channel; off São Luiz, northeast Brazil. At 505–3,186 ft (154 to 971 m) deep, although some found in shallow water.

HABITAT
Groups of cirri along a curved stalk anchor the lily on hard substratum.

BEHAVIOR
Individual arms wave rapidly up and down to prevent the settlement of other organisms or undesired particles on the crown. Can also move from place to place by crawling over substratum using their arms. Relocation may be stimulated by the presence of other individuals or other organisms as possible predators.

FEEDING ECOLOGY AND DIET
Feeds through the parabolic filtration fan posture.

REPRODUCTIVE BIOLOGY
Nothing is known.

CONSERVATION STATUS
Not listed by the IUCN.

SIGNIFICANCE TO HUMANS
None known. ◆

Resources

Books

Ausich, William I. "Origin of Crinoids." In *Echinoderm Research 1998*, edited by Candia Carnevali and F. Bonasoro. Rotterdam: Balkema, 1999.

Hess, Hans, William I. Ausich, Carlton E. Brett, and Michel J. Simms, eds. *Fossil Crinoids*. Cambridge: Cambridge University Press, 1999.

Littlewood, D. Tim J., Andrew B. Smith, K. A. Clough, and Roland H. Emson. "Five Classes of Echinoderm and One School of Thought." In *Echinoderms: San Francisco*, edited by R. Mooi and M. Telford. Rotterdam: Balkema, 1998.

Simms, Mike J. "The Phylogeny of Post-Palaeozoic Crinoids." In *Echinoderm Phylogeny and Evolutionary Biology*, edited by C. R. C. Paul and A. B. Smith. Oxford: Clarendon Press, 1988.

Periodicals

Ameziane, Nadia, and Michel Roux. "Biodiversity and Historical Biogeography of Stalked Crinoids (Echinodermata) in the Deep Sea." *Biodiversity and Conservation* 6 (1997): 1557–1570.

Ausich, William I. "Early Phylogeny and Subclass Division of the Crinoidea (Phylum Echinodermata)." *Journal of Paleontology* 72 (1998): 499–510.

Ausich, William I., and Thomas W. Kammer. "The Study of Crinoids During the 20th Century and the Challenges of the 21st Century." *Journal of Paleontology* 75 (2001): 1161–1173.

Baumiller, Tomasz K., Michael LaBarbera, and Jeremy D. Woodley. "Ecology and Functional Morphology of the Isocrinid *Cenocrinus asterius* (Linnaeus) (Echinodermata: Crinoidea): In Situ and Laboratory Experiments and Observations." *Bulletin of Marine Science* 48 (1991): 731–748.

Fabricius, Katharina E. "Spatial Patterns in Shallow-Water Crinoid Communities on the Central Great Barrier Reef." *Australian Journal of Marine and Freshwater Research* 45 (1994): 1225–1236.

Guensburg, Thomas E., and James Sprinkle. "Earliest Crinoids: New Evidence for the Origin of the Dominant Paleozoic Echinoderms." *Geology* 29 (2001): 131–134.

Holland, Nicholas D., J. Rudi Strickler, and A. B. Leonard. "Particle Interception, Transport and Rejection by the

Resources

Feather Star *Oligometra serripina* (Echinodermata: Crinoidea), Studied by Frame Analysis of Videotapes." *Marine Biology* 93 (1986): 111–126.

Lahaye, M. C., and Michel Jangoux. "Functional Morphology of the Podia and Ambulacral Grooves of the Comatulid Crinoid *Antedon bifida* (Echinodermata)." *Marine Biology* 86 (1985): 307–318.

MacCord, Fábio S., and Luiz F. Duarte, L. "Dispersion in Populations of *Tropiometra carinata* (Crinoidea: Comatulida) in the São Sebastião Channel, São Paulo State, Brazil." *Estuarine Coastal and Shelf Science* 54 (2002): 219–225.

Macurda, Jr., Donald B., and David L Meyer. "Feeding Posture of Modern Stalked Crinoids." *Nature* 247 (1974): 394–396.

McClintock, James B., Bill J. Baker, Tomasz K. Baumiller, and Charles G. Messing. "Lack of Chemical Defense in Two Species of Stalked Crinoids: Support for the Predation Hypothesis for Mesozoic Bathymetric Restriction." *Journal of Experimental Marine Biology and Ecology* 232 (1999): 1–7.

McEdward, Larry R., and Benjamin G. Miner. "Larval and Life-Cycle Pattern in Echinoderms." *Canadian Journal of Zoology* 79 (2001): 1125–1170.

Messing, Charles G., M. Christine RoseSmyth, Stuart R. Mailer, and John E. Miller. "Relocation Movement in a Stalked Crinoid (Echinodermata)." *Bulletin of Marine Science* 42 (1988): 480–487.

Meyer, David L. "Distribution and Living Habits of Comatulid Crinoids Near Discovery Bay, Jamaica." *Bulletin of Marine Science* 23 (1973): 244–259.

———. "Feeding Behavior and Ecology of Shallow-Water Unstalked Crinoids (Echinodermata) in the Caribbean Sea." *Marine Biology* 22 (1973): 105–129.

Meyer, David L., Charles G. Messing, and Donald B. Macurda, Jr. "Zoogeography of Tropical Western Atlantic Crinoidea (Echinodermata)." *Bulletin of Marine Science* 28 (1978): 412–441.

Nichols, David. "Evidence for a Sacrificial Response to Predation in the Reproductive Strategy of the Comatulid Crinoid *Antedon bifida* from the English Channel." *Oceanologica Acta* 19 (1996): 237–240.

———. "Reproductive Seasonality in the Comatulid Crinoid *Antedon bifida* (Pennant) from the English Channel." *Philosophical Transactions of the Royal Society of London B* 343 (1994): 113–134.

Vail, Lyle. "Diel Patterns of Emergence of Crinoids (Echinodermata) from Within a Reef at Lizard Island, Great Barrier Reef, Australia." *Marine Biology* 93 (1987): 551–560.

———. "Reproduction in Five Species of Crinoids at Lizard Island, Great Barrier Reef." *Marine Biology* 95 (1987): 431–446.

———. "Arm Growth and Regeneration in *Oligometra serripina* (Carpenter) (Echinodermata: Crinoidea) at Lizard Island, Great Barrier Reef." *Journal of Experimental Marine Biology and Ecology* 130 (1989): 189–204.

Young, Craig M., and Roland H. Emson. "Rapid Arm Movements in Stalked Crinoids." *Biological Bulletin* 188 (1995): 89–97.

Fábio Sá MacCord, MSc

Asteroidea

(Sea stars)

Phylum Echinodermata
Class Asteroidea
Number of families 35

Thumbnail description
Conspicuous and successful bottom-dwelling animals that can survive without food for months and feed on almost every type of marine organism encountered on the seabed; they range in size from 0.4 in (1 cm) in diameter to more than 3 ft (91 cm) across and inhabit virtually every latitude and ocean depths

Photo: Chocolate-chip sea star (*Protoreaster nodosus*). (©Shedd Aquarium. Photo by Patrice Ceisel. Reproduced by permission.)

Evolution and systematics

The class Asteroidea is a highly diverse group comprised of seven orders, 35 families, and an estimated 1,600 known living species, although their precise phylogenetic relationship and hence classification still proves challenging to taxonomists.

Asteroids belong to a major group of other bottom-dwelling animals called echinoderms. Collectively this group includes echinoids (sea urchins), holothurians (sea cucumbers), crinoids (feather stars), and ophiuriods (brittle stars), the latter group closely resembling sea stars. All echinoderms share similar pentamerous radial symmetry and spiny skin characteristics, although sea stars differ slightly because they have five or more arms large enough to contain space for digestive and reproductive glands. Another group of animals thought to belong to echinoderms are concentricycloids, or sea daisies. These small disc-shaped animals discovered in the abyssal seas off New Zealand and Bahamas in the late 1980s are considered an evolutionary forerunner to asteroids.

Sea stars have an ancient linage that shows embryologically they are not too distantly related to the phylum Chordata (back-boned animals). The fossil record places a form of asteroid over 300 millions years before the dinosaurs, sharing a common ancestry with ophiuroids, yet within 50 million years of their appearance they became clearly differentiated. Their evolutionary path has included some bizarre taxa that have been hard to classify, yet this successful group has persisted and remain ecologically important to many marine communities worldwide.

Physical characteristics

Sea stars vary considerably in size, shape, and color, even within the same populations. Their diverse forms reflect evolutionary adaptation to the cosmopolitan habitats they occupy. Despite this diversity they all share similar physical characteristics. All are star-shaped (stellate) with a central body or disc that has symmetrically projecting arms with rows of tube feet running along the lower surface of a V-shaped furrow called the ambulacral groove. Typically, the number of arms is five, but some species such as the coral-eating crown-of-thorns *Acanthaster planci* can have up to 30. Their size ranges from the 0.4 in (1 cm) arms of the cushion star *Patiriella parvivipara*, which gives the appearance of a nearly spherical body, to the long skinny arms of *Novodinia antillensis*, which span almost 3 ft (91 cm) in diameter. In most species, the arm tips carry an optic cushion of red-pigmented and light-sensitive cells that sense changes in the prevailing environment.

The skeleton of a sea star consists of small calcium carbonate plates called ossicles. These are often studded and spiny, and provide a firm but flexible skeleton of connective tissue. Flexibility enables a variety of postures to be adopted without muscular effort, thus providing an effective means to capture and handle prey and allow individuals to closely follow irregular substrates in search of food. Alternatively, their flexibility can enable sea stars to upright themselves if overturned.

The surface of a sea star looks and feels rough because of the numerous small and transparent sacs called papulae that cover the body, which provide a respiratory surface for exchanging oxygen. The upper and lower body surfaces also contain pincer-like structures called pedicellariae, which come in a variety of forms from simple modified spines to highly specialized opposing hooks. Their function is to rid areas around the papulae of small organisms and debris, and in some species capture prey by detecting their presence. These are

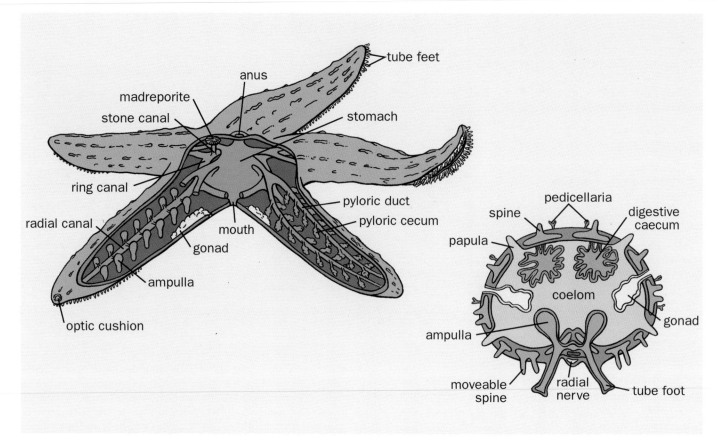

Sea star anatomy. (Illustration by Patricia Ferrer)

usually small fish or shrimp-like crustaceans on which the sea star feeds. The shape of pedicellariae is an important characteristic for asteroid taxonomy.

Sea stars have an unusual way of moving. Water is taken in through the madreportite, a small, perforated plate on the upper surface of the disc, and into the water vascular system, a canal of tubes connected to the tube feet. Following muscular contraction water is directed under pressure to the tube feet, which then extend under its force. Movement is achieved through a coordinated stepping motion where, on muscle contraction, the feet adhere to the sediment surface, pushing the individual forward. Depending on the species, tube feet have suckers that help stick the sea star to hard surfaces or assist in prying open the shells of its mollusk prey. Besides involvement in prey capture; tube feet also have a respiratory function. Species with more than five arms and reproduce asexually will have numerous madreportites.

Distribution

The greatest diversity of sea stars occur in coastal regions, although as a group, they are well represented globally from the Antarctic, Pacific, Atlantic, and Indian Oceans. They inhabit wave-exposed inter-tidal zones of coastal waters to the calm sandy pavements of the deepest oceans. The Benthopectinidae family of sea stars, for example, lives exclusively in the deep-sea of the Atlantic and Pacific Oceans, whereas

the species *Odontaster validus*, which belongs to the family Ganeriidae, are found only in the Antarctic Ocean. Perhaps the most well-known and ubiquitous group of sea stars belongs to the order Forcipulatida. This group includes the genus *Asterias*, a veracious predator of mussels and oysters in many coastal waters around the world. In the Far East, the species *Asterias amurensis* has extended its normal range into New Zealand and Australia, where as a non-native species it has caused extensive economical and ecological damage to the shellfish industry.

Habitat

As a group, sea stars live in virtually every habitat found in the sea, ranging from tidal pools, rocky shores, sea grass and kelp beds, beneath rock rubble, on coral reefs, sand, and mud. In some species a broad and flattened body may act as a snowshoe when foraging on very soft mud. In the upper shore, they are periodically exposed by the retreating tide, resulting in extended periods of desiccation. The only refuge is cover in moist crevices beneath rocks. By contrast, in the deep sea at depth greater than 29,530 ft (9,000 m) they are found inhabiting sandy bottoms and steep cliffs.

They are prominent seafloor predators. Perhaps their success and influence comes from a unique combination of attributes. These include indeterminate growth, a morphology and digestive system generalized enough to capture, handle, and ingest many different prey types and sizes, and a sensory ability sophisticated enough to respond quickly to the presence of prey and changes in the prevailing environment. Moreover, their flexible bodies and suckered tube feet enable them to adhere firmly to the seabed whilst manipulating prey, thus enabling them to survive in high stress environments by withstanding the full force of crashing waves.

Behavior

Sea stars have a "central nervous system," or diffuse nerve net, but lack anything identified as a brain. Despite this, they are sophisticated enough to adapt to change based on previous experiences (conditioning), whereby behavior that is persistently unsuccessful, usually a feeding one, is stopped.

They are not considered social animals, yet many species tend to aggregate or swarm in large numbers during certain times of the year. These events tend to be triggered during spawning periods, feeding frenzies, or seasonal migrations to deeper waters offshore. Some sea stars show avoidance behavior to other species or attraction towards members of the opposite sex. Feeding is perhaps the most common cause of aggregation, where sea stars can appear in thousands to prey on mussels, oysters, or coral.

The daily activity patterns in many sea stars are synchronized to changes in light intensity, usually around dawn and dusk. Such activity may help to avoid predators and coincide sea star foraging activity with the activity of their preferred prey. In others such as *Astropecten irregularis*, daily activity patterns are synchronized to periods of slack water on a high and low tide when velocities are low enough to optimize foraging success.

Feeding ecology and diet

Sea stars are carnivorous, preying on sponges, shellfish, crabs, corals, worms, and even on other echinoderms. Most are generalists, feeding on anything that is too slow to escape, such as mussels and clams, whilst others are specialized feeders preying exclusively on sponges, corals, bivalves, or algae. Prey is located by the chemical odors emanating from its waste products or by small movements that betray its presence when detected by a sea star. Food preferences can change depending on availability of prey, which change geographically and seasonally. Even weather conditions in temperate species and reproductive state (usually during gonad growth) affects dietary requirements.

Feeding strategies can be divided into those that are scavengers, feeding mainly on decaying fish and invertebrates; those that are deposit feeders, filling their stomachs with mud from which they extract microscopic organisms and organic matter; and those that are suspension feeders, filtering prey and food particles from the water (e.g., *Novodinia antillensis*).

Depending on the species, sea stars have two different feeding methods. Intra-oral feeders ingest their prey into their stomach alive, sometimes distending or rupturing their disc in the process. The burrowing sand star *Astropecten irregularis*, for example, can swallow hundreds of live juvenile mollusks during one foraging period. In some cases prey such as clams

The bat star (*Patiria miniata*) is found along the Pacific coast of North America. (©Shedd Aquarium. Photo by Patrice Ceisel. Reproduced by permission.)

and snails resist digestion by keeping their valves or operculum plate tightly closed, forcing sea stars to take weeks to digest them. Extra-oral feeders devour their prey (usually oysters and mussels) by pulling the shells apart using their tube feet and arms. Digestion occurs once the sea star's stomach is everted through its mouth and brought into direct contact with soft tissue. Often they take advantage of imperfections in the seal of the prey's shell and squeeze their stomach into 0.1 mm-wide gaps. Some animals such as clams, worms, and crustaceans avoid predation by co-existing as commensals. The worm *Acholoë astericola*, for example, lives within the arm grooves of its host, the burrowing starfish *Astropecten irregularis*, but when threatened the worm seeks refuge inside the sea star's stomach. In coral eaters, such as the crown-of-thorns *Acanthaster planci*, the stomach is applied directly to the coral and a patch digested. Sea stars can go without food for months. *Pisaster ochraceus*, for example, can survive for 18 months without eating, losing an estimate 35% of its body weight.

They have few predators as adults due to their armored spiny skeleton and rigid nature. In less heavily armored and juvenile sea stars, protection from predators comes from having a cryptic coloration. Other defensives include toxic spines or skin (e.g., *Crossaster papposus* and *Acanthaster planci*) and predator avoidance by burrowing beneath the sediment surface (e.g., *Astropecten irregularis* and *Anseropoda placenta*). Some crabs, fish, birds, and other echinoderms are known to prey on sea stars. Usually, they feed on arm tips, as their calcified bodies are difficult to eat and not very nutritious.

Reproductive biology

Most sea stars have separate sexes with no visible differences between them. Internally, each arm contains a pair of gonads that become almost filled with eggs or sperm, depending on the sex, at the time of breeding. The majority of species are broadcast spawners where eggs and sperm are released into the water column to be fertilized. To increase the chances of fertilization, sea stars aggregate when they are ready to spawn. These events usually rely on environmental cues, such as day length, to coordinate timing and may use chemical signals to indicate readiness. The crown-of-thorns sea star, for example, releases a potent chemical into the water column to attract the opposite sex. Fertilized eggs rapidly develop into free-living bipinnaria and later brachiolaria larvae that are planktonic. Eventually, they undergo metamorphosis and settle on the seabed to grow into adults. This type of reproductive strategy is known as indirect-development.

Some sea stars brood their young, where females hold their fertilized eggs in a brood space under the arm (e.g., *Asterina phylactica*), in the stomach (e.g., *Leptasterias hexactis*), or incubate them in the gonads (e.g., *Patiriella parvivipara*). In the last two cases, young develop internally and escape through small openings the female's body wall called gonopores. Many brooding sea stars inhabit polar and deep-sea regions. Some brooding sea stars, however, produce unguarded egg masses that they attach to the seabed (*Asterina gibbosa*).

Asexual reproduction is another method of development that involves either fission or regeneration of entire animal from arm parts. Almost a dozen species divide through their disc, producing clones with identical genetic makeup (e.g., *Linckia laevigata*). Seven species are known to voluntarily pinch off one or more arms (autotomous asexual reproduction) that subsequently regenerate a complete new disc and arms; these species tend to be very small. Even sexually reproducing animals can show asexual characteristics at different stages of their life cycle. For example, larvae can pinch off body structures capable of growing into another independent feeding larvae.

Conservation status

No asteroid is listed in the IUCN Red List of Threatened Species. Some species are protected, however, at local levels, particularly in tropical destinations where souvenir hunters have lead to a decline in numbers. In the Caribbean, for example, the sea star *Oreaster reticulatus* has protection.

Significance to humans

Various sea stars cause significant ecological and commercial impact, particularly to harvested shellfish throughout the world. On the North Atlantic coasts, *Asterias forbesi* feed intensely on oysters, mussels, and scallops, with sea star aggregations causing massive damage to shell fisheries. In New Zealand and Australia the accidental introduction of *Asterias amurensis* has caused extensive damage to both commercial fisheries and endemic communities. Their arrival is thought to be from ballast water discharged before docking by visiting ships and has triggered a nationwide strategy to halt their progress. Coral reefs often fall victim to the destructive feeding power of some sea stars that invade these globally important habitats. In Indo-Pacific regions *Acanthaster planci* is an infamous coralivore of coral reefs, causing devastating infestations and major management problems.

However, sea stars do have some commercial value. In Denmark, *Asterias* are used as an ingredient in fish meal, which is fed to poultry. The ancient Indians of British Columbia and the Egyptians used them as fertilizer. Some companies collect sea stars for biological supplies to schools and collectors. Their multi-rayed image is emblematic of the sea, making their dried bodies a valuable commodity to the souvenir trade.

1. Crown-of-thorns (*Acanthaster planci*); 2. Cushion star (*Oreaster reticulatus*); 3. Sand star (*Astropecten irregularis*); 4. Sunflower star (*Pycnopodia helianthoides*); 5. Northern Pacific sea star (*Asterias amurensis*). (Illustration by Barbara Duperron)

1. Blue starfish (*Linckia laevigata*); 2. Cushion star (*Odontaster validus*); 3. Cushion star (*Patiriella parvivipara*); 4. Velcro sea star (*Novodinia antillensis*); 5. Ocher star (*Pisaster ochraceus*). (Illustration by Barbara Duperron)

Species accounts

Velcro sea star
Novodinia antillensis

ORDER
Brisingida

FAMILY
Brisingidae

TAXONOMY
Novodinia antillensis Rowe, 1989.

OTHER COMMON NAMES
None known.

PHYSICAL CHARACTERISTICS
The sea star has between 10 and 14 arms with rows of spines and teeth-like pedicellaria. Arms are long and thin. Red brick coloration.

DISTRIBUTION
Atlantic Ocean, West Indies down to depths of 1,970–2,625 ft (600–800 m).

HABITAT
Found attached to hard substratum with steeply sloping rocky surfaces; under cliffs. Prefers areas where current speeds are relatively strong. Often associated with large semi-sedentary filter-feeding animals such as large sponges, sea fans, and stony corals.

BEHAVIOR
Semi-sedentary. Spiny arms and pedicellaria act like Velcro® by sticking the sea star to virtually any surface.

FEEDING ECOLOGY AND DIET
An opportunistic suspension-feeder. Characteristic arm posture creates a basket-like appearance as arms extend into the water column and their tips curl inwards over its mouth, providing maximum exposure to currents. Food is captured as it becomes impinged on the array of arm spines and hook-like structures adapted to piercing and gripping objects. Feeds on planktonic crustaceans such as copepods, mysids, and amphipods. Remain relatively inactive whilst in the feeding posture, but slowly bend their arms to envelope captured prey.

REPRODUCTIVE BIOLOGY
Little known about its life-history. Sexual reproduction.

CONSERVATION STATUS
Not listed by the IUCN.

SIGNIFICANCE TO HUMANS
None known. ◆

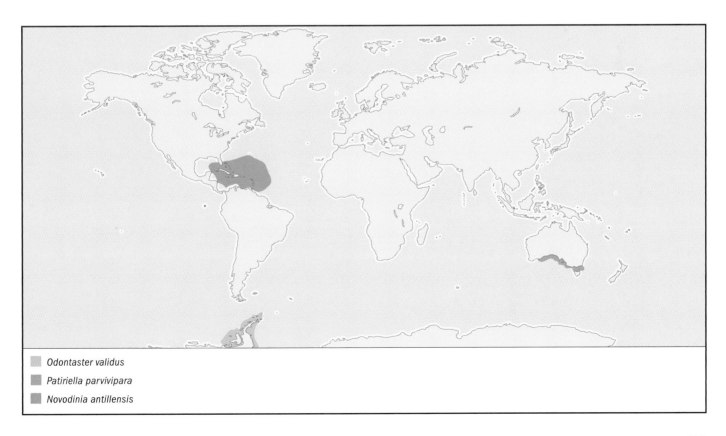

Odontaster validus

Patiriella parvivipara

Novodinia antillensis

Northern Pacific sea star
Asterias amurensis

ORDER
Forcipulatida

FAMILY
Asteriidae

TAXONOMY
Asterias amurensis Lutken, 1871.

OTHER COMMON NAMES
English: Flatbottom sea star, Amur sea star.

PHYSICAL CHARACTERISTICS
Diameter of 16–20 in (40–50 cm), with five arms that are distinctly turned up at the tips. Colors can include rosy brown, ochre and yellowish brown, red, and purple. The underside is very flat. Skin covered with numerous unevenly arranged small spines with jagged ends.

DISTRIBUTION
Far East, Russia, Korea, Japan, China, Alaska (north and south of the Alaska Peninsula), and ranges from British Columbia, Canada, and the northern Pacific down to a depth below 820 ft (250 m).

HABITAT
Found in shallow water on sheltered coasts. It can tolerate a range of temperatures (45°F [7°C] and 72°F [22°C]) and salini-

ties, which is unusual in many sea stars; hence is also found living in estuaries. Found on sandy, mud and rock sediments, among stones and algae thickets.

BEHAVIOR
Forms dense spawning aggregations, where the females have been observed lifting themselves above ground on their rays and release the eggs between the arms while the male sea star crawls beneath. Polychaete *Actonoe* has a symbiotic relationship with the sea stars and serves to clean its surface of unwanted microorganisms.

FEEDING ECOLOGY AND DIET
A generalist feeder. Diet includes scallops, oysters, mussels, shrimp, and even other echinoderms. Juvenile king crab *Paralithodes* shelter between its arms, presumably for protection against predators. Feeds by using its tube feet and arms to pull apart the shells of its prey before everting its stomach.

REPRODUCTIVE BIOLOGY
Sexual reproduction. Spawning geographically variable; in Russia, June–July and September, and in Australia, July–October. Estimated 20 million eggs are released, and develop into free-living larvae.

CONSERVATION STATUS
Not listed by the IUCN.

SIGNIFICANCE TO HUMANS
Accidentally introduced into southeastern Australia and Tasmania, causing extensive commercial and ecological damage. ◆

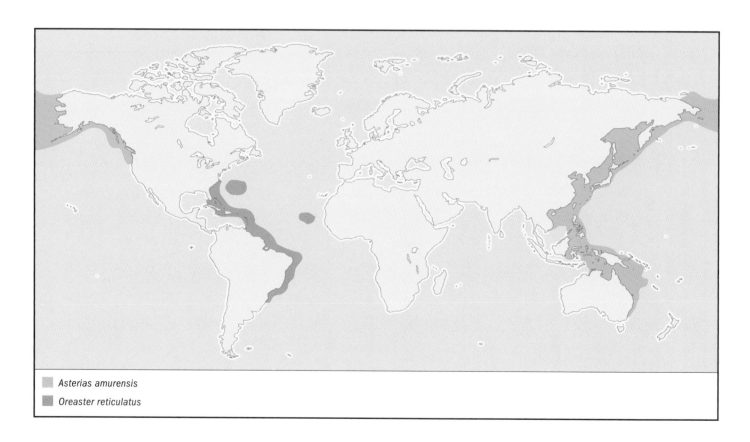

Asterias amurensis

Oreaster reticulatus

Ocher star
Pisaster ochraceus

ORDER
Forcipulatida

FAMILY
Asteriidae

TAXONOMY
Pisaster ochraceus Brandt, 1835.

OTHER COMMON NAMES
English: Purple sea star.

PHYSICAL CHARACTERISTICS
Large central disc with stout tapering arms, varying from four to seven, but usually five. Size variable but can reach 11 in (28 cm). Commonly yellow, orange, brown, and purple in color. Body covered with numerous small white spines.

DISTRIBUTION
Pacific coast from Alaska to California and down to a depth of 328 ft (100 m).

HABITAT
Intertidal rocky shores exposed to strong wave action; predator of kelp forests. Also found inhabiting tide pools at low tide.

BEHAVIOR
Keystone predator because it has impact on its marine community that is disproportionately large. Can withstand 50 hours exposed to air if among moist algae.

FEEDING ECOLOGY AND DIET
Feeds mainly on the mussel *Mytilus californinus*, although can feed on other bivalves, snails, limpets, and chitons. Uses tube feet to pull apart shells and everts stomach to digest soft tissue. Few predators, but some are eaten by sea otters and gulls.

REPRODUCTIVE BIOLOGY
Sexual reproduction, shedding eggs and sperm into water column. Spawn between April and May. Free-swimming larvae that feed on small planktonic organisms until they settle out on rocks. Can regenerate arms.

CONSERVATION STATUS
Not listed by the IUCN.

SIGNIFICANCE TO HUMANS
None known. ◆

Sunflower star
Pycnopodia helianthoides

ORDER
Forcipulatida

FAMILY
Asteriidae

TAXONOMY
Pycnopodia helianthoides Brandt, 1835.

▨ *Pisaster ochraceus*
▨ *Linckia laevigata*

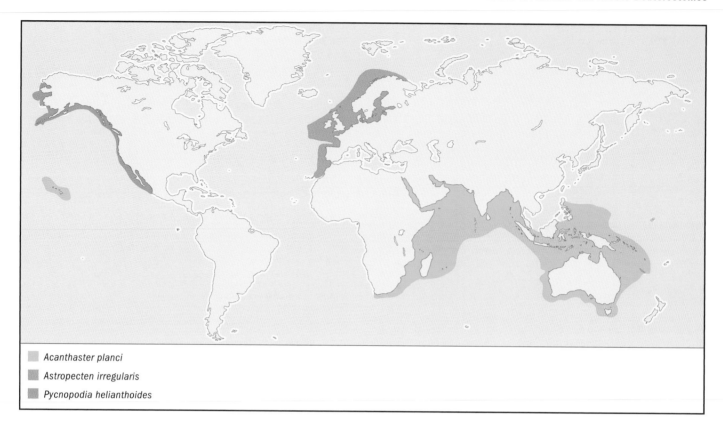

Acanthaster planci

Astropecten irregularis

Pycnopodia helianthoides

OTHER COMMON NAMES
English: Sun star, twenty-rayed star.

PHYSICAL CHARACTERISTICS
Adults usually have between 10 and 24 arms, while juveniles have only 5. One of the largest and heaviest sea stars; sizes can range between a radius of 16 in (40 cm) and 35 in (90 cm). Color variable from pink, purple, brown, red, orange, or yellow. A broad central disc and armed with over 15,000 tube feet. Skeleton has few ossicles, so the species has a soft and flexible body wall ideal for stretching mouth to accommodate large prey.

DISTRIBUTION
Northeast Pacific coastal waters. Found inhabiting the intertidal and subtidal zones from Alaska to California down to a depth of 1,640 ft (500 m).

HABITAT
Commonly found in dense seaweeds in low intertidal zones on rocky shores because their fragile bodies need the support of surrounding water.

BEHAVIOR
Solitary. A fast-moving predator that can reach speeds of 5 ft (1.6 m) per minute. When two individuals meet, they display aggressive or combative behavior.

FEEDING ECOLOGY AND DIET
Feeds on bivalves, polychaetes, chitons, snails, crabs, sea cucumbers, sea urchins (e.g., *Strongylocentrotus purpuratus*), sand dollars, sea stars (e.g., *Leptasterias*), and dead or dying squid when seasonally available. Uses sucker feet when capturing prey and swallows whole, although has the ability to partly evert stomach. Main predator is the king crab *Paralithodes*.

REPRODUCTIVE BIOLOGY
Sexual reproduction. Shed eggs and sperm into water column. Spawn from March to July, peaking in May and June. Planktonic larvae stage lasts between 2 and 10 weeks. Have the ability to regenerate arms.

CONSERVATION STATUS
Not listed by the IUCN.

SIGNIFICANCE TO HUMANS
None known. ◆

Sand star
Astropecten irregularis

ORDER
Paxillosida

FAMILY
Astropectinidae

TAXONOMY
Asterias irregularis Pennant, 1777.

OTHER COMMON NAMES
English: Burrowing starfish.

PHYSICAL CHARACTERISTICS
Varies in size, but commonly between 2 in (5 cm) and 4 in (10 cm), but up to 8 in (20 cm) in deep waters. Pale violet to yellowish color with five relatively short and tapering arms that form stiff and distinct angles. It has well-developed upper and

lower marginal plates fringed with small spines. The tube feet are pointed and sucker-less.

DISTRIBUTION
Geographical range extends from Norway to Morocco and found sub-tidally between 16 ft (5 m) and 3,281 ft (1,000 m).

HABITAT
Inhabits a variety of different substratum ranging from coarse gravel to fine mud, although it is more commonly found in sandy substrata. Usually buried either partially or completely within the sediment.

BEHAVIOR
Migrates offshore into deeper water during the winter months to avoid cooling seawater temperatures and being dislodged by strong onshore storm surges. They show quadri-diurnal pattern of locomotory activity that coincide with periods of slack water during high and low tidal cycles, enabling prey buried in sediment to be detected more easily. It has a commensal worm *Acholoe squamosa*, which freely enters the stomach and lives in the ambulacral grooves.

FEEDING ECOLOGY AND DIET
Intra-oral feeder, prey are excavated from the sediment and swallowed whole. Voracious predators of mollusks, particularly the clam *Spisula subtruncata*, polychaetes, crustaceans, and other echinoderms. It has a limited olfactory ability and relies on detecting prey by touch.

REPRODUCTIVE BIOLOGY
Sexual reproduction. Spawning occurs between May and July following a rise in seawater temperature. Prior to spawning sea stars aggregate together to reproduce.

CONSERVATION STATUS
Not listed by the IUCN.

SIGNIFICANCE TO HUMANS
Arm damage is used as an indicator of bottom trawling impact. ◆

Crown-of-thorns
Acanthaster planci

ORDER
Valvatida

FAMILY
Acanthasteridae

TAXONOMY
Acanthaster planci Linnaeus, 1758.

OTHER COMMON NAMES
None known.

PHYSICAL CHARACTERISTICS
Size over 16 in (40 cm) in diameter with between 10 and 30 arms covered in dense thorn-like spines, which are mildly venomous; can inflict painful wounds that are slow to heal. Red and green coloration with reddish tips to spines. Juveniles are cryptic in color. Tube feet can function in gas exchange and feeding.

DISTRIBUTION
Pacific and Indian coral reefs, particularly associated with reefs in Hawaii, Australia, the Red Sea, India, and South Africa.

HABITAT
Adults found on open sand and feed among coral, whilst juveniles tend to hide among the coral, under rocks, and coral rubble.

BEHAVIOR
Sedentary dwellers of reef habitats. Large numbers may suddenly appear feeding on coral and then disappear.

FEEDING ECOLOGY AND DIET
Solitarily, generally feeds at night. A voracious predator of hard corals. Digests food by everting its stomach over coral, releasing a digestive enzyme and then absorbing liquefied tissue. Can survive without food for six months and feed on an estimated 3.1 mi² (8 km²) of coral per year, leaving behind dead coral skeletons.

REPRODUCTIVE BIOLOGY
Sexual reproduction. Planktonic larvae undergo bipinnaria and brachiolaria development. Regenerates broken arms to form another individual.

CONSERVATION STATUS
Not listed by the IUCN.

SIGNIFICANCE TO HUMANS
Have caused widespread damage to coral reefs in the Indo-Pacific Ocean, Red Sea, and Australia's Great Barrier Reef. Toxic spines capable of stinging humans, inflicting pain at site of sting and causing nausea. ◆

Cushion star
Patiriella parvivipara

ORDER
Valvatida

FAMILY
Asterinidae

TAXONOMY
Patiriella parvivipara Dartnall, 1972.

OTHER COMMON NAMES
None known.

PHYSICAL CHARACTERISTICS
One of the world's smallest known sea stars, measuring up to 0.4 in (1 cm) in diameter with stout arms. They are conspicuous yellow-orange color. Morphologically, they are similar to a co-occurring species *Patiriella exigua*.

DISTRIBUTION
Among sea stars, this species has the most restricted distribution. Currently found only within the coastal waters of southern Australia.

HABITAT
In either sheltered or exposed shores, usually under small boulders. At low tide, they remain covered with a few centimeters of water, although occasionally they are completely exposed.

BEHAVIOR
Slow-moving and spends most of their time beneath the underside of boulders to avoid predators and desiccation at low tide.

FEEDING ECOLOGY AND DIET
Opportunistic feeder, consuming essentially algal growth and detritus, although small epifaunal organisms and decaying animals are also eaten.

REPRODUCTIVE BIOLOGY
Unusual life-history. It is simultaneous hermaphrodite (self-fertilizing), has intragonadal fertilization, and incubates its young in the gonads. The strategy is to produce few eggs and small amounts of sperm at any one time. The advantage is a higher survival rate of offspring compared to the more usual strategy of broadcasting species. Cannibalism by juveniles feeding on other juveniles is common in this species. Most juveniles crawl away from the parent when sufficient size is reached. Emergence of juveniles appears to be influenced by temperature increases during the summer months.

CONSERVATION STATUS
Not listed by the IUCN.

SIGNIFICANCE TO HUMANS
None known. ◆

Cushion star
Odontaster validus

ORDER
Valvatida

FAMILY
Odontasteridae

TAXONOMY
Odontaster validus Koehler, 1906.

OTHER COMMON NAMES
None known.

PHYSICAL CHARACTERISTICS
Broad central disc with five short arms tapering to a blunt tip. Can reach a size of 5.5 in (14 cm) in diameter and adopts a characteristic position with arm tips slightly raised. Colors variable, ranging from dark brown, purple, purple-red, orange, red-orange, red, brick red, dark carmine, and pink. It may have light-colored arm tips with yellowish white under surface.

DISTRIBUTION
Found throughout Antarctica and the Antarctic Peninsula, South Shetland Islands, South Orkney Islands, South Sandwich Islands, South Georgia Island, Shag Rocks, Marion and Prince Edward Islands, and Bouvet Island at depths down to 2,950 ft (900 m).

HABITAT
Commonly found inhabiting the shallow shelf waters of Antarctica, usually occurring between 49 ft (15 m) and 660 ft (200 m) depths.

BEHAVIOR
Attack large prey in gangs (e.g., the sea urchin *Sterechinus neumayeri* and sea star *Acodontaster conspicuus*). Recognizes chemical odor of individuals from the same species during feeding, minimizing the risk of cannibalism.

FEEDING ECOLOGY AND DIET
An omnivore, capable of filter-feeding and eating a varied diet, including detritus, Weddell seal feces, diatoms, algae, crustaceans, mollusks, hydroids, bryozoans, sponges, polychaetes, and sea urchins. Everts stomach to feed. Predator is another sea star *Macroptychaster accrescens* and anemone *Urticinopsis antarcticus*.

REPRODUCTIVE BIOLOGY
Sexual reproduction. Broadcast spawning. Larvae feed on bacteria and algae, and have exceptionally low metabolic rates, which are ideal for long-term survival. Slow-growing, taking up to nine years to reach normal adult size.

CONSERVATION STATUS
Not listed by the IUCN.

SIGNIFICANCE TO HUMANS
None known. ◆

Blue starfish
Linckia laevigata

ORDER
Valvatida

FAMILY
Ophidiasteridae

TAXONOMY
Asterias laevigata (Linnaeus, 1758).

OTHER COMMON NAMES
None known.

PHYSICAL CHARACTERISTICS
Usually five arms with a body diameter that can reach 12 in (30 cm). Adults have brilliant blue coloration. Juveniles are blue-green, purplish with dark spots. The genus *Linckia* has many color morphs, making it difficult to identify species.

DISTRIBUTION
Common in shallow waters of Indo-Pacific Oceans. In particular, eastern Africa to Hawaii and the South Pacific Islands to Japan.

HABITAT
Adults found along coral gravel substrates of reef terraces in direct sunlight, sandy sediments, and under rocks.

BEHAVIOR
Adults characteristically knock over coral when foraging. Hides during the day in coral and rocky crevices. Juveniles may sometimes aggregate in large numbers under coral and rock. Has a commensal shrimp *Periclimenes cornutus*.

FEEDING ECOLOGY AND DIET
Non-selective grazer, feeding on detritus, debris, and small organisms. Everts stomach to feed on prey. Predator is the Triton trumpet snail *Charonia tritonis*.

REPRODUCTIVE BIOLOGY
Little known about their natural history. Asexual reproduction.

CONSERVATION STATUS
Not listed by the IUCN.

SIGNIFICANCE TO HUMANS
Used in home aquaria and are the most commonly imported sea star. ◆

Cushion star
Oreaster reticulatus

ORDER
Valvatida

FAMILY
Oreasteridae

TAXONOMY
Asterias reticulata (Linnaeus, 1758).

OTHER COMMON NAMES
English: West Indian sea star.

PHYSICAL CHARACTERISTICS
A robust animal that can reach 20 in (50 cm) in radius. A massive inflated central disc that supports five short, slightly tapering arms. Highly variable in color from juveniles, which are mottled green, brown, tan, and gray, to adults, which are yellow, brown, or orange. All have a beige or cream lower surface. Upper surface has thick, heavy plates forming a reticulate pattern. The plates carry numerous prominent tubercles with bluntly rounded tips.

DISTRIBUTION
Found on both sides of the Atlantic, ranging from North Carolina, Bermuda, south to Florida, the Bahamas, Brazil, and the Cape Verde Islands off western Africa.

HABITAT
Prefers shallow, quiet waters of reef flats, sea grass, lagoons, and mangrove channels. Often found among sea grass beds (e.g., *Thalassia testudinum* and *Halodule wrightii*) and on sand flats associated with these sea grasses.

BEHAVIOR
Solitary. Visible during day on sand and sea grass patches. Sluggish and large sea star with an estimated speed of 5–13 in (12–33 cm) per minute.

FEEDING ECOLOGY AND DIET
Feeds mainly on microorganisms and particulate matter associated with sand, sea grass, and algal substrates. Stomach everted for feeding.

REPRODUCTIVE BIOLOGY
Sexual reproduction. Broadcast fertilized eggs into the water column. Larvae undergo planktonic stage before settling to the seafloor. Can regenerate arms.

CONSERVATION STATUS
Not listed by the IUCN, but considered rare. Vulnerable to humans. Locally protected in the Caribbean.

SIGNIFICANCE TO HUMANS
Exploited for its ornamental value and sold as souvenirs as dried or garishly painted objects. ◆

Resources

Books

Birkeland, C. "The Influence of Echinoderms on Coral-reef Communities." In *Echinoderm Studies*, edited by Michael Jangoux and John Lawrence. Rotterdam, The Netherlands: A. A. Balkema, 1989.

Blake, B. D. "Asteroidea: Functional Morphology, Classification and Phylogeny." In *Echinoderm Studies*, edited by Michael Jangoux and John Lawrence. Rotterdam, The Netherlands: A. A. Balkema, 1989.

Clark, A. M., and M. E. Downey. *Starfish of the Atlantic.* London: Chapman and Hall, 1992.

Ebert, T. A. "Recruitment in Echinoderms." In *Echinoderm Studies*, edited by Michael Jangoux and John Lawrence. Rotterdam, The Netherlands: A. A. Balkema, 1983.

Eugene, N. K. *Marine Invertebrates of the Pacific Northwest.* Seattle: University of Washington Press, 1996.

Feder, H., and A. M. Christensen. "Aspects of Asteroid Biology." In *Echinoderm Studies*, edited by Michael Jangoux and John Lawrence. Rotterdam, The Netherlands: A. A. Balkema, 1983.

Freeman, S. M., C. A. Richardson, and R. Seed. "Seasonal Abundance, Prey Selection and Locomotory Activity Patterns of *Astropecten irregularis*." In *Echinoderm Research. Proceedings of the 5th European Conference on Echinoderms*, edited by Maria D.C. Carnevali and Francesco Bonasoro. Rotterdam, The Netherlands: A. A. Balkema, 1999.

Hendler, G., J. E. Miller, D. L. Pawson, and P. M. Kier. *Sea Stars, Sea Urchins, and Allies: Echinoderms of Florida and the Caribbean.* Washington, DC, and London: The Smithsonian Institution Press, 1995.

Jangoux, M. "Digestive Systems: Asteroidea." In *Echinoderm Nutrition*, edited by Michael Jangoux and John Lawrence. Rotterdam, The Netherlands: A. A. Balkema, 1982.

———. "Food and Feeding Mechanism: Asteroidea." In *Echinoderm Nutrition*, edited by Michael Jangoux and John Lawrence. Rotterdam, The Netherlands: A. A. Balkema, 1982.

Janies, D., and R. Mooi. "*Xyloplax* Is an Asteroid." In *Echinoderm Research. Proceedings of the 5th European Conference on Echinoderms*, edited by Maria D.C. Carnevali and Francesco Bonasoro. Rotterdam, The Netherlands: A. A. Balkema, 1999.

Kozloff, E. N. *Seashore Life of the Northern Pacific Coast.* Seattle: University of Washington Press, 1993.

Lerman, M. *Marine Biology: Environment, Diversity, and Ecology.* Redwood City, CA: Cummings Publishing Company, 1986.

Menge, B. "Effects of Feeding on the Environment: Asteoidea." In *Echinoderm Nutrition*, edited by Michael Jangoux and John Lawrence. Rotterdam, The Netherlands: A. A. Balkema, 1982.

O'Clair, R. M., and E. O. O'Clair. *Southeast Alaska's Rocky Shores, Animals.* Auke Bay, AK: Plant Press, 1998.

Resources

Picton, B. E. *EA Field Guide to the Shallow-water Echinoderms of the British Isles.* London: Immel Publishing Ltd., 1993.

Sloan, N. A., and A. C. Campbell. "Perception of Food." In *Echinoderm Nutrition*, edited by Michael Jangoux and John Lawrence. Rotterdam, Netherlands: A. A. Balkema, 1982.

Periodicals

Blake, B. D. "Adaptive Zones of the Class Asteroidea (Echinodermata)." *Bulletin of Marine Science* 46 (1990): 701–718.

———. "A Classification and Phylogeny of Post-Paleozoic Sea Stars (Asteroidea: Echinodermata)." *Journal of Natural History* 21 (1987): 481–528.

Bryne, M. "Reproduction of Sympatric Populations of *Patiriella guunii, P. calcar* and *P. exigua* in New South Wales, Asterinid Sea Stars with Direct Development." *Marine Biology* 114 (1992): 297–316.

———. "Viviparity and Intragonadal Cannibalism in the Diminutive Sea Stars *Patiriella vivipara* and *P. parvivipara* (family Asterinidae)." *Marine Biology* 125 (1996): 551–567.

Bryne, M., and A. Cerra. "Evolution of Intragonadal Development in the Diminutive Asterinid Sea Star *Patiriella vivipara* and *P. parvivipara* with an Overview of Development in the Asterinidae." *Biological Bulletin* 191 (1996): 17–26.

Bryne, M., M. G. Morrice, and B. Wolf. "Introduction of the Northern Pacific Asteroid *Asterias amurenis* to Tasmania: Reproduction and Current Distribution." *Marine Biology* 127 (1997): 73–685.

Ebert, T. A. "The Consequences of Broadcasting, Brooding, and Asexual Reproduction in Echinoderm Metapopulations." *Oceanologica Acta* 19 (1996): 217–226.

Emson, R. H., and I. C. Wilkie. "Fission and Autotomy in Echinoderms." *Oceanography Biological Annual Review* 18 (1973): 389–438.

Emson, R. H., and C. M. Young. "Feeding Mechanism of the Brisingid Starfish *Novodinia antillensis*." *Marine Biology* 118 (1994): 433–442.

Freeman, S. M., C. A. Richardson, and R. Seed. "The Distribution and Occurrence of *Acholoe squamosa* (Polychaeta: Polynoidae), a Commensal with the Burrowing Starfish *Astropecten irregularis* (Echinodermata: Asteroidea)." *Estuarine, Coastal and Shelf Science* 47 (1998): 107–118.

———. "Seasonal Abundance, Spatial Distribution, Spawning and Growth of *Astropecten irregularis* (Echinodermata: Asteroidea)." *Estuarine, Coastal Shelf Science* 53 (2001): 39–49.

Kaiser, M. J. "Starfish Damage as Indicator of Trawling Intensity." *Marine Ecology Progress Series* 134 (1996): 303–307.

Keesing, J. K., and J. S. Lucas. "Field Measurement of Feeding and Movement Rates of the Crown-of-thorns Starfish *Acanthaster planci*." *Journal of Experimental Marine Biology and Ecology* 156 (1992): 89–104.

Kidawa, A. "Antarctic Starfish, *Odontaster validus*, Distinguish Between Fed and Starved Conspecifics." *Polar Biology* 24 (2001): 408–410.

Lawrence, J. M., M. Byrne, L. Harris, B. Keegan, S. M. Freeman, and B. C. Cowell. "Sublethal Arm Loss in *Asterias amurensis, A. rubens, A. vulgaris* and *A. forbesi* (Echinodermata: Asteroidea)." *Vie et Milieu* 49 (1999): 69–73.

Lawrence, J. M., and J. Herrera. "Stress and Deviant Reproduction in Echinoderms." *Zoological Studies* 39 (2000): 151–171.

Laxton, J. H. "A Preliminary Study of the Biology and Ecology of the Blue Starfish *Linckia laevigata* on the Australian Great Barrier Reef and an Interpretation of Its Role in the Coral Reef Ecosystem." *Biological Journal of the Linnean Society* 6 (1974): 47–64.

Littlewood, D. T. J. "Echinoderm Class Relationships Revisited." In *Echinoderm Research*, edited by R. H. Emson, A. B.Smith, and A. C. Cambell. Rotterdam, Netherlands: A.A. Balkema, 1995.

McClintock, J. B. "Trophic Biology of Antarctic Shallow-water Echinoderms." *Marine Ecology Progress Series* 111 (1994): 191–202.

Menge, B. A., E. L. Berlow, S. A. Blanchette, S. A. Navarrete, and S. B. Yamada. "The Keystone Species Concept: Variation in Interaction Strength in a Rocky Intertidal Habitat." *Ecological Monographs* 64 (1994): 249–286.

Mosig, J. "Pacific Sea Star Looms as Threat to Aquaculture." *Austasia Aquaculture* 4 (1998): 57–58.

Paine, R. T. "A Note on Trophic Complexity and Community Stability." *American Naturalist* 103 (1969): 91–93.

Pawson, D. L. "Some Aspects of the Biology of Deep-sea Echinoderms." *Thalassia Jugoslavica* 12 (1976): 287–293.

Pearse, J. S. "Reproductive Periodicities in Several Contrasting Populations of *Odontaster validus* Koehler, a Common Antarctic Asteroid." *Biology of the Antarctic Seas 2, Antarctic Research Series* 5 (1965): 39–85.

Scheibling, R. E. "Dynamics and Feeding Activity of High Density Aggregations of *Oreaster reticulatus* (L.) (Echnodermata: Asteroidea) in a Sand Patch Habitat." *Marine Ecology Progress Series* 2 (1980): 321–327.

———. "Feeding Habits of *Oreaster reticulatus* (Echnodermata: Asteroidea)." *Bulletin of Marine Science* 32 (1982): 504–510.

Sloan, N. A. "Aspects of the Feeding Biology of Asteroids." *Marine Biology Annual Review* 18 (1980): 57–124.

Tyler, P. A., S. L. Pain, and J. D. Gage. "The Reproductive Biology of the Deep-sea Asteroid *Bathybiaster vexillifer*." *Journal of the Marine Biological Association of the United Kingdom* 62 (1982): 57–69.

Zann, L., J. Brodie, C. Berryman, and M. Naqasima. "Recruitment, Ecology, Growth and Behavior of Juvenile *Acanthaster planci* (L.) (Echinodermata: Asteroidea)." *Bulletin of Marine Science* 41 (1987): 56–575.

Other

Freeman, S. M. "The Ecology of *Astropecten irregularis* and Its Potential Role as a Benthic Predator in Structuring a Soft-sediment Community." *PhD Thesis*. School of Ocean Sciences, University of North Wales, Bangor, UK, 1999.

Steven Mark Freeman, PhD

Concentricycloidea
(Sea daisies)

Phylum Echinodermata

Class Concentricycloidea

Number of families 1

Thumbnail description

Small disc-shaped echinoderms with a circular skeleton and water vascular system

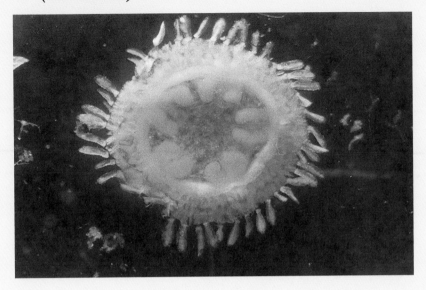

Photo: Medusiform sea daisy (*Xyloplax medusiformis*), dorsal side. (Photo by Alan N. Baker. Reproduced by permission.)

Evolution and systematics

The unusual echinoderm class Concentricycloidea (sea daisies) was erected for the genus and species *Xyloplax medusiformis* by Baker, et al., (1986) because of its radical departure in morphology from any other known living echinoderm (starfishes, brittle stars, sea urchins, sea cucumbers, and sea lilies). Concentrically arranged skeletal structures and single series of tube feet arranged in a circle were novel in living echinoderms. Given these special features, it was not considered possible, on the basis of morphology, to accommodate the genus in any existing echinoderm class. Rowe, et al., (1988) did consider, however, that *Xyloplax* was derived from the Asteroidea (starfishes), possibly from an ancestral precursor to the order Valvatida. In 1988, a second species, *X. turnerae*, was discovered which confirmed the unique morphology of the class.

The morphological differences between each of the existing orders of the Asteroidea are quite minor compared with the radical rearrangement of the axial skeleton, water vascular system, and the form of the spermatozoa in the genus *Xyloplax*. So, where to place this enigmatic taxon within the Echinodermata?

Recently attempts have been made to clarify the morphology and examine the evolutionary history of *Xyloplax* using cladistic and phylogenetic analyses of DNA sequences.

A case has been made for the Concentricyloidea to be included in the class Asteroidea based on a form of cladistic analysis of rDNA from one species, *X. turnerae*. Despite regarding *Xyloplax* as "morphologically enigmatic," the analysis placed *Xyloplax* as a sister taxon of *Rathbunaster*, a morphologically completely unrelated genus of the starfish order Forcipulatida. The sequencing has not been subsequently replicated in *X. turnerae*, and no DNA has yet been successfully sequenced from *X. medusiformis*. Also, there is a question regarding the effect of initial preservation of some of the samples of *X. turnerae* in formalin, which usually precludes or confuses accurate DNA analysis. There is an urgent need to apply standard and repeatable phylogenetic analysis of rDNA control region sequences to samples from both *X. turnerae* and *X. medusiformis*, and to as wide a range of examples of the Asteroidea as possible, to further elucidate the systematic position of this morphologically highly divergent taxon.

If the Concentricycloidea is to be regarded as a taxon at some level within the Asteroidea, then a re-defining of that class to accommodate the two *Xyloplax* species is essential. This would also require a revision of the higher systematics of the somasteroids and ophiuroids. Until more fresh material of *Xyloplax* is collected and examined both morphologically and genetically, and comparisons are made with a much larger number of taxa, the class Concentricycloidea must remain distinct to accommodate the extraordinary unique morphology of the two species of *Xyloplax*.

The class Concentricycloidea contains one order, Peripodida, and one family, Xyloplacidae.

Physical characteristics

Sea daisies have a small, discoidal, radially symmetrical echinoderm. They are circular or slightly sub-pentagonal in outline, without arms, and weakly inflated. A sea daisy has a delicately plated dorsal surface which delimits the disc-like body with a single peripheral series of flattish, oar blade-like spines. The ventral surface is dominated by a skeletal mouth frame in the form of a ring of ossicles that supports either a thin velum or a peristome opening into a shallow stomach. There are 10 large gonads, and a single peripheral ring of tube feet.

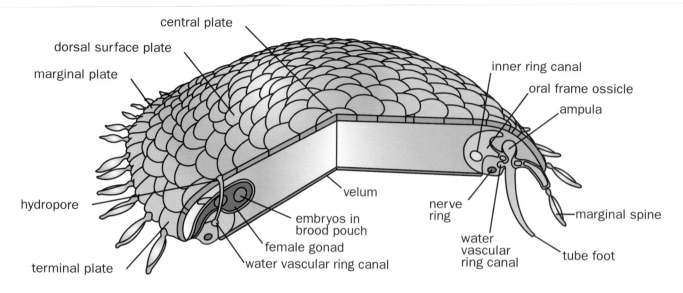

Section through a female medusiform sea daisy

Sea daisy anatomy. (Illustration by Laura Pabst)

They range in size from 0.03 to 0.54 in (0.9–13.6 mm) in diameter, including the marginal spines. They are translucent, with underlying skeletal structure and gonads visible.

Sea daisies have an outer water vascular ring system with lateral canals, ampullae and podia, not arranged in five radiating arms. An inner ring was originally thought to be also part of the water vascular system as it contained coelomocytes similar to those in the water vascular system but it may be part of the haemal system; however, this is not yet proven. The reproductive system is unique amongst echinoderms. The sexes are separate, and the gonoducts of each pair of 10 gonads are fused to form a single, interradial duct leading to a gonopore that opens in the females at the body margin, and in the males, slightly beyond the margin as the penial projection, which is most likely used for copulation. The highly filiform nature of the spermatozoa also suggests that these extremely modified cells do not contact the external environment, and that copulation takes place.

Distribution
Sea daisies are found only off New Zealand and the Bahamas.

Habitat
Sea daisies are found only on sunken wood below 3,280 ft (1,000 m).

Behavior
Not much is known. But in order to move to new habitats, the sea daisies may move by pulsating medusoid actions of the velum or stomach, or drift "parachute-like" in the demersal plankton.

Feeding ecology and diet
Not much is known about the diet of sea daisies, but it is suspected to be composed of bacteria and dissolved organic material on sunken wood, and detritus, and possibly micromollusks.

In the case of *X. medusiformis*, which lacks a stomach, direct absorption must occur through the velum and general body tissues. Shell fragments found in *X. turnerae*, which possesses a primitive stomach, may indicate utilization of whole food.

Reproductive biology
The sexes are separate in Concentricycloids, and the disclike body of the species, and the radial arrangement of gonopores (female) and penial projections (male), makes it mechanically possible, although somewhat unlikely, for one female to be serviced by five males, or alternatively, five females could surround one male, or an alternate male/female chain of reproductively active individuals might occur.

Oocytes are ovoid and become increasingly yolk-filled with growth. The largest develop towards the blind end of the gonad, and measure up to 0.00709 in (180 µm) in length.

Spermatozoa are streamlined, elongate cells over 0.00106 in (27 µm) long and composed of a finely tapered segmented astrodome, tail attachment area, free flagellum, nucleus, and a single, elongate mitochondrion posterior to the nucleus.

The two *Xyloplax* species are dioecious and sexually dimorphic: males are smaller than females. Different reproductive strategies are adopted by each species: *X. medusiformis* is an intra-ovarian, non-placental, viviparous species, and *X. turnerae* is presumed to be capable of depositing fertilized eggs externally. In *X. medusiformis*, reproduction is asynchronous between the 10 gonads—not all gonads show equal development of oocytes. The developing juveniles are clearly recognizable as cone-shaped or flattened individuals with developing tube feet, a well-developed ectoneural nerve, and a vestigial gut (which is lost in immediate prenatal specimens).

No larval stages have been observed, but it is likely that they exist in *X. turnerae*, as developmental stages have not been observed in the ovaries of that species.

Conservation status

Xyloplax medusiformis is known only from nine specimens taken off the coast of New Zealand. It is therefore thought to be endemic to that region. *Xyloplax turnerae* is known from several hundred specimens collected off the Bahamas; it may

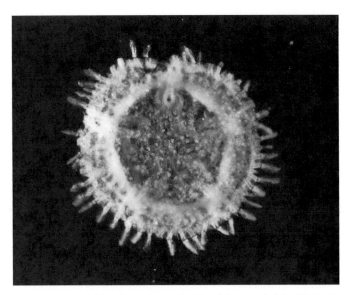

Medusiform sea daisy (*Xyloplax medusiformis*), ventral side. (Photo by Alan N. Baker. Reproduced by permission.)

well be endemic to the tropical west Atlantic Ocean. Neither species is known to be threatened.

Significance to humans

None known.

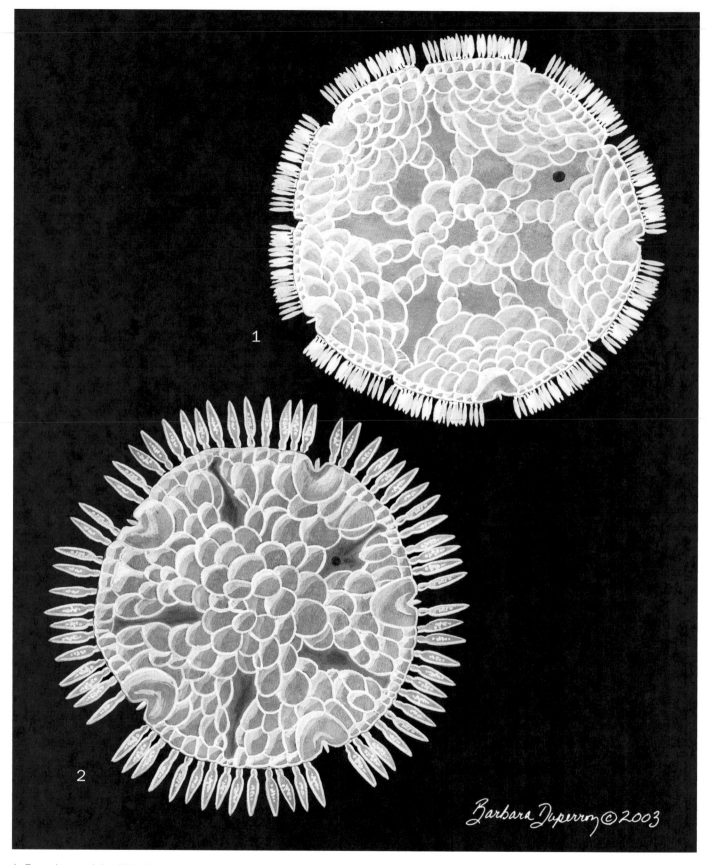

1. Turner's sea daisy (*Xyloplax turnerae*); 2. Medusiform sea daisy (*Xyloplax medusiformis*). (Illustration by Barbara Duperron)

Species accounts

Medusiform sea daisy
Xyloplax medusiformis

ORDER
Peripodida

FAMILY
Xyloplacidae

TAXONOMY
Xyloplax medusiformis Baker, Rowe, and Clark, 1986.

OTHER COMMON NAMES
None known.

PHYSICAL CHARACTERISTICS
Body is disc-like, slightly inflated, up to 0.31 in (7.8 mm) body diameter or 0.35 in (9 mm) including peripheral spines. Abactinal (dorsal) plates imbricate outwards towards the margin of the body. The peripheral spines are of one size group—length 40–75 μm; 10–30 mouth frame ossicles present; no stomach, instead an oral velum is supported by the mouth frame ossicles.

DISTRIBUTION
East and west coasts of New Zealand.

HABITAT
Found only on sunken wood in deep water (below 3,470 ft [1,057 m]).

BEHAVIOR
Nothing is known.

FEEDING ECOLOGY AND DIET
Food (dissolved organic material) absorbed through velum. Bacteria may be held associated with ventral surface and may release radicals that would dissolve through the velum. Organic materials would be released by decomposing sunken wood which also would be subject to boring organisms (i.e., mollusks), resulting in detritus being made available to the sea daisies.

REPRODUCTIVE BIOLOGY
Dioecious and sexually dimorphic, reproduction is non-placental, viviparous, and intra-ovarian.

CONSERVATION STATUS
Not listed by the IUCN.

SIGNIFICANCE TO HUMANS
None known. ◆

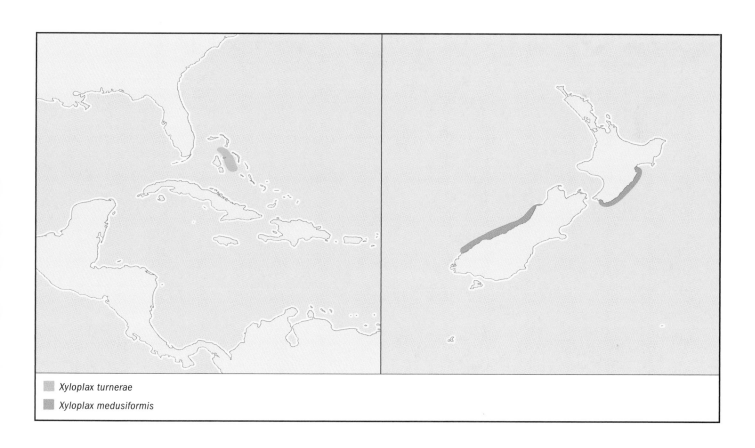

■ *Xyloplax turnerae*
■ *Xyloplax medusiformis*

Turner's sea daisy
Xyloplax turnerae

ORDER
Peripodida

FAMILY
Xyloplacidae

TAXONOMY
Xyloplax turnerae Rowe, Baker, and Clark, 1988.

OTHER COMMON NAMES
None known.

PHYSICAL CHARACTERISTICS
Body disc-like, slightly inflated, up to 0.47 in (12 mm) body diameter or 0.54 in (13.6 mm) including marginal spines. Abactinal (dorsal) plates imbricate towards centrodorsal plate, marginal spines in two sizes: 0.00354–0.00591 in (90–150 µm) and 0.00118–0.0157 in (300–400 µm). A shallow, blind sac-like stomach with a central mouth is present.

DISTRIBUTION
Tongue of the Ocean, Bahamas, Atlantic Ocean.

HABITAT
Attached to wood panels placed in 6,780 ft (2,066 m) for collection of molluskan spat.

BEHAVIOR
Nothing is known.

FEEDING ECOLOGY AND DIET
Detritus and shell fragments have been found in the stomach of this species, suggesting utilization of whole food items.

REPRODUCTIVE BIOLOGY
Vitellogenic oocytes, but no developmental stages have been observed in the ovaries, indicating oviposition and possible larval stages. Larvae or yolky eggs could drift long distances in the demersal plankton assisted by sea floor ocean currents.

CONSERVATION STATUS
Not listed by the IUCN.

SIGNIFICANCE TO HUMANS
None known. ◆

Resources

Books

Janies, D., and R. Mooi. "*Xyloplax* Is an Asteroid." In *Echinoderm Research*, edited by M. Candia Carniveli and F. Bonasoro. Rotterdam, The Netherlands: A .A. Balkema, 1998.

Pearse, V. B., and J. S. Pearse. "Echinoderm Phylogeny and the Place of Concentricycloids." In *Echinoderms Through Time*, edited by A. Guille, J. P. Feral, and M. Roux. Rotterdam, The Netherlands: A. A. Balkema, 1994.

Smith, A. B. "To Group or Not to Group: The Taxonomic Position of *Xyloplax*. In *Echinoderm Biology*, edited by R. D. Burke, P. V. Mladenov, P. Lambert, and R. L. Parsely. Rotterdam, The Netherlands: A. A. Balkema, 1998.

Periodicals

Baker, A. N., F. W. E. Rowe, and H. E. S. Clark. "A New Class of Echinodermata from New Zealand." *Nature* 321, no. 6073 (1986): 862–864.

Healy, J. M., F. W. E. Rowe, and D. T. Anderson. "Spermatozoa and Spermiogenesis in *Xyloplax* (Class Concentricycloidea): A New Type of Spermatozoon in the Echinodermata." *Zoologica Scripta* 17, no. 3 (1988): 297–310.

Rowe, F. W. E., A. N. Baker, and H. E. S. Clark. "The Morphology, Development and Taxonomic Status of *Xyloplax* Baker, Rowe, and Clark (1986) (Echinodermata: Concentricycloidea), with Description of a New Species." *Proceedings of the Royal Society of London* B 233 (1988): 431–459.

A. N. Baker, PhD

Ophiuroidea

(Brittle and basket stars)

Phylum Echinodermata

Class Ophiuroidea

Number of families 16

Thumbnail description
Small- to medium-size echinoderms with a flattened disk often covered with a series of scales, granules and small spines; usually five long thin (in comparison with the disk) articulated arms that break off easily; a row of papillae (small nipple-shaped structures) known as the arm comb near the base of each arm; lower surface of the disk containing a central mouth leading to the stomach pouch; mouth divided by five jaws

Photo: The snake star (*Pectinura maculata*) is an unusually large and mobile species. It is carnivorous, and has scales, not bristles. (Photo by ©David Hall/Photo Researchers, Inc. Reproduced by permission.)

Evolution and systematics

The fossil record of the ophiuroids extends back some 500 million years to the early Ordovician period. They are the most diverse class of echinoderms, with some 250 described genera and 2,000 species. The most recent phylogeny and upper-level taxonomy places extant ophiuroids into two subclasses: the Oegophiuridea with only one family, the Ophiocanopidae; and the Ophiuridea, which is divided into two distinct orders—the Euryalida, including the basket stars with branched arms, and the more familiar Ophiurida or brittle stars. The phylogenetic tree for ophiuroids suggest that they underwent a major diversification in the Triassic and early Jurassic periods; that is, about 200–250 million years ago. The classification of the ophiuroids, however, remains a subject of debate. The placement of the unusual *Ophiocanops fugiens*, thought to be a "living fossil," into the subclass Oegophiuridea, is especially interesting, since this subclass was thought to have become extinct after the late Carboniferous period. The families are as follows: Ophiocanopidae; Asteronychidae; Gorgonocephalidae; Asteroschematidae; Euryalidae; Ophiomyxidae; Hemieuryalidae; Ophiuridae; Amphiuridae; Ophiothricidae; Ophiactidae; Ophionereididae; Ophiocomidae; Ophichitonidae; Ophiodermatidae; Ophiolepididae.

Physical characteristics

Ophiuroids have long slender flexible arms that are sharply separated from the disk. The common name "brittle star" refers to the fact that the arms of many species are easily broken off. Locomotion involves the entire arm; movement is made possible by an internal skeleton that supports the arm. The skeletal portions of the arm are made of calcium carbonate and look like vertebrae; they are called vertebral ossi-

cles. Five arms are usual, but a few species of ophiurid ophiuroids have six or seven, and euryalid ophiuroids may have as many as 20 major additional arms.

Species with branched arms are restricted to the families Gorgonocephalidae and Euryalidae and are called basket stars. The basket stars are the largest ophiuroids; *Gorgonocephalus stimpsoni* can measure up to 27.5 in (70 cm) in arm length with a disk diameter of 5.63 in (14.3 cm). The branching of the arms in basket stars is repeated in the formation of smaller and smaller units in a fernlike manner.

The upper surface of the disk in ophiurid ophiuroids is covered with a series of scales. In contrast, euryalid ophiuroids are characterized by the presence of a covering on the disk without any large plates underneath. The mouth on the lower surface of the disk is framed by five jaws bearing spinelike teeth and papillae. The mouth leads directly to the saclike stomach, which ends in a blind pouch since the organism lacks intestines and an anus.

The oral side of the disk contains the bursal slits on each side of the base of each arm. These slits are openings for the respiratory bursae, which are specialized sac- or pouchlike formations in the body wall that serve for gas exchange. The gonads are attached to the walls of the bursae; in a few species, however, the gonads extend into the coelom (body cavity) of the arms. The number of gonads varies.

Distribution

Ophiuroids range from the poles to the tropics and from intertidal to abyssal plains as well as in deep sea trenches. The deepest recordings were taken from the Bourgainville Trench at 26,270 ft (8,006 m).

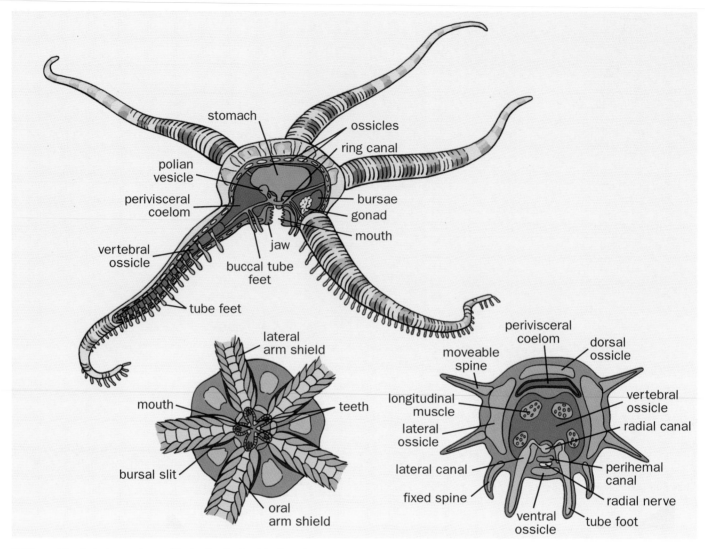

Brittle and basket star anatomy. (Illustration by Patricia Ferrer)

Habitat

Ophiuroids are found in all marine habitats. There are burrowing ophiuroids, notably members of the family Amphiuridae, that live on muddy or sandy sea bottoms, as well as species that live either exposed or cryptic (hidden) on all types of ocean substrates. In addition, some ophiuroids live on or in association with such other organisms as algae, sponges, and corals. Many are filter-feeding species, such as the basket stars that occupy habitats swept by currents. Like echinoderms in general, ophiuroids cannot tolerate low levels of salt concentration in the water; a few species, however, including *Ophiura albida* and *Ophiophragmus filegranus*, can adapt to brackish water and live in estuaries with salt concentrations as low as 10 ppt (ocean water is usually 34–35 ppt).

Behavior

Many ophiuroids move by powerful arm strokes that lift their disks and thrust them forward. Some species may even swim short distances in this manner. Other species, particu-larly those with very long arms, move with a slithering type of motion that pulls them along the substrate. Because of this motion, they are sometimes referred to as "serpent stars" or "snake stars." Only a few species use their tube feet for moving about. Ophiuroids are good climbers; filter-feeding species may cling to rock outcrops or to such other filter feeders as gorgonians and corals in order to position themselves in stronger currents above the sea floor.

Ophiuroids show a range of responses to light intensity, ranging from rapid escape, feeding at night, or color change to behavior that indicates relative indifference to light. For example, *Ophiocoma wendtii* is a highly photosensitive species that changes color from dark brown during the day to banded gray and black during the night; it moves rapidly away from shadows by crawling into crevices. It has been shown in this species that the arm plates function as compound eyes, focusing light through nerve bundles on the dorsal arm plates.

The best-known adaptation to partial predation in motile animals is probably autotomy, or the self-amputation of ap-

Brittle star (*Ophiothrix* sp.) scavenges on Alcyonarian coral at night. (Photo by ©Jeff Rotman/Photo Researchers, Inc. Reproduced by permission.)

pendages. Ophiuroids may lose arms and even disks in some amphiurid species. The brittle star's ability to shed parts of its arms has to do with the mechanical properties of the ligaments between the vertebral ossicles of the arms. Stimulation of the nerves at an arm joint causes this particular tissue—mutable collagenous tissue or MCT—to disintegrate, and the arm breaks off. The lost body parts are subsequently regenerated. Regeneration of missing arms may take a few months in warm tropical water but more than a year in cool temperate water.

Several species of ophiuroids are bioluminescent. It has been shown that the nocturnally active *Ophiopsila riisei* and *Ophiopsila californica* deter predators by luminescent flashes from their arms, and that their bioluminescence also functions as an aposematic (warning) signal to discourage crustacean predators.

Feeding ecology and diet

Most ophiuroid species are able to feed on more than one type of food and use more than one feeding method. Researchers recognize two major feeding groups, however. There is a carnivorous group of predators and carrion feeders that usually have short arm spines and tube feet, and seize their food in one of their arm loops. Species in the second major group feed mainly on small particles from the substrate or suspended in the water. The arm spines and tube feet in this group are relatively long, and are the principal food-trapping organs. It has also been shown that some species, such as the amphiurid *Microphiopholis gracillima*, are able to take up dissolved organic material from the water. Basket stars, however, appear to be specialized predatory suspension feeders that mainly take zooplankton, or live animal food, from the water column. Basket stars capture their prey by a rapid flexing movement of the arm that encircles the prey. The prey is then gripped by tiny sharp hooks arranged along the distant part of the arms.

Reproductive biology

Most ophiuroids have separate sexes. When the gametes are mature they are discharged into the bursae and expelled into the water through the bursal slits. The eggs are usually small with a low yolk content and a diameter of 0.0039–0.07 in (0.1–0.18 mm). Following fertilization, the eggs develop into a ciliated blastula which then matures further into the characteristic ophiopluteus larvae with arms, a mouth, and an anus. The ophiopluteus larva is planktonic; the larval stage

A juvenile spiny brittle star (*Ophiocoma paucigranulata*) on a wide mesh sea fan (*Gorgonia mariae*). (Photo by ©Mary Beth Angelo/Photo Researchers, Inc. Reproduced by permission.)

lasts two to five weeks, during which the larvae feed on such microscopic particles as diatoms, dinoflagellates and other small flagellates. Subsequently the larval body goes through a remarkable metamorphosis into a young brittle star. The body flattens, the larval gut closes, and the anus disappears.

Other species develop via a vittelaria larva, which originates from eggs with a high yolk content. This larva does not feed; it develops into a young brittle star within approximately 12 days. There are also species that undergo direct development; the larval stages are suppressed and a juvenile brittle star emerges directly from the gastrula stage.

There are approximately 60 hermaphroditic species of ophiuroids that practice brood care. The eggs are retained and developed in the maternal body; in most cases, the bursae serve as brood chambers in which the young brittle stars also may be fed by the mother, as in *Amphipholis squamata*.

Lastly, there are about 45 species of brittle stars with a mixed life history that includes both sexual reproduction and clonal proliferation by fission. These brittle stars have six arms, and spontaneously divide into two halves. Each half then subsequently regenerates three new arms to form two new complete individuals. Although asexual species are uncommon in this class, they can be very successful, range widely, and occur in high densities in certain habitats. For most of these species, the asexual mode is associated with small body size, while sexual maturity is usually attained after an individual reaches a certain size.

Conservation status

Such conservation organizations as the IUCN do not list any species of ophiuroids as endangered.

Significance to humans

Ophiuroids are of no direct significance to humans except for purposes of scientific research.

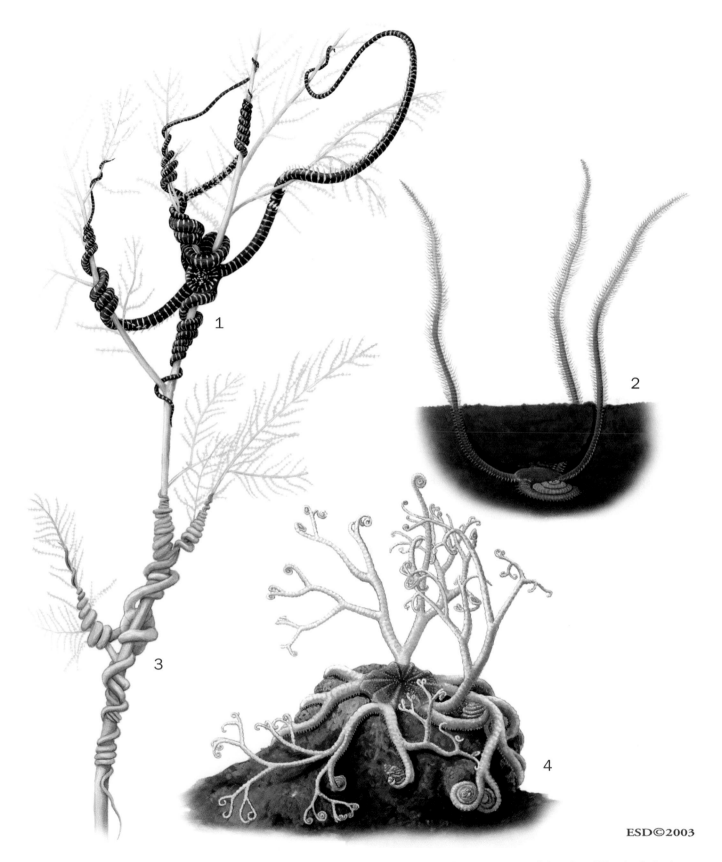

1. Nighttime feeding posture of snake star striped morph (*Astrobrachion constrictum*); 2. Feeding posture of *Amphiura filiformis*; 3. Snake star (*A. constrictum*) yellow morph coiled up during the day; 4. Northern basket star (*Gorgonocephalus arcticus*). (Illustration by Emily Damstra)

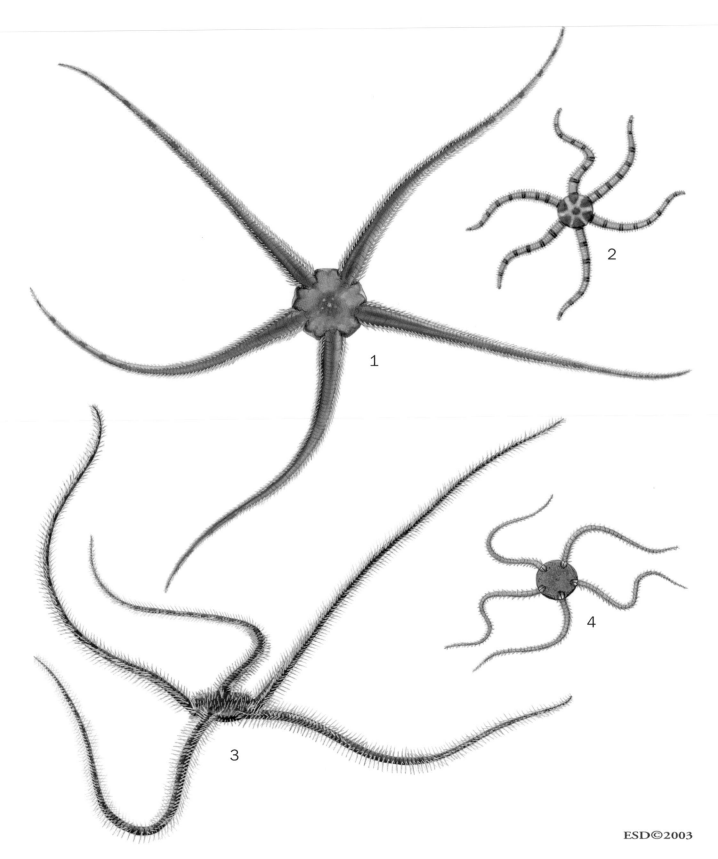

1. *Ophiura ophiura*; 2. Tropical brittle star (*Ophiactis savignyi*); 3. Common brittle star (*Ophiothrix fragilis*); 4. Dwarf brittle star (*Amphipholis squamata*). (Illustration by Emily Damstra)

Species accounts

Snake star
Astrobrachion constrictum

ORDER
Euryalida

FAMILY
Asteroschematidae

TAXONOMY
Astrobrachion constrictum Farquhar, 1900.

OTHER COMMON NAMES
None known.

PHYSICAL CHARACTERISTICS
A large brittle star with a disk as much as 0.9 in (23 mm) in diameter. The arms are long, about 13.7 in (350 mm) and broken near the tip. A soft, smooth skin covers both the disk and the arms. The spines are shorter than the arm width; the distal spines are flattened, turning into compound hooks with 2–4 lateral teeth. Colors range from uniform bright yellow or deep red to almost black. Dark red or black individuals may have white spots or stripes.

DISTRIBUTION
The species is found throughout New Zealand, southeastern Australia, and New Caledonia.

HABITAT
Astrobrachion constrictum lives in a mutualistic relationship with the black coral *Anthipates* spp. These species are normally found together in relatively deep water (164–590 ft or 50–180 m) along the continental shelf but also in shallower water (up to 33 ft or 10 m deep) in the fjords of southwestern New Zealand. The black corals attach themselves to hard substrates and the snake stars coil around the branches of the coral.

BEHAVIOR
Astrobrachion constrictum is a nocturnal feeder. During the day it remains tightly coiled around its host. The relationship with the black coral is probably mutualistic; in the process of feeding by gathering mucus from the coral, the snake star also protects its host from organisms and detritus that would otherwise smother the colony. Snake stars show a strong preference for black corals; if they are removed, they move relatively quickly toward a black coral colony, probably by sensing chemical signals from the coral.

FEEDING ECOLOGY AND DIET
Snake stars feed during the night by stretching 2–3 arms out in the water column to snare drifting plankton with their tube feet and arm spines. They also wipe the branches of the coral in order to feed on the plankton and detritus collected on the coral.

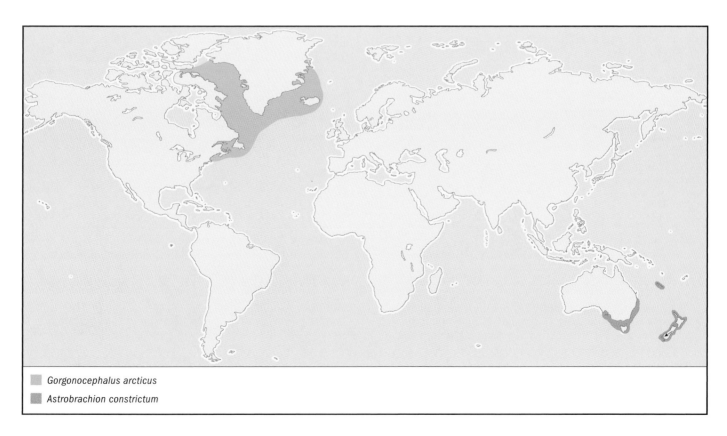

Gorgonocephalus arcticus
Astrobrachion constrictum

REPRODUCTIVE BIOLOGY

A distinctive feature of this family is the position of the gonads within the arms. One shallow-water study of the fjords of southern New Zealand reported that these brittle stars have a prolonged reproductive period, possibly even continuous, as found in many deep-sea ophiuroids. The oocytes are relatively large (0.015–0.019 in or 400–500 μm), indicating that development is lecitotrophic (larvae do not feed), either direct or abbreviated. A larval stage, however, is not known.

CONSERVATION STATUS

Not listed by the IUCN.

SIGNIFICANCE TO HUMANS

None known. ◆

Northern basket star

Gorgonocephalus arcticus

ORDER

Euryalida

FAMILY

Gorgonocephalidae

TAXONOMY

Gorgonocephalus arcticus Leach, 1819.

OTHER COMMON NAMES

German: Gorgonenhaupt; Norwegian: Medusahode.

PHYSICAL CHARACTERISTICS

A large basket star with a disk 4.01 in (102 mm) in diameter, with highly branched long (14.1 in or 360 mm) arms. The disk is pentagonal and bare, leathery with five pairs of ridges radiating from the center to the sides of the arms. The five sturdy arms divide into two branches near the disk and then redivide equally five or more times. The arm joints have short hooked spines. Color varies from yellowish brown to darker brown with lighter arms.

DISTRIBUTION

North Atlantic, from the arctic region to Cape Cod in the northeastern United States.

HABITAT

On rocky bottoms swept by currents, often clinging to gorgonians at a depth of 164–4,921 ft (5–1,500 m).

BEHAVIOR

Northern basket stars are adapted to live in strong ocean currents and seek out positions high up in the water column in order to spread out their feeding fans in the form of a concave dish facing the current. They can hold out their arms in a stiff position for long periods of time. There is morphological evidence that the mutable collagenous tissues (MCT) of basket stars may be important in maintaining their stiff fans. The use of MCT lowers the rate of energy consumption in comparison to using muscles for the same purpose. In contrast to shallow-water basket stars that are strictly nocturnal, *Gorgonocephalus arcticus* uses its feeding fan during the day.

FEEDING ECOLOGY AND DIET

Gorgonocephalus arcticus is a predatory suspension feeder that captures zooplankton, primarily such macroscopic crustaceans as krill, by rapidly coiling and bending its arms. Rings of sharp

hooks on the arms assist the basket star in capturing food by sticking to the prey.

REPRODUCTIVE BIOLOGY

Not much is known about the reproduction of gorgonocephalids. The genus has a large number of gonads, as many as 1000 per individual. The closely related *Gorgonocephalus caput-medusae* lays its eggs free in water; however, a larval form is not known. Juveniles of *Gorgonocephalus* spp. are also known to live on the soft coral *Gersemia rubiformis*.

CONSERVATION STATUS

Not listed by the IUCN.

SIGNIFICANCE TO HUMANS

None known. ◆

Dwarf brittle star

Amphipholis squamata

ORDER

Ophiurida

FAMILY

Amphiuridae

TAXONOMY

Amphipholis squamata Delle Chiaje, 1828, Mediterranean, Naples, Italy.

OTHER COMMON NAMES

English: Scaly brittle star, long-armed brittle star; French: Ophiure écailleuse; German: Schuppiger Schlangenstern; Norwegian: Overgslangestjerne.

PHYSICAL CHARACTERISTICS

The disk is very small, attaining a maximum diameter of only 0.19 in (5 mm); it is bluish or gray in color, circular in shape, and covered with scales. The arms are relatively short and thin. The radial plates are half-moon shaped and conspicuous. Dwarf brittle stars come in a range of colors from orange, dark brown or beige to black and gray. Previously these color varieties were thought to belong to sibling species, but genetic analyses do not support this hypothesis.

DISTRIBUTION

Amphipholis squamata has a world-wide distribution; it is found in all oceans of the world, including subarctic and subantarctic waters. This pattern of distribution is interesting; one would expect the organism to be restricted to certain regions since it is a brooder and supposedly a poor disperser. Genetic analyses indicate, however, that sporadic long-distance dispersal does in fact occur, probably through passive transport (drifting or rafting on macroalgae).

HABITAT

Dwarf brittle stars are found in the mid- and lower littoral zones among algae, bryozoans, and similar organisms; and sublittorally in waters several hundred meters deep. These brittle stars live mainly under stones but also on sandy surfaces.

BEHAVIOR

Amphipholis squamata is a good climber that also uses its tube feet when it moves. The arms are extremely flexible in a verti-

Amphipholis squamata

cal direction. If the brittle star is dislodged from its substrate, it coils its arms over its disk and sinks rapidly to the bottom, probably to avoid being exposed to predators. When disturbed it produces light that appears as spots along each arm. Its bioluminescence is attributed to specific photocytes under the control of ganglia, or groups of nerve cells.

FEEDING ECOLOGY AND DIET
Amphipholis squamata feeds on deposits left on sediment and particles suspended in the water. It uses its tube feet to wipe off particles from its sticky spines. The stomach content includes a variety of such items as unicellular algae, small gastropods, foraminiferans and amphipod limbs. *Amphipholis* has sometimes been observed to feed on dead fish. It can absorb dissolved free amino acids from sea water, primarily through symbiosis with bacteria living under its cuticle.

REPRODUCTIVE BIOLOGY
The dwarf brittle star is a simultaneous hermaphroditic viviparous brooder that can brood several embryos at different stages of development in each bursa at the same time. The eggs are small (0.0039 in or 100 µm), which suggests that the larvae must obtain nutrition from the parent within the bursae. Breeding occurs throughout the year. It has also been shown in the laboratory that *Amphipholis* can reproduce in isolation. This finding suggests that it can reproduce by self-fertilization or possibly by parthenogenesis.

CONSERVATION STATUS
Not listed by the IUCN.

SIGNIFICANCE TO HUMANS
None known. ◆

No common name
Amphiura filiformis

ORDER
Ophiurida

FAMILY
Amphiuridae

TAXONOMY
Amphiura filiformis Müller, 1776, Scandinavia, probably southern Norway.

OTHER COMMON NAMES
None known.

PHYSICAL CHARACTERISTICS
Disk up to 0.39 in (10 mm) in diameter, reddish gray in color, with very long fine arms. The dorsal side is covered with fine scales but the underside is bare.

DISTRIBUTION
Northeastern Atlantic from Iceland and western Norway to the Iberian peninsula; Mediterranean.

HABITAT
Buries itself in fine muddy silt or sand at depths of 15–3,600 ft (5–1,200 m). Very common in the North Sea; populations may be as dense as 4,000 individuals per 10.8 ft^2(1 m^2).

BEHAVIOR
When *Amphiura* is burrowing, it uses its tube feet to move material away from itself and virtually sinks into the substrate.

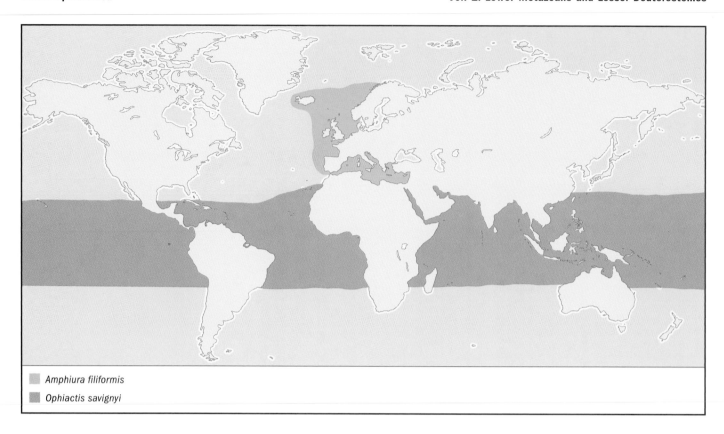

Amphiura filiformis

Ophiactis savignyi

The animal then uses wavelike movements of its arms and digging movements of the tube feet to hollow out a burrow lined with mucus and equipped with channels to the surface of the sediment. For ventilation, it waves one or more of its arms to allow oxygen-rich water to enter the burrow. *Amphiura filiformis* is selectively pursued by some flatfishes like dab (*Limanda limanda*) and frequently sheds its arms when attacked. It is able to sense the presence of predators attacking other members of its species in the surrounding water and will then pull its arms back into the burrow.

FEEDING ECOLOGY AND DIET
Amphiura filiformis is primarily a passive suspension feeder that collects small organic particles on its long sticky tube feet, which are covered with numerous papillae. In moving water, it raises its arms about an inch (2.5 cm) above the seafloor. The particles that it collects are packed together with mucus into a bolus (soft rounded mass) and carried along its arm to its mouth beneath the sediment surface. This species may also collect particles from the surface of the sediment or feed on larger dead organisms.

REPRODUCTIVE BIOLOGY
Amphiura filiformis has separate sexes and a seasonal developmental pattern with a spawning period that extends from late spring to late summer. When spawning, it emerges from its burrow and uses its arms to lift its disk about 0.39 in (1 cm) above the bottom. Its life history includes an ophiopluteus larva.

CONSERVATION STATUS
Not listed by the IUCN.

SIGNIFICANCE TO HUMANS
None known. ◆

Tropical brittle star
Ophiactis savignyi

ORDER
Ophiurida

FAMILY
Ophiactidae

TAXONOMY
Ophiactis savignyi Müller and Troschel, 1842, Egypt, presumably Red Sea.

OTHER COMMON NAMES
English: Savigny's brittle star.

PHYSICAL CHARACTERISTICS
A very small brittle star with a disk only 0.039–0.11 in (1–3 mm) in diameter. It usually has six arms, but since it is fissiparous some individuals that have not completed regeneration may have as few as two arms. *Ophiactis* has a conspicuous color pattern of green with darker markings and large bare radial shields.

DISTRIBUTION
Found in all tropical seas. Studies of its genetic pattern indicate that it has recently spread to the Atlantic Ocean from the Indo-Pacific, probably in ballast water or in fouling communities coming through the Panama Canal on ships.

HABITAT
Ophiactis savignyi is often associated with algal turfs and sponges but also hides in reef crevices.

BEHAVIOR

Ophiactis savignyi is cryptic and colonizes mainly algae or sponges; it probably receives shelter from predators among these organisms. It grows larger when it lives among sponges in comparison to algal turf, and may benefit from particles gathered by the sponge colonies from the water they take in. *Ophiactis* can reach high densities, sometimes more than 280 per deciliter of sponge.

FEEDING ECOLOGY AND DIET

The diet and feeding patterns of *Ophiactis savignyi* have not been studied as of 2003. Ophiactids, however, belong to a group that feeds mainly on small particles from the substrate or suspended in the water.

REPRODUCTIVE BIOLOGY

Ophiactis practices both sexual and asexual reproduction by fission. During fission, the mutable collagenous tissue of the disk softens, forms a groove, and tears irregularly across the disk. The two halves then regenerate into two new six-armed individuals. A single clone can in this way colonize a sponge or patch of alga. Several investigated populations seem to be maintained by asexual reproduction only. Sexual reproduction takes place when the brittle star has reached a size larger than 0.11 in (3 mm). The larva is a typical planktotrophic ophiopluteus.

CONSERVATION STATUS

Not listed by the IUCN.

SIGNIFICANCE TO HUMANS

None known. ◆

Ophiura ophiura

Ophiothrix fragilis

Common brittle star

Ophiothrix fragilis

ORDER

Ophiurida

FAMILY

Ophiothricidae

TAXONOMY

Ophiothrix fragilis Abildgaard, 1789, Denmark.

OTHER COMMON NAMES

Dutch: Gewone brokkelster; German: Zerbrechlicher; Schlangenstern; Norwegian: Håstjerne.

PHYSICAL CHARACTERISTICS

A large brittle star with a disk up to 0.78 in (20 mm) in diameter. The arms are long, colorfully banded and very bristly with seven glassy spines on each segment. Its color varies; it may be patterned in red, yellow, orange, violet, gray or brown.

DISTRIBUTION

Widely distributed in the eastern Atlantic from northern Norway to the Cape of Good Hope; also common in the Mediterranean.

HABITAT

Occurs from the lower shore to about 1,640 ft (500 m) on hard sea bottoms swept by currents; or beneath boulders, in crevices, or on substrates among sessile organisms. It may also form dense groups or aggregations (brittle star beds) on bare sand or shell sediments, with as many as 2000 individuals per square meter in habitats with few predators.

BEHAVIOR

Predation pressure controls whether *Ophiothrix fragilis* hides in crevices or under rocks; that is, whether it is cryptic or lives in the open on the substrate. The formation of groups or aggregations on bare substrates helps individuals to maintain their position in strong currents. Under these conditions individual brittle stars interlock their supporting arms while lowering their disks onto the substrate. Lowering frictional resistance to the current helps to keep them stable. Juveniles seem to prefer to settle on or seek out adults; they can be seen clinging to the spines and bodies of adult individuals. They also settle on such other suspension feeders as sponges.

FEEDING ECOLOGY AND DIET

Ophiothrix fragilis is mainly a passive suspension feeder; like *Amphiura filiformis*, it collects organic particles on long and sticky tube feet covered with numerous papillae. It raises its arms about 3 in (7–8 cm) above the substrate; the collected particles are formed into a bolus and carried along the arm to the mouth.

REPRODUCTIVE BIOLOGY

The sexes are separate, with a seasonal developmental pattern. The ophiopluteus larva can be found in the northwestern Mediterranean from March to October, with the peak settlement period in June. The time from fertilization to metamorphosis is about 26 days. The ophiopluteus has unusually long arms, indicating that it is capable of dispersing over long distances.

CONSERVATION STATUS

Not listed by the IUCN.

SIGNIFICANCE TO HUMANS

None known. ◆

No common name
Ophiura ophiura

ORDER
Ophiurida

FAMILY
Ophiuridae

TAXONOMY
Ophiura ophiura Linnaeus, 1758 "the sea," (probably North Sea).

OTHER COMMON NAMES
French: Ophiure commune; German: Gemusterter Schlangenstern; Norwegian: grå slangestjerne.

PHYSICAL CHARACTERISTICS
Active brittle star with a large disk as much as 1.38 in (35 mm) in diameter and relatively short arms. The dorsal side of the disk is gray-brown or sandy orange and patterned with plates. The arm combs are well developed, with 20–30 papillae in each comb.

DISTRIBUTION
Northeastern Atlantic from northern Norway to Madeira; Mediterranean.

HABITAT
Lives on the surface of soft ocean bottoms at depths of 6.5–660 ft (2–200 m).

BEHAVIOR
Ophiura ophiura moves rapidly in a rowing motion when it is disturbed. Occasionally it burrows shallowly in sediment. At rest, it assumes a position with its arms curved upwards at their tips, possibly to help it detect food from chemical cues in the water current.

FEEDING ECOLOGY AND DIET
Ophiura ophiura is an active omnivorous predatory brittle star that seizes its prey in one of its arm loops and then pounces on the food. It ingests a variety of small organisms that live on sediment, including mollusks, echinoderms, crustaceans and polychaetes; in addition it feeds on sediment, detritus, and benthic diatoms.

REPRODUCTIVE BIOLOGY
Ophiura ophiura has separate sexes. Its developmental pattern is seasonal, with a long spawning period that lasts from spring to late summer. Its life history includes an ophiopluteus larva.

CONSERVATION STATUS
Not listed by the IUCN.

SIGNIFICANCE TO HUMANS
None known. ◆

Resources

Books

Hyman, L. H. *The Invertebrates.* Vol. IV, *Echinodermata.* New York: The Maple Press Company, 1955.

Jangoux, M., and J. M. Lawrence. *Echinoderm Nutrition.* Rotterdam: A. A. Balkema, 1982.

Lawrence, J. M. *A Functional Biology of Echinoderms.* Baltimore: Johns Hopkins University Press, 1987.

McKnight, D. G. *The Marine Fauna of New Zealand: Basket Stars and Snake Stars (Echinodermata: Ophiuroidea: Euryalinida).* Wellington: NIWA (National Institute of Water and Atmospheric Research), 2000.

Mladenov, P. V., and R. D. Burke. "Echinodermata: Asexual Propagation." In *Asexual Propagation and Reproductive Strategies.* New Delhi, Bombay, and Calcutta: R. G. Oxford & IBH Publishing Co. Pvt. Ltd., 1994.

Picton, B. E. *A Field Guide to the Shallow-Water Echinoderms of the British Isles.* London: Immel Publishing, 1993.

Periodicals

Aizenberg, J., A. Tkachenko, S. Weiner, L. Addadi, and G. W. Hendler. "Calcitic Microlenses as Part of the Photoreceptor System in Brittlestars." *Nature* 412 (2001): 819–822.

Allen, J. R. "Suspension Feeding in the Brittle-Star *Ophiothrix fragilis*: Efficiency of Particulate Retention and Implications for the Use of Encounter Rate Models." *Marine Biology* 132 (1998): 383–390.

Emson, R. H., P. V. Mladenov, and K. Barrow. "The Feeding Mechanism of the Basket Star *Gorgonocephalus arcticus.*" *Canadian Journal of Zoology* 69 (1991): 449–455.

Mladenov, P. V. "Environmental Factors Influencing Asexual Reproductive Processes in Echinoderms." *Oecologica Acta* 19, no. 3–4 (1996): 227–235.

Rosenberg, R. and E. Selander. "Alarm signal response in the brittle star *Amphiura filiformis.*" *Marine Biology* 136 (2000): 43–48.

Roy, M. S., and R. Sponer. "Evidence of Human-Mediated Invasion of the Tropical Western Atlantic by the World's Most Common Brittlestar." *Proceedings of the Royal Society of London, Series B: Biological Sciences* 269, no. 1495 (2002): 1017–1023.

Sköld, M. "Escape Responses in Four Epibenthic Brittle Stars (Ophiuroidea: Echinodermata)." *Ophelia* 49 (1998): 163–179.

Sköld, M., L-O Loo, and R. Rosenberg. "Production, Dynamics and Demography of an *Amphiura filiformis* Population." *Marine Ecology Progress Series* 103 (1994): 81–90.

Smith, A. B., G. L. J. Paterson, and B. Lafay. "Ophiuroid Phylogeny and Higher Taxonomy: Morphological, Molecular and Palaeontological Perspectives." *Zoological Journal of the Linnean Society* 114 (1995): 213–243.

Sponer, R., and M. S. Roy. "Phylogeographic Analysis of the Brooding Brittle Star *Amphipholis squamata* (Echinodermata) Along the Coast of New Zealand Reveals High Cryptic Genetic Variation and Cryptic Dispersal Potential." *Evolution* 56, no. 10 (2002): 1954–1967.

Sponer, R., and M. S. Roy. "Large Genetic Distances Within a Population of *Amphipholis squamata* (Echinodermata;

Resources

Ophiuroidea) Do Not Support Color Varieties as Sibling Species." *Marine Ecology Progress Series* 219 (2001): 169–175.

Stewart, B. "Can a Snake Star Earn Its Keep? Feeding and Cleaning Behavior in *Astrobrachion constrictum* (Farquhar) (Echinodermata: Ophiuroidea), a Euryalid Brittle-star Living in Association with the Black Coral, *Antipathes fiordensis* (Grange, 1990)." *Journal of Experimental Marine Biology and Ecology* 221 (1998): 173–189.

Stewart, B., and P. V. Mladenov. "Reproductive Periodicity in the Euryalinid Snake Star *Astrobrachion constrictum* in a New Zealand Fiord." *Marine Biology* 123 (1995): 543–553.

Turon, X, M. Codina, I. Tarjuelo, M. J. Uriz, and M. A. Becerro. "Mass Recruitment of *Ophiothrix fragilis* (Ophiuroidea) on Sponges: Settlement Patterns and Post-Settlement Dynamics." *Marine Ecology Progress Series* 200 (2000): 201–212.

Tyler, P. A. "Seasonal Variation and Ecology of Gametogenesis in the Genus *Ophiura* (Ophiuroidea: Echinodermata) from the Bristol Channel." *Journal of Experimental Marine Biology and Ecology* 30 (1977): 185–197.

Mattias Sköld, PhD

Echinoidea
(Sea urchins and sand dollars)

Phylum Echinodermata
Class Echinoidea
Number of families 46

Thumbnail description
Ubiquitous, spine-covered animals that often live beneath the sand surface or hide out in rocky crevices and sea grass beds

Photo: Red sea urchin (*Strongylocentrotus franciscanus*) (©Shedd Aquarium Photo by Patrice Ceisel. Reproduced by permission.)

Evolution and systematics

The class is divided into the subclass Perischoechinoidea, which has only one order and contains the most taxonomically primitive members of the class (e.g., pencil urchins), and the subclass Euechinoidea, which has 19 orders and contains the "true" echinoids. These are divided into two broad groups based on body shape: the regular urchins, which contain sea urchins, and the irregular urchins, which include both heart urchins and sand dollars. Collectively, there are 46 families and about 900 known species.

Echinoids are members of a much larger group of marine animals called echinoderms. These include asteroids (sea stars), ophiuroids (brittle stars), crinoids (feather stars), concentricycloids (sea daisies), and holothurians (sea cucumbers). The last group has the closest fossil linkage to echinoids than any other echinoderm class. The fossil record shows divergence from sea stars nearly 450 million years ago in the late Ordovician period, the oldest of these being the Perischoechinoidea. Regular and irregular urchins may have first evolved in the lower Jurassic, but these early forms remain poorly understood. In particular, the fossil genus *Loriolella* has an intermediate morphology, showing both irregular and regular characteristics. However, it is believed that during the Cretaceous period, sea level rise may have accounted for a dramatic increase in their diversity. Today, urchins play an important ecological role in many marine communities globally.

Physical characteristics

Echinoids come in a variety of different shapes, sizes, and color, but like all echinoderms, they have pentamerous radial symmetry (with the exception of irregular echinoids that have bilateral symmetry) and spiny skin. One of the smallest recorded urchins, *Echinocyamus scaber*, has a test diameter of only 0.2 in (0.5 cm), whereas the largest, *Sperosoma giganteum*, a deep-water species, has a test diameter up to 15 in (38 cm).

Echinoids have a hard calcareous shell made up of a skeleton of tightly packed or fused plates called a test. The skeleton is designed to prevent cracks from spreading if damaged. All urchins are covered with moveable spines that articulate like a ball and socket joint. However, their morphology is highly variable between species. Some range from the thick blunt spines of the pencil urchin, *Eucidaris tribuloides*, to the long poisonous spines of *Diadema antillarum*. Spines are primarily used for locomotion and defense against predators, although some species cover them with shell fragments, algae, or encrusting organisms to camouflage themselves from visual predators or, alternatively, to provide shade from direct sunlight.

The echinoid test has tiny pincer-like structures called pedicellariae that, in some species, are poisonous. These species tend to have fewer spines and use their pedicellariae in defense, while others use them to clear away settling microorganisms, unwanted parasites, and detritus. Sometimes they are visible to the naked eye, e.g., the stalked globular-shaped pedicellariae of the West Indian sea egg, *Lytechinus variegatus*.

As in other echinoderms, echinoids posses rows of tube feet that go from the very top of the anus down to the top of the mouth and they use them to trap food such as detritus and algae, assist in locomotion, prey capture, adhere to substrata and kelp stipes, and even respiration. On the dorsal (upper) surface of the test, a small sieve-like or perforated plate, madreportite, allows seawater into the water vascular system,

A color reconstruction of a fossil echinoderm, a blastoid. Fossil blastoids have been found that show evidence of a striped pattern of pigmentation. (Illustration by Emily Damstra)

which is a canal of tubes running through the body to the tube feet. Following muscular contraction, seawater is drawn into the canal under pressure and directed at the tube feet, which then extend under its force. In most species, the anus is located on the dorsal surface. In some species the distinctive coloration of the anus is used by biologists and naturalist to identify differences between species. *Diadema steosum*, for example, has an orange ring around its anus, which is the only visible distinction from *D. savignyi*, which has an iridescent blue ring.

In regular urchins, the mouth is located on the aboral (opposite to the mouth) surface and consists of an array of very tough calcium carbonate plates embedded in tissue and five teeth arranged symmetrically. The entire feeding apparatus has a characteristic pattern has given rise to the name "Aris-

totle's lantern," a term used because of the resemblance to a five-sided Greek lantern and because urchins and sand dollars were a favorite animal of Aristotle. Some species such as the rock boring urchin, *Echinostrephus aciculatus*, use the teeth to rasp away at rock to form a cup-shaped hiding place. Urchins have a well-developed digestive system because they process a large amount of indigestible material such as sand, stones, and plant matter.

The test shape of regular urchins tends to be globular and pentaradiate, whereas that of irregular urchins tends to have a more flattened, oval, and bilateral symmetry. Their test shape generally reflects their mode of life. Irregular urchins, for example, are flattened for burrowing efficiently into sediment, and are covered in very short spines to aid feeding when beneath the seabed surface. Moreover, some species (e.g., sand dollar, *Leodia sexiesperforata*) have a series of slits or perforations called lunules. These may prevent it from being washed out of the sediment by the pressure of a passing wave surge. The internal organization of irregular urchins is generally similar to regular echinoids. The exception is that regular urchins have five pairs of gonads, whereas irregular urchins have between two and five.

Distribution

Echinoids live within the intertidal and subtidal waters of the Antarctic, Pacific, Atlantic, and Indian Oceans, although they are most abundant between 32.8 ft and 164 ft (10–50 m). Deep-water species can inhabit depths below 16,400 ft (5,000 m). By contrast, some species such as the short-spined sea urchin, *Anthocidaris crassispina*, can survive being periodically exposed during deep wave surges or left behind in shallow rock pools at low tide.

Habitat

Regular sea urchins inhabit a broad range of environments from wave exposed rocky outcrops, crevices within rocks, rock pools, coral reefs, sandy lagoons to sea grass beds and kelp forests. The Antarctic species, *Sterechinus neumayeri*, has an energy-efficient metabolism that is physiologically adapted to freezing and food-deprived environments. Echinoids in warmer waters, where food is plentiful, have many predators and attempts to elude them involve highly evolved cryptic coloration to blending in with the background. The pencil-like spines of the pencil urchin, *Eucidaris*, for example, encourage encrusting algae to settle, making it virtually invisible during the day. This is not the case in the Galápagos Islands, where spines are not covered with epiphytes or epizotes, and the *Eucidaris* are completely exposed. Urchins inhabiting sea grass beds adopt a slightly different avoidance strategy; for example, some species cover their test with shell fragments, algae, and other types of debris to provide appropriate camouflage.

Heart urchins and sand dollars have a different mode of existence to sea urchins. Their flattened test is perfectly suited to life beneath the sediment surface. This instantly provides cover from visually orientated predators. Most species occupy sandy habitats and avoid muddy or silt-dominated sediments. Others such as the purple-heart urchin *Spatangus purpureus* prefer coarse gravel habitats.

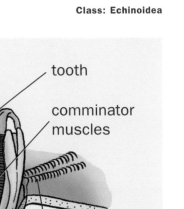

The feeding complex (Aristotle's lantern) of a sea urchin. A. Retracted; B. Protracted. (Illustration by Emily Damstra)

Behavior

Most urchins are nocturnal foragers, although some locomotory activity can coincide with tidal changing if they occur during the day. Generally, these are small movements to compensate for change in seawater level associated with an outgoing or incoming tide. On wave-exposed shores, urchins can aggregate together to interlock spines, thus reducing the risk of being dislodged by a strong wave surge. Other aggregation patterns occur in defensive moves directed at predators, whereas other patterns usually signify feeding aggregations.

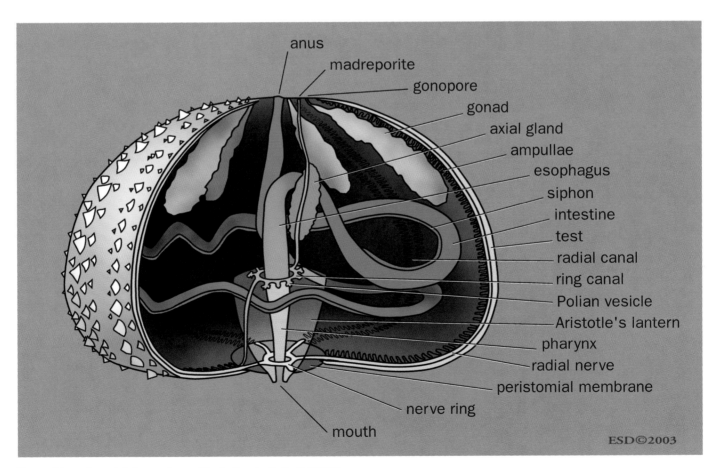

Sea urchin and sand dollar anatomy. (Illustration by Emily Damstra)

Green sea urchins (*Strongylocentrotus droebachiensis*) feeding on crab. (Photo by ©Andrew J. Martinez/Photo Researchers, Inc. Reproduced by permission.)

A toxic sea urchin (*Tripneustes gratilla*) covers itself with debris as camouflage. (Photo by ©A. Flowers & L. Newman/Photo Researchers, Inc. Reproduced by permission.)

Sand dollars burrow into the sediment by moving at an angle to the surface, and heart urchins burrow by excavating allowing the animal to sink vertically into the sediment. Forward movement is then achieved by using its powerful laterally orientated spines.

Feeding ecology and diet

Sea urchins feed in a variety of different ways. Regular urchins use their teeth to bite and rasp encrusting algae, sea grasses, and seaweeds (herbivores), while others feed on sessile organisms, carrion, and detritus (ominvores). Foraging for food usually occurs at night to avoid predators. Most sea urchins are slow moving.

Echinoids are formidable grazers and are capable of significantly altering the marine community in which they live. Following a mass mortality of sea urchins (often caused by a parasitic disease), there is a rapid growth in algae that can have devastating impacts to coral reefs. In many shallow marine communities, such species are recognized as keystone because their impact is disproportionately large. Little is known about deep-sea urchins other than they are believed to have similar feeding habitats to shallow-water species.

Irregular urchins are primarily deposit feeders, processing fine organic matter that settles on the seafloor. Heart urchins process sand in bulk to feed on organic particles trapped in sediment. Some species draw down surface detritus into their burrows, via a respiratory funnel, using their tube feet. Sand dollars, however, are more efficient as they sieve fine sediment particles while burrowing. Their spines are dense enough to prevent sand grains from falling through, yet fine enough to allow particles to drop out onto strings of mucus before being transported to the mouth. Typically, sand dollars feed on small diatoms and organic matter that accumulate in sand. Some species can become suspension feeders by lifting the test obliquely out of the sediment (e.g., *Dendraster excentricus*) to face the prevailing water currents. The tube feet of some sand dollars are specialized for deposit feeding, whereas some heart urchins have tube feet around the mouth specialized in picking up organic-rich detritus from sediment (selective deposit feeding).

Predators of echinoids include sea otters, sea stars, crabs, eels, lobsters, and fishes such as wrasses, wolffish, American plaice, butterflyfish, porcupine fish, and triggerfish. In coral habitats, the urchin, *Diadema*, is consumed by triggerfish, which blow jets of water at it in an attempt to turn the test upside down. American plaice feed exclusively on sand dollars, which in most cases only live for a few years. By contrast, heart urchins have fewer predators and can survive for up to 12 years. Some deep-sea echinoids have poisonous spines to defend against predators.

Reproductive biology

All echinoids have separate sexes, which cannot be distinguished by their external appearance. Only Antarctic urchins show a tendency to brood their young. The sexual dimorphism of these brooding heart urchins is generally easy to see. The "marsupial" is pronounced in females; present but not conspicuous in males. Generally, females are broadcast spawners, releasing millions of eggs into the water column to be fertilized by male sperm. Sea urchin gonads are connected to small openings in the body wall called gonopores, through which eggs and sperm are released. The resultant larvae undergo a multistage planktonic phase before settlement and metamorphosis. Most have a characteristic larvae development, passing through a free-swimming stage called echinopluteus. Larvae are bilaterally symmetric and show no signs of pentaradiate symmetry, a characteristic of sea urchins. Metamorphosis gives rise to an adult body form either before or after settling on the sea floor. The duration between planktonic development and settlement is species- and geographically dependent.

Very few echinoids are brooders. The heart urchin, *Abatus cordatus*, broods its young to increase survival in freezing temperatures and food-deprived environments. These species omit a planktonic stage (lecithotrophs), a strategy to produce a small number of offspring that are larger in size.

Unlike sea stars, urchins are less likely to regenerate, although they will grow lost spines, tube feet, and repair holes in test.

A sand dollar (*Echinarachnius parma*) on its side for feeding. (Photo by ©Andrew J. Martinez/Photo Researchers, Inc. Reproduced by permission.)

Conservation status

No echinoid is listed on the IUCN Red List of Threatened Species. However, *Diadema* is protected by the state of Florida.

Significance to humans

Some echinoids are commercially valuable. In the United States, for example, red sea urchin, *Strongylocentrotus franciscanus*, purple sea urchin, *S. purpuratus*, and green urchin, *S. droebachiensis*, are harvested for their roe. In Japan, urchin eggs and their reproductive organs are eaten as a delicacy. In European waters, overexploitation of *Paracentrotus lividus* has resulted in habitat destruction and decline in population numbers. In areas where predators of urchins have been overfished, urchin numbers can explode, causing devastating impacts to the marine community. In the Caribbean, for example, urchins have been responsible for 90% of the bioerosion of coral reefs. Ironically, their efficiency at consuming unwanted algae and detritus has made them popular animals for aquariums.

The sea urchin, *Diadema*, is perhaps the most well known for its long spines that easily puncture human skin, leaving deep and painful wounds. The spine tips easily break off under the skin and are almost impossible to remove.

Urchins' dried and empty tests are a familiar component of strandlines on beaches and provide a valuable commodity for souvenir hunters.

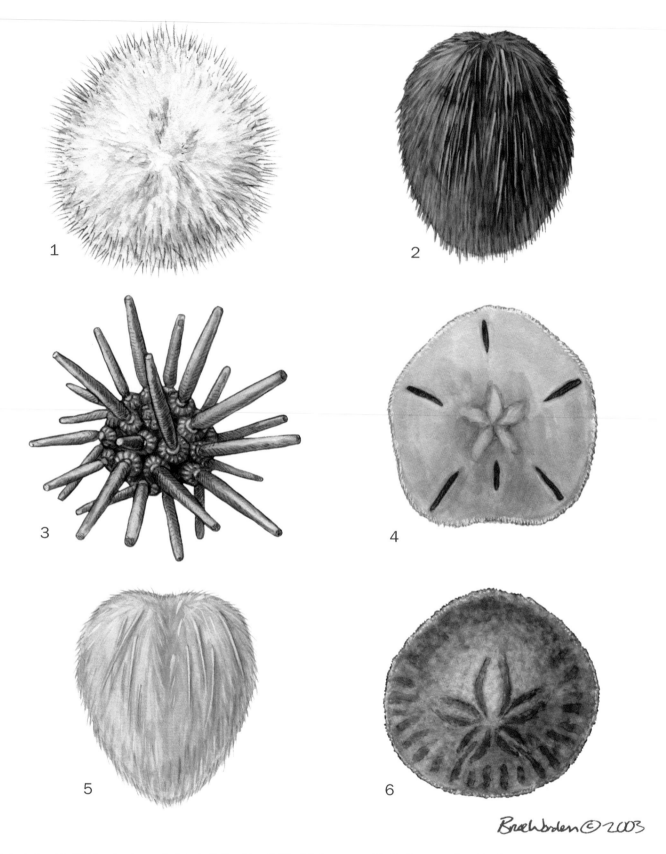

1. Rock boring urchin (*Echinostrephus aciculatus*); 2. Sea biscuit (*Plagiobrissus grandis*); 3. Slate-pencil urchin (*Eucidaris tribuloides*); 4. Six key-hole sand dollar (*Leodia sexiesperforata*); 5. Heart urchin (*Abatus cordatus*); 6. Western sand dollar (*Dendraster excentricus*). (Illustration by Bruce Worden)

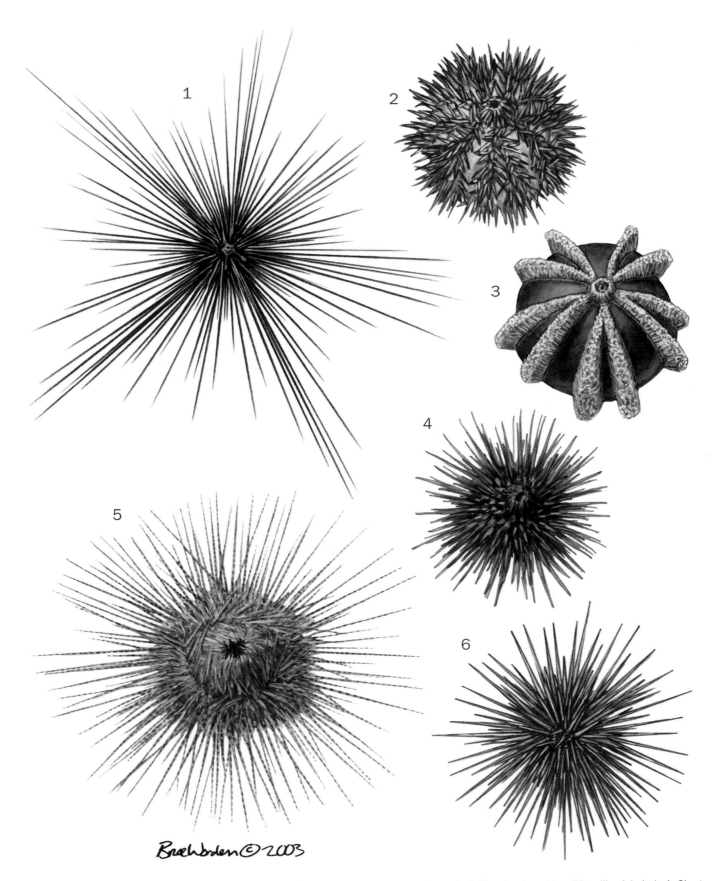

1. Long-spined sea urchin (*Diadema savignyi*); 2. West Indian sea egg (*Lytechinus variegatus*); 3. Tuxedo pincushion (*Mespilia globulus*); 4. Short-spined sea urchin (*Anthocidaris crassispina*); 5. Magnificent urchin (*Astropyga magnifica*); 6. Pea urchin (*Echinocyamus pusillus*). (Illustration by Bruce Worden)

Species accounts

Slate-pencil urchin
Eucidaris tribuloides

ORDER
Cidariida

FAMILY
Cidaridae

TAXONOMY
Eucidaris tribuloides Lamarck, 1816.

OTHER COMMON NAMES
English: Mine urchin, club urchin.

PHYSICAL CHARACTERISTICS
Size can reach 3.1 in (8 cm) in diameter, but with spines up to 5.1 in (13 cm). Coloration variable, from brown, red-brown to sometimes mottled brown-red with grayish white spine. Spines vary in color because of the attachment of encrusting organisms, particularly algae such as coralline. Solid brown cylindrical spines attached to a globular test.

DISTRIBUTION
Commonly found in less than 165 ft (50 m) of water, although can occur to a depth of 2,265 ft (800 m). Inhabits the coastal waters of North Carolina to Brazil.

HABITAT
Hides among sea grass beds such as turtle grass and in rocky crevices amongst coral, under rocks, and coral rubble. Usually found in lagoon areas. Spine length related to habitat. In high wave-exposed areas, spines are usually short.

BEHAVIOR
Forages for food at night. In coral reef habitats, spines used to defend against predators and as a mechanism to wedge itself in a crevice. In sea grass beds, it does not cover itself with algae and detritus as camouflage, as do other urchins. Instead, it allows encrusting organisms to settle and grow on spines; a very slow-moving urchin.

FEEDING ECOLOGY AND DIET
Omnivorous, feeding on algae and small invertebrates like sea squirts, sponges, bryozoans, algae, gastropods, and bivalves. Although *E. tribuloides* has not been shown to be destructive to corals, a close relative that shares similar feeding habits, *E. galapagensis*, is destructive.

REPRODUCTIVE BIOLOGY
Sexual reproduction; eggs and sperm are shed into the water column where fertilized eggs develop into free-living and transparent larvae with reddish eyespots before metamorphosing into juveniles and settling on the seabed. Spawning period linked to day-length and lunar cycle, but actual event varies geographically.

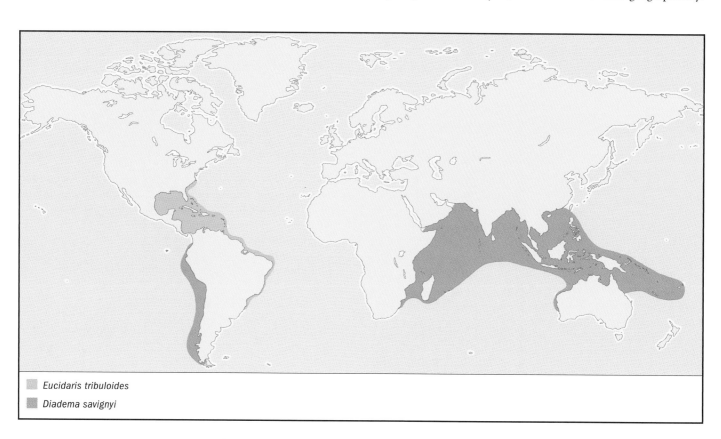

Eucidaris tribuloides
Diadema savignyi

CONSERVATION STATUS
Not listed by the IUCN.

SIGNIFICANCE TO HUMANS
None known. ◆

Magnificent urchin
Astropyga magnifica

ORDER
Diadematida

FAMILY
Diadematidae

TAXONOMY
Astropyga magnifica Clark, 1934.

OTHER COMMON NAMES
None known.

PHYSICAL CHARACTERISTICS
The test diameter can reach 7.8 in (20 cm) with brilliant coloration of golden yellow and iridescent blue spots. Spines are long and banded with reddish brown and yellowish white. Areas of test are bare.

DISTRIBUTION
Tropical coastal waters of the western Atlantic to the northeast South America down to about 295 ft (90 m).

HABITAT
Sandy bottoms, shell sand, and limestone outcrops on coral reefs.

BEHAVIOR
A fast-moving urchin owing to its ability to walk on its spines. Patchy distribution, but some are known to aggregate. When threatened by predator, it clumps its spines in a defense position. Similar biology to the sea urchin, *Diadema antillarum*. Species associated with a number of different types of commensal animals such as small fish and shrimps.

FEEDING ECOLOGY AND DIET
Grazer, feeding on algae and occasionally small invertebrates.

REPRODUCTIVE BIOLOGY
Sexual reproduction. Sexes are separate. Eggs and sperm are shed into the water column, where they are fertilized and develop into free-living larvae.

CONSERVATION STATUS
Not listed by the IUCN.

SIGNIFICANCE TO HUMANS
None known. ◆

Long-spined sea urchin
Diadema savignyi

ORDER
Diadematida

FAMILY
Diadematidae

TAXONOMY
Diadema savignyi Michelin, 1845.

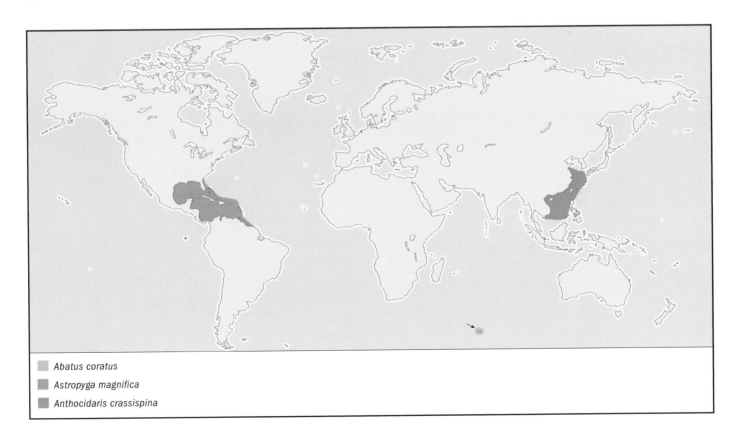

■ *Abatus coratus*
■ *Astropyga magnifica*
■ *Anthocidaris crassispina*

OTHER COMMON NAMES
English: Blue eye urchin, black long-spined urchin.

PHYSICAL CHARACTERISTICS
Relatively flattened, but oval test that can reach 3.1 in (8 cm) in diameter. Distinctive long and slender black or white spines, which vary in length from 1.9–3.9 in (5–10 cm). Test and spines are fragile. It has a distinctive iridescent blue ring around anus. Often mistaken for *D. steosum*, which has an orange ring. Darker animals tend to exist in the open on sand, whereas paler colored animals live in crevices or in turbid waters.

DISTRIBUTION
A ubiquitous shallow water species found throughout the Indo-Pacific and South Pacific Oceans to a depth of 1,312 ft (400 m).

HABITAT
Usually found inhabiting sheltered areas of coral reefs on rocky substrata or sandy lagoons, but occasionally occur in sea grass beds.

BEHAVIOR
An active animal that is highly responsive to changes in light intensity. Often the test color will change according to changes in light. High densities may provide protection against predators. A variety of commensals associated with urchins, living among the spines, ranging from small shrimps and mysids to juvenile fish.

FEEDING ECOLOGY AND DIET
Voracious grazer of algal turf. Following unexpected mass mortalities, growth of algae is usually rapid and detrimental to many coral reefs globally. Tend to hide during the day in rocky crevices, but forage for food at night. Has many predators, mostly fish, mollusks, and crustaceans.

REPRODUCTIVE BIOLOGY
Sexual reproduction; depending on geographical location, known to aggregate during spawning. Release of eggs and sperm into the water column coincides with the lunar cycle usually between summer and early winter.

CONSERVATION STATUS
Not listed by the IUCN.

SIGNIFICANCE TO HUMANS
Considered ecologically important to many coral reef ecosystems. Spines are poisonous and easily puncture human skin, often leaving infection. ◆

PHYSICAL CHARACTERISTICS
Size variable, but commonly occurring between 0.8 and 2 in (2–5 cm) in diameter. Have black to dark violet coloration, with short spines.

DISTRIBUTION
Commonly found inhabiting the lower intertidal and shallow subtidal areas throughout the coastal waters of southern China, where it is an important benthic grazer of many algae-dominated communities.

HABITAT
Rocky shores are their principle habitat, where they appear to be well adapted to living in crevices and rock pools, especially on wave-exposed shores. Geographical distribution is limited by saline conditions; urchin densities tend to be lower on the oceanic southern and eastern coasts of Asia. When found on sand, their spines are usually covered in algae and shell fragments to camouflage them from predators.

BEHAVIOR
Foraging activities are predominately nocturnal, although movements during the day correlate with changes in seawater depth during the tidal cycle.

FEEDING ECOLOGY AND DIET
Grazer, feeding mainly on encrusting algae (e.g., *Corallina*, *Sargassum*, *Colpomenia*, and *Ulva*), although the diet broadly reflects their seasonal availability. During the summer months, when the availability of prey is generally low, the diet is dominated by *Corallina*. Occasionally, other small organisms are consumed, such as bivalves and crabs.

REPRODUCTIVE BIOLOGY
Sexual reproduction; spawning usually occurs between July and September, depending on seawater temperature. Broadcast fertilized eggs into the water column. Larvae undergo planktonic stage before settling to the seafloor.

CONSERVATION STATUS
Not listed by the IUCN.

SIGNIFICANCE TO HUMANS
Commercially harvested for roe in the coastal waters of Asia. Also used as an indicator species for toxic chemicals in the marine environment. The occurrence of an indicator species is usually associated with impacts from anthropogenic activity, such as high concentrations of chemicals released from outfalls, or industrial spill run off from the land into the sea. ◆

Short-spined sea urchin
Anthocidaris crassispina

ORDER
Echinida

FAMILY
Echinometridae

TAXONOMY
Anthocidaris crassispina Agassiz, 1863.

OTHER COMMON NAMES
English: Far Eastern violet sea urchin, purple sea urchin, black urchin, decorator urchin.

Rock boring urchin
Echinostrephus aciculatus

ORDER
Echinida

FAMILY
Echinometridae

TAXONOMY
Echinostrephus aciculatus Agassiz, 1863.

OTHER COMMON NAMES
English: Needle-spined urchin, reef boring sea-hedgehog.

PHYSICAL CHARACTERISTICS
Flattened and circular test up to 3.1 in (8 cm) in diameter. Has black to reddish purple coloration with rigid needle-like spines.

DISTRIBUTION
Indo-Pacific coastal waters to a depth of 165 ft (50 m).

HABITAT
Lives in rock and coral.

BEHAVIOR
Bores into rock to create living-chamber. Uses spines on its upper surface as defense against predators. As they grow, individuals can become trapped in chamber.

FEEDING ECOLOGY AND DIET
Dependent on catching drifting algae that passes across mouth of burrow. Uses tube feet to capture algae.

REPRODUCTIVE BIOLOGY
Sexual reproduction. Sexes are separate. Eggs and sperm are shed into the water column, where they are fertilized and develop into free-living larvae.

CONSERVATION STATUS
Not listed by the IUCN.

SIGNIFICANCE TO HUMANS
None known. ◆

Pea urchin
Echinocyamus pusillus

ORDER
Laganina

FAMILY
Fibulariidae

TAXONOMY
Echinocyamus pusillus Muller, 1776.

OTHER COMMON NAMES
English: Dwarf sea urchin.

PHYSICAL CHARACTERISTICS
Among one of the smallest urchins, its tiny oval test measures up to a maximum 0.6 in (1.5 cm) in length. Usually gray-green to bright green in color with very short spines, which give the animal a velvet texture to touch.

DISTRIBUTION
European coastal waters to a depth of 165 ft (50 m).

HABITAT
Lives buried in coarse gravel, sandy sediment, and among *Zosteria* and maerl beds.

BEHAVIOR
Little known.

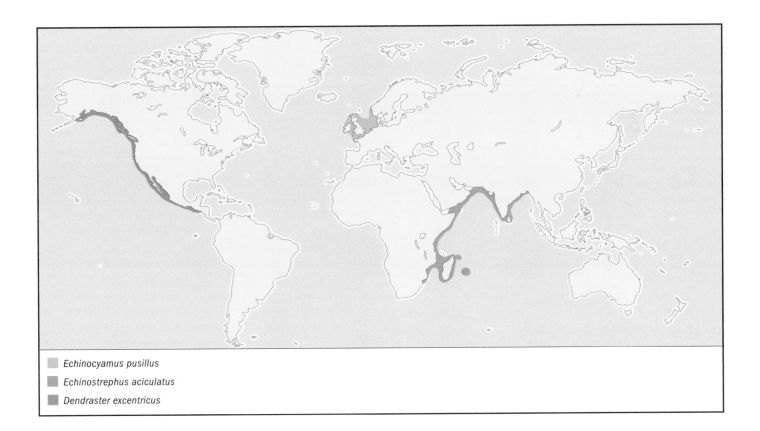

■ *Echinocyamus pusillus*

■ *Echinostrephus aciculatus*

■ *Dendraster excentricus*

FEEDING ECOLOGY AND DIET
Primarily deposit feeder, but also preys on foraminiferans and grazes on individual sediment grains for its organic coating.

REPRODUCTIVE BIOLOGY
Sexual reproduction. Sexes are separate. Eggs and sperm are shed into the water column, where they are fertilized and develop into free-living larvae.

CONSERVATION STATUS
Not listed by the IUCN.

SIGNIFICANCE TO HUMANS
None known. ◆

Western sand dollar
Dendraster excentricus

ORDER
Scutellina

FAMILY
Dendrasteridae

TAXONOMY
Dendraster excentricus Eschscholtz, 1831.

OTHER COMMON NAMES
English: Eccentric sand dollar.

PHYSICAL CHARACTERISTICS
Rigid test measuring up to 3.5 in (9 cm) and covered with moveable spines. Pale gray-lavender to dark purplish black coloration; characteristic pentaradiate or petal-shaped pattern tube feet on upper surface of test.

DISTRIBUTION
Northeastern coasts of the Pacific ocean from southern Alaska to Mexico. Found in depths between 130–295 ft (40–90 m).

HABITAT
Inhabits sandy bottoms within sheltered bays, lagoons, and open coastal areas. Commonly found in dense aggregations forming a carpet of animals.

BEHAVIOR
Occurs in large numbers on the seabed. These high densities are thought to help influence the nature of near-bed currents to assist in feeding.

FEEDING ECOLOGY AND DIET
The only echinoid to suspension feed, trapping suspended organic particles as they drift in passing water currents. Mucus strands assist in trapping organic matter. Juveniles ingest sand when feeding to act as ballast. Stand in the sand obliquely when feeding and use their tube feet for respiration and catching food.

REPRODUCTIVE BIOLOGY
Sexual reproduction; spawning period between July and August. Adults may feed on larvae, but not their eggs because of a protective coating. Settling larvae have a greater protection from predators if they settle within existing sand dollar beds. Estimated lifespan is up to 15 years.

CONSERVATION STATUS
Not listed by the IUCN.

SIGNIFICANCE TO HUMANS
Selected by beachcombers for their aesthetic value. ◆

Six keyhole sand dollar
Leodia sexiesperforata

ORDER
Scutellina

FAMILY
Mellitidae

TAXONOMY
Leodia sexiesperforata Leske, 1778.

OTHER COMMON NAMES
English: Sand dollar.

PHYSICAL CHARACTERISTICS
Thin and flattened disc with six characteristic slot-like holes and distinctive pentaradiate or petal-like pattern tube feet on upper surface. Diameter is up to 3.9 in (10 cm). Yellow to light brown in color.

DISTRIBUTION
Occurs along coastal waters of the western Atlantic from North Carolina to Uruguay. Subtidally to a depth of 200 ft (60 m).

HABITAT
Commonly found inhabiting open sandy areas clear of algae.

BEHAVIOR
Burrows vertically into sand to several inches (centimeters) below surface. Species is host to the crab, *Dissodactylus crinitichelis*.

FEEDING ECOLOGY AND DIET
Feeds on organic particles such as algae and detritus. Uses mucus strands to collected food. Known predators include the triggerfish (*Balistes capriscus*) and helmet conch (*Cassis tuberosa*).

REPRODUCTIVE BIOLOGY
Sexual reproduction; spawning tends to occur during rainy season between late summer and autumn.

CONSERVATION STATUS
Not listed by the IUCN.

SIGNIFICANCE TO HUMANS
Collected by beachcombers for their aesthetic value. ◆

Sea biscuit
Plagiobrissus grandis

ORDER
Spatangida

FAMILY
Brissidae

TAXONOMY
Plagiobrissus grandis Gmelin, 1788.

■ *Plagiobrissus grandis*
■ *Leodia sexiesperforata*

OTHER COMMON NAMES
English: Heart urchin, long-spined sea biscuit, great red-footed urchin.

PHYSICAL CHARACTERISTICS
One of the largest irregular urchins; longest dimension is their oval-shaped test that can reach 9.8 in (25 cm) in length. Colors vary from yellow or tan to reddish brown. Test covered with medium-sized spines, although some on the upper surface are up to 3.9 in (10 cm) long. Flat underneath with five rows of tube feet in ambulacral groove.

DISTRIBUTION
Found in shallow waters from Florida to the West Indies and Brazil. Common at depths less than 164 ft (50 m), but can occur in depths greater than 655 ft (200 m).

HABITAT
Inhabits sandy coastal lagoons.

BEHAVIOR
Burrows into the sediment. Its long spines are used to defend itself from predators when buried. Often associated with the commensal crab, *Dissodactylus primitivus*.

FEEDING ECOLOGY AND DIET
Deposit feeder; feeding is suspected to occur both night and day. Main predators are the large helmet conchs, *Cassis tuberosa* and *C. madagascariensis spinella*.

REPRODUCTIVE BIOLOGY
Sexual reproduction. Sexes are separate. Eggs and sperm are released into the water column, where they are fertilized and develop into free-living larvae.

CONSERVATION STATUS
Not listed by the IUCN.

SIGNIFICANCE TO HUMANS
Known to have poisonous spines. ◆

Heart urchin
Abatus cordatus

ORDER
Spatangida

FAMILY
Schizasteridae

TAXONOMY
Abatus cordatus Verrill, 1876.

OTHER COMMON NAMES
None known.

PHYSICAL CHARACTERISTICS
Diameter can reach 2.3 in (6 cm). Coloration usually brownish yellow. Oval shaped with dense coat of fine spines that keep sediment clear of test during burrowing activity.

DISTRIBUTION
Limited distribution. Endemic to the Kerguelen Islands, Antarctica.

HABITAT
Inhabits shallow inlets and bays, living buried in fine sand usually protected from wave swell.

BEHAVIOR
Little known, but lives within dense populations.

FEEDING ECOLOGY AND DIET
Little known but other species belonging to the genus *Abatus* are deposit feeders, usually gathering detritus with their tube feet.

REPRODUCTIVE BIOLOGY
Females brood their young in dorsal pouch. The strategy is to produce few eggs and small amounts of sperm at any one time. The advantage is a higher survival rate of offspring compared to the more usual strategy of broadcasting species.

CONSERVATION STATUS
Not listed by the IUCN.

SIGNIFICANCE TO HUMANS
None known. ◆

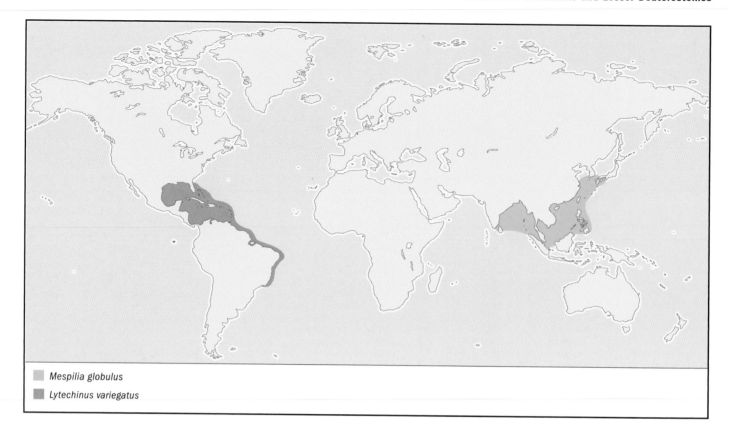

Mespilia globulus

Lytechinus variegatus

Tuxedo pincushion urchin
Mespilia globulus

ORDER
Temnopleurida

FAMILY
Temnopleuridae

TAXONOMY
Mespilia globules Linnaeus, 1758.

OTHER COMMON NAMES
English: Royal urchin, sphere urchin, globular sea urchin, ball sea-hedgehog.

PHYSICAL CHARACTERISTICS
Test can reach 2.3 in (6 cm) in diameter and has five to 10 broad bands of bright blue coloration with bands of reddish-brown spines, which give it a very striking appearance.

DISTRIBUTION
Coastal waters of China, Japan, Indian Ocean, and the Philippines.

HABITAT
During the day will hide among rocks, rubble, and in crevices associated with coral reefs.

BEHAVIOR
Solitary animal. Camouflages test with shell fragments and algae held on with fine cilia hairs between spines.

FEEDING ECOLOGY AND DIET
Herbivore, feeding on micro-algae and encrusting coralline. Forages for food at night.

REPRODUCTIVE BIOLOGY
Sexual reproduction.

CONSERVATION STATUS
Not listed by the IUCN.

SIGNIFICANCE TO HUMANS
Used in home aquaria. ◆

West Indian sea egg
Lytechinus variegatus

ORDER
Temnopleurida

FAMILY
Toxopneustidae

TAXONOMY
Lytechinus variegates Lamarck, 1816.

OTHER COMMON NAMES
English: Short-spined urchin.

PHYSICAL CHARACTERISTICS
Large, up to 3.9 in (10 cm) in diameter with spherical test that appears greenish with green-purple spines. Stalked pedicellariae are white or pink and visible to the naked eye.

DISTRIBUTION

Found in coastal lagoons from the southwestern Atlantic to the Caribbean and Brazil. Subtidal to a depth of 1,090 ft (250 m).

HABITAT

Found inhabiting sea grass beds (*Thalassia*), sandy bottoms, and among rocks. Avoids turbid waters containing high concentrations of suspended silts.

BEHAVIOR

Covers its test with shell fragments, rubble, grass blades, and algae for camouflage or protection against strong sunlight.

FEEDING ECOLOGY AND DIET

Herbivore, feeding on turtle grass (*Thalassia*). Predators include reef fish, helmet conch (*Cassis tuberosa*), and shore birds.

REPRODUCTIVE BIOLOGY

Sexual reproduction; spawning season between spring and summer and occurs within the lunar cycle. Metamorphosis of settling young tends to occur in the intertidal regions.

CONSERVATION STATUS

Not listed by the IUCN.

SIGNIFICANCE TO HUMANS

Roe may be consumed locally. ◆

Resources

Books

Anzalone, L. "Loriolella, and the Transition from Regular to Irregular Echinoids." In *Echinoderm Research, Proceedings of the 5th European Conference on Echinoderms*, edited by Maria D. C. Carnevali and Francesco Bonasoro. Rotterdam: A. A. Balkema, 1999.

Birkeland, C. "The Influence of Echinoderms on Coral-reef Communities." In *Echinoderm Studies*, edited by Michael Jangoux and John Lawrence. Rotterdam: A. A. Balkema, 1989.

Ebert, T. A. "Recruitment in Echinoderms." In *Echinoderm Studies*, edited by Michael Jangoux and John Lawrence. Rotterdam: A. A. Balkema, 1983.

Eugene, N. K. *Marine Invertebrates of the Pacific Northwest.* Seattle: University of Washington Press, 1996.

Hendler, G., J. E. Miller, D. L. Pawson, and P. M. Kier. *Sea Stars, Sea Urchins, and Allies: Echinoderms of Florida and the Caribbean.* Washington, DC, and London: The Smithsonian Institution Press, 1995.

Kozloff, E. N. *Seashore Life of the Northern Pacific Coast.* Seattle: University of Washington Press, 1993.

Lawrence, J. *A Functional Biology of Echinoderms.* London and Sydney: Croom Helm Press, 1987.

Lerman, M. *Marine Biology: Environment, Diversity, and Ecology.* San Francisco: Cummings Publishing Company, 1986.

Littlewood, D. T. J. "Echinoderm Class Relationships Revisited." In *Echinoderm Research*, edited by H. Roland, A. Emson, B. Smith, and A. C. Campbell. Rotterdam: A. A. Balkema, 1995.

O'Clair, R. M., and E. O. O'Clair. *Southeast Alaska's Rocky Shores, Animals.* Auke Bay, AK: Plant Press, 1998.

Picton, B. E. *A Field Guide to the Shallow-water Echinoderms of the British Isles.* London: Immel Publishing Ltd, 1993.

Sloan, N. A., and A. C. Campbell. "Perception of Food." In *Echinoderm Nutrition*, edited by Michael Jangoux and John Lawrence. Rotterdam: A. A. Balkema, 1982.

Periodicals

Chiappone, M., D. W. Swanson, and S. L. Miller. "Density, Spatial Distribution and Size Structure of Sea Urchins in Florida Keys Coral Reef and Hard-bottom Habitats." *Marine Ecology Progress Series* 235 (2002): 117–126.

Chiu, S. T. "*Anthocidaris crassispina* (Echinodermata: Echinoidea) Grazing Epibenthic Macroalgae in Hong Kong." *Asian Marine Biology* (1988): 1–79.

Drummond, A. E. "Reproduction of the Sea Urchins *Echinometra mathaei* and *Diadema savignyi* on the South African Eastern Coast." *Marine Freshwater Research* 46 (1995): 751–755.

Ebert, T. A. "The Consequences of Broadcasting, Brooding, and Asexual Reproduction in Echinoderm Metapopulations." *Oceanologica Acta* 19 (1996): 217–226.

Emson, R. H., and I. C. Wilkie. "Fission and Autotomy in Echinoderms." *Oceanography Biological Annual Review* 18 (1973): 389–438.

Freeman, S. M. "Size-dependent Distribution and Diurnal Rythmicity Patterns in the Short-spined Sea Urchin *Anthocidaris crassispina*." *Estuarine, Coastal Shelf Science* (2003).

Ghiold, J. "Observations on the Clypeasteroid *Echinocyamus pusillus* (Muller)." *Journal of Experimental Marine Biology and Ecology* 61 (1982): 57–74.

Kobayashi, N. "Fertilized Sea Urchin Eggs as an Indicatory Material for Marine Pollution Bioassay, Preliminary Experiments." *Marine Biology Laboratory* (1971): 379–406.

McClintock, J. B. "Trophic Biology of Antarctic Shallow-water Echinoderms." *Marine Ecology Progress Series* 111 (1994): 191–202.

Mooi, R. "Sand Dollars of the Genus *Dendraster* (Echinoidea: Clypeasteroidea): Phylogenic Systematics, Heterochrony, and Distribution of Extant Species." *Bulletin of Marine Science* 61 (1997): 343–375.

Lawrence, J. M., and J. Herrera. "Stress and Deviant Reproduction in Echinoderms." *Zoological Studies* 39 (2000): 151–171.

Paine, R. T. "A Note on Trophic Complexity and Community Stability." *American Naturalist* 103 (1969): 91–93.

Pawson, D. L. "Some Aspects of the Biology of Deep-sea Echinoderms." *Thalassia Jugoslavica* 12 (1976): 287–293.

Poulin, E., and J. P. Feral. "Pattern of Spatial Distribution of a Brood-protecting Schizasterid Echinoid, *Abatus cordatus*,

Endemic to the Kerguelen Islands." *Marine Ecology Progress Series* 118 (1995): 179–186.

Smith, A. B. "The Stereom Microstructure of the Echinoid Test." *Special Papers in Palaeontology* (1981): 1–85.

Telford, M., A. S. Harold, and R. Mooi. "Feeding Structures, Behavior and Microhabitat of *Echinocyamus pusillus* (Echinoidea: Clypeasteoidea)." *Biological Bulletins* 165 (1983): 745–757.

Timlo, P. L. "Sand Dollars as Suspension Feeders: A New Description of Feeding in *Dendraster excentricus*." *Biological Bulletin* 151 (1976): 247–259.

Other

"Classification of the Extant Echinodermata." [July 15, 2003]. <http://www.calacademy.org/research/izg/echinoderm/classify.htm>.

Steven Mark Freeman, PhD

Holothuroidea

(Sea cucumbers)

Phylum Echinodermata

Class Holothuroidea

Number of families 25

Thumbnail description
Worm-like echinoderms with a mouth surrounded by feeding tentacles and often a reduced skeleton of microscopic ossicles

Photo: A sea cucumber (*Bohadschia argus*) expelling cuverian tubules for defense. (Photo by © A. Flowers & L. Newman/Photo Researchers, Inc. Reproduced by permission.)

Evolution and systematics

As for most soft-bodied animals, the fossil history of holothuroids, or sea cucumbers, is threadbare. Only 19 species have been described from body fossils, although one form, an *Achistrum* species from the Middle Pennsylvanian Mazon Creek Formation in North America, is known from study of several thousand, often quite well-preserved, specimens. Most ancient species are known from study of isolated fossils of their ossicles. These microscopic skeletal elements, found in the body walls and internal organs of most taxa, are an important feature in defining extinct and living species. However, because ossicle form varies even within a single animal, most fossil ossicles are classified as paraspecies on the basis of unique morphological features. Largely on the basis of this record of ossicles as well as the few known body fossils, approximately 12 of the 25 living families of holothuroids have been found to have ancient representatives.

The earliest undisputed fossils of holothuroids are of isolated ossicles from the Middle Silurian Period circa 425 million years ago (mya). Plate ossicles attributed to holothuroids are known from the Ordovician Period 450 mya, but their identity as holothuroid is uncertain because they resemble the plates of other echinoderms. Holothuroids appear to have evolved perhaps 480 mya from a poorly known group of extinct burrowing echinoderms called *ophiocystioids*, which resembled spineless sea urchins with a reduced number of large, plated tube feet. The oldest described body fossil is of *Palaeocucumaria hunsrueckiana* from the Lower Devonian Period 395 mya. This species is unique among known holothuroids in having plated tentacles, a feature that suggests in part a link to the ophiocystioids. Holothuroids continued to diversify during the Paleozoic Era, when members assigned to the orders Apodida, Elasipodida, Dendrochirotida, and Dacytlochirotida first appeared.

Holothuroids were decimated 250 mya by the end-Permian mass extinction event, as were nearly all other marine organisms. During this time, other classes of echinoderms either became extinct or were reduced to representatives from one or two genera. Holothuroids, however, survived as several divergent groups, perhaps aided by a deep-water, burrowing, or detritus-feeding lifestyle. By the Middle Jurassic Period, approximately 180 mya, holothuroids had diversified considerably, and several new important groups arose, including the living orders Aspidochirotida and Molpadiida, as well as the family Synaptidae within Apodida. In addition, several still-living taxa, now known only from deep water, disappeared from the fossil record. This disappearance may indicate that these organisms invaded the deep sea around that time. By the Jurassic, the family Achistridae, one of the most successful holothuroid groups from the late Paleozoic, had again become a dominant component of the fossil fauna. It met its demise in the Lower Cretaceous 140 mya. Holothuroid fossils younger than 65 million years old are surprisingly scarce, possibly because of a lack of collecting effort. Still, several groups at the family and subfamily levels appear to have arisen during this time, including the family Molpadiidae and two subfamilies within Synaptidae—the Leptosynaptinae and Synaptinae.

The class name Holothuroidea is derived from the term *holothourion* used by Aristotle to describe an animal that was only "slightly different from the sponges," "without feeling," and "unattached," but "plant-like." This puzzling description resulted in use of the latinized term *Holothuria*, and its variants from at least the early sixteenth century until the late eighteenth century for siphonophore jellyfish, sea squirts (Tunicata), and priapulid worms, in addition to what we now call holothuroids. Linnaeus originally referred most holothuroids to *Fistularia*, a name preoccupied by a fish and

A sea apple (*Pseudocolochirus violaceus*) with its feeding tentacles extended. (Photo by ©A. Flowers & L. Newman/Photo Researchers, Inc. Reproduced by permission.)

subsequently abandoned. However, by the nineteenth century, *Holothuria* referred almost exclusively to certain holothuroids. Nevertheless, the historic uncertainty in ascription engendered several alternative names for the entire class that gained some currency. Arguments over the provenance of the term continued into the twentieth century until *Holothuria* was formally assigned to the eponymous class of echinoderms in 1924.

Living holothuroids are divided into six orders. Ordinal assignment is based largely on the form of the calcareous ring and tentacles as well as the presence of certain organs, such as respiratory trees and the muscles that retract the oral region. Each order is described according to its taxonomic diversity and major diagnostic features:

APODIDA. Footless sea cucumbers. The order contains approximately 269 species in 32 genera and three families. Tentacles are digitate, pinnate, or, in some small species, simple. Respiratory trees are absent. Tube feet are completely absent. The calcareous ring is low and bandlike, without posterior projections. The body wall is very thin and often

transparent. These sea cucumber are found in both shallow and deep water.

ELASIPODIDA. Deep-sea sea cucumbers. The order contains approximately 141 species in 24 genera and five families. Tentacles are shield shaped and used in shoveling sediment. Respiratory trees are present. The calcareous ring is without posterior projections. With the exception of one family, Deimatidae, the body wall is soft to gelatinous. All forms live in deep water.

ASPIDOCHIROTIDA. Shield-tentacle sea cucumbers. There are approximately 340 species in 35 genera and three families in this order. Tentacles are shield shaped, that is, flattened and pad-like. Respiratory trees are present. The calcareous ring is without posterior projections. The body wall is generally soft and pliant. Most forms live in shallow water, although one family is restricted to the deep sea.

MOLPADIIDA. Rat-tailed sea cucumbers. Approximately 95 species compose 11 genera and four families in this order. Tentacles are digitate to simple. Respiratory trees are present. The calcareous ring may have short posterior projections. The body wall is generally soft and pliant. Most forms live in relatively shallow water, although one family is restricted to the deep sea.

DENDROCHIROTIDA. Suspension-feeding sea cucumbers. The order contains approximately 550 species in 90 genera and seven families. Tentacles are highly branched. Respiratory trees are present. Some members have a calcareous ring composed of numerous small pieces or have long posterior extensions. These animals have muscles for retracting the oral introvert. In a few species, the body is hardened from enlarged plate-like ossicles and is U shaped. These sea cucumbers live either attached to hard bottoms or burrow in soft sediment. Most species live in shallow water.

DACTYLOCHIROTIDA. U-shaped sea cucumbers. The order contains approximately 35 species in seven genera and three families. Tentacles are simple or have a few small digits. Respiratory trees are present. The calcareous ring is without posterior projections. These sea cucumbers have muscles for retracting the oral introvert. All members have a rigid body encased in enlarged flattened ossicles. The body usually is U shaped. All members live burrowed in soft sediment. Most live in deep water.

Compared with that of the other four living classes of echinoderms, the phylogeny of holothuroids was poorly known. These animals lack the integrated skeleton that provides an extensive fossil record and do not have the numerous morphological characteristics of other groups of echinoderms. The first speculations about evolutionary relations appeared in a tree figured in 1868 by the German zoologist Carl Semper. Several of Semper's suggestions have been corroborated with formal comparative analyses of morphological features and DNA sequences. The morphological work shows that apodans (members of Apodida) branched off quite early from the other holothuroids, which are united by the presence of hemal vessels and tube feet on the body wall. Among the latter forms, the elasipodans diverged next. Holothuroids in the remaining sister group are united by the

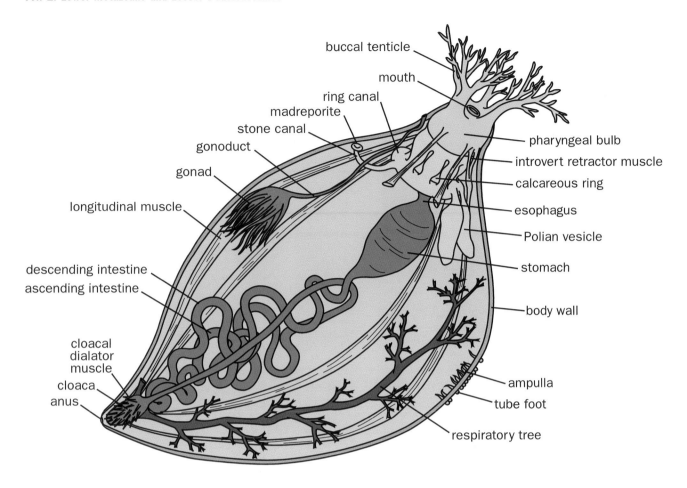

Sea cucumber anatomy. (Illustration by Laura Pabst)

presence of respiratory trees. This group radiated in the Upper Triassic to Lower Jurassic into the aspidochirotes and a group united by several characteristics, including posterior projections on the calcareous ring. The latter group diverged into the molpadiians and another lineage, comprising the dendrochirotes and dactylochirotes, the members of which have a retractible oral region called an introvert. Evidence from DNA sequences is largely congruent with the morphological data. However, several important points of disagreement remain: The arrangement of family-level branches within the dendrochirotes is poorly supported, and several families with hardened skeletons may turn out to be distantly related or subsumed within soft-bodied groups. The family-level arrangement within Elasipodida is still uncertain. In addition, the aspidochirote family Synallactidae may comprise two or more independently evolved lineages. Finally, the phylogenetic affinities of the enigmatic families Eupyrgidae and Gephyrothuriidae are unknown.

Physical characteristics

Most sea cucumbers are soft bodied and worm- or slug-like. Some tropical species have thick, muscular body walls, whereas many deep-sea forms are gelatinous and transparent. Most species are perhaps 20 in (15 cm) long, although some apodans are as small as a few millimeters or, in one species, *Synapta maculata*, more than 118 in (3 m) long. Another species in the aspidochirote Stichopodidae, *Thelenota anax*, may weigh more than 11 lb (5 kg). Although the description "echinoderm worms" is apt, some holothuroids deviate notably from a vermiform appearance. Several burrowing forms have a foreshortened dorsum, giving them an inflated U-shaped appearance. The limit of this trend is seen in the flask-shaped members of Rhopalodinidae, the mouth and anus of which are adjacent atop a long narrow stalk. Varied development of the papillae—modified tube feet—also contributes to a diversity of form. The numerous enlarged papillae in some Stichopodidae and Synallactidae give them a prickly appearance. The elasipodans Deimatidae and Elpidiidae may have elongate ventrolaterally positioned papillae that serve as "legs," locomotor structures for raising them above, and for negotiating, soft deep-sea sediments. Most holothuroids are dark colored or, in burrowing forms, pale gray to white. In contrast, many shallow-water tropical taxa are brightly colored, being green, red, orange, or yellow. Deep-sea species often are transparent or have a violet to pinkish cast.

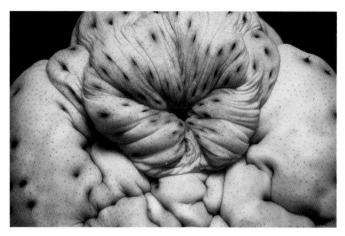

Close-up view of the mouth of a holothurodian (sea cucumber). (Photo by Bill Wood. Bruce Coleman, Inc. Reproduced by permission.)

The skeleton of holothuroids is considerably modified from that of most other echinoderms. In 90 percent of living species, the skeleton of the body wall is reduced to microscopic ossicles approximately 0.0004–0.04 in (10 µm–1 mm) long. The great variety in shape makes ossicles of considerable taxonomic importance. Ossicles are classified as rods, rosettes, crosses, buttons, tables, and wheels and anchors, among other shapes. In some dendrochirotes, such as the family Psolidae, ossicles may be secondarily enlarged and plate-like so that the animal is rigid. Another important skeletal feature, one definitive for Holothuroidea and used in higher-level taxonomy, is an internal calcareous ring that encircles the pharynx or throat. This ring serves as an attachment surface for muscles operating the oral tentacles and the anterior ends of other muscles that contract the body longitudinally.

As in other echinoderms, the holothurian water vascular system powering the tube feet consists of an anterior ring canal around the pharynx from which arise long canals running posteriorly. Despite their similarity to the radial canals of other echinoderms, the latter structures arise embryologically in a quite different manner. For this reason these canals in holothurians have been recently renamed longitudinal canals. In holothuroids, the larval structures that would in other echinoderms extend away from the mouth and form the radial canals instead become the five primary oral tentacles. This circlet of oral tentacles, from five to more than 20 in number, is another definitive feature of Holothuroidea. The tentacles may be simple, digitate (with finger-like projections), pinnate (feather-like), or peltate (flattened and shield-like). In most echinoderms, the water vascular system exchanges water with the environment through a sieve plate, or madrepore, that opens externally. In most holothuroids, however, with the notable exception of elasipodans and some molpadiians, the madrepore is internal and opens into the body cavity or coelom.

Sea cucumbers, with the exception of members of Elasipodida and Apodida, have respiratory trees used in gas exchange. These structures are paired, heavily branched tubes inside the body cavity that attach to the rectum. These structures allow a type of breathing called cloacal breathing also present in an unrelated group, the echiuran worms. In many species from the mostly tropical family Holothuriidae, numerous cuvierian tubules insert at the base of the respiratory trees. These tubules apparently serve as defensive structures in most species that have them. Members of the aspidochirote Holothuriidae and Stichopodidae and the molpadiian Molpadiidae and Caudinidae have a rete mirable, a well-developed dorsal plexus of hemal vessels over the left respiratory tree, which facilitates gas exchange. Ciliated funnels, cups, or vibratile urns are small, numerous organs arranged along the insertion of the intestinal mesenteries into the body walls of the apodans Chiridotidae and Synaptidae. Cup interiors are ciliated and appear to function in removing foreign particles from the coelomic fluid. Statocysts are small organs of presumed importance in balance and are arranged along the anterior radial nerves of apodan families, elasipodan Elpidiidae, and some molpadiians. Some species in the apodan family Synaptidae have "eyes" called ocelli, or optic cups. These structures are small patches of pigmented cells that enclose photosensitive cells at the base of the tentacles.

Distribution

Occurs worldwide from the equator to polar regions at all ocean depths. Latitudinal variation in taxonomic composition is pronounced, even at the level of orders. The shallow-water tropics to warm temperate regions are the most diverse and are dominated by members of the aspidochirote families Holothuriidae and Stichopodidae. The diversity of these families peaks on coral reefs, where 20 species per 2.5 acres (1 hectare) is not uncommon. Dendrochirotes live here as well but become a dominant part of the holothuroid fauna only in shallow cool-temperate to polar seas. Whereas the apodan Synaptidae, subfamily Synaptinae, are primarily found in the tropics, the diversity of Apodida generally increases away from the equator. Molpadiidae also are primarily found at higher latitudes or in deeper water. In the tropics, the Caribbean fauna is distinct from that of the Indo-Pacific region. The waters around southern Africa and New Zealand harbor numerous unusual endemic forms.

Marked taxonomic variation in depth also occurs. The aspidochirote Holothuriidae, in addition to being an essentially tropical family, is primarily a shallow-water group. Most holothuroid habitat, however, is in the deep sea. Many families of holothuroids have at least some deep-sea members. Most Dactylochirotida live at depth. All species of the aspidochirote Synallactidae and Gephyrothuriidae, apodan Myriotrochidae, and the order Elasipodida are found in the deep sea. Among the Elasipodida, species in Laetmogonidae live primarily at bathyl depths (3,000–6,000 ft [915–1,830 m]), whereas those in Psychropotidae and the *Peniagone* species in Elpidiidae characterize abyssal depths (6,000–18,000 ft [1,830–5,490 m]). The region below approximately 18,000 ft (5,490 m) comprises only approximately 1% of the area of the ocean floor, and a noticeable decrease in species diversity occurs. These depths consist of geologically less stable and inclined substrata in oceanic trenches that extend to 36,000 ft (10,970 m). Nevertheless, in these regions holothuroids dominate the benthic fauna in terms of weight of living organisms. Although nearly all holothuroids are restricted to particular depth ranges, a few species are remarkably indiscriminate. For example, *Elpidia glacialis* live in waters as shallow as 230 ft (70 m) in northern Europe to as deep as 33,000 ft (10,058 m).

Habitat

Holothuroids are found throughout the marine realm. They may be briefly exposed at low tide or occur in large aggregations on the deepest ocean bottom. Other species are limited to wave-hammered reef crests and rocky shorelines. Many species, particularly those in Aspidochirotida and Elasipodida, are epibenthic, living atop either hard or soft substrata. Others in Dactylochirotida, Apodida, and Molpadiida primarily burrow in sediment. There also are several swimming species, which may venture miles above the sea floor, making Holothuroidea the only class of echinoderms with pelagic members.

Behavior

With few exceptions, holothuroids are very slow-moving animals. Many aspidochirotes rear up and extend their anterior ends into the water column when spawning. Other species writhe violently or inflate when they encounter a predator. Some, mostly deep-sea taxa have adaptations for swimming, such as a flattened body or fringes of webbed papillae that can be undulated rhythmically. Epibenthic taxa wander in an apparently random manner as they feed. Many tropical species are nocturnal, living in crevices or under the sand during the day. Others, usually large species, live permanently exposed in shallow water. This lifestyle may be aided by the presence of toxins in the body wall that deter predation by fishes. The juvenile of one aspidochirote species, *Pearsonothuria graeffei*, appears to mimic the bright coloration of a toxic species of nudibranch gastropod. One species of tiny apodan lives attached to deep-sea fishes. Many species in the aspidochirote Holothuriidae have cuvierian tubules for use in defense. These structures are expelled through the anus, whereupon they expand dramatically in length and become sticky, entangling or deterring would-be predators, such as crabs and gastropods. Disturbance of some holothuroids can cause them to eviscerate. Dendrochirotes eviscerate anteriorly by detaching their tentacle crown. Conversely, many aspidochirotes eviscerate through the anus. The eviscerated animals usually live and regrow the expelled organs.

Feeding ecology and diet

Holothuroids are either deposit feeders or suspension feeders. Approximately 33 percent of species are suspension feeders, nearly all of them within Dendrochirotida. This group has richly branched tentacles that are lightly coated in mucus and extend into currents to capture algae, planktonic animals, or organic matter. Food is captured passively on mucus-coated sites on the tentacles or mechanically. The tentacles are brought into the mouth one at a time and are wiped

clean by contracting muscles encircling the pharynx. While they are being withdrawn from the mouth, the tentacles may be reprovisioned with mucus by small glands in the pharynx. The pelagic elasipodan *Pelagothuria* captures settling floc with a circumoral funnel of webbed papillae. Most holothuroids, however, feed on bacteria, algae, or detritus in surficial deposits. The variety of ways in which holothuroids feed is reflected in the diversity of tentacle form. Epibenthic aspidochirotes and elasipodans shovel in, or mop up, sediment with peltate tentacles. Synaptid apodans lash the surface with plume-like tentacles. Other apodans, as well as dactylochirotes and molpadiians, are burrowers and have digitate tentacles that probably aid ingestion of the surrounding sediment. One enigmatic molpadiian, *Ceraplectana*, has short tentacles in sclerous, claw-like sheaths of unknown function.

Numerous animals, including sharks, rays, other large fishes, crabs, gastropods, sea stars, and marine mammals such as walrus, occasionally feed on adult holothuroids, However, holothuroids are a regular part of the diets of only a few fishes and sea stars. Only gastropods in the genus *Tonna* appear to specialize on holothuroids. These large snails engulf holothuroids with an extensible proboscis or, if the holothuroid is too large, rasp out circular sections of the body wall of prey. Harpacticoid copepods are voracious predators of holothuroid larvae in culture and may therefore also be important predators in the wild.

While sea cucumbers have few specialist predators, they host numerous types of commensals and parasites. These include species living on the body surface, tentacles, and in the anal opening such as portunid and pinnotherid crabs, palaemonid shrimp, polynoid polycheate worms, and flattened polyctene comb jellies (Ctenophora). Eulimid snails (Gastropoda) burrow into the body wall. Turbellarian flatworms and eel-like carapodid pearlfishes may live within the body cavity. Pearlfishes feed externally, but seek refuge in the sea cucumber during the day by entering through the anus and a tear in the rectal wall. Juvenile pearlfishes may feed on the sea cucumber gonad.

Reproductive biology

Holothuroids are unique among echinoderms in having a single, anteriorly positioned gonad and gonoduct leading to a dorsal gonopore. Spawning usually is annual, occurring in the spring or summer. Some species may have a second, usually smaller, autumnal spawning event. Species can have separate sexes or be hermaphroditic. At least one deep-sea species appears to be pair forming. Among broadcast spawners, eggs and sperm are released into the water column, where fertilization and development of the larvae take place. Among brooding species, however, females gather the eggs with their tentacles as they emerge and retain them ventrally or in special pouches. A few species brood their larvae within the body cavity. Brooding is most common among littoral, cold-water species, whereas tropical taxa are almost entirely broadcast spawners, as are, apparently, many deep-sea forms. As with other echinoderms, development is largely either indirect or direct. Indirectly developing species pass through a distinct planktonic and feeding larval phase, the auricularia, before metamorphosing into a barrel-shaped doliolaria and settling as a miniature adult called a pentacula. During metamorphosis, the bilaterally arranged organ systems are reorganized into the pentamerous adult body plan. In direct development, radical metamorphosis to adult morphologic features does not occur but development proceeds directly from a nonfeeding vitellaria larva. The larva may be planktonic or not and is provisioned with lipid stores. Some tropical species may also reproduce asexually as adults by transverse binary fission.

Conservation status

No species is listed by the IUCN or under the CITES convention. Nevertheless, drastic declines in local populations have been caused by commercial overharvesting. Dried and processed holothuroids, called *beche-de-mer* or *trepang*, are sold as a gourmet food item in Asian markets, where they form the basis of a multimillion dollar industry. Demand for beche-de-mer is increasing, and overfishing is a threat in many areas. The most valuable species are slow-growing, long-lived tropical forms in shallow water, which are easily harvested. Buyers often move into unregulated areas, where lack of management programs allows unsustainable exploitation. Several areas, such as the Galapagos Islands, Fiji, Sulawesi, Solomon Islands, and Cook Islands, have been overfished, and recovery is slow. Anecdotal evidence suggests that the current lack of commercially valuable species around some islands is due to overharvesting there before World War II. Regulation of harvesting in other areas, such as northern Australia and western North America, has led to long-term, stable fisheries.

Significance to humans

Holothuroids are a food item in several Asian and Pacific Island countries. The widespread use of holothuroids as food and medicine in Asia extends to at least the late sixteenth century, when detailed Chinese and European accounts of commerce first began mentioning trade in beche-de-mer. This long-term, domestic familiarity with holothuroids in the region is reflected in a small role for the animal in northern Asian culture as an object of poetry and popular cartoons. Several thousand individuals of colorful tropical species are harvested annually as part of the worldwide marine aquarium trade. Holothuroids are of minor medical significance because the potent dermal toxins of some species cause severe contact dermatitis in some people. These same toxins are of commercial interest because of their pharmacological properties. Compounds extracted from holothuroids exhibit antimicrobial, anticoagulating, tumor-inhibiting, and antiinflammatory activity. Other compounds are potent respiratory toxins in vertebrates. This feature is used by fishers in the Pacific Islands, who use abraded or chopped holothuroids to poison fishes and force octopuses from their lairs. The sticky cuvierian tubules also are spread over coral cuts to stem bleeding.

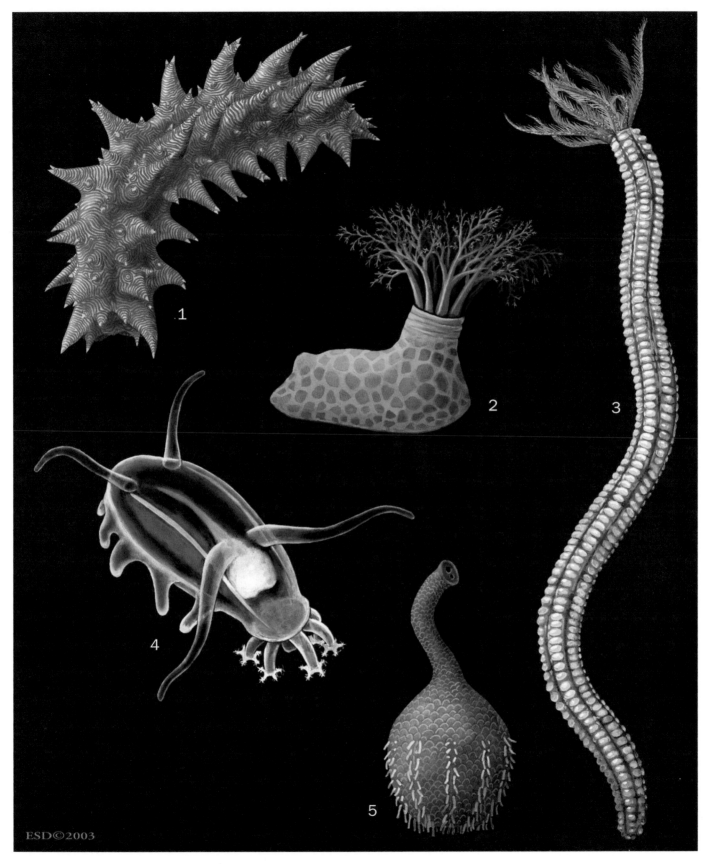

1. Candy cane sea cucumber (*Thelenota rubralineata*); 2. Slipper sea cucumber (*Psolus chitinoides*); 3. Giant medusan worm (*Synapta maculata*); 4. Sea pig (*Scotoplanes globosa*); 5. Flask-shaped sea cucumber (*Rhopalodina lageniformis*). (Illustration by Emily Damstra)

1. Pelagic sea cucumber (*Pelagothuria natatrix*); 2. Sea apple (*Pseudocolochirus violaceus*); 3. Hydrothermal vent sea cucumber (*Chiridota hydrothermica*); 4. Tiger's tail sea cucumber (*Holothuria [Thymiosycia] thomasi*); 5. Rat-tailed sea cucumber (*Molpadia oolitica*). (Illustration by Emily Damstra)

Species accounts

Hydrothermal vent sea cucumber
Chiridota hydrothermica

ORDER
Apodida

FAMILY
Chiridotidae

TAXONOMY
Chiridota hydrothermica Smirnov, 2000, Manus Basin at 8,622 ft (2,628 m) depth (3°6.63′S, 50°21.62′E).

OTHER COMMON NAMES
None known.

PHYSICAL CHARACTERISTICS
Grayish brown, cylindrical holothuroid with thin, translucent body wall to more than 10 in (25.4 cm) long. Tentacles incompletely surround mouth, leaving a ventral gap. Tips of tentacles have round, flat lobes and scalloped edges. Ossicles are typical of the genus *Chiridota*. Wheel ossicles are concentrated in body-wall papillae and branched rods in the tentacles.

DISTRIBUTION
Manus and Fiji basins in equatorial western Pacific and southeast Pacific rise near the Galápagos Islands between 6,600 and 8,500 ft (2,010–2,590 m) depth.

HABITAT
Observed only within the immediate vicinity of active hydrothermal vents, sometimes at the base of black smoking chimneys on bare rock or atop communities of sessile vent organisms.

BEHAVIOR
Often seen with posterior end hidden in a crevice and in small aggregations of up to three individuals per square foot (0.1 m²).

FEEDING ECOLOGY AND DIET
Suspension and deposit feeder. Feeds on suspended matter by raising the anterior end upward and spreading its tentacles but also has been seen feeding on benthic material.

REPRODUCTIVE BIOLOGY
Gonad is composed of clusters of short tubercles to 0.6 in (1.5 cm) long. Nothing else concerning reproduction is known.

CONSERVATION STATUS
Not listed by the IUCN or under the CITES convention.

SIGNIFICANCE TO HUMANS
None known. ◆

Giant medusan worm
Synapta maculata

ORDER
Apodida

FAMILY
Synaptidae

TAXONOMY
Holothuria maculata (Chamisso and Eysenhardt, 1821), Marshall Islands, Micronesia. Three subspecies recognized.

OTHER COMMON NAMES
German: Wurmseegurke.

PHYSICAL CHARACTERISTICS
An unmistakable species, the longest sea cucumber with a maximum length of 10 ft (3 m), although most animals reach only approximately 3–5 ft (1.0–1.5 m) long. This sea cucumber is serpentine and a mottled light and dark brown. There are 20-40 tentacles, which are feather-like. Ossicles are anchors and oblong perforated plates as well as tiny rosettes and rough rods. The sharp tines of the anchor ossicles protrude from the body wall so that when handled the animal seems to stick to the hands.

DISTRIBUTION
Tropical western Indian Ocean to central Pacific Ocean.

HABITAT
Coral reefs and adjacent sand flats in subtidal areas to approximately 40 ft (12 m) depth.

BEHAVIOR
A common species that is active during the day, the giant medusan worm moves slowly by peristalsis, using the posteriorly recurved anchor ossicles protruding from its body wall to gain a purchase on the substrate. When attacked by its principal predator, the gastropod *Tonna perdix*, the giant medusan worm may allow the snail to tear off the posterior most portion of its body without any apparent ill effect.

FEEDING ECOLOGY AND DIET
The giant medusan worm is a deposit feeder. It feeds by lashing its feathery tentacles over the sediment, rocks, and sea grass blades.

REPRODUCTIVE BIOLOGY
Like several other members of Synaptidae, the giant medusan worm is hermaphroditic. Eggs are less than 0.004 in (0.1 mm) in diameter. The animal is a broadcast spawner with a feeding auricularia larva that lives as plankton until it metamorphoses and settles to the bottom as a juvenile.

CONSERVATION STATUS
Not listed by the IUCN or under the CITES convention.

SIGNIFICANCE TO HUMANS
None known. ◆

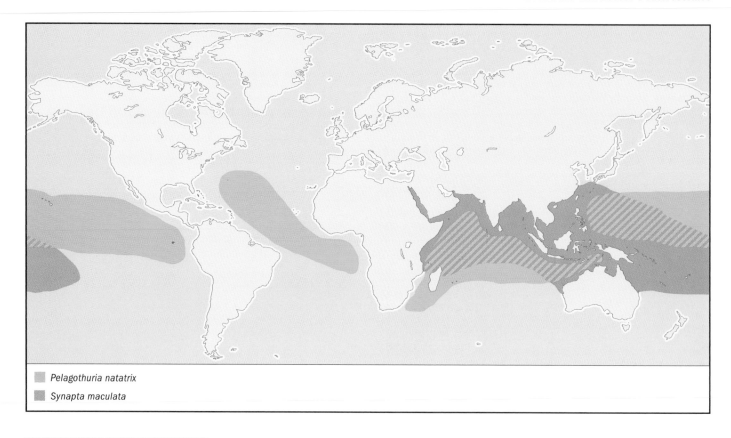

Pelagothuria natatrix

Synapta maculata

Tiger's tail sea cucumber
Holothuria (Thymiosycia) thomasi

ORDER
Aspidochirotida

FAMILY
Holothuriidae

TAXONOMY
Holothuria (Thymiosycia) thomasi Pawson and Caycedo, 1980, Colombia.

OTHER COMMON NAMES
None known.

PHYSICAL CHARACTERISTICS
Large holothuroid to 6.5 ft (2 m) long. Mottled light brown with white-tipped papillae and 20 usually light-colored peltate tentacles. Ventral side lighter than dorsum with scattered tube feet. Skin is thick but soft. The large scattered papillae on the dorsal surface give this sea cucumber a somewhat shaggy appearance. Ossicles in the body wall are ellipsoid buttons with two longitudinal rows of holes in pairs. Tower ossicles usually have a squat spire and a square base ringed with 12 holes.

DISTRIBUTION
Most of the Caribbean Sea from the Bahamas to Colombia, eastward to Panama and Mexico.

HABITAT
Steep forereefs with living corals from 10–100 ft (3–30 m) depth.

BEHAVIOR
Hides in reef crevices during the day when not feeding and at most extends only the anterior end to feed. When disturbed,

the tiger's tail sea cucumber swells its posterior end to prevent dislodgment and retracts quickly into its shelter.

FEEDING ECOLOGY AND DIET
A nocturnal deposit feeder capable of ingesting large pieces of coral rubble.

REPRODUCTIVE BIOLOGY
Nothing is known. However, gonads of specimens found in the Virgin Islands appeared ripe in July and consisted of numerous elongate tubules. Other species from the subgenus *Thymiosycia* have a feeding, planktonic auricularia larva.

CONSERVATION STATUS
Not listed by the IUCN or under the CITES convention.

SIGNIFICANCE TO HUMANS
None known. ◆

Candy cane sea cucumber
Thelenota rubralineata

ORDER
Aspidochirotida

FAMILY
Stichopodidae

TAXONOMY
Thelenota rubralineata Massin and Lane, 1991, Madang, Papua New Guinea.

OTHER COMMON NAMES
None known.

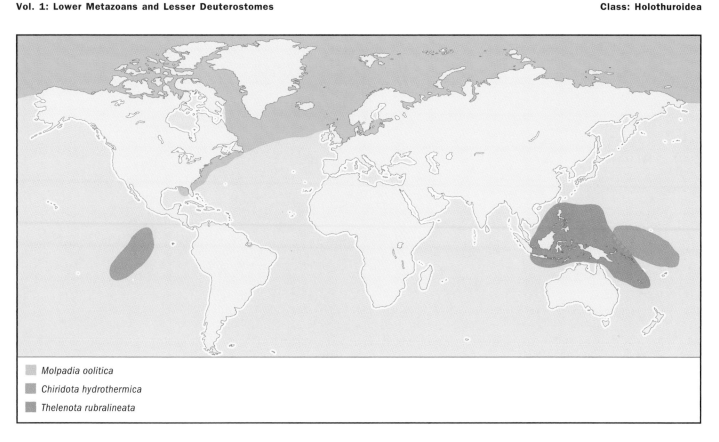

Molpadia oolitica

Chiridota hydrothermica

Thelenota rubralineata

PHYSICAL CHARACTERISTICS
A large and colorful sea cucumber to 20 in (51 cm) long and trapezoidal in cross section, the candy cane sea cucumber has a pronounced, flattened ventral sole crowded with tube feet. The dorsal side bears numerous large, pointed papillae. The candy cane sea cucumber has a unique crimson herringbone-like pattern of stripes on a white background and bears approximately 20 dull red, peltate tentacles.

DISTRIBUTION
New Guinea, Indonesia, Philippines, Sulawesi, Solomon Islands, New Caledonia, and Guam.

HABITAT
Forereef slope on sand patches from 20 ft (6 m) to at least 200 ft (60 m) depth.

BEHAVIOR
Very little is known about the biology of this rare sea cucumber. It crawls exposed on the reef during the day and night and curls up by bringing its anterior and posterior ends together when disturbed.

FEEDING ECOLOGY AND DIET
A deposit feeder that ingests reef sediments.

REPRODUCTIVE BIOLOGY
Nothing is known. Probably similar to the reproductive biology of the other two members of the genus, which broadcast spawn and have indirectly developing larvae.

CONSERVATION STATUS
Not listed by the IUCN or under the CITES convention.

SIGNIFICANCE TO HUMANS
Harvested incidentally in small numbers with other commercially valuable holothuroids used in the beche-de-mer industry. ◆

Flask-shaped sea cucumber
Rhopalodina lageniformis

ORDER
Dactylochirotida

FAMILY
Rhopalidinidae

TAXONOMY
Rhopalodina lageniformis Gray, 1853, Congo.

OTHER COMMON NAMES
None known.

PHYSICAL CHARACTERISTICS
Unusual flask-shaped holothuroid to 4 in (10 cm) long. Body covered in plates. Mouth and anus adjacent atop a slender stalk above a globose body. Fifteen to 25 digitate tentacles in two concentric whorls. The doubled-over body gives the appearance of 10 radii along the body, unlike the canonical five of other echinoderms. The radii do not cross the ventral pole of the body. Ossicles are small knobby towers. Cruciform plates are present at the ventral pole. Tube foot plates have an elongate roughened end.

DISTRIBUTION
Atlantic Ocean along western coast of Africa from Senegal to Cabinda.

HABITAT
Coastal mud bottoms at 7–20 ft (2–6 m) depth.

BEHAVIOR
Remains burrowed in mud with only its mouth and anus exposed.

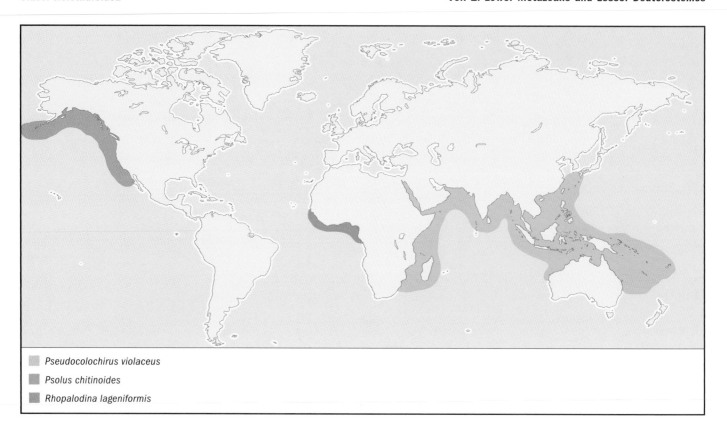

Pseudocolochirus violaceus

Psolus chitinoides

Rhopalodina lageniformis

FEEDING ECOLOGY AND DIET
Nothing is known. The tentacle structure suggests this sea cucumber is a deposit feeder.

REPRODUCTIVE BIOLOGY
Nothing is known.

CONSERVATION STATUS
Not listed by the IUCN or under the CITES convention.

SIGNIFICANCE TO HUMANS
None known. ◆

Sea apple
Pseudocolochirus violaceus

ORDER
Dendrochirotida

FAMILY
Cucumariidae

TAXONOMY
Colochirus violaceus (Théel, 1886), Philippines. Synonyms include *P. axiologus*, *P. arae*, *P. bicolor*, and *P. tricolor*.

OTHER COMMON NAMES
French: Pomme de mer; German: Seeapfel.

PHYSICAL CHARACTERISTICS
A large, colorful species to 7 in (18 cm) long. Color is variable, often purple. There are three ventral, longitudinal rows of tube feet. The dorsal side has two rows of papillae and small scat-

tered papillae. The body is curved in life so that the mouth and anus point upward. The 10 tentacles are bushy purple to red with white tips. Body-wall ossicles are rounded, smooth plates with a few holes and occasionally are absent in large animals.

DISTRIBUTION
Indian Ocean to western Pacific Ocean in continental and continental-island areas, such as Fiji through Indonesia north to southern Japan, south to Australia and Lord Howe Islands but absent from true oceanic islands. India west to Red Sea, Madagascar, and South Africa.

HABITAT
Hard substrates, including coral reefs, to 40 ft (12 m) depth in areas with currents or upwellings.

BEHAVIOR
Lives partly concealed to fully exposed with tentacles expanded, even during the day.

FEEDING ECOLOGY AND DIET
The sea apple is a suspension feeder. It can feed continuously, capturing large phytoplankton with outstretched arborescent tentacles lightly coated in mucus.

REPRODUCTIVE BIOLOGY
Sexes are separate. Females are distinguished by having a gonopore atop a single unadorned tube a few millimeters long. The counterpart structure in males is tipped in numerous papillae. Males and females release gametes into the water column, where fertilization and development of the larvae take place.

CONSERVATION STATUS
Not listed by the IUCN or under the CITES convention.

SIGNIFICANCE TO HUMANS
Sea apples are taken in moderate numbers for the marine aquarium trade but are of only minor economic significance, because only approximately 1,000 of these animals are imported annually into North American and European countries. ◆

Slipper sea cucumber
Psolus chitinoides

ORDER
Dendrochirotida

FAMILY
Psolidae

TAXONOMY
Psolus chitinoides H. L. Clark, 1901, Puget Sound, Washington, United States.

OTHER COMMON NAMES
English: Armored sea cucumber, pedal sea cucumber.

PHYSICAL CHARACTERISTICS
A yellow to pinkish orange and ovoid sea cucumber to 3 in (8 cm) long. Both mouth and anus are upturned. The dorsum is arched and covered in large, flat plates. The bottom is a soft, flattened sole with tube feet concentrated around its perimeter and scattered down the center. The 10 white-tipped, red tentacles are extensively branched. In addition to the large dorsal plates, the ventral wall of this species has smaller flat and oval ossicles with closely spaced holes and, in larger ossicles, knobs or a central reticulated mound.

DISTRIBUTION
Pacific coast of North America from the Aleutian Islands south to central Baja California.

HABITAT
Most common in intertidal areas such as rocky shorelines, but occurs from 0–800 ft (0–244 m) on hard, inclined surfaces swept by current.

BEHAVIOR
The slipper sea cucumber firmly attaches itself to smooth rocks but can use its tube feet to creep along slowly. When the sea cucumber is positioned, the body often becomes covered in debris or other organisms, leaving only bright red tentacles in view.

FEEDING ECOLOGY AND DIET
The slipper sea cucumber feeds in a manner very similar to that of the sea apple, by extending its bushy tentacles into the current to capture passing particles of food.

REPRODUCTIVE BIOLOGY
Sexes are separate. Spawning occurs from March until May by release of gametes into the water column, where fertilization takes place. Males aid the dispersal of sperm into the water column by waving a tentacle across the gonopore. Eggs are red and approximately 0.02 in (600 μm) in diameter. The larva does not feed while in the plankton but is provisioned with lipid stores, which see the larva through development.

CONSERVATION STATUS
Not listed by the IUCN or under the CITES convention.

SIGNIFICANCE TO HUMANS
None known. ◆

Sea pig
Scotoplanes globosa

ORDER
Elasipodida

FAMILY
Elpidiidae

TAXONOMY
Elpidia globosa Theél, 1879, Western Pacific Ocean below 2,000 ft (610 m).

OTHER COMMON NAMES
English: Sea cow.

PHYSICAL CHARACTERISTICS
Transparent, rounded sea cucumber 2–4 in long, with 10 tentacles and a small number of large papillae. The dorsal papillae are of two widely spaced antenna-like pairs. The other papillae are arranged in a row around the edge of the somewhat flattened ventrum. The tentacles are discoid with marginal lobes. Ossicles in the body wall are smooth to spiny rods and smaller C-shaped rods nearly identical to those found in some demospongean sponges.

DISTRIBUTION
Nearly cosmopolitan, although apparently absent from the North Atlantic Ocean and the westernmost Pacific Ocean from Central and South America.

HABITAT
Deep ocean bottoms from 1,800 to 2,400 ft (550–730 m). Lives at the shallow end of its bathymetric range at higher latitudes and colder water.

BEHAVIOR
Moves above the sediment with the aid of long, locomotory papillae. Sea pigs aggregate, forming large "herds," sometimes in response to the presence local accumulations of relatively fine sediment. This aggregating behavior, the leg-like papillae, and curved dorsal papillae have earned these animals the alternative name "sea cow."

FEEDING ECOLOGY AND DIET
Feeds on fine surface sediment on the deep ocean bottom by pushing material into the mouth by means of tentacles with flattened ends. On most specimens a sediment-filled gut is easily seen through the thin body wall.

REPRODUCTIVE BIOLOGY
Very little is known about the reproduction of deep-sea holothuroids, including sea pigs. Maximum egg size is approximately 0.008 in (0.2 mm) in diameter, a characteristic that suggests the larvae are nonfeeding. The ripe gonad may be visible through the body wall near the anterior end.

CONSERVATION STATUS
Not listed by the IUCN or under the CITES convention.

SIGNIFICANCE TO HUMANS
None known. ◆

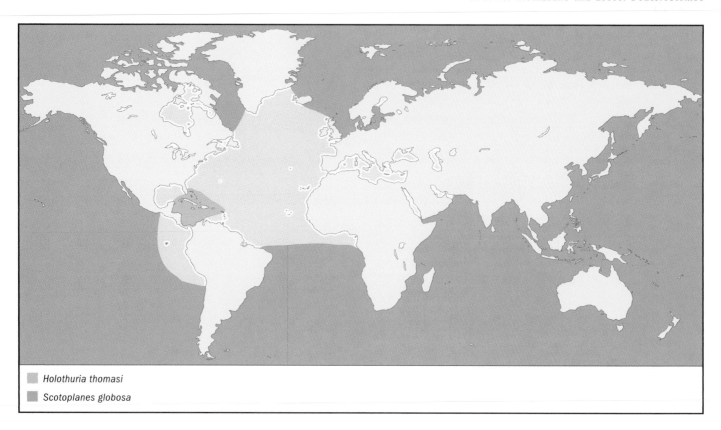

Holothuria thomasi

Scotoplanes globosa

Pelagic sea cucumber
Pelagothuria natatrix

ORDER
Elasipodida

FAMILY
Pelagothuriidae

TAXONOMY
Pelagothuria natatrix Ludwig, 1894, Gulf of Panama, Eastern
Pacific Ocean below 2,000 ft (610 m) depth.

OTHER COMMON NAMES
None known.

PHYSICAL CHARACTERISTICS
A bizarre, transparent sea cucumber 2–3 in (5–8 cm) long with
12–16 webbed papillae exceeding the body's length and forming a veil around the mouth. This sea cucumber is slender and
pale pinkish purple. The tentacles are two-pronged and number approximately 15. Ossicles and a calcareous ring are absent.

DISTRIBUTION
Western Indian Ocean to eastern Pacific Ocean and Atlantic
Ocean at subtropical to tropical latitudes.

HABITAT
Open ocean from near surface waters to at least 10,000 ft
(3,050 m) depth.

BEHAVIOR
The pelagic sea cucumber often is encountered hanging motionless and drifting. It swims by flapping its veil posteriorly and
gliding. The veil collapses and is pulled inward and upward, and
the stroke is repeated. Because the veil is incompletely webbed,
the sea cucumber moves at an angle to its body axis.

FEEDING ECOLOGY AND DIET
The pelagic sea cucumber is a suspension feeder. It concen-
trates descending organic matter in mid-water by hanging vertically in the water column and spreading its extensive veil of
webbed papillae into an inverted cone. Gut contents suggest
that this species never ingests bottom sediment.

REPRODUCTIVE BIOLOGY
Spawning and a larval form have not been reported for the
pelagic sea cucumber. Postlarval juveniles are pelagic and have
midventral tube feet that are lost during growth. The ripe gonad sometimes is visible through the body wall as a white or
yellowish tuft.

CONSERVATION STATUS
Not listed by the IUCN or under the CITES convention.

SIGNIFICANCE TO HUMANS
None known. ◆

Rat-tailed sea cucumber
Molpadia oolitica

ORDER
Molpadiida

FAMILY
Molpadiidae

TAXONOMY
Chiridota oolitica (Pourtalès, 1851), Florida, United States.

OTHER COMMON NAMES
None known.

PHYSICAL CHARACTERISTICS
A brownish gray to reddish black, sausage-shaped sea cucumber to 6 in (15 cm) long with 15 digitate tentacles and a small, sometimes indistinct "tail." The body surface is smooth. The tail and oral region usually are somewhat lighter in color than the rest of the body. Body color is age related; older individuals often are darker as a result of the presence in the body wall of microscopic phosphatic bodies that replace the calcareous ossicles as the animal ages. Ossicles in the body wall are few or absent in larger animals. When present, ossicles are more common in the tail in the form of three-spired towers with a tripartite to elongate base.

DISTRIBUTION
Eastern Gulf of Mexico and western Atlantic coast from Florida to Labrador, west to the North Sea and the Arctic Ocean.

HABITAT
Soft sand, mud or silt in subtidal areas to beyond the continental slope to at least 1,500 ft (457 m) in depth.

BEHAVIOR
A burrowing sea cucumber that can occur at densities of several animals per square foot (0.1 m²) in some areas. These populations considerably modify bottom topography and the vertical grain-size distribution of sediment. The relatively stable fecal mounds become home to clams, amphipod crustaceans, and tube-dwelling worms.

FEEDING ECOLOGY AND DIET
Ingests and processes large amounts of fine sediment in a conveyor-belt manner while positioned vertically, oral end down, in the substrate. Feeding is apparently selective for the finer fraction of sediment, which passes through the gut without being pelletized. The anus lies at the surface, where evacuated material forms a large, wide mound.

REPRODUCTIVE BIOLOGY
Nothing is known. Other species of *Molpadia*, such as *M. intermedia* and *M. blakei*, have large, yolky ova and a nonfeeding, planktonic vitellaria larva.

CONSERVATION STATUS
Not listed by the IUCN or under the CITES convention.

SIGNIFICANCE TO HUMANS
None known. ◆

Resources

Books

Féral, J.-P., and G. Cherbonnier. "Les holothuries." In *Guide des étoiles de mer, oursins e autres échinodermes du lagon de Nouvelle-Calédonie*, edited by Guille, A., P. Laboute, and J.-L. Menou. Paris: ORSTOM, 1986.

Hendler, G., J. E. Miller, D. L. Pawson, and P. M. Kier. *Sea Stars, Sea Urchins, and Allies: Echinoderms of Florida and the Caribbean*. Washington, DC: Smithsonian Institution Press, 1995.

Lambert, P. *Sea Cucumbers of British Columbia, Southeast Alaska and Puget Sound*. Vancouver: University of British Columbia Press, 1997.

Picton, B. E., and R. H. Johnson. *A Field Guide to the Shallow-Water Echinoderms of the British Isles*. London: Immel, 1993.

Smiley, S. "Holothuroidea." In *Microscopic Anatomy of Invertebrates*, Vol 14, *Echinodermata*, edited by F. W. Harrison and F.-S. Chia. New York: Wiley-Liss, 1994.

Smiley, S., F. S. McEuen, S. Chaffee, and S. Krishnan. "Echinodermata: Holothuroidea." In *Reproduction of Marine Invertebrates*, Vol 6, edited by A. C. Giese, J. S. Pearse, and V. B. Pearse. Pacific Grove, CA: Boxwood Press, 1991.

Periodicals

Gilliland, P. M. "The Skeletal Morphology, Systematics and Evolutionary History of Holothurians." *Special Papers in Palaeontology* 47 (1993): 1–147.

Kerr, A. M., and J. Kim. "Phylogeny of Holothuroidea (Echinodermata) Inferred from Morphology." *Zoological Journal of the Linnean Society* 133 (2001): 63–81.

Miller, J. E., and D. L. Pawson. "Swimming Sea Cucumbers (Echinodermata: Holothuroidea): A Survey, with Analysis of Swimming Behavior in Four Bathyal Species." *Smithsonian Contributions to the Marine Sciences* 35 (1990): 1–18.

Other

Kerr, A. M. "Holothuroidea: Sea Cucumbers." *Tree of Life*. 1 Dec. 2000 [14 July 2003]. <http://tolweb.org/tree?group= Holothuroidea&contgroup;=Echinodermata>.

Alexander M. Kerr, PhD

Chaetognatha
(Arrow worms)

Phylum Chaetognatha
Number of families 6

Thumbnail description
Long, narrow worms that have transparent bodies and that are known for the hooks that they use to catch prey

Photo: Arrow worms are a dominant part of plankton. They prey on copepods, crustaceans, and larval fish, as well as other chaetognaths. (Photo by Animals Animals ©Peter Parks, OSF. Reproduced by permission.)

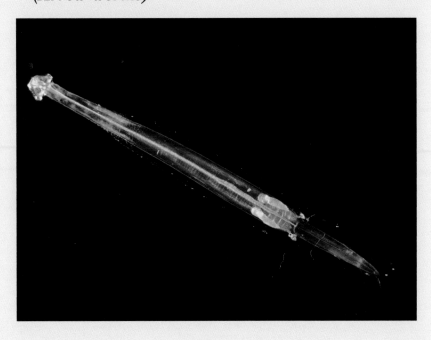

Evolution and systematics

The phylum Chaetognatha is thought to be an old group with a long and separate path of development. It is not clear what the nearest relatives are, but evidence provided by developmental and molecular studies show that this group may be an early split-off of the metazoan phyla. It is a paradox that there are so few species, because the group has been so successful in increasing its number of individuals and maintaining widespread distribution. It is a soft-bodied group and several researchers (Owre and Bayer, 1962, and Conway-Morris, 1977) have raised doubts concerning fossils of certain species (e.g., *Amiskwia sagittiformis*) that have been recorded. *A. sagittiformis* has raised particular doubts because it has no substantive structures that are usually present in a chaetognath, such as hooks or teeth; more over it has a gut that runs to the end of the tail unlike any chaetognath.

Two orders are usually recognized, Phragmophora and Aphragmophora. The order Phragmophora contains three families (Heterokrohniidae [three genera], Spadellidae [four genera], Eukrohniidae [one genus]), and the order Aphragmophora also contains three families (Sagittidae [one genus], Pterosagittidae [one genus], Krohnittidae [one genus]). The genus *Sagitta* (Sagittidae) is split into 11 genera by some authors, but because of the uncertain nature of some of the groupings and the need for further molecular evidence to determine the true nature of these genera, the name *Sagitta* will be used in this chapter. There are about 100 species under discussion but there may be as many as 200 valid species, many of which may be benthic.

Physical characteristics

The animals are bilaterally symmetric with a head, trunk, and tail. They are transparent: the internal organs can clearly be seen in living specimens and in specimens that have been well preserved in formalin. The body cavity is filled with fluid that is surrounded by muscles and tough membrane; a multilayered epidermis covers the outer layer of the body. The head has a complex musculature that supports the grasping spines or hooks that are the most obvious and recognizable features of this animal. There also is a mouth, one or two rows of teeth, and eyes. The eyes have photoreceptive cells, and most species also have spots of pigment in their eyes.

The nervous system consists of six ganglia that are present in the head, a superficial cerebral ganglion, and two lateral ganglia at either side of the esophagus. The cerebral ganglion is connected to a ganglion present on the ventral side of the trunk; paired nerves extend from the ventral ganglion toward the tail. Very little is known about the functional roles of the different parts of the nervous system.

The animals detect the movement of prey with ciliary tufts present on the body. The body wall is folded in the neck region, which forms a hood that can be folded over the head to allow the animals to swim more smoothly. In some species there is a layer of loose tissue present in the collarette region of the neck. Chaetognatha have tail fins and one or two pairs of fins on the sides of the body; all fins are supported by fin rays.

The alimentary structure consists of the mouth, esophagus, gut (which can have paired diverticula), and anus. The anus is located just before the septum that separates the trunk and the tail. The ovaries are situated at the posterior end of the trunk in an opening present on both sides of the intestine, just before the septum. The tail contains the testes. Ripe sperm is stored in sperm packages within the seminal vesicles that project from both sides of the tail.

The warm-water species are generally smaller than the cold-water species, with adults between 0.12–5.91 in (3–150 mm) in size. The largest species, *Sagitta gazellae*, lives in Antarctic waters and reaches a size of 2.76 in (70 mm). The benthic species are the smallest arrow worms, with *Spadella cephaloptera* reaching maturity at 0.12 in (3 mm).

Distribution

Large numbers of Chaetognatha live in every region of the ocean. About half of the species are planktonic. Neritic (inhabiting shallow waters near coastlines) and open-ocean species are in the pelagic realm, the latter being confined to the waters above the continental shelf, which is about 656 ft (200 m) deep. Oceanic species are widely distributed, but a very common distribution pattern is tropical-subtropical from 40°N–40°S in all three oceans; the species *Pterosagitta draco* follows this line of distribution. More tropical species, such as *Sagitta regularis*, are confined to roughly 30°N–30°S and are usually seen only in the Indo-Pacific Ocean. Distributions are much more restricted for neritic species, such as *Sagitta setosa*, because of the topography and variability of the locations. The width of a shelf, outflow of rivers, or local upwellings can have a tremendous impact on neritic distribution.

The largest concentration of individuals is found in the epipelagic layer shallower than 656 ft (200 m). Arrow worms that live in surface waters, such as *Sagitta enflata*, are often transparent, which helps them avoid predators such as fishes. There are fewer individuals but more species found in the mesopelagic layer from 656–3,280 ft (200–1,000 m). The guts of species in the mesopelagic layers, such as *Eukrohnia fowleri*, are very often yellow or red in color, which they obtain by consuming prey of the same colors. Species from deeper waters, such as *Sagitta planctonis*, are more muscular and less transparent.

The mid-water species usually perform diurnal vertical migration. They swim to the surface at night to feed and sink to deeper layers during the day. There also is an ontogenic vertical distribution: juveniles live higher up in the water column than the adults. This results in more species being present in the upper layers at night than during the day. In general, the very deep-living species are thought to be globally distributed. However, the high cost of sampling these waters has prevented extensive surveys. As a result, very little is known about the deep bathypelagic and benthopelagic layers.

Neritic species are adapted to more variation in environmental conditions. They show more restricted areas of distribution than the oceanic species. Because they can tolerate a wider range of environmental conditions, it has been possible to keep specimens for a limited amount of time in an aquarium to study feeding and swimming behavior. However, scientists have so far been unable to keep specimens alive in an aquarium setting for more than one generation.

Benthic species live attached to objects on the substrate such as sea grass or rocks. The distributions of most benthic species are very restricted; some species have only been found in the area of their type locality. *Spadella cephaloptera* is presumed to be globally distributed, but it may be a species complex.

Habitat

Chaetognaths are strictly marine and can live in every part of the ocean. The benthic Spadellas are very small and adhere to rocks or other objects by using special adhesive papillae, but they are also able to swim short distances.

Behavior

The pelagic species position themselves in water in a slightly oblique position. If their side bristles detect movement, a quick sweep of their tail allows them to swiftly dart off in the direction of prey. They are able to swim across short distances very quickly. Their ability to swim is probably dependent on their body muscles, but the precise manner of how it is done is still not known. The fins are thin, transparent structures supported by fin rays and do not contain muscle fiber. Fins enlarge body width and enhance the buoyancy of the animal. In some species the fins contain gelatinous material. Fins do not play an active role in swimming.

Feeding ecology and diet

Chaetognaths are carnivores and swim very actively in order to catch prey. They detect their prey by sensing movement with the ciliar tufts on their body. The chaetognath uses its hooks to grab prey, which is usually a small copepod whose nicely rounded shape makes it easy for the chaetognath to swallow. However, arrow worms can feed on anything that is of a certain size, including fish larvae, other chaetognath species, and even phytoplankton. The function of the teeth is not known. Arrow worms appear to immobilize their prey after capture. Tetrodotoxin, which is probably synthesized by the bacterium *Vibrio algynoliticus*, has been isolated from the head of several chaetognath species. How chaetognaths acquire this bacterium is still unknown, but tetrodoxin can cause immobilization. However, direct observations of feeding chaetognaths are very scarce, and it is not known how many species of chaetognaths may have or use the venom.

Digestion is rapid, and it is very rare that arrow worms are found with full guts. It is estimated that chaetognaths consume between two and 50 prey in a day, the latter number coming from laboratory experiments to establish the worms' food saturation point. Chaetognaths play an important part in the food chain. Estimates have shown that they comprise 30% of the biomass of copepods. Chaetognaths are mainly eaten by fishes, but they are also prey for larger carnivorous animals. Chaetognaths do acquire parasites from prey but not very frequently; the most common parasites ingested by chaetognaths are trematodes, nematodes, and cestodes (helminths), although the latter occurs very infrequently. There is no host-specifc parasite known in chaetognaths, which is remarkable for such an old group.

Reproductive biology

Arrow worms are hermaphrodites; male reproductive organs are in the tail, and female organs are in the trunk. They are protandric, meaning that the sperm develops first and the eggs follow later. The ovaries are present in the lower part of the trunk. Openings just before the septum divide the trunk and tail on either side of the gut. Sperm is deposited on the body and swims toward the openings, which they enter; the openings are also used for egg laying. The maximum length of mature ovaries varies with the species. The ovaries can be very long. In species such as *Pterosagitta draco*, when the ovary is full of eggs it can extend all the way to the neck. However, in *Sagitta enflata*, the ovaries are very short.

The open-ocean species are adapted to a large space and as such have not been cultured. This makes it difficult to study behavior and life cycles. For example, knowing whether they cross-fertilize or fertilize themselves would be important in studying genetic variation, but scientists have not yet determined the method of fertilization for these species. What is known about the reproductive behavior of chaetognaths comes from observations of *Spadella cephaloptera*. This species has been held in aquaria for observation studies. The sperm package of this species is deposited on the body of another individual of the same species. The sperm moves into the ovaries and is stored there to fertilize the eggs. The fertilized eggs are then released into the sea water. In deep-water species such as *Eukrohnia fowleri*, brood sacs hang from the ovary openings.

Arrow worms do not have larval stages. Small chaetognaths hatch from the eggs and then continue to grow. The animals require a temperature that is high enough to allow eggs to develop. At higher temperatures, individuals mature in a shorter period of time than they do in cool water. If the temperature is too low, they use energy in order to grow and reach a greater length. For example, the temperature of the water in which it grows greatly affects the size of *Sagitta tasmanica*; mature individuals range from 0.31 to 0.67 in (8–17 mm) in size. When the animals are transported to water masses that are too cold for them to reproduce, they grow to be giant in size, but they never reach a mature state and are referred to as sterile expatriates. In cold water like the Canadian Arctic, they may take two years to mature, while in tropical waters their complete life span may be as little as six weeks.

Conservation status

No Chaetognatha species is listed by the IUCN Red List. Planktonic species have huge populations with very large distributions and as such, are not expected to die out. There is

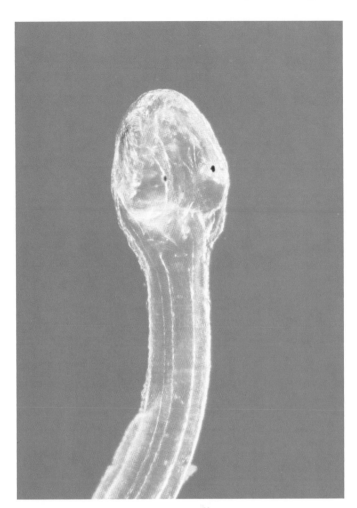

Close up of the head of an arrow worm seen in the Great Barrier Reef. (Photo by Animals Animals ©Peter Parks, OSF. Reproduced by permission.)

no other animal on the planet with such a large distribution area, and species worldwide are not threatened. However, climate changes have been known to cause distribution areas to shift and local populations of species are able to appear and disappear. For example, we can expect populations of warm-water species in the North Atlantic to begin moving northwards if ocean temperatures continue to increase, and it is possible that the arctic-subarctic cold-water species *Sagitta elegans* could disappear from the North Sea, which is its southernmost area of distribution.

Significance to humans

There is no direct significance between humans and chaetognaths. However, chaetognataha are an important part of the marine food chain, and they are used as indicators to determine the level of water from the Atlantic Ocean moving into the North Sea. Arrow worms may play a role in the oceanic distribution of tetradotoxin-producing bacteria.

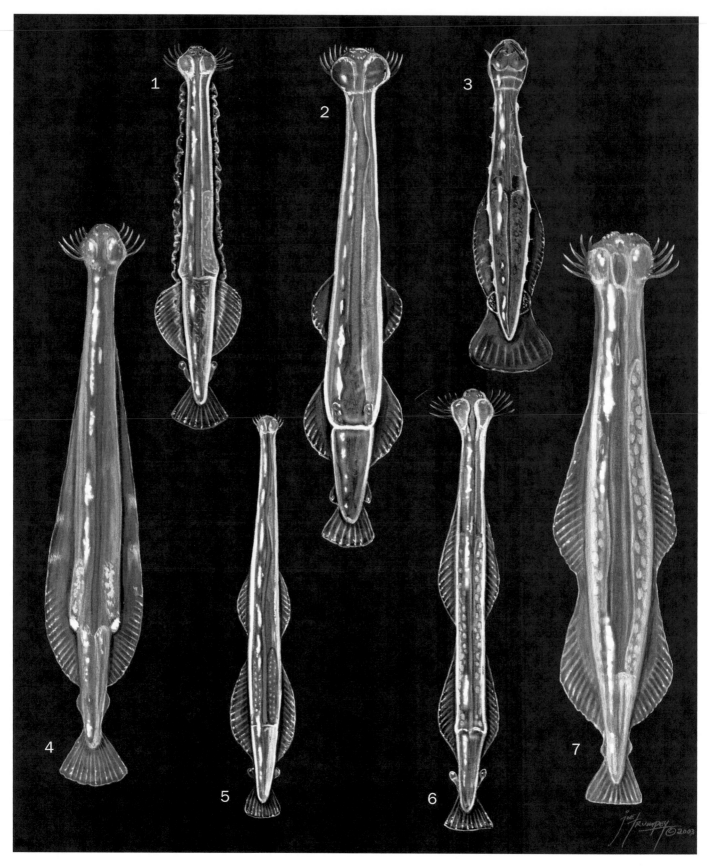

1. *Pterosagitta draco*; 2. *Sagitta enflata*; 3. *Spadella cephaloptera*; 4. *Eukrohnia fowleri*; 5. *Sagitta setosa*; 6. *Sagitta bipunctata*; 7. *Sagitta planctonis*. (Illustration by Joseph E. Trumpey)

Species accounts

No common name
Pterosagitta draco

ORDER
Aphragmophora

FAMILY
Pterosagittidae

TAXONOMY
Sagitta draco (Krohn, 1853), Mediterranean Sea off Messina.

OTHER COMMON NAMES
None known.

PHYSICAL CHARACTERISTICS
Individuals have 8–10 hooks, 6–10 anterior teeth, and 8–18 posterior teeth. Maximum adult body length is 0.43 in (11 mm), and the relative tail length is 38–45% of total body size. The body is firm, broad, and opaque. One pair of fins is present on the tail segment; fins are completely rayed and round. Head is fairly large. There is a large, broad collarette along the trunk and part of the tail. There are no gut diverticula. Eyes are small with T-shaped pigment spots. Seminal vesicles are present on the head and trunk, close to tail fin, and touch lateral fins. Ovaries are long, reaching the neck region; ova are large.

DISTRIBUTION
Epipelagic, occurring between 40°N and 40°S in all three oceans.

HABITAT
Lives in the pelagic realm of tropical and subtropical oceans, at depths of 3–85 ft (10–300 m).

BEHAVIOR
Shows no evidence of diurnal vertical distribution. Rapid darting movements are made over short distances to catch prey.

FEEDING ECOLOGY AND DIET
Prefers small copepods. Feeds at night after moving to superficial layers.

REPRODUCTIVE BIOLOGY
Reproduction is dependent on water temperature and occurs several times a year. Hermaphroditic, and sperm is stored in the oviducts waiting for the eggs to mature. Fertilized eggs are released in sea water. There is no larval stage.

CONSERVATION STATUS
Not listed by IUCN.

SIGNIFICANCE TO HUMANS
None known. ◆

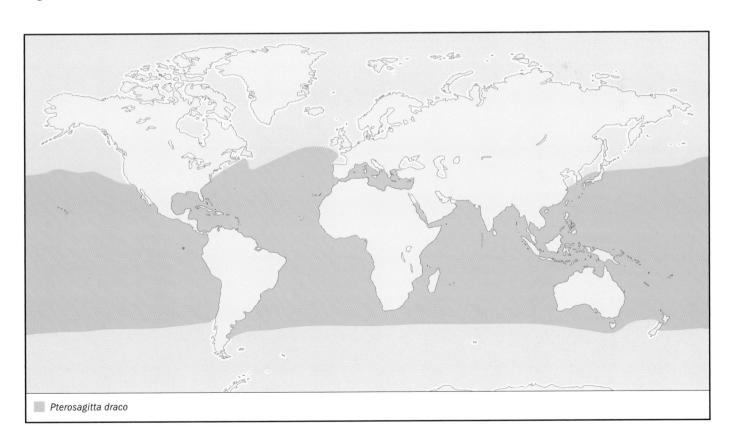

Pterosagitta draco

No common name
Sagitta bipunctata

ORDER
Aphragmophora

FAMILY
Sagittidae

TAXONOMY
Sagitta bipunctata Quoy and Gaimard, 1827, Strait of Gibraltar.

OTHER COMMON NAMES
None known.

PHYSICAL CHARACTERISTICS
Individuals have 8–10 hooks, 5–8 anterior teeth, and 8–16 posterior teeth. Maximum adult body length is 0.75 in (19 mm), with a relative tail length of 22–29% of total body size. The body is firm and opaque. There are two pairs of lateral fins and no fin bridge. Anterior fins are medium in length, rounded, and completely rayed. Posterior fins are medium in length, completely rayed, and triangular in shape. This species does not have a collarette, nor does it have gut diverticula. Eyes are small, with T-shaped pigment spots. Seminal vesicles are present in the head and trunk and touch the tail fin, although they are well separated from posterior fins. Ovaries are medium in length and reach the region of the ventral ganglion; ova are large.

DISTRIBUTION
Epipelagic, occurring between 40°N and 40°S in all three oceans.

HABITAT
A pelagic, oceanic species that lives in tropical and subtropical regions at depths of 0–328 ft (0–100 m).

BEHAVIOR
One of the few species in the northwest Atlantic that lives in the upper 164 ft (50 m) of the water column. Shows evidence of diurnal vertical distribution, with population numbers being most abundant at depths of 0–33 ft (0–10 m) at night and 33–82 ft (10–25m) in the daytime. Rapid darting movements are made over short distances to catch prey.

FEEDING ECOLOGY AND DIET
Prefers small copepods.

REPRODUCTIVE BIOLOGY
Reproduces one or more times a year depending on water temperature. Hermaphroditic, and sperm is stored in the oviducts waiting for the eggs to mature. Fertilized eggs are released in the sea water. There is no larval stage.

CONSERVATION STATUS
Not listed by the IUCN.

SIGNIFICANCE TO HUMANS
The first chaetognath species ever described. ◆

No common name
Sagitta enflata

ORDER
Aphragmophora

FAMILY
Sagittidae

TAXONOMY
Sagitta enflata Grassi, 1881, Bay of Naples.

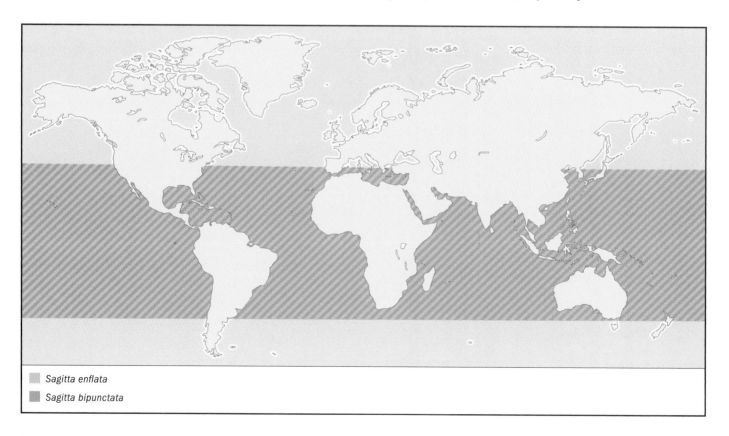

  Sagitta enflata

  Sagitta bipunctata

OTHER COMMON NAMES
None known.

PHYSICAL CHARACTERISTICS
Individuals have 8–10 hooks, 4–10 anterior teeth, and 4–15 posterior teeth. Maximum adult body length is 0.98 in (25 mm), and the relative tail length is quite short at 14–17% of total body size. The body is flaccid and transparent. There are two pairs of lateral fins and no fin bridge. Anterior fins are very short, partially rayed, and round. Posterior fins are short, partially rayed, and round. There is no collarette, nor are there gut diverticula. Eyes are small, with star-shaped pigment spots. Seminal vesicles are round, touch the tail fin, and are well separated from the posterior fins. Ovaries are small, reaching to the middle of the post fins; ova are large.

DISTRIBUTION
Epipelagic and cosmopolitan, occurring from 40°N to 40°S.

HABITAT
Oceanic and pelagic species that lives in tropical and subtropical waters at depths of 0–985 ft (0–300 m). Reported to tolerate different degrees of salinity and may occur in inshore waters.

BEHAVIOR
Shows evidence of diurnal vertical distribution. Juveniles occur at higher levels of the water column than do the adults. Rapid darting movements are made over short distances to catch prey.

FEEDING ECOLOGY AND DIET
Prefers small copepods. Feeds at night after moving to superficial layers.

REPRODUCTIVE BIOLOGY
Reproduces several times a year, but the number of times is dependent on water temperature. Hermaphroditic, and sperm is stored in the oviducts waiting for eggs to mature. Fertilized eggs are released in the sea water. There is no larval stage.

CONSERVATION STATUS
Not listed by the IUCN.

SIGNIFICANCE TO HUMANS
None known. ◆

No common name
Sagitta planctonis

ORDER
Aphragmophora

FAMILY
Sagittidae

TAXONOMY
Sagitta planctonis Steinhaus, 1896, south Equatorial Current, Atlantic Ocean.

OTHER COMMON NAMES
None known.

PHYSICAL CHARACTERISTICS
Individuals have 8–11 hooks, 6–9 anterior teeth, and 10–14 posterior teeth. Maximum adult body length is 1.46 in (37 mm), and relative tail length is 19–21% of total body size. The body is large, firm, and opaque. There are two pairs of lateral fins and a fin bridge. Anterior fins are long, angular, partially rayed, and reach the middle of the ventral ganglion. Posterior fins are long, angular, and partially rayed. There is a large col-

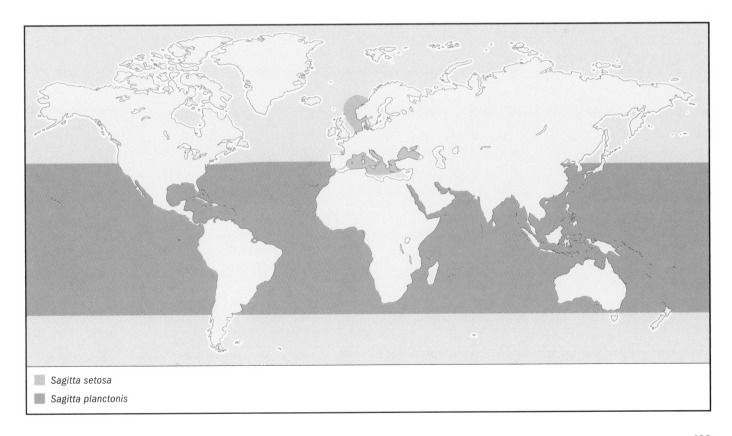

■ *Sagitta setosa*
■ *Sagitta planctonis*

larette and gut diverticula. Eyes are small with T-shaped pigment spots. Seminal vesicles are conical and touch posterior fins. Ovaries are very broad and long, reaching the neck region; ova are fairly small

DISTRIBUTION
Shallow-mesopelagic and cosmopolitan, occurring between 40°N and 40°S.

HABITAT
Occurs in the pelagic realm of tropical and subtropical regions of the ocean, at depths of 328–1,640 ft (100–500 m).

BEHAVIOR
Shows evidence of diurnal vertical distribution. Juveniles occur at higher levels in the water column than the adults. Rapid darting movements are made over short distances to catch prey.

FEEDING ECOLOGY AND DIET
Prefers small copepods. Feeds at night after moving to superficial layers.

REPRODUCTIVE BIOLOGY
Reproduces once or twice a year, depending on water temperature. Hermaphroditic, and sperm is stored in the oviducts waiting for the eggs to mature. Fertilized eggs are released in sea water. There is no larval stage.

CONSERVATION STATUS
Not listed by the IUCN.

SIGNIFICANCE TO HUMANS
None known. ◆

No common name
Sagitta setosa

ORDER
Aphragmophora

FAMILY
Sagittidae

TAXONOMY
Sagitta setosa Müller, 1847, North Sea.

OTHER COMMON NAMES
None known.

PHYSICAL CHARACTERISTICS
Individuals have 8–9 hooks, 6–8 anterior teeth, and 10–16 posterior teeth. Maximum adult body length is 0.55 in (14 mm), and the relative tail length is 16–25% of total body size. The body is small, narrow, and transparent. There are two pairs of lateral fins and no fin bridge. Anterior fins are relatively short, completely rayed, and round. Posterior fins are short, completely rayed, and round. The collarette is small or not present, and there are no gut diverticula. Eyes have star-shaped pigment spots. Seminal vesicles are present in the head and trunk, touch the tail fin, and are somewhat separated from the posterior fins. Ovaries are short and ova are small. Recent molecular

evidence has revealed substantial genetic differences between populations from the North and Mediterranean seas.

DISTRIBUTION
Neritic, occurring in the North, Mediterranean, and Black seas.

HABITAT
Lives in shallow pelagic waters of the Mediterranean, Black, and North seas, especially where the continental shelf is relatively wide. Reported to tolerate different degrees of salinity and may inhabit inshore waters and estuaries. Occurs in the western part of the Baltic Sea as long as the salinity is not too low.

BEHAVIOR
Nothing is known.

FEEDING ECOLOGY AND DIET
Prefers small copepods. Feeds at night after moving to superficial layers.

REPRODUCTIVE BIOLOGY
Reproduces once a year, during the summer months with a peak in late August. Hermaphroditic, and sperm is stored in the oviducts waiting for the eggs to mature. Fertilized eggs are released in the sea water. There is no larval stage.

CONSERVATION STATUS
Not listed by the IUCN.

SIGNIFICANCE TO HUMANS
Used as an indicator species to follow the movement of water masses. ◆

No common name
Eukrohnia fowleri

ORDER
Phragmphora

FAMILY
Eukrohniidae

TAXONOMY
Eukrohnia fowleri Ritter-Zahony, 1909, southern Indian Ocean.

OTHER COMMON NAMES
None known.

PHYSICAL CHARACTERISTICS
Individuals have 8–14 hooks and 2–31 posterior teeth. There are no anterior teeth. Maximum adult body length is 1.57 in (40 mm), and the relative tail length is 22–27% of total body size. The body is firm, broad, and opaque. Transversal muscles are present in the anterior part of the trunk. The neck is narrower than the region at the septum between trunk and tail segment. Pair of long, lateral fins are located on both the trunk and tail and are partially rayed. This species does not have a collarette, nor does it have gut diverticula. The head is small, with big oval eyes that have diamond-shaped pigment spots. Seminal vesicles are oval and rounded and do not touch the tail fin or the lateral fins. Ovaries are short and broad, and the ova are large.

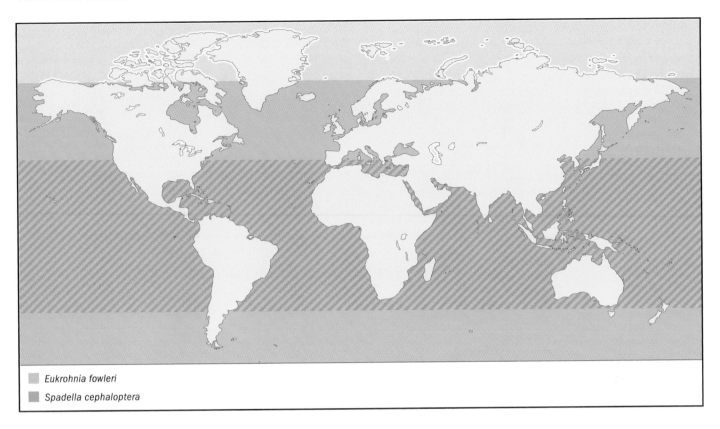

■ *Eukrohnia fowleri*

■ *Spadella cephaloptera*

DISTRIBUTION
Deep-mesopelagic, occurring between 1,640 and 4,920 ft (500–1,500 m). Circumglobal, from 70°N to 70°S in all three oceans.

HABITAT
Oceanic species that lives in the pelagic realm in all oceans except the Arctic and Antarctic.

BEHAVIOR
Nothing is known.

FEEDING ECOLOGY AND DIET
Prefers small copepods.

REPRODUCTIVE BIOLOGY
Reproduction is not seasonally determined. Growth and reproduction are very slow, and the fertilized eggs are brooded in a sac that hangs out of the ovaries.

CONSERVATION STATUS
Not listed by the IUCN.

SIGNIFICANCE TO HUMANS
None known. ◆

No common name
Spadella cephaloptera

ORDER
Phragmphora

FAMILY
Spadelliidae

TAXONOMY
Sagitta cephaloptera Busch, 1851 Northeast Atlantic, Orkney Islands.

OTHER COMMON NAMES
None known.

PHYSICAL CHARACTERISTICS
Individuals have 7–10 hooks, 3–5 anterior teeth, and 2–4 posterior teeth. Maximum adult body length is 0.22 in (5.5 mm), and the relative tail length is 49–58% of total body size. The body is firm, broad, and opaque. Transversal muscles are present in the trunk. A pair of short lateral fins is located almost entirely on the tail segment and is connected to the tail fin by rays. The species has a collarette, small gut diverticula, and a prominent head. Eyes contain pigment spots. Seminal vesicles are oval or rounded, touching both the tail fin and lateral fins. Ovaries are short and broad. The ova are large and are sometimes loose in the trunk cavity. There are glands and adhesive papillae on the ventral side of the tail.

DISTRIBUTION
Neritic and benthic. Occurs between 0 and 82 ft (0–25 m), from about 40°N to 40°S.

HABITAT
Adheres to objects such as rocks or sea grass in benthic coastal areas. It can swim for short distances.

BEHAVIOR
Lives in benthic coastal layers. Benthic species are the only chaetognatha that can be observed in aquariums for long periods of time. Observations about behavior are known from these benthic species.

FEEDING ECOLOGY AND DIET
Prefers small copepods.

REPRODUCTIVE BIOLOGY
Two mature individuals with full seminal vesicles maneuver close to each other to mate. One will deposit a sperm package or spermatophore on the dorsal neck region of the other individual. Then the sperm moves toward the genital opening and is stored until eggs are mature. The eggs are fertilized internally and then released into the water.

CONSERVATION STATUS
Not listed by the IUCN.

SIGNIFICANCE TO HUMANS
None known. ◆

Resources

Books

Bone, Q., H. Kapp, and A. C. Pierrot-Bults, eds. *The Biology of Chaetognaths.* Oxford, New York, Tokyo: Oxford University Press, 1991.

Conway Morris, S. *Fossil Priapulid Worms.* London: Palaeontological Association, 1977; distributed by B.H. Blackwell.

Periodicals

Schram, F. R. "Pseudocoelomates and a nemertine from the Illinois Pennsylvanian." *Journal of Palaeontology* 47 (1973): 985–989.

Annelies C. Pierrot-Bults, PhD

Hemichordata
(Hemichordates)

Phylum Hemichordata
Number of families 8

Thumbnail description
Small wormlike marine animals that live individually or in colonies, depending on the species

Photo: Tornaria larva of an *Enteropneusta* worm. (Photo by Animals Animals ©Peter Parks, OSF. Reproduced by permission.)

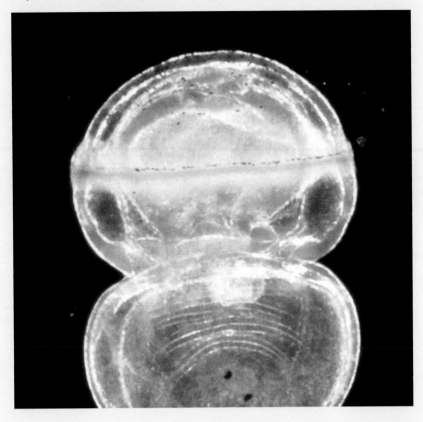

Evolution and systematics

The extant members of the phylum Hemichordata (formerly called Stomochordata) number about 92 and are typically divided into three classes:

- Enteropneusta or acorn worms: four families;

- Pterobranchia or pterobranchs: three families; and

- Planctosphaeroidea: one family.

Hemichordates are closely related to the chordates, sharing many chordate or chordate-like features. In the enteropneusts, these features include lateral openings, which are pharyngeal gill pores that connect the pharynx or airway to the exterior; and a stomochord that is somewhat analogous to the chordate notochord, although it has been described as "no more than an anterior extension of the buccal cavity." Despite these similarities, genetic studies, morphological comparisons of the larvae and anatomy of the nervous system, and biochemical evidence have recently indicated that hemichordates may be more closely related to echinoderms (starfishes and sea urchins) than to chordates.

In addition, recent genetic analyses of the extant hemichordates suggest that the pterobranchs are not plesiomor-

phic (similar in form) as previously thought, but have actually evolved from a forerunner to the enteropneusts. Under this arrangement, the lineage of the Ptychoderidae family within the enteropneusts split off first in the course of evolution, followed by a later separation of the Harrimaniidae family of the enteropneusts and the pterobranchs.

In addition to these three extant classes of Hemichordata, there is an extinct class, Graptolithina, known from fossils found in rocks dating from the Ordovician and Silurian periods (505–410 million years ago).

Physical characteristics

Hemichordates have two major body plans. One of the most notable characteristics of the enteropneustan hemichordates is their three-part body plan, which includes a protosome, or anterior proboscis (sometimes called a pre-oral lobe); followed by a mesosome or collar; and finally a metasome, or trunk. Cilia, which are present over all body areas, play roles in locomotion and in distributing the proteinaceous mucus secreted by the acorn worms. The largest species is the enteropneust *Balanoglossus gigas* of Brazil, an acorn worm that reaches 4.9 ft (1.5 m) in length and lives in long burrows stretching over more than 9.8 ft (3 m).

A Hawaiian acorn worm (*Ptychodera flava*) near the Lizard Islands, northern Great Barrier Reef. (Photo by © A. Flowers & L. Newman/Photo Researchers, Inc. Reproduced by permission.)

Pterobranchs have a three-part body plan like the enteropneusts, but with a shorter, shield-shaped proboscis and a more complex collar. In some species, the collar has tentacled arms. Pterobranchs form colonies, often with individuals attached by so-called stolons or stems. The individual animals are called zooids and are quite small, typically less than 0.04 in (1 mm) long. Groups form and live within a coenecium, which is a network of proteinaceous tubes built with secretions from each animal's proboscis.

The class Planctosphaeroidea has only one species, *Planctosphaera pelagica*, and it is known only from its larvae. Although several times larger at 0.3–1 in (8–25 mm) long, the almost-spherical, transparent *P. pelagica* larva is otherwise quite similar to enteropneust tornaria, having a gelatinous body covered with cilia. Unlike tornaria, however, the epidermis of *P. pelagica* has two deep invaginations (pouchlike formations) as well as numerous glands that secrete mucus.

Distribution

Enteropneusts, pterobranchs, and planctosphaeroids occur in oceans throughout the world. In general, the acorn worms live in shallower areas and the pterobranchs in deeper waters. The single species known from the class Planctosphaeroidea is found in both the Atlantic and Pacific Oceans at depths between 246 ft (75 m) and about 3,280 ft (1,000 m).

Habitat

Habitat varies by class. Adult acorn worms are typically found in either intertidal or shallow marine areas, although they are occasionally found in deeper water. They generally inhabit burrows in the sea bottom but also live sometimes in the sand inside shells, under rocks, in thick seaweed, or between root tangles. Adult pterobranchs are colonial forms that live in secreted tubular coenecia, and the planctosphaerids are planktonic.

Behavior

The acorn worms are solitary animals that are generally found sheltered in burrows, under rocks, or in thick vegetation. The burrowing species, like *Balanoglossus clavigerus*, use their proboscis primarily to fashion U-shaped burrows. They line the burrow walls with epidermal secretions that provide added strength. Each end of the burrow lies at the surface of the sea bottom and the remainder of the "U" is underground. One end is a cone-shaped depression in the sand bottom, and the other can be identified by a several-inch-tall pile of worm castings a short distance away. Besides this main burrow, *Balanoglossus* also employs a few side tunnels. Frequently, acorn worms will stretch their proboscis and collar out of the tunnel, but they spend the bulk of their time underground. When threatened, acorn worms respond by expanding the proboscis, effectively anchoring the animal in its burrow or tangle of vegetation while withdrawing the rest of the body. Studies of phototaxis (movement toward or away from a light source) reveal that illumination stimulates some species, like *Saccoglossus ruber*, to burrow deeper.

Because of the burrowing nature of most hemichordates, little is known about the reproductive and other behaviors of many species.

Feeding ecology and diet

Hemichordates may be either suspension- or sediment-feeders. The latter, like *Balanoglossus clavigerus*, take in sediment and obtain nutrients from the organic matter contained

Acorn worms, such as *Balanoglossus australiensis* are named for their acorn shaped head. They dig a U-shaped burrow, and live there with only the head sticking out of one opening, eating bacteria, micro-algae, and detritus. (Photo by Animals Animals ©Joyce & Frank Burek. Reproduced by permission.)

in it. The suspension-feeding adult hemichordates, as well as the filter-feeding tornaria larvae, gather their meals by generating currents with the cilia located on their bodies and drawing in organic matter. There is some uncertainty about the role their mucus plays in prey capture. Some scientists believe food sticks to the mucus-covered proboscis, and the cilia then beat in a pattern that draws the mucus and the food together to the mouth at the bottom of the proboscis. Researchers studying such species as *Rhabdopleura normani*, on the other hand, have found that normal feeding does not involve mucus; instead, the organism relies on the cilia to change direction in movements called local reversals and thereby directs food particles to its mouth.

Indirect-developing species have free-swimming tornaria larvae that live on plankton for weeks to months. Some species, like *Saccoglossus horsti*, have free-swimming larvae that obtain all their nutrition from their yolk, and within a few days take up the sessile (permanently attached) lifestyle. Studies of *P. pelagica* larvae indicate that their mucus may facilitate feeding, although the details are unclear and several alternative hypotheses have been suggested for the mucus.

Reproductive biology

Enteropneusts normally reproduce sexually via external fertilization, and develop either directly or via tornaria larvae. The indirect developers, including *Balanoglossus* and *Ptychodera* species, are in the majority. These species develop from egg to planktonic tornaria larva to adult form. The tornaria larvae eventually become sessile, with the burrow-dwellers developing tails behind the anus that they use to anchor themselves in their mucus-lined tunnels. Direct developers, on the other hand, hatch into adult animals, bypassing the planktonic phase. An example is *Saccoglossus kowalevskii*. Enteropneusts are also known to reproduce asexually by fragmentation of the adult's body, but this mode of reproduction is uncommon. Typically, the females lay up to 3,000 eggs at a time, and the males release sperm that appear to find the eggs by following chemical cues. Reproduction in many species is cyclical. *Saccoglossus horsti*, for example, breeds in late spring to midsummer. Water temperature and tides appear to affect reproductive timing in hemichordates.

The pterobranchs reproduce via short-lived larvae in a fashion similar to the enteropneusts, but more often resort to reproduction by asexual budding. Many, perhaps all, of the hemichordates are able to regenerate portions of their trunks.

Conservation status

No hemichordates are listed as threatened by the IUCN.

Significance to humans

The hemichordates are perhaps most important to humans for the information they can provide about the origin of chordates, deuterostomes, and bilateral animals.

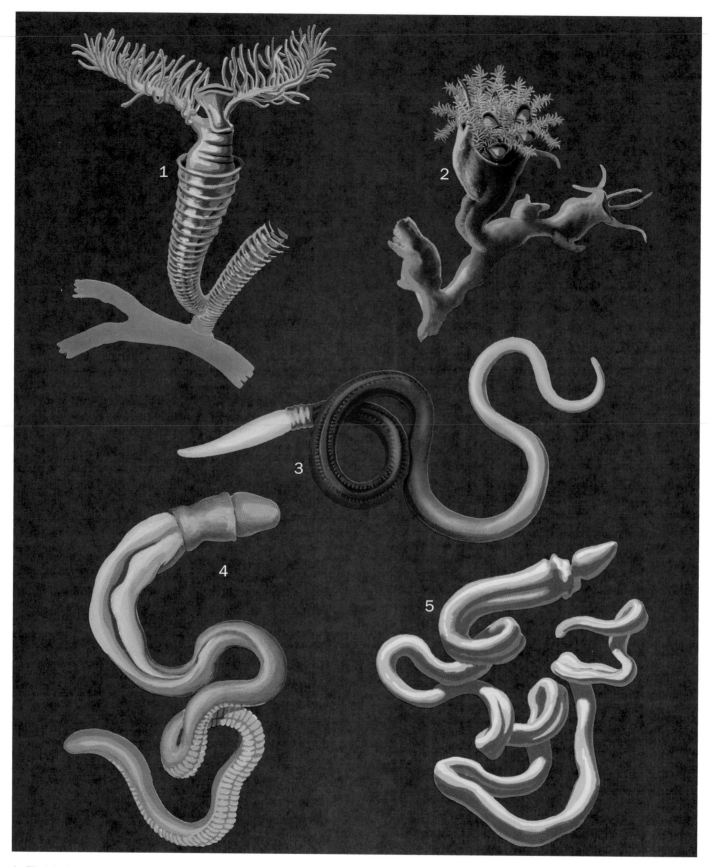

1. *Rhabdopleura normani*; 2. *Cephalodiscus gracilis*; 3. *Saccoglossus kowalevskii*; 4. Hawaiian acorn worm (*Ptychodera flava*); 5. Spaghetti worm (*Saxipendium coronatum*). (Illustration by John Megahan)

Species accounts

No common name
Cephalodiscus gracilis

ORDER
Cephalodiscida

FAMILY
Cephalodiscidae

TAXONOMY
Cephalodiscus gracilis M'Intosh, 1882, Straits of Magellan.

OTHER COMMON NAMES
None known.

PHYSICAL CHARACTERISTICS
Cephalodiscus gracilis is a three-part animal with a cephalic shield, complex collar, and trunk. The collar has two rows of up to five arms each. The arms have ciliated tentacles that reach 0.2–0.3 in (5–7 mm) in length.

DISTRIBUTION
Originally found at the southern tip of South America, they have since been found in Atlantic and Indo-Pacific waters.

HABITAT
These marine hemichordates live in hollow coenecia, but are able to travel freely inside each coenecium and on its tubes. They are typically found in waters less than 65.6 ft (20 m) deep.

BEHAVIOR
Little is known about the behavior of this species.

FEEDING ECOLOGY AND DIET
These organisms are filter feeders, using their cilia to create currents to direct the food. Their tentacles also flick food into the food canal. Their diet consists of organic food particles.

REPRODUCTIVE BIOLOGY
Embryos are brooded in the coenecium tubes. *Cephalodiscus* has a free-swimming larval stage followed by a enteropneust-like adult form. Adults eventually take on the characteristic U-shape.

CONSERVATION STATUS
Not listed by the IUCN.

SIGNIFICANCE TO HUMANS
None known. ◆

No common name
Saccoglossus kowalevskii

ORDER
No order designation

FAMILY
Harrimaniidae

TAXONOMY
Saccoglossus kowalevskii Agassiz 1873.

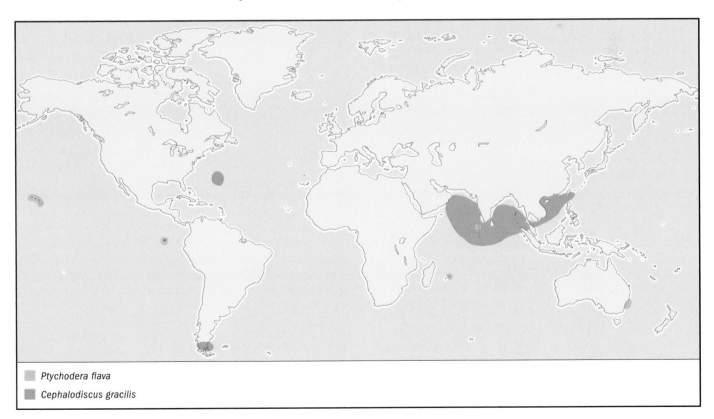

Ptychodera flava
Cephalodiscus gracilis

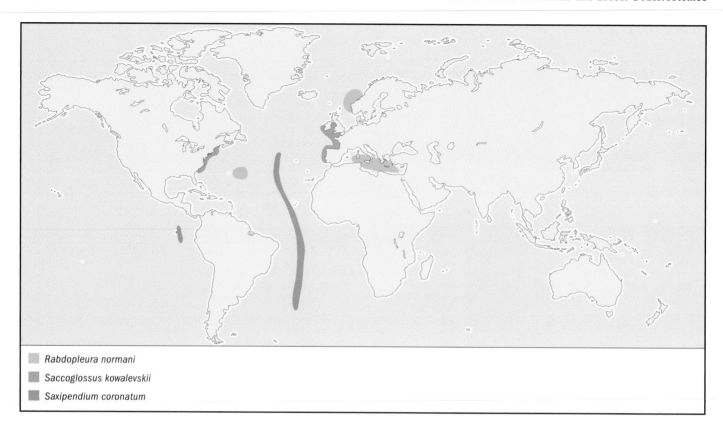

Rabdopleura normani

Saccoglossus kowalevskii

Saxipendium coronatum

OTHER COMMON NAMES
None known.

PHYSICAL CHARACTERISTICS
Adults are wormlike animals with a pointed proboscis, followed by a short collar and a long trunk. They range 4.0–5.9 in (10–15 cm) long. The proboscis is generally whitish yellow or light pink, the collar is orange, and the trunk can be pink to orange-brown.

DISTRIBUTION
North Atlantic, off the coast of Europe and the United States from Massachusetts to South Carolina.

HABITAT
Intertidal zones, typically sandy or silty areas.

BEHAVIOR
When threatened, an individual of this and other species of the genus *Saccoglossus* will swell its proboscis to serve as a holdfast in its burrow while drawing in the rest of its body. A mostly subterranean animal, *Saccoglossus* is most often seen with its proboscis poking out from its sea-bottom burrow. Its reproductive activity appears to be affected by seawater temperature, with spawning stimulated when temperatures shift from 80.6 to 71.6°F (27 to 22°C).

FEEDING ECOLOGY AND DIET
Saccoglossus kowalevskii gathers food from the uppermost layer of sediment in its immediate surroundings. Food items may include bacteria, microalgae and diatoms that live in the sediment, and dissolved or particulate organic material in the water. Each day, individuals of this species may take in as much as 300 times their body weight in sediment. Research in-

dicates that this acorn worm is induced to feed by the presence of microalgae and diatom populations as well as other food sources that are high in chlorophyll.

REPRODUCTIVE BIOLOGY
Fertilization is external. Females release mucus-bound eggs into the water, males respond by releasing sperm, and fertilization occurs in the sea water. The eggs, which average about 0.02 in (0.4 mm) in diameter, hatch in seven days into wormlike young that begin a sessile lifestyle immediately. *Saccoglossus kowalevskii* is a direct-developing species that hatches from eggs into the adult form without the planktonic larval stage common to many other hemichordates.

CONSERVATION STATUS
Not listed by IUCN.

SIGNIFICANCE TO HUMANS
None known. ◆

Hawaiian acorn worm
Ptychodera flava

ORDER
No order designation

FAMILY
Ptychoderidae

TAXONOMY
Ptychodera flava Eschscholtz, 1825.

OTHER COMMON NAMES
None known.

PHYSICAL CHARACTERISTICS
The adult is a yellowish brown animal with a small cone-shaped proboscis, short collar, and long trunk.

DISTRIBUTION
Pacific Ocean, particularly near Hawaii and Japan.

HABITAT
Coastal marine waters.

BEHAVIOR
Spawning times are related to sea water temperatures. In Hawaii, these acorn worms spawn in late November to early December, usually around 6 P.M. or dusk.

FEEDING ECOLOGY AND DIET
The diet of *Ptychodera flava* consists of organic food particles.

REPRODUCTIVE BIOLOGY
Fertilization, which is external, occurs when females release mucus-bound eggs into the water, and the males respond by releasing sperm. Fertilization occurs in the sea water. The eggs, which measure about 0.00433–0.00472 in (110–20 µm) in diameter, hatch into tornaria larvae in about two days. Tornaria larvae have a keyhole shape with a sphere above a broadened bell-shaped bottom. Two eyespots are present on the dorsal (upper) surface.

CONSERVATION STATUS
Not listed by IUCN.

SIGNIFICANCE TO HUMANS
None known. ◆

No common name
Rhabdopleurida normani

ORDER
Rhalodopleurida

FAMILY
Rhabdopleuridae

TAXONOMY
Rhabdopleura normani Allman, 1869, Norway.

OTHER COMMON NAMES
None known.

PHYSICAL CHARACTERISTICS
Adults have a cephalic shield, complex collar, and trunk. They can reach 0.1 in (3 mm) long, but typically are about 0.04 (1 mm) in length. There are two tentacled arms rising from the collar. The tentacles may reach 0.06 in (1.5 mm) in length.

DISTRIBUTION
Atlantic and Arctic Oceans, Mediterranean Sea.

HABITAT
These marine hemichordates are sessile colonial organisms found in shallow water up to 33 ft (10 m) deep, clinging to the bottoms or other protected nooks and crannies of rocks, corals, and other hard underwater surfaces. They live in and are con-fined to a coenecium because of their attachment to a continuous organic stem or stolon.

BEHAVIOR
Adult individuals are known as zooids. They live in separate, translucent tubes within a coenecium built out of secretions from their cephalic shields. The zooids are attached to a stolon and are unable to leave the coenecium. The larvae, on the other hand, use their cilia for swimming.

FEEDING ECOLOGY AND DIET
Individual adult *R. normani* worms feed by poking their proboscis, or cephalic shields, out of the coenecium and extending their two ciliated "arms" into the water. Although it has been suggested that other hemichordates use the mucus they secrete to transport food items, recent research indicates that individuals of this suspension-feeding species beat their cilia to draw food particles toward the mouth and do not rely on mucus to trap their food.

REPRODUCTIVE BIOLOGY
R. normani engages in sexual and asexual reproduction. Asexual budding occurs at the stolon, which runs along the bottoms of the tubes and forms connections among the zooids. Sexual reproduction occurs throughout the year. Females secrete tubes that are coiled at the base. The coils serve as brood chambers for one to seven eggs, which are deposited at different times, and therefore are at different developmental stages. The eggs are yolky, a creamy yellow in color, and about 0.0078 in (200 µm) in diameter. Each larva hatches and grows to a size of 0.016–0.018 in (400–450 µm). At that point, it swims past the other eggs and exits the tube. The larva continues swimming in search of its own place to settle on the bottom.

CONSERVATION STATUS
Not listed by IUCN.

SIGNIFICANCE TO HUMANS
None known. ◆

Spaghetti worm
Saxipendium coronatum

ORDER
No order designation

FAMILY
Saxipendiidae

TAXONOMY
Saxipendium coronatum Woodwick & Sensenbaugh, 1985, near "Rose Garden" geothermal vent, Galápagos Rift, at a depth of 8130 ft (2478 m).

OTHER COMMON NAMES
Italian: Verme tentacolato.

PHYSICAL CHARACTERISTICS
A long thin yellowish-white hemichordate, this species can reach 6.6–9.8 ft (2–3 m) in length. The proboscis is tapered to a soft point toward its front, and the collar is short.

DISTRIBUTION
Near the Galápagos Islands.

HABITAT
Found near hydrothermal vents in the deep sea, loosely attached to rocks.

BEHAVIOR
Little is known about the behavior of this species, but individuals are typically seen in "tortuous coils, wrapped upon themselves and welded in mucus."

FEEDING ECOLOGY AND DIET
Little is known about the diet of *S. coronatum*.

REPRODUCTIVE BIOLOGY
Fertilization is external; otherwise, little is known about this species' mode of reproduction.

CONSERVATION STATUS
Not listed by the IUCN.

SIGNIFICANCE TO HUMANS
None known. ◆

Resources

Books

Barrington, E. J. W. *The Biology of Hemichordata and Protochordata.* San Francisco: W. H. Freeman and Co., 1965.

Periodicals

Cameron, C. B., J. R. Garey, and B. J. Swalla. "Evolution of the Chordate Body Plan: New Insights from Phylogenetic Analyses of Deuterostome Phyla." *Proceedings of the National Academy of Sciences* 97, no. 9 (April 25, 2000): 4469–4474.

Halanych, K. M. "Suspension Feeding by the Lophophore-Like Apparatus of the Pterobranch Hemichordata *Rhabdopleura normani*." *Biology Bulletin* 185 (December 1993): 417–427.

Hart, M. W., R. L. Miller, and L. P. Madin. "Form and Feeding of a Living *Planctosphaera pelagica* (Phylum Hemichordata)." *Marine Biology* 120 (1994): 521–533.

Lester, S. M. "Ultrastructure of Adult Gonads and Development and Structure of the Larva of *Rhabdopleura normani* (Hemichordata: Pterobranchia)." *Acta Zoologica (Stockholm)* 69, no. 2 (1988): 95–109.

Organizations

British & Irish Graptolite Group. c/o Dr. A. W. A. Rushton, The Natural History Museum, Cromwell Rd., South Kensington, London, SW7 5BD United Kingdom. Web site: <http://www.graptolites.co.uk/>

The Graptolite Working Group of the International Palaeontological Association. c/o Dr. Charles E. Mitchell, Department of Geology, State University of New York at Buffalo, Buffalo, NY 14260-3050 United States. E-mail: cem@acsu.buffalo.edu Web site: <http://www.geology .buffalo.edu/gwg/index.htm>

Other

"Chris Cameron's Homepage." (15 July 2003). <http://cluster3 .biosci.utexas.edu/faculty/cameronc/CBC.htm>.

"Hemichordate Phylogeny." University of Washington Faculty Web Server. (15 July 2003). <http://faculty.washington.edu/ bjswalla/Hemichordata/hemichordata.html>.

Leslie Ann Mertz, PhD

Ascidiacea

(Sea squirts)

Phylum Chordata

Class Ascidiacea

Number of families 24

Thumbnail description
Benthic, solitary, and colonial species whose adults are sessile, almost exclusively fixed, firmly attached, or lying free on the sea floor; they usually filter feed; larvae are free swimming

Photo: These tunicates (*Phallusia julinea*) have a larval stage with notochord, forerunner of the vertebrate spine. (Photo by ©David Hall/Photo Researchers, Inc. Reproduced by permission.)

Evolution and systematics

There are very few fossil organisms that can be interpreted as ascidians: ascidians have no hard skeleton elements or calcareous shells, making it unlikely that their soft bodies can be fossilized. The single known fossil species (from the Pliocene), *Cystodytes incrassatus*, belongs to the widely distributed recent genus *Cystodytes*, characterized by the presence of relatively large (up to 0.04 in, or 1 mm, in diameter) spicules of discoid form, allowing easy identification. Another, much more ancient (about 300 million years old) form, *Jaekelocarpus oklahomensis*, has been interpreted as a tunicate, but recent ascidian taxonomists do not support this view. Thus, in the absence of paleontological data, phylogenetic relationships of ascidians can be understood mainly on the basis of the morphological characteristics of the recent forms.

The nature and relationships of ascidians were not understood for a long time, although these common marine animals were known even to Aristotle, more than 2,300 years ago. Earlier authors, including Carl Linnaeus, placed colonial ascidians in Zoophytes, a compound group that contained many different unrelated taxa, while the solitary ascidians were regarded as members of the phylum Mollusca. This view was based mostly on a wrong opinion, that the tunic, or the test, covering the body of adult ascidians is a modification of the calcareous shell of mollusks; furthermore, solitary ascidi-

ans and bivalve mollusks both are filter feeders and have two siphons, intruding and extruding. In 1816 J. C. Savigny first recognized the common nature of the solitary and colonial ascidians, as well as some pelagic forms, and in the same year J. B. Lamarck created the group Tunicata. He believed that it was a distinct class between Alcyonaria and Vermes, although other authors still treated this group as a class Molluscoides of Mollusca. From 1867 to 1872 the Russian embryologist A. Kowalevsky published several works on larval development and morphological characteristics of ascidians and was the first to recognize the close relationship between ascidians and chordates. Kowalevsky showed that the development of ascidians was similar to that of lancelets, a small group of primitive chordate animals now placed in the subphylum Cephaolchrodata of the phylum Chordata.

There are three important features linking ascidians and other tunicates with the chordate animals. First there is the presence of a notochord, a rod of specialized cells in the tail of ascidian larvae. The notochord disappears during the metamorphosis from free-swimming larva to sessile adult ascidian. In lancelets it is well developed in adults. In adult vertebrates the chorda dorsalis is wholly retained only in some fishes or is represented by remnants between the vertebrae. Second, a dorsal hollow nerve chord also is present only in ascidian larvae and disappears in adults. It lies dorsal to the notochord,

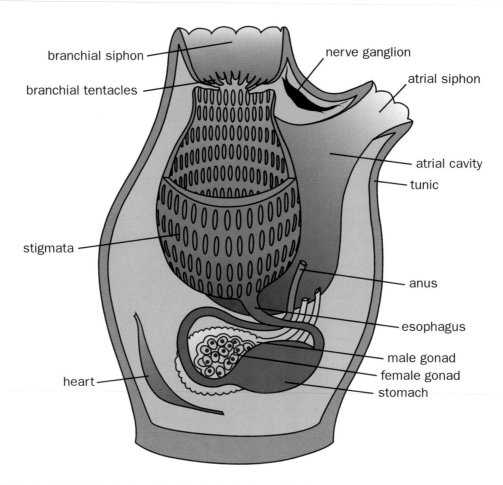

Tunicate anatomy. Tunicates are invertebrate chordates. (Illustration by Laura Pabst)

as in chordates, whereas it is always ventral in other phyla of invertebrate animals. Third, pharynx perforations, or gill slits, known in Ascidiacea as "stigmata," are present in all chordates, at least during embryonic development, but such structures are never present in invertebrate animals.

Class Ascidiacea belongs to the subphylum Tunicata and is divided into two orders: Enterogona (with two suborders, Aplousobranchia and Phlebobranchia) and Pleurogona (with one suborder, Stolidobranchia). The definition of the orders is based entirely on embryonic characters and reflects the origin of the atrial cavity—from a single or paired dorsal invaginations. For practical purposes, taxonomists use mostly suborders, saying that the ascidians are divided into three main groups, Aplousobranchia, Phlebobranchia, and Stolidobranchia. This subdivision is based mostly on the morphological features of adults, in particular, the structure of the branchial sac, the position of the gut, and some other features. This subdivision does not reflect colonial organization: Stolidobranchia and Phlebobranchia contain mostly solitary species but also include colonial species, and most Aplousobranchia are colonial, but some species are solitary.

There are 24 families, and, at present, about 185 (from about 240 described) genera are treated as valid. The number

of valid species is more difficult to count; the estimated number is 2,500–2,800. The interesting peculiarity of the ascidian classification is the relatively small number of genera in relation to the number of species. More than half of all known species belong to 10 very large genera listed here in decreasing order of number of species: *Aplidium*, *Didemnum*, *Molgula*, *Polycarpa*, *Ascidia*, *Styela*, *Eudistoma*, *Pyura*, *Cnemidocarpa*, and *Synoicum*. Most genera are well separated and can be identified easily, but species identification is often much more difficult. In general, the current ascidian classification seems to be close to the ideal, or "natural," system; that is, it appears to reflect the real phylogenetic relationships between taxa.

Physical characteristics

Usually, ascidians are described as sessile organisms with sac-like bodies firmly attached by the posterior end to the substratum. In reality, the shape of different ascidian species may be very diverse, and sometimes it is even difficult to recognize them as belonging to Ascidiacea. Although they grow side by side, is hard to imagine that the large, upright, cylindrical, solitary *Ciona* species belong to the same order as the *Didemnum*, which look like thin, encrusting stones and resemble a sponge colony.

Painted tunicates (*Clavelina picta*) in Saba, Dutch Antilles. (Photo by ©Andrew J. Martinez/Photo Researchers, Inc. Reproduced by permission.)

The body is always covered with the test, a protective layer of cellulose-like material secreted by the epithelium. The test may be clear and often brightly colored, as in the spectacular Pacific species *Halocynthia aurantium*, or it may be covered by various kinds of spines, as in the northern *Boltenia echinata*; it even may contain calcareous spicules, as in snow-white *Bathypera ovoida*. Among species living on a sandy or muddy bottom, the test often has long, thin outgrows, or test hairs, and sometimes is covered by a dense layer of sand grains, making the specimen cryptic, like many species of the family Molgulidae.

The body muscles typically are arranged in transverse and longitudinal bands. Sometimes they are numerous, thick, and strong, forming a solid muscular wall; in many species, however, the muscles are represented by only a few thin fibers. There are two openings on the body: an oral or branchial opening and an atrial opening. These openings may be sessile or set on the ends of siphons of various lengths, the atrial siphon always dorsal to the branchial siphon. The oral siphon leads to the voluminous pharynx, called in ascidians the "branchial sac," The branchial sac occupies most the space in the body of solitary ascidians. The space inside the branchial sac is the branchial cavity, and the space between the branchial sac and the body wall is the atrial cavity. The atrial cavity opens to the exterior through the atrial siphon, and the branchial and atrial cavities are filled with seawater. The seawater, with food particles and oxygen, is drawn into the branchial sac through the branchial siphon and a circle of oral tentacles and passes from the branchial to the atrial cavity through numerous perforations in the wall of the branchial sac, where food items are sieved; then filtered water moves out from the body through the atrial siphon.

Perforations of the wall of the branchial sac generally are small and have ciliated margins called stigmata, The shape of the stigmata is an important taxonomic character: they may be straight and arranged in transverse or sometimes longitu-

dinal rows, or they may be spiral figures, sometimes forming high funnels protruding into the branchial cavity. The wall of the branchial sac has transverse and longitudinal branchial vessels crossing each other and forming rectangular meshes. In some deepwater ascidians true stigmata are absent, the wall of the branchial sac is reduced, and the branchial sac is represented only by wide rectangular meshes formed by crossing longitudinal and transverse branchial vessels. The branchial sac is bilaterally symmetrical, with its mid-dorsal line marked by a fold termed the "dorsal lamina." The mid-ventral line has an endostyle, a groove lined with ciliated glandular epithelium secreting mucus. The mucus constantly moves from the endostyle to the dorsal lamina and then, with the filtered food particles, to the esophagus, to which the bottom of the branchial sac opens. The esophagus typically is short and always is much narrower than the branchial sac. It leads to the stomach and then to the intestine. The intestine makes a loop and opens into the atrial cavity.

The position of the gut loop in relation to the branchial sac is an important taxonomic character of the suborder and family levels: the gut loop may be on the left or, in one family, on the right side of the branchial sac; between the branchial sac and the body wall; or, mostly in colonial ascidians, under the branchial sac. Gonads are hermaphroditic and situated in the gut loop, under the gut, or on the body wall on the sides of the branchial sac. Gonoducts always open into the atrial cavity, and only in one highly specialized genus do they penetrate the test and open directly to the exterior. The nervous system, as in all sessile animals, is simple and represented by an elongated ganglion situated on the dorsal side between the siphons and a few nerves running from it. The heart is a thin-walled tubular organ on the ventral side of the body or, in colonial ascidians, in the bottom of elongated zooids.

In the case of colonial ascidians, the test forms the so-called common test, a mass in which individuals, called "zooids," are completely or partly embedded. The general plan of the structure of the zooids is the same as in solitary ascidians, but details may differ significantly. The body may be undivided or divided into two (thorax and abdomen) or three (thorax, abdomen, and post-abdomen) regions. The thorax contains the branchial sac, the gut loop is in the abdomen, and the gonads are either in the abdomen or the post-abdomen. Branchial siphons of all zooids in a colony open directly to the exterior, but atrial apertures in many species open into the cloacal cavity within the common test. The colony may contain single or several isolated cloacal cavities, each exposed to the exterior through one or several openings. Zooids connected with one cloacal cavity form a system. The shape of the systems varies; zooids may be arranged in circular systems around the single cloacal opening in the center, or they may form long double rows along cloacal canals. The form of the systems usually can be recognized easily on living colonies; the systems make a characteristic pattern on the surface of the colony, which, in certain cases, helps to identify the species.

The size of most ascidians varies from 0.04 to 0.4 in (1–10 mm), rarely 0.6 in (15 cm). Certain species, however, are much larger; in favorable conditions some solitary species may reach

Sea pig (*Halocynthia pyriformis*) nestled into the rocks in the Gulf of Maine. (Photo by ©Andrew J. Martinez/Photo Researchers, Inc. Reproduced by permission.)

19.7 in (50 cm) in height, and thin, encrusting colonies of certain species of didemnids grow to 9.8 sq ft (3 sq m). The largest species is the colonial Antarctic *Distaplia cylindrica*; its long, sausage-shaped colonies may be up to 23 ft (7 m) in length and 3 in (8 cm) in diameter. The smallest species is the deepwater *Minipera pedunculata*, whose diameter is only 0.02 in (0.5 mm).

Many ascidians are brightly colored, usually red, brown, and yellow or, rarely, blue. The coloring of colonial ascidians is especially diverse: that of the zooids and the common test may be different, resulting in complex, often very beautiful patterns on the colony surface. Some species, such as *Clavelina*, have a transparent body, often marked with variously colored spots and lines.

Many ascidians have a peculiar feature: they are able to accumulate vanadium. More primitive species tend to have higher levels of vanadium in their tissues.

Distribution

Ascidians are distributed widely in all oceans and seas, from the intertidal zone to the abyssal depths. In exceptional cases, species can survive in brackish or freshwater. Several species have expanded outside their original range and propagated all over the world. For example, *Molgula manhattensis*, naturally distributed on the Atlantic coast of North America from Maine to Louisiana, was found in large numbers on the Pacific coast in San Francisco Bay in the 1950s, and later it was introduced into Japan on ships' hulls. In 1975 it was recorded in Australia, and in 2000 it was recorded in large numbers on the Russian coast of the Sea of Japan.

Habitat

Ascidians cover a wide range of benthic habitats. Most species require a hard substratum and firmly attach themselves to rocks, stones, shells, algae, and so on. Other species are adapted for living on soft, muddy or sandy bottoms; such species are not firmly attached but often have thin hairlike outgrows on the test, anchoring them in the mud. There is a group of several interesting interstitial species: these rarely recorded minute species live between gravel grains; some of them are not fixed and can move. Ascidians occur from the intertidal zone to abyssal depths and have been recorded in all deepwater trenches; the maximal recorded depth is 27,560 ft (8,400 m). Most species inhabit shallow waters, some of them, such as the thin calcified colonies of certain didemnids, can survive on the open shores under strong wave actions.

Behavior

Ascidians are sessile, usually firmly fixed to the substratum. In response to an external stimulus, they can only slowly contract the body and close the siphons. Some of the very few interstitial species can actively move between gravel particles. Very slow movement of the colony also has been recorded in one or two tropical colonial species. Abyssal *Situla* and *Megalodicopia* species can quickly close their enormously large, bilobed branchial siphons in attempt to catch small swimming invertebrates, such as copepods.

Feeding ecology and diet

Almost all ascidians are filter feeders. The cilia lining the stigmata on the pharyngeal wall constantly pump water with suspended organic particles through the branchial sac. This feeding method is called active filtration: an ascidian expends energy to create a steady flow of water through the branchial sac. With the filtration membrane, a constantly moving mucous sheet on the inner surface of the branchial sac, ascidians are able to capture very fine organic particles including bacteria and phytoplankton. The species living in water with higher concentrations of suspended particles usually have more complex and dense branchial sacs; the stigmata in these species typically are small and numerous, and high branchial folds increase the branchial surface.

At great depths, water contains very little organic matter. Ascidians having the usual type of branchial sac, with ciliated stigmata, are very small in size there. Some deepwater species, such as *Culeolus*, have lost true ciliated stigmata; the branchial sac in such species resembles a loose net made by the crossing vessels, without tissue between them. These branchial sacs have less resistance to water flow than do the dense filters of shallow-water ascidians. These species typically have very large, widely opened branchial siphons oriented to the water current, allowing so-called passive filtration: water passes through the branchial sac without any muscular action or beating of the cilia, thus saving energy. Such highly adapted abyssal species often are large. Ascidians of the family Octacnemidae have a mixed diet. Some of them apparently can catch small swimming invertebrates with their large, bilobed branchial siphons.

Reproductive biology

All ascidians are hermaphroditic. All colonial species of the suborder Aplousobranchia are viviparous; in these species ova are fertilized internally and developed within the parent colony, in the atrial cavity of zooids or in special outgrowths of the zooid body wall, termed brood pooches, or in the common test of the colony. Viviparous species release swimming larvae. Solitary species are either viviparous or oviparous, releasing ova. In viviparous solitary species, larvae are incubated in the atrial cavity. Ascidian larvae are called tadpole larvae. Larvae have an oval trunk, usually less than 0.04 in (1 mm) long, and a tail that is longer than the trunk. The larval trunk contains the larval organs, including the cerebral vesicle with a light receptor and statocyte and adhesive papillae on the anterior end, and a rudiment of the future adult. In many colonial species this rudimentary ascidian within the larval trunk is rather well developed and resembles a zooid of the adult ascidian. Sometimes, as in colonial *Diplosoma*, the larval trunk may contain two or more future zooids. Ascidian larvae never feed; they swim for a short time and then attach to the substratum and start metamorphosis. During metamorphosis all larval organs degenerate, and an ascidian develops from the rudiment located in the larval trunk. In the case of solitary ascidians only one specimen develops from each larva, whereas in colonial ascidians individuals (zooids) replicate themselves by asexual reproduction (budding) and form a colony.

Conservation status

No species of Ascidiacea are listed by the IUCN.

Significance to humans

Ascidians have no direct and important significance to humans. Like many other marine organisms, they are a potential source for bioactive chemicals that may be used in pharmacology. Also, several species are consumed as food in some countries, for example, Japan. Some species may cause problems in fish farming operations, growing on the bottoms of ships and negatively affecting operations such as mussel and oyster cultivation.

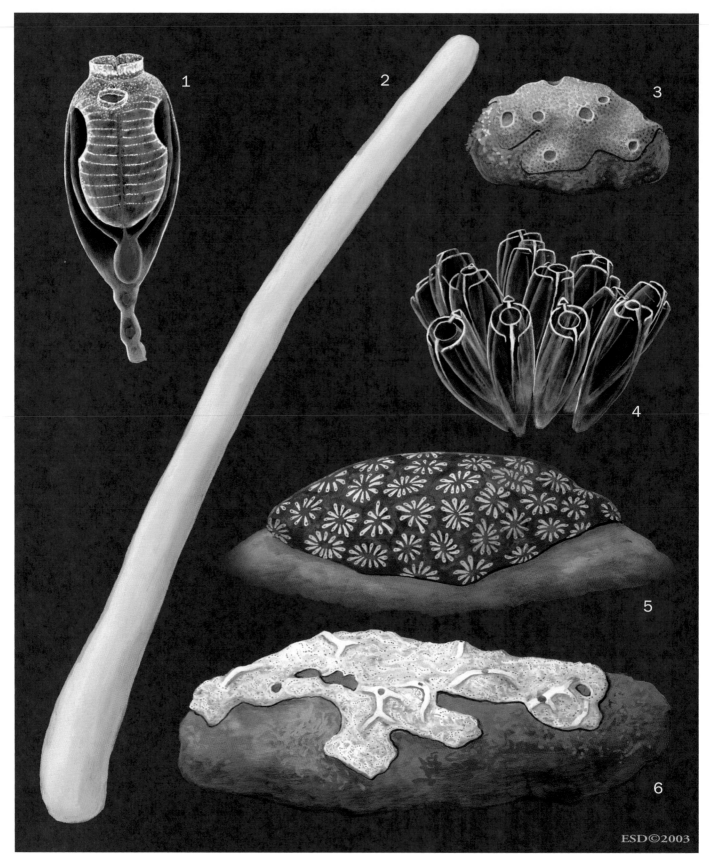

1. *Clavelina dellavallei*; 2. *Distaplia cylindrica*; 3. *Didemnum commune*; 4. *Clavelina lepadiformis*; 5. *Botryllus schlosseri*; 6. *Didemnum studeri*. (Illustration by Emily Damstra)

1. *Culeolus likae*; 2. *Pelonaia corrugata*; 3. Sea pig (*Halocynthia pyriformis*); 4. *Ciona intestinalis*; 5. *Corella parallelogramma*; 6. *Hartmeyeria triangularis*; 7. *Chelyosoma orientale*; 8. *Octacnemus kottae*; 9. *Situla pelliculosa*. (Illustration by Emily Damstra)

Species accounts

No common name
Ciona intestinalis

ORDER
Enterogona

FAMILY
Cionidae

TAXONOMY
Ascidia intestinalis Linnaeus, 1767, Europe.

OTHER COMMON NAMES
None known.

PHYSICAL CHARACTERISTICS
Cylindrical body up to 5.9 in (15 cm) in length, attached by posterior end. Test soft and translucent, colorless. Siphons are prominent and terminal, the branchial siphon marked by eight and the atrial siphon by six red pigment spots on the margin.

DISTRIBUTION
One of the most widely distributed, almost cosmopolitan ascidian species, especially abundant along the coasts of northern Europe. Recorded also in the Mediterranean Sea, along the At-lantic coast of North America and parts of the Atlantic and Pacific coasts of South America, California, Hawaii, South Africa, Australia, New Zealand, and Japan.

HABITAT
Occurs on rocks, stones, shells, and algae at depths from 0 to about 1,640 ft (0–500 m).

BEHAVIOR
Solitary species, often forming large populations of many crowded specimens. Sessile, attached, immobile species.

FEEDING ECOLOGY AND DIET
Filter feeder.

REPRODUCTIVE BIOLOGY
Ova released directly into seawater through the atrial siphon. Larvae are not incubated in the atrial cavity.

CONSERVATION STATUS
Not listed by the IUCN.

SIGNIFICANCE TO HUMANS
Often used as a laboratory specimen for various general types of biological research, in particular, cell biology and embryology. ◆

Culeolus likae

Ciona intestinalis

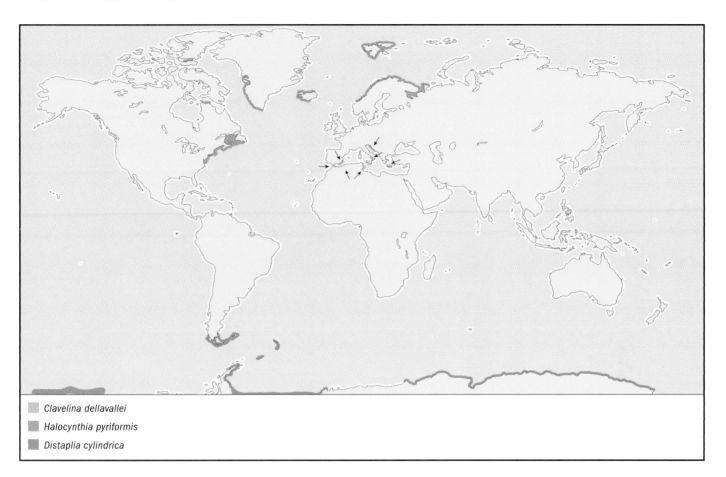

Clavelina dellavallei
Halocynthia pyriformis
Distaplia cylindrica

No common name

Clavelina dellavallei

ORDER
Enterogona

FAMILY
Clavelinidae

TAXONOMY
Bradiclavella dellavallei Zirpolo, 1925, Naples Gulf, Italy.

OTHER COMMON NAMES
None known.

PHYSICAL CHARACTERISTICS
Colonial species. Colony consists of a few, sometimes only one, upright zooids connected by basal stolons. Zooids are club-shaped, up to 2.4 in (6 cm) in length, and taper basally to a thin stalk. There are two large, plain-edged terminal siphons. Test is completely transparent, and all internal organs are clearly visible through it. The only opaque organ is the small gut loop; it is situated under the branchial sac and has deep blue coloring. Two vertical blue lines are seen inside the branchial sac, corresponding to the dorsal lamina and the endostyle.

DISTRIBUTION
This species is known only from the Mediterranean Sea.

HABITAT
Rocky bottoms at depths from 33 to 295 ft (10–90 m).

BEHAVIOR
Sessile, attached, immobile species.

FEEDING ECOLOGY AND DIET
Filter feeder.

REPRODUCTIVE BIOLOGY
Embryos are incubated in the atrial cavity on the right side of the branchial sac; they are very numerous, up to 50 in each zooid.

CONSERVATION STATUS
Not listed by the IUCN.

SIGNIFICANCE TO HUMANS
None known. ◆

No common name

Clavelina lepadiformis

ORDER
Enterogona

FAMILY
Clavelinidae

TAXONOMY
Ascidia lepadiformis Mueller, 1776, Norwegian fjords.

Clavelina lepadiformis
Pelonaia corrugata
Situla pelliculosa

OTHER COMMON NAMES
None known.

PHYSICAL CHARACTERISTICS
Colonial species. Colony consists of numerous upright (up to 1.2 in, or 3 cm, in height) zooids united only by their bases. The species can be recognized easily by its color pattern: completely transparent zooids have longitudinal white lines along the endostyle and dorsal lamina, white rings on the siphons and on the upper edge of the branchial sac, and a short white strip between the siphons.

DISTRIBUTION
Europe from western Norway to Spain and the Mediterranean Sea. Local populations are found in Morocco, Madeira, and the Azores. Introduced to South Africa, probably via the bottoms of ships.

HABITAT
Rocky bottom at depths from 0 to 164 ft (0–50 m).

BEHAVIOR
Sessile, attached, immobile species.

FEEDING ECOLOGY AND DIET
Filter feeder.

REPRODUCTIVE BIOLOGY
Larvae incubated in the atrial cavity.

CONSERVATION STATUS
Not listed by the IUCN.

SIGNIFICANCE TO HUMANS
None known. ◆

No common name
Chelyosoma orientale

ORDER
Enterogona

FAMILY
Corellidae

TAXONOMY
Chelyosoma orientale Redikorzev, 1911, Tatarsky Strait, Sea of Japan.

OTHER COMMON NAMES
None known.

PHYSICAL CHARACTERISTICS
Solitary species. The external appearance very characteristic: the low, wide body forms a flat disk, up to 3.9 in (10 cm) in diameter, covered by horny plates that have concentric lines of growth. Tunic is very hard, brown. Two sessile siphons on the disk, each siphon surrounded by six triangular plates. When growing on the sandy bottom, the species forms long, thick, root-shaped outgrows firmly anchoring the specimen to the substratum, but these outgrows are not present in specimens attached to stones.

DISTRIBUTION
Northwestern Pacific from the north part of the Sea of Japan to the Bering Strait and Chukchi Sea.

HABITAT
Occurs at depths from 66 to more than 984 ft (20–300 m) on stony, sandy, or muddy bottoms.

■ Botryllus schlosseri
■ Chelyosoma orientale

BEHAVIOR
Sessile, attached, immobile species.

FEEDING ECOLOGY AND DIET
Filter feeder.

REPRODUCTIVE BIOLOGY
Nothing is known, except that this species does not incubate larvae.

CONSERVATION STATUS
Not listed by the IUCN.

SIGNIFICANCE TO HUMANS
None known. ◆

No common name
Corella parallelogramma

ORDER
Enterogona

FAMILY
Corellidae

TAXONOMY
Ascidia parallelogramma Mueller, 1776, Oslo Fjord, Norway.

OTHER COMMON NAMES
None known.

PHYSICAL CHARACTERISTICS
Solitary species. Body reaches 2 in (5 cm) in height and is nearly rectangular in outline and attached by the posterior end. Test transparent and colorless; internal organs clearly seen through it. Red lines, making a characteristic pattern on living specimens, are body muscles; they are better developed on the left side of the body.

DISTRIBUTION
Norway, Sweden, North Sea, United Kingdom, and Mediterranean Sea.

HABITAT
On hard substratum (shells, stones, and algae) at depths from 0 to about 656 ft (0–200 m).

BEHAVIOR
Sessile, attached, immobile species.

FEEDING ECOLOGY AND DIET
Filter feeder.

REPRODUCTIVE BIOLOGY
Larvae are not incubated in atrial cavity.

CONSERVATION STATUS
Not listed by the IUCN.

SIGNIFICANCE TO HUMANS
None known. ◆

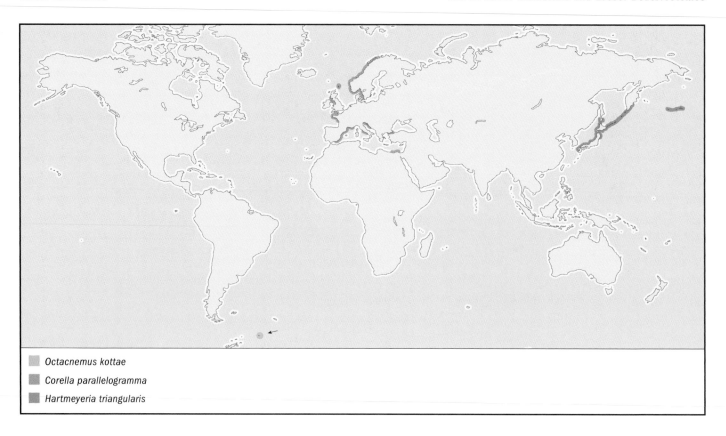

Octacnemus kottae

Corella parallelogramma

Hartmeyeria triangularis

No common name
Didemnum commune

ORDER
Enterogona

FAMILY
Didemnidae

TAXONOMY
Leptoclinum commune Della Valle, 1877, Naples Gulf, Italy.

OTHER COMMON NAMES
None known.

PHYSICAL CHARACTERISTICS
Colonial species with small (up to 1.2 in, or 3 cm, in diameter), red, encrusting colonies. Colony contains numerous minute calcareous spicules and feels hard to the touch. Very small zooids are completely embedded in the common test; numerous small openings of branchial siphons and several larger cloacal openings are seen on the colony surface.

DISTRIBUTION
European coasts of the Mediterranean Sea.

HABITAT
Rocky bottoms at depths from 0 to 230 ft (0–70 m). Common below intertidal zone, crevices, and caves exposed to strong wave actions.

BEHAVIOR
Sessile, attached, immobile species.

FEEDING ECOLOGY AND DIET
Filter feeder.

REPRODUCTIVE BIOLOGY
Larvae found in the colony from April to October.

CONSERVATION STATUS
Not listed by the IUCN.

SIGNIFICANCE TO HUMANS
None known. ◆

No common name
Didemnum studeri

ORDER
Enterogona

FAMILY
Didemnidae

TAXONOMY
Didemnum studeri Hartmeyer, 1911, Kerguelen Islands.

OTHER COMMON NAMES
None known.

PHYSICAL CHARACTERISTICS
Colonial species forming thin stone colonies on rocks, which may reach 0.4 ft (1 m) or more in diameter, while the thickness is always less than 3.3 in (5 mm). Colony contains minute calcareous spicules and is white.

DISTRIBUTION
Circumpolar in the subantarctic. Especially common in the Strait of Magellan and surrounding islands.

■ *Didemnum commune*
■ *Didemnum studeri*

HABITAT

Grows on stones and rocks and often found on roots or stems of algae. Occurs at depths from 16 to 5,554 ft (5–1,693 m) but chiefly in shallow waters.

BEHAVIOR

Sessile, attached, immobile species.

FEEDING ECOLOGY AND DIET

Filter feeder.

REPRODUCTIVE BIOLOGY

Larvae incubate in colony in September.

CONSERVATION STATUS

Not listed by the IUCN.

SIGNIFICANCE TO HUMANS

None known. ◆

No common name

Distaplia cylindrica

ORDER

Enterogona

FAMILY

Holozoidae

TAXONOMY

Holozoa cylindrica Lesson, 1830, Magellan Strait.

OTHER COMMON NAMES

None known.

PHYSICAL CHARACTERISTICS

Colonial species with very long, sausage-shaped colonies. This species is the largest known ascidian; the colonies may reach 23 ft (7 m) in length and 3 in (8 cm) in diameter. Colony is white or yellowish and soft to the touch; it is attached by one end to rocks and grows vertically. As for most ascidians, individual zooids are small and located only in the superficial layer of the colony, where they form numerous small, oval systems.

DISTRIBUTION

Completely circumpolar in the Antarctic.

HABITAT

Rocky bottoms at depths from 66 to 1,444 ft (20–440 m). Sometimes colonies are found floating on the surface, but such detached colonies always have degenerated zooids.

BEHAVIOR

Sessile, attached, immobile species.

FEEDING ECOLOGY AND DIET

Filter feeder.

REPRODUCTIVE BIOLOGY

Hermaphrodite, but zooids in distinct parts of the colony may have different sexual status: some zooids have better developed male gonads, whereas others have better developed female gonads. Zooids with well-developed gonads are found in colonies in December and January; fully developed tailed larvae appear from January to April. Larvae are incubated in brood pooches within parent colony.

CONSERVATION STATUS
Not listed by the IUCN.

SIGNIFICANCE TO HUMANS
None known. ◆

No common name
Octacnemus kottae

ORDER
Enterogona

FAMILY
Octacnemidae

TAXONOMY
Octacnemus kottae Sanamyan and Sanamyan, 2002, west of
South Sandwich Trench, Atlantic.

OTHER COMMON NAMES
None known.

PHYSICAL CHARACTERISTICS
Solitary species. This transparent abyssal species resembles enlarged zooids of Octocorallia, rather than ascidians. It has a
distinct cylindrical peduncle about 1.2 in (3 cm) long and a
flattened oral disk with a crown of eight triangular lobes that
have well-developed musculature and numerous lateral pinnules.

DISTRIBUTION
Although it is recorded only from the southern Atlantic, near
the South Sandwich Islands, the species probably has a wider
distribution, because several other records of *Octacnemus* may
belong to this species.

HABITAT
Soft, muddy bottoms at depths of 12,140–12,830 ft
(3,700–3,910 m).

BEHAVIOR
Nothing is known.

FEEDING ECOLOGY AND DIET
Nothing is known about how this and other species of *Octacnemus* feed. Unlike many other ascidians, this species probably is
not a filter feeder. In the gut of other *Octacnemus* species, fragments of copepods, polychaetes, and diatoms have been found.

REPRODUCTIVE BIOLOGY
Nothing is known.

CONSERVATION STATUS
Not listed by the IUCN.

SIGNIFICANCE TO HUMANS
None known. ◆

No common name
Situla pelliculosa

ORDER
Enterogona

FAMILY
Octacnemidae

TAXONOMY
Situla pelliculosa Vinogradova, 1969, Kurile-Kamchatka Trench,
northwestern Pacific.

OTHER COMMON NAMES
None known.

PHYSICAL CHARACTERISTICS
This species is one of the most unusual ascidians. Large, up to
13.4 in (34 cm) in height. Upright, very soft body attached to
small stones by a short, thick stem. Branchial orifice is enormously large and opens wide, exposing all internal organs.
Atrial orifice is small and inconspicuous. Branchial sac is modified to thin, flat membrane; all organs are in small nucleus in
the center of this branchial membrane.

DISTRIBUTION
Kurile-Kamchatka Trench, northwestern Pacific.

HABITAT
Abyssal species recorded at depths from 16,400 to 27,560 ft
(5,000–8,400 m).

BEHAVIOR
Sessile species.

FEEDING ECOLOGY AND DIET
The species probably has a mixed diet that includes small
plankton invertebrates, such as copepods, and fine organic particles obtained by passive filtration.

REPRODUCTIVE BIOLOGY
Nothing is known.

CONSERVATION STATUS
Not listed by the IUCN.

SIGNIFICANCE TO HUMANS
None known. ◆

No common name
Culeolus likae

ORDER
Pleurogona

FAMILY
Pyuridae

TAXONOMY
Culeolus likae Sanamyan and Sanamyan, 2002, Scotia Sea between South Georgia and Falkland Islands.

OTHER COMMON NAMES
None known.

PHYSICAL CHARACTERISTICS
Solitary species. Barrel-shaped body, 3.1 in (8 cm) in length, is supported on thin and very long, up to 33.5 in (85 cm), peduncle. Peduncle breaks basally into small tufts of delicate rhizoids anchoring the specimen in the mud. Peduncle inserted into anterior end of the body, on ventral side of large and widely opened branchial siphon. Atrial siphon is smaller and situated on the opposite end of the body. Soft and transparent body has characteristic posteroventral crest consisting of three high lamellae.

DISTRIBUTION
Southern Atlantic: recorded between South Georgia and South Falkland Islands and in the Argentine basin.

HABITAT
Abyssal species recorded at 15,301–18,474 ft (4,664–5,631 m).

BEHAVIOR
Immobile species fixed on a stalk.

FEEDING ECOLOGY AND DIET
Passive filter feeder. Large branchial siphon with peduncle inserted near it, oriented in such a way that the bottom current passes through the branchial sac.

REPRODUCTIVE BIOLOGY
Nothing is known.

CONSERVATION STATUS
Not listed by the IUCN.

SIGNIFICANCE TO HUMANS
None known. ◆

Sea pig
Halocynthia pyriformis

ORDER
Pleurogona

FAMILY
Pyuridae

TAXONOMY
Ascidia pyriformis Rathke, 1806, Bergen, Norway.

OTHER COMMON NAMES
None known.

PHYSICAL CHARACTERISTICS
Solitary species. Red or purple barrel-shaped body, up to 3.9 in (10 cm) in height, attached by the posterior end to hard substratum. Two well-developed siphons are positioned on the anterior end of body and covered by thin spines. Similar, but smaller spines cover the whole test.

DISTRIBUTION
Arctic North America and north Atlantic coasts of North America; Greenland; Iceland; Spitsbergen, Norway; White Sea; Barents Sea.

HABITAT
Rocky or stony bottoms at depths from 0 to about 656 ft (0–200 m).

BEHAVIOR
Sessile, attached, immobile species.

FEEDING ECOLOGY AND DIET
Filter feeder.

REPRODUCTIVE BIOLOGY
Larvae are not incubated in the atrial cavity.

CONSERVATION STATUS
Not listed by the IUCN.

SIGNIFICANCE TO HUMANS
None known. ◆

No common name
Hartmeyeria triangularis

ORDER
Pleurogona

FAMILY
Pyuridae

TAXONOMY
Hartmeyeria triangularis Ritter, 1913, Kyska Harbor, Aleutian Islands, Alaska, United States.

OTHER COMMON NAMES
None known.

PHYSICAL CHARACTERISTICS
Solitary species. Small, not exceeding 0.4 in (1 cm) in diameter; triangular and slightly flattened laterally, with body covered by thick layer of sand grains. Two small red siphons are set on opposite ends of upper body and covered by minute spines. Thin and relatively long peduncle anchors the specimen in the sand.

DISTRIBUTION
North Pacific from Alaska and Aleutian Islands to Kamchatka and further south the middle of Japan.

HABITAT
Lives on sandy bottoms at depths from 23 to 98 ft (7–30 m). Apparently cannot survive if the sand contains large amounts of mud.

BEHAVIOR
Immobile species.

FEEDING ECOLOGY AND DIET
Filter feeder.

REPRODUCTIVE BIOLOGY
Nothing is known.

CONSERVATION STATUS
Not listed by the IUCN.

SIGNIFICANCE TO HUMANS
None known. ◆

No common name
Botryllus schlosseri

ORDER
Pleurogona

FAMILY
Styelidae

TAXONOMY
Alcyonium schlosseri Pallas, 1766, Falmouth, English Channel.

OTHER COMMON NAMES
None known.

PHYSICAL CHARACTERISTICS
Colonial species. Colonies usually are flat, encrusting stones or algae, or sometimes lobed. Typically not exceeding 3.9 in (10 cm) in diameter. Zooids completely embedded in common test and arranged in circular systems. Each system has single central, common cloacal opening surrounded by a circle of branchial openings for each zooid. The test is soft and slimy, generally dark brown. Each zooid has a white or yellow strip on the dorsal side between the orifices; whole colony has a beautiful stellate color pattern. The color pattern, however, may vary significantly.

DISTRIBUTION
Common on northern European coasts and the Mediterranean Sea. Recorded on Atlantic coasts of North America, South Africa, Australia, New Zealand, and Japan. There is an opinion, however, that Pacific records belong to another similar, but distinct species.

HABITAT
Shallow-water species may grow on all kinds of hard substratum: stones, shells, and algae.

BEHAVIOR
Sessile, attached, immobile species.

FEEDING ECOLOGY AND DIET
Filter feeder.

REPRODUCTIVE BIOLOGY
Egg fertilization and larvae development occur within parent colony.

CONSERVATION STATUS
Not listed by the IUCN.

SIGNIFICANCE TO HUMANS
None known. ◆

No common name
Pelonaia corrugata

ORDER
Pleurogona

FAMILY
Styelidae

TAXONOMY
Pelonaia corrugata Forbes and Goodsir, 1841, Firth of Clyde, Scotland.

OTHER COMMON NAMES
None known.

PHYSICAL CHARACTERISTICS
Solitary species with elongated (up to 3.9 in, or 10 cm long) vermiform body resembling a sipunculid rather than an ascidian. Test is brown and leathery, covered by short, thin hairs. Branchial and atrial apertures are on small sessile siphons close to each other on the anterior, narrower end of the body.

DISTRIBUTION
Widely distributed in cold and temperate waters of Northern Hemisphere, in the North Atlantic and North Pacific, probably circumpolar in the Arctic.

HABITAT
Lives unattached in soft mud or sand at depths from 6.6 to 656 ft (2 to 200 m).

BEHAVIOR
Immobile species.

FEEDING ECOLOGY AND DIET
Filter feeder.

REPRODUCTIVE BIOLOGY
Nothing is known.

CONSERVATION STATUS
Not listed by the IUCN.

SIGNIFICANCE TO HUMANS
None known. ◆

Resources

Periodicals
Kott, P. "The Australian Ascidiacea." Part 1, "Phlebobranchia and Stolidobranchia." *Memoirs of the Queensland Museum* 23 (1985): 1–440.

———. "The Australian Ascidiacea." Part 2, "Aplousobranchia (1)." *Memoirs of the Queensland Museum* 29, no. 1 (1990): 1–226.

———. "The Australian Ascidiacea." Part 3, "Aplousobranchia (2)." *Memoirs of the Queensland Museum* 32, no. 2 (1992): 375–620.

———. "The Australian Ascidiacea." Part 4, "Aplousobranchia (3), Didemnidae." *Memoirs of the Queensland Museum* 47, no. 1 (2001): 1–407.

Van Name, W. G. "The North and South American Ascidians." *Bulletin of the American Museum of Natural History* 84 (1945): 1–476.

Karen Sanamyan, PhD

Thaliacea

(Salps)

Phylum Chordata

Class Thaliacea

Number of families 5

Thumbnail description
Marine, holoplanktonic organisms with cylindrical bodies that have openings on both ends

Photo: An open water solitary salp (*Lasis* sp.), seen only in the deep sea. (Photo by ©OSF/Animals Animals. Reproduced by permission.)

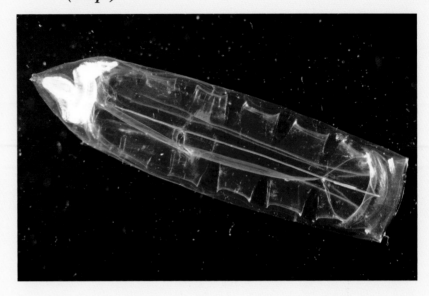

Evolution and systematics

The class Thaliacea is made up of three orders, five families, and 73 species. The three orders are Pyrosomatida, with one family and 10 species; Doliolida, with three families and 23 species; and Salpida, with one family and 40 species. Recent work indicates the thaliaceans are an artificial group because each of the orders evidently arose from a different group of benthic tunicates (class Ascidiacea). Evidence for this polyphyly is in comparative embryology and anatomy, where each of the orders show their ascidiacean ancestry to greater or lesser degrees. Their similar body forms are an example of convergent evolution in which animals look alike despite their differing phylogeny. Molecular work being done on thaliaceans should help determine relationships more precisely; consequently, this group may be dismantled and reorganized into separate groups in the future. Salps are related to the benthic tunicate class Ascidiacea.

Physical characteristics

Salps have a cylindrical body, called the test, that is mostly clear with openings called siphons on both ends. One opening is the oral and the other the cloacal siphon. Their body walls appear jelly-like and are therefore grouped as gelatinous zooplankton. Varying numbers of muscle bands are embedded within the gelatinous wall of the test. They range in size from 0.2–7.8 in (5 mm–20 cm), with some colonial forms many feet (meters) in length

Distribution

Thaliaceans are found in tropical and temperate seas. They tend to be found in the epipelagic zone from the surface to 656 ft (200 m), but some have been found in the mesopelagic zone from 656 to 3,280 ft (200 to 1,000 m).

Habitat

Thaliaceans tend to be more abundant in phytoplankton-rich surface waters of tropical and temperate seas. They are free swimming and holoplanktonic (at the mercy of ocean currents during their whole lives). Thaliaceans are often found with hyperiid amphipods living on the outside or even on the inside of their bodies in a symbiotic relationship.

Behavior

They move through the water by either ciliary action, as in the pyrosomatids, or by contraction of muscles embedded in the tunic, as in the doliolids and salps moving water into the oral siphon and out the cloacal siphon.

At times, because of their fast generation time, large numbers of thaliaceans can make up the largest proportion of gelatinous zooplankton.

Feeding ecology and diet

Thaliaceans are all filter feeders upon phytoplankton. Water flows into the oral siphon through a mucous sheet covering a basket and out the cloacal siphon. Phytoplankton and other small organisms are caught on the mucus. The mucus is then moved to the mouth and eaten. Because of their large numbers at certain times of the year, thaliaceans are important consumers of phytoplankton and make a large impact on the flux of carbon because of their production of massive amounts of fecal material.

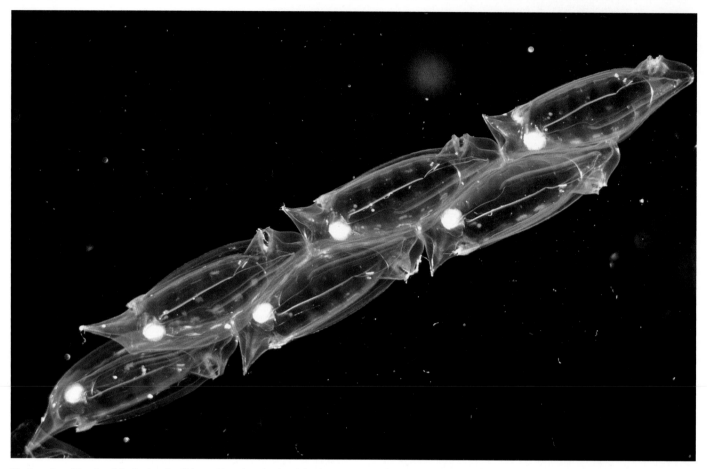

Chain salps (*Urochordata iasis atlantic*) are abundant in the Atlantic and Indian Oceans. (Photo by ©OSF/Animals Animals. Reproduced by permission.)

Reproductive biology

Alternation of generations enables thaliaceans to colonize seas quickly, taking advantage of phytoplankton blooms resulting in large swarms. One of the generations is asexual with vegetative growth resulting in the sexual stage. The other generation is sexual, which produces the egg and sperm that fuse to make the asexual stage.

Thaliaceans show different strategies to ensure successful reproduction. Pyrosomatids develop eggs that, when fertilized, develop to hatching inside the zooid. Doliolids retain the tadpole larval stage of their benthic tunicate ancestors during development. Salpids retain both asexual and sexual stages and true embryonic connection (via a placenta) to the parent during development.

Conservation status

Although there is much more to be learned about the ecology of thaliaceans, none of them is known to be in danger of extinction. No species is listed by the IUCN.

Significance to humans

At times, some thaliaceans can become so abundant that they can outcompete other herbivores at the base of the oceanic food web, such as copepods. Copepods are important in the diet of young fish species that humans rely on for food. In this way, the thaliaceans have a negative impact on human fishing activities.

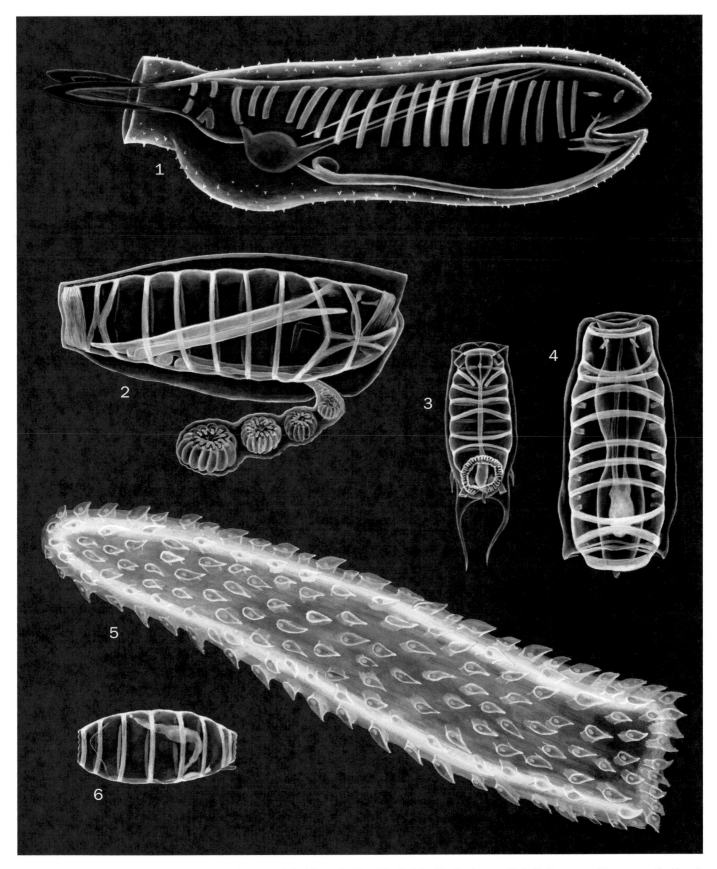

1. Salp (*Thetys vagina*); 2. Salp (*Cyclosalpa affinis*); 3. Salp (*Salpa fusiformis*); 4. Salp (*Thalia democratica*); 5. Pyrosome (*Pyrosoma atlanticum*); 6. Doliolid (*Dolioletta gegenbauri*). (Illustration by Emily Damstra)

Species accounts

Doliolid
Dolioletta gegenbauri

ORDER
Doliolida

FAMILY
Doliolidae

TAXONOMY
Dolioletta gegenbauri (Uljanin, 1884).

OTHER COMMON NAMES
None known.

PHYSICAL CHARACTERISTICS
Barrel-shaped, clear, gelatinous body about 0.35 in (9 mm) in length with eight prominent muscle bands in the asexual stage and the 0.47-in (12-mm) sexual stage has nine muscle bands.

DISTRIBUTION
Semi-cosmopolitan in cool waters of both hemispheres. (Specific distribution map not available.)

HABITAT
Little is known about vertical distribution, but commonly found in oceanic surface waters.

BEHAVIOR
Known for their jumpy swimming pattern when disturbed. Swims by contracting muscle bands present in the gelatinous body.

FEEDING ECOLOGY AND DIET
Can feed on a wide variety of food particle sizes ranging from bacteria 0.0002 in (5 µm) to large diatoms >0.00394 in (>100 µm). Can be very important herbivores, completely clearing the waters they inhabit in about two days.

REPRODUCTIVE BIOLOGY
The complicated lifecycle includes an asexual stage (sometimes called a nurse) during which a tail will grow. Many feeding stages bud on the tail, when released they grow a tail on which the sexual stage grows. The hermaphroditic sexual stage releases fertilized eggs that develop into tadpole larvae lacking the characteristic ascidiacean tail; the larvae quickly grows into the asexual nurse stage. Although complicated, the lifecycle helps doliolids respond to blooms of phytoplankton, resulting in blooms of their own, reported to be hundreds of square miles (kilometers) in the open ocean.

CONSERVATION STATUS
Not listed by the IUCN.

SIGNIFICANCE TO HUMANS
None known. ◆

Pyrosome
Pyrosoma atlanticum

ORDER
Pyrosomatida

FAMILY
Pyrosomatidae

TAXONOMY
Pyrosoma atlanticum (Peron, 1804), Mediterranean Sea.

OTHER COMMON NAMES
None known.

PHYSICAL CHARACTERISTICS
Individuals called zooids are about 0.33 in (8.5 mm) and embedded in a thick, clear tubular test that can reach 23.6 in (60 cm) in length. Colonies are pink or yellowish pink. The oral siphons of the zooids are on the outer surface of the tube and the cloacal siphon point to the inside. Water comes in the oral siphon and empties into the common opening inside of the tube. One end is closed and the water exits out a common opening at the other end of the colony, propelling it through the water. Each of the zooids is brilliantly bioluminescent and this is the reason for the scientific name of pyrosoma, "fire body."

DISTRIBUTION
Semi-cosmopolitan in temperate to tropical waters, it is the most common pyrosome. (Specific distribution map not available.)

HABITAT
Commonly seen in surface waters, but daily vertically migrates more than 2,460 ft (750 m).

BEHAVIOR
Known for the brilliant luminescence of each zooid, communication among zooids results in waves of light along the entire colony.

FEEDING ECOLOGY AND DIET
Water passes into each zooid through the oral siphon, into a mucous filter where phytoplankton is filtered out for food. Pyrosomes can form huge swarms that can result in significant quantities of fecal pellets. The sinking pellets can be very important in carbon input into the depths.

REPRODUCTIVE BIOLOGY
Zooids are hermaphrodites, possessing both testis and ovary. In each zooid of the colony a single egg is fertilized, which grows to a four-zooid stage that leaves the parent to start a new colony asexually by budding. This form of reproduction can result in huge swarms of pyrosome colonies that are dependent upon abundant phytoplankton for food.

CONSERVATION STATUS
Not listed by the IUCN.

SIGNIFICANCE TO HUMANS
The brilliant light displays given off from colonies of pyrosomes have bewildered and fascinated sailors for generations.◆

Salp

Cyclosalpa affinis

ORDER
Salpida

FAMILY
Salpidae

TAXONOMY
Cyclosalpa affinis (Chamisso, 1819).

OTHER COMMON NAMES
None known.

PHYSICAL CHARACTERISTICS
Both stages possess a clear, barrel-shaped, gelatinous body with circular muscle bands embedded in the body wall. Solitary (asexual) stage 2.9 in (74 mm) in length with seven muscle bands and a thick test. Aggregate form (sexual) is irregularly shaped, about 1.7 in (4.5 cm) long with four muscle bands. The aggregates are connected to form a chain of linked circles.

DISTRIBUTION
Temperate and tropical oceanic waters. (Specific distribution map not available.)

HABITAT
Most commonly found in surface of offshore waters.

BEHAVIOR
The only salp to form chains of linked circles.

FEEDING ECOLOGY AND DIET
Known for the ability to filter small food particles like phytoplankton with 100% efficiency.

REPRODUCTIVE BIOLOGY
The fertilized egg grows into the asexual stage inside the sexual stage, connected to the parent by a placenta used for norishment. Once released the sexual stage grows a tail where the sexual stage buds in chains of linked circles. Each sexual stage contains testis and ovary, which produce sperm and egg to produce the next generation of the asexual stage. They lack the tadpole larval stage.

CONSERVATION STATUS
Not listed by the IUCN.

SIGNIFICANCE TO HUMANS
None known. ◆

Salp

Salpa fusiformis

ORDER
Salpida

FAMILY
Salpidae

TAXONOMY
Salpa fusiformis (Cuvier, 1804).

OTHER COMMON NAMES
None known.

PHYSICAL CHARACTERISTICS
Both stages possess a clear, barrel-shaped, gelatinous body with circular muscle bands embedded in the body wall. Solitary (asexual) stage 0.39–1.9 in (1–5 cm) in length with nine muscle bands. Aggregate form (sexual) 0.39–1.5 in (1–4 cm) long with six muscle bands and can be found solitary after breaking off long chain. The body shape of the solitary (asexual) stage is symmetrical; it is asymmetrical for the aggregate (sexual) stage, with short outerior and posterior projections. Gut is the only prominently colored part of the body.

DISTRIBUTION
Semi-cosmopolitan in temperate to tropical waters, mostly oceanic with occasional near-shore swarms. (Specific distribution map not available.)

HABITAT
Most commonly found in surface waters at night. Can make daily vertical migrations of up to 1,640 ft (500 m).

BEHAVIOR
Long chains of aggregate forms can move through the water surprisingly quickly, up to 1.85 in (4.7 cm) per second.

FEEDING ECOLOGY AND DIET
Important filter feeders on phytoplankton. Known for fast growth rates.

REPRODUCTIVE BIOLOGY
Asexual stage produces a tail onto which bud two rows of the sexual stage that stay connected together, forming long chains several feet (meters) in length. The sexual stage can break away from the chain and swim along independently. The sexual stage bears fertilized eggs that are connected to a kind of placenta.

CONSERVATION STATUS
Not listed by the IUCN.

SIGNIFICANCE TO HUMANS
None known. ◆

Salp

Thalia democratica

ORDER
Salpida

FAMILY
Salpidae

TAXONOMY
Thalia democratica (Forskal, 1775).

OTHER COMMON NAMES
None known.

PHYSICAL CHARACTERISTICS
The asexual stage is about 0.47 in (12 mm) long with a pair of lateral posterior projections. The sexual stage is about 0.23 in (6 mm). The test is thick and has five prominent muscle bands.

DISTRIBUTION
Semi-cosmopolitan in temperate to tropical waters. (Specific distribution map not available.)

HABITAT
Usually found in surface waters.

BEHAVIOR
Swim actively through the water by contraction of the muscles in the test.

FEEDING ECOLOGY AND DIET
Important filter feeders on phytoplankton. Recorded growth rates among the fastest of any multicellular organism (up to 70% length increase per hour).

REPRODUCTIVE BIOLOGY
Asexual stage produces a tail onto which bud rows of the sexual stage that stay connected together, forming long chains several feet (meters) in length. They lack the tadpole larval stage, the egg develops directly into the asexual stage. Generation times can be as fast as a few days, resulting in huge swarms covering hundreds of square miles (kilometers) of open ocean.

CONSERVATION STATUS
Not listed by the IUCN.

SIGNIFICANCE TO HUMANS
None known. ◆

Salp
Thetys vagina

ORDER
Salpida

FAMILY
Salpidae

TAXONOMY
Thetys vagina (Tilesius, 1802).

OTHER COMMON NAMES
None known.

PHYSICAL CHARACTERISTICS
Solitary stage has a very thick test, 20 weakly developed muscle bands, two prominent posterior green lateral projections, and can reach up to 13 in (33 cm) in length, making it the largest of all salps. Aggregate stage up to 10 in (25 cm) and five muscle bands. Prominent gut mass is often red-pigmented in both stages.

DISTRIBUTION
Semi-cosmopolitan in temperate to tropical waters. (Specific distribution map not available.)

HABITAT
Commonly found in oceanic surface waters down to 490 ft (150 m).

BEHAVIOR
Can be very prominent member of oceanic and near-shore surface waters. The large aggregate stage stays connected, forming very long chains many feet in length.

FEEDING ECOLOGY AND DIET
Feed on phytoplankton.

REPRODUCTIVE BIOLOGY
The fertilized egg is connected to the inside of the sexual stage by a placenta; it swims out only when it has developed. It grows into the asexual stage, which grows a tail where the sexual stages bud in two rows. Each sexual stage contains testis and ovary, which produce sperm and egg to produce the next generation of asexual stage.

CONSERVATION STATUS
Not listed by the IUCN.

SIGNIFICANCE TO HUMANS
None known. ◆

Resources

Books
Berrill, N. J. *The Tunicata: With an Account of the British Species.* London: The Ray Society, 1950.

Brusca, R. C., and G. L. Brusca. *Invertebrates.* Sunderland, MA: Sinauer Associates, Inc., 1990.

Bone, Q. *Biology of Pelagic Tunicates.* New York: Oxford University Press, 1997.

Esnal, G. B., "Pyrosomatida." In *South Atlantic Zooplankton,* volume 2, edited by D. Boltovskoy. Leiden, The Netherlands: Backhuys, 1999.

Esnal, G. B., and M. C. Daponte. "Doliolida." In *South Atlantic Zooplankton,* volume 2, edited by D. Boltovskoy. Leiden, The Netherlands: Backhuys, 1999.

———. "Salpida." In *South Atlantic Zooplankton,* volume 2, edited by D. Boltovskoy. Leiden, The Netherlands: Backhuys, 1999.

Wrobel, D., and C. Mills. *Pacific Coast Pelagic Invertebrates.* Monterey, CA: Sea Challengers and Monterey Bay Aquarium, 1999.

Yamaji, I. *Illustrations of the Marine Plankton of Japan.* Osaka: Hoikusha Publishing Co., 1976.

Periodicals
Madin, L. P., and G. R. Harbison. "The Associations of Amphipoda Hyperiidea with Gelatinous Zooplankton. I. Associations with Salpidae." *Deep-Sea Research* 24 (1977): 449–463.

Other
<http://www.jellieszone.com>. [July 15, 2003.]

eurapp/target/imagecol/col.html>.

<http://bonita.mbnms.nos.noaa.gov/sitechar/pelagic.html>.

Michael S. Schaadt, MS

Appendicularia
(Larvaceans)

Phylum Chordata

Class Appendicularia

Number of families 3

Thumbnail description
Transparent, generally small organisms that produce a complex external mucous net (the house) to collect tiny planktonic food and that retain the larval tadpole stage as sexually mature adults

Photo: An appendicularian (*Oikopleura labradoriensis*) seen in the Gulf of Maine, USA. (Photo by ©Andrew J. Martinez/Photo Researchers, Inc. Reproduced by permission.)

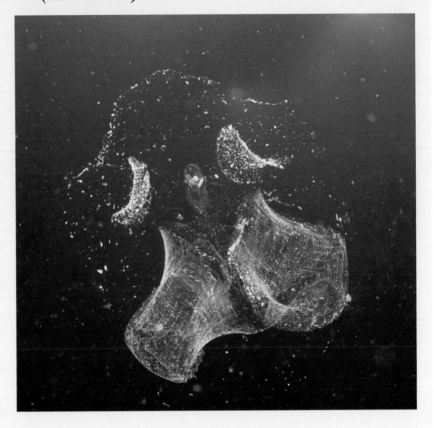

Evolution and systematics

The class Appendicularia is made up of one order, three families, and 64 species. Appendicularians are related to benthic tunicates and are commonly called larvaceans because they retain the larval tadpole stage as sexually mature adults. These animals are transparent and lack the outside covering, or tunic, of their benthic relatives. The body is composed of a trunk containing most of the internal organs and of a tail with a notochord running down the middle. The trunk secretes a mucous house, which may enclose the animal, as in *Oikopleura* and *Bathochordaeus* species, or enclose only the tail, as in *Fritillaria* species. The body and house appear jelly-like and are therefore grouped as gelatinous zooplankton. Larvaceans are small, ranging from a 0.04 in (1-mm) body length with a 0.2 in (4 mm) house to *Oikopleura* to 1 in (25 mm) body length with a house more than 6.6 ft (2 m) in diameter for the giant larvacean *Bathochordaeus*.

Physical characteristics

Body length ranges from 0.04 in (1 mm) for *Oikopleura dioica* to 1 in (25 mm) for *Bathochordaeus charon*.

Distribution

Cosmopolitan in ocean currents.

Habitat

Appendicularians are pelagic and live in oceanic and near-shore waters.

Behavior

Appendicularians move their tails rhythmically inside the house to produce a current that filters tiny food particles and to move the house through the water. If the filters become clogged or something bumps the house, the appendicularian abandons the house through a mucous trap door in the posterior of the house. The beginnings of a new house lie on the trunk of the body, and the animal inflates the new house and flips inside.

Some appendicularians have bioluminescent granules embedded in the house wall. It is thought that predators of the animal may eat an empty house that is flashing light while the

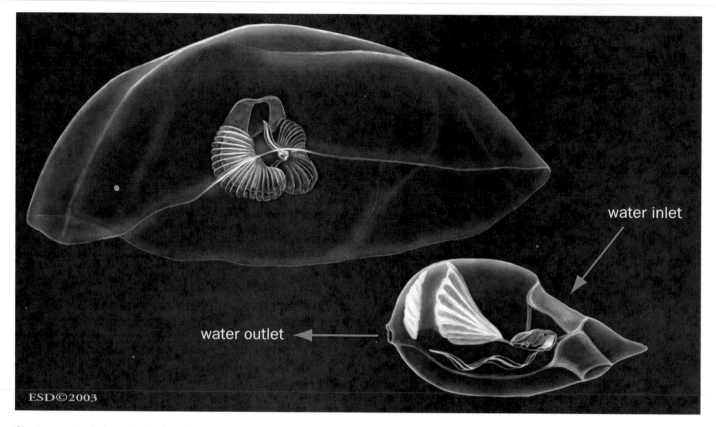

water inlet

water outlet

ESD©2003

Giant appendicularian with feeding filters and mucous sheet (top). Mucous house of *Oikopleura* (bottom). (Illustration by Emily Damstra)

original house builder swims away to make another house. Surface waters of some coastal bays and harbors can look brilliantly bioluminescent from large aggregations of appendicularians.

Feeding ecology and diet

All appendicularians are filter feeders with an amazing filtration apparatus. The mucous house has two prefilters embedded in the wall and an inner filter connected to the animal's mouth. The filters are made of strands of mucus that allow only the smallest food particles <0.0004 in (<1 µm) into the tube leading to the mouth.

Appendicularians can be important prey for many animals, including larval and adult fish. At least one fish important to humans, the anchovy, relies heavily on appendicularians for food.

Reproductive biology

Only sexual reproduction occurs in appendicularians. All except *Oikopleura dioica* are hermaphrodites. Gametes are

shed directly into the surrounding water. Hermaphrodites release sperm, and the eggs burst out of the body wall, a process that results in the death of the animal. If phytoplankton numbers are high, production of a large number of eggs and a fast generation time result in rapid development of blooms of appendicularians, sometimes in a matter of a few days.

Conservation status

Although there is much to be learned about the ecology of appendicularians, none of these species is known to be in danger of extinction. No species is listed by the IUCN.

Significance to humans

Because of potentially high population numbers, fast generation times, and ability to clear waters of phytoplankton, appendicularians can have considerable effect on the food web of the pelagic environment and thus on the numbers of fish important to humans.

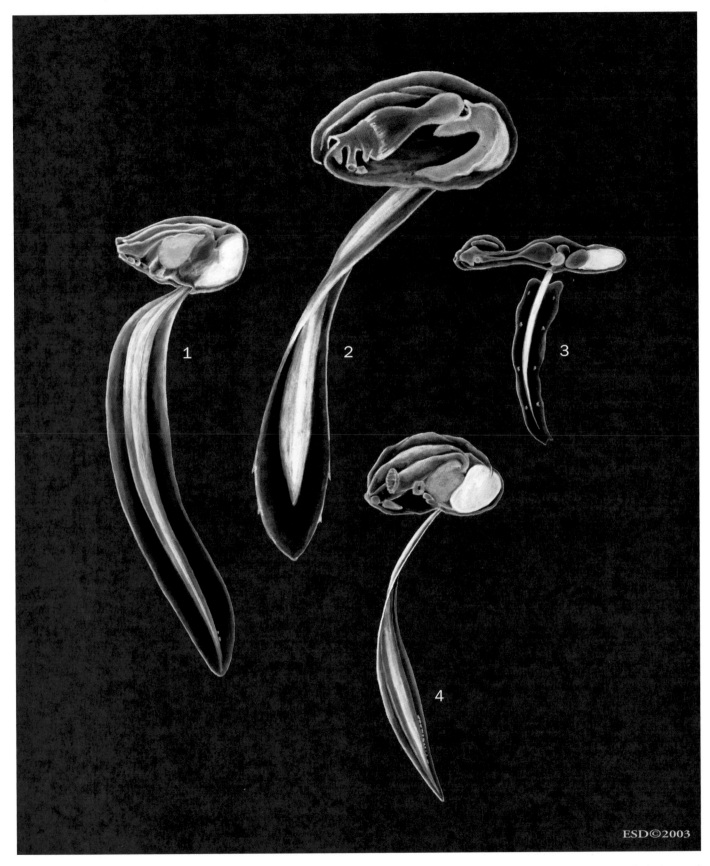

1. *Oikopleura dioica*; 2. Giant larvacean (*Bathochordaeus charon*); 3. *Fritillaria borealis*; 4. *Oikopleura labradoriensis*. (Illustration by Emily Damstra)

Species accounts

No common name
Fritillaria borealis

ORDER
Copelata

FAMILY
Fritillaridae

TAXONOMY
Fritillaria borealis (Lohmann, 1896), Irminger Sea.

OTHER COMMON NAMES
None known.

PHYSICAL CHARACTERISTICS
Body length to 0.2 in (4.5 mm) (trunk, 0.06 in [1.5 mm]; tail, 0.1 in [3 mm]). Distinguished from other appendicularians by connection of the tail at the midpoint of the trunk. The tail is the only part that is entirely inside the very small house.

DISTRIBUTION
Cosmopolitan in epipelagic zone (less than 656 ft [200 m]). (Specific distribution map not available.)

HABITAT
One of the most common appendicularians worldwide.

BEHAVIOR
The house is small and encloses only the tail and part of the trunk. When the tail stops moving, the house collapses on the body. Once the animal floats away, the tail starts moving again, inflating the house. Filter feeding continues in a new area.

FEEDING ECOLOGY AND DIET
Filters only the smallest particles out of the water (less than 0.0004 in [1 μm]).

REPRODUCTIVE BIOLOGY
These hermaphrodites release sperm first then release eggs by rupturing their body wall, a process that results in the death of the animal.

CONSERVATION STATUS
Not listed by the IUCN.

SIGNIFICANCE TO HUMANS
None known. ◆

Giant larvacean
Bathochordaeus charon

ORDER
Copelata

FAMILY
Oikopleuridae

TAXONOMY
Bathochordaeus charon (Chun, 1900), Cape of Good Hope.

OTHER COMMON NAMES
None known.

PHYSICAL CHARACTERISTICS
Largest of all appendicularians, reaching a length of 1 in (25 mm). Short, broad tail with clearly visible notochord embedded. The gut and gonads are easily seen in the large conspicuous trunk. The house is a series of elaborate tubes and filters used for feeding. A roughly ovoid sheet of mucus that can reach 6.6 ft (2 m) in diameter is continuously secreted above the house for protection and for catching sinking food particles.

DISTRIBUTION
Cosmopolitan in mesopelagic zone (328–1,640 ft [100–500 m]) (Specific distribution map not available.)

HABITAT
Pelagic in deeper waters (328–1,640 ft [100–500 m]).

BEHAVIOR
The tail beats slowly inside the house producing a current that moves water and food particles into the filters. The giant larvacean discards its mucous sheet and house when disturbed or when food particles clog the filters. Because abandoned mucous sheets and houses are laden with food, detritus-feeding zooplankton, and smaller microbes, they become important contributors of organic materials to the deep ocean as they sink to the botoom.

FEEDING ECOLOGY AND DIET
The large mucous sheet secreted above the house catches sinking particles, which are strained and directed to the inner filter of the house. The house filters further strain food particles of the proper size onto a mucous sheet, which is eaten.

REPRODUCTIVE BIOLOGY
Little is known except that these animals are hermaphrodites.

CONSERVATION STATUS
Not listed by the IUCN.

SIGNIFICANCE TO HUMANS
None known. ◆

No common name
Oikopleura dioica

ORDER
Copelata

FAMILY
Oikopleuridae

TAXONOMY
Oikopleura dioica (Fol, 1872), Mediterranean Sea.

OTHER COMMON NAMES
None known.

PHYSICAL CHARACTERISTICS

Smallest of all appendicularians with a body length to 0.1 in (3 mm) (trunk, 0.04 in [1 mm]; tail, 0.08 in [2 mm]). House approximately 0.2 in (4 mm) long.

DISTRIBUTION

Cosmopolitan in tropical and temperate waters in epipelagic zone (less than 656 ft [200 m]). (Specific distribution map not available.)

HABITAT

Can be abundant in coastal surface waters and is one of the most common appendicularians worldwide.

BEHAVIOR

When the filters become clogged, the house is abandoned, and a new house is inflated. Studies have shown that on average a new house can be made every 4 hours.

FEEDING ECOLOGY AND DIET

Filters only the smallest particles out of the water (less than 0.0004 in [1 µm]). Some fish, such as the blacksmith *Chromis bipunctata* in California kelp forests, are known for feeding almost exclusively on *Oikopleura dioica*.

REPRODUCTIVE BIOLOGY

Separate male and female forms release their sperm and eggs into the surrounding water.

CONSERVATION STATUS

Not listed by the IUCN.

SIGNIFICANCE TO HUMANS

None known. ◆

No common name
Oikopleura labradoriensis

ORDER
Copelata

FAMILY
Oikopleuridae

TAXONOMY
Oikopleura labradoriensis (Lohmann, 1892), Labrador Current.

OTHER COMMON NAMES
None known.

PHYSICAL CHARACTERISTICS
Body length to 0.24 in (6 mm) (trunk, 0.08 in [2 mm]; tail, 0.16 [4 mm]). House is approximately 0.4 in (9 mm).

DISTRIBUTION
Cosmopolitan in temperate to cold waters, in epipelagic zone (less than 656 ft [200 m]).

HABITAT
Can be abundant in cooler offshore and coastal surface waters.

BEHAVIOR
These animals have bioluminescent granules embedded in the walls of the house that may help confuse predators, who eat abandoned houses rather than animals that have abandoned the houses.

FEEDING ECOLOGY AND DIET
Filters only the smallest particles out of the water (less than 0.0004 in [1 µm]).

REPRODUCTIVE BIOLOGY
These hermaphrodites release sperm first then release eggs by rupturing the body wall, a process that results in the death of the animal.

CONSERVATION STATUS
Not listed by the IUCN.

SIGNIFICANCE TO HUMANS
None known. ◆

Resources

Books
Bone, Q. *Biology of Pelagic Tunicates.* New York: Oxford University Press, 1997.

Esnal, G. B. "Appendicularia." In *South Atlantic Zooplankton,* Vol. 2, edited by D. Boltovskoy. Leiden, The Netherlands: Backhuys, 1999.

Wrobel, D., and C. Mills. *Pacific Coast Pelagic Invertebrates.* Monterey, CA: Sea Challengers and Monterey Bay Aquarium, 1999.

Yamaji, I. *Illustrations of the Marine Plankton of Japan.* Osaka: Hoikusha, 1976.

Periodicals
Hamner, W. M., and B. H. Robison. "In Situ Observations of Giant Appendicularians in Monterey Bay." *Deep-Sea Research* 39 (1992): 1299–1313.

Silver, M. W., S. L. Coale, C. H. Pilskaln, D. R. Steinberg. "Giant Aggregates: Importance as Microbial Centers and Agents of Material Flux in the Mesopelagic Zone." *Limnology and Oceanography* 43 (1998): 498–507.

Other
"The JelliesZone." (15 July 2003). <www.jellieszone.com>.

<http://bonita.mbnms.nos.noaa.gov/sitechar/pelagic.html>

Michael S. Schaadt, MS

Sorberacea

(Sorberaceans)

Phylum Chordata

Class Sorberacea

Number of families 1

Thumbnail description
Small group of exclusively deep-water Tunicata comprising solitary benthic species with a strongly reduced branchial sac

Illustration: *Oligotrema sandersi.* (Illustration by John Megahan)

Evolution and systematics

The class Sorberacea was erected in 1975 by the French ascidiologists Claude and Françoise Monniot and Françoise Gail for several highly adapted deepwater species belonging to four genera (*Oligotrema, Hexacrobylus, Gasterascidia,* and *Sorbera*) united in the family Hexacrobylidae. The characters used to distinguish the Sorberacea from class Ascidiacea were the absence of the branchial sac, the presence of dorsal nervous cord, a very superficial position of the nervous ganglion, the histology of the digestive tract, and some other features.

The taxonomic position of Hexacrobylidae was not certain, but the family was most often considered as related to the ascidian family Molgulidae, or placed into a separate order, Aspiraculata, within the Ascidiacea. However, other taxonomists, including the Australian ascidiologist Patricia Kott, doubted whether the Hexacrobylidae should be removed from the class Ascidiacea. Kott considered that the characters used to distinguish Sorberacea and Ascidiacea could not be confirmed as true differences between Hexacrobylidae and the rest of the Ascidiacea. In particular, although the branchial sac is strongly reduced in Hexacrobylidae, it is not absent, and the neural complex in Hexacrobylidae is in the same position beneath the epidermis as in other ascidians. In addition, the structure and histology of the digestive tract varies in other ascidiaceans.

There are also problems in defining genera and species in the group. Of the four nominal genera mentioned above, *Gasterascidia, Sorbera,* and *Hexacrobylus* are now treated as synonyms of *Oligotrema,* and all species assigned previously to *Hexacrobylus* (apart from the type species of this genus) now belong to *Asajirus.* Thus the family Hexacrobylidae contains only two valid genera: *Oligotrema* and *Asajirus. Oligotrema* contains five species, and *Asajirus* (formerly known as *Hexacrobylus*), contains 12 nominal species, six of which are most probably synonyms of widely distributed *Asajirus indicus.*

Physical characteristics

All species of sorberaceans are solitary. Body shape is rather constant, *Asajirus* species almost always have an oval or egg-shaped body, and *Oligotrema* species are usually slightly more elongated and the siphons are situated on opposite ends of the body and directed away from each other. The test, a protective layer secreted by epithelium, is covered by short hair-like processes with adhering sand, mud particles, or, more often, tests of foraminiferans. The cloacal siphon is always small and inconspicuous. The branchial siphon is large, situated on the anterior end or on a side of the body, and surrounded by six large lobes. The opening of the branchial siphon leads to the buccal cavity lined with the test; the buccal cavity is homologous to the branchial siphon of other ascidians. As in all tunicates, The pharynx is perforated, but unlike in other ascidians, is small and has only a few openings leading to a system of thin-walled chambers that open into the atrial cavity by ciliated stigmata. Some species have a very large globular stomach, which occupies most of the internal space of the animal. All species have a kidney, a large thin-walled excretory organ, similar to the kidney of ascidians in the family Molgulidae (suggesting a possible relationship with this family).

All Hexacrobylidae are relatively small, from less than 1 in (3 mm) up to 2.4 in (6 cm), but usually they are usually no more than 1.4 in (1 cm) in greatest dimension.

Distribution

Sorberaceans are widely distributed in bathyal and abyssal depths in all oceans, but records are not numerous. They have not been recorded in the high Arctic and North Pacific Oceans, but deepwater tunicates in these regions have been less intensively sampled.

Habitat

Most specimens of sorberaceans were collected in deep waters at 3,280–16,400 ft (1,000–5,000 m). However, in 1903 a single specimen of *Oligotrema psammites* was recorded at 308 ft (94 m), and in 1990 another specimen of the same species was recorded at 886 ft (270 m). The deepest record is of a single specimen of *Asajirus indicus* at 25,853 ft (7,880 m); only four other tunicates have been recorded at such great depths. The bottom at abyssal depths is almost always soft and covered by thick layer of mud, sorberaceans live unattached on the mud.

Behavior

Sorberaceans do not survive capture, and no one has directly observed their behavior. Although they are not firmly attached to the substratum, it is unlikely that they are mobile, as their body covered by hairs with attached sediments suggests that they are anchored in the mud and cannot actively move. Their muscular branchial siphon has fingerlike lobes on the end, suggesting that it can operate as a hand to capture small invertebrates.

Feeding ecology and diet

Unlike most other benthic tunicates sorberaceans are carnivorous and not filter feeders. The specimens examined had an almost-empty digestive tract, so they are certainly not detritus-feeders, otherwise the digestive tract would be full of sediments. Possibly they can actively search and capture small moving invertebrates using their muscular (and sometimes eversible) branchial siphon. The large globular stomachs of some species suggest that the prey may be relatively large, and small copepods, isopods, ostracods, nematodes, and polychaetes were found in the stomachs of some specimens.

Reproductive biology

Almost nothing is known about sorberacean reproduction. As with all other benthic tunicates, they are hermaphrodites. All species are apparently oviparous, and larvae are not known. The youngest recorded specimens, 0.02 in (0.5 mm) in diameter, already resembled adult forms.

Conservation status

No species of Sorberacea are listed by the IUCN.

Significance to humans

None known.

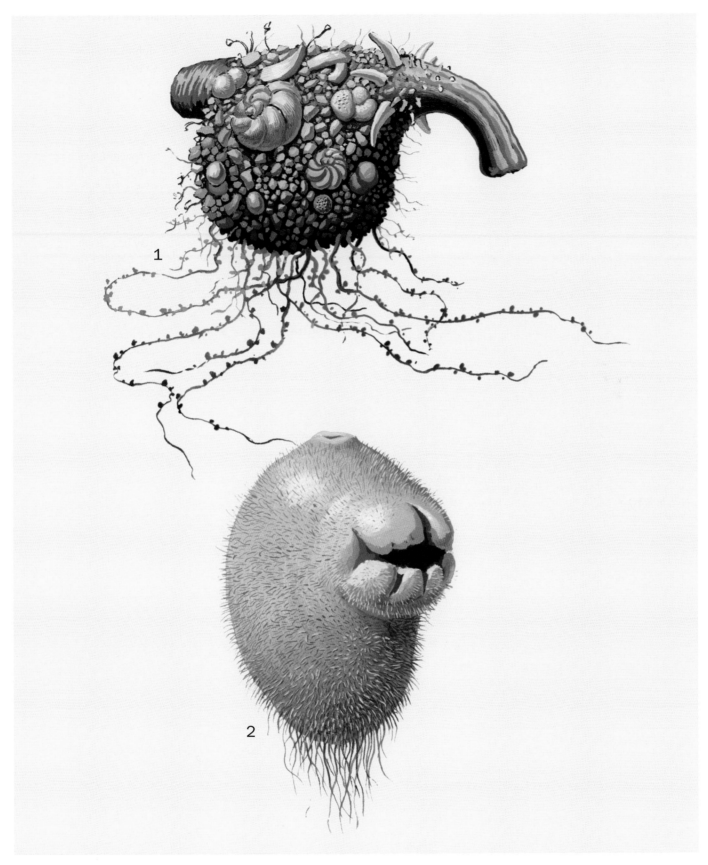

1. *Oligotrema sandersi*; 2. *Asajirus indicus*. (Illustration by John Megahan)

Species accounts

No common name
Asajirus indicus

FAMILY
Hexacrobylidae

TAXONOMY
Hexacrobylus indicus Oka, 1913, Sri Lanka, Indian Ocean.

OTHER COMMON NAMES
None known.

PHYSICAL CHARACTERISTICS
Body oval, not usually exceeding 0.8 in (2 cm) high. Atrial siphon relatively long, but it is internal, and thus can be seen only on dissected specimens after test is removed. On surface of test, only small inconspicuous atrial opening is present on top of body. Branchial siphon is large and muscular. On preserved specimens, siphon is internal, lying wholly within the body, but it is not known whether this is normal in living specimens. Branchial siphon opens on anterior part of the body by wide slit surrounded by six wide lobes. Body coated by fine short hairs that are significantly longer on the posterior end of body, suggesting specimens sit vertically at the bottom.

DISTRIBUTION
All oceans except central Arctic and North and East Pacific Oceans.

HABITAT
Lives unattached on soft muddy bottom of continental slopes and abyssal plains and trenches; recorded from 1,970 to 24,540 ft (600 to 7,480 m) deep.

BEHAVIOR
Nothing is known.

FEEDING ECOLOGY AND DIET
Carnivorous, including small invertebrates.

REPRODUCTIVE BIOLOGY
Nothing is known.

CONSERVATION STATUS
Not listed by the IUCN.

SIGNIFICANCE TO HUMANS
None known. ◆

Asajirus indicus
Oligotrema sandersi

No common name
Oligotrema sandersi

FAMILY
Hexacrobylidae

TAXONOMY
Gasterascidia sandersi Monniot and Monniot, 1968, northwest Atlantic Ocean.

OTHER COMMON NAMES
None known.

PHYSICAL CHARACTERISTICS
Small, not exceeding 0.2 in (5 mm). Oval body covered by sparse hairs with attached mud particles and foraminiferan tests. Two prominent siphons on opposite sides of body. Highly muscular, completely eversible branchial siphon longer than atrial siphon, has six fingerlike lobes. Stomach very large, occupies most of body.

DISTRIBUTION
Widely distributed in Atlantic Ocean, not recorded in other oceans.

HABITAT
Abyssal; recorded at about 4,920–16,400 ft (1,500–5,000 m) deep.

BEHAVIOR
Nothing is known.

FEEDING ECOLOGY AND DIET
Carnivorous, including small invertebrates.

REPRODUCTIVE BIOLOGY
Nothing is known.

CONSERVATION STATUS
Not listed by the IUCN.

SIGNIFICANCE TO HUMANS
None known. ◆

Resources

Periodicals

Monniot, C., and F. Monniot. "Nouvelles Sorberacea (Tunicata) profondes de l'Atlantique Sud et l'Océan Indien." *Cahiers de Biologie Marine* 25 (1984): 197–215.

———. "Revision of the Class Sorberacea (Benthic Tunicates) with Descriptions of Seven New Species." *Zoological Journal of the Linnean Society* 99, no. 3 (1990): 239–290.

Monniot, C., F. Monniot, and F. Gaill. "Les Sorberacea: une nouvelle classe de tuniciers." *Archives de Zoologie expérimentale et générale* 116, no. 1 (1975): 77–122.

Kott, P. "The Australian Ascidiacea." *Memoirs of the Queensland Museum* 32, no. 2 (1992): 621–655.

———. "The Family Hexacrobylidae Seeliger, 1906 (Ascidiacea, Tunicata)." *Memoirs of the Queensland Museum* 27, no. 2 (1989): 517–534.

Karen Sanamyan, PhD

Cephalochordata
(Lancelets)

Phylum Chordata

Class Cephalochordata

Number of families 1

Thumbnail description
Exclusively marine species with slender, fish-like shape, tapered at both ends; easily recognized by the externally visible v-shaped lines that separate the iterated muscle blocks and by the oral cirri that guard the mouth opening against unwarranted particles

Photo: Lancelet (*Amphioxus lanceolatus*) larva head. Lancelets have no heart but do have gills. (Photo by Animals Animals ©P. Parks, OSF. Reproduced by permission.)

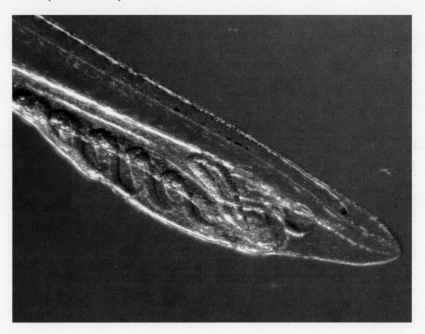

Evolution and systematics

Cephalochordates (also called acraniates or lancelets) possess key features that characterize them as chordates. Those characters include the notochord, dorsal nerve cord, branchial basket, and a post anal tail with fins. It is generally believed that cephalochordates are the closest living relatives to the vertebrates. The two groups share unique traits such as the organization of the main body musculature into separate segments and the organization of the circulatory system. Another feature, which unites cephalochordates and vertebrates, is a caecum that is probably homologous to the vertebrate liver. Thus, the morphological evidence is strong. In fact it is so strong that it was used as evidence against phylogenetic conclusions, which were drawn solely from DNA sequences when a molecular study placed the cephalochordates as relatives of echinoderms, not even within the chordates.

The potential for fossilization of cephalochordates is limited because the animals are soft bodied. Even their main skeletal element, the notochord, consists of specialized muscle tissue. Nevertheless several fossils have been described that could have affinities to cephalochordates.

Pikaia gracilens from the Canadian Burgess Shale (Cambrian, ca. 530 million years old) is a promising candidate for consideration as a cephalochordate fossil. It has the typical tapered body form at both ends, and it has been argued that it possessed a notochord. However, *Pikaia* also shows characters not found in extant cephalochordates. In some fossils the lines, interpreted as myocommata, are w-shaped rather reminiscent of vertebrates than of the simpler v-shape of cephalochordates. Also, the anterior end of some fossil remnants displays tentacles, a feature not found in living cephalochordates, but known, for example, from hagfishes.

The nature of several other fossils (e.g., *Cathaymyrus diadexus* [Cambrian], *Lagenocystis pyramidalis* [Ordovician], *Palaeobranchiostoma hamatotergum* [Permian]) remains more equivocal. An instructive example is the fossil *Yunnannozoon lividum* (Cambrian). Originally reconstructed as a cephalochordate, it was re-described as a worm-like hemichordate only to be interpreted as a sea pen (phylum Cnidaria) shortly thereafter.

Evolutionary relationships among cephalochordate species are unclear. However, it is thought that the genus *Epigonichthys* is more derived compared to *Branchiostoma*.

The class Cephalochordata (some experts now categorize it as a subphylum) contains one order, Amphioxiformes, and one family, Branchiostomatidae.

Physical characteristics

As the closest living relatives of vertebrates, cephalochordates occupy an eminent phylogenetic position. They are invaluable in reconstructing the evolutionary step from invertebrate to vertebrate along the lineage that leads to mammals and humans. Therefore, cephalochordates are of great interest to numerous researchers and consequently their anatomy is well known.

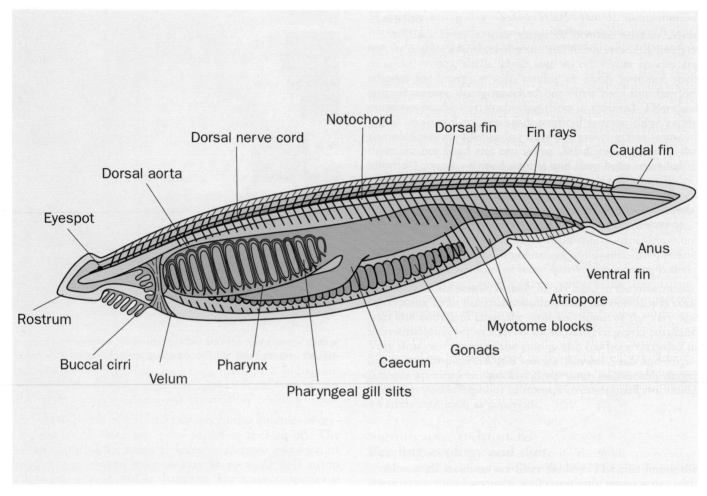

Lancelet anatomy. (Illustration by Christina St. Clair)

A median fin is present along the entire dorsal length of the animals, extending at the anterior end into a short rostral fin and at the posterior end a tail fin. The ventral fin runs between anus and atriopore. In front of the atriopore paired fin-like folds, the metapleural folds, extend to the anterior region of the pharynx. The dorsal and ventral fins contain so-called fin chambers enclosing coelomic cavities and fin rays that consist of a substance that could serve as an energy store.

The epidermis consists of a single-celled layer, a primitive feature compared to the multi-cellular epidermis in vertebrates. It contains gland cells that secrete a thin sheet of mucus covering the surface of the animal. Different types of sensory cells are distributed in the epidermis, especially in the rostral region.

The dorsal nerve cord of cephalochordates has a central canal that is enlarged at the anterior end of the cord. This enlarged region is known as the cerebral vesicle and is homologous to parts of the vertebrate brain as indicated by morphological similarities (infundibular cells, Reissner's fiber) and similarities in the expression patterns of developmental genes. A peculiarity of the cephalochordate neural cord is the presence of numerous pigmented ocelli, the so-called pigment-cup ocelli. In addition, several more structures in the neural cord are probably light sensitive, e.g., the Joseph cell receptors and the lamellar body in the cerebral vesicle. A photoreceptor in the anterior cerebral vesicle of larvae shows significant similarities to the paired eyes of vertebrates and could be homologous to them. Outside the central nervous system one finds receptors (e.g., corpuscules de Quatrefage), especially in the epidermis of the buccal cirri, velar tentacles, the atrium, and along the metapleural folds.

The notochord extends from the anterior to the posterior end. It is a skeletal element that is the antagonist of the lateral muscle cells that attach to it. Curiously, it is itself an active tissue. It consists of specialized muscle cells and is innervated from the nerve cord. Vacuoles within and between the cells act as a hydroskeleton and help in adjusting the rigidity of the notochord. The notochord is the main skeletal element. Aside from it, in the pharynx, between the gill slits, the branchial bars contain skeletal, cartilaginous rods. In addition, similar rods support the anterior cirri.

A pattern of repeated V-shaped lines is visible on the side of a cephalochordate. It is similar to the pattern seen in a fish fillet, though simpler, and of course, minute. This pattern is caused by the connective tissue that separates individual muscle segments. The innervation of the muscle cells is quite different from the vertebrates. There are no ventral spinal nerves that innervate the cells as in vertebrates. Instead the muscle cells themselves form long thin extensions that approach the nerve cord and are innervated directly there.

It has long been believed that cephalochordates possessed excretory cells very similar to polychaetes. Electron microscopy showed that the similarity is superficial. The excretory cells of cephalochordates are unique in the animal kingdom though in evolutionary terms they are probably derived from the so-called podocytes that are found in other deuterostomes. Excretory organs (nephridia; sing. nephridium) are found serially along the dorsal part of the pharynx. They are associated with blood vessels and function in a similar way as the vertebrate kidney. In addition to these nephridia a nephridium is situated just in front of the mouth (Hatschek's nephridium).

The general arrangement of blood vessels is similar to vertebrates, but the vessels are not lined by cells as in vertebrates. The region where a heart would be expected is contractile. In notable similarity with vertebrates a portal vein connects a capillary system around the posterior intestine with a capillary system around the hepatic caecum.

A pharynx with gill slits is common to all chordates: tunicates, cephalochordates, and the more primitive members of the vertebrates. In cephalochordates a vestibule lies in front of the pharynx that is guarded by the cirri. The vestibule contains ciliated tracts (wheel organ), an excretory organ (Hatschek's nephridium), and a shallow groove (Hatschek's pit) that is probably homologous to the adenohypophysis in vertebrates. The pharynx proper begins with the mouth opening that is surrounded by velar tentacles. The pharynx is perforated by up to 200 gill slits. Along its ventral midline runs a groove, the endostyle that produces mucus and thyroxine, indicating homology to the thyroid in vertebrates. The gill slits do not open into the open water, but the whole pharynx is covered by the outer body wall. In this way a space around the pharynx is created, which is called the atrium. The atrium opens in the ventral midline through the atriopore anterior to the anus.

The pharynx leads into the short esophagus and from there the gut extends straight posteriorly and opens through the anus. Ventrally, at about the junction of the pharynx and esophagus a blind ending caecum projects anterior. It is called the hepatic caecum and is probably homologous to the vertebrate liver.

Depending on species, adult animals range in size 0.4–3.2 in (1–8 cm) in total length. The color of all species is whitish to a creamy yellow, sometimes with a tint of pink. The mucus covering the epidermis can reflect iridescently.

For identification and taxonomic purposes the so-called myotome (muscle block) formula is used. Three numbers signify the number of myotomes anterior to the atriopore, between atriopore and anus, and posterior to the anus. In

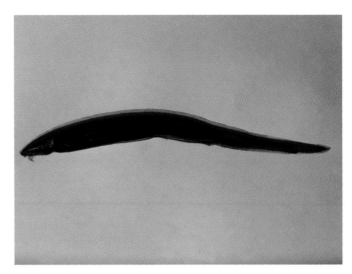

A lancelet (*Amphioxus*) is seldom more than 3 inches (8 cm) long; it resembles a slender fish without head definition or eyes. (Photo by Animals Animals ©E. R. Degginger. Reproduced by permission.)

addition, the number of fin chambers in the dorsal fin serves as a diagnostic character. *Epigonichthys* differs from *Branchiostoma* in that it possesses only one series of gonads on the right side of the body, whereas the gonads are paired in *Branchiostoma*.

Distribution

Cephalochordates are known from continental shelves of the tropic and temperate regions worldwide.

Habitat

Cephalochordates are semi-sessile filter feeders. They can swim vigorously, forward and backward, yet they live most of their life buried halfway in the substrate. Cephalochordates are found in shallow coastal areas, where they inhabit sandy bottoms. They prefer stable, well-reutilated, and smooth-textured grains. Larvae are planktonic and may drift over long distances before colonizing a suitable habitat.

Behavior

Cephalochordates can be found in high densities of over 9,000 animals per square meter. Sexes in cephalochordates are separate, and males and females are equally abundant within a population. Recruitment of larval stages influences the age structure in a population, which in turn depends on the currents.

Depending on the coarseness of the habitat, cephalochordates assume different feeding positions. In coarse sediment they bury their body entirely within the substrate. The preferred position is with their anterior end exposed to the open water. In fine sediment they lie on the substrate.

Feeding ecology and diet

Cephalochordates feed on plankton with plant cells, diatoms being the most important. The entire branchial basket is an adaptation for feeding with the endostyle continuously producing a mucus-net that is transported over the gill slits by cilia and capturing particles out of the water. The food particles together with the mucus net are rolled into a mass and passed on into the esophagus and the posterior intestine where they are digested.

Reproductive biology

Eggs are about 0.00394 in (100 µm) in diameter and contain numerous yolk droplets. Spermatozoa are simple and typical for marine invertebrate species with external fertilization.

Eggs and sperm are released into the water. How spawning is synchronized is not known. In subtropical populations the reproductive period stretches over several months and individuals spawn repeatedly. In temperate areas the reproductive period is shorter.

Total, radial cleavage is followed by the formation of a blastula, then gastrulation and neurulation ensue, and a peculiarly asymmetric larva is formed. The large mouth is situated on the left side. A single row of gill slits forms and the larvae start feeding immediately. More gill slits are added and later an additional row of gill slits forms on the right side. When about 12–15 pairs of gill slits are formed the larvae sink to the bottom and metamorphose into juveniles. This metamorphosis mainly involves the formation of the atrium by an outgrowth of the epidermis that surrounds the pharynx. Simultaneously the gill slits are divided and new gill slits are added. From now on the animals only grow in size, adding additional gill slits and segments while developing the gonads to full maturity. The larval period in the plankton lasts from several weeks up to a few months. In the following year the animals become sexually mature. Offshore plankton samples yielded relatively large cephalochordates that resembled larval stages but had developed rudimentary gonads. These were described under the genus name *Amphioxides*. However, it is more likely that these forms are merely larval forms of typically benthic species that did not encounter a suitable substrate rather than representing a genus with a different life history.

Conservation status

No cephalochordate species is regarded as threatened.

Significance to humans

It is believed that the "crocodile fish" that occurs in southern Chinese mythology is the cephalochordate species *Branchiostoma belcheri*. The mythical hero Han Yü killed many crocodiles, some of which escaped mortally wounded to the Xiamen region before they died. The maggots that emerged from the crocodile carcasses gradually changed into cephalochordates. This myth originated in the Tan Dynasty (A.D. 616–905).

In southern China local fishermen, deploying traditional techniques, fish for *Branchiostoma belcheri*. The catch is used for human consumption. *Branchiostoma lanceolatum* occurs as an appetizer on a gala dinner menu in honor of the influential German zoologist Ernst Haeckel. The greatest importance of cephalochordates to humans, however, is cultural: they occupy a prominent place as a linchpin for the theory of evolution.

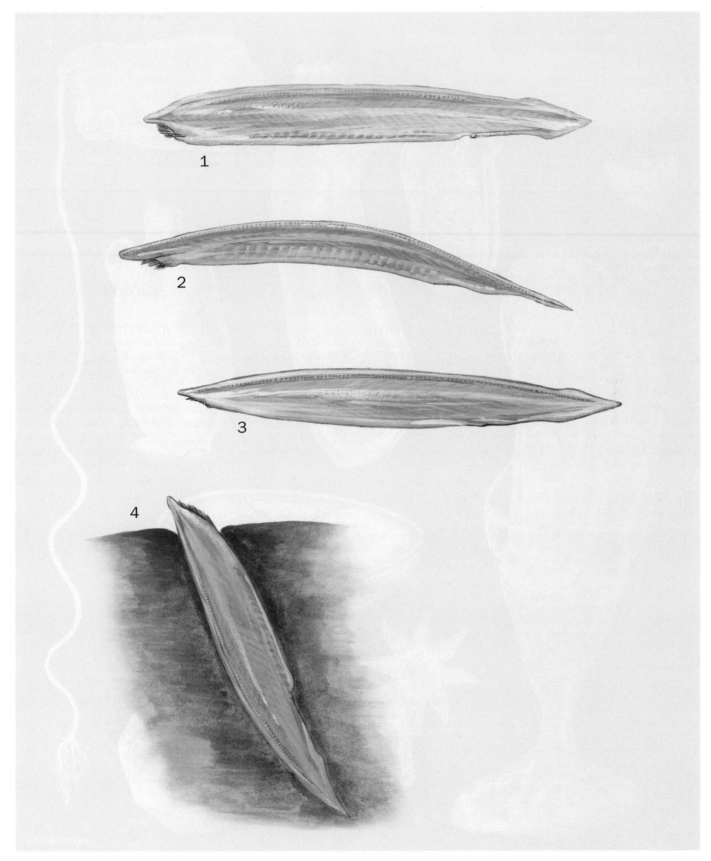

1. *Epigonichthys cultellus*; 2. Bahama lancelet (*Epigonichthys lucayanus*) 3. Smalltail lancelet (*Branchiostoma belcheri*); 4. European lancelet (*Branchiostoma lanceolatum*). (Illustration by Michelle Meneghini)

Species accounts

Smalltail lancelet
Branchiostoma belcheri

ORDER
Amphioxiformes

FAMILY
Branchiostomatidae

TAXONOMY
Branchiostoma belcheri Gray, 1847, Borneo. Possibly three sub-species: *B. belcheri belcheri, B. belcheri japonicus, B. belcheri tsingtauense.*

OTHER COMMON NAMES
(Listed names were originally coined for the European species *Branchiostoma lanceolatum*. Because of the morphological uniformity of cephalochordates they are usually used for all species.) English: Amphioxus, lancelet; French: Amphioxus, lancelet; German: Amphioxus, Lanzettfischchen; Spanish: Anfioxo, pez lanceta, lanceta.

PHYSICAL CHARACTERISTICS
Sixty-five (61–69) myotomes: 38 (34–43) preatriopore, 17 (14–21) atriopore to anus, 10 (8–12) postanal. Two hundred ninety (222–360) dorsal fin chambers. Reaches up to 2.8 in (70 mm) in total length.

DISTRIBUTION
A widespread species found from southern Japan, along the Chinese coast, the Philippines, the northern Australian coast, to Madagascar, and southeastern Africa.

HABITAT
Sandy bottoms from 3.28–82.02 ft (1–25 m) in depth.

BEHAVIOR
Seems to avoid mud.

FEEDING ECOLOGY AND DIET
Consumes microscopic plankton, mainly diatoms.

REPRODUCTIVE BIOLOGY
Reproductive season in Amoy (China) lasts from May to July. Second reproductive period seems to occur in December.

CONSERVATION STATUS
Not threatened. For cultural reasons, *Branchiostoma belcheri* is protected in the "Xiamen Rare Marine Creatures Conservation Area."

SIGNIFICANCE TO HUMANS
In some villages at the southern coast of China the majority of fishermen were fishing for *Branchiostoma belcheri*. Although the importance of this specialized and traditional fishery has declined, it is still viable in the area around Xiamen. The animals

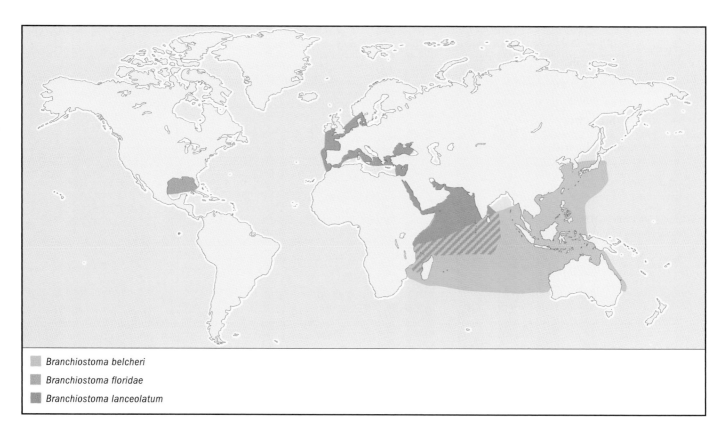

Branchiostoma belcheri
Branchiostoma floridae
Branchiostoma lanceolatum

are intended for human consumption. Some animals also are fished for academic purposes and support active research programs around the world. ◆

Florida lancelet
Branchiostoma floridae

ORDER
Amphioxiformes

FAMILY
Branchiostomatidae

TAXONOMY
Branchiostoma floridae Hubbs, 1922, Tampa Bay, Florida.

OTHER COMMON NAMES
(Listed names were originally coined for the European species *Branchiostoma lanceolatum*. Because of the morphological uniformity of cephalochordates they are usually used for all species.) English: Amphioxus, lancelet; French: Amphioxus, lancelet; German: Amphioxus, Lanzettfischchen; Spanish: Anfioxo, pez lanceta, lanceta.

PHYSICAL CHARACTERISTICS
Sixty (56–64) myotomes: 36 (33–40) preatriopore, 16 (14–17) atriopore to anus, 8 (6–11) postanal. Two hundred eighty-six (206–349) dorsal fin chambers. Reaches up to 2.3 in (59 mm) in total length.

DISTRIBUTION
Gulf of Mexico from southwestern Florida westward to Texas. Maybe also along Central American coasts from Mexico to Venezuela.

HABITAT
Sandy bottoms from 1.64–98.42 ft (0.5–30 m) in depth.

BEHAVIOR
Probably same as for family.

FEEDING ECOLOGY AND DIET
Probably same as for family.

REPRODUCTIVE BIOLOGY
Reproductive period occurs in Tampa Bay, Florida, from late spring to late summer. Individuals can spawn, ripen, and spawn again in the same breeding season.

CONSERVATION STATUS
Not threatened.

SIGNIFICANCE TO HUMANS
Branchiostoma floridae occurs in very shallow water where it is accessible without boats. A population in Tampa Bay, Florida, has recently become a major source especially for molecular studies of embryonic development. ◆

European lancelet
Branchiostoma lanceolatum

ORDER
Amphioxiformes

FAMILY
Branchiostomatidae

TAXONOMY
Branchiostoma lanceolatum Pallas, 1774, Cornwall, England.

OTHER COMMON NAMES
English: Amphioxus, lancelet; French: Amphioxus, lancelet; German: Amphioxus, Lanzettfischchen; Spanish: Anfioxo, pez lanceta, lanceta.

PHYSICAL CHARACTERISTICS
Sixty-one (59–65) myotomes: 36 (34–38) preatriopore, 14 (11–17) atriopore to anus, 11 (10–14) postanal. Two hundred twelve (183–288) dorsal fin chambers. Reaches up to 2.4 in (60 mm) in total length.

DISTRIBUTION
Most widespread species in the genus. Reported from the North Sea to the Mediterranean and the Black Sea. Also found in the Suez Canal, off the East Africa coast, Oman, India, and Sri Lanka. Specimens from the Solomon Islands also may belong to this species.

HABITAT
Sandy bottoms from 3.28–196.85 ft (1–60 m) in depth.

BEHAVIOR
Probably same as for family.

FEEDING ECOLOGY AND DIET
Semi-sessile filter feeders. Feed on plankton organisms, especially diatoms, but foraminiferans, radiolarians, and cladocerans were also found in content analyses of the gut.

REPRODUCTIVE BIOLOGY
Reproductive period starts in April in the Mediterranean, and in June/July in the North Sea. Spawning seems to occur in the evening; spermatozoa are shed before the eggs.

CONSERVATION STATUS
Not threatened.

SIGNIFICANCE TO HUMANS
Several European marine biological stations supply this species to educational and research institutions. ◆

No common name
Epigonichthys cultellus

ORDER
Amphioxiformes

FAMILY
Branchiostomatidae

TAXONOMY
Epigonichthys cultellus Peters, 1877, Moreton Bay, Australia.

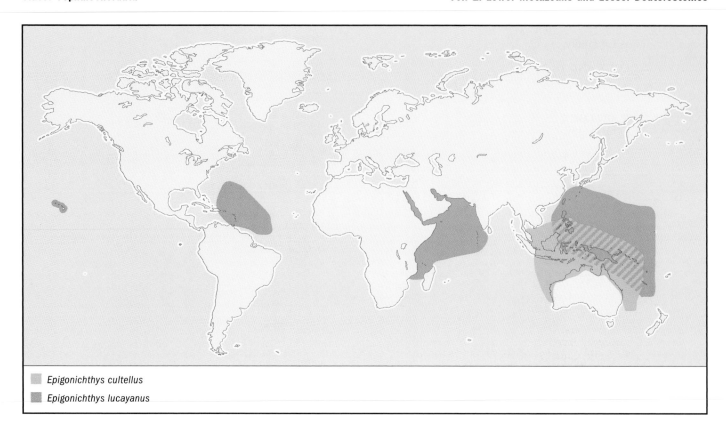

Epigonichthys cultellus

Epigonichthys lucayanus

OTHER COMMON NAMES
(Listed names were originally coined for the European species *Branchiostoma lanceolatum*. Because of the morphological uniformity of cephalochordates they are usually used for all species.)
English: Amphioxus, lancelet; French: Amphioxus, lancelet; German: Amphioxus, Lanzettfischchen; Spanish: Anfioxo, pez lanceta, lanceta.

PHYSICAL CHARACTERISTICS
Gonads on right side of body only. Fifty-one (48–52) myotomes: 31 (29–36) preatriopore, 15 (13–18) atriopore to anus, 9 (9–13) postanal. Two hundred twenty (180–254) dorsal fin chambers. Reaches up to 1.6 in (40 mm) in total length.

DISTRIBUTION
Solomon Islands, Queensland, Indonesia, Philippines, western Australia, Thailand, Sri Lanka, and Tanzania.

HABITAT
Sand and mud from about 13.12–65.6 ft (4–20 m).

BEHAVIOR
Probably same as for family.

FEEDING ECOLOGY AND DIET
Probably same as for family.

REPRODUCTIVE BIOLOGY
Probably same as for family.

CONSERVATION STATUS
Not threatened.

SIGNIFICANCE TO HUMANS
None known. ◆

Bahama lancelet
Epigonichthys lucayanus

ORDER
Amphioxiformes

FAMILY
Branchiostomatidae

TAXONOMY
Epigonichthys lucayanus Andrews, 1893, Alice Town, North Bimini, Bahamas.

OTHER COMMON NAMES
(Listed names were originally coined for the European species *Branchiostoma lanceolatum*. Because of the morphological uniformity of cephalochordates they are usually used for all species.)
English: Amphioxus, lancelet; French: Amphioxus, lancelet; German: Amphioxus, Lanzettfischchen; Spanish: Anfioxo, pez lanceta, lanceta.

PHYSICAL CHARACTERISTICS
Gonads on right side of body only. Sixty (55–62) myotomes: 37 (35–52) preatriopore, 16 (13–18) atriopore to anus, 8 (6–9) postanal. Three hundred seven (167–484) dorsal fin chambers. Reaches up to 0.83 in (21 mm) in total length.

DISTRIBUTION
Tropical and subtropical regions of Atlantic, Pacific, and Indian Oceans.

HABITAT
Coral sands to mud flats, from 0–65.61 ft (0–20 m) in depth (up to 3,001.97 ft [915 m] for larvae).

BEHAVIOR
Probably same as for family.

FEEDING ECOLOGY AND DIET
Semi-sessile filter feeder. Feeds on plankton organisms, especially diatoms and coccolithophorids.

REPRODUCTIVE BIOLOGY
Probably same as for family.

CONSERVATION STATUS
Not considered threatened.

SIGNIFICANCE TO HUMANS
None known. ◆

Resources

Books

Gans, C., N. Kemp, and S. Poss, eds. *The Lancelets (Cephalochordata): A New Look at Some Old Beasts. The Results of a Workshop.* Volume 42, *Israel Journal of Zoology.* Jerusalem: Laser Pages, 1996.

Gee, Henry. *Before the Backbone.* London, Weinheim, New York, Tokyo, Melbourne, Madras: Chapman & Hall, 1996.

Harrison, Frederick W., and Edward E. Ruppert, eds. *Microscopic Anatomy of Invertebrates. Vol. 15. Cephalochordata (Acrania).* New York, Chichester, Weinheim, Brisbane, Singapore, Toronto: Wiley-Liss, 1997.

Periodicals

Stokes, M. Dale, and Nicholas D. Holland. "The Lancelet." *American Scientist* 86, no. 6 (1998): 552–560.

Other

Xiamen Rare Marine Creatures Conservation Areas. [3 July 2003]. <http://ois.xmu.edu.cn/cbcm/english/resource/save/save04.htm>

Thomas Günther Stach, PhD

For further reading

Abramson, C. I. *Invertebrate Learning: A Laboratory Manual and Source Book*. Washington, DC: American Psychological Association, 1990.

———. *A Primer of Invertebrate Learning: The Behavioral Perspective*. Washington, DC: American Psychological Association, 1994.

Abramson, C. I., and I. S. Aquino. *A Scanning Electron Microscopy Atlas of the Africanized Honey Bee* (Apis mellifera): *Photographs for the General Public*. Campina Grande, PB, Brazil: Arte Express, 2002.

Abramson, C. I., Z. P. Shuranova, and Y. M. Burmistrov, eds. *Russian Contributions to Invertebrate Behavior*. Westport, CT: Praeger, 1996.

Adrianov A. V., and V. V. Malakhov. *Kinorhyncha: Structure, Development, Phylogeny and Classification*. Moscow: Nauka Publishing, 1994.

———. *Cephalorhyncha of the World Oceans*. Moscow: KMK Scientific Press Ltd., 1999.

Arai, Mary N. *A Functional Biology of Scyphozoa*. London: Chapman and Hall, 1997.

Barrington, E. J. W. *The Biology of Hemichordata and Protochordata*. San Francisco: W. H. Freeman and Co., 1965.

Baslow, Morris H. *Marine Pharmacology*. Baltimore: Williams and Wilkins, 1969.

Bergquist, P. R. *Sponges*.Berkeley: University of California Press, 1978.

Berrill, N. J. *The Tunicata: With an Account of the British Species*. London: The Ray Society, 1950.

Bird, Alan F. *The Structure of Nematodes*. New York: Academic Press, 1971.

Bone, Q. *Biology of Pelagic Tunicates*. New York: Oxford University Press, 1997.

Bone, Q., H. Kapp, and A. C. Pierrot-Bults, eds. *The Biology of Chaetognaths*. Oxford, New York, Tokyo: Oxford University Press, 1991.

Bouillon, J., F. Boero, F. Cicogna, and P. F. S. Cornelius, eds. *Modern Trends in the Systematics, Ecology and Evolution of Hydroids and Hydromedusae*. Oxford: Clarendon Press, 1987.

Bouillon, J., C. Gravili, F. Pages, J. M. Gili, and N. Boero. *An Introduction to Hydrozoa*. In press.

Brooks, D. R., and D. A. McLennan. *Parascript: Parasites and the Language of Evolution*. Washington, DC, and London: Smithsonian Institution Press, 1993.

Brusca, R. C., and G. J. Brusca. *Invertebrates*. New York: Sinauer Associates Inc., 1990, 2nd ed., 2003.

Cannon, L. R. G. *Turbellaria of the World—A Guide to Families & Genera*. South Brisbane, Australia: Queensland Museum. 1986.

Cheng, Thomas. *General Parasitology*, 2nd ed. Orlando, FL: Academic Press, 1986.

Chitwood, B. G., and M. B. Chitwood. *Introduction to Nematology*. Baltimore: University Park Press, 1950.

Clark, A. M., and M. E. Downey. *Starfish of the Atlantic*. London: Chapman and Hall, 1992.

Combes, C. *Parasitism: The Ecology and Evolution of Intimate Interactions*. Chicago and London: University of Chicago Press, 2001.

Conn, David Bruce. *Atlas of Invertebrate Reproduction and Development*, 2nd ed. New York: Wiley-Liss, 2000.

Conway Morris, S. *Fossil Priapulid worms*. London: Palaeontological Association, 1977; distributed by B.H. Blackwell.

Croll, Neil A., and Bernard E. Matthews. *Biology of Nematodes*. Glasgow and London: Blackie and Son Limited, 1977.

Crompton, D. W. T., and Brent B. Nickol. *Biology of the Acanthocephala*. Cambridge: Cambridge University Press, 1985.

Cunningham, W. P., and M. A. Cunningham. *Principles of Environmental Science: Inquiry and Applications*, 1st ed. New York: McGraw-Hill, 2002.

Doss, Mildred A. *Index Catalogue of Medical and Veterinary Zoology: Trematoda*, Parts 1–8. Washington, DC: U. S. Government Printing Office, 1966.

Eugene, N. K. *Marine Invertebrates of the Pacific Northwest.* Seattle: University of Washington Press, 1996.

Fabricius, Katharina, and Philip Alderslade. *Soft Corals and Sea Fans.* Townsville, Australia: Australian Institute of Marine Science, 2001.

Farrand, John, ed. *The Audubon Society Encyclopedia of Animal Life.* New York: Clarkson N. Potter, 1982.

Friese, U. Erich. *Sea Anemones: As a Hobby.* Neptune City, NJ: T. F. H. Publications, Inc., 1993.

Fusetani, Nobuhiro, ed. *Drugs from the Sea.* Basel: Karger, 2000.

Gee, Henry. *Before the Backbone.* London, Weinheim, New York, Tokyo, Melbourne, Madras: Chapman and Hall, 1996.

Gibson, Ray. *Nemerteans. Synopses of the British Fauna.* Dorchester, U. K.: The Linnean Society of London, 1994.

Gibson, Ray, Janet Moore, and Per Sundberg. *Advances in Nemertean Biology.* Dordrecht, The Netherlands: Kluwer Academic Publishers, 1993.

Gilbert, Scott F., and Anne M. Raunio, eds. *Embryology: Constructing the Organism.* Sunderland, MA: Sinauer Associates, 1997.

Hendler, G., J. E. Miller, D. L. Pawson, and P. M. Kier. *Sea Stars, Sea Urchins, and Allies: Echinoderms of Florida and the Caribbean.* Washington, DC, and London: The Smithsonian Institution Press, 1995.

Hess, Hans, William I. Ausich, Carlton E. Brett, and Michel J. Simms, eds. *Fossil Crinoids.* Cambridge: Cambridge University Press, 1999.

Hooper, J. N. A., and R. W. M. Van Soest, eds. *Systema Porifera. Guide to the Supraspecific Classification of Sponges and Spongiomorphs (Porifera).* New York: Plenum, 2002.

Hyman, L. H.*The Invertebrates: Acanthocephala, Aschelminthes, and Entoprocta. The Pseudocoelomate Bilateria,* vol. 3. New York: McGraw-Hill, 1951.

Jangoux, M., and J. M. Lawrence. *Echinoderm Nutrition.* Rotterdam, The Netherlands: A. A. Balkema, 1982.

Jefferies, R. P. S. *The Ancestry of the Vertebrates.* London: British Museum of Natural History, 1986.

Kearn, G. C. *Parasitism and the Platyhelminths.* New York: Chapman and Hall, 1998.

Kennedy, M. W., and W. Harnett, eds. *Parasitic Nematodes: Molecular Biology, Biochemistry and Immunology.* Oxon, U.K.: CABI Publishing, 2001.

Khalil, L. F., A. Jones, and R. A. Bray. *Keys to the Cestode Parasites of Vertebrates.* Wallingford: CAB International, 1994.

Koste, Walter. *Rotatoria: Die Rädertiere Mitteleuropas.* Stuttgart, Germany: Gebrüder Borntraeger, 1978.

Kozloff, E. N. *Seashore Life of the Northern Pacific Coast.* Seattle: University of Washington Press, 1993.

Lambert, P. *Sea Cucumbers of British Columbia, Southeast Alaska and Puget Sound.* Vancouver, Canada: University of British Columbia Press, 1997.

Lawrence, J. *A Functional Biology of Echinoderms.* London and Sydney: Croom Helm Press, 1987.

Lerman, M. *Marine Biology: Environment, Diversity, and Ecology.* Redwood City, CA: Cummings Publishing Company, 1986.

Lévi, C., ed. *Sponges of the New Caledonian Lagoon.* Paris: ORSTOM, 1998.

Levin, Simon Asher, ed. *Encyclopedia of Biodiversity.* San Diego, CA: Academic Press, 2001.

Littlewood, D. T. J., and R. A. Bray, eds. *Interrelationships of the Platyhelminthes.* London: Taylor and Francis, 2001.

Lutz, P. E. *Invertebrate Zoology.* Menlo Park, CA: Benjamin/Cummings Publishing Company, Inc, 1986.

Maggenti, Armand. *General Nematology.* New York: Springer-Verlag, 1981.

Malakhov, V. V., ed. by W. Duane Hope, and trans. by George V. Bentz. *Nematodes: Structure, Development, Classification, and Phylogeny.* Washington, DC, and London: Smithsonian Institution Press, 1994.

Margulis, Lynn. *Symbiosis in Cell Evolution,* 2nd ed. New York: W. H. Freeman, 1993.

Matsumoto, G. I. *Phylum Ctenophora (Orders Lobata, Cestida, Beroida, Cydippida, and Thalassocalycida): Functional Morphology, Locomotion, and Natural History.* PhD diss., University of California, Los Angeles, 1990.

McKnight, D. G. *The Marine Fauna of New Zealand: Basket Stars and Snake Stars (Echinodermata: Ophiuroidea: Euryalinida).* Wellington, Australia: NIWA (National Institute of Water and Atmospheric Research), 2000.

Mehlhorn, Heinz, ed. *Encyclopedic Reference of Parasitology: Diseases, Treatment, Therapy,* 2nd ed. New York: Springer, 2001.

Moss, David, and Graham Ackers, eds. *The UCS Sponge Guide.* Ross-on-Wye, U.K.: The Underwater Conservation Society, 1982.

Muller, Ralph. *Worms and Human Diseases.* Cambridge, MA: CABI Publishing, 2002.

Nielson, C. *Animal Evolution: Interrelationships of the Living Phyla.* Oxford: Oxford University Press, 2001.

Nybbaken, J. W. *Marine Biology: An Ecological Approach.* New York: Harper and Row, 1982.

O'Clair, R. M., and E. O. O'Clair. *Southeast Alaska's Rocky Shores, Animals.* Auke Bay, AK: Plant Press, 1998.

Olsen, O. Wilford. *Animal Parasites: Their Biology and Life Cycles.* Minneapolis, MN: Burgess Publishing Co., 1967.

Paracer, Surindar, and Vernon Ahmadjian. *An Introduction to Biological Associations,* 2nd ed. New York: Oxford University Press, 2000.

Parker, Sybil P., ed. *Synopsis and Classification of Living Organisms.* New York: McGraw-Hill Book Company, 1982.

Picton, B. E. *A Field Guide to the Shallow-water Echinoderms of the British Isles.* London: Immel Publishing Ltd, 1993.

Pietra, Francesco. *Biodiversity and Natural Product Diversity.* London: Pergamon, 2002.

Poinar, George O. Jr. *The Natural History of Nematodes.* Englewood Cliffs, NJ: Prentice-Hall, Inc., 1983.

Preston-Mafham, R., and K. Preston-Mafham. *The Encyclopedia of Land Invertebrate Behavior.* Cambridge, MA: The MIT Press, 1993.

Prudhoe, S. *A Monograph on Polyclad Turbellaria.* New York: Oxford University Press, 1985.

Quammen, D. *The Song of the Dodo: Island Biogeography in an Age of Extinctions.* New York: Scribner, 1996.

Roberts, L. S., and J. Janovy. *Gerald D. Schmidt and Larry S. Roberts' Foundations of Parasitology,* 6th ed. Boston: McGraw-Hill Co., 2000.

Romoser, W. S., and J. G. Stoffolano, Jr. *The Science of Entomology.* 3rd ed. Dubuque, IA: W. C. Brown Publishers, 1994.

Ruppert, E. E., and R. D. Barnes. *Invertebrate Zoology,* 6th ed. Forth Worth, TX: Saunders College Publishing, 1994.

Schell, Stewart C. *How to Know the Trematodes.* Dubuque, IA: William C. Brown Co., Publishers, 1970.

Scheuer, Paul J. *Chemistry of Marine Natural Products.* New York: Academic Press, 1973.

Schmidt, G. D. *CRC Handbook of Tapeworm Identification.* Boca Raton, FL: CRC Press, Inc., 1986.

Schmidt-Rhaesa, Andreas. *Süsswasserfauna von Mitteleuropa: Nematomorpha.* Stuttgart, Germany: Gustav Fisher Verlag, 1997.

Shick, J. Malcolm. *A Functional Biology of Sea Anemones.* New York: Chapman and Hall, 1991.

Strathmann, Megumi F., ed. *Reproduction and Development of Marine Invertebrates of the Northern Pacific Coast.* Seattle: University of Washington Press, 1987.

Veron, John E. N. *Corals of the World.* Townsville, Australia: Australian Institute of Marine Science, 2000.

Wallace, Alfred Russell. *The Geographical Distribution of Animals, with a Study of the Relations of Living and Extinct Faunas as Elucidating Past Changes of the Earth's Surface.* New York: Harper, 1876.

Wallace, Carden C. *Staghorn Corals of the World.* Collingwood, Australia: CSIRO, 1999.

Wharton, David A. *A Functional Biology of Nematodes.* London: Croom Helm, 1986.

Wilson, E. O. *The Diversity of Life.* New York: Harvard University Press, 1992.

Wilson, W. H., Stephen A. Sticker, and George L. Shinn, eds. *Reproduction and Development of Marine Invertebrates.* Baltimore: Johns Hopkins University Press, 1994.

Wrobel, D., and C. Mills. *Pacific Coast Pelagic Invertebrates.* Monterey, CA: Sea Challengers and Monterey Bay Aquarium, 1999.

Yamaji, I. *Illustrations of the Marine Plankton of Japan.* Osaka: Hoikusha, 1976.

Young, Craig M., M. A. Sewell, and Mary E. Rice, eds. *Atlas of Marine Invertebrate Larvae.* San Diego, CA: Academic Press, 2002.

Zaitsev, Yu, and V. Mamaev. *Marine Biological Diversity in the Black Sea: A Study of Change and Decline.* New York: United Nations Publications, 1997.

Zelinka, K. *Monographie der Echinodera.* Leipzig: Engelman, 1928.

• • • • •

Organizations

Agricultural Research Service
<http://www.ars.usda.gov/is/AR/archive/may01/
worms0501.htm>

American Society of Parasitologists
<http://asp.unl.edu>

American Zoo and Aquarium Association
8403 Colesville Road, Suite 710
Silver Spring, MD 20910 USA
<http://www.aza.org>

Australian Regional Association of Zoological Parks and Aquaria
PO Box 20
Mosman, NSW 2088
Australia
Phone: 61 (2) 9978-4797
Fax: 61 (2) 9978-4761
<http://www.arazpa.org>

British and Irish Graptolite Group
c/o Dr. A. W. A. Rushton
The Natural History Museum, Cromwell Rd.
London SW7 5BD United Kingdom
<http://www.graptolites.co.uk/>

European Association of Zoos and Aquaria
PO Box 20164
1000 HD Amsterdam
The Netherlands
<http://www.eaza.net>

Helminthological Society of Washington
Allen Richards, Ricksettsial Disease Department
Naval Medical Research Center
503 Robert Grant Ave.
Silver Spring, MD 20910-7500 USA

Hydrozoan Society: Dedicated to the Study of Hydrozoan Biology
<http://www.ucmp.berkeley.edu/agc/HS/>

International Society of Endocytobiology
<http://www.endocytobiology.org/>

International Symbiosis Society
<http://www.ma.psu.edu/lkh1/iss/>

Monterey Bay Aquarium Research Institute
7700 Sandholdt Road
Moss Landing, CA 95039 USA
Phone: (831) 775-1700
Fax: (831) 775-1620
E-mail: mage@mbari.org
<http://www.mbari.org>

Species Survival Commission, IUCN—The World Conservation Union
Rue Mauverney 28
Gland CH-1196 Switzerland
Phone: +41-22-999-0152
Fax: +41-22-999-0015
E-mail: ssc@iucn.org
<http://www.iucn.org>

The Graptolite Working Group of the International Palaeontological Association
c/o Dr. Charles E. Mitchell, Department of Geology
State University of New York at Buffalo
Buffalo, NY 14260-3050 USA
E-mail: cem@acsu.buffalo.edu
<http://www.geology.buffalo.edu/gwg/index.htm>

Xiamen Rare Marine Creatures Conservation Areas
<http://ois.xmu.edu.cn/cbcm/english/resource/save/save04
.htm>

Dr. Fritz Dieterlen
Zoological Research Institute,
A. Koenig Museum
Bonn, Germany

Dr. Rolf Dircksen
Professor, Pedagogical Institute
Bielefeld, Germany

Josef Donner
Instructor of Biology
Katzelsdorf, Austria

Dr. Jean Dorst
Professor, National Museum of
Natural History
Paris, France

Dr. Gerti Dücker
Professor and Chief Curator,
Zoological Institute, University
of Münster
Münster, Germany

Dr. Michael Dzwillo
Zoological Institute and Museum,
University of Hamburg
Hamburg, Germany

Dr. Irenäus Eibl-Eibesfeldt
Professor and Director, Institute of
Human Ethology, Max Planck
Institute for Behavioral Physiology
Percha/Starnberg, Germany

Dr. Martin Eisentraut
Professor and Director,
Zoological Research Institute and
A. Koenig Museum
Bonn, Germany

Dr. Eberhard Ernst
Swiss Tropical Institute
Basel, Switzerland

R. D. Etchecopar
Director, National Museum of
Natural History
Paris, France

Dr. R. A. Falla
Director, Dominion Museum
Wellington, New Zealand

Dr. Hubert Fechter
Curator, Lower Animals, Zoological
Collection of the State of Bavaria
Munich, Germany

Dr. Walter Fiedler
Docent, University of Vienna, and
Director, Schönbrunn Zoo
Vienna, Austria

Wolfgang Fischer
Inspector of Animals, Animal Park
Berlin, Germany

Dr. C. A. Fleming
Geological Survey Department of
Scientific and Industrial Research
Lower Hutt, New Zealand

Dr. Hans Frädrich
Zoological Garden
Berlin, Germany

Dr. Hans-Albrecht Freye
Professor and Director, Biological
Institute of the Medical School
Halle a.d.S., Germany

Günther E. Freytag
Former Director, Reptile and
Amphibian Collection, Museum of
Cultural History in Magdeburg
Berlin, Germany

Dr. Herbert Friedmann
Director, Los Angeles County
Museum of Natural History
Los Angeles, California, U.S.A.

Dr. H. Friedrich
Professor, Overseas Museum
Bremen, Germany

Dr. Jan Frijlink
Zoological Laboratory, University
of Amsterdam
Amsterdam, The Netherlands

Dr. H.C. Karl Von Frisch
Professor Emeritus and former
Director, Zoological Institute,
University of Munich
Munich, Germany

Dr. H. J. Frith
C.S.I.R.O. Research Institute
Canberra, Australia

Dr. Ion E. Fuhn
Academy of the Roumanian Socialist
Republic, Trajan Savulescu Institute
of Biology
Bucharest, Romania

Dr. Carl Gans
Professor, Department of Biology,
State University of New York
at Buffalo
Buffalo, New York, U.S.A.

Dr. Rudolf Geigy
Professor and Director,
Swiss Tropical Institute
Basel, Switzerland

Dr. Jacques Gery
St. Genies, France

Dr. Wolfgang Gewalt
Director, Animal Park
Duisburg, Germany

Dr. H.C. Viktor Goerttler
Professor Emeritus, University of Jena
Jena, Germany

Dr. Friedrich Goethe
Director, Institute of Ornithology,
Heligoland Ornithological Station
Wilhelmshaven, Germany

Dr. Ulrich F. Gruber
Herpetological Section,
Zoological Research Institute and
A. Koenig Museum
Bonn, Germany

Dr. H. R. Haefelfinger
Museum of Natural History
Basel, Switzerland

Dr. Theodor Haltenorth
Director, Mammalology, Zoological
Collection of the State of Bavaria
Munich, Germany

Barbara Harrisson
Sarawak Museum, Kuching, Borneo
Ithaca, New York, U.S.A.

Dr. Francois Haverschmidt
President, High Court (retired)
Paramaribo, Suriname

Dr. Heinz Heck
Director, Catskill Game Farm
Catskill, New York, U.S.A.

Dr. Lutz Heck
Professor (retired), and Director,
Zoological Garden, Berlin
Wiesbaden, Germany

Dr. H.C. Heini Hediger
Director, Zoological Garder
Zurich, Switzerland

Dr. Dietrich Heinemann
Director, Zoological Garden, Münster
Dörnigheim, Germany

Dr. Helmut Hemmer
Institute for Physiological Zoology,
University of Mainz
Mainz, Germany

Dr. W. G. Heptner
Professor, Zoological Museum,
University of Moscow
Moscow, Russia

Dr. Konrad Herter
Professor Emeritus and Director
(retired), Zoological Institute,
Free University of Berlin
Berlin, Germany

Dr. Hans Rudolf Heusser
Zoological Museum,
University of Zurich
Zurich, Switzerland

Dr. Emil Otto Höhn
Associate Professor of Physiology,
University of Alberta
Edmonton, Canada

Dr. W. Hohorst
Professor and Director,
Parasitological Institute, Farbwerke
Hoechst A.G.
Frankfurt-Höchst, Germany

Dr. Folkhart Hückinghaus
Director, Senckenbergische Anatomy,
University of Frankfurt a.M.
Frankfurt a.M., Germany

Francois Hüe
National Museum of Natural History
Paris, France

Dr. K. Immelmann
Professor, Zoological Institute,
Technical University
of Braunschweig
Braunschweig, Germany

Dr. Junichiro Itani
Kyoto University
Kyoto, Japan

Dr. Richard F. Johnston
Professor of Zoology,
University of Kansas
Lawrence, Kansas, U.S.A.

Otto Jost
Oberstudienrat,
Freiherr-vom-Stein Gymnasium
Fulda, Germany

Dr. Paul Kähsbauer
Curator, Fishes, Museum of
Natural History
Vienna, Austria

Dr. Ludwig Karbe
Zoological State Institute
and Museum
Hamburg, Germany

Dr. N. N. Kartaschew
Docent, Department of Biology,
Lomonossow State University
Moscow, Russia

Dr. Werner Kästle
Oberstudienrat, Gisela Gymnasium
Munich, Germany

Dr. Reinhard Kaufmann
Field Station of the Tropical Institute,
Justus Liebig University,
Giessen, Germany
Santa Marta, Colombia

Dr. Masao Kawai
Primate Research Institute,
Kyoto University
Kyoto, Japan

Dr. Ernst F. Kilian
Professor, Giessen University and
Catedratico Universidad Austral,
Valdivia-Chile
Giessen, Germany

Dr. Ragnar Kinzelbach
Institute for General Zoology,
University of Mainz
Mainz, Germany

Dr. Heinrich Kirchner
Landwirtschaftsrat (retired)
Bad Oldesloe, Germany

Dr. Rosl Kirchshofer
Zoological Garden,
University of Frankfort a.M.
Frankfurt a.M., Germany

Dr. Wolfgang Klausewitz
Curator, Senckenberg Nature
Museum and Research Institute
Frankfurt a.M., Germany

Dr. Konrad Klemmer
Curator, Senckenberg Nature
Museum and Research Institute
Frankfurt a.M., Germany

Dr. Erich Klinghammer
Laboratory of Ethology,
Purdue University
Lafayette, Indiana, U.S.A.

Dr. Heinz-Georg Klös
Professor and Director,
Zoological Garden
Berlin, Germany

Ursula Klös
Zoological Garden
Berlin, Germany

Dr. Otto Koehler
Professor Emeritus, Zoological
Institute, University of Freiburg
Freiburg i. BR., Germany

Dr. Kurt Kolar
Institute of Ethology, Austrian
Academy of Sciences
Vienna, Austria

Dr. Claus König
State Ornithological Station of
Baden-Württemberg
Ludwigsburg, Germany

Dr. Adriaan Kortlandt
Zoological Laboratory,
University of Amsterdam
Amsterdam, The Netherlands

Dr. Helmut Kraft
Professor and Scientific Councillor,
Medical Animal Clinic,
University of Munich
Munich, Germany

Dr. Helmut Kramer
Zoological Research Institute and
A. Koenig Museum
Bonn, Germany

Dr. Franz Krapp
Zoological Institute,
University of Freiburg
Freiburg, Switzerland

Dr. Otto Kraus
Professor, University of Hamburg,
and Director, Zoological Institute
and Museum
Hamburg, Germany

Dr. Hans Krieg
Professor and First Director (retired),
Scientific Collections of the State
of Bavaria
Munich, Germany

Dr. Heinrich Kühl
Federal Research Institute for
Fisheries, Cuxhaven Laboratory
Cuxhaven, Germany

Dr. Oskar Kuhn
Professor, formerly University
Halle/Saale
Munich, Germany

Dr. Hans Kumerloeve
First Director (retired), State
Scientific Museum, Vienna
Munich, Germany

Dr. Nagamichi Kuroda
Yamashina Ornithological Institute,
Shibuya-Ku
Tokyo, Japan

Dr. Fred Kurt
Zoological Museum of
Zurich University,
Smithsonian Elephant Survey
Colombo, Ceylon

Dr. Werner Ladiges
Professor and Chief Curator,
Zoological Institute and Museum,
University of Hamburg
Hamburg, Germany

Leslie Laidlaw
Department of Animal Sciences,
Purdue University
Lafayette, Indiana, U.S.A.

Dr. Ernst M. Lang
Director, Zoological Garden
Basel, Switzerland

Dr. Alfredo Langguth
Department of Zoology,
Faculty of Humanities and Sciences,
University of the Republic
Montevideo, Uruguay

Leo Lehtonen
Science Writer
Helsinki, Finland

Bernd Leisler
Second Zoological Institute,
University of Vienna
Vienna, Austria

Dr. Kurt Lillelund
Professor and Director, Institute for
Hydrobiology and Fishery Sciences,
University of Hamburg
Hamburg, Germany

R. Liversidge
Alexander MacGregor Memorial
Museum
Kimberley, South Africa

Dr. Konrad Lorenz
Professor and Director, Max Planck
Institute for Behavioral Physiology
Seewiesen/Obb., Germany

Dr. Martin Lühmann
Federal Research Institute for the
Breeding of Small Animals
Celle, Germany

Dr. Johannes Lüttschwager
Oberstudienrat (retired)
Heidelberg, Germany

Dr. Wolfgang Makatsch
Bautzen, Germany

Dr. Hubert Markl
Professor and Director,
Zoological Institute, Technical
University of Darmstadt
Darmstadt, Germany

Basil J. Marlow, B.SC. (Hons)
Curator, Australian Museum
Sydney, Australia

Dr. Theodor Mebs
Instructor of Biology
Weissenhaus/Ostsee, Germany

Dr. Gerlof Fokko Mees
Curator of Birds,
Rijks Museum of Natural History
Leiden, The Netherlands

Hermann Meinken
Director,
Fish Identification Institute, V.D.A.
Bremen, Germany

Dr. Wilhelm Meise
Chief Curator, Zoological Institute
and Museum, University of Hamburg
Hamburg, Germany

Dr. Joachim Messtorff
Field Station of the Federal Fisheries
Research Institute
Bremerhaven, Germany

Dr. Marian Mlynarski
Professor, Polish Academy of
Sciences, Institute for Systematic and
Experimental Zoology
Cracow, Poland

Dr. Walburga Moeller
Nature Museum
Hamburg, Germany

Dr. H.C. Erna Mohr
Curator (retired), Zoological State
Institute and Museum
Hamburg, Germany

Dr. Karl-Heinz Moll
Waren/Müritz, Germany

Dr. Detlev Müller-Using
Professor, Institute for Game
Management, University of Göttingen
Hannoversch-Münden, Germany

Werner Münster
Instructor of Biology
Ebersbach, Germany

Dr. Joachim Münzing
Altona Museum
Hamburg, Germany
Dr. Wilbert Neugebauer
Wilhelma Zoo
Stuttgart-Bad Cannstatt, Germany

Dr. Ian Newton
Senior Scientific Officer,
The Nature Conservancy
Edinburgh, Scotland

Dr. Jürgen Nicolai
Max Planck Institute for
Behavioral Physiology
Seewiesen/Obb., Germany

Dr. Günther Niethammer
Professor, Zoological Research
Institute and A. Koenig Museum
Bonn, Germany

Dr. Bernhard Nievergelt
Zoological Museum,
University of Zurich
Zurich, Switzerland

Dr. C. C. Olrog
Institut Miguel Lillo San Miguel
de Tucuman
Tucuman, Argentina

Alwin Pedersen
Mammal Research and aRctic Explorer
Holte, Denmark

Dr. Dieter Stefan Peters
Nature Museum and Senckenberg
Research Institute
Frankfurt a.M., Germany

Dr. Nicolaus Peters
Scientific Councillor and Docent,
Institute of Hydrobiology and
Fisheries, University of Hamburg
Hamburg, Germany

Dr. Hans-Günter Petzold
Assistant Director, Zoological Garden
Berlin, Germany

Dr. Rudolf Piechocki
Docent, Zoological Institute,
University of Halle
Halle a.d.S., Germany

Dr. Ivo Poglayen-Neuwall
Director, Zoological Garden
Louisville, Kentucky, U.S.A.

Dr. Egon Popp
Zoological Collection of the State
of Bavaria
Munich, Germany

Dr. H.C. Adolf Portmann
Professor Emeritus, Zoological
Institute, University of Basel
Basel, Switzerland

Hans Psenner
Professor and Director, Alpine Zoo
Innsbruck, Austria

Dr. Heinz-Siburd Raethel
Oberveterinärrat
Berlin, Germany

Dr. Urs H. Rahm
Professor, Museum of Natural History
Basel, Switzerland

Dr. Werner Rathmayer
Biology Institute,
University of Konstanz
Konstanz, Germany

Walter Reinhard
Biologist
Baden-Baden, Germany

Dr. H. H. Reinsch
Federal Fisheries Research Institute
Bremerhaven, Germany

Dr. Bernhard Rensch
Professor Emeritus, Zoological
Institute, University of Münster
Münster, Germany

Dr. Vernon Reynolds
Docent, Department of Sociology,
University of Bristol
Bristol, England

Dr. Rupert Riedl
Professor, Department of Zoology,
University of North Carolina
Chapel Hill, North Carolina, U.S.A.

Dr. Peter Rietschel
Professor (retired), Zoological
Institute, University of Frankfurt a.M.
Frankfurt a.M., Germany

Dr. Siegfried Rietschel
Docent, University of Frankfurt;
Curator, Nature Museum and
Research Institute Senckenberg
Frankfurt a.M., Germany

Herbert Ringleben
Institute of Ornithology, Heligoland
Ornithological Station
Wilhelmshaven, Germany

Dr. K. Rohde
Institute for General Zoology,
Ruhr University
Bochum, Germany

Dr. Peter Röben
Academic Councillor, Zoological
Institute, Heidelberg University
Heidelberg, Germany

Dr. Anton E. M. De Roo
Royal Museum of Central Africa
Tervuren, South Africa

Dr. Hubert Saint Girons
Research Director, Center for
National Scientific Research
Brunoy (Essonne), France

Dr. Luitfried Von Salvini-Plawen
First Zoological Institute,
University of Vienna
Vienna, Austria

Dr. Kurt Sanft
Oberstudienrat, Diesterweg-Gymnasium
Berlin, Germany

Dr. E. G. Franz Sauer
Professor, Zoological Research
Institute and A. Koenig Museum,
University of Bonn
Bonn, Germany

Dr. Eleonore M. Sauer
Zoological Research Institute and
A. Koenig Museum, University of Bonn
Bonn, Germany

Dr. Ernst Schäfer
Curator, State Museum of
Lower Saxony
Hannover, Germany

Dr. Friedrich Schaller
Professor and Chairman,
First Zoological Institute,
University of Vienna
Vienna, Austria

Dr. George B. Schaller
Serengeti Research Institute,
Michael Grzimek Laboratory
Seronera, Tanzania

Dr. Georg Scheer
Chief Curator and Director,
Zoological Institute,
State Museum of Hesse
Darmstadt, Germany

Dr. Christoph Scherpner
Zoological Garden
Frankfurt a.M., Germany

Dr. Herbert Schifter
Bird Collection,
Museum of Natural History
Vienna, Austria

Dr. Marco Schnitter
Zoological Museum,
Zurich University
Zurich, Switzerland

Dr. Kurt Schubert
Federal Fisheries Research Institute
Hamburg, Germany

Eugen Schuhmacher
Director, Animals Films, I.U.C.N.
Munich, Germany

Dr. Thomas Schultze-Westrum
Zoological Institute,
University of Munich
Munich, Germany

Dr. Ernst Schüt
Professor and Director (retired),
State Museum of Natural History
Stuttgart, Germany

Dr. Lester L. Short Jr.
Associate Curator, American Museum
of Natural History
New York, New York, U.S.A.

Dr. Helmut Sick
National Museum
Rio de Janeiro, Brazil

Dr. Alexander F. Skutch
Professor of Ornithology,
University of Costa Rica
San Isidro del General, Costa Rica

Dr. Everhard J. Slijper
Professor, Zoological Laboratory,
University of Amsterdam
Amsterdam, The Netherlands

Bertram E. Smythies
Curator (retired), Division of Forestry
Management, Sarawak-Malaysia
Estepona, Spain

Dr. Kenneth E. Stager
Chief Curator, Los Angeles County
Museum of Natural History
Los Angeles, California, U.S.A.

Dr. H.C. Georg H.W. Stein
Professor, Curator of Mammals,
Institute of Zoology and Zoological
Museum, Humboldt University
Berlin, Germany

Dr. Joachim Steinbacher
Curator, Nature Museum and
Senckenberg Research Institute
Frankfurt a.M., Germany

Dr. Bernard Stonehouse
Canterbury University
Christchurch, New Zealand

Dr. Richard Zur Strassen
Curator, Nature Museum and
Senckenberg Research Institute
Frankfurt a.M., Germany

Dr. Adelheid Studer-Thiersch
Zoological Garden
Basel, Switzerland

Dr. Ernst Sutter
Museum of Natural History
Basel, Switzerland

Dr. Fritz Terofal
Director, Fish Collection, Zoological
Collection of the State of Bavaria
Munich, Germany

Dr. G. F. Van Tets
Wildlife Research
Canberra, Australia

Ellen Thaler-Kottek
Institute of Zoology, University
of Innsbruck
Innsbruck, Austria

Dr. Erich Thenius
Professor and Director, Institute of
Paleontolgy, University of Vienna
Vienna, Austria

Dr. Niko Tinbergen
Professor of Animal Behavior,
Department of Zoology,
Oxford University
Oxford, England

Alexander Tsurikov
Lecturer, University of Munich
Munich, Germany

Dr. Wolfgang Villwock
Zoological Institute and Museum,
University of Hamburg
Hamburg, Germany

Zdenek Vogel
Director,
Suchdol Herpetological Station
Prague, Czechoslovakia

Dieter Vogt
Schorndorf, Germany

Dr. Jiri Volf
Zoological Garden
Prague, Czechoslovakia
Otto Wadewitz
Leipzig, Germany

Dr. Helmut O. Wagner
Director (retired),
Overseas Museum, Bremen
Mexico City, Mexico

Dr. Fritz Walther
Professor, Texas A & M University
College Station, Texas, U.S.A.

John Warham
Zoology Department,
Canterbury University
Christchurch, New Zealand

Dr. Sherwood L. Washburn
University of California at Berkeley
Berkeley, California, U.S.A.

Eberhard Wawra
First Zoological Institute,
University of Vienna
Vienna, Austria

Dr. Ingrid Weigel
Zoological Collection of the State
of Bavaria
Munich, Germany

Dr. B. Weischer
Institute of Nematode Research,
Federal Biological Institute
Münster/Westfalen, Germany

Herbert Wendt
Author, Natural History
Baden-Baden, Germany

Dr. Heinz Wermuth
Chief Curator,
State Nature Museum, Stuttgart
Ludwigsburg, Germany

Dr. Wolfgang Von Westernhagen
Preetz/Holstein, Germany

Dr. Alexander Wetmore
United States National Museum,
Smithsonian Institution
Washington, D.C., U.S.A.

Dr. Dietrich E. Wilcke
Röttgen, Germany

Dr. Helmut Wilkens
Professor and Director,
Institute of Anatomy,
School of Veterinary Medicine
Hannover, Germany

Dr. Michael L. Wolfe
Utah, U.S.A.

Hans Edmund Wolters
Zoological Research Institute and
A. Koenig Museum
Bonn, Germany

Dr. Arnfrid Wünschmann
Research Associate, Zoological Garden
Berlin, Germany

Dr. Walter Wüst
Instructor, Wilhelms Gymnasium
Munich, Germany

Dr. Heinz Wundt
Zoological Collection of the State
of Bavaria
Munich, Germany

Dr. Claus-Dieter Zander
Zoological Institute and Museum,
University of Hamburg
Hamburg, Germany

Dr. Fritz Zumpt
Director, Entomology and
Parasitology, South African Institute
for Medical Research
Johannesburg, South Africa

Dr. Richard L. Zusi
Curator of Birds,
United States National Museum,
Smithsonian Institution
Washington, D.C., U.S.A.

Glossary

4d cell—Mesentoblast; a blastomere cell that results from zygotes that have spiral cleavage divisions, and contains an unidentified cytoplasmic factor that causes the cell and its progeny to form mesoderm.

Abdomen—The posterior of the main body divisions.

Abyssal—Of, or relating to, the deepest regions of the ocean.

Acanthor—First larval stage of acanthocephalans.

Aciculum—Small needlelike structure resembling a rod that supports the divisions of the parapodium.

Acoelomate—An organism, particularly an invertebrate, lacking a coelom that is characterized by bilateral symmetry.

Actinotrocha—Tentacle-like ciliated larva of phoronids.

Aestivation—A period of dormancy that is entered into when conditions are not favorable, particularly during very warm or very dry seasons.

Aflagellate—An organism that lacks a flagella.

Agamete—Nucleus within the plasmodium that divides mitotically and gives rise to a sexual adult.

Ametabolous—Development in which little or no external metamorphic changes are noticeable in the larval to adult transition.

Amphiblastula—Free-swimming larval stage of sponges.

Anal—Relating to or being close to the anus.

Anamorphic—Development in which only part of the adult segments are present in recently hatched young.

Ancestrula—Zooid that develops from an egg.

Antibiosis—A provocative association between organisms that is detrimental, inhibitive, and preventative to one or more of them but produces a metabolic product in another.

Aphotic zone—Region of the ocean where no sun light reaches and exists in complete darkness.

Apical field—An area inside the circumapical band of rotifers that is devoid of cilia.

Arboreal—An organism that lives in, on, or among trees.

Ascidiologists—Scientists who study the Ascidiacea.

Auricularia—Primary larval stage in holothuroid development.

Autapomorphy—A derived trait unique to a taxonomic group.

Benthic—An organism that lives on the bottom of the ocean floor.

Bipinnaria—Free-swimming larval stage of asteroids.

Bivoltine—The production of two broods or generations in a season or year.

Blastomere—Zygote cleavage divisions that result in a cell.

Blastopore—The first opening of the early digestive tract.

Blastula—Sphere of blastomeres.

Brachiolaria—Second stage of asteroid larva.

Brood—When the care of eggs takes place outside or inside of the mother's body for at least the early part of development.

Buccal cavity—A cavity that is present within the mouth.

Bud—The development of new progeny cells or new outgrowth.

Caudal—Referring or pertaining to the posterior end of the body.

Cephalic—Referring or pertaining to the anterior end of the body.

Cephalothorax—The body region that consists of the head and thoracic segments.

Chelicera—Pair of appendages present in the anterior body of arachnids.

Chorion—The shell or covering of an egg.

Cilia—Outgrowth present on the cell surface that is short and produces a lashing movement capable of creating locomotion.

Cleavage—The process of cellular divisions in a fertilized egg that changes it from a single-celled zygote into a multicellular embryo.

Cloaca—Chamber into which the intestinal and urogenital tracts discharge.

Cnidocytes—Prey-capture and defensive cells unique to cnidarians.

Coelom—The epithelium-lined space between the body wall and the digestive tract.

Colony—Body composed of zooids that share resources.

Commensalism—Symbiotic relationship between two or more species in which no group is injured, and at least one group benefits.

Commercial fishery—The industry of catching a certain species for sale.

Communal—Cooperation between females of one species in production and building, but not in caring for the brood.

Conspecific—Belonging to the same species.

Coracidium—Ciliated free-swimming stage of cestode.

Cosmopolitan—Occurring throughout most of the world.

Cuticle—The noncellular outer layers of the body.

Cydippid—Free-swimming ctenophore larva.

Cyphonarutes—Planktonic larva of some nonbrooding gymnolaemate bryozoans.

Definitive host—See Primary host.

Demersal—Aquatic animals that live near, are deposited on, or sink to the bottom of the sea.

Dentate—Having teeth, or structures that function as teeth.

Denticles—Teeth, or structures that function as, or are derived from, teeth.

Deposit feeders—Animals that feed upon matter that has settled on the substrate.

Desmosomes—Structures involved in cellular adhesion.

Detritus—Fragments of plant, animal, or waste remnants.

Deuterostome—Division of the animal kingdom that includes animals that are bilaterally symmetrical, have indeterminate cleavage and a mouth that does not arise from the blastopore.

Diapause—A period of time in which development is suspended or arrested and the body is dormant.

Dioecious—Organisms that have male reproductive organs in one individual and female in another.

Diverticulum—A pouch or sac on a hollow structure or organ.

Doliolaria—Barrel-shaped larval stage.

Ecdysis—Molting or shedding of the exoskeleton.

Ectoparasite—A parasite that lives on the outside of a host.

Endemic—Belonging to or from a particular geographical region.

Endocuticle—The innermost layer of the cuticle.

Endoparasite—A parasite that lives inside the body of its host.

Endosymbiont—Symbiotic relationship in which a symbiont dwells within the body of its symbiotic partner.

Epicuticle—The surface layers of the cuticle.

Epiphragm—Temporary mucus door over the aperture (opening) that hardens to seal the snail inside.

Epithelium—Membranous tissue that covers the body or lines a body cavity or tube.

Epizoic—An animal or plant that lives on another animal or plant.

Estuary—A semi-enclosed body of water that is diluted by freshwater input and has an open connection to the sea. Typically, there is a mixing of sea and fresh water, and the influx of nutrients from both sources results in high productivity.

Eukaryote—A cell with a nucleus that contains DNA; or an organism made up of such cells.

Eurybathic—An animal that occurs in a wide range of depths.

Euryhaline—An animal that occurs in a wide variety of salinities.

Eurythermic—An animal that occurs in a wide range of temperatures.

Eversible—Capable of being turned inside out.

Exocuticle—Hard and darkened layer of the cuticle lying between the endocuticle and epicuticle.

Exoskeleton—The external plates of the body wall.

Fibrillae—Small filaments, hairs, or fibers.

Fishery—The industry of catching fish, crustaceans, mollusks or other aquatic animals for commercial, recreational, subsistence or aesthetic purposes.

Fusiform—Having a shape that tapers toward each end.

Gametogenesis—Production of gametes (sex cells).

Ganglion—A nerve tissue mass containing cell bodies of neurons external to the brain or spinal cord.

Girdle—Outer mantle of the polyplacophoran that is thick and stiff, extending out from the shell plate.

Glycocalyx—Protein and carbohydrate surface coat in cells.

Gonochoric—An animal with separate sexes.

Gonopore—Reproductive aperture or pore present in the genital area.

Gonozooid—A reproductive zooid of a hydroid.

Gynandromorph—An individual that exhibits both male and female characteristics.

Hematophagous—A group that feeds or subsides on blood.

Hemitransparent—Half or partially transparent.

Hermaphrodite—An organism that has both male and female sexual organs.

Heterothermic springs—Springs (of water) that may freeze in the winter.

Heterotroph—An organism that is unable to produce its own food, but must obtain its nutrition by feeding on other organisms.

Higgins larva—Loriciferan larval stage.

Holoplankton—An animal that lives in plankton all of its life.

Homothermic springs—Those with a constant temperature throughout the year.

Host—The organism in or on which a parasite lives.

Hyaline—Transparent, clear, and colorless.

Hydromedusa—Medusa of the hydrozoans.

Hyperparasite—A parasitic organism whose host is another parasite.

Infauna—An animal that lives among sediment.

Inquiline—Animal that lives in the nests or abode of another species.

Integument—A layer of skin, membrane, or cuticle that envelops an organism or one of its parts.

Intermediate host—Host for the larval stage of a parasitic organism.

Intracytoplasmic—Located within or taking place within a cell's cytoplasm.

Kinesis—A movement that lacks directional orientation and depends upon the intensity of stimulation.

Lamina—Thin, parallel plates of soft vascular sensitive tissue.

Larva—An immature development stage.

Larviparous—Eggs brooded within the female that are later released as larvae.

Lecithotrophic—Larvae that do not feed, but rather derive nutrition from yolk.

Lorica—Specialized girdle-like structure made of a set of hardened parts that protect the body, named for the segmented corselet of armor worn by Roman soldiers.

Lumen—Cavity of a tubular organ.

Mandible—The jaw.

Manubria—Tube that bears the mouth and hangs down from the subumbrella or medusae.

Maxilla—One of two components of the mouth immediately behind the mandibles.

Medusae—Well-developed cnidarian that is gelatinous and free-swimming.

Meiosis—Cellular process that results in the number of chromosomes in gamete-producing cells (usually sex cells) being reduced to one half.

Mesoderm—Tissue derived from the three primary embryonic germ layers, and the source of many bodily tissues and structures.

Metachronous—Using coordinated waves, as in bands of cilia beating metachronously.

Metamorphosis—A change in physical form or substance.

Miracidium—Free-swimming first larva of trematodes that is ciliated.

Mitosis—A process that takes place in the nucleus of a dividing cell that results in the formation of two new nuclei having the same number of chromosomes as the parent nucleus.

Moult—The shedding of the exoskeleton.

Mutualism—Symbiotic relationship in which both members of the relationship benefit.

Myoepithelial—Cells of the epithelium.

Nematocyst—Stingers of cnidarians.

Neoblasts—Undifferentiated cells that form the blastema, which precedes regeneration in planarians.

Neritic—An organism that inhabits the region of shallow water adjoining the seacoast.

Nocturnal—An organism that is active mostly at night.

Obligate ectoparasites—External parasites that cannot complete their cycle when removed from their host.

Oocyte—The egg before it has reached maturation.

Ootheca—The cover or case that surrounds a mass of eggs.

Oozoid—The zooid that develops from the fertilized egg of urochordates.

Ophiopleuteus—Planktonic larva of echinoids (urchins).

Oral lamella—Oral membrane or layer.

Oviparous—An organism that lays eggs.

Ovipositor—The apparatus through which the female lays eggs.

Ovoviviparous—An organism that produces young that hatch out of their egg while still within their mother.

Paedomorphosis—Retention of juvenile characteristics by adults.

Paraphyletic—A group that contains some of the descendants of a common ancestor.

Parapodium—Appendage present on annelids that resembles a paddle.

Parasite—An organism that lives in or on the body of another living organism, feeding off of its host.

Parenchymula—Larval sponge.

Parthenogenetic—Development of an egg without fertilization.

Pelagic—Organisms that live in the open sea, above the ocean floor.

Pelagosphera—Second planktotrophic larva of sipunculans.

Pentaradial symmetry—Five-part radial symmetry, as seen in echinoderms.

Petancula—Stage of metamorphosis for holothuroids.

Phoresy—Nonparasitic relationship between two organisms in which one uses the other as a means of transportation.

Photokinesis—Activity induced by the presence of light.

Photophore—Cell or group of cells that produce light.

Phytophagous—An organism that solely feeds upon plants.

Pilidium—Free-swimming, planktotrophic larva of heteronemerteans.

Pinnules—Small branches.

Planktotrophic larvae—Larvae that feed during their planktonic phase.

Planulae—Larval cnidarians.

Plasmodium—A life cycle stage in which several young organisms join to form a mass of protoplasm.

Plerocercoid—Last larval lifestage of tapeworms.

Polyembryony—The production of several embryos from a single egg.

Polyp—Cnidarian form that is sessile.

Polyphagous—An organism that consumes a variety of foods.

Polyphyletic—A group that does not contain the most recent ancestor of the organisms.

Positively phototactic—Movement toward light.

Predaceous—An organism that preys on other organisms.

Predator—An animal that attacks and feeds on other animals.

Primary host—An organism that acts as the host for an adult stage of a parasite. Also called definitive host.

Protandric hermaphrodites—Animals hatch as males and later develop into females.

Protonephridia—Ciliated excretory tube that is specialized for filtration.

Protonymph—The second instar of a mite.

Protostome—Bilateral metazoans characterized by determinate and spiral cleavage, the formation of a mouth and anus directly from the blastopore, and the formation of the coelom by the embryonic mesoderm having split.

Pseudovipositor—Terminal abdominal segment of females from which eggs are layed.

Radial symmetry—The exact arrangement of parts or organs around a central axis.

Ramate—An animal or organism with branches.

Raptorial—An organism that has specially adapted the ability to seize and grasp prey.

Rhagon—Stage of development in demosponge larva.

Rostrum— The beak, snout, spine, proboscis, or anterior median prolongation of the carapace or head of an organism.

Saprophytic—An organism that lives on dead or decaying organic matter.

Scalids—Sets of complex spines that allow the organism to move, capture food, or sense changes in its environment.

Scyphistoma—Scyphozoan polyp.

Sclerites—Thick layer of the exoskeleton.

Segment—A rings or subdivisions of the body.

Sensu stricto—In the "strict sense."

Seta—A bristle.

Spermatophore—Packet of sperm that is usually transferred from one individual to another during mating.

Spicule—A slender, pointed structure.

Spiral cleavage—Cleavage pattern in which spindles or places are oblique to the axis of the egg.

Spiralians—Animal groups that show spiral cleavage patterns.

Spirocyst—Adhesive threads present on Cnidarians that capture prey and attach to immobile objects.

Stereom—A unique skeletal microstructure (a network of interconnected holes) in echinoderms.

Stock—A biologically distinct and interbreeding population within a species of aquatic animals.

Stoloniferous—An organism that bears or develops a branch from its base to produce new plants from buds, or an extension of the body wall that develops buds giving rise to new zooids.

Stomochord—A hollow pouch of the gut found in the proboscis of hemichordates.

Strobilation—Asexual reproduction by division into body segments.

Subsistence fishery—A fishery in which the harvested resource is used directly by the fisher.

Symbiont—An organism living in a close relationship with another organism.

Symbiosis—An intimate association, union, or living arragement between two dissimilar organisms in which at least one of the organisms is dependent upon the other.

Synanthropic—Associated with human habitation.

Synapomorphy—A derived trait shared by two or more taxonomic groups.

Syncytial—Multinucleate mass of cytoplasm resulting from the fusion of cells.

Taxis—Reflex movement by an organism in relation to a source of stimulation.

Tegument—Outer, nonciliated layer of the body wall of platyhelminth parasites.

Test—Shell-like encasement or skeleton.

Totipotency—The ability of a cell to differentiate into any type of body cell.

Triploblastic—Embryos with three germ layers.

Triradial symmetry—Three-part radial symmetry, as seen in the life stages of some echinoderms.

Trochophore—Larva that has a girdle ring of cilia.

Troglophilous—An organism that lives in caves.

Unci—Hooked anatomical structure.

Uncinus—Miniature hooked anatomical structure.

Univoltine—A group that produces only one generation per year.

Velum—Shelf present under the umbrella of most hydromedusae, or a ciliated growth with which larva swim.

Vermiform larva—A legless, worm-like larva without a well-developed head.

Vibrissae—A pair of large bristles that is present just above the mouth in some organisms.

Vitellarium—Part of the ovary that produces yolk-filled nurse cells.

Viviparous—An organism that produces live young.

Zoea—Second to last larval stage of many crustaceans.

Zooid—Individual invertebrate that reproduces nonsexually by budding or splitting, especially one that lives in a colony in which each member is joined to others by living material, for example, a coral.

Zooplankton—Free-swimming, microscopic planktonic animals present in lakes and oceans.

Lower metazoans and lesser deuterostomes order list

Porifera [Phylum]
  Calcarea [Class]
    Baeriida [Order]
    Clathrinida
    Leucosoleniida
    Lithonida
    Murrayonida

  Hexactinellida [Class]
    Amphidiscosida [Order]
    Aulocalycoida
    Hexactinosida
    Lychniscosida
    Lyssacinosida

  Demospongiae [Class]
    Agelasida [Order]
    Astrophorida
    Chondrosida
    Dendroceratida
    Dictyoceratida
    Hadromerida
    Halichondrida
    Halisarcida
    Haplosclerida
    Homosclerophorida
    Poecilosclerida
    Spirophorida
    Verongida
    Verticillitida

Placozoa [Phylum]
  No order designations

Monoblastozoa [Phylum]
  No order designations

Rhombozoa [Phylum]
  Dicyemida [Order]
  Heterocyemida

Orthonectida [Phylum]
  Orthonectida [Order]

Cnidaria [Phylum]
  Anthozoa [Class]
    Actiniaria [Order]

Alcyonacea
Antipatharia
Cerianthia
Corallimorpharia
Helioporacea
Pennatulacea
Scleractinia
Zoanthidea

Hydrozoa [Class]
  Actinulida [Order]
  Capitata
  Conica
  Cystonectae
  Filifera
  Laingiomedusae
  Limnomedusae
  Moerisiida
  Narcomedusa
  Proboscoida
  Physonectae
  Polypodiozoa
  Trachymedusa

Cubozoa [Class]
  Cubomedusae [Order]

Scyphozoa [Class]
  Coronatae [Order]
  Rhizostomeae
  Semaeostomeae
  Stauromedusae

Ctenophora [Phylum]
  Beroida [Order]
  Cestida
  Cydippida
  Ganeshida
  Lobata
  Platyctenida
  Thalassocalycida

Platyhelminthes [Phylum]
  Acoela [Class]
    No order designations

Turbellaria [Class]
  Catenulida [Order]
  Haplopharyngida
  Lecithoepitheliata
  Macrostomida
  Polycladida
  Prolecithophora
  Proplicastomata
  Proseriata
  Rhabdocoela
  Tricladida

Trematoda [Class]
  Aspidogastrida [Order]
  Azygiida
  Echinostomida
  Opisthorchiida
  Plagiorchiida
  Strigeatida

Monogenea [Class]
  Monopisthocotylea [Order]
  Polyopisthocotylea

Cestoda [Class]
  Amphilinidea [Order]
  Caryophyllidea
  Cyclophyllidea
  Diphyllidea
  Gyrocotylidea
  Haplobothriidea
  Lecanicephalidea
  Litobothriidea
  Nippotaeniidea
  Proteocephalidea
  Pseudophyllidea
  Spathebothriidea
  Tetrabothriidea
  Tetraphyllidea
  Trypanorhyncha

Nemertea [Phylum]
  Anopla [Class]
    Heteronemertea [Order]
    Palaeonemertea

Enopla [Class]
Hoplonemertea [Order]

Rotifera [Phylum]
Bdelloida [Order]
Collothecacea
Flosculariacea
Ploimida
Seisonida

Gastrotricha [Phylum]
Chaetonotida [Order]
Macrodasyida

Kinorhyncha [Phylum]
Cyclorhagida [Order]
Homalorhagida

Nematoda [Phylum]
Adenophorea [Class]
Araeolaimida [Order]
Chromadorida
Desmodorida
Desmoscolecida
Dorylaimida
Enoplida
Isolaimida
Mermithida
Monhysterida
Mononchida
Muspiceida
Stichosomida
Trichocephalida

Secernentea [Class]
Aphelenchida [Order]
Ascaridida
Camallanida
Diplogasterida
Rhabditida
Spirurida
Strongylida
Tylenchida

Nematomorpha [Phylum]
Gordioidea [Order]
Nectonematoidea

Acanthocephala [Phylum]
Apororhynchida [Order]
Echinorhynchida
Gigantorhynchida

Gyracanthocephala
Moniliformida
Neoechinorhynchida
Oligacanthorhynchida
Polymorphida

Entoprocta [Phylum]
Coloniales [Order]
Solitaria

Micrognathozoa [Phylum]
Limnognathida [Order]

Gnathostomulida [Phylum]
Bursovaginoidea [Order]
Filospermoidea

Priapulida [Phylum]
Halicryptomorphida [Order]
Priapulimorphida
Seticoronarida

Loricifera [Phylum]
Nanaloricida [Order]

Cycliophora [Phylum]
Symbiida [Order]

Echinodermata [Phylum]
Crinoidea [Class]
Bourgueticrinida [Order]
Comatulida
Cyrtocrinida
Isocrinida
Millericrinida

Asteroidea [Class]
Brisingida [Order]
Forcipulatida
Notomyotida
Paxillosida
Spinulosida
Valvatida
Velatida

Concentricycloidea [Class]
Peripodida [Order]

Ophiuroidea [Class]
Euryalida [Order]
Oegophiurida
Ophiurida

Echinoidea [Class]
Arbacioida [Order]
Cassiduloida
Cidaroida
Clypeasteroida
Diadematoida
Echinoida
Echinothuroida
Holasteroida
Holectypoida
Pedinoida
Phymosomatoida
Salenoida
Spatangoida
Temnopleurida

Holothuroidea [Class]
Apodida [Order]
Aspidochirotida
Dactylochirotida
Dendrochirotida
Elasipodida
Molpadiida

Chaetognatha [Phylum]
Aphragmophora [Order]
Phragmophora

Hemichordata [Phylum]
Cephalodiscida [Order]
Rhabdopleurida

Chordata [Phylum]
Urochordata [Subphylum]
Ascidiacea [Class]
Enterogona [Order]
Pleurogona

Thaliacea [Class]
Doliolida [Order]
Pyrosomatida
Salpida

Appendicularia [Class]
Copelata [Order]

Sorberacea [Class]
Aspiraculatav [Order]

Cephalochordata [Subphylum]
Amphioxiformes [Order]

LOWER METAZOANS AND LESSER DEUTEROSTOMES ORDER LIST

• • • • •

A brief geologic history of animal life

A note about geologic time scales: A cursory look will reveal that the timing of various geological periods differs among textbooks. Is one right and the others wrong? Not necessarily. Scientists use different methods to estimate geological time—methods with a precision sometimes measured in tens of millions of years. There is, however, a general agreement on the magnitude and relative timing associated with modern time scales. The closer in geological time one comes to the present, the more accurate science can be—and sometimes the more disagreement there seems to be. The following account was compiled using the more widely accepted boundaries from a diverse selection of reputable scientific resources.

Geologic time scale

Era	Period	Epoch	Dates	Life forms
Proterozoic			2,500-544 mya*	First single-celled organisms, simple plants, and invertebrates (such as algae, amoebas, and jellyfish)
Paleozoic	Cambrian		544-490 mya	First crustaceans, mollusks, sponges, nautiloids, and annelids (worms)
	Ordovician		490-438 mya	Trilobites dominant. Also first fungi, jawless vertebrates, starfishes, sea scorpions, and urchins
	Silurian		438-408 mya	First terrestrial plants, sharks, and bony fishes
	Devonian		408-360 mya	First insects, arachnids (scorpions), and tetrapods
	Carboniferous	Mississippian	360-325 mya	Amphibians abundant. Also first spiders, land snails
		Pennsylvanian	325-286 mya	First reptiles and synapsids
	Permian		286-248 mya	Reptiles abundant. Extinction of trilobytes. Most modern insect orders
Mesozoic	Triassic		248-205 mya	Diversification of reptiles: turtles, crocodiles, therapsids (mammal-like reptiles), first dinosaurs, first flies
	Jurassic		205-145 mya	Insects abundant, dinosaurs dominant in later stage. First mammals, lizards, frogs, and birds
	Cretaceous		145-65 mya	First snakes and modern fish. Extinction of dinosaurs and ammonites, rise and fall of toothed birds
Cenozoic	Tertiary	Paleocene	65-55.5 mya	Diversification of mammals
		Eocene	55.5-33.7 mya	First horses, whales, monkeys, and leafminer insects
		Oligocene	33.7-23.8 mya	Diversification of birds. First anthropoids (higher primates)
		Miocene	23.8-5.6 mya	First hominids
		Pliocene	5.6-1.8 mya	First australopithecines
	Quaternary	Pleistocene	1.8 mya-8,000 ya	Mammoths, mastodons, and Neanderthals
		Holocene	8,000 ya-present	First modern humans

*Millions of years ago (mya)

• • • • •

Index

Bold page numbers indicate the primary discussion of a topic; page numbers in italics indicate illustrations.

A

Abalones, 1:51
Abatus cordatus. See Heart urchins
Abiotic environments, 1:24
Acanthaster planci. See Crown-of-thorns
Acanthocephala. *See* Thorny headed worms
Achistridae, 1:417
Achistrum spp., 1:417
Acoelomorpha, 1:11, 1:179
Acoels, 1:5, 1:7, 1:11, **1:179–183**, 1:*180*
Acoleidae, 1:226
Acorn worms, 1:13, 1:22, 1:*443*, 1:*444*, 1:445, 1:*446*, 1:*447*, 1:448–449
Acraniates. *See* Lancelets
Acrobothriidae, 1:225
Acropora millepora, 1:*110*, 1:*117*
Acropora palmata. See Elkhorn corals
Actiniaria. *See* Sea anemones
Actinulidae, 1:123
Adenophorea. *See* Roundworms
Aequorea spp., 1:127
Aequorea victoria, 1:49, 1:129, 1:*131*, 1:137
Aequorin, 1:47
Aerosols, 1:52
   See also Conservation status
Aethomerus spp. *See* Longhorn beetles
African river blindness nematodes, 1:35–36, 1:*296*, 1:*299*
Agelasida, 1:77, 1:79
Aglantha digitale, 1:*130*, 1:145
Aglaophenia picardi, 1:25
Aglaophenia pluma, 1:25, 1:*131*, 1:137–138
Aglapheniids, 1:129
Alaria spp., 1:199
Alcyonacea, 1:103
Alcyonaria. *See* Octocorallia
Algae, 1:*9*, 1:32–33
Alternation of generations. *See* Reproductive duality
Amabiliidae, 1:226
American diamond girdle wearers, 1:345, 1:*346*, 1:*347*, 1:349
American tube dwelling anemones, 1:*109*, 1:*111*, 1:115
Amiskwia sagittiformis, 1:433
Amphibdella spp., 1:214
Amphilina foliacea, 1:231, 1:*232*, 1:234–235
Amphilinidae, 1:225
Amphilinidea, 1:225, 1:226, 1:227, 1:228, 1:230, 1:231
Amphioxiformes. *See* Lancelets
Amphioxus spp. *See* Lancelets
Amphioxus lanceolatus. See Lancelets

Amphipholis squamata. See Dwarf brittle stars
Amphiporus spp., 1:253
Amphiura filiformis, 1:*391*, 1:395–396
Amphiuridae, 1:387, 1:388
Amplexidiscus fenestrafer. See Elephant ear polyps
Amur sea stars. *See* Northern Pacific sea stars
Anaea butterfly caterpillars, 1:42
Ancylostoma caninum. See Dog hookworms
Ancylostoma duodenale, 1:35
Anemones, **1:103–116**, 1:*109*, 1:*110*
   behavior, 1:29–30, 1:32, 1:40, 1:42, 1:106
   clownfish and, 1:33
   conservation status, 1:107
   distribution, 1:105
   evolution, 1:103
   feeding ecology, 1:27–28, 1:106–107
   habitats, 1:105
   humans and, 1:107–108
   physical characteristics, 1:103–*105*
   reproduction, 1:107
   species of, 1:111–116
   taxonomy, 1:103
Anemotaxis, 1:40
Angiostrongylus cantonensis. See Rat lungworms
Anoplans, 1:11, **1:245–251**, 1:*248*
Anoplocephalidae, 1:226, 1:231
Anseropoda placenta, 1:369
Antedon bifida. See Rosy feather stars
Anteroporidae, 1:226
Anthocidaris crassispina. See Short-spined sea urchins
Anthomedusae, 1:123, 1:124–125
Anthopleura xanthogrammica. See Giant green anemones
Anthosigmella spp., 1:29
Anthozoa, **1:103–122**, 1:*109*, 1:*110*
   behavior, 1:106
   conservation status, 1:107
   distribution, 1:105
   evolution, 1:103
   feeding ecology, 1:106–107
   habitats, 1:24, 1:25, 1:105
   humans and, 1:107–108
   physical characteristics, 1:103–*105*
   reproduction, 1:107
   species of, 1:*111*–121
   taxonomy, 1:10, 1:103
Anthracomedusa turnbulli, 1:147
Anticancer drugs, 1:44–46, 1:49
   See also Humans
Antigomonidae, 1:275
Antipatharia, 1:103, 1:104, 1:105, 1:106, 1:108
Antipathella fiordensis. See Black corals

Antiponemertes allisonae, 1:255
Aphragmophora, 1:433
Aphrocallistes vastus. See Cloud sponges
Aplidium spp., 1:452
Aplidium albicans, 1:46
Aploparaksis spp., 1:229
Aplousobranchia, 1:452, 1:455
Aplydine, 1:46
Aplysina archeri. See Stove-pipe sponges
Aplysina cauliformis. See Row pore rope sponges
Aplysinellidae, 1:79
Apodida, 1:417, 1:418, 1:420, 1:421
Apolemia uvaria, 1:127, 1:*133*, 1:144
Appendicularia. *See* Larvaceans
Aquaculture
   for drug production, 1:45
   vs. fisheries, 1:51–52
Ara-A (arabinosyl adenine), 1:44
Ara-C (arabinosyl cytosine), 1:44
Arabinosyl adenine (Ara-A), 1:44
Arabinosyl cytosine (Ara-C), 1:44
Archaea, 1:49
Archaeocyathans, 1:10
Archeocytes, 1:21
Archiacanthocephala, 1:311
Archigetes iowensis, 1:230
Archigetes limnodrili, 1:230
Archigetes sieboldi, 1:230
Arenicola grubei, 1:40
Argonemertes hillii, 1:255
Aristotle, 1:417, 1:451
Armored sea cucumbers. *See* Slipper sea cucumbers
Arrow worms, 1:9, **1:433–442**, 1:*435*, 1:*436*
   behavior, 1:42, 1:434
   conservation status, 1:435
   distribution, 1:434
   evolution, 1:433
   feeding ecology, 1:434–435
   habitats, 1:434
   humans and, 1:435
   physical characteristics, 1:433–434
   reproduction, 1:22, 1:435
   species of, 1:*437*–442
   taxonomy, 1:9, 1:13, 1:433
Asajirus spp., 1:481
Asajirus indicus, 1:479, 1:480, 1:*481*, 1:*482*
Asbestopluma spp., 1:10
Asbestopluma hypogea. See Carnivorous sponges
Ascaris lumbricoides. See Maw-worms
Aschelminthes, 1:9
Ascidia spp., 1:29, 1:45–46, 1:48, 1:452
Ascidiacea. *See* Sea squirts

INDEX

INDEX

INDEX